Subcellular Biochemistry

Volume 72

Series editor
J. Robin Harris
University of Mainz, Mainz, Germany

The book series SUBCELLULAR BIOCHEMISTRY is a renowned and well recognized forum for disseminating advances of emerging topics in Cell Biology and related subjects. All volumes are edited by established scientists and the individual chapters are written by experts on the relevant topic. The individual chapters of each volume are fully citable and indexed in Medline/Pubmed to ensure maximum visibility of the work.

More information about this series at http://www.springer.com/series/6515

Kazuyuki Akasaka • Hitoshi Matsuki
Editors

High Pressure Bioscience

Basic Concepts, Applications and Frontiers

Springer

Editors
Kazuyuki Akasaka
Graduate School of Life and Environmental Sciences
Kyoto Prefectural University
Kyoto, Japan

Hitoshi Matsuki
Department of Life System
Institute of Technology and Science
Tokushima University
Tokushima, Japan

ISSN 0306-0225
Subcellular Biochemistry
ISBN 978-94-017-9917-1 ISBN 978-94-017-9918-8 (eBook)
DOI 10.1007/978-94-017-9918-8

Library of Congress Control Number: 2015945028

Springer Dordrecht Heidelberg New York London

Printed on acid-free paper

Springer Science+Business Media B.V. Dordrecht is part of Springer Science+Business Media (www.springer.com)

Preface

One century ago, the coagulation of egg white by applying pressure of 700 MPa showed for the first time that pressure can denature a protein (P. W. Bridgman, J Biol Chem 19:511–512, 1914). A half century later, the protein denaturation phenomenon began to be understood on *P–T* plane as kinetic and thermodynamic processes (Suzuki, Rev Phys Chem Jpn 29:91–98, 1960; Hawley, Biochemistry 10:2436–2442, 1971), while independently the protein-folding phenomenon was established as a thermodynamic process (Anfinsen, Proc Natl Acad Sci USA 47:1309–1314, 1961, Nobel prize 1972). Since then, up to the present, stable folded structures of proteins and nucleic acids have been disclosed in mass quantities in crystal as well as in solution, with a hope to understand their function based on structures. However, it is now clear that generally the knowledge on these stable, and basically single, folded structures alone does not give direct answers to their function, but the knowledge on the multiple and dynamic reality of their structures is crucial. It is on this historical background that methods capable of disclosing the multiple-conformational and dynamic reality of bio-macromolecules, notably high-pressure NMR, X-ray crystallography/scattering, and others with pressure as least "invasive" perturbation, have been developed in recent years. These methods are rapidly disclosing knowledge on the hidden reality of proteins that is crucial to in-depth understanding not only of protein structures and function but also of most life phenomena such as motility, physiology/disease, and environmental adaptation/evolution, as well as to the strategic and efficient use of pressure in agricultural and medical applications in the years to come. In other words, pressure is no longer an "odd" and "foreign" parameter to life. Instead, it is a crucial variable parameter with which one can disclose the secret of nature in its very basic atomic design and, furthermore, a manageable external parameter with which one can select and enhance some important molecular events in nature to our practical needs. In short, we are currently facing a new era of high-pressure bioscience, where advanced knowledge on bio-macromolecular structure is to be shared by all people interested, to deepen our understanding of basic life phenomena and to expand the utility of pressure variable to a wide area of bioscience and biotechnology. It is in this timing and concept, marking the centennial anniversary of the first report on pressure denaturation of proteins, that we plan this monograph covering from the basics to the application with pressure as the key variable.

The 33 chapters of this monograph are all original contributions by experts in the field. They are divided into nine parts, each of which consists of three to four chapters, which are arranged in an as much as unified manner to provide readers with a comprehensive view on the entire field of research from the basics (in Parts I through III) to the application (in Parts VI through VIII) and both (in Parts IV to V). This monograph is intended also to serve as a first comprehensive textbook on bio-macromolecular events under pressure and their possible roles in macroscopic life phenomena as well as in application to high-pressure biotechnology, for a wide audience including students and researchers in both basic and applied biosciences.

Kyoto, Japan Kazuyuki Akasaka
Tokushima, Japan Hitoshi Matsuki

Contents

Contributors

Fumiyoshi Abe Department of Chemistry and Biological Science, College of Science and Engineering, Aoyama Gakuin University, Sagamihara, Japan

Institute of Biogeosciences, Japan Agency for Marine-Earth Science and Technology (JAMSTEC), Yokosuka, Japan

Kazuyuki Akasaka High Pressure Protein Research Center, Institute of Advanced Technology, Kinki University, Kinokawa, Japan

Present Affiliation: Graduate School of Life and Environmental Sciences, Kyoto Prefectural University, Kyoto, Japan

Yoriko Atomi 204 Research Center for Science and Technology, Tokyo University of Agriculture and Technology, Tokyo, Japan

Shana P.C. Barroso Instituto de Bioquímica Médica Leopoldo de Meis, Universidade Federal do Rio de Janeiro, Rio de Janeiro, Brazil

Tigran V. Chalikian Department of Pharmaceutical Sciences, Leslie Dan Faculty of Pharmacy, University of Toronto, Toronto, Canada

Mariano Dellarole Centre de Biochimie Structurale, UMR UM1&UM2/5048 CNRS/1054 INSERM, Montpellier, France

Anne-Claire Dhaussy CRISMAT, ENSICAEN, Caen, France

Carlos H. Dumard Instituto de Bioquímica Médica Leopoldo de Meis, Universidade Federal do Rio de Janeiro, Rio de Janeiro, Brazil

Instituto Nacional de Ciência e Tecnologia de Biologia Estrutural e Bioimagem, Rio de Janeiro, Brazil

Tetsuro Fujisawa Department of Chemistry and Biomolecular Science, Gifu University, Gifu, Japan

Kunihiko Gekko Hiroshima Synchrotron Radiation Center, Hiroshima University, Higashi-Hiroshima, Japan

Eric Girard University of Grenoble Alpes, IBS, Grenoble, France

CNRS, IBS, Grenoble, France

CEA, DSV, IBS, Grenoble, France

Andre M.O. Gomes Instituto de Bioquímica Médica Leopoldo de Meis, Universidade Federal do Rio de Janeiro, Rio de Janeiro, Brazil

Instituto Nacional de Ciência e Tecnologia de Biologia Estrutural e Bioimagem, Rio de Janeiro, Brazil

Fumio Hirata College of Life Sciences, Ritsumeikan University, Kusatsu, Japan

Eri Hoshino Research Institute, Echigo Seika Co., Ltd., Niigata, Japan

Jun Hoshino Research Institute, Echigo Seika Co., Ltd., Niigata, Japan

Hitoshi Iwahashi Department of Applied Life Science, Faculty of Applied Biological Science, Gifu University, Gifu, Japan

Hidemi Izumi Faculty of Biology-Oriented Science and Technology, Kinki University, Kinokawa, Japan

Hans Robert Kalbitzer Institute of Biophysics and physical Biochemistry, University of Regensburg, Regensburg, Germany

Yuji O. Kamatari Life Science Research Center, Gifu University, Gifu, Japan

Chiaki Kato Institute of Biogeosciences, Japan Agency for Marine-Earth Science and Technology (JAMSTEC), Yokosuka, Japan

Mariko Kawamura Research Institute, Echigo Seika Co., Ltd., Niigata, Japan

Ryo Kitahara College of Pharmaceutical Sciences, Ritsumeikan University, Kusatsu, Japan

Atsushi Kobayashi Research Institute, Echigo Seika Co., Ltd., Niigata, Japan

Takeshi Kobayashi Department of Physiology, Nagoya University Graduate School of Medicine, Nagoya, Japan

Christian A. Lenz Lehrstuhl für Technische Mikrobiologie, Technische Universität München, Freising, Germany

Hua Li National Center for Protein Science Shanghai, State Key Laboratory of Molecular Biology, Institute of Biochemistry and Cell Biology, Shanghai Institutes for Biological Sciences, Chinese Academy of Sciences, Shanghai, China

Akihiro Maeno High Pressure Protein Research Center, Institute of Advanced Technology, Kinki University, Kinokawa, Japan

Hitoshi Matsuki Department of Life System, Institute of Technology and Science, Tokushima University, Tokushima, Japan

Ygara S. Mendes Instituto de Bioquímica Médica Leopoldo de Meis, Universidade Federal do Rio de Janeiro, Rio de Janeiro, Brazil

Tadayuki Nishiumi Department of Applied Biological Chemistry, Faculty of Agriculture, Niigata University, Niigata, Japan

Masayoshi Nishiyama The Hakubi Center for Advanced Research/Institute for Integrated Cell-Material Sciences, Kyoto University, Kyoto, Japan

Philippe M. Oger Laboratoire de Géologie de Lyon, UMR CNRS 5276, Ecole Normale Supérieure de Lyon, Lyon, France

Miyuki Ogino Department of Applied Biological Chemistry, Faculty of Agriculture, Niigata University, Niigata, Japan

Eri Ohara Research Institute, Echigo Seika Co., Ltd., Niigata, Japan

Eiji Ohmae Department of Mathematical and Life Sciences, Graduate School of Science, Hiroshima University, Higashi-Hiroshima, Japan

Andréa C. Oliveira Instituto de Bioquímica Médica Leopoldo de Meis, Universidade Federal do Rio de Janeiro, Rio de Janeiro, Brazil

Instituto Nacional de Ciência e Tecnologia de Biologia Estrutural e Bioimagem, Rio de Janeiro, Brazil

Fumihisa Ono Department of Applied Science, Okayama University of Science, Okayama, Japan

Department of Chemistry, University College London, London, UK

Julien Roche Centre de Biochimie Structurale, UMR UM1&UM2/5048 CNRS/1054 INSERM, Montpellier, France

Christian Roumestand Centre de Biochimie Structurale, UMR UM1&UM2/5048 CNRS/1054 INSERM, Montpellier, France

Catherine A. Royer Department of Biology, Rensselaer Polytechnic Institute, Troy, NY, USA

Patricia S. Santos Instituto de Bioquímica Médica Leopoldo de Meis, Universidade Federal do Rio de Janeiro, Rio de Janeiro, Brazil

Instituto Nacional de Ciência e Tecnologia de Biologia Estrutural e Bioimagem, Rio de Janeiro, Brazil

Yasuyuki Sawada Department of Physiology, Nagoya University Graduate School of Medicine, Nagoya, Japan

Toru Shigematsu Faculty of Applied Life Sciences, Niigata University of Pharmacy and Applied Life Sciences (NUPALS), Niigata, Japan

Jerson L. Silva Instituto de Bioquímica Médica Leopoldo de Meis, Universidade Federal do Rio de Janeiro, Rio de Janeiro, Brazil

Instituto Nacional de Ciência e Tecnologia de Biologia Estrutural e Bioimagem, Rio de Janeiro, Brazil

László Smeller Department of Biophysics and Radiation Biology, Semmelweis University, Budapest, Hungary

Masahiro Sokabe Mechanobiology Laboratory, Nagoya University Graduate School of Medicine, Nagoya, Japan

Mechanobiology Institute, National University of Singapore, Singapore, Singapore

Keizo Suzuki Emeritus Professor, Department of Chemistry, College of Science and Engineering, Ritsumeikan University, Kusatsu, Japan

Hideki Tachibana Department of Biotechnological Science, School of Biology-Oriented Science and Technology, and High Pressure Protein Research Center, Institute of Advanced Technology, Kinki University, Kinokawa, Japan

Rudi F. Vogel Lehrstuhl für Technische Mikrobiologie, Technische Universität München, Freising, Germany

Nobuhisa Watanabe Synchrotron Radiation Research Center, Nagoya University, Nagoya, Japan

Mike P. Williamson Department of Molecular Biology and Biotechnology, University of Sheffield, Sheffield, UK

Roland Winter Physical Chemistry I – Biophysical Chemistry, TU Dortmund University, Dortmund, Germany

Akira Yamazaki Research Institute, Echigo Seika Co., Ltd., Niigata, Japan

Nomenclature and Abbreviations

Nomenclature

α Thermal expansion coefficient ($=1/V(\partial V/\partial T)_p$) or thermal expansivity ($= (\partial V/\partial T)_p$)
B_2 Second virial coefficient
β_s Adiabatic compressibility coefficient ($= -1/V(\partial V/\partial p)_s$)
β_T Isothermal compressibility coefficient ($= -1/V(\partial V/\partial p)_T$)
c_i Molar concentration of component i
C_p Heat capacity at constant pressure
c_p Partial molar or specific heat capacity at constant pressure
C_v Heat capacity at constant volume
D Denatured state
ϕ Apparent molar or specific volume
F Helmholtz free energy
G Gibbs free energy
H Enthalpy
I Intermediate state, peak intensity
h_i Partial molar or specific enthalpy of component i
K Equilibrium constant
k Rate constant
k_B Boltzmann constant
κ_s Adiabatic compression ($= -(\partial V/\partial p)_s$)
κ_T Isothermal compression ($= -(\partial V/\partial p)_T$)
μ_i Chemical potential of component i
N Native state
N_i Number concentration of component i
R Gas constant
S Entropy
s_i Partial molar or specific entropy of component i
τ_i Relaxation time of component i

U	Unfolded state
ρ	Density of solution
V	Total volume, molar volume
v_i	Partial molar or specific volume of component i
‡	Activated state (superscript symbol)

Abbreviations

AsA	Ascorbic acid
BPTI	Bovine pancreatic trypsin inhibitor
BSA	Bovine serum albumin
CCW	Counterclockwise
CD	Circular dichroism
CFU	Colony-forming unit
CLE	Cortex lytic enzyme
CpMV	Cowpea mosaic virus
CW	Clockwise
DAC	Diamond anvil cell
DAPI	4′,6-Diamidino-2-phenylindole
DHFR	Dihydrofolate reductase
DLVO	Derjaguin, Landau, Verwey, Overbeek
DPA	Dipicolinic acid (pyridine-2,6-dicarboxylic acid)
DPH	1,6-Diphenyl-1,3,5-hexatriene
ECM	Extracellular matrix
ER	Endoplasmic reticulum
FA	Fatty acid, focal adhesion
FPGM	Freeze-pressure generation method
GABA	Gamma-aminobutyric acid
GAP	GTPase-activating protein
GD	Gramicidin D
GEF	Guanine nucleotide exchange factor
HEWL	Hen egg white lysozyme
HHP	High hydrostatic pressure
Hi-Pit	High-pressure-induced transformation
HIV	Human immunodeficiency virus
HPMX	High-pressure macromolecular crystallography
HPT	High-pressure thermal processing
HSQC	Heteronuclear single-quantum correlation
IPMDH	3-Isopropylmalate dehydrogenase
LLPT	Liquid–liquid phase transition
MAP	Modified atmosphere packaging
MD	Molecular dynamics
MG	Molten globule

MIP	Minimum interdigitation pressure
MSA	Mean spherical approximation
MSC	Mechanosensitive ion channel
MT	Microtubule
MUFA	Monounsaturated fatty acid
nGR	Germinant receptors responding to nutrient
NMR	Nuclear magnetic resonance
NOE	Nuclear overhauser effect
ORAC	Oxygen radical scavenging capacity
OVM	Ovomucoid
PATP	Pressure-assisted thermal processing
PC	Phosphatidylcholine
PE	Phosphatidylethanolamine
PEG	Polyethylene glycol
PG	Phosphatidylglycerol, peptidoglycan
PUFA	Polyunsaturated fatty acid
RalGDS	Ral guanine dissociation stimulator
Ras	Rat sarcoma
RISM	Reference interaction site model
RNase	Ribonuclease
SASP	Small acid-soluble protein
SAXS	Small-angle X-ray scattering
SF	Actin cytoskeleton stress fiber
sHSP	Small heat shock protein
SNase	Staphylococcal nuclease
SOD	Superoxide dismutase
TMAO	Trimethylamine-*N*-oxide
TRFAM	Time-resolved fluorescence anisotropy measurement
TSE	Transition state ensemble
UOx	Urate oxidase
WT	Wild type

Part I
Why and How Proteins Denature Under Pressure?

Editors' Foreword of Part I

Ever since the first observation of pressure coagulation of egg white a century ago (Bridgman, J Biol Chem 19:511–512, 1914), the question why and how pressure (and temperature) coagulates proteins remained for a long time. In early days of 50s and 60s Suzuki and coworkers challenged this problem, and found that it follows a first-order kinetics with respect to the protein concentration and mapped the rate on the *P-T* plane (Suzuki, Rev Phys Chem Jpn 29:91–98, 1960).

In this part, Chaps. 1 and 2 are the reviews of the attempts in early days and of the actual expressions of the unfolding phenomena as two-state equilibrium, while Chaps. 3 and 4 give two typical contemporary views on the mechanisms of pressure denaturation of proteins. Chapter 1 is a contribution from Prof. Keizo Suzuki who himself is a living historical figure, who tells how people, including himself, approached the Bridgman's finding to describe it as physicochemical phenomena between 40s and 60s when three-dimensional structures of proteins are almost unknown. Suzuki analyzed, for the first time, the protein unfolding/coagulation reaction as a kinetic process on *P-T* plane. Chapter 2 by Smeller gives a comprehensive survey of the works that followed in 70s up to present, treating the protein unfolding (denaturation) event as an two-state equilibrium process, with the Gibbs energy difference to the second order to *P* and *T*, giving fundamental thermodynamic quantities, partial molar volume, compressibility and expansibility of a protein in solution. A number of actual examples are given, with deviations for specific cases from the simple two-state approach. Chapters 3 and 4 give two representative current views on the molecular mechanism of pressure denaturation (unfolding) of proteins by two leading scientists in the field, each describing the general origin for the volume difference between the folded state and the unfolded state of a protein in solution. In Chap. 3, the answer is sought systematically in the differential volumes of various structural components of the polypeptide chain between the folded state and the unfolded state, the volume of the latter itself is pressure-dependent, while in Chap. 4 Royer shows convincingly that a void volume, a space unoccupied by the protein atoms, is crucial in determining the volume change upon unfolding. Readers will find that, in the frontier of pressure bioscience, the long-standing fundamental question why and how proteins denature under pressure is still under debate.

Chapter 1	Early Days of Pressure Denaturation Studies of Proteins	Keizo Suzuki
Chapter 2	Protein Denaturation on *p-T* Axes – Thermodynamics and Analysis	László Smeller
Chapter 3	Driving Forces in Pressure-Induced Protein Transitions	Tigran V. Chalikian
Chapter 4	Why and How Does Pressure Unfold Proteins?	Catherine A. Royer

Chapter 1
Early Days of Pressure Denaturation Studies of Proteins

Keizo Suzuki

Abstract The denaturation of protein by pressure has been generally well known since the findings of the perfect coagulation of egg white by a pressure of 7,000 atm within 30 min by Bridgman (J Biol Chem 19:511–512, 1914), and Kiyama and Yanagimoto (Rev Phys Chem Jpn 21:41–43, 1951) confirmed that the coagulation occurs above 3,880 kg cm^{-2}. Grant et al. (Science 94:616, 1941) and Suzuki and Kitamura (Abstracts of 30th annual meeting of Japanese Biochemical Society, 1957) found that SH groups are detected at the compressed sample of ovalbumin. On the other hand, Johnson and Campbell (J Cell Comp Physiol 26:43–49, 1945), Tongur (Kolloid Zhur 11:274–279, 1949; Biokhimiya 17:495–503, 1952) and Suzuki et al. (Mem Res Inst Sci Eng Ritsumeikan Univ 3:1–4, 1958) reported that the thermal denaturation of proteins is retarded in a few examples by the low pressure of about 1,000 atm. Before 1960, the studies of denaturation under high pressure were, however, rare and almost qualitative compared with those by heat, acid, urea and so on, so that there was no theory for the influence of hydrostatic pressure on the mechanism of denaturation. Here I review how I started experiments and analysis on pressure denaturation of proteins in early days of 1950s and 1960s in my laboratory and others.

Keywords Globular protein • Kinetics • Pressure denaturation • Pressure inactivation • Renaturation

1.1 Introduction

It seems that there are three common interesting points on the protein denaturation. They are: (1) Although the denaturation phenomena were very complex and characteristic, they could be derived from the changes of structure of proteins. Therefore, the studies of denaturation gave the important key to make clear the

K. Suzuki (✉)
Emeritus Professor, Department of Chemistry, College of Science and Engineering, Ritsumeikan University, 1-1-1, Nojihigashi, Kusatsu, Shiga 525-8577, Japan
e-mail: hideyuki@kit.ac.jp

K. Akasaka, H. Matsuki (eds.), *High Pressure Bioscience*,
Subcellular Biochemistry 72, DOI 10.1007/978-94-017-9918-8_1

structure of protein. (2) This phenomena were probably in the course of the process that the living matter goes to the nonliving matter, and therefore much were paid from the wide survey of biology. (3) The well-known facts that the temperature coefficients of the rate of thermal denaturation were generally very large compared with those of the usual chemical reactions, for example in ovalbumin the rate increased by a factor of several hundreds with the temperature rise of 10 °C, that the temperature coefficient was negative in the rate of urea denaturation, and that the activation entropy was negative in the acid denaturation of hemoglobin, and so on, were very interesting from the stand point of chemical kinetics (Eyring and Stearn 1939). The above mentioned interests on the denaturation phenomena were common to the studies of the denaturation under pressure, and especially the third interest from the kinetics became the decisive motive to take up this research under pressure, because kinetic treatments were seldom in the field of high pressure.

1.2 The Kinetics of Protein Denaturation by High Pressure: Ovalbumin and Carbonylhemoglobin

The characteristics in the pressure denaturation studies of proteins obtained at early days up to the beginning of about 1960 were summarized by Suzuki (1960). The outlines of this article were reported at the First Congress of the Japan Society of High Pressure Science and Technology held at Kyoto University, on November, 1959. The thermodynamic functions in the activation process of denaturation were calculated from the relations of reaction rates using the results of ovalbumin (Suzuki 1958) which is a most typical globular protein, and carbonylhemoglobin (Suzuki and Kitamura 1960a) which is the representative of chromoprotein to discuss the mechanism of protein denaturation under pressure.

The schematic layout of the compressing apparatus is shown in Fig. 1.1 (Suzuki 1958). This apparatus was used to compress samples at pressures up to 10,000 kg cm^{-2}, and the properties of the compressed samples were measured after releasing pressure. A sample solution charged in a polyvinylchloride sack was placed in the high-pressure chamber A, which was then filled up with pressure transmitting medium (water). The pressure vessel was held in the oil-press B, and the sample was hydrostatically compressed. The pressure in chamber A was estimated from the product of the pressure gauge C reading by the ratio of the cross-sectional area of piston J_2 to that of J_1. The reaction temperature was kept constant by circulating water from the thermostat through the jacket I surrounding the vessel, but owing to large heat capacity and conductance of the apparatus, the temperature in the vessel was often found to be appreciably different from the thermostat. So the temperature in the vessel was measured directly by the thermistor K in a small side-hole of the chamber.

As the measure to examine the denaturation process, after a definite time under compression, the sack containing the test solution prepared at the isoelectric point was taken out and the sample was filtered immediately. Then the protein

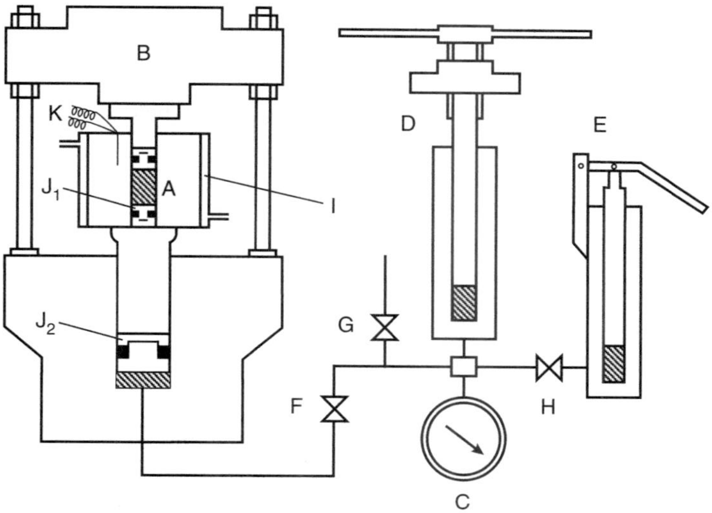

Fig. 1.1 Schematic layout of apparatus: (*A*) high pressure chamber, (*B*) compression apparatus, (*C*) Bourdon type pressure gauge, (*D*) injection pump, (*E*) hand pump, (*F, G, H*) high pressure valves, (*I*) water jacket, (J_1, J_2) self-tightening packing (Bridgman type), (*K*) thermistor (Suzuki 1957, 1958)

concentration of the filtrate containing the native protein was determined by the colorimetric measurement of biuret reaction, corrected by the micro-Kjeldahl analysis of nitrogen by means of an electron- tube-photometer.

Ovalbumin was prepared from hen's egg white by the method of Sörensen and Höyrup (1915), recrystallized three times, dialyzed against water until free from ammonium sulfate, and stored as the stock solution in a refrigerator. This stock was diluted with water and acetate buffer of pH 4.8 to a given protein concentration (usually 0.1 M), and used as a test solution. It was found that the rate of denaturation is very sensitive to the freshness of ovalbumin, even if stored in a refrigerator, so that newly prepared ovalbumin was used within a week. Carbonylhemoglobin; carbon monoxide gas was bubbled through oxyhemoglobin solution to saturation. Oxyhemoglobin; fresh bovine red cells well airated and repeatedly washed with isotonic saline solution, were hemolyzed by addition of distilled water, and then the stroma was removed by centrifuging. The solution thus obtained was stored as a stock solution in ice box of 0 °C and used within 12 h.

The characteristic points are summarized as follows. The rate of denaturation was essentially of the first order with regard to the protein concentration through the whole range of experiment. In Fig. 1.2 the points with a given rate constant (0.1 min^{-1}) are plotted against temperature and pressure in both cases of albumin and hemoglobin (Suzuki 1960). It was found from the figure that the results resembles each other in both proteins, that is, the figures are composed of three parts: AB, BC and CD respectively. It is to be especially noted that the rate of denaturation has the negative temperature coefficient in the part of AB. The behavior

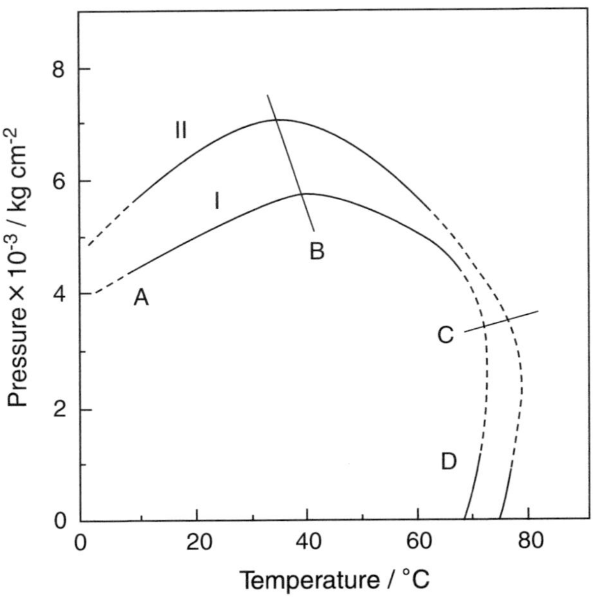

Fig. 1.2 Variations of rate constants of protein denaturation in relation to temperature and pressure. Lines are drawn connecting the points with a rate constant of 0.1 min^{-1}. *Line I*: ovalbumin at pH 4.8. *Line II*: carbonylhemoglobin at pH 6.8 (Suzuki 1960)

of hemoglobin was remarkably different from that of albumin in the reversibility of coagulation. That is, in ovalbumin the coagulated protein caused by pressure did not change in its nature after releasing pressure and standing for some time, while in hemoglobin the coagulation once formed became soluble again. Such a difference offered a key to discuss the structure of both proteins.

The thermodynamic functions in the activation process of denaturation were calculated from the rate constant obtained. The numerical values of thermodynamic functions calculated are omitted here. Considering the signs of the values of $\Delta H^{\ddagger}$, $\Delta S^{\ddagger}$ and $\Delta V^{\ddagger}$, results were to be classified into three regions as shown in Table 1.1. It was to be noted from the table that $\Delta H^{\ddagger}$ changes the sign around 40 °C, and $\Delta V^{\ddagger}$ has the negative value except the case where the retardation in thermal denaturation occurs. $\Delta S^{\ddagger}$ in the region II seemed to change the sign according to the conditions, namely it was supposed $\Delta S^{\ddagger}$ may tend to become negative at the higher pressure and the lower temperature.

The rate of denaturation was of the first order respect to the protein concentration through all over the regions, and so it was found that the intramolecular process governs the rate-determining step. In the region III (>60 °C, <3,000 kg cm^{-2} as shown in Table 1.1) thermal denaturation was retarded by pressure. The values of $\Delta H^{\ddagger}$, $\Delta S^{\ddagger}$ and $\Delta V^{\ddagger}$ were all positive and the very large value of $\Delta S^{\ddagger}$ at the atmospheric pressure decreased with the rise of pressure in this region. It was generally considered to explain the extraordinary large value $\Delta S^{\ddagger}$ in thermal

Table 1.1 Classification into three regions based on the signs of the thermodynamic functions (Suzuki 1960)

	Temperature (°C)	Pressure ($kg\ cm^{-2}$)	$\Delta H^{\ddagger}$	$\Delta S^{\ddagger}$	$\Delta V^{\ddagger}$
I	<30	>4,000	−	−	−
II	>40	>3,000	+	±	−
III	>60	<3,000	+	+	+

denaturation that the disorganization of protein molecule or the unfolding of peptide chains occurs because of the breaking of the subsidiary bridge caused by the thermal vibration. Therefore, the behavior in the region III was to be understood if it was taken into consideration that such a disorderliness of configuration is retarded by pressure. Dilatometer measurements by Heymann (1936) indicated an expansion of 80.3 cm^{-3} per mole (at 80 °C) in the thermal denaturation of ovalbumin at the isoelectric point, and so it was also estimated from the principle of Le Chaterier that pressure behaves to suppress the thermal denaturation. And it was interesting to make a comparison between the values of $\Delta V^{\ddagger}$: 24 cm^{-3} per mole (at 70 °C) obtained in Table 1.1 and the above 80.3 cm^{-3} per mole.

In the region I (<30 °C, >4,000 kg cm^{-2} as shown in Table 1.1) where pressure denaturation only occurs, it was noteworthy that all the values of $\Delta H^{\ddagger}$, $\Delta S^{\ddagger}$ and $\Delta V^{\ddagger}$ become negative. Now, the following mechanism on pressure denaturation was proposed, considering the values of these thermodynamic functions and the facts that the rate was of the first order with regard to the protein concentration, and so on;

$$P + nH_2O \rightleftarrows P(H_2O)_n \tag{1.1}$$

$$P(H_2O)_n \rightarrow P_D \tag{1.2}$$

$$mP_D \rightarrow (P_D)_m \tag{1.3}$$

where P is the native protein; P_D is the denatured protein; $(P_D)_m$ is the coagulated protein; n is the number of the water molecules combined with the protein; m is the number of the denatured protein which forms coagulation. $P(H_2O)_n$ is the protein hydrate which exists always in the equilibrium relation of (1.1) under high pressure and is constructed with the rigid structure having much more water molecules adhered (or frozen) to protein. The rate determining step is the process of (1.2). As the formation of $P(H_2O)_n$ is exothermic and accompanied with the volume decrease, the equilibrium of (1.1) tends to move much more toward the right hand if the temperature is lowered and the pressure increases. Accordingly, the forming rate of denatured protein becomes fast at such conditions (1.3) is the associating process of denatured protein. It was deduced in following way that the rate of denaturation is of the first order with respect to the protein concentration;

$$\mathrm{d}\,[\mathrm{P_D}]\,/\mathrm{d}t = k'_2\,[\mathrm{P(H_2O)}_n] = k'_2 K[\mathrm{H_2O}]^n\,[P] = k'\,[P]\,,$$
$$k' = k'_2 K\,[\mathrm{H_2O}]^n = \text{constant} \tag{1.4}$$

where k_2' is the rate constant in (1.2); K is the equilibrium constant in (1.1); k' is the over-all rate constant which is obtained actually from the experiment.

The problem that $\Delta H^{\ddagger}$, $\Delta S^{\ddagger}$ and $\Delta V^{\ddagger}$ which are deduced from the rate constant k' are negative was considered. Now, it is irrational in itself that activation enthalpy and activation entropy become negative in an elementary process. Therefore, $\Delta H_2{}^{\ddagger}$ and $\Delta S^{\ddagger}$ of $P(H_2O)_n$ in the process of only (1.2) should take positive values respectively. And it was reasonable to assume that the activation volume, $\Delta V_2{}^{\ddagger}$ takes also positive value, because the denatured protein is in the unfolding state. These problems were to be explained in the following way. It is readily known that the relations of $\Delta H_2{}^{\ddagger} = \Delta H^{\ddagger} - \Delta H$, $\Delta S_2{}^{\ddagger} = \Delta S^{\ddagger} - \Delta S$ and $\Delta V_2{}^{\ddagger} = \Delta V^{\ddagger} - \Delta V$ are established, where ΔH, ΔS and ΔV are quantities in (1.1) respectively. Accordingly, $\Delta H_2{}^{\ddagger}$ and $\Delta S_2{}^{\ddagger}$ may be positive, even if the formation of $P(H_2O)_n$ is in exothermic (moreover, $\Delta H > \Delta H^{\ddagger}$) and $P(H_2O)_n$ is in the lower state of entropy (moreover, $\Delta S > \Delta S^{\ddagger}$), because of the rigidness of the structure, according to the denaturation mechanism above described. And it was reasonable to assume such properties of $P(H_2O)_n$.

In the case of $\Delta V_2{}^{\ddagger}$ the same elucidation was applied from the mechanism that the process of (1.1) is accompanied with the volume decreases. It was clear that the volume usually decreases in the solvation, but in this case it was necessary to consider with what mechanism much more plenty of water is adhered to protein and then the volume decreases. The first was such a case that owing to the increase of ionization of the dissociating groups in protein caused by pressure, water molecules are attracted to the ionizing groups by the electro-striction, and then the total volume decreases. Though the volume decrease was about 10 cm^{-3} per mole of ionizing group (Johnson et al. 1954), it seemed this volume change becomes very large in macro-molecules such as proteins, for this volume change is additive. Therefore the increase of pressure favors the process of ionization, and large influences of pressure are expected on the volume contraction of the system.

As the second mechanism, the functions of the water molecule modified by pressure (Johnson et al. 1954; Cuddeback et al. 1953), were to be considered. Owing to its characteristic actions, the water molecule modified by pressure behaves in the analogous way to urea molecule in the urea denaturation in which it was supposed that a great number of urea are adhered to protein (Suzuki 1960; Hopkins 1930), and therefore the total volume decreases, because much more water molecules are adhered (or frozen) to protein molecule.

It seemed to be difficult to decide uniquely which mechanism is more favorable, and which mechanism is more predominant if both mechanisms may participate in the process at the same time. It was, however, supposed that the former mechanism may be predominant at relatively higher temperature, and latter at relatively lower temperature. But from the experimental results obtained by Drickamer et al.

(Cuddeback et al. 1953)[1] on self-diffusion in water and the necessity that $P(H_2O)_n$ must be more thoroughly lower state in enthalpy and entropy to explain the negative values of $\Delta H^{\ddagger}$ and $\Delta V^{\ddagger}$ as above mentioned, the latter mechanism in which is considered the functions of the water molecule modified by pressure seemed to be more favorable than the former. It was, however, common to both mechanisms to have to acknowledge the existence of the singular protein hydrate in the intermediate process to denaturation, though the structures will be not necessarily the same. It was supposed that in the region III (>60 °C, >3,000 kg cm^{-2} as shown in Table 1.1) both denaturation by pressure and heat occur at the same time.

1.3 Reversibility of Pressure Denatured Proteins

We have studied the denaturation of some globular proteins by high-pressure treatment (Suzuki 1957, 1958, 1960; Suzuki and Suzuki 1962; Suzuki and Kitamura 1960a, b; Suzuki et al. 1963a, b, c), and found that the characteristics of denaturation are considerably different in each protein, especially in the case of renaturation. For instance, pressure-coagulated ovalbumin does not dissolve by releasing the pressure. On the other hand, pressure-coagulated serum albumin dissolves completely and quickly. The denatured oxyhemoglobin is intermediate between these two proteins, that is, it is restored partially and slowly after the release of pressure. The ease of renaturation may be a more profound reflection of the molecular configuration of protein than the ease of denaturation, because these three proteins begin to denature almost at the same pressure, 4,000–5,000 kg cm^{-2}. Therefore, it appeared interesting to continue the investigation of pressure denaturation on another protein, paying special attention to the ease of renaturation.

The pressure denaturation of proteins had been studied by several workers, but all results published were obtained on solutions after the pressure was released. However, it has been known that some denatured proteins return to the original state when the denaturant is removed as above mentioned. If the rate of recovery from the denatured state is slow, it may be possible to deduce properties of the protein in the denatured state from the data obtained on the solution after releasing the pressure. On the other hand, if the rate of recovery is fast, or if the recovery from the denatured state takes place instantly when the pressure is released, the measurement must be made directly on the solution under pressure. We measured the properties of pressure-treated horse serum albumin (HSA) solution after releasing the pressure (Suzuki and Kitamura 1961). In that experiment no precipitation was observed even if the isoelectric solution of HSA was compressed up to 10,000 kg cm^{-2}. It was

[1] It was found from the experimental results of self-diffusion in water at 25 °C that the tetrahedrally coordinated structure of the usual water is broken down with the increase of pressure, and a stabilized new structure is formed around 4,000 atm. It was very interesting to note that the pressure denaturation begins at the same pressure region (Cuddeback et al. 1953).

not certain, however, the protein does not denature by compression, or whether the pressure denaturation of serum albumin is reversible.

An additional reason for this investigation was the similarity of the influence of high pressure and of urea on the denaturation process. It was observed that the activation volume and activation entropy of the denaturation reaction are negative in both cases (Suzuki 1960; Simpson and Kauzmann 1953). Moreover, it has been well known that the dielectric constant of solvent water at 5,000 kg cm^{-2} is the same order as that of an aqueous solution of 8 M urea (Scaife 1955; Wyman 1933). These similarities between the effects of pressure and of urea suggested to us that the pressure denaturation and urea denaturation might be similar in character. If this is true, the pressure-denatured protein might be soluble under pressure and coagulate only when the pressure is released, since the protein dissolves in highly concentrated urea solution and precipitates only when solution is dialyzed. These facts required us to overcome the various difficulties in the experimental technique and observe the turbidity of the solution directly under high pressure.

The outlines of this work were presented at the First Congress of the Biophysical Society of Japan held at Osaka, on May 1–3, 1962. It was first necessary to construct a high pressure apparatus with optical windows that could withstand a pressure of several thousand kilograms per square centimeter. The apparatus was designed and constructed after Fishman and Drickamer (1956) as shown in Fig. 1.3. The overall dimensions of the apparatus were 32 cm in height, 13.6 cm in diameter. The light of a tungsten lamp struck a photoelectric cell after passing through a glass filter and the protein solution. The light intensity was measured by an ammeter.

Ovalbumin and HSA were crystallized with ammonium sulfate from hen's egg white and horse blood plasma, respectively. The substances were recrystallized three times, and then dialyzed against running water. The pH of the protein solution was adjusted to the isoelectric point, pH 4.8, with acetate buffer solution, the final concentration being 0.1 M.

The results on ovalbumin are shown in Figs. 1.4 and 1.5. The pressure was increased in step up to 6,000 and 4,500 kg/cm^{-2}, respectively, and then decreased in steps down to atmospheric pressure. The turbidity of the solution was measured at each step. In these figures, the reading of the ammeter which corresponds to the light intensity is plotted against time. It was seen that the intensity of the light decreases sharply when the protein is compressed over 4,000 kg cm^{-2}, indicating that the denatured protein aggregates to give a turbid solution. It was also seen that the ovalubumin solution does not only remain turbid after the release of pressure, but that the molecules continue to aggregate at low pressure (see Figs. 1.4 and 1.5) though this additional reaction may be due to the slowness of the process. In other words, the pressure denaturation of ovalbumin was irreversible.

The result on HSA is shown in Fig. 1.6. It was found that the light intensity begins to decrease rapidly at a pressure of 5,000 kg cm^{-2} and quickly returns to that of the original solution if the pressure is decreased below 4,000 kg cm^{-2}. This result showed that the denatured protein molecules aggregate to give a turbid solution under a pressure above 5,000 kg cm^{-2} as in the case of ovalbumin, but the turbidity

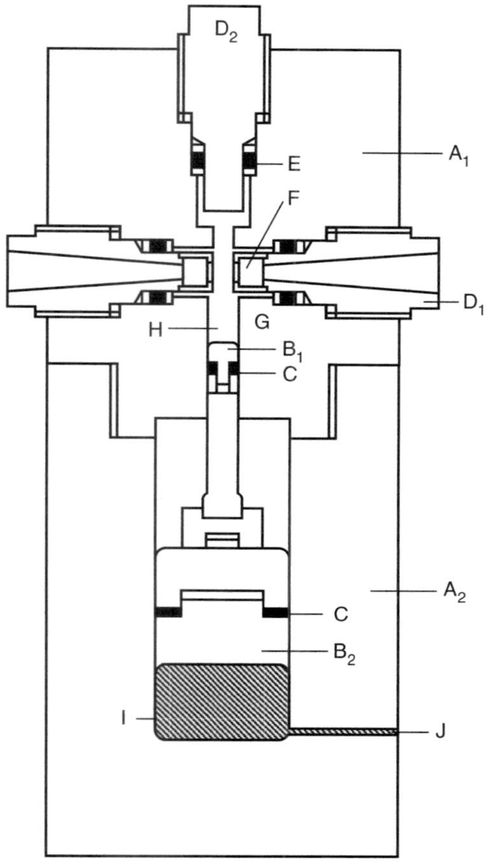

Fig. 1.3 High-pressure apparatus with optical windows: (A_1) cylinder of high pressure part (AISI-4340, H_{RC} 48–50), (A_2) cylinder of low-pressure part (mild steel), (B_1) piston of Bridgman type in high-pressure part (SKD-3, H_{RC} 58–60), (B_2) piston of Bridgman type in low-pressure part (mild steel), (*C*) rubber packing, (D_1) plug with optical windows (SKD-3, H_{RC} 50), (D_2) plug (SKD-3, H_{RC} 50), (*E*) packing composed of rings of copper, lead, and steel, (*F*) optical window (12 mm, $\phi \times 10$ mm, sapphire), (*G*) cap to support optical window (18–8 N-Cr), (*H*) chamber to charge sample solution, (*I*) pressure transmitting medium (machine oil), (*J*) to oil-injection pump (Suzuki et al. 1963c)

of the solution disappears absolutely when the pressure is released. The behavior of turbidity of serum albumin was quite different from that of ovalbumin, that is, the pressure denaturation of serum albumin was reversible.

The measurements of the turbidity of HSA and ovalbumin solutions suggested that at pH 4.8 HSA molecules apparently denature under pressure above several 1,000 km/cm^{-2}, and moreover the aggregation of ovalbumin molecules denatured by high pressure is reversible by releasing pressure. Therefore, the pressure denaturation of HSA could only be studied if the measurement was carried out directly under pressure. The expectation that the denatured molecules might be soluble under pressure, based on an analogy to the urea denaturation, proved to be false. At the isoelectric point, the denatured molecules of ovalbumin and HSA are soluble in urea. It's more beautiful and fresh than the picture of molecular dynamics ion, while they are insoluble under high pressure. However, the reversibility of the denaturation of serum albumin and the irreversibility of ovalbumin were similar in urea denaturation (Suzuki et al. 1963c) and pressure denaturation. Probably, this analogy was not caused by the similarity of denaturants, but by the properties of the proteins themselves.

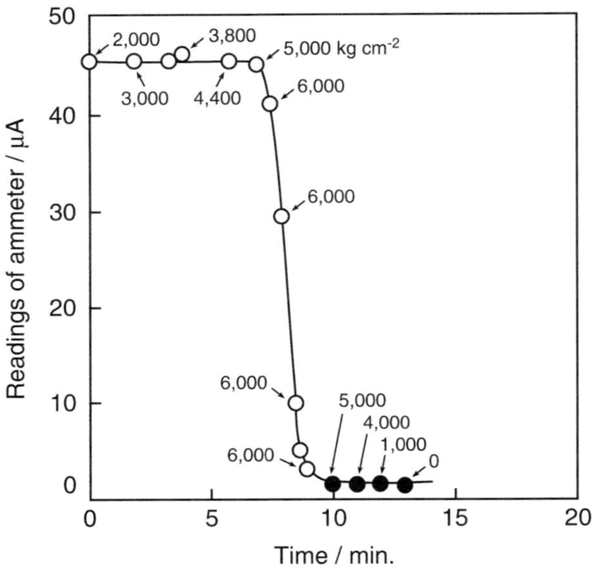

Fig. 1.4 Change of light intensity passing through the ovalbumin solution on progressing pressure ovalbumin upward -○- and downward -●- at 32 °C, pH 4.8 (0.1 M acetate buffer), protein concentration 4 mg ml^{-1} (Suzuki et al. 1963c)

1.4 Pressure Renaturation of Coagulated Proteins: γ-Globulin

Johnson and Campbell (1946) studied the effect of pressure on the heat denaturation of serum γ-globulin: however, the pressure denaturation of this protein had not been investigated. Tongur and Kasatochkin (1954) reported the interesting result that heat-denatured γ-globulin was renatured by compression at a moderately high pressure of about 1,000 kg cm^{-2}. The denaturation and the renaturation of bovine serum γ-globulin under high pressure using solubility as a measure of denaturation are described below.

Bovine serum γ-globulin was purchased from Boehringwerke, Germany. According to the electrophoretic analysis of pH 7.6 solution, its purity was about 90 % and two other fractions except γ-globulin were observed. Isoelectric protein solution (0.01 M phosphate buffer of pH 6.0, protein concentration 0.5–1.0 %) in 0.25 M sodium chloride was used in all denaturation studies, except in the experiment of pH. Pressure-treated samples used for renaturation experiments were prepared by compression of the isoelectric γ-globulin solution, to yield a suspension of the denatured protein. The pressure-treated sample was used as rapidly as possible.

Two kinds of pressure apparatus were used: one for experiments of denaturation (Fig. 1.1), and the other for experiments of renatration (Fig. 1.7). With these instruments the analysis of denaturation was done after the release of pressure.

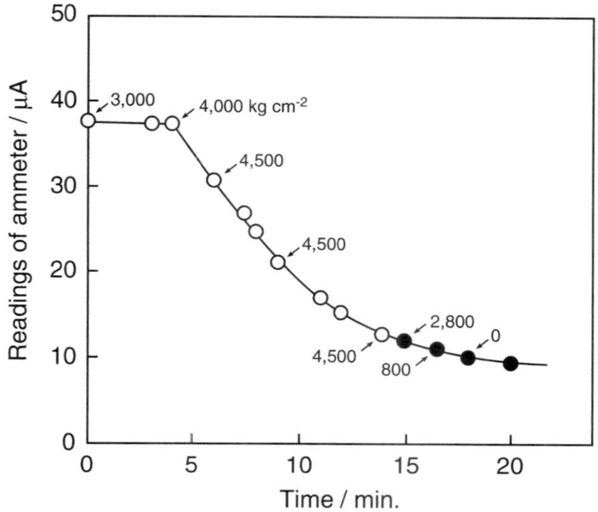

Fig. 1.5 Change of light intensity passing through the ovalbumin solution on progressing pressure upward -O- and downward -●- at 32 °C, pH 4.8 (0.1 M acetate buffer), protein concentration 5 mg ml^{-1} (Suzuki et al. 1963c)

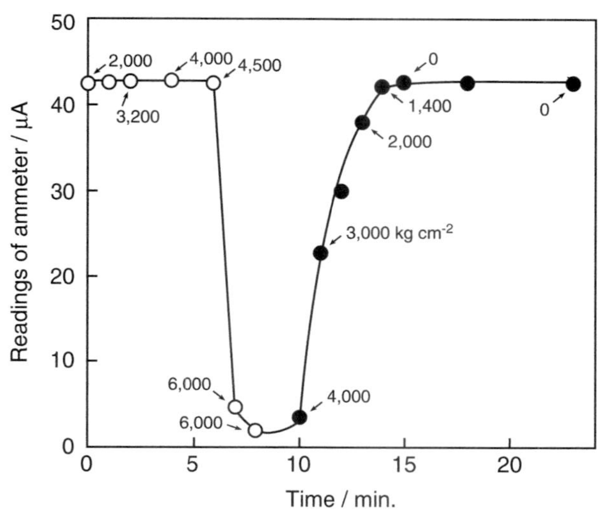

Fig. 1.6 Change of light intensity passing through the horse serum albumin solution on progressing pressure upward -O- and downward -●- at 31 °C, pH 4.8 (0.1 M acetate buffer), protein concentration 6 mg ml^{-1} (Suzuki et al. 1963c)

As an index of denaturation, the change in solubility at the isoelectric point was measured following the pressure treatment. After filtration of the coagulated protein, the quantity of soluble protein remaining in the filtrate was colorimetrically

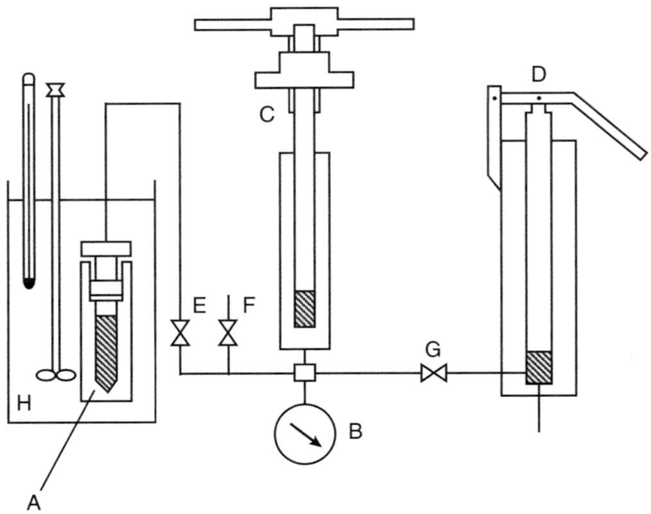

Fig. 1.7 Schematic layout of moderate pressure apparatus: (*A*) reaction vessel, (*B*) Bourdon-type pressure gauge, (*C*) injection pump, (*D*) hand pump, (*E*, *F*, *G*) high pressure valves, (*H*) thermostat (Miyagawa et al. 1964)

measured at the wave length of 530 nm by the biuret reaction with a spectrophotometer. Other physico-chemical properties of a protein solution except solubility were examined in order to observe the differences between the soluble protein after the pressure treatment and the native protein. In addition the differences between the completely renatured protein, (from solubility), and the native protein were studied. The ultraviolet absorption spectrum was obtained with a spectrophotometer, and the optical rotation was measured with a polarimeter, The viscosity measurement was done with an Ostwald type viscometer, and the susceptibility to enzymatic proteolysis was examined by the colorimetric proteolysis of the digested products at 660 nm using Folin reagent. The outlines of this study were summarized as follows.

The denaturation and the renaturation of bovine serum γ-globulin under high pressure were studied using solubility as a measure of denaturation. The coagulation of isoelectric γ-globulin was initiated by compression around 4,000 kg cm^{-2} and completed by compression for 5 min at 8,000 kg cm^{-2}. The depression effect of pressure on the heat denaturation at 70 °C had a maximum at about 2,000 kg cm^{-2}. The temperature coefficient of pressure denaturation was positive up to 50 °C. The rate of pressure coagulation was more rapid when the pH of the protein solution (pH 5–10) was lower, and the initial protein concentration was higher.

The most striking result differing from ovalbumin and hemoglobin was the fact that the amount of soluble protein in this final state depended upon the magnitude of pressure and temperature. Some physico-chemical properties of the soluble protein were different from those of the native protein, although the differences were small.

It might be assumed that there was an equilibrium between soluble protein and coagulated protein, although the reverse process (dissolution) was not so readily established.

The dissolution of partially pressure-coagulated γ-globulin was remarkably accelerated by recompression at moderately high pressure, and the rate increased with temperature up to 40 °C. Some physico-chemical properties of completely dissolved protein were assumed not to be the same as those of the native protein. It was not possible to cause the completely pressure-coagulated sample (100 % denatured) to go into solution. However, partially pressure-coagulated γ-globulin dissolved more rapidly than heat- coagulated γ-globulin.

1.5 Pressure Inactivation of an Enzyme: α-Amylase

The inactivation of enzymes by high pressure was first investigated in 1932 by Basset and Macheboeuf (1932). Afterwards, there appeared some elaborate works on purified enzyme by Matthews et al. (1940), and Curl and Jansen (1950a, b). Kinetic treatments, however, were very few, and the detail of pressure inactivation was still remained to be clarified.

It is generally accepted that the biological activity of enzyme-protein arises from the active site located at some part of the conformation of the molecule. Since the protein denaturation is a phenomenon due to the change in conformation of the molecule, it seemed to be interesting to see if the kinetics of pressure inactivation of enzyme would be similar to those of pressure denaturation of globular proteins above mentioned (Suzuki 1960), and to be worthwhile to investigate the loss in biological activity and changes in some other properties by the denaturation of enzyme molecule in parallel. On the other hand, activation volume in the activation process of enzymes by heat and acid etc. can be obtained by examining the effects of a relatively low pressure, which cannot cause the denaturation, on the rates of their inactivation process, and then those values may afford some information on the changes of the conformation by these denaturants which cause the change in biological activity.

The outlines of pressure inactivation of various kind of enzymes investigated by us in early days were reported at the 1st Congress of the Biophysical Society of Japan held at Osaka, on May 1–3, 1962. Our first publication on this field "Studies on the Kinetics of Inactivation of α-amylase of *Bacillus subtilis* under High Pressure" (Suzuki and Kitamura 1963) was summarized as follows.

1. The change in biological activity and the change in susceptibility to enzymatic proteolysis of α-amylase of *Bacillus subtilis* by the pressure up to 10,000 kg cm^{-2} were investigated.
2. The rate of inactivation becomes progressive at the higher pressure than about 5,000 kg cm^{-2} at pH 5.8, and is the faster in the more acidic solution if examined in the range between pH 4.8 and 8.0. Calcium ion markedly inhibits the pressure inactivation.

3. The rate of pressure inactivation is of first order. The values of activation volume and activation entropy were negative as well as in the pressure denaturation of some globular proteins.
4. The change in biological activity and the change in susceptibility to enzymatic proteolysis coincides with each other within experimental error.
5. In the view of the results in 3 and 4, especially in 4, it is confirmed that the inactivation of this enzyme by the pressure treatment must arise from denaturation of the protein molecule, that is, the change of conformation of the protein molecule as a whole.
6. The moderate pressure below 1,000 kg/cm^{-2} retards the heat inactivation but accelerates acid inactivation, that is, the activation volume in the former is positive, and that in the latter is negative.

Other examples were reported by us in the early days of 1960 (Miyagawa (1965), Miyagawa and Suzuki (1963a, b), Miyagawa and Suzuki (1964), Miyagawa et al. (1964)).

References

Basset J, Macheboeuf MA (1932) Etudes aur les effets biologiques des ultra-pressions: Résistance des bactéries, des diastases et des toxines aux pressions très élevées. C R Acad Sci 195:1431–1433

Cuddeback RB, Koeller RC, Drickamer HG (1953) The effect of pressure on diffusion in water and in sulfate solutions. J Chem Phys 21:589

Curl AL, Jansen EF (1950a) Effect of high pressures on trypsin and chymotrypsin. J Biol Chem 184:45–54

Curl AL, Jansen EF (1950b) The effect of high pressures on pepsin and chymotrypsinogen. J Biol Chem 185:713–723

Eyring H, Stearn AE (1939) The application of the theory of absolute reaction rates to proteins. Chem Rev 24:253–270

Fishman E, Drickamer HG (1956) Equipment for high pressure optical and spectroscopic studies. Anal Chem 28:804–806

Heymann E (1936) Dilatometric investigations on the heat-denaturation of proteins. I. Biochem J 30:127–131

Hopkins FG (1930) Denaturation of proteins by urea and related substances. Nature 126:328–330, 383–384

Johnson FH, Campbell DH (1946) Pressure and protein denaturation. J Biol Chem 163:689–698

Johnson FH, Eyring H, Polissar MJ (1954) The kinetic basis of molecular biology. Wiley, New York, p 303

Matthews JE, Dow RB, Anderson AE (1940) The effects of high pressure on the activity of pepsin and rennin. J Biol Chem 135:697–705

Miyagawa K (1965) Studies on Taka-amylase A under high pressure: III. Some physicochemical properties of pressure-denatured amylase A. Arch Biochem Biophys 110:381–387

Miyagawa K, Suzuki K (1963a) Pressure inactivation of enzyme: some kinetic aspects of pressure inactivation of trypsin. Rev Phys Chem Jpn 32:43–50

Miyagawa K, Suzuki K (1963b) Pressure inactivation of enzyme: some kinetic aspects of pressure inactivation of chymotrypsin. Rev Phys Chem Jpn 32:51–56

Miyagawa K, Suzuki K (1964) Studies on Taka-amylase A under high pressure: 1. Some kinetic aspects of pressure inactivation of Taka-amylase A. Arch Biochem Biophys 105:297–302

Miyagawa K, Sannoe K, Suzuki K (1964) Studies on Taka-amylase A under high pressure treatment: II. Recovery of enzymic activity of pressure inactivated Taka-amylase A and its enhancement by retreatment at moderate pressure. Arch Biochem Biophys 106:467–474

Scaife BKP (1955) Isothermal pressure dependence of the dielectric properties of eugenol, glycerol and water. Proc Phys Soc B68:790–792

Simpson RB, Kauzmann W (1953) The kinetics of protein denaturation. I. The behavior of the optical rotation of ovalbumin in urea solutions. J Am Chem Soc 75:5139–5152

Sörensen SPL, Höyrup M (1915) On the preparation of egg-albumin solutions of well-defined composition, and on the analytical methods used. C R Trav Lab Carlsberg 12:12–67

Suzuki K (1957) Denaturation of egg albumin under high pressure. Mem Res Inst Sci Eng Ritsumeikan Univ 2:19–25

Suzuki K (1958) Studies on the denaturation of egg albumin under high pressure. Rev Phys Chem Jpn 28:24–30

Suzuki K (1960) Studies on the kinetics of protein denaturation under high pressure. Rev Phys Chem Jpn 29:91–98

Suzuki K, Kitamura K (1960a) Denaturation of hemoglobin under high pressure, II. Rev Phys Chem Jpn 29:86–90

Suzuki K, Kitamura K (1960b) Denaturation of hemoglobin under high pressure, I. Rev Phys Chem Jpn 29:81–85

Suzuki K, Kitamura K (1961) Presented at the 14th meeting of Japanese Chemical Society. Tokyo, April 1961

Suzuki K, Kitamura K (1963) Inactivation of enzyme under high pressure: studies on the kinetics of inactivation of α-amylase of bacillus subtilis under high pressure. J Biochem 54:214–219

Suzuki C, Suzuki K (1962) The protein denaturation by high pressure: changes of optical rotation and susceptibility to enzymic proteolysis with ovalbumin denatured by pressure. J Biochem 52:67–71

Suzuki C, Kitamura K, Suzuki K, Osugi J (1963a) The protein denaturation under high pressure: horse serum albumin. Rev Phys Chem Jpn 32:30–36

Suzuki C, Kitamura K, Suzuki K, Osugi J (1963b) The protein denaturation under high pressure: effects of pH and some substances on the pressure denaturation of ovalbumin solution. Rev Phys Chem Jpn 32:37–42

Suzuki K, Miyosawa Y, Suzuki C (1963c) Protein denaturation by high pressure. Measurements of turbidity of isoelectric ovalbumin and horse serum albumin under high pressure. Arch Biochem Biophys 101:225–228

Tongur VS, Kasatochkin VI (1954) Protein regeneration under pressure. The kinetics and thermodynamics of the process. Trudy Vsesoyuz Obshchestva Fiziologov, Biokhimikov i Farmakologov Akad Nauk SSSR 2:166–176

Wyman J (1933) Dielectric constants: ethanol-diethyl ether and urea-water solutions between 0 and 50 °. J Am Chem Soc 55:4116–4121

Chapter 2
Protein Denaturation on *p-T* Axes – Thermodynamics and Analysis

László Smeller

Abstract Proteins are essential players in the vast majority of molecular level life processes. Since their structure is in most cases substantial for their correct function, study of their structural changes attracted great interest in the past decades. The three dimensional structure of proteins is influenced by several factors including temperature, pH, presence of chaotropic and cosmotropic agents, or presence of denaturants. Although pressure is an equally important thermodynamic parameter as temperature, pressure studies are considerably less frequent in the literature, probably due to the technical difficulties associated to the pressure studies.

Although the first steps in the high-pressure protein study have been done 100 years ago with Bridgman's ground breaking work, the field was silent until the modern spectroscopic techniques allowed the characterization of the protein structural changes, while the protein was under pressure.

Recently a number of proteins were studied under pressure, and complete pressure-temperature phase diagrams were determined for several of them. This review summarizes the thermodynamic background of the typical elliptic *p-T* phase diagram, its limitations and the possible reasons for deviations of the experimental diagrams from the theoretical one. Finally we show some examples of experimentally determined pressure-temperature phase diagrams.

Keywords High pressure • Intrinsically disordered protein • Lysozyme • Myoglobin • Parvalbumin • Phase diagram • Ribonuclease • Rv3221c • Staphylococcal nuclease • Titin

2.1 Introduction

Proteins are one of the major constituents of the living matter. Since their structure is in most cases essential for their correct function, study of protein structure and dynamics attracted a lot of interest in the past decades. It is clear that the

L. Smeller (✉)
Department of Biophysics and Radiation Biology, Semmelweis University, Budapest, Hungary
e-mail: smeller.laszlo@med.semmelweis-univ.hu

K. Akasaka, H. Matsuki (eds.), *High Pressure Bioscience*,
Subcellular Biochemistry 72, DOI 10.1007/978-94-017-9918-8_2

three-dimensional structure of proteins is influenced by several physicochemical factors including temperature, pH, presence of chaotropic and cosmotropic agents, or presence of denaturants. All these are widely studied, but pressure, although it is another important thermodynamic parameter, is used in a significantly smaller amount of scientific publications. None of the physicists or chemists would treat pressure as less significant parameter than temperature, but the technical difficulties of pressure experiments could have prevented many scientists from the pressure studies.

Although Bridgman (1914), made the first steps in the high-pressure protein study 100 years ago the field was silent for a half of a century, and the research in the field started again with the seminal works of Kauzmann, Taniguchi and Suzuki (Suzuki 1960; Zipp and Walter 1973; Taniguchi and Suzuki 1983). The development and wide spread of the physical-biophysical methods usable to assay protein structure gave a renaissance to high-pressure protein research. In the meantime several practical applications appeared: pressure treatment of biological systems, like microorganisms and complex food systems were studied and some of these studies led to the development of products which were successfully commercialized (Sasagawa et al. 2005; Yamakura et al. 2005; Rastogi et al. 2007; Yaldagard et al. 2008; Knorr et al. 2011). Simultaneously more and more proteins were experimentally investigated and characterized under pressure as part of basic research projects (Brandts et al. 1970; Hawley 1971; Panick et al. 1999; Smeller et al. 1999; Lassalle et al. 2000; Meersman et al. 2002, 2005; Smeller 2002, 2009; Maeno et al. 2009; Somkuti et al. 2012, 2013a, b).

These achievements on both pure and applied scientific level required the clear thermodynamic description of the underlying processes, among them the description of the protein denaturation under effect of various thermodynamic parameters like pressure and temperature. The first attempt to give a simple but consistent picture was presented by Hawley describing the unfolding *p-T* diagram of chymotrypsin (Hawley 1971). He also analyzed the data obtained on ribonuclease-A by Brandts (Brandts et al. 1970). Since then a number of proteins were studied under pressure, and the complete *p-T* phase diagrams have been determined for several of them (Zipp and Walter 1973; Panick et al. 1999; Lassalle et al. 2000; Meersman et al. 2005; Smeller 2002; Maeno et al. 2009; Somkuti et al. 2011, 2012, 2013a). This review summarizes the present knowledge about the *p-T* phase diagram of protein unfolding.

2.2 The Thermodynamic Picture

A typical protein consisting of about 100 residues has several hundred degrees of freedom due to possible rotation around two bonds in each residue. This results in an astronomically high number of possible conformations while generally only few of them are biologically active. Finding these conformations can be described by the folding funnel, which clearly represents the nonrandom, free energy driven nature of the folding process (Fig. 2.1) (Dill and Chan 1977; Bryngelson et al. 1995).

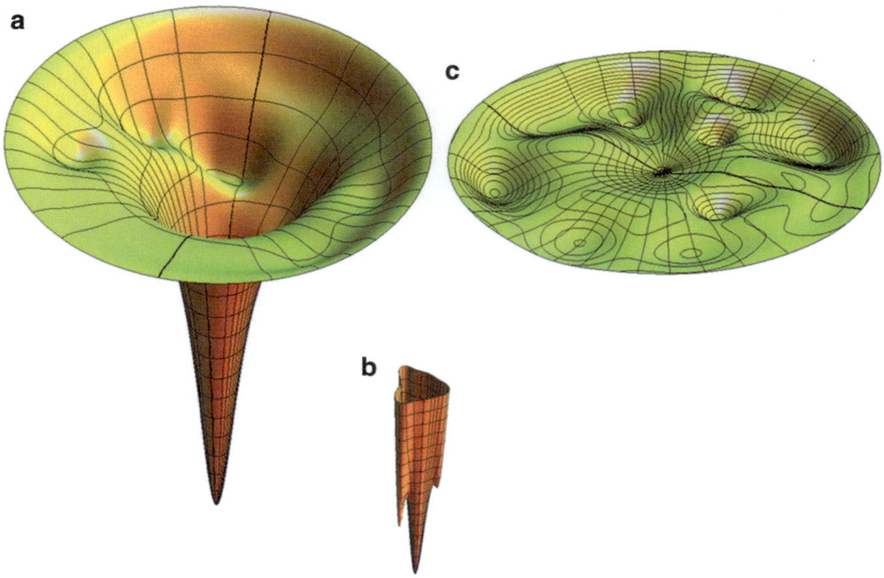

Fig. 2.1 The folding funnel which explains the directed nature of the folding process. (**a**) under conditions where the native state of the protein is thermodynamically stable, (**b**) the enlarged view of the bottom of the funnel shows several low lying conformations, (**c**) the same energy landscape under denaturing conditions

The protein folding is a complex phenomenon (Osvath et al. 2003, 2009; Gruebele 2009). Although some proteins fold co-translationally or quickly after being synthesized, there are some which face the danger of misfolding, if they are not assisted by chaperone. The folding funnel can have a very rugged surface allowing the accumulation of several long lived intermediate structures. Folding of the so called downhill folder proteins is fast and simple; no partially folded states appear on the folding pathway (Liu and Gruebele 2008). In this case the folding and the unfolding of the protein can be treated as simple two state processes:

$$N \leftrightarrow U$$

where N and U refer to the native and unfolded states respectively.

2.2.1 *Thermodynamic Description of the Two State Folding: The Elliptic Phase Diagram*

If we can treat the folding-unfolding process as a simple two-state one, the thermodynamic description is based on the Gibbs free energy difference between the native and unfolded states (Smeller 2002; Hawley 1971):

$$\Delta G = G_U - G_N \tag{2.1}$$

The Gibbs free energies of all states of the protein are pressure and temperature dependent. Treatment of these dependencies can be simplified by a Taylor expansion of the Gibbs free energy. Choosing an appropriate reference point (T_0, p_0), and including up to second order terms, we obtain:

$$\begin{aligned}\Delta G = \Delta G_0 &+ \left(\frac{\partial \Delta G}{\partial T}\right)_0 (T - T_0) + \frac{1}{2}\left(\frac{\partial^2 \Delta G}{\partial T^2}\right)_0 (T - T_0)^2 \\ &+ \left(\frac{\partial \Delta G}{\partial p}\right)_0 (p - p_0) + \frac{1}{2}\left(\frac{\partial^2 \Delta G}{\partial p^2}\right)_0 (p - p_0)^2 \\ &+ \left(\frac{\partial^2 \Delta G}{\partial T \partial p}\right)_0 (T - T_0)(p - p_0)\end{aligned} \tag{2.2}$$

where $\Delta G_0 = \Delta G(T_0, p_0)$, and all the derivatives have to be calculated at the reference point (T_0, p_0). The derivatives of ΔG can be written using thermodynamic parameters, like volume (V), compression (κ), entropy (S), isobaric heat capacity (C_p) and thermal expansivity (α), according to the following thermodynamic definitions:

$$\begin{aligned} -\left(\frac{\partial \Delta G}{\partial T}\right)_p = \Delta S \qquad \left(\frac{\partial \Delta G}{\partial p}\right)_T = \Delta V \\ -\left(\frac{\partial^2 \Delta G}{\partial T^2}\right)_p = \frac{\Delta C_p}{T} \qquad -\left(\frac{\partial^2 \Delta G}{\partial p^2}\right)_T = \Delta \kappa \qquad \left(\frac{\partial^2 \Delta G}{\partial T \partial p}\right) = \Delta \alpha \end{aligned} \tag{2.3}$$

The free energy difference can be now written as:

$$\begin{aligned}\Delta G = \Delta G_0 - \Delta S_0 (T - T_0) - \frac{\Delta C_\mathrm{p}}{2T_0}(T - T_0)^2 \\ + \Delta V_0 (p - p_0) - \frac{\Delta \kappa}{2}(p - p_0)^2 + \Delta \alpha (T - T_0)(p - p_0) + \ldots\end{aligned} \tag{2.4}$$

where Δ refers to the difference of the value in unfolded and native state (similarly to Eq. 2.1). The subscript 0 indicates that the value has to be taken at the reference point (T_0, p_0). The truncation of the series at the second order terms means that we neglect the temperature and pressure dependence of the second order derivatives, i.e. C_p, κ and α are independent of T and p. This is why they do not have a zero index.

Compressibility is defined by:

$$\beta = -\frac{1}{V}\left(\frac{\partial V}{\partial p}\right)_T \tag{2.5}$$

It has to be noted, that an expression similar to the right hand side of Eq. 2.4 can be obtained for ΔG if one starts to integrate the changes in ΔG starting from the

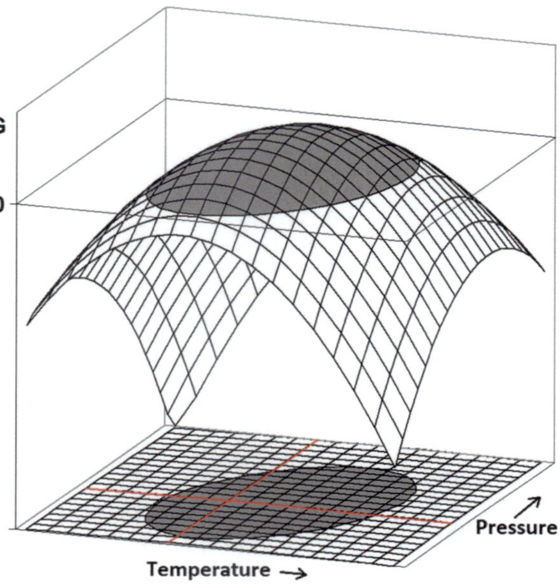

Fig. 2.2 Gibbs free energy of unfolding as a function of pressure and temperature. The *thick ellipse* indicates the transition curve where DG = 0. Inside the ellipse the native, outside the unfolded state is stable

(T_0, p_0) point until any arbitrary (T, p) point. The only difference is in the third term of the right hand side of Eq. 2.4, which is given by:

$$-\Delta C_p \left[T \left(\ln \frac{T}{T_0} - 1 \right) + T_0 \right] \tag{2.6}$$

This can be approximated by the third term of Eq. 2.4 if $(T - T_0)/T_0 << 1$.

From mathematical point of view Eq. 2.4 describes a dome-shaped surface as function of T and p as shown in Fig. 2.2. The boundary of the native phase can be obtained by:

$$\Delta G\,(T_t, p_t) = 0 \tag{2.7}$$

where (T_t, p_t) is an arbitrary point on the boundary. Crossing this transition boundary one can observe unfolding or folding of the protein. Combining Eqs. 2.4 and 2.7 one gets the quadratic equation for a conic section. The curve can be ellipse, hyperbola or parabola, depending on the discriminant of the equation. In the observed cases of proteins, the boundary of the native state was always an ellipse. The requirement for the elliptic shape is:

$$\Delta\alpha^2 - \Delta\kappa\Delta C_p/T_0 < 0 \tag{2.8}$$

Further requirement to have a real (and not imaginary) ellipse is:

$$-Det/\left(\Delta\kappa + C_p/T_0\right) < 0 \tag{2.9}$$

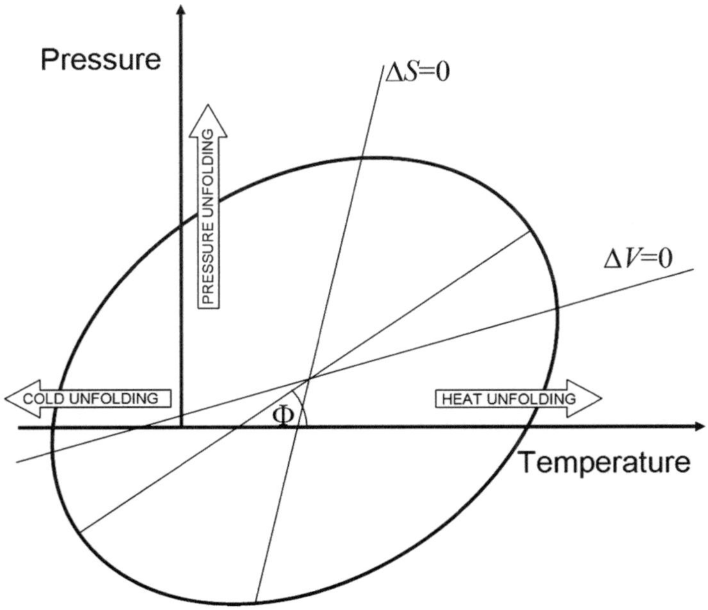

Fig. 2.3 The elliptic phase diagram

where:

$$Det = \begin{vmatrix} -\frac{\Delta C_p}{2T_0} & \frac{\Delta\alpha}{2} & -\frac{\Delta S_0}{2} \\ \frac{\Delta\alpha}{2} & -\frac{\Delta\kappa}{2} & \frac{\Delta V_0}{2} \\ -\frac{\Delta S_0}{2} & \frac{\Delta V_0}{2} & \Delta G_0 \end{vmatrix}$$

These requirements seem to be fulfilled for the protein solutions studied so far. The experimental transition points can be well fitted with an elliptic curve in a number of published cases (Brandts et al. 1970; Hawley 1971; Panick et al. 1999; Lassalle et al. 2000; Maeno et al. 2009; Somkuti et al. 2013a). If the $\Delta G(T, p)$ function follows a dome shape, like the one in Fig. 2.2, the elliptic boundary surrounds the region, where the protein is native, while outside of the elliptic region it is unfolded (see Fig. 2.3).

The central point (T_c, p_c) of the ellipse is given by the solution of the following system of linear equations:

$$\begin{aligned} -\frac{\Delta C_p}{T_0}T_c + \Delta\alpha p_c - \Delta S_0 = 0 \\ \Delta\alpha T_c - \Delta\kappa p_c + \Delta V_0 = 0 \end{aligned} \tag{2.10}$$

More interestingly, the angle Φ between the temperature axis and the principal axis of the ellipse (Fig. 2.3) is given by the following equation:

$$tg2\Phi = -\frac{2\Delta\alpha}{\Delta C_p/T_0 - \Delta\kappa} \tag{2.11}$$

This equation emphasizes the role of the thermal expansivity ($\Delta\alpha$) in the determination of the direction of the ellipse.

The elliptic diagram in Fig. 2.3 makes it clear that the usually observed heat unfolding and also the pressure and the cold unfolding are only three special cases from the infinitely many possibilities to cross the elliptic boundary and unfold the protein. A number of papers tried to explore and compare these three specific unfolding ways. Cold unfolding of proteins was described and its thermodynamics has been analyzed in details by Privalov (1990, 1997). It was however observed only for few proteins, since the cold unfolding temperature often falls below the freezing point of water. This problem was overcome by high pressure experimentalists, by using a subdenaturing pressure, where the freezing point of the water could be lowered by ca. 20 °C (at 200 MPa). This trick is called pressure assisted cold unfolding. Cold denaturations of ribonuclease A (Zhang et al. 1995) myoglobin (Meersman et al. 2002, 2005) and ubiquitin (Kitahara and Akasaka 2003) were measured using this method.

An important question is, whether the unfolded states are different if the unfolding path is different? Are there distinct heat-denatured, pressure denatured and cold denatured states? A clear proof would be if one could observe a phase transition when the cold or pressure denatured protein is heated above the heat denaturation temperature under sufficiently high pressure to keep it in the unfolded state, and subsequently it is cooled to the atmospheric pressure (reaching the heat denaturation conditions) without crossing the elliptic boundary. No such transition was ever detected. This does not necessarily mean the identity of these unfolded states, it indicates that the transition between them is smooth if it exists. We can recall a well-known example from the physical chemistry of water. It is possible to go smoothly from water to vapor (without phase transition) if the path goes above the critical point. Similar smooth conformational drift can also happen within the unfolded ensemble of polypeptide chains in the denatured phase.

From a thermodynamic point of view the driving force in case of the cold, heat and pressure denaturations is different. In Fig. 2.3 the lines for $\Delta V = 0$ and $\Delta S = 0$ are also shown. Their position on the phase diagram makes clear that the pressure and cold unfolding are driven mainly by ΔV while in case of heat unfolding ΔS plays the most important role. This was confirmed by NMR experiments on Ribonuclease A, where the protection factors for the H/D exchange were similar for the cold and the pressure unfolded states. These states were also found to be more compact than the heat unfolded structure (Nash et al. 1996).

The above theory gives a very general description. It does not consider any specific feature of the protein, thus any two-state system which fulfills Eq. 2.8 will show an elliptic phase diagram. There are examples in the literature of the liquid crystals for similar elliptic phases, which are called there as "re-entrant" phases (Cladis 1988).

2.2.2 *Limitations of the Simple Thermodynamic Theory*

The elliptic shape of the phase diagram is the consequence of the fact that we stopped the series expansion at the second order terms. This means physically, that the coefficients of the second order terms (i.e. ΔC_p, $-\ \Delta\kappa$, $\Delta\alpha$) are temperature and pressure independent. If this is not true, the third order terms proportional to $(T - T_0)^3$, $(T - T_0)^2(p - p_0)$, $(T - T_0)(p - p_0)^2$ and $(p - p_0)^3$ have to be taken into account. Several experimental works found pressure or temperature dependence of the above coefficients, e.g. pressure dependence of C_p was reported in the case of ribonuclease A (Yamaguchi et al. 1995).

In our earlier work we pointed out that the third order terms do not cause significant alteration from the elliptic shape, if the T and p dependencies of ΔC_p, $-\ \Delta\kappa$ and $\Delta\alpha$ are not too large (Smeller 2002; Smeller and Heremans 1997). This could be the reason for the success of the above theory, since the small deviations do not influence the fitting of the experimental values with the elliptic boundary.

Further significant limitation to the use of the elliptic phase diagram could be the existence of intermediate states on the folding pathway. This results in the existence of metastable states: in certain regions of the phase diagram two or even more structures can appear depending on the history of the sample.

Another important issue, which is not taken into account in the above simple theory, is the appearance of the intermolecular interactions in the denatured and/or intermediate states. This can lead to misfolding or to formation of amorphous or fibrous aggregates, depending on the physicochemical environment of the polypeptide chain. The fibrous aggregates are of interest in the medical research, since such insoluble protein aggregates are associated with a number of neurodegenerative diseases, like Alzheimer's disease, type II diabetes and the transmissible spongiform encephalopathies (Stefani and Dobson 2003; Meersman and Dobson 2006). Amorphous aggregates are however everyday features in the food industry or in the kitchen. The simplest example for this is the heat denaturation, where a majority of the proteins forms amorphous aggregates. It has been shown that the amorphous and the fibrous aggregates have drastically different pressure stability (Dirix et al. 2005). Since aggregation is a slow and multistep process (Smeller et al. 2004, 2006, 2008), its description using methods of the equilibrium thermodynamics is very difficult.

2.3 Experimentally Determined Phase Diagrams

The first important step in this field was done by Hawley (1971), who introduced the elliptic phase diagram and experimentally verified his concept on chymotrypsinogen. He also used the data measured by Brandts (Brandts et al. 1970) on ribonuclease, to support his elliptic theory. Since that a number of pressure-temperature phase diagrams were recorded. Here we discuss some examples, without attempting to be exhaustive.

2.3.1 Myoglobin

The phase diagram of myoglobin was first studied by Zipp and Kauzmann (Zipp and Kauzmann 1973). They investigated sperm whale metmyoglobin. The boundary of the folded state was represented by curves similar to ellipses. The size of the *p-T* region of the native state was strongly dependent on the pH of the solution. Also a remarkable observation is that the experiments were very difficult to perform in the high-temperature and high-pressure region of the phase diagram. Here the system showed long relaxation times and it was very difficult to reach equilibrium. Additionally, in the pH range of 5–9 they reported precipitation of the sample especially at high temperatures.

Zipp and Kauzmann used absorption spectroscopy to detect the phase transitions, where they measured the signal of the prosthetic group namely the absorption of the porphyrin ring, which means the information they obtained reported only indirectly the structure of the polypeptide chain.

Infrared studies on myoglobin were performed in Heremans's laboratory. The amide I band of the infrared spectrum (between 1,600 and 1,700 cm^{-1}) is directly characteristic for the secondary structure of the protein (Susi and Byler 1986; Smeller et al. 1995), while the intensity of the amide II band (around 1,550 cm^{-1}) shows the hydrogen deuterium exchange and thus it reports the loosening of the tertiary structure.

Meersman et al. used infrared spectroscopy to compare the cold, heat and pressure unfolded states of metmyoglobin from horse heart (Meersman et al. 2002). In a successive work they measured the complete *p-T* phase diagram of myoglobin (Meersman et al. 2005). Infrared measurements require high protein concentration, which makes the aggregation of the protein at high temperature and low pressure unavoidable. Since the aggregation is irreversible, the aggregated phase remains stable (or metastable) after returning to ambient conditions. Under these circumstances a purely equilibrium phase diagram cannot be constructed. In certain regions of the *p-T* diagram two or even three metastable states can coexist, depending on the history of the sample. Aggregation is disfavored by pressure, therefore aggregation does not happen in pressure denaturation experiments, but folding intermediate states can form during the release of the pressure (Smeller et al. 1999). These intermediates can either refold or aggregate, depending on the temperature. In many cases three states can be observed at ambient conditions: native, aggregated, folding intermediate (Smeller 2002). The high-temperature behavior of the aggregated phase was investigated in detail also by the group of Heremans (Meersman et al. 2005). Interestingly, they found, that the high-temperature aggregates can be

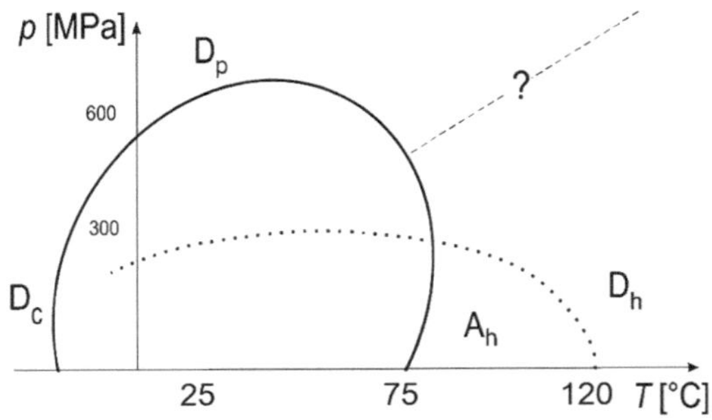

Fig. 2.4 Phase diagram of myoglobin (Reprinted with permission from Meersman et al. (2005), Copyright 2005 John Wiley and Sons)

dissociated not only by pressure as it was thought before, but they also dissociate at very high temperatures, well above 100 °C. Similar results were obtained for lysozyme under reducing conditions (Meersman and Heremans 2003). The hypothesis emerged that a similar elliptic phase boundary could be constructed for the aggregated phase, as it is known for the native one (Meersman et al. 2005) (Fig. 2.4). This hypothesis has never been tested experimentally in detail, although it seems quite plausible. The difficulties lie again in the very slow kinetics of the aggregation.

2.3.2 *Lysozyme*

A number of experimental works were published on the pressure and temperature unfolding of lysozyme (Sasahara et al. 2001; Refaee et al. 2003; Vogtt and Winter 2005; Maeno et al. 2009; Smeller 2009). Most often the hen egg white (HEW) lysozyme was investigated, since it is commercially available. Lysozyme is quite pressure stable under neutral pH, the midpoint of the pressure unfolding is 680 MPa at 25 °C (Smeller et al. 2006).

Akasaka's group determined the detailed phase diagram of HEW lysozyme at pH 2 using ^{1}H-NMR and Trp fluorescence spectroscopies (Maeno et al. 2009). Figure 2.5 shows the pressure-temperature diagram obtained by them. As it is clear from the figure, lowering the pH reduces pressure stability to 450 MPa, which was obtained at 5 °C. They analyzed their data using a similar thermodynamic analysis to the one described in Eq. 2.4. They neglected, however, the difference of the compressibility between the native and unfolded state based on an earlier work (Royer 2006). This assumption makes the elliptic barrier similar to a parabolic one, which describes the experimental points in this case quite well. The maximum position of the fluorescence emission was fitted by a sigmoid curve, which describes the two-state transition. The NMR results followed the same model quite well in the pressure range, where NMR spectroscopy was available. Lysozyme denaturation at low pH was shown in this paper to behave as a two-state process in the whole pressure-temperature region studied (Maeno et al. 2009).

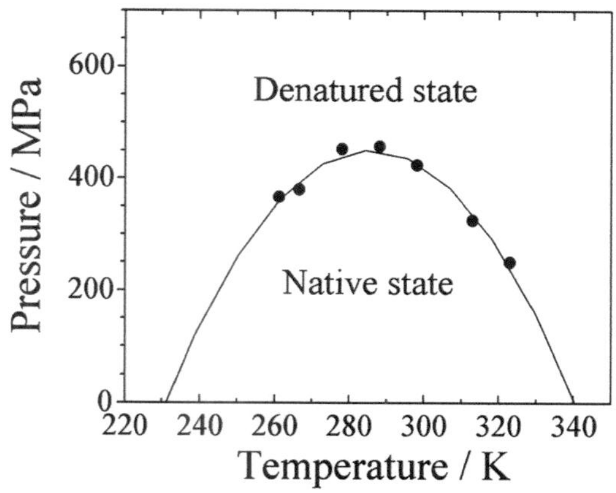

Fig. 2.5 Phase diagram of lysozyme at pH 2 (Reprinted from BIOPHYSICS Vol. 5 (pp. 1–9, 2009) (Maeno et al. 2009), with permission)

FTIR spectroscopic experiments were performed by Heremans's group at neutral pH (Smeller et al. 2006). They suggested the presence of folding intermediates, which can be considerably populated on the refolding path, similarly to the case of myoglobin mentioned above (Smeller 2002). Our later experiments showed the existence of molten globule states in the pressure-temperature phase diagram of lysozyme. The formation of molten globule precedes the temperature and also the pressure unfolding (Smeller 2009). Deviation from the two-state unfolding was found in complex environment, where the effect of mild ($\leq$100 MPa) pressure on denaturant (GdnCl) induced unfolding of lysozyme at pH 4 was investigated (Sasahara et al. 2001).

Winter's group studied the heat denatured and the pressure assisted cold denaturated state of HEW lysozyme using COSY proton NMR (Vogtt and Winter 2005). They measured the hydrogen-deuterium exchange rate and determined the protection factors for several amino acid residues. The exchange kinetics upon heat treatment was found to be a two-step process with an initial slow exchange followed by a fast one. This was interpreted by the protection of the slow exchanging transient state. Such effect was not found at cold denaturation.

Since lysozyme is quite stable its pressure region, which causes no unfolding, but only elastic deformations, is broad and can be conveniently studied (Refaee et al. 2003). The effect of subdenaturing pressure gives valuable information about the structure of the folding energy landscape. Pressure stabilizes selectively the conformations with smaller volumes. These conformational states are present under physiological conditions, but they cannot be observed due to their low thermodynamic probability in ambient conditions. Subdenaturating pressures can reveal these conformations (Akasaka et al. 2013).

Besides HEW lysozyme, the T4 lysozyme was also investigated extensively. A number of mutants were prepared in Matthews's laboratory with cavities of different size inside the protein (Ando et al. 2008). The cavity formation influenced the volume change (ΔV) and the unfolding pressure too. The main conclusion from the pressure unfolding studies on the cavity forming mutants was that water penetration into the protein interior is more consistent with the results, rather than transfer of the hydrophobic residues to the water during the pressure denaturation (Ando et al. 2008).

2.3.3 Ribonuclease

Ribonuclease A is one of the first proteins whose phase diagram was determined (Brandts et al. 1970; Hawley 1971). Since that, many pressure experiments were performed on different types of ribonucleases (Yamaguchi et al. 1995; Nash et al. 1996; Yamasaki et al. 1998; Ribo et al. 2006; Zhai and Winter 2013). Jonas's group compared the cold, heat and pressure denaturation of ribonuclease using NMR spectroscopy (Zhang et al. 1995).

Recently the phase diagram of ribonuclease A was reinvestigated to study the effect of molecular crowding on the phase diagram (Zhai and Winter 2013). Dextran was used as crowding. Molecular crowding caused by 30 % dextran stabilized the folded state of the protein against both pressure and heat denaturation (Fig. 2.6) (Zhai and Winter 2013). Understanding the effect of molecular crowding is important, since the cellular environment is highly crowded.

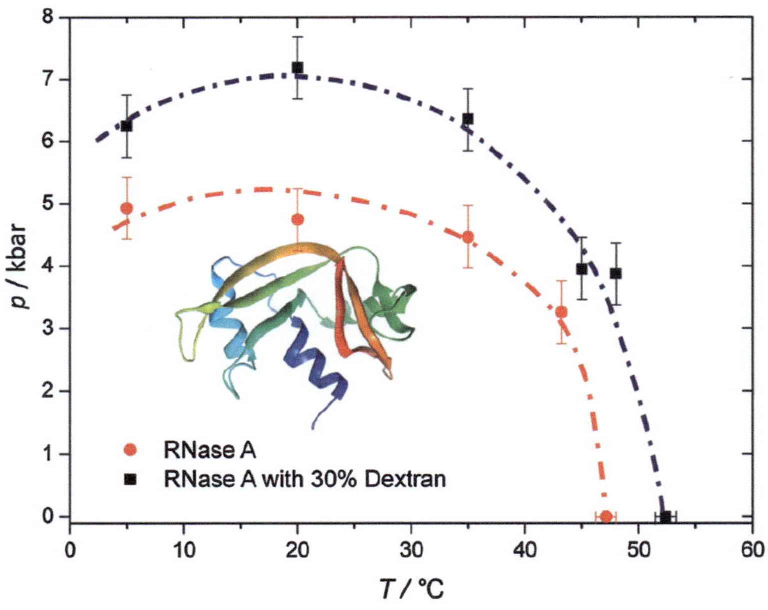

Fig. 2.6 Phase diagram of Ribonuclease A with and without crowding agent (dextran) (Reprinted with permission from Zhai and Winter (2013), Copyright 2013 John Wiley and Sons)

2.3.4 Staphylococcal Nuclease (SNase)

Staphylococcal nuclease is a small 17.5 kDa protein containing 149 residues and no disulfide bonds. According to the crystal structure roughly one quarter of the structure is helical and another quarter of the residues belong to beta structure (Hynes and Fox 1991). It has a relatively large number of ionizable side groups. SNase has one single tryptophan residue, which makes it ideal for fluorescence studies. A number of high pressure works were performed on SNase and on its mutants, to explore its pressure behavior and to describe the role of hydration and cavities in the pressure unfolding, and to study the effect of different cosolvents (Stites et al. 1991; Frye and Royer 1998; Panick et al. 1999; Herberhold et al. 2004; Kitahara et al. 2011; Roche et al. 2012; Zhai and Winter 2013).

The pressure-temperature phase diagram of SNase was determined by the groups of Winter, Royer and Akasaka, using several experimental techniques, like tryptophan fluorescence, FTIR spectroscopy, small-angle X-ray scattering (Panick et al. 1999) and NMR spectroscopy (Lassalle et al. 2000). The results allowed to create the *p-T* phase diagram of the protein at pH 5.5 (Fig. 2.7). The phase boundary resembles the elliptic shape, although fitting to the Eq. 2.4 was not done. The tryptophan fluorescence experiments gave slightly lower denaturation pressures, which can probably be explained by the terminal position of the tryptophan residue. The V66A mutant (Stites et al. 1991) was also investigated, and both the pressure and the temperature stability of the mutant was found smaller than those of the wild type.

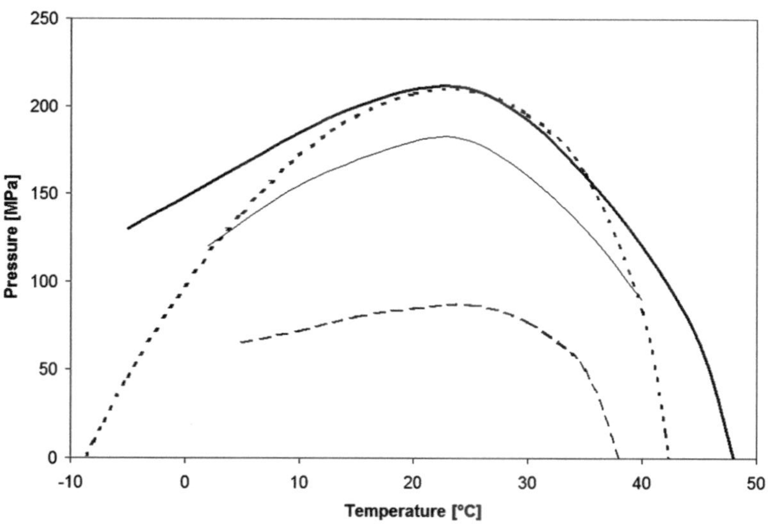

Fig. 2.7 Phase diagram of SNase (Adapted with permission from (Panick et al. 1999), Copyright 1999 American Chemical Society). *Thick line*: wild type SNase measured by FTIR and SAXS; *thin line*: wild type measured by fluorescence; *broken line*: V66A SNase by FTIR. The *dotted curve* was calculated from the parameters in Lassalle et al. (2000)

Table 2.1 Fitted parameters of Eq. 2.4

Parameter	Unit	Chimotrypsinogen[a]	Ribonuclease[a]	Snase[b]
ΔG_0	kJ mol^{-1}	10.6	10.5	13.18
ΔS_0	kJ (mol · K)$^{-1}$	−0.95	0.052	0.32
ΔV_0	ml mol^{-1}	−14.3	−48.6	−41.9
ΔC_p	kJ (mol · K)$^{-1}$	15.9	7.11	13.12
$\Delta\kappa$	ml (MPa · mol)$^{-1}$	0.296	0.195	0.2
$\Delta\alpha$	ml (mol · K)$^{-1}$	1.32	0.252	1.33
T_0	°C	0	0	24
p_0	MPa	0.1	0.1	0.1
pH		2.07	2.0	5.3

[a]From Hawley (1971)
[b]From Lassalle et al. (2000)

Akasaka's group determined the parameters of the Eq. 2.4 from a series of pressure experiments at different temperatures. They obtained the unfolding curves of Snase by following the ε protons of His8 and His124 (Lassalle et al. 2000). The parameters obtained for the elliptic fit are listed in the Table 2.1. The ellipse constructed from their parameters is also drawn in Fig. 2.7.

Several mutants of Snase were expressed and characterized. The most interesting of them were the mutations in which residue 66 was exchanged. This residue is buried in the hydrophobic core. The V66K mutation was found to affect the plasticity and stability of the protein (Kitahara et al. 2011).

Another important branch of investigations focused on the effect of chaotropic and cosmotropic cosolvents on the pressure and temperature stability. While glycerol, sorbitol, sucrose and K_2SO_4, increased the stability against pressure unfolding, $CaCl_2$ and urea destabilized the protein. The pressure stabilization and the temperature stabilization effects correlated well with each other (Herberhold et al. 2004).

2.3.5 *The Rv3221c Protein from Mycobacterium Tuberculosis*

The Rv3221c protein can be found in *Mycobacterium tuberculosis* (Cole et al. 1998). Its specific function is not clear yet, but its importance can be hypothesized from the fact that this protein is highly conserved in *Mycobacteria* (Cole et al. 2001; Kumar et al. 2008). In our experiments the protein showed mainly beta structure, which unfolded at both high temperature and high pressure giving a *p-T* diagram with a nice elliptic shape (Somkuti et al. 2013a). The midpoint of the pressure unfolding was 530 MPa (at 30 °C), while the heat unfolding happened at 65 °C (around atmospheric pressure). The native state binds biotin, which presumably stabilizes the native conformation. Biotin dissociates in the pressure-unfolded state, and only part of the polypeptide chains rebind biotin after the release of the pressure.

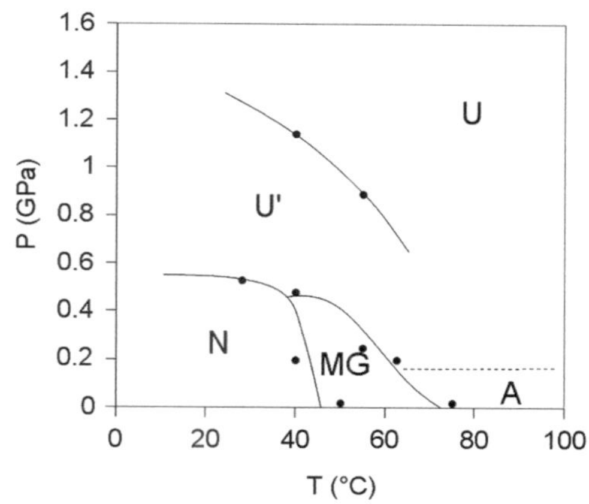

Fig. 2.8 Phase diagram of cod parvalbumin (Gad m1.01) (Reprinted with permission from Somkuti et al. (2012), Copyright 2012 American Chemical Society)

2.3.6 *Parvalbumin*

We have chosen parvalbumin for our *p-T* studies, because it is the main allergen protein in fish (Ma et al. 2008). Exploration of the phase diagram of allergen proteins can provide information about possible pressure inactivation (Somkuti and Smeller 2013). Parvalbumin is a small protein consisting of 109 residues. It has a mostly helical structure. The helices form so called EF-hand motives, which can accommodate two Ca^{2+}-binding sites. Usually the presence of Ca^{2+} ions makes the unfolding far more complex than a two state process. The phase diagram of cod parvalbumin is also quite exotic (Fig. 2.8) (Somkuti et al. 2012). Although the overall elliptic shape can be recognized on the diagram, a number of states appear besides the usual folded and unfolded states. The protein adopts a molten globule state at temperatures above 50 °C at ambient pressure. This structure unfolds and immediately aggregates if the temperature reaches 75 °C. Pressure unfolding also shows two steps. A residual structure remains in the partially unfolded state above 500 MPa. Very high pressure and elevated temperature is needed to reach the completely unfolded state. This phase diagram is a remarkable example for the complexity of the unfolding process, and for the appearance of the folding intermediates, if additional stabilizing factors are present, like the two Ca^{2+}-binding sites of parvalbumin.

2.3.7 *IG27 of Titin*

Titin is the largest protein in nature, with a molar weight exceeding 3 MDa (Labeit et al. 1990). It is responsible for the passive force generated upon stretching the muscle (Trombitas et al. 1998). Titin structure is dominated by two structurally different types of elements. Over 90 % of its molecular weight is made up of immunoglobulin

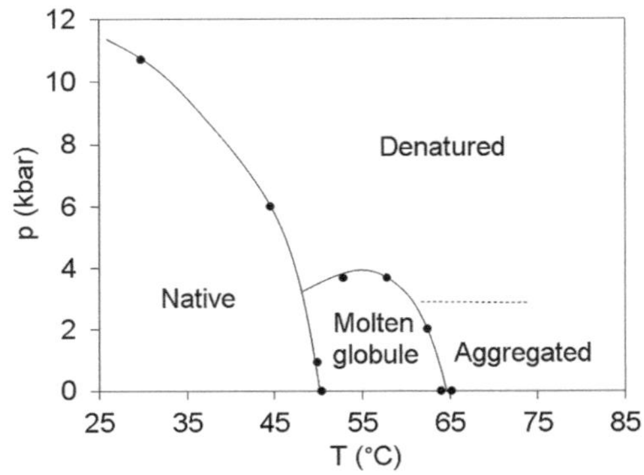

Fig. 2.9 Phase diagram of I27 domain of titin (Reprinted with permission from Somkuti et al. (2013b), Copyright 2013 Elsevier)

and fibronectin III domains. The remaining 10 % is a disordered segment (called PEVK domain) (Labeit et al. 1990; Labeit and Kolmerer 1995). The most studied immunoglobulin domain is the I27, which has a beta sandwich fold and contains a single tryptophan residue in the middle, making it ideal for the folding studies by both FTIR and fluorescence spectroscopy. The pressure-temperature phase diagram of I27 was determined using FTIR and tryptophan fluorescence spectroscopies in our laboratory (Fig. 2.9) (Somkuti et al. 2013b). The most remarkable feature of the diagram is the appearance of the molten globule state at slightly elevated temperature in the low-pressure region. The native → molten globule transition could be clearly detected by tryptophan fluorescence. At 50 °C the compact structure of the protein opens, and the polarity of the interior changes. Simultaneously the infrared spectrum shows the completion of the hydrogen-deuterium exchange, showing the loosening of the tertiary structure. The secondary structure, however, disappears only above 65 °C, where the protein aggregates immediately after the unfolding.

The pressure unfolding was found to be a two-state transition, without formation of any intermediates or partially folded states.

2.3.8 *Intrinsically Disordered Proteins (IDPs)*

Recently a big family of IDPs has been described, which do not follow the classical structure–function paradigm (Tompa 2009). They show a disordered conformation under physiological conditions. Different types of IDPs exist: some will be ordered upon binding to their target proteins, but a distinct category of them acts as entropic spring. The PEVK domain of titin is a typical example for such proteins.

From thermodynamic point of view it is an important question, whether the elliptic folded phase of these proteins is just shifted far from the physiological conditions on the *p*-*T* diagram, or it is completely absent. The latter case means that ΔG is always negative, for all (p,T) points of the pressure-temperature phase diagram, i.e. the inequality of Eq. 2.9 is not fulfilled. Visually this means that the whole dome-shaped curve of Fig. 2.2 is under the zero level of ΔG.

This area is mainly uncovered yet, since pressure works on IDPs are practically absent from the literature. The PEVK domain of titin was investigated by FTIR spectroscopy under various pH and temperature conditions (Somkuti et al. 2013b). Since these experiments did not reveal any region where the polypeptide chain would adopt any ordered structure, the second situation discussed above is more probable for PEVK, which means that this protein domain has always negative ΔG.

2.4 Concluding Remarks

The conventional Hawley theory leads to an elliptic boundary in the pressure-temperature plane for the native state. The theory is based on a two-state assumption. It can treat only situations in which only the native and the unfolded states are populated significantly. The model can describe a number of simple proteins in a dilute solution. This model fails to take into account the intermolecular interactions, and the consequent aggregation processes. If molten globule or partially folded intermediates can appear in certain ranges of the pressure–temperature plane, the phase diagram can be more complex. The complexity of the phase diagram becomes particularly apparent, if there are several factors which stabilize the protein, e.g. when binding of ions plays stabilizing role.

Since proteins in their native environment are tightly surrounded by other macromolecules, intermolecular interactions, and their effect on the native state, and also their effect on the phase diagram are of particular interest.

Acknowledgement This work was supported by the Hungarian Research Fund project OTKA 77730. The author is very grateful to Sz. Osvath for reading the manuscript and for the fruitful discussions.

References

Akasaka K, Kitahara R, Kamatari YO (2013) Exploring the folding energy landscape with pressure. Arch Biochem Biophys 53:110–115. doi:10.1016/j.abb.2012.11.016

Ando N, Barstow B, Baase WA, Fields A, Matthews BW, Gruner SM (2008) Structural and thermodynamic characterization of T4 lysozyme mutants and the contribution of internal cavities to pressure denaturation. Biochemistry 47:11097–11109. doi:10.1021/bi801287m

Brandts JF, Oliveira RJ, Westort C (1970) Thermodynamics of protein denaturation. Effect of pressure on the denaturation of ribonuclease A. Biochemistry 17:1038–1047

Bridgman PW (1914) The coagulation of albumen by pressure. J Biol Chem 19:511–512

Bryngelson JD, Onuchic JN, Socci ND, Wolynes PG (1995) Funnels, pathways, and the energy landscape of protein folding: a synthesis. Proteins Struct Funct Genet 21:167–195. doi:10.1002/prot.340210302

Cladis PE (1988) A 100 year perspective of the reentrant nematic phase. Mol Cryst Liq Cryst 165:85–121. doi:10.1080/00268948808082197

Cole ST, Brosch R, Parkhill J, Garnier T, Churcher C, Harris D, Gordon SV, Eiglmeier K, Gas S, Barry CE, Tekaia F, Badcock K, Basham D, Brown D, Chillingworth T, Connor R, Davies R, Devlin K, Feltwell T, Gentles S, Hamlin N, Holroyd S, Hornby T, Jagels K, Krogh A, McLean J, Moule S, Murphy L, Oliver K, Osborne J, Quail MA, Rajandream MA, Rogers J, Rutter S, Seeger K, Skelton J, Squares R, Squares S, Sulston JE, Taylor K, Whitehead S, Barrell BG (1998) Deciphering the biology of Mycobacterium tuberculosis from the complete genome sequence. Nature 393(6685):537–544

Cole ST, Eiglmeier K, Parkhill J, James KD, Thomson NR, Wheeler PR, Honore N, Garnier T, Churcher C, Harris D, Mungall K, Basham D, Brown D, Chillingworth T, Connor R, Davies RM, Devlin K, Duthoy S, Feltwell T, Fraser A, Hamlin N, Holroyd S, Hornsby T, Jagels K, Lacroix C, Maclean J, Moule S, Murphy L, Oliver K, Quail MA, Rajandream MA, Rutherford KM, Rutter S, Seeger K, Simon S, Simmonds M, Skelton J, Squares R, Squares S, Stevens K, Taylor K, Whitehead S, Woodward JR, Barrell BG (2001) Massive gene decay in the leprosy bacillus. Nature 409(6823):1007–1011. doi:10.1038/35059006

Dill KA, Chan HS (1977) From Levinthal to pathways to funnels. Nature Struct Biol 4:10–19

Dirix C, Meersman F, MacPhee CE, Dobson CM, Heremans K (2005) High hydrostatic pressure dissociates early aggregates of TTR105-115, but not the mature amyloid fibrils. J Mol Biol 347:903–909. doi:10.1016/j.jmb.2005.01.073

Frye KJ, Royer CA (1998) Probing the contribution of internal cavities to the volume change of protein unfolding under pressure. Protein Sci 7:2217–2222

Gruebele M (2009) Protein dynamics: from molecules, to interactions, to biology. Int J Mol Sci 10:1360–1368. doi:10.3390/ijms10031360

Hawley SA (1971) Reversible pressure-temperature denaturation of chymotrypsinogen. Biochemistry 10:2436–2442. doi:10.1021/bi00789a002

Herberhold H, Royer CA, Winter R (2004) Effects of chaotropic and kosmotropic cosolvents on the pressure-induced unfolding and denaturation of proteins: an FT-IR study on staphylococcal nuclease. Biochemistry 43:3336–3345. doi:10.1021/bi036106z

Hynes TR, Fox RO (1991) The crystal structure of staphylococcal nuclease refined at 1.7 Å resolution. Proteins 10:92–105. doi:10.1002/prot.340100203

Kitahara R, Akasaka K (2003) Close identity of a pressure-stabilized intermediate with a kinetic intermediate in protein folding. Proc Natl Acad Sci U S A 100:3167–3172. doi:10.1073/pnas.0630309100

Kitahara R, Hata K, Maeno A, Akasaka K, Chimenti MS, Garcia-Moreno B, Schroer MA, Jeworrek C, Tolan M, Winter R, Roche J, Roumestand C, de Guillen KM, Royer CA (2011) Structural plasticity of staphylococcal nuclease probed by perturbation with pressure and pH. Proteins 79:1293–1305. doi:10.1002/prot.22966

Knorr D, Froehling A, Jaeger H, Reineke K, Schlueter O, Schoessler K (2011) Emerging technologies in food processing. Ann Rev Food Sci Technol 2:203–235

Kumar N, Shukla S, Kumar S, Suryawanshi A, Chaudhry U, Ramachandran S, Maiti S (2008) Intrinsically disordered protein from a pathogenic mesophile Mycobacterium tuberculosis adopts structured conformation at high temperature. Proteins 71:1123–1133. doi:10.1002/prot.21798

Labeit S, Kolmerer B (1995) Titins – giant proteins in charge of muscle ultrastructure and elasticity. Science 270:293–296. doi:10.1126/science.270.5234.293

Labeit S, Barlow DP, Gautel M, Gibson T, Holt J, Hsieh CL, Francke U, Leonard K, Wardale J, Whiting A, Trinick J (1990) A regular pattern of 2 types of 100-residue motif in the sequence of titin. Nature 345:273–276. doi:10.1038/345273a0

Lassalle MW, Yamada H, Akasaka K (2000) The pressure-temperature free energy-landscape of staphylococcal nuclease monitored by H-1 NMR. J Mol Biol 298:293–302. doi:10.1006/jmbi.2000.3659

Liu F, Gruebele M (2008) Downhill dynamics and the molecular rate of protein folding. Chem Phys Lett 461:1–8. doi:10.1016/j.cplett.2008.04.075

Ma Y, Griesmeier U, Susani M, Radauer C, Briza P, Erler A, Bublin M, Alessandri S, Himly M, Vazquez-Cortes S, de Arellano IRR, Vassilopoulou E, Saxoni-Papageorgiou P, Knulst AC, Fernandez-Rivas M, Hoffmann-Sommergruber K, Breiteneder H (2008) Comparison of natural and recombinant forms of the major fish allergen parvalbumin from cod and carp. Mol Nutr Food Res 52:S196–S207. doi:10.1002/mnfr.200700284

Maeno A, Matsuuo H, Akasaka K (2009) The pressure-temperature phase diagram of hen lysozyme at low pH. Biophys 5:1–9. doi:10.2142/biophysics.5.1

Meersman F, Dobson CM (2006) Probing the pressure-temperature stability of amyloid fibrils provides new insights into their molecular properties. Biochim Biophys Acta 1764:452–460. doi:10.1016/j.bbapap.2005.10.021

Meersman F, Heremans K (2003) Temperature-induced dissociation of protein aggregates: accessing the denatured state. Biochemistry 42:14234–14241. doi:10.1021/bi035623e

Meersman F, Smeller L, Heremans K (2002) Comparative Fourier transform infrared spectroscopy study of cold-, pressure-, and heat-induced unfolding and aggregation of myoglobin. Biophys J 82:2635–2644

Meersman F, Smeller L, Heremans K (2005) Extending the pressure-temperature state diagram of myoglobin. Helv Chim Acta 88:546–556

Nash D, Lee BS, Jonas J (1996) Hydrogen-exchange kinetics in the cold denatured state of ribonuclease A. Biochim Biophys Acta 1297:40–48. doi:10.1016/0167-4838(96)00085-4

Osvath S, Sabelko JJ, Gruebele M (2003) Tuning the heterogeneous early folding dynamics of phosphoglycerate kinase. J Mol Biol 333:187–199. doi:10.1016/j.jmb.2003.08.011

Osvath S, Quynh LM, Smeller L (2009) Thermodynamics and kinetics of the pressure unfolding of phosphoglycerate kinase. Biochemistry 48:10146–10150. doi:10.1021/bi900922f

Panick G, Vidugiris GJA, Malessa R, Rapp G, Winter R, Royer CA (1999) Exploring the temperature-pressure phase diagram of staphylococcal nuclease. Biochemistry 38:4157–4164

Privalov PL (1990) Cold denaturation of proteins. Biophys J 57:A26

Privalov PL (1997) Thermodynamics of protein folding. J Chem Thermodyn 29:447–474. doi:10.1006/jcht.1996.0178

Rastogi NK, Raghavarao KSMS, Balasubramaniam VM, Niranjan K, Knorr D (2007) Opportunities and challenges in high pressure processing of foods. Crit Rev Food Sci Nutr 47:69–112. doi:10.1080/10408390600626420

Refaee M, Tezuka T, Akasaka K, Williamson MP (2003) Pressure-dependent changes in the solution structure of hen egg-white lysozyme. J Mol Biol 327:857–865. doi:10.1016/s0022-2836(03)00209-2

Ribo M, Font J, Benito A, Torrent J, Lange R, Vilanova M (2006) Pressure as a tool to study protein-unfolding/refolding processes: the case of ribonuclease A. Biochim Biophys Acta 1764:461–469. doi:10.1016/j.bbapap.2005.11.011

Roche J, Dellarole M, Caro JA, Guca E, Norberto DR, Yang Y, Garcia AE, Roumestand C, Garcia-Moreno B, Royer CA (2012) Remodeling of the folding free energy landscape of staphylococcal nuclease by cavity-creating mutations. Biochemistry 51:9535–9546. doi:10.1021/bi301071z

Royer CA (2006) Probing protein folding and conformational transitions with fluorescence. Chem Rev 106:1769–1784. doi:10.1021/cr0404390

Sasagawa A, Gomi M, Ohura K, Yamazaki A, Yamada A (2005) Production of Miso based on Koji prepared from mixed different grains using high-pressure treatment. J Japanese Soc Food Sci Technol-Nippon Shokuhin Kagaku Kogaku Kaishi 52:485–490

Sasahara K, Sakurai M, Nitta K (2001) Pressure effect on denaturant-induced unfolding of hen egg white lysozyme. Proteins- Struct Funct Genet 44:180–187. doi:10.1002/Prot.1083

Smeller L (2002) Pressure-temperature phase diagrams of biomolecules. Biochim Biophys Acta 1595:11–29

Smeller L (2009) Evidence for metastable states of lysozyme revealed by high pressure FTIR spectroscopy. Biophys J 96:388a. doi:http://dx.doi.org/10.1016/j.bpj.2008.12.2900

Smeller L, Heremans K (1997) Some thermodynamic and kinetic consequences of the phase diagram of protein denaturation. In: Heremans K (ed) High pressure research in bioscience and biotechnology. Leuven University Press, Leuven, pp 55–58

Smeller L, Goossens K, Heremans K (1995) Determination of the secondary structure of proteins at high pressure. Vib Spectrosc 8:199–203

Smeller L, Rubens P, Heremans K (1999) Pressure effect on the temperature-induced unfolding and tendency to aggregate of myoglobin. Biochemistry 38:3816–3820

Smeller L, Fidy J, Heremans K (2004) Protein folding, unfolding and aggregation. Pressure induced intermediate states on the refolding pathway of horseradish peroxidase. J Phys 16:S1053–S1058

Smeller L, Meersman F, Heremans K (2006) Refolding studies using pressure: the folding landscape of lysozyme in the pressure-temperature plane. Biochim Biophys Acta 1764:497–505. doi:10.1016/j.bbapap.2006.01.016

Smeller L, Meersman F, Heremans K (2008) Stable misfolded states of human serum albumin revealed by high-pressure infrared spectroscopic studies. Eur Biophys J 37:1127–1132

Somkuti J, Smeller L (2013) High pressure effects on allergen food proteins. Biophys Chem 183:19–29. doi:10.1016/j.bpc.2013.06.009

Somkuti J, Houska M, Smeller L (2011) Pressure and temperature stability of the main apple allergen Mal d1. Eur Biophys J 40:143–151. doi:10.1007/s00249-010-0633-8

Somkuti J, Bublin M, Breiteneder H, Smeller L (2012) Pressure-temperature stability, Ca^{2+} binding, and pressure-temperature phase diagram of cod parvalbumin: Gad m 1. Biochemistry 51:5903–5911. doi:dx.doi.org/10.1021/bi300403h

Somkuti J, Jain S, Ramachandran S, Smeller L (2013a) Folding-unfolding transitions of Rv3221c on the pressure-temperature plane. High Pressure Res 33:250–257. doi:10.1080/08957959.2013.780055

Somkuti J, Martonfalvi Z, Kellermayer MSZ, Smeller L (2013b) Different pressure-temperature behavior of the structured and unstructured regions of titin. Biochim Biophys Acta 1834:112–118. doi:10.1016/j.bbapap.2012.10.001

Stefani M, Dobson CM (2003) Protein aggregation and aggregate toxicity: new insights into protein folding, misfolding diseases and biological evolution. J Mol Med 81:678–699. doi:10.1007/s00109-003-0464-5

Stites WE, Gittis AG, Lattman EE, Shortle D (1991) In a staphylococcal nuclease mutant the side-chain of a lysine replacing valine-66 is fully buried in the hydrophobic core. J Mol Biol 221:7–14. doi:10.1016/0022-2836(91)80195-z

Susi H, Byler DM (1986) Resolution-enhanced Fourier-transform infrared-spectroscopy of enzymes. Methods Enzymol 130:290–311

Suzuki K (1960) Studies on the kinetics of protein denaturation under high pressure. Rev Phys Chem Jpn 29:91–98

Taniguchi Y, Suzuki K (1983) Studies of polymer effects under pressure. Part 7. Pressure inactivation of alpha-chymotrypsin. J Phys Chem 87:5185–5193. doi:10.1021/J150643a025

Tompa P (2009) Structure and function of intrinsically disordered proteins. Chapman and Hall/CRC, Boca Raton

Trombitas K, Greaser M, Labeit S, Jin JP, Kellermayer M, Helmes M, Granzier H (1998) Titin extensibility in situ: entropic elasticity of permanently folded and permanently unfolded molecular segments. J Cell Biol 140:853–859. doi:10.1083/jcb.140.4.853

Vogtt K, Winter R (2005) Pressure-assisted cold denaturation of hen egg white lysozyme: the influence of co-solvents probed by hydrogen exchange nuclear magnetic resonance. Brazilian J Med Biol Res 38:1185–1193. doi:10.1590/s0100-879x2005000800005

Yaldagard M, Mortazavi SA, Tabatabaie F (2008) The principles of ultra high pressure technology and its application in food processing/preservation: a review of microbiological and quality aspects. Afr J Biotechnol 7:2739–2767

Yamaguchi T, Yamada H, Akasaka K (1995) Thermodynamics of unfolding of ribonuclease A under high pressure. A study by proton NMR. J Mol Biol 250:689–694. doi:10.1006/jmbi.1995.0408

Yamakura M, Haraguchi K, Okadome H, Suzuki K, Tran UT, Horigane KA, Yoshida M, Homma S, Sasagawa A, Yamazaki A, Ohtsubo K (2005) Effects of soaking and high-pressure treatment on the qualities of cooked rice. J Appl Glycosci 52:85–93

Yamasaki K, Taniguchi Y, Takeda N, Nakano K, Yamasaki T, Kanaya S, Oobatake M (1998) Pressure-denatured state of Escherichia coli ribonuclease HI as monitored by Fourier transform infrared and NMR spectroscopy. Biochemistry 37:18001–18009. doi:10.1021/bi981046w

Zhai Y, Winter R (2013) Effect of molecular crowding on the temperature-pressure stability diagram of ribonuclease A. ChemPhysChem 14:386–393. doi:10.1002/cphc.201200767

Zhang J, Peng XD, Jonas A, Jonas J (1995) NMR-study of the cold, heat, and pressure unfolding of ribonuclease-A. Biochemistry 34:8631–8641. doi:10.1021/bi00027a012

Zipp A, Kauzmann W (1973) Pressure denaturation of metmyoglobin. Biochemistry 12: 4217–4228

Chapter 3
Driving Forces in Pressure-Induced Protein Transitions

Tigran V. Chalikian

Abstract The molecular mechanisms underlying pressure-induced protein denaturation can be analyzed based on the pressure-dependent differences in the apparent volume occupied by amino acids inside the protein and when exposed to water in an unfolded conformation. This chapter presents a volumetric analysis of the peptide group and the 20 naturally occurring amino acid side chains in the interior of the native state, the micelle-like interior of the pressure-induced denatured state, and in the unfolded conformation modeled by low-molecular analogs of proteins. The transfer of a peptide group from the protein interior to water becomes increasingly favorable as pressure increases. This observation classifies solvation of peptide groups as a major driving force in pressure-induced protein denaturation. Polar side chains do not appear to exhibit significant pressure-dependent changes in their preference for the protein interior or solvent. The transfer of nonpolar side chains from the protein interior to water becomes more unfavorable as pressure increases. An inference can be drawn that a sizeable population of nonpolar side chains remains buried inside a solvent-inaccessible core of the pressure-induced denatured state. At elevated pressures this core, owing to the absence of structural constraints, may become packed almost as tightly as the interior of the native state. The presence and partial disappearance of large intraglobular voids is another driving force facilitating pressure-induced protein denaturation. Volumetric data presented here have implications for the kinetics of protein folding and shed light on the nature of the folding transition state ensembles.

Keywords Hydration • Kinetics • Protein transitions • Pressure • Thermodynamics

---

T.V. Chalikian (✉)
Department of Pharmaceutical Sciences, Leslie Dan Faculty of Pharmacy, University of Toronto, Toronto, ON M5S 3 M2, Canada
e-mail: chalikan@phm.utoronto.ca

K. Akasaka, H. Matsuki (eds.), *High Pressure Bioscience*,
Subcellular Biochemistry 72, DOI 10.1007/978-94-017-9918-8_3

3.1 Introduction

The conformational preferences of proteins are governed by a subtle balance of intramolecular and intermolecular (solute-solvent) interactions (Shakhnovich 2006; Dill 1990; Dill et al. 2008; Dill and MacCallum 2012). An increase in pressure shifts the balance of such interactions bringing about an increase or a decrease in protein stability, although, in most (but not all) empirical observations, proteins tend to denature at elevated pressures. According to Le Châtelier's principle, an increase in pressure shifts a chemical equilibrium toward the state with a smaller volume. Near-zero changes in volume accompanying protein transitions arise from a complex interplay between sizeable contributions reflecting the intramolecular architecture of the protein, its hydration, and thermal volume (Chalikian and Breslauer 1996; Chalikian and Filfil 2003; Chalikian 2003; Ben-Naim 2012). Although there have been attempts to ascribe the transition change in protein volume to a single contribution (Roche et al. 2012), the real picture is more complex depending on and reflecting the pressure-induced shift in a myriad of molecular interactions and their related volumetric contributions (Chalikian and Breslauer 1996; Chalikian and Filfil 2003; Ben-Naim 2012; Kharakoz 1997). Theoretical deconvolution of the transition volume for each individual protein into contributions is an extremely complex endeavor that would require the detailed knowledge of the structural features of the protein in its native state and the ensemble of pressure-induced denatured conformations as well as the microscopic volumetric properties of each water molecule around solvent-exposed protein group. On the other hand, empirical insights into the nature of the forces governing the pressure-dependent conformational preferences of a protein can be gained from comparison of the apparent volumes occupied by amino acid side chains and peptide bond inside the protein and when exposed to solvent.

In this respect, Gerstein and colleagues have evaluated the Voronoi volumes of the 20 naturally existing amino acid residues buried inside the protein and compared these data with the volumes occupied by the same residues in zwitterionic amino acids and, in some cases, tripeptides in aqueous solution (Gerstein et al. 1994, 1995; Gerstein and Chothia 1996; Tsai and Gerstein 2002). These results make a good starting point for estimating changes in volume associated with solvent-exposure of previously buried residues at atmospheric pressure. However, to analyze the conformational response of a protein to elevated pressure one must examine, in addition to volume, the compressibility and the pressure derivatives of the compressibility of the individual amino acids. The importance of this notion is emphasized by simulations by Kharakoz (1997). Specifically, the magnitude and the sign of the difference in volume between conformational states of a protein may change with pressure, depending on the relative values of the partial compressibility and the pressure derivative of the partial compressibility of populated protein states (Kharakoz 1997).

This chapter describes results of a recent work in which we have evaluated and compared the volume occupied by each of the 20 naturally occurring amino acids in water with its volume inside the native and pressure-induced denatured states of a protein as a function of pressure (Chalikian and Macgregor 2009). This study explores and outlines the propensities of various groups to opt for the inside or outside of the protein as pressure increases independent of the number of actual states involved in pressure-induced protein transitions. Our analysis suggests that pressure-induced protein denaturation is a result of a complex and subtle interplay between the "outward" drive of peptide groups, the "inward" drive of hydrophobic side chains, and the compaction of domains consisting of nonpolar residues.

3.2 Thermodynamic Approach

The change in free energy associated with exposing a buried amino acid residue to solvent is given by the relationship:

$$\Delta G(P) = \int_{P0}^{P} \Delta V(P)\,\mathrm{d}P \tag{3.1}$$

where $\Delta V(P)$ is the pressure-dependent difference in the apparent molar volume occupied by an amino acid residue in its buried and solvent-exposed states; and P_0 is the reference pressure conventionally taken equal to atmospheric pressure.

A Taylor expansion of $\Delta V(P)$ yields the approximation:

$$\begin{aligned}\Delta V(P) = \Delta V - \Delta K_T\,(P - P_0) - (1/2)\,(\partial \Delta K_T/\partial P)_T(P - P_0)^2 \\ - (1/6)\left(\partial^2 \Delta K_T/\partial P^2\right)_T(P - P_0)^3 - \ldots\end{aligned} \tag{3.2}$$

where $\Delta V = V_S - V_M$, $\Delta K_T = K_S - K_M$, and the subscripts M and S refer to the residue inside the protein and fully exposed to solvent, respectively. Using Eq. 3.2, we can rewrite Eq. 3.1:

$$\begin{aligned}\Delta G(P) = \Delta V\,(P - P_0) - (1/2)\,\Delta K_T(P - P_0)^2 - (1/6)\,(\partial \Delta K_T/\partial P)_T(P - P_0)^3 \\ - (1/24)\left(\partial^2 \Delta K_T/\partial P^2\right)_T(P - P_0)^4 - \ldots\end{aligned} \tag{3.3}$$

The pressure range over which Eqs. 3.2 and 3.3 are applicable increases with the number of terms in the approximation. While partial molar volumes and compressibilities are known for many proteins and their low-molecular weight analogs, the pressure derivatives of the compressibility have been measured only for a few compounds (Sarvazyan and Chalikian 1989; Chalikian et al. 1994b; Hedwig et al. 2008). Therefore, Eqs. 3.2 and 3.3 are often limited to the first two terms (containing ΔV and ΔK_T). To estimate the error introduced by this approximation, we have calculated the pressure dependences of the partial molar volumes of zwitterionic Gly and Ala [for which $(\partial K_M/\partial P)_T$ has been measured (Chalikian et al. 1994b)] with and without the third term of the expansion. For these two compounds, the difference between calculated the values of V(P) based on these two assumptions reaches ~25 % at 1.5 kbar, which we have taken as the upper limit of the acceptability for the numerical analysis and discussion presented below. This is a suitable range for such studies given that many proteins show a tendency to denature or begin to denature at around 1 kbar. Whenever possible, we will use the $(\partial K^{\circ}_T/\partial P)_T$ terms in Eqs. 3.2 and 3.3. However, omission of these terms will not compromise the veracity of the conclusions reached (especially, the qualitative ones).

3.3 Volume of Amino Acid Residues Inside the Native Protein Interior

The average Voronoi volumes, V_M, of the 20 naturally occurring amino acids have been evaluated for 130 globular proteins based on their X-ray crystallographic structures (Gerstein et al. 1994, 1995; Tsai et al. 1999; Tsai and Gerstein 2002; Harpaz et al. 1994). In the Voronoi tessellation, the volume ascribed to each atom is the volume of its corresponding Voronoi cell which is defined by the smallest polyhedron formed by the planes bisecting the lines connecting the central atom with all of its neighbors (Fig. 3.1) (Tsai and Gerstein 2002; Poupon 2004; Richards 1985). Each point within a polyhedron is closer to its central atom than to any other atom. The intrinsic volume of a protein is equal to the sum of the volumes of the Voronoi cells of all of its constituent atoms (there is no free volume left inside the protein once the Voronoi tessellation is complete). The Voronoi method provides a good estimate of the apparent volume occupied by a buried (non-surface) atom where each polyhedron is fully closed and bounded by faces defined by well-positioned neighboring atoms (Harpaz et al. 1994; Liang and Dill 2001). Results are less straightforward when the method is applied to calculating the volume of a surface atom with no well-defined neighbors. The Voronoi volume of each of the 20 amino acid residues buried deeply inside a native protein is remarkably constant and weakly depends on the specific protein (Gerstein et al. 1994; Tsai et al. 1999; Tsai and Gerstein 2002; Harpaz et al. 1994).

The second column in Table 3.1 shows the Voronoi volumes, V_M, of glycine residue (–CH_2CONH-) and the 19 amino acid side chains buried inside the interior of the native conformation presented by Gerstein et al. (Tsai and Gerstein 2002). The V_M of a specific residue consists of its van der Waals volume plus the volume of the void space around it. The latter reflects the packing defects of the native protein conformation. The collective standard deviation of the Voronoi volumes of individual amino acid residues in Table 3.1 is 3.4 ± 0.7 % (Tsai and Gerstein 2002). The observed deviations reflect the presence and the actual size of intrinsic voids (packing defects) within individual proteins normalized per specific amino acid type.

Fig. 3.1 Graphic presentation of Voronoi polyhedra constructed for atoms 1 and 2

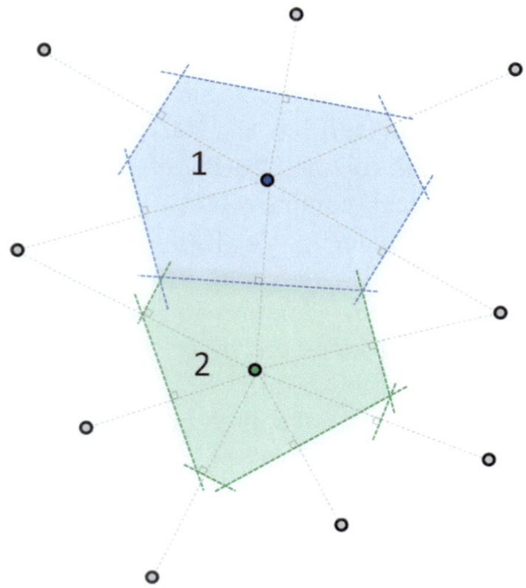

Table 3.1 Volumetric parameters for the glycine residue and the 19 naturally occurring amino acid side chains at 25 °C

AA	V_M	$K_M \times 10^4$	$(\partial K_M/\partial P)_T \times 10^7$	V_S	$K_S \times 10^4$	$V_M - V_S$
	$cm^3\ mol^{-1}$	$cm^3\ mol^{-1}\ bar^{-1}$	$cm^3\ mol^{-1}\ bar^{-2}$	$cm^3\ mol^{-1}$	$cm^3\ mol^{-1}\ bar^{-1}$	$cm^3\ mol^{-1}$
-CH_2CONH-	38.4 ± 1.6	9.6	−2.6	37.4 ± 0.1	−1.1 ± 0.5	+1.0
Ala	15.4 ± 2.1	3.9	−1.1	17.0 ± 0.2	2.0 ± 0.8	−1.6
Val	44.8 ± 2.9	11.2	−3.1	47.7 ± 0.6	0.6 ± 0.4	−2.9
Leu	59.8 ± 3.5	15.0	−4.1	65.4 ± 0.2	0.4 ± 0.6	−5.6
Ile	59.7 ± 3.2	14.9	−4.1	63.8 ± 0.1	−0.8 ± 0.4	−4.1
Pro	34.6 ± 2.2	8.7	−2.4	35.6 ± 0.1	−4.0 ± 0.6	−1.0
Phe	76.5 ± 2.9	19.1	−5.3	79.9 ± 0.3	2.4 ± 1.2	−3.4
Trp	97.9 ± 3.2	24.5	−6.7	101.9 ± 0.2	3.8 ± 0.6	−4.0
Met	61.4 ± 3.3	15.4	−4.2	62.8 ± 0.1	−1.4 ± 0.4	−1.4
Cys	29.5 ± 3.3	7.4	−2.0	29.7 ± 0.1	−4.9 ± 0.6	−0.2
Tyr	78.8 ± 3.0	19.7	−5.4	82.2 ± 0.2	−2.2 ± 0.6	−3.4
Ser	17.9 ± 2.3	4.5	−1.2	17.1 ± 0.1	−1.3 ± 0.6	+0.8
Thr	33.6 ± 2.5	8.4	−2.3	33.1 ± 0.1	−2.2 ± 0.6	+0.5
Asn	35.3 ± 2.8	8.8	−2.4	34.0 ± 0.2	−3.3 ± 0.6	+1.3
Gln	50.0 ± 2.6	12.5	−3.4	50.8 ± 0.1	−1.0 ± 0.4	−0.8
Asp	30.5 ± 2.3	7.6	−2.1	31.7 ± 0.1	0 ± 0.4	−1.2
Glu	45.2 ± 2.6	11.3	−3.1	47.7 ± 0.7	1.9 ± 0.8	−2.5
His	56.4 ± 2.5	14.1	−3.9	57.0 ± 0.6	1.8 ± 0.8	−0.6
Lys	61.0 ± 4.2	15.3	−4.2	70.1 ± 0.4	−2.4 ± 0.6	−9.1
Arg	76.2 ± 2.8	19.1	−5.2	67.4 ± 0.3	−5.1 ± 0.6	+8.8

As pressure increases, the intrinsic volume of the protein (while still in its native conformation) decreases according to Eq. 3.2 (Kundrot and Richards 1987, 1988; Katrusiak and Dauter 1996; Akasaka et al. 1997, 1999; Li et al. 1998; Kamatari et al. 2001). To use Eq. 3.2 for evaluating the pressure-dependent volume, V_M(P), of a buried amino acid residue, one needs to know its apparent compressibility, $K_M = \beta_M V_M$, and the pressure derivative of the latter, $(\partial K_M/\partial P)_T = V_M[(\partial \beta_{TM}/\partial P)_T - \beta_M^2]$, where $\beta_M = -\ (V_M)^{-1}(\partial V_M/\partial P)_T$ is the local coefficient of isothermal compressibility of the interior of a native protein. Compression of a native protein is not uniform; generally, α-helices and loops are more compressible than β-strands (Kundrot and Richards 1987; Akasaka et al. 1999; Gekko and Hasegawa 1986). However, a uniform compression of the protein interior can be assumed as a first approximation, *i. e.* uniform values of β_M and $(\partial \beta_{TM}/\partial P)_T$ can be ascribed to all buried amino acid residues and their constituent groups. Experiment-based estimates of the average coefficient of isothermal compressibility of the interior of a native protein and its pressure derivative are $\beta_M = 25 \times 10^{-6}$ bar^{-1} and $(\partial \beta_M/\partial P)_T = -6.25 \times 10^{-9}$ bar^{-2}, respectively (Chalikian 2003; Taulier and Chalikian 2002; Kharakoz 2000). Table 3.1 presents our evaluated values of $K_M = \beta_M V_M$ and $(\partial K_M/\partial P)_T = V_M[(\partial \beta_{TM}/\partial P)_T - \beta_M^2]$ of the glycine residue ($-CH_2CONH$-) and the 19 amino acid side chains. With these results, one can use Eq. 3.2 to calculate the pressure dependent volumes of the glycine residue and each of the amino acid side chains.

3.4 Free Energies of Solvent Exposure of Glycine Residue and Peptide Group

There is an ambiguity concerning the volumetric properties of a solvent-exposed glycine residue depending on the identity of the specific model compounds used to derive the group numbers of the glycine residue (Chalikian et al. 1994a; Hakin et al. 2000; Hedwig 2006). In our analysis, we use the volume and compressibility contributions of the -CH_2CONH- group derived from oligoglycine-based studies (Chalikian et al. 1994a). At 25 °C, the volume and compressibility contributions of a glycine residue are 37.4 ± 0.1 cm^3 mol^{-1} and $-(1.1 \pm 0.5) \times 10^{-4}$ cm^3 mol^{-1} bar^{-1}, respectively (Chalikian et al. 1994a). The pressure dependences of the adiabatic and isothermal compressibility contributions, K_S, of a glycine residue have been measured for Ala(Gly)$_n$ oligopeptides at pressures up to 1 kbar (Hedwig et al. 2008). The pressure derivative of the compressibility of a glycine residue, $(\partial K_S/\partial P)_T$, is equal to $(8.3 \pm 1.5) \times 10^{-7}$ cm^3 mol^{-1} bar^{-2} as can be calculated using the exponential approximation of the pressure dependence of K_S (Hedwig et al. 2008).

We do not discriminate between the partial molar adiabatic and isothermal compressibilities of amino acids in solution. The difference between the partial molar adiabatic and isothermal compressibilities of a solute is proportional to the square of the coefficient of thermal expansion, α_0, of the solvent and inversely proportional to its specific heat capacity, c_{P0} (Gucker et al. 1950; Desnoyers and Philip 1972; Blandamer et al. 2001). In aqueous solutions with small α_0 and large c_{P0}, the difference between the partial molar adiabatic and isothermal compressibilities of a solute is not large, being on the order of 10–15 % or less. This difference falls within uncertainties of our analysis.

A change in free energy, ΔG, accompanying solvent exposure of a glycine residue can be calculated from Eq. 3.3. The results of these calculations are plotted in Fig. 3.2a. Inspection of this plot reveals that, in contrast to most of the side chains (see below), a glycine residue exhibits a steady decrease in ΔG with increasing pressure. The value of ΔG decreases by 21 cal mol^{-1} as pressure increases from atmospheric to 1.5 kbar. Assuming additivity, the free energy of solvent exposure of the peptide group, –CONH–, can be evaluated by subtracting the ΔG contribution of a methylene group, $-CH_2-$, from that of a glycine residue, $-CH_2CONH-$. The Voronoi volume, V_M, of a fully buried methylene group is 15.1 ± 0.2 cm^{3}mol^{-1} as can be determined from the data on glycine and amino acid residues with aliphatic side chains (Gerstein et al. 1994; Tsai et al. 1999; Tsai and Gerstein 2002). The adiabatic compressibility, K_M, and the pressure dependence of the compressibility, $(\partial K_M/\partial P)_T$, of a buried methylene group are $(3.8 \pm 0.1) \times 10^{-4}$ cm^3 mol^{-1} bar^{-1} and $-(1.1 \pm 0.1) \times 10^{-7}$ cm^3 mol^{-1} bar^{-2}, respectively. On the other hand, the contributions of a solvent-exposed methylene group to volume, V_S, and adiabatic compressibility, K_S, are 15.7 ± 0.2 cm^3 mol^{-1} and $-(1.9 \pm 0.2) \times 10^{-4}$ cm^3 mol^{-1} bar^{-1}, respectively (Kharakoz 1989, 1991; Chalikian et al. 1993). The pressure slope of the compressibility of a solvent-exposed methylene group, $(\partial K_s/\partial P)_T$, can be roughly approximated by the average incremental change of the pressure slope of the partial molar adiabatic compressibility of zwitterionic glycine, alanine, valine, leucine, and isoleucine normalized per aliphatic carbon (Sarvazyan and Chalikian 1989). This approximation yields a value of $(\partial K_s/\partial P)_T$ of $(4.6 \pm 0.3) \times 10^{-7}$ cm^3 mol^{-1} bar^{-2}.

Figure 3.2a presents the calculated pressure dependences of the protein interior-to-solvent transfer free energy for a methylene and a peptide groups (glycine residue minus methylene group). As is seen from Fig. 3.2a, the transfer of a methylene group from the protein interior to solvent becomes more unfavorable as pressure increases (an increase in ΔG is 29 cal mol^{-1} when pressure increases from atmospheric to 1.5 kbar). In contrast, the transfer free energy of the peptide group decreases by 51 cal mol^{-1} upon an increase in pressure to 1.5 kbar. Thus, solvation of a peptide group becomes increasingly favorable at elevated pressures thereby providing a major driving force for pressure-induced denaturation of a globular protein.

A note of caution is in order here. Our results suggest that solvent exposure of the peptide group from the interior of a native protein is accompanied by a negative change in volume and, hence, facilitates pressure-induced protein denaturation. The negative sign of the transfer volume reflects the difference in the apparent volumes occupied by an "average" peptide group inside the native protein and when solvent-exposed. The situation is different for an isolated α-helical oligopeptide.

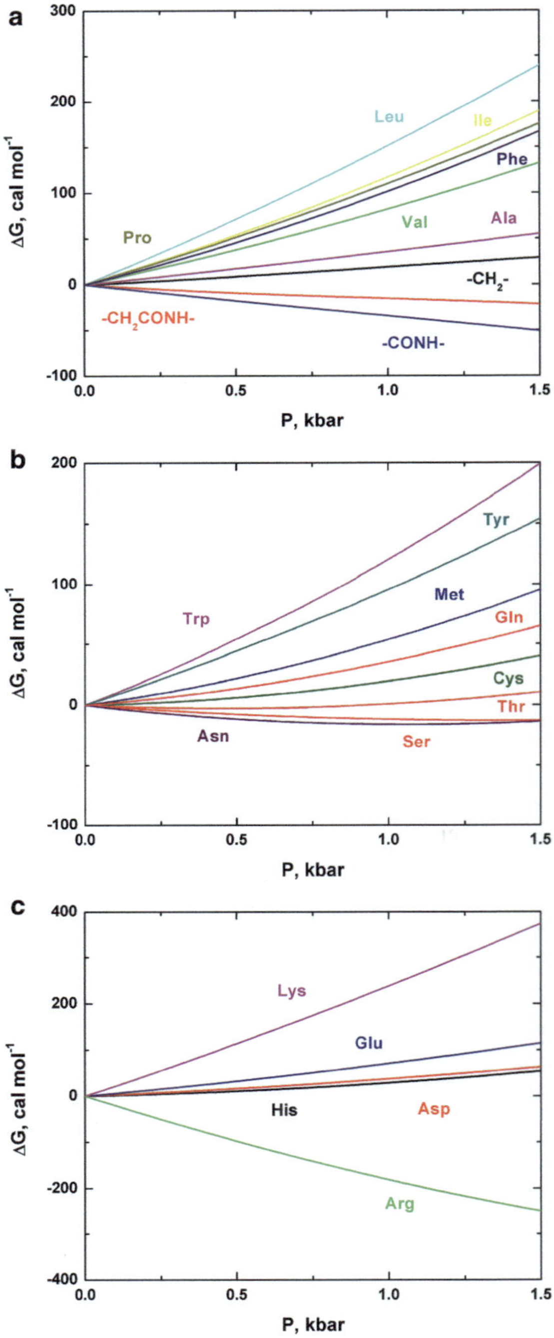

Fig. 3.2 (**a**) Pressure dependences of the protein interior-to-solvent transfer free energies for a glycine residue ($-CH_2CONH$-), a methylene group ($-CH_2$-), a peptide group ($-CONH$-), and the 6 naturally occurring amino acids. (**b**) Pressure dependences of the protein interior-to-solvent transfer free energies for the 8 naturally occurring amino acids. (**c**) Pressure dependences of the protein interior-to-solvent transfer free energies for the 5 naturally occurring amino acids

The stability of α-helical peptides increases with increasing pressure, although their helix-to-coil transitions are not accompanied by any significant change in solvent-accessible surface area (Takekiyo et al. 2005; Imamura and Kato 2009; Neumaier et al. 2013). This observation suggests that, in contrast to an "average" peptide group inside the protein, peptide groups engaged in intra-helical hydrogen bonds in an isolated peptide occupy a smaller volume than when forming hydrogen bonds with water. The differential behavior of a peptide group inside a protein and in an isolated α-helical peptide should be kept in mind when extrapolations are made towards understanding the nature of the pressure-induced unfolded conformation of a protein. The latter conformation, while being devoid of the tight packing characteristic of the native protein, may still display a population of isolated α-helices.

3.5 Free Energies of Solvent Exposure of Side Chains

For evaluating the pressure-dependent volumes of the 19 amino acid side chains in solution, we use partial molar volume and adiabatic compressibility data on *N*-acetyl amino acid amides, uncharged analogs of zwitterionic amino acids (Lee et al. 2008). The values of V_S and K_S for the amino acid side chains at 25 °C are presented in Table 3.1. The pressure dependences of the partial molar adiabatic compressibilities of the *N*-acetyl amino acid amides and, hence, the compressibility contributions of their side chains in solution are not known. Therefore, the Taylor expansion of the pressure-induced changes in the apparent volume is limited to the two first terms in Eqs. 3.2 and 3.3. This approximation does not alter significantly the qualitative picture derived from such calculations for pressures not exceeding ~1.5 kbar.

The sign of the difference between V_S and V_M (shown in the last column of Table 3.1) suggests that solvent exposure of the side chains of Ser, Thr, Asn, or Arg results in a decrease in volume at atmospheric pressure. All other side chains exhibit a smaller apparent molar volume inside the protein relative to their fully solvent-exposed configuration and, therefore, on average, contribute positively to a change in volume associated with protein unfolding at atmospheric pressure. Note, however, that, although the Voronoi volumes, V_M, of buried amino acids appear to be essentially constant, the observed deviations ranging from $\pm$ 1.6 cm^3 mol^{-1} to $\pm$ 4.2 cm^3 mol^{-1} have a magnitude similar to the difference $\Delta V = (V_S - V_M)$. The deviations in V_M reflect the local packing of specific amino acids, which, in turn, reflects the voids or packing defects of individual proteins. The cumulative effect of such deviations may prove sufficiently large to alter the magnitude and even the sign of a change in volume ($\Delta V = V_S - V_M$) accompanying solvent exposure of individual amino acid residues.

Equation 3.3 can be used in conjunction with data in Table 3.1 to calculate pressure-dependent changes in protein-to-solvent transfer free energy for each amino acid side chain. Figure 3.2 (panels a, b, and c) presents the results of these calculations. Inspection of Fig. 3.2a–c reveals that, at elevated pressures, all side chains, with the exception of Ser, Asn, and Arg, exhibit unfavorable changes in the protein-to-solvent transfer free energy, ΔG. The changes are especially unfavorable for side chains containing sizeable hydrophobic moieties. For example, an increase in pressure from atmospheric to 1.5 kbar causes an increase in ΔG of 55, 150, 167, 189, and 240 cal mol^{-1} for Ala, Val, Phe, Ile, and Leu, respectively. The same pressure change causes Ser and Asn to exhibit a small negative change in ΔG of -13 cal mol^{-1}. Uncharged Arg is the only amino acid side chain which manifests a significant favorable change in the transfer free energy; $\Delta\Delta G$ reaches -250 cal mol^{-1} at 1.5 kbar.

Our derivations were made for uncharged forms of the Asp, Glu, His, Lys, and Arg side chains. When charged, their solvent volumes will become smaller due to electrostriction (Rasper and Kauzmann 1962; Chalikian et al. 1992, 1998). This will bring about a decrease in the value of ΔV in Eq. 3.3, even changing its sign from positive to negative, thereby making solvent exposure of charged residues an increasingly favorable event with increasing pressure. Contrary to common perception, there are many charged residues buried inside a protein (Kajander et al. 2000).

3.6 Water-Inaccessible Structural Regions in the Pressure-Induced Denatured State

Our analysis suggests that solvent-exposure of peptide groups is a major factor facilitating the denaturation of proteins at elevated pressures, whereas the majority of side chains (especially nonpolar ones) show a tendency to remain water-inaccessible in the pressure-induced denatured conformation. To reconcile the two opposing tendencies, we propose that the pressure-induced denatured state represents an ensemble of non-native conformations with a large portion of peptide groups exposed to solvent and sizeable cluster(s) of nonpolar side chains. One can further hypothesize that the clusters lack the rigid arrangement characteristic of the native conformation and can be modeled by the nonpolar interior of a micelle. A methylene group can be used as a representative functionality to explore the energetics of the switch from the native state to the pressure-induced denatured state. The volume and compressibility contributions of the -CH_2- groups in micelles have been determined from the incremental change in partial molar volume and adiabatic compressibility upon an increase in the number of -CH_2- groups in alkyltrimethylammonium bromides above their critical micelle concentration (Kudryashov et al. 1998). At 25 °C, the volume and compressibility contributions of a -CH_2- group inside the micelle are 16.3 ± 0.1 cm^3 mol^{-1} and $(9.7 \pm 0.3) \times 10^{-4}$ cm^3 mol^{-1} bar^{-1}, respectively (Kudryashov et al. 1998).

Figure 3.3 shows the pressure dependences of the apparent volume exhibited by a methylene group in the native protein interior and a micelle calculated from Eq. 3.2. There are no data on the pressure derivative of the intrinsic compressibility

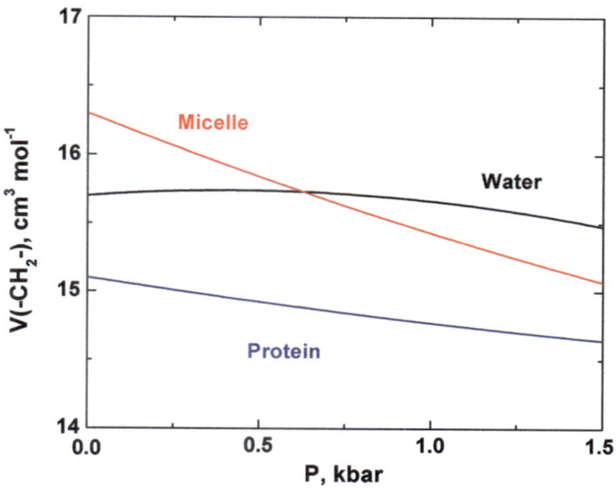

Fig. 3.3 Pressure dependences of the apparent volume occupied by a methylene group ($-CH_2$-) in the native protein interior, a micelle, and solution

of a micelle; therefore, the calculation was carried out with the approximation that $(\partial K_M/\partial P)_T$ equals zero. For comparison, Fig. 3.3 also presents the similar dependence for a methylene group in solution. The latter exhibits an initial increase in volume up to ~0.5 kbar followed by a decrease at higher pressures. This result is in qualitative agreement with MD simulations of the pressure dependence of the partial molar volume of methane in water (Moghaddam and Chan 2007). Inspection of Fig. 3.3 reveals that, while at atmospheric pressure the apparent volume occupied by a methylene group in a micelle is higher than that in a native protein, the difference rapidly diminishes with increasing pressure due to the higher intrinsic compressibility of a micelle. Although, within the pressure range we have analyzed, the difference does not diminish to zero, the trend, if continued, will eventually result in a tighter packing of nonpolar groups inside the micelle-like interior of the denatured state compared to the native protein interior. It is evident that the exact value of the threshold pressure will depend on the specific protein.

Equation 3.3 can be used to calculate a change in free energy accompanying the transfer of a -CH_2- group from the native protein interior to the micelle-like environment of the pressure-induced denatured conformation. Figure 3.4 graphically presents results of these calculations along with the pressure dependence of the free energy of the transfer of a methylene group from a micelle to water. The latter are consistent with the published data on the pressure dependences of the critical micelle concentrations of ionic and nonionic surfactants, including sodium alkylsulfates, alkyltrimethylammonium bromides, and *n*-alkyl polyoxyethylene ethers (Hara et al. 1989; Lesemann et al. 1998; Baden et al. 2002; Kaneshina et al. 1974; Brun et al. 1978; Tanaka et al. 1973, 1974). The critical micelle concentration of these surfactants exhibits an initial increase with pressure followed by a decrease with a maximum at around 1 kbar. This behavior is in satisfactory agreement with the pressure dependence of the free energy change we have calculated for the transfer of methylene groups from micelle to water which shows a maximum of ΔG at

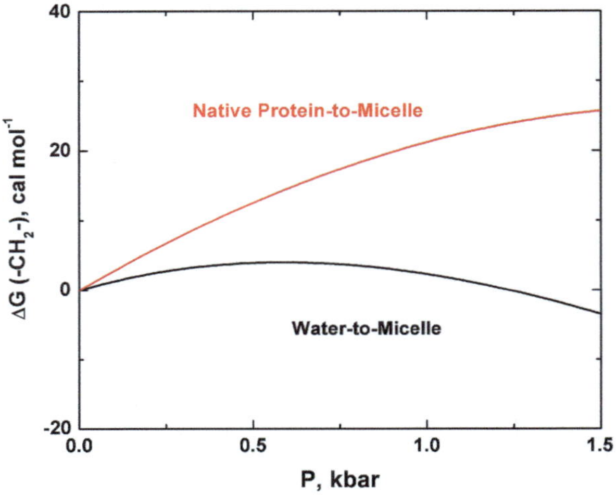

Fig. 3.4 Pressure-dependent changes in free energy accompanying the transfer of a methylene group ($-CH_2$-) from native protein to micelle and form micelle to water

around 0.6 kbar. This agreement lends credence to our analysis and the volumetric parameters used.

Inspection of Figs. 3.3 and 3.4 reveals a number of important observations that shed light on the role of nonpolar side chains in protein denaturation at elevated pressures. Between atmospheric pressure and 1.5 kbar, a -CH_2- group in a micelle, and by extension, the interior of the pressure-induced denatured state, occupies a larger volume than that in the interior of a native protein. However, due to a higher compressibility of a micelle, the unfavorable change in free energy accompanying the transfer of a -CH_2- group from a native protein to micelle (as a model of the solvent-inaccessible core of the pressure-induced denatured state) has a parabolic shape that exhibits a maximum at around 1.7 kbar and decreasing at higher pressures. If this tendency persists, transfer free energy will change its sign from positive to negative at $\sim$3 kbar. An important qualitative conclusion that ensues from this result is that, at elevated pressures, nonpolar amino acid side chains inside the core of a protein will tend to lose their rigid, native-like arrangement (becoming micelle-like) while predominantly remaining water-inaccessible.

3.7 Emerging Model of Pressure-Induced Denaturation of Proteins

The results and analyses presented above are consistent with a process in which proteins denature at high pressures by converting into a relatively compact state which preserves a sizeable core and clusters of predominantly nonpolar side chains while a significant number of previously buried peptide groups show a tendency to become accessible to solvent. Solvent exposure of peptide groups becomes increasingly favorable at higher pressures thereby contributing to and facilitating pressure-induced denaturation of proteins. As suggested by experimental studies of isolated α-helical oligopeptides under elevated pressures, the pressure-induced unfolded state of a protein may display an α-helical content (Takekiyo et al. 2005; Imamura and Kato 2009; Neumaier et al. 2013). In general, the pressure-induced unfolded state may structurally resemble the molten globule state which, in fact, has been observed in pressure-induced denaturation of apomyoglobin (Kitahara et al. 2002).

On the other hand, solvent exposure of nonpolar groups becomes increasingly unfavorable. Consequently, the pressure-induced unfolded state involves unstructured, solvent-inaccessible domains predominantly consisting of nonpolar side-chains. We propose that, to a good approximation, the volumetric properties of these solvent-inaccessible domains can be modeled by those of the micelle interior.

At low pressures, the packing of the side chains inside the micelle-like domains is less tight than that inside the native protein. Consequently, at low pressures, the native arrangement of amino acid side chains is favored over the loose, unstructured interior of the domains of a denatured protein. As pressure increases, the opposite is true. Due to the higher compressibility of the micelle-like interior of the denatured state, its packing becomes tighter (the volume occupied by specific side chains decreases) with the ensuing favorable change in free energy accompanying destruction of the native-like arrangement of amino acid residues inside the water-inaccessible domains of the protein at elevated pressures. One may speculate that, at elevated pressures, a few water molecules may be found within the micelle-like interior of the pressure-induced denatured conformation as has been suggested by computer simulations (Hummer et al. 1998, 2000; Ghosh et al. 2001). Such an event may make the formation of the unstructured hydrophobic domains at elevated pressures even more favorable thereby further shifting equilibrium toward the ensemble of pressure-induced denatured states. While this explanation and the related analysis may provide a good qualitative understanding of the molecular processes involving nonpolar side chains, further studies are required to establish to what extent the volumetric properties of the interior of the pressure-induced denatured conformation can be quantitatively modeled by those of a micelle.

The presence of "additional" molecular-size packing defects around specific amino acid residues (such as large intraglobular cavities) will increase their actual Voronoi volumes. The reduction of such intraglobular voids in proteins will make the free energy of transfer of individual amino acids to water or the interior of the solvent-inaccessible domains of the pressure denatured state more favorable. Clearly, this represents another factor facilitating pressure-induced protein denaturation (Roche et al. 2012; Liang and Dill 2001; Rashin et al. 1986; Matthews and Liu 2009). Alternatively, a tightly packed protein with a lower than average amount of intraglobular voids (leading to a negative deviation in the Voronoi volume, V_M, of buried amino acid residues) will prove unusually pressure-resistant.

In a recent study, changes in volume, ΔV, accompanying pressure-induced denaturation of a number of cavity forming mutants of staphylococcal nuclease have been measured (Roche et al. 2012). It has been found that ΔV, generally, correlates with the cavity volume introduced by a mutation in which leucine, isoleucine, valine, or phenylalanine is replaced by alanine (Roche et al. 2012). This result is not unexpected since the mutants are practically identical to each other with the exception of the mutated amino acid. Consequently, all other contributions to ΔV (intramolecular and hydration) are very similar if not identical. Hence, the measured transition volumes, ΔV, should correlate to the volume of the cavity produced by the mutation. This result underscores the notion that intraglobular voids contribute to ΔV. However, by no means, it can be interpreted as suggesting that intraglobular voids are the only or the most important factor determining ΔV and, hence, governing the pressure stability of a globular protein. The simplistic nature of the analysis in which pressure-induced protein denaturation is considered to solely arise from the volume of intraglobular cavities has been emphasized in a recent theoretical study of Ben-Naim (2012).

3.8 Implications for the Kinetics of Folding/Unfolding Transitions

Pressure-jump measurements of the folding transitions of staphylococcal nuclease and *trp* repressor have shown the activation volumes for these processes to be large and positive (Vidugiris et al. 1995; Desai et al. 1999; Panick et al. 1999; Woenckhaus et al. 2001; Brun et al. 2006; Winter et al. 2007). A larger volume of the folding transition state ensemble (TSE) compared to both the native and denatured states appears to be the volumetric signature of folding TSE. This experimental observation is consistent with a mechanism in which the transition from the native to the pressure-induced denatured conformation proceeds through a sequence of transition states with at least some of them being characterized by significant solvent exposure of nonpolar side chains. As follows from the discussion above, this state should exhibit a greater volume than the native or denatured states resulting in a positive activation volume. This agrees with the emerging view that folding TSE is populated by conformations with native-like topologies manifesting a highly hydrated hydrophobic core (Cheung et al. 2002; Fernandez-Escamilla et al. 2004; Levy and Onuchic 2006). The final state of the folding transition should involve cooperative expulsion of water molecules from the hydrophobic core region (or regions) resulting in a dry and well-packed protein exhibiting a smaller volume.

3.9 Conclusion

Analysis of the pressure-dependent changes in volume and free energy accompanying the transfer of the peptide group and the 20 naturally occurring amino acid side chains between the native state, the micelle-like interior of the pressure-induced denatured state, and the unfolded state characterizes the transfer of peptide groups from the protein interior to water as a major force driving pressure-induced protein denaturation. Polar side chains do not appear to exhibit any significant pressure-dependent changes in their preference for the protein interior or solvent. The transfer of nonpolar side chains from the protein interior to water becomes more unfavorable as pressure increases. One can, therefore, conclude that a sizeable fraction of nonpolar side chains remain buried inside relatively unstructured, solvent-inaccessible clusters that arise in the pressure-induced denatured state. At elevated pressures these clusters may become packed almost as tightly as the interior of the native state. The presence and partial disappearance of "additional" intraglobular voids (or packing defects) provides yet another driving force facilitating pressure-induced denaturation of individual proteins. These results, in conjunction with the observation of a positive activation volume for the folding reaction, are consistent with the picture in which TSE contains a significant population of highly hydrated conformations with nonpolar groups brought in contact with water.

References

Akasaka K, Tezuka T, Yamada H (1997) Pressure-induced changes in the folded structure of lysozyme. J Mol Biol 271:671–678

Akasaka K, Li H, Yamada H, Li RH, Thoresen T, Woodward CK (1999) Pressure response of protein backbone structure. Pressure-induced amide N-15 chemical shifts in BPTI. Protein Sci 8:1946–1953

Baden N, Kajimoto O, Hara K (2002) High-pressure studies on aggregation number of surfactant micelles using the fluorescence quenching method. J Phys Chem B 106:8621–8624

Ben-Naim A (2012) Theoretical aspects of pressure and solute denaturation of proteins: a Kirkwood-buff-theory approach. J Chem Phys 137:235102

Blandamer MJ, Davis MI, Douheret G, Reis JCR (2001) Apparent molar isentropic compressions and expansions of solutions. Chem Soc Rev 30:8–15

Brun TS, Hoiland H, Vikingstad E (1978) Partial molal volumes and isentropic partial molal compressibilities of surface active agents in aqueous solution. J Colloid Interface Sci 63:89–96

Brun L, Isom DG, Velu P, Garcia-Moreno B, Royer CA (2006) Hydration of the folding transition state ensemble of a protein. Biochemistry 45:3473–3480

Chalikian TV (2003) Volumetric properties of proteins. Annu Rev Biophys Biomol Struct 32:207–235

Chalikian TV, Breslauer KJ (1996) On volume changes accompanying conformational transitions of biopolymers. Biopolymers 39:619–626

Chalikian TV, Filfil R (2003) How large are the volume changes accompanying protein transitions and binding? Biophys Chem 104:489–499

Chalikian TV, Macgregor RB Jr (2009) Origins of pressure-induced protein transitions. J Mol Biol 394:834–842

Chalikian TV, Kharakoz DP, Sarvazyan AP, Cain CA, Mcgough RJ, Pogosova IV, Gareginian TN (1992) Ultrasonic study of proton transfer reactions in aqueous solutions of amino acids. J Phys Chem 96:876–883

Chalikian TV, Sarvazyan AP, Breslauer KJ (1993) Partial molar volumes, expansibilities, and compressibilities of α, ω-aminocarboxylic acids in aqueous solutions between 18 and 55°C. J Phys Chem 97:13017–13026

Chalikian TV, Sarvazyan AP, Funck T, Breslauer KJ (1994a) Partial molar volumes, expansibilities, and compressibilities of oligoglycines in aqueous solutions at 18–55°C. Biopolymers 34:541–553

Chalikian TV, Sarvazyan AP, Funck T, Cain CA, Breslauer KJ (1994b) Partial molar characteristics of glycine and alanine in aqueous solutions at high pressures calculated from ultrasonic velocity data. J Phys Chem 98:321–328
Chalikian TV, Gindikin VS, Breslauer KJ (1998) Hydration of diglycyl tripeptides with nonpolar side chains: a volumetric study. Biophys Chem 75:57–71
Cheung MS, Garcia AE, Onuchic JN (2002) Protein folding mediated by solvation: water expulsion and formation of the hydrophobic core occur after the structural collapse. Proc Natl Acad Sci U S A 99:685–690
Desai G, Panick G, Zein M, Winter R, Royer CA (1999) Pressure-jump studies of the folding unfolding of trp repressor. J Mol Biol 288:461–475
Desnoyers JE, Philip PR (1972) Isothermal compressibilities of aqueous solutions of tetraalkylammonium bromides. Can J Chem 50:1094–1096
Dill KA (1990) Dominant forces in protein folding. Biochemistry 29:7133–7155
Dill KA, MacCallum JL (2012) The protein-folding problem, 50 years on. Science 338:1042–1046
Dill KA, Ozkan SB, Shell MS, Weikl TR (2008) The protein folding problem. Annu Rev Biophys 37:289–316
Fernandez-Escamilla AM, Cheung MS, Vega MC, Wilmanns M, Onuchic JN, Serrano L (2004) Solvation in protein folding analysis: combination of theoretical and experimental approaches. Proc Natl Acad Sci U S A 101:2834–2839
Gekko K, Hasegawa Y (1986) Compressibility-structure relationship of globular proteins. Biochemistry 25:6563–6571
Gerstein M, Chothia C (1996) Packing at the protein-water interface. Proc Natl Acad Sci U S A 93:10167–10172
Gerstein M, Sonnhammer ELL, Chothia C (1994) Volume changes in protein evolution. J Mol Biol 236:1067–1078
Gerstein M, Tsai J, Levitt M (1995) The volume of atoms on the protein surface calculated from simulation, using Voronoi polyhedral. J Mol Biol 249:955–966
Ghosh T, Garcia AE, Garde S (2001) Molecular dynamics simulations of pressure effects on hydrophobic interactions. J Am Chem Soc 123:10997–11003
Gucker FT, Lamb FW, Marsh GA, Haag RM (1950) The adiabatic compressibility of aqueous solutions of some simple amino acids and their uncharged isomers at 25°C. J Am Chem Soc 72:310–317
Hakin AW, Hoiland H, Hedwig GR (2000) Volumetric properties of some oligopeptides in aqueous solution: partial molar expansibilities and isothermal compressibilities at 298.15 K for the peptides of sequence Ala(gly)(n), $n = 1$–4. Phys Chem Chem Phys 2:4850–4857
Hara K, Suzuki H, Takisawa N (1989) High-pressure studies of a fluorescence probe for the critical micelle concentration in sodium dodecyl sulfate. J Phys Chem 93:3710–3713
Harpaz Y, Gerstein M, Chothia C (1994) Volume changes on protein-folding. Structure 2:641–649
Hedwig GR (2006) Isentropic and isothermal compressibilities of the backbone glycyl group of proteins in aqueous solution. Biophys Chem 124:35–42
Hedwig GR, Hogseth E, Hoiland H (2008) Volumetric properties of the glycyl group of proteins in aqueous solution at high pressures. Phys Chem Chem Phys 10:884–897
Hummer G, Garde S, Garcia AE, Paulaitis ME, Pratt LR (1998) The pressure dependence of hydrophobic interactions is consistent with the observed pressure denaturation of proteins. Proc Natl Acad Sci U S A 95:1552–1555
Hummer G, Garde S, Garcia AE, Pratt LR (2000) New perspectives on hydrophobic effects. Chem Phys 258:349–370
Imamura H, Kato M (2009) Effect of pressure on helix-coil transition of an alanine-based peptide: an FTIR study. Proteins 75:911–918
Kajander T, Kahn PC, Passila SH, Cohen DC, Lehtio L, Adolfsen W, Warwicker J, Schell U, Goldman A (2000) Buried charged surface in proteins. Structure 8:1203–1214
Kamatari YO, Yamada H, Akasaka K, Jones JA, Dobson CM, Smith LJ (2001) Response of native and denatured hen lysozyme to high pressure studied by $^{15}N/^{1}H$ NMR spectroscopy. Eur J Biochem 268:1782–1793

Kaneshina S, Tanaka M, Tomida T, Matuura R (1974) Micelle formation of sodium alkylsulfate under high pressures. J Colloid Interface Sci 48:450–460
Katrusiak A, Dauter Z (1996) Compressibility of lysozyme protein crystals by X-ray diffraction. Acta Cryst Sect D 52:607–608
Kharakoz DP (1989) Volumetric properties of proteins and their analogs in diluted water solutions. 1. Partial volumes of amino acids at 15–55°C. Biophys Chem 34:115–125
Kharakoz DP (1991) Volumetric properties of proteins and their analogs in diluted water solutions. 2. Partial adiabatic compressibilities of amino acids at 15–70°C. J Phys Chem 95:5634–5642
Kharakoz DP (1997) Partial volumes and compressibilities of extended polypeptide chains in aqueous solution: additivity scheme and implication of protein unfolding at normal and high pressure. Biochemistry 36:10276–10285
Kharakoz DP (2000) Protein compressibility, dynamics, and pressure. Biophys J 79:511–525
Kitahara R, Yamada H, Akasaka K, Wright PE (2002) High pressure NMR reveals that apomyoglobin is an equilibrium mixture from the native to the unfolded. J Mol Biol 320:311–319
Kudryashov E, Kapustina T, Morrissey S, Buckin V, Dawson K (1998) The compressibility of alkyltrimethylammonium bromide micelles. J Colloid Interface Sci 203:59–68
Kundrot CE, Richards FM (1987) Crystal structure of hen egg white lysozyme at a hydrostatic pressure of 1000 atmospheres. J Mol Biol 193:157–170
Kundrot CE, Richards FM (1988) Effect of hydrostatic pressure on the solvent in crystals of hen egg white lysozyme. J Mol Biol 200:401–410
Lee S, Tikhomirova A, Shalvardjian N, Chalikian TV (2008) Partial molar volumes and adiabatic compressibilities of unfolded protein states. Biophys Chem 134:185–199
Lesemann M, Thirumoorthy K, Kim YJ, Jonas J, Paulaitis ME (1998) Pressure dependence of the critical micelle concentration of a nonionic surfactant in water studied by H-1-NMR. Langmuir 14:5339–5341
Levy Y, Onuchic JN (2006) Water mediation in protein folding and molecular recognition. Annu Rev Biophys Biomol Struct 35:389–415
Li H, Yamada H, Akasaka K (1998) Effect of pressure on individual hydrogen bonds in proteins, basic pancreatic trypsin inhibitor. Biochemistry 37:1167–1173
Liang J, Dill KA (2001) Are proteins well-packed? Biophys J 81:751–766
Matthews BW, Liu L (2009) A review about nothing: are apolar cavities in proteins really empty? Protein Sci 18:494–502
Moghaddam MS, Chan HS (2007) Pressure and temperature dependence of hydrophobic hydration: volumetric, compressibility, and thermodynamic signatures. J Chem Phys 126:114507
Neumaier S, Buttner M, Bachmann A, Kiefhaber T (2013) Transition state and ground state properties of the helix-coil transition in peptides deduced from high-pressure studies. Proc Natl Acad Sci U S A 110:20988–20993
Panick G, Vidugiris GJA, Malessa R, Rapp G, Winter R, Royer CA (1999) Exploring the temperature-pressure phase diagram of staphylococcal nuclease. Biochemistry 38:4157–4164
Poupon A (2004) Voronoi and Voronoi-related tessellations in studies of protein structure and interaction. Curr Opin Struct Biol 14:233–241
Rashin AA, Iofin M, Honig B (1986) Internal cavities and buried waters in globular proteins. Biochemistry 25:3619–3625
Rasper J, Kauzmann W (1962) Volume changes in protein reactions. 1. Ionization reactions of proteins. J Am Chem Soc 84:1771–1777
Richards FM (1985) Calculation of molecular volumes and areas for structures of known geometry. Methods Enzymol 115:440–464
Roche J, Caro JA, Norberto DR, Barthe P, Roumestand C, Schlessman JL, Garcia AE, Garcia-Moreno B, Royer CA (2012) Cavities determine the pressure unfolding of proteins. Proc Natl Acad Sci U S A 109:6945–6950
Sarvazyan AP, Chalikian TV (1989) Relationship between nonlinear acoustic properties and thermodynamic characteristics of solutions of biological substances. In Ultrasonic international 89, Butterworth, pp 704–710

Shakhnovich E (2006) Protein folding thermodynamics and dynamics: where physics, chemistry, and biology meet. Chem Rev 106:1559–1588

Takekiyo T, Shimizu A, Kato M, Taniguchi Y (2005) Pressure-tuning FTIR spectroscopic study on the helix-coil transition of ala-rich oligopeptide in aqueous solution. Biochim Biophys Acta 1750:1–4

Tanaka M, Kaneshina S, Tomida T, Noda K, Aoki K (1973) Effect of pressure on solubilities of ionic surfactants in water. J Colloid Interface Sci 44:525–531

Tanaka M, Kaneshina S, Kaoru SN, Okajima T, Tomida T (1974) Partial molal volumes of surfactant and its homologous salts under high pressure. J Colloid Interface Sci 46:132–138

Taulier N, Chalikian TV (2002) Compressibility of protein transitions. Biochim Biophys Acta 1595:48–70

Tsai J, Gerstein M (2002) Calculations of protein volumes: sensitivity analysis and parameter database. Bioinformatics 18:985–995

Tsai J, Taylor R, Chothia C, Gerstein M (1999) The packing density in proteins: standard radii and volumes. J Mol Biol 290:253–266

Vidugiris GJA, Markley JL, Royer CA (1995) Evidence for a molten globule-like transition state in protein folding from determination of activation volumes. Biochemistry 34:4909–4912

Winter R, Lopes D, Grudzielanek S, Vogtt K (2007) Towards an understanding of the temperature/pressure configurational and free-energy landscape of biomolecules. J Non-Equilib Thermodyn 32:41–97

Woenckhaus J, Kohling R, Thiyagarajan P, Littrell KC, Seifert S, Royer CA, Winter R (2001) Pressure-jump small-angle x-ray scattering detected kinetics of staphylococcal nuclease folding. Biophys J 80:1518–1523

Chapter 4
Why and How Does Pressure Unfold Proteins?

Catherine A. Royer

Abstract This year, 2014, marks the 100th anniversary of the first publication reporting the denaturation of proteins by high hydrostatic pressure (Bridgman 1914). Since that time a large and recently increasing number of studies of pressure effects on protein stability have been published, yet the mechanism for the action of pressure on proteins remains subject to considerable debate. This review will present an overview from this author's perspective of where this debate stands today.

Keywords Cavity • Cavity-containing variant • Conformational fluctuation • Pressure unfolding • Water penetration

4.1 Introduction

At and around room temperature, the application of pressure to protein solutions often, but not always, results in their partial or complete unfolding, as assessed by the spectroscopic parameter used in the experiment. Most often the observable has been the fluorescence of intrinsic tryptophan or tyrosine residues, but visible and IR absorption measurements have been used as well. Recently, high-pressure NMR measurements have become not only possible, as in the past, but now rather straightforward. It is important to note at the start of this discussion, that while protein solutions are compressible, the *difference* in compressibility between folded and unfolded states is quite small (Seemann et al. 2001). Hence, most of the pressure unfolding profiles reported in the literature can be fit to a simple two-state transition, assuming that the relative stability of the folded state with respect to the unfolded state changes linearly with pressure. This linear and constant coefficient corresponds to the volume change upon unfolding, ΔV_u, which is negative. Therefore, pressure acts simply to stabilize the lower volume unfolded state relative to the higher volume folded state, thus shifting the equilibrium towards unfolding. For a protein of a given stability, the ease with which pressure does this depends upon the magnitude of the

C.A. Royer (✉)
Department of Biology, Rensselaer Polytechnic Institute, Troy, NY 12180, USA
e-mail: royerc@rpi.edu

K. Akasaka, H. Matsuki (eds.), *High Pressure Bioscience*,
Subcellular Biochemistry 72, DOI 10.1007/978-94-017-9918-8_4

difference in volume between these two states. It is important to remember that this difference in volume is inherent to the protein system. The unfolded state presents a lower molar volume than the folded state, even at atmospheric pressure. Pressure simply favors the state that occupies the smaller volume.

For the discussion that follows we must consider the magnitude of this difference in volume between folded and unfolded proteins. Measured values tend to cluster around 50–100 ml/mol, although, somewhat smaller and somewhat larger values have been reported. Generally speaking these values correspond to between 0.5 % and 2 % of the protein's molar volume. One must be cautious therefore in the quest to determine the physical basis for ΔV_u. It is a very small number. Be that as it may, the values for the volume change upon protein unfolding are clearly large enough such that pressure generally unfolds proteins, and as noted above, many studies have been carried out taking advantage of this fact.

If a protein is not observed by any method to unfold when subjected to pressure, this can arise for one (or both) of two reasons. Either the volume difference between the folded and unfolded states is negligibly small, and hence there is no intrinsic effect of pressure possible, or alternatively, the protein is simply too stable, and even with a volume change of reasonable magnitude, unfolding cannot be observed in the pressure range accessible to liquid high pressure studies, maximally ~1,500 MPa, but more typically 300–400 MPa. Often, destabilizing the folded state relative to the unfolded state by changing the pH, adding a small amount of chemical denaturant such as urea or guanidinium hydrochloride, or making a destabilizing single site mutation to the protein sequence will bring the transition into the accessible range, allowing the volume change of unfolding to be determined.

Another complicating factor in pressure effects on proteins is that they are highly temperature dependent. Since the seminal experiments by Kauzmann, Hawley and Brandts in the 1970s (Brandts et al. 1970; Hawley 1971; Zipp and Kauzmann 1973), many researchers have confirmed that the pressure-temperature phase diagram of protein stability (assuming two-state transitions) can be described by an ellipse (Meersman et al. 2006). In addition to the difference in heat capacity between folded and unfolded states, the most important coefficient defining the shape of this ellipse in the p-T plane is the difference in thermal expansivity between the folded and unfolded states, the difference in compressibility being rather small as noted above (Seemann et al. 2001). The expansivity term arises because the thermal expansion of the molar volume of unfolded states of proteins is generally larger than the expansion of their folded states, for reasons that are not understood. Because of this difference in thermal expansion, the magnitude of ΔV_u is a strong and decreasing function of temperature, resulting in the observation that the pressure effect on protein stability is largest at lower temperature, and decreases linearly (and even can change sign) with increasing temperature. This brings another cautionary note to the present discussion. It is difficult to discuss the underlying basis for the volume change of unfolding because its magnitude is temperature dependent. It is not inherently obvious that comparison of ΔV_u values for different proteins at an arbitrary reference temperature represents a truly meaningful exercise if the magnitude of their thermal expansion is very different.

Bearing in mind the caveats listed above, we turn to consideration of the different factors contributing to the molar volume of folded proteins in solution, and then to differences in these factors when the protein unfolds. The total volume (not the molar volume) of a protein solution includes the total volume of the protein molecules, V_{prot}, and the total volume of the water molecules, V_{water}.

$$V_T = V_{prot} + V_{water} \tag{4.1}$$

For the moment we discuss the total volumes rather than the molar volumes. The total volume of the protein molecules corresponds to the sum of the van der Waals volume of the atoms of the protein and the void volume resulting from imperfect packing of the atoms in the folded structure.

$$V_{prot} = V_{vdW} + V_{void} \tag{4.2}$$

The total volume of the water molecules can be defined as the sum of the total volume of the bulk water molecules and the total volume of the water molecules that interact with the protein surface. These two "categories" of water molecules may have different density.

$$V_{water} = V_{bulk} + V_{hyd} \tag{4.3}$$

Moreover, the volume of the water molecules interacting with the protein surface can be further categorized into water molecules interacting with polar or charged moieties on the protein surface and water molecules interacting with apolar moieties on the protein surface.

$$V_{hyd} = V_{polar} + V_{apolar} \tag{4.4}$$

When the protein unfolds, two things occur. First, the internal void volume decreases because the packing defects present in the ensemble of configurations populated in the unfolded state are fewer than in the folded structure. The sign of this effect is not subject to debate, although its magnitude, relative to the total change in molar volume of unfolding, is not known. The packing density of proteins moieties inside proteins is limited by hard sphere considerations, but more importantly perhaps, by the polymeric nature of the chain and the energetic constraints imposed by the proteins' three dimensional structure. Secondly, upon unfolding a certain number of water molecules change categories, moving from the bulk to interaction with the newly exposed protein surface, which in turn, scales linearly with the size of the protein.

The contribution to ΔV_u of changes in the density of water of hydration is the subject of much debate, not only concerning its magnitude, but also its sign. Since Frank and Evans (Frank and Evans 1945) coined the term "ice-like" clathrate structures to explain the loss in entropy observed in hydrophobic hydration, debate in the field has raged concerning the structure (and hence the density or volume) of water molecules interacting with hydrophobic groups. A multitude of theoretical and experimental studies have been published, which will not be reviewed here. Evidence from pressure perturbation calorimetry suggests that at low temperature the density of water around hydrophobic solutes is lower than the bulk, while around polar compounds it is higher, as expected. These differences tend to fade, however, as temperature increases (Lin et al. 2002). Despite the ongoing debate, there does appear to be a consensus that the differences in water structure around small hydrophobic moieties with respect to bulk water are small (see for example, (Huang and Chandler 2000; Gallagher and Sharp 2003; Galamba 2013; Galamba 2014)). Water density around charged, and to a lesser extent polar residues is generally considered to be slightly higher than bulk, due to electrostriction (Lepori and Gianni 2000; Lin et al. 2002).

From a different perspective, one can think of the volume change of unfolding as resulting uniquely from differences in empty volume in the unfolded as opposed to folded state, as defined by scaled particle theory. According to scaled particle theory (see for example (Lepori and Gianni 2000)) the void or empty volume of small molecules in solution is defined as the portion of the cavity required in the solvent to dissolve the particle that exceeds the van der Waals volume. This empty volume can be quite large, equal to the van der Waals volume for some molecules and scales linearly with this latter. It is a function of the hard sphere diameters of the solute and the solvent, and also includes a term called the interaction volume which involves differences in hydrating water density with respect to the bulk. At standard temperature and pressure, the solvent interaction volume term is always ≤ 0; it is ~ 0 for hydrocarbons (in agreement with the papers cited above concerning the small hydrophobic hydration volume effect) and negative for polar and charged groups due to electrostriction (Lepori and Gianni 2000). For example, the van der Waals volume of CH_3COO^- is 31 ml mol^{-1}, whereas, the measured molar volume in water at STP is 46 ml mol^{-1}. This difference is largely due to the hard sphere empty volume. However, the electrostriction (solvent shrinkage) of the COO^- group is estimated to be -6.3 ml mol^{-1}. Without this effect, the molar volume would be even larger, ~ 52 ml mol^{-1}. Thus, the volume difference upon unfolding of proteins can be thought of as arising from packing density differences for each of the protein's moieties in the solvent with respect to the protein interior. In any case, given the small size of the total change in molar volume of proteins upon unfolding, and the large number of different types of moieties, not to mention the polymeric nature of the chain and apparent temperature effects, prediction of volume changes based on sequence and/or structure appears to this investigator to be an extremely difficult exercise. Rather than attempt to predict the change in volume upon unfolding of proteins based on the sum of the contributing factors, we set out to perturb model protein systems by site specific mutagenesis and then to measure the effect of these perturbations on the volume change of unfolding.

4.2 Testing the Effect of Changes in Exposed Protein Size and Accessible Surface Area

To test the significance of hydration of protein surface area upon unfolding, we chose a model repeat protein system, the Ankyrin domain of the Notch Receptor (Nank), and a series of domain deletion variants whose chemical denaturation had been well documented (Bradley and Barrick 2006; Street et al. 2007; Barrick et al. 2008; Tripp and Barrick 2008; Street and Barrick 2009). We reasoned that if hydration of exposed surface area represented a significant contribution to the magnitude of ΔV_u, then the observed ΔV_u should decrease in magnitude as the number of repeats decreased. We measured unfolding by pressure via the fluorescence from the single tryptophan residue present in the central region of the WT protein (Rouget et al. 2010) and for several variants with C- and N-terminal deletions of 1, 2 or 3 repeats out of a total of seven in the WT protein (Rouget et al. 2011). Rather than a decrease in the magnitude of the pressure effect, deletion of one or two repeats either from the N or C-terminal domains (D1, D2 or D6, D7) resulted in a significant increase in the magnitude of ΔV_u (Fig. 4.1). The variant

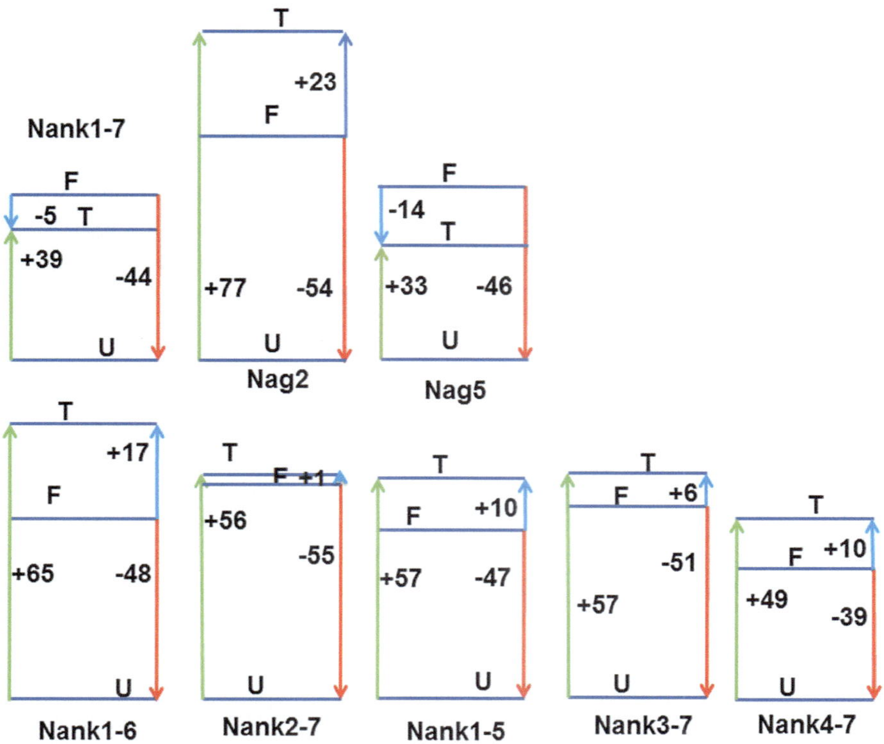

Fig. 4.1 Volume diagrams of the Nank deletion variants obtained from the fluorescence detected (Reprinted with permission from (Rouget et al. 2011). Copyright 2011 American Chemical Society)

with only 4/7 repeats, 40 % smaller than the WT protein exhibited a ΔV_u value only 11 % smaller. Moreover, single amino acid substitutions of alanine with glycine had very different effects. The mutation in domain two increased the magnitude of ΔV_u by over 20 % (a larger effect than deleting 40 % of the protein), but had almost no effect in domain five. Note as well (Fig. 4.1) the interesting effects of the domain deletions and single site substitutions on the activation volume for unfolding, which is negative for the full-length protein and becomes positive for many of the variants, placing the transition state higher in volume than the folded state. These results revealed that for Nank, the determinants for the volume change of unfolding were concentrated in the central repeats, four and five of the protein, and not homogeneously distributed throughout the structure, as would be expected for surface area. Moreover, the increases in the pressure effect upon deletion of the terminal repeats, suggests an expansion of the structure when these structural constraints are released by deletion. In the supplemental material for the Nank deletion work (Rouget et al. 2011), we reported that for a series of proteins for which the volume changes had been reported in the literature at approximately the same temperature (20 or 21 C) (Frye and Royer 1998; Mei et al. 1999; Mohana-Borges et al. 1999; Sasahara and Nitta 1999; Herberhold and Winter 2002; Jacob et al. 2002; Kitahara et al. 2002), there was a rather poor correlation between the magnitude of ΔV_u and the number of residues, the unfolded state accessible surface area, the m-value or the molecular volume (Fig. 4.2). However, reasonable correlations were found when one considered the packing efficiency represented either as the size of the largest cavity or the ratio of the void volume with respect to van der Waals volume. Indeed, molecular dynamics simulations of the WT Nank protein revealed the existence of solvent excluded cavities in the central repeats of the protein, the region harboring the determinants for the value of ΔV_u (Fig. 4.3). These results, demonstrating a lack of correlation of the volume change with the protein size, provide very strong evidence against the notion that hydration of exposed surface area upon protein unfolding represents a major contributing factor to the magnitude of the pressure effect.

4.3 Testing the Effect of Internal Solvent-Excluded Void Volume

In addition to suggesting a probable lack of significance for solvent density differences upon hydration of exposed protein surfaces, the above results strongly suggested that internal solvent-excluded void volume could make a significant contribution to the magnitude of ΔV_u. We sought to test this hypothesis by constructing variants of a model protein system with internal voids, via site-directed mutagenesis. We had attempted such a study previously (Frye and Royer 1998), but the strong destabilization of the protein by the cavity creating mutations limited the structural information we could obtain, and hence the conclusions which could be drawn. To overcome this drawback we created single site substitutions of bulky

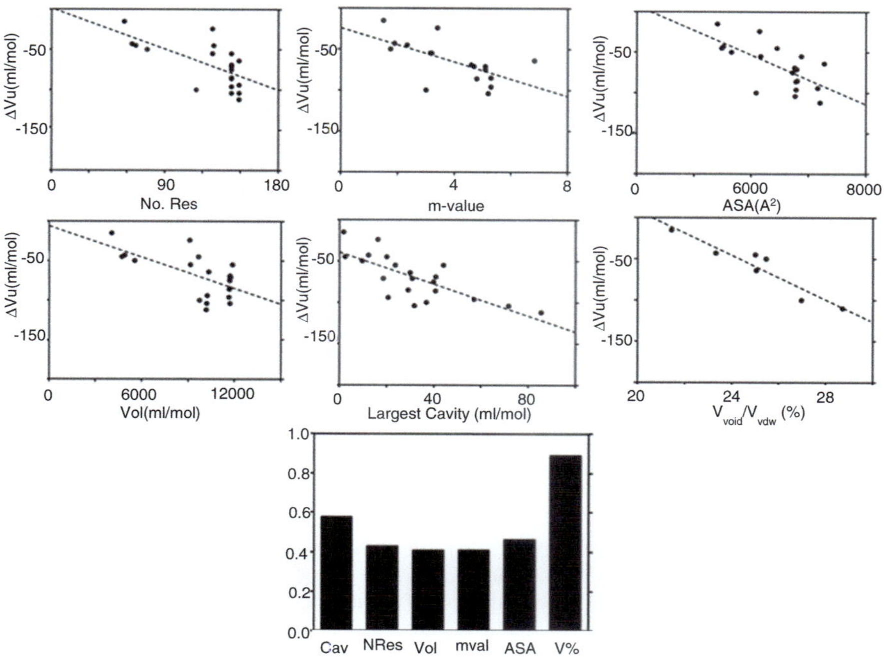

Fig. 4.2 Correlations between volume change of unfolding at 20 or 21 C and various structural parameters for different proteins. Values are taken from published results (Frye and Royer 1998; Mei et al. 1999; Sasahara and Nitta 1999; Herberhold and Winter 2002; Jacob et al. 2002, Kitahara et al. 2002) (Reprinted with permission from (Rouget et al. 2011). Copyright 2011 American Chemical Society)

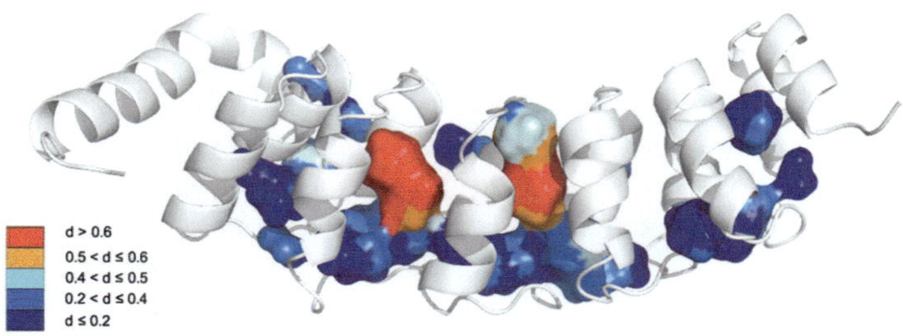

Fig. 4.3 Example of cavity calculations on Nank1-7 full-length construct. Representations of the internal cavities detected using the McVol algorithm (Till and Ullmann 2010) and a probe size of 1.1 Å for a representative configuration from simulations of Nank1-7. Cavities are colored blue to red for the distance from their center to the nearest water molecule (Reprinted with permission from (Rouget et al. 2011). Copyright 2011 American Chemical Society)

amino acids (L, I, V) by alanine in a hyperstable variant of staphylococcal nuclease (SNase) called $\Delta+$PHS (Roche et al. 2012). This variant is deleted for a flexible loop (the Δ) and bears three substitutions in the C-terminus (P117G, H124L and S128A), all of which contribute to the increased stability ($\sim$12 kcal mol^{-1} compared

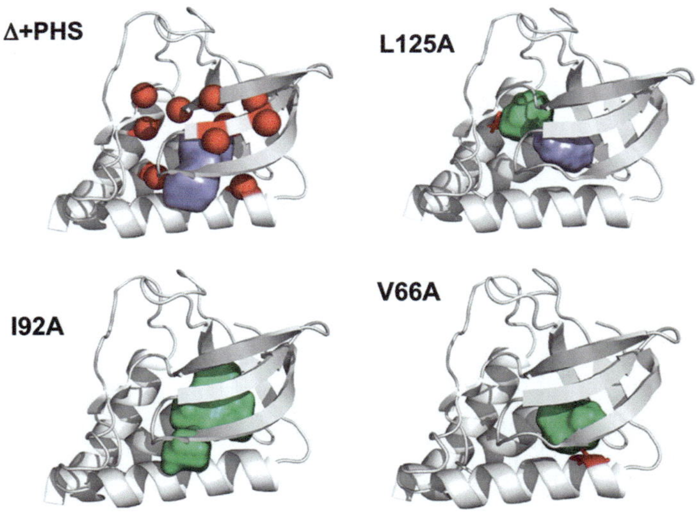

Fig. 4.4 Ribbon diagram of the structure of top left, the reference protein, $\Delta+$PHS (pdb code: 3BCD), with the positions of the substitutions in red. The central cavity of the reference structure is in purple. The other three are ribbon diagrams of the three dimensional structures of three of the alanine containing variants all in the $\Delta+$PHS background, from their crystal structures, I92A (pdb code: 3MEH), L125A (pdb code: 3OSO) and V66A (pdb code: 3NQT). Cavities due to the mutation are shown in green, the mutated residue is shown in red sticks. Cavities were detected using the McVol algorithm (Till and Ullmann 2010) and a probe size of 1.1 Å (Reprinted with permission from (Roche et al. 2012). Copyright (2012) National Academy of Sciences, USA)

to ~5 kcal mol^{-1} for the WT protein). We determined the three dimensional structures of 10 variants with the alanine substitutions (Roche et al. 2012). All of the variants presented a cavity at the expected position, but there was very little change in their structures other than at the cavity position. While they showed lower stability than the reference protein, they all remained significantly more stable than the true WT SNase. We measured the pressure-induced unfolding of all the variants by fluorescence of the single tryptophan residue at position 140. High pressure 2-D NMR (^{15}N-^{1}H HSQC) was used as well, to follow the unfolding of the $\Delta+$PHS reference protein and three of the variants (shown in Fig. 4.4). All the cavity containing variants exhibited a larger magnitude for ΔV_{u} compared to the reference protein (Fig. 4.5a, b), but the actual value depended upon the position of the mutation. The largest change was observed for the I92A substitution, deep in the protein's hydrophobic core. The magnitude of the volume change for this variant doubled with respect to that of the reference protein. NMR measurements revealed the same percentage increase for the absolute value of ΔV_{u} for the different variant tested (Fig. 4.5c). The average ΔV_{u} values obtained from the NMR studies were somewhat larger than those obtained from fluorescence, likely because the tryptophan undergoes some hydration in the pre-transition pressure range. These experiments demonstrate clearly that modification of the amount of internal void

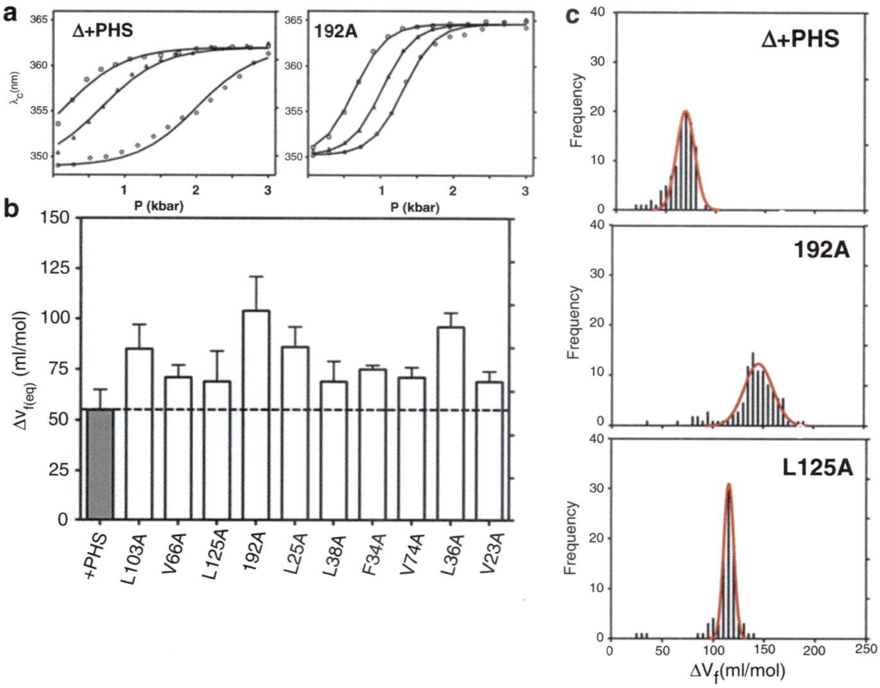

Fig. 4.5 Results of the pressure unfolding of the cavity bearing variants of SNase obtained (**a** and **b**) from fluorescence measurements and (**c**) from the equilibrium pressure dependent ^{15}N-^{1}H HSQC NMR measurements of the indicated variants (Reprinted with permission from (Roche et al. 2012). Copyright (2012) National Academy of Sciences, USA)

volume by single amino acid substitutions leads to significant differences (up to two-fold) in the magnitude of the pressure effect. Taken together the results on the cavity-containing variants and the previous study on the Nank deletion variants strongly support the notion that solvent-excluded void volume represents the major contributing factor to magnitude of the pressure effect on protein stability.

4.4 Consequences of the Role of Internal Solvent Excluded Void

A mechanism for pressure-induced unfolding of protein based primarily on the existence and amount of solvent excluded void volume in the folded state presents some interesting advantages that can be exploited in the study of protein folding. Indeed, unlike denaturant induced unfolding, which depends upon a homogeneously distributed property of the *unfolded state*, namely the solvent accessible surface area, which itself scales with the size of the protein, pressure acts due to a structure and sequence specific property of the *folded state*, namely its packing density, which is inhomogeneously distributed throughout the structure. Hence, chemical denaturants act homogeneously across the structure, whereas pressure does not. This mechanism of action of pressure provides a unique tool for dissecting cooperative interaction networks in proteins. For example, if two domains in a protein fold cooperatively, yet only one of the domains presents significant packing defects, then pressure will induce unfolding of both domains in a cooperative manner. In contrast, if there is little cooperativity between domains, even if they have the same stability, pressure will cause only the cavity containing domain to unfold, leaving the other intact. In such a manner, intermediate species can be more easily brought to light in pressure experiments. We note also the exquisite detail available from unfolding studies on the SNase variants using pressure coupled to multidimensional NMR, with over 100 observable parameters, individual site-specific HSQC resonances. We were able to exploit this site-specific structural information to describe folding intermediates (structurally and energetically) for several of the variants (Roche et al. 2012) (See also Roche et al. in Chap. 13).

Another advantage of pressure, touched upon only briefly in describing the Nank results, is the fact that the activation volume for folding is large and positive for many proteins (see the volume diagrams in Fig. 4.1). Based on our conclusions concerning the mechanism of the pressure effect, this means that the transition states contain significant solvent-excluded void volume. A large positive activation volume for folding results in significant decreases in the folding rate constant as pressure increases. This can slow folding considerably. For example, above 1 kbar, the V66A variant of $\Delta + \mathrm{PHS}$ SNase requires nearly 24 h to relax after each pressure jump ($\sim$200 bar jump magnitude) (Fig. 4.6). (See also Roche et al. in Chap. 13). Even relatively rapid kinetics can be slowed sufficiently for multi-dimensional NMR using appropriate experiments, such as ^{15}N-^{1}H SOFAST HMQC experiments (Schanda et al. 2005; Schanda et al. 2006; Gal et al. 2009) (Fig. 4.6). These slow relaxation times allow one to "watch" by NMR every part of the protein as it unfolds or refolds (one can jump pressure in either direction). Using this approach with the SNase cavity containing variants (Roche et al. 2013) revealed significantly different folding rates and routes for two of the variants, L125A and I92A, that exhibited the same overall thermodynamics stability for their folded states. We showed that introduction of a cavity between subdomains at position 125 had little effect on folding pathways, although rates for both folding and unfolding increased significantly. The protein folded via the same pathway as the WT and $\Delta + \mathrm{PHS}$ variants. In contrast, placing a cavity deep in the core of the protein with the I92A substitution resulted in a highly perturbed and complex folding landscape with multiple intermediates and parallel routes exhibiting a very different and distributed transition state ensemble. Thus although, the apparent thermodynamic signature of these two substitutions is similar they had very different effects on the ruggedness of the protein folding landscape.

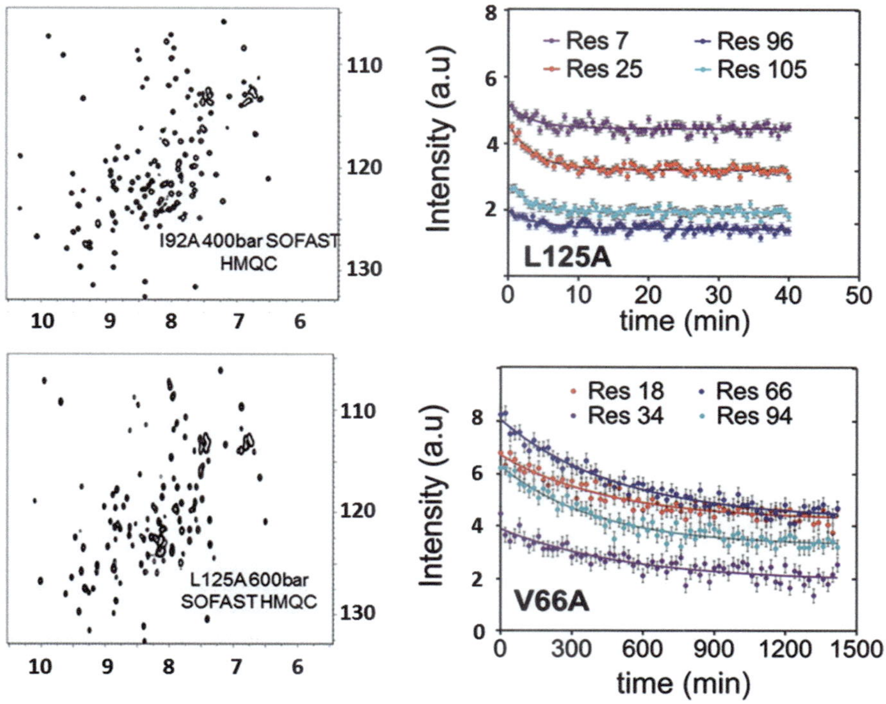

Fig. 4.6 Pressure-jump multi-dimensional NMR experiments of SNase variants. Left: ^{15}N-^{1}H HMQC SOFAST (Schanda et al. 2005; Schanda et al. 2006; Gal et al. 2009) spectra for two of the variants under pressure as indicated. Right: relaxation profiles of HMQC peak intensity vs time for selected residues, as indicated, from two of the variants, L125A at 1,200 bar and 0.85 M guanidinium hydrochloride and V66A at 1,200 bar and 1.1 M guanidinium hydrochloride (Reprinted with permission from (Roche et al. 2013). Copyright 2013 American Chemical Society)

4.5 Conclusions

While we have not succeeded or even attempted to parameterize the magnitude of the volume change of unfolding of proteins, responsible for the pressure effect on their stability, we have, using designed perturbations of well-studied model systems, demonstrated two important facts. The first is that perturbations that significantly change the amount of accessible surface area in the *unfolded* state have little or no effect on the magnitude of ΔV_u. Second, changes that introduce internal void volume in the *folded* protein structure lead to significant increases in the magnitude of ΔV_u. This identifies solvent-excluded void volume as the major contributing factor to pressure-induced unfolding of proteins. We have shown this unique mechanism of pressure effects to be particularly useful in the mapping of protein free energy landscapes.

References

Barrick D, Ferreiro DU, Komives EA (2008) Folding landscapes of ankyrin repeat proteins: experiments meet theory. Curr Opin Struct Biol 18:27–34

Bradley CM, Barrick D (2006) The notch ankyrin domain folds via a discrete, centralized pathway. Structure 14:1303–1312

Brandts JF, Oliveira RJ, Westort C (1970) Thermodynamics of protein denaturation. Effect of pressure on the denaturation of ribonuclease A. Biochemistry 9:1038–1047

Bridgman PW (1914) The coagulation of albumin by pressure. J Biol Chem 19:511–512

Frank HS, Evans MW (1945) Free volume and entropy in condensed systems. III. Entropy in binary liquid mixtures; partial molal entropy in dilute solutions; structure and thermodynamics in aqueous electrolytes. J Chem Phys 13:507–532

Frye KJ, Royer CA (1998) Probing the contribution of internal cavities to the volume change of protein unfolding under pressure. Protein Sci 7:2217–2222

Gal M, Kern T, Schanda P, Frydman L, Brutscher B (2009) An improved ultrafast 2D NMR experiment: towards atom-resolved real-time studies of protein kinetics at multi-Hz rates. J Biomol NMR 43:1–10

Galamba N (2013) Water's structure around hydrophobic solutes and the iceberg model. J Phys Chem B 117:2153–2159

Galamba N (2014) Water tetrahedrons, hydrogen-bond dynamics, and the orientational mobility of water around hydrophobic solutes. J Phys Chem B 118:4169–4176

Gallagher KR, Sharp KA (2003) A new angle on heat capacity changes in hydrophobic solvation. J Am Chem Soc 125:9853–9860

Hawley SA (1971) Reversible pressure–temperature denaturation of chymotrypsinogen. Biochemistry 10:2436–2442

Herberhold H, Winter R (2002) Temperature- and pressure-induced unfolding and refolding of ubiquitin: a static and kinetic Fourier transform infrared spectroscopy study. Biochemistry 41:2396–2401

Huang DM, Chandler D (2000) Temperature and length scale dependence of hydrophobic effects and their possible implications for protein folding. Proc Natl Acad Sci U S A 97:8324–8327

Jacob MH, Saudan C, Holtermann G, Martin A, Perl D, Merbach AE, Schmid FX (2002) Water contributes actively to the rapid crossing of a protein unfolding barrier. J Mol Biol 318:837–845

Kitahara R, Royer C, Yamada H, Boyer M, Saldana JL, Akasaka K, Roumestand C (2002) Equilibrium and pressure-jump relaxation studies of the conformational transitions of P13MTCP1. J Mol Biol 320:609–628

Lepori L, Gianni P (2000) Partial molar volumes of ionic and nonionic organic solutes in water: a simple additivity scheme based on the intrinsic volume approach. J Solut Chem 29:405–447

Lin LN, Brandts JF, Brandts JM, Plotnikov V (2002) Determination of the volumetric properties of proteins and other solutes using pressure perturbation calorimetry. Anal Biochem 302:144–160

Meersman F, Smeller L, Heremans K (2006) Protein stability and dynamics in the pressure-temperature plane. Biochim Biophys Acta 1764:346–354

Mei G, Di VA, Campeggi FM, Gilardi G, Rosato N, De MF, Finazzi-Agro A (1999) The effect of pressure and guanidine hydrochloride on azurins mutated in the hydrophobic core. Eur J Biochem 265:619–626

Mohana-Borges R, Silva JL, Ruiz-Sanz J, de Prat-Gay G (1999) Folding of a pressure-denatured model protein. Proc Natl Acad Sci U S A 96:7888–7893

Roche J, Caro JA, Norberto DR, Barthe P, Roumestand C, Schlessman JL, Garcia AE, Garcia-Moreno BE, Royer CA (2012) Cavities determine the pressure unfolding of proteins. Proc Natl Acad Sci U S A 109:6945–6950

Roche J, Dellarole M, Caro JA, Norberto DR, Garcia AE, Garcia-Moreno EB, Roumestand C, Royer CA (2013) Effect of internal cavities on folding rates and routes revealed by real-time pressure-jump NMR spectroscopy. J Am Chem Soc 135:14610–14618

Rouget JB, Schroer MA, Jeworrek C, Puhse M, Saldana JL, Bessin Y, Tolan M, Barrick D, Winter R, Royer CA (2010) Unique features of the folding landscape of a repeat protein revealed by pressure perturbation. Biophys J 98:2712–2721

Rouget JB, Aksel T, Roche J, Saldana JL, Garcia AE, Barrick D, Royer CA (2011) Size and sequence and the volume change of protein folding. J Am Chem Soc 133:6020–6027

Sasahara K, Nitta K (1999) Pressure-induced unfolding of lysozyme in aqueous guanidinium chloride solution. Protein Sci 8:1469–1474

Schanda P, Kupce E, Brutscher B (2005) SOFAST-HMQC experiments for recording two-dimensional heteronuclear correlation spectra of proteins within a few seconds. J Biomol NMR 33:199–211

Schanda P, Forge V, Brutscher B (2006) HET-SOFAST NMR for fast detection of structural compactness and heterogeneity along polypeptide chains. Magn Reson Chem 44:S177–S184

Seemann H, Winter R, Royer CA (2001) Volume, expansivity and isothermal compressibility changes associated with temperature and pressure unfolding of Staphylococcal nuclease. J Mol Biol 307:1091–1102

Street TO, Barrick D (2009) Predicting repeat protein folding kinetics from an experimentally determined folding energy landscape. Protein Sci 18:58–68

Street TO, Bradley CM, Barrick D (2007) Predicting coupling limits from an experimentally determined energy landscape. Proc Natl Acad Sci U S A 104:4907–4912

Till MS, Ullmann GM (2010) McVol – a program for calculating protein volumes and identifying cavities by a Monte Carlo algorithm. J Mol Model 16:419–429

Tripp KW, Barrick D (2008) Rerouting the folding pathway of the Notch ankyrin domain by reshaping the energy landscape. J Am Chem Soc 130:5681–5688

Zipp A, Kauzmann W (1973) Pressure denaturation of metmyoglobin. Biochemistry 12:4217–4228

Part II
Volume, Compressibility, Fluctuation and Interaction in Proteins

Editors' Foreword of Part II

The partial specific (or molar) volume, expansibility, and compressibility of a protein, introduced as empirical parameters in Chap. 2, are fundamental thermodynamic quantities to characterize protein structures in solution. In Chap. 5, Gekko starts with the definitions of these volumetric quantities, and describes method of measurements and implications to protein structure and dynamics. In Chap. 6, Williamson succeeded in expressing non-uniform compression of folded protein structure by pressure as changes in atomic coordinates, based on the NMR chemical shift changes. Chapter 7 by Hirata is devoted to show the essential role of solvent water on structural fluctuations of proteins, based purely on statistical mechanical calculations developed by him and coworkers. The theory also showed that the structural changes observed at high pressure is the magnification of the conformational fluctuation at ambient pressure. In Chap. 8, the topic is intermolecular interactions of proteins that would become important in the phenomena like protein aggregation, crystallization, and phase behavior in general. Here Winter combines an experimental approach by small-angle X-ray scattering and a liquid-state theoretical approach to understand the peculiarity of dense lysozyme solutions as a function of temperature and pressure as well as of salts and osmolytes.

Chapter 5	Volume and Compressibility of Proteins	Kunihiko Gekko
Chapter 6	Pressure-Dependent Conformation and Fluctuation in Folded Protein Molecules	Mike P. Williamson
Chapter 7	Water Turns the "Non-biological" Fluctuation of Protein into "Biological" One	Fumio Hirata
Chapter 8	Pressure Effects on the Intermolecular Interaction Potential of Condensed Protein Solutions	Roland Winter

Chapter 5
Volume and Compressibility of Proteins

Kunihiko Gekko

Abstract The partial specific (or molar) volume, expansibility, and compressibility of a protein are fundamental thermodynamic quantities for characterizing its structure in solution. We review the definitions, measurements, and implications of these volumetric quantities in relation to protein structural biology. The partial specific volumes under constant molality (isomolal) and chemical potential (isopotential) conditions of the cosolvent (multicomponent systems) are explained in terms of preferential solvent interactions relevant to the solubility and stability of proteins. The partial expansibility is briefly discussed in terms of the effects of temperature on protein–solvent interactions (hydration) and internal packing defects (cavities). We discuss the compressibility–structure–function relationships of proteins based on analyses of the correlations between the partial adiabatic compressibilities and the structures or functions of various globular proteins (including mutants), focusing on the roles of the internal cavities in structural fluctuations. The volume and compressibility changes associated with various conformational transitions are also discussed in terms of the changes in hydration and cavities in order to elucidate the nonnative structures and the transition mechanisms, especially those associated with pressure denaturation.

Keywords Cavity • Compressibility • Expansibility • Hydration • Partial volume • Protein • Volume fluctuation

5.1 Introduction

A huge amount of thermodynamic data of proteins has been accumulated using various techniques (e.g., calorimetry, densimetry, and spectroscopy), as compiled in many books (e.g., Hinz 1986). Thermodynamic data refer to macroscopic quantities, but they are indispensable for understanding the structural stability, dynamics, and function of a protein. Among the thermodynamic quantities, the partial specific (or

K. Gekko (✉)
Hiroshima Synchrotron Radiation Center, Hiroshima University,
Higashi-Hiroshima 739-0046, Japan
e-mail: gekko@hiroshima-u.ac.jp

K. Akasaka, H. Matsuki (eds.), *High Pressure Bioscience*,
Subcellular Biochemistry 72, DOI 10.1007/978-94-017-9918-8_5

molar) volume is a unique parameter for intuitively grasping the overall structure of a protein in solution, although unlike X-ray crystallography and NMR spectroscopy it does not provide atomic resolution. The partial specific volume involves the contributions of protein–solvent interactions (hydration) and internal packing defects (cavities) in addition to the constituent atomic volume, both of which depend on the solvent composition, temperature, and pressure. The partial specific volumes under constant molality (isomolal) and chemical potential (isopotential) conditions of a cosolvent (multicomponent systems) are different due to the preferential binding or exclusion of the cosolvent, which is a decisive factor for the protein solubility and stability (Casassa and Eisenberg 1964; Durchschlag 1986; Timasheff 1995). The temperature and pressure derivatives of the partial volume correspond to the partial expansibility and compressibility, respectively, which are characteristic of the protein structure since they modify both the cavities and hydration. Especially, the partial compressibility sensitively depends on the protein structure due to the counteracting effects of pressure on the cavities and hydration. Most compressibility studies have been carried out with adiabatic compressibility because measuring the isothermal compressibility of a protein is technically difficult. These volumetric data of proteins have been comprehensively discussed in many reviews (Kupke 1973; Durchschlag 1986; Gekko 1991; Chalikian 2003; Schweiker and Makhatadze 2009).

Interpretation of the volumetric data of proteins is still tentative because of the complicated contributions of cavities and hydration. There are arguments about the differences between the adiabatic and isothermal compressibilities of proteins from both theoretical and methodological viewpoints (Nölting 1995; Pinfield and Povey 1997; Gekko et al. 2009). Nevertheless, these volumetric data give new insights into protein structure and dynamics that cannot be obtained by other spectroscopic techniques. The compressibility is of particular interest to biophysical chemists because it is linked to volume fluctuation (Cooper 1976). Analyses of the correlation between the adiabatic compressibility and structure of proteins have revealed some of the factors that control the compressibility (Gekko and Noguchi 1979; Gekko and Hasegawa 1986). The effects of ligand binding and mutations on adiabatic compressibility demonstrate the important role of cavities in structural fluctuations and function (Kamiyama and Gekko 2000; Gekko 2002; Gekko et al. 2004; Son et al. 2012). On the other hand, changes in partial volume, expansibility, and compressibility upon a conformational transition have been measured in order to characterize the nonnative protein structures and the transition mechanism in terms of cavities and hydration (Tamura and Gekko 1995; Taulier and Chalikian 2002). Volumetric data give useful information especially for analyzing pressure denaturation, which is complicated due to the temperature–pressure cross-factor and the variable water structure. Thus, in accordance with the recent development of the techniques for accurate sound-velocity and density measurements, the compressibility data are becoming increasingly important in protein structural biology.

This chapter first provides an overview of the definitions, measurements, and implications of the partial volume, expansibility, and compressibility of a protein. The partial specific volumes in multicomponent solutions are explained in terms of the preferential solvent interaction in relation to protein solubility and stability.

The difference between the adiabatic and isothermal compressibilities of a protein is discussed from the theoretical and methodological viewpoints. We then discuss the compressibility–structure–function relationships of a protein based on analyses of the correlation between the adiabatic compressibility data and the structures or functions of various globular proteins, including mutants. Finally, we survey the volume and compressibility changes due to various conformational transitions so as to characterize the structures of nonnative proteins and the transition mechanisms, focusing on changes in cavities and hydration.

5.2 Partial Specific (or Molar) Volume

5.2.1 Definitions and Measurements

The intrinsic volume of a protein molecule can be determined from its X-ray crystal structure using any of several computational algorithms (Richards 1977; Chalikian et al. 1996). In solution, the volume relevant to thermodynamics is the partial specific (or molar) volume, which is defined as the change in total volume per unit mass upon adding an infinitesimal amount of protein at constant temperature, pressure, and masses in grams (or in moles) of all the other components. The partial specific volume involves not only the volume of the protein itself, but also the contribution of protein–solvent interactions (hydration). In general, a protein solution is a multicomponent system containing the protein itself, water, salts, and other substances. However, it is more convenient to reduce any such systems to a binary one comprising the protein as the first component and the second component being the solvent containing water, salts, and other substances. At constant temperature T and pressure P, the total volume V of such a binary solution comprising g_1 g of solvent and g_2 g of protein is expressed as

$$V = g_1 v_1 + g_2 v_2 \tag{5.1}$$

where v_1 and v_2 are the partial specific volumes of the solvent and the protein, respectively, as defined by

$$v_1 = (\partial V/\partial g_1)_{T,P,g2} \tag{5.2}$$

$$v_2 = (\partial V/\partial g_2)_{T,P,g1} \tag{5.3}$$

In practice, the apparent specific volume ϕ_2 is defined instead of the partial specific volume, by assuming the specific volume of the solvent in solution to be the same as that of the solvent in a pure phase, $v_1{}^{\text{o}}$:

$$V = g_1 v_1^{\text{o}} + g_2 \phi_2 \tag{5.4}$$

The apparent and partial specific volumes are interconnected as follows:

$$v_2 = \phi_2 + g_2(\partial\phi_2/\partial g_2)_{g1} \tag{5.5}$$

From this equation it is apparent that ϕ_2 equals v_2 at infinite dilution ($g_2 = 0$) or when ϕ_2 has no concentration dependence; that is, $(\partial\phi_2/\partial g_2)_{g1} = 0$. The partial molar volume V_2 is v_2 multiplied by the molecular weight of the protein.

The value of ϕ_2 can be determined experimentally by measuring the density ρ (in units of grams per milliliter) of the protein solution as a function of its concentration c in grams per milliliter (Kupke 1973):

$$\phi_2 = (1/c)\ [1 - (\rho - c)/\rho_o] \tag{5.6}$$

where ρ_o is the density of the solvent used to prepare the protein solution (in this case the molalities of the solvent components are constant). Various techniques including pycnometry, dilatometry, the magnetic float method, and the mechanical oscillator technique have been used to measure the density (Kupke 1973), but most density data are currently obtained by the mechanical oscillator technique (digital density meter) because of its accuracy and convenience (Kratky et al. 1973). Various types of instruments using this technique are manufactured by the Anton Paar Company in Graz, Austria. This technique can be used to make accurate density measurements of a small sample of a protein solution (a minimum of about 0.7 cm^3) with a high precision of $\pm 1.5 \times 10^{-6}$ g cm^{-3}, allowing for ϕ_2 values to be determined to three significant figures (e.g., 0.735 cm^3 g^{-1}). The procedures for the density measurement of a protein are described in detail elsewhere (Kupke 1973; Lee et al. 1979; Durchschlag 1986).

In general, ϕ_2 is dependent on the protein concentration due to intermolecular interactions, and so the value of ϕ_2 at infinite dilution (denoted as $\phi_2{}^o$) is used for the volumetric analysis of a protein molecule in solution. It is apparent from Eq. 5.5 that $\phi_2{}^o$ equals the value of v_2 at infinite dilution (denoted as $v_2{}^o$). The $v_2{}^o$ (or $\phi_2{}^o$) value measured for a constant molality of solvent components is the true thermodynamic volume of a protein. However, measuring $v_2{}^o$ (or $\phi_2{}^o$) often requires difficult sample preparation, and hence most experiments usually involve solutions that have been subjected to equilibrium dialysis against the solvent. At equilibrium, the chemical potentials of diffusible components are equal between the protein solution and dialyzate on the two sides of the dialysis membrane, but their molalities are not necessarily identical mainly due to the Donnan effects and the preferential binding of solvent components to the protein. Therefore, the partial (or apparent) specific volume obtained using the ρ_o value of dialyzate, which is denoted as $v'_2{}^o$ (or $\phi'_2{}^o$), is not necessarily identical to $v_2{}^o$ (or $\phi_2{}^o$); however, the difference between $v_2{}^o$ and $v'_2{}^o$ ($\phi_2{}^o$ and $\phi'_2{}^o$) is negligibly small in a two-component system and a dilute buffer system.

Books that have compiled large sets of $v_2{}^o$ or $\phi_2{}^o$ ($v'_2{}^o$ or $\phi'_2{}^o$) data of proteins are available (Fasman 1976; Durchschlag 1986). In an aqueous solution or a dilute buffer system at room temperature, the $v_2{}^o$ values of native (folded) proteins fall in the range from 0.69 to 0.76 cm^3 g^{-1}, with most exhibiting values in the range from 0.72 to 0.75 cm^3 g^{-1}. These values are within the range of the partial specific volumes of amino acids, and hence the $v_2°$ value of a protein has often been estimated from its amino acid composition by assuming additivity of the partial specific volumes (Cohn and Edsall 1943; Zamyatnin 1984). However, it should be noted that the additivity principle does not strictly apply to proteins, with the partial specific volume being a thermodynamic quantity that depends on the solvent composition, temperature, and pressure.

5.2.2 *Effects of the Cosolvent (Preferential Solvent Interactions)*

In the presence of the cosolvent at a high concentration (typically 0.3–10 M), the solvent composition in the vicinity of the protein molecule is different from the bulk solvent composition, with the magnitude of the difference varying with the affinity of the cosolvent to the protein (Fig. 5.1). An excess of cosolvent over the bulk-solvent composition indicates preferential binding and a deficiency of cosolvent in the vicinity of the protein indicates preferential hydration or preferential exclusion of the cosolvent.

The preferential solvent interaction includes all thermodynamic contributions from the chemical characteristics (ionic, polar, and nonpolar residues) on the protein surface, steric exclusion of the cosolvent (excluded volume effect), surface tension,

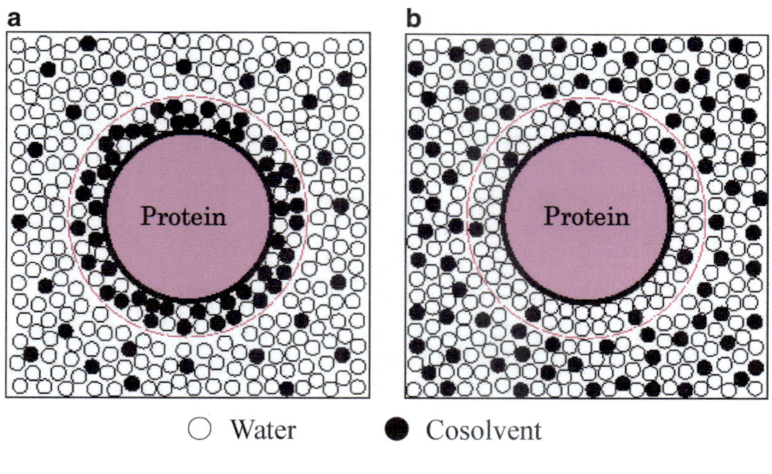

Fig. 5.1 Schematics of preferential binding (**a**) and preferential hydration (**b**) of a protein in cosolvent systems (Reproduced from Na and Timasheff (1981))

modified water structure, and others, and then any thermodynamic properties are affected by the preferential solvent interaction. In densitometry, the preferential solvent interaction manifests itself in the difference between the apparent specific volumes measured under isomolal ($\phi_2{}^{\mathrm{o}}$) and isopotential ($\phi'_2{}^{\mathrm{o}}$) conditions of solvent components as follows (Casassa and Eisenberg 1964; Lee et al. 1979; Timasheff 1995):

$$\phi_2^{\mathrm{o}} - \phi_2'^{\mathrm{o}} = \xi_1\,(1/\rho_{\mathrm{o}} - v_1) = \xi_3\,(1/\rho_{\mathrm{o}} - v_3) \tag{5.7}$$

where v_1 and v_3 are the partial specific volumes of water (component 1) and cosolvent (component 3), respectively, and ξ_1 and ξ_3 are the preferential interaction parameters defined as the number of grams of water and cosolvent, respectively, that must be added per gram of protein (component 2) to maintain components 1 and 3 at constant chemical potential (μ): $\xi_3 = (\partial g_3/\partial g_2)_{T,P,\mu 3}$ or $\xi_1 = (\partial g_1/\partial g_2)_{T,P,\mu 1}$. Here g_i is the concentration of component *i* expressed in grams of *i* per gram of the principal solvent (water). A positive value of ξ_3 indicates preferential binding, and a negative value (or positive value of ξ_1) indicates preferential hydration. According to Eq. 5.7, the ξ_1 and ξ_3 values can be determined from the measurements of $\phi_2{}^{\mathrm{o}}$ and $\phi'_2{}^{\mathrm{o}}$ of the protein (detailed experimental procedures are described in Lee et al. 1979). ξ_3 and ξ_1 can be converted into the preferential interaction parameters in molar units [$\psi_3 = (\partial m_3/\partial m_2)_{T,P,\mu 3}$ and $\psi_1 = (\partial m_1/\partial m_2)_{T,P,\mu 1}$, respectively] using the molality *m* of each component. ξ_3 is related to the amounts of cosolvent (A_3) and water (A_1) in grams bound per gram of protein, and ψ_3 is related to the amounts of cosolvent (B_3) and water (B_1) in moles bound per mole of protein:

$$\xi_3 = -g_3\xi_1 = A_3 - g_3 A_1$$

or

$$\psi_3 = -\,(m_3/m_1)\,\psi_1 = B_3 - (m_3/m_1)\,B_1 \tag{5.8}$$

Table 5.1 lists the values of $\phi_2{}^{\mathrm{o}}$, $\phi'_2{}^{\mathrm{o}}$, and ξ_3 of two proteins in various cosolvent systems. In general, preferential binding is observed for strong denaturants such as guanidine hydrochloride (GuHCl) and alcohols, and preferential hydration is observed for stabilization and crystallization reagents such as polyols and salts. Preferential hydration systems are of practical importance in biochemistry and food sciences (Gekko and Morikawa 1981; Gekko and Timasheff 1981).

The values of ψ_3 and ψ_1 are directly linked to the chemical potential of the protein (μ_2) in solution (Timasheff 1995). When ψ_3 is positive, the value of μ_2 in the cosolvent system is lower than its value in water, indicating that the interaction of the cosolvent with the protein is thermodynamically favorable to increase the solubility and solvent-accessible surface area of the protein, thereby facilitating unfolding and dissociation. When ψ_3 is negative, μ_2 increases in the cosolvent system, leading to decreases in the solubility and solvent-accessible surface area of the protein, thereby facilitating crystallization, folding,and association. The sign

Table 5.1 Apparent specific volumes of proteins at isomolal ($\phi_2{}^o$) and isopotential ($\phi_2{}'^o$) conditions of cosolvents, and preferential interaction parameters (ξ_3)

Solvent	$\phi_2{}^o$ ($cm^3\ g^{-1}$)	$\phi_2{}'^o$ ($cm^3\ g^{-1}$)	ξ_3 ($g\ g^{-1}$)
Bovine serum albumin			
Water (25 °C)	0.735	0.735	0
1 *M* Na_2SO_4 (pH 5.6, 20 °C)[a]	0.735	0.788	−0.074
30 % glycerol (25 °C)[b]	0.729	0.744	−0.101
30 % sorbitol (25 °C)[b]	0.738	0.761	−0.092
20 % propylene glycol (pH 2, 25 °C)[c]	0.740	0.726	0.211
1 *M* GdHCl (20 °C)[d]	0.735	0.729	0.025
Chymotrypsinogen A			
30 % glycerol (pH 5.8, 20 °C)[e]	0.727	0.745	−0.123
1 *M* sucrose (1 m*M* HCl, 20 °C)[f]	0.736	0.773	−0.138
6 *M* GdHCl (20 °C)[g]	0.729	0.712	0.15

[a]Arakawa and Timasheff (1982), [b]Gekko and Morikawa (1981), [c]Gekko and Koga (1984), [d]Arakawa and Timasheff (1984), [e]Gekko and Timasheff (1981), [f]Lee and Timasheff (1981), [g]Lee and Timasheff (1974)

of ψ_3 may change at a critical cosolvent concentration or pH where the relative magnitudes of $\phi_2{}^o$ and $\phi'_2{}^o$ are reversed (Eq. 5.7). Although measurements are rare, the $\phi_2{}^o$ and $\phi'_2{}^o$ values at various temperatures represent important information on the mechanisms of preferential solvent interaction and protein stabilization (Gekko and Morikawa 1981). Directly measuring both $\phi_2{}^o$ and $\phi'_2{}^o$ under high pressure is technically difficult, but the possible contribution of preferential solvent interaction must be considered in studies of multicomponent protein solutions under high pressure (e.g., pressure denaturation in a urea solution).

According to the Wyman linkage relation (Wyman and Gill 1990; Timasheff 1995), the dependence of the cosolvent concentration (in activity unit a_3) on the equilibrium constant K for the equilibrium reaction X $\leftrightarrow$ Y is related to the difference in ψ_3 between the two states, $\Delta\psi_3$:

$$d\ \ln K/d\ \ln a_3 = \Delta\psi_3 = \psi_3^{\mathrm{Y}} - \psi_3^{\mathrm{X}} = \left(B_3^{\mathrm{Y}} - B_3^{\mathrm{X}}\right) - (m_3/m_1)\ \left(B_1^{\mathrm{Y}} - B_3^{\mathrm{X}}\right) \tag{5.9}$$

At a low cosolvent concentration, where m_3/m_1 is negligible, $\Delta\psi_3$ or d ln K/d ln a_3 is determined only by the difference in B_3 between states X and Y, while it is determined by changes in both B_3 and B_1 at a high cosolvent concentration. $\Delta\psi_3$ is generally positive for strong denaturants (e.g., urea and GuHCl) due to $0 < \psi_3{}^{\mathrm{X}} < \psi_3{}^{\mathrm{Y}}$, while it is negative for stabilizing cosolvents (polyols and sugars) due to $\psi_3{}^{\mathrm{Y}} < \psi_3{}^{\mathrm{X}} < 0$. Thus, the partial specific volumes in multicomponent solutions are important not only in determining the volumetric quantities of the solution, but also in yielding thermodynamic information on the phase- and chemical-equilibrium phenomena of proteins.

5.2.3 *Contributions of Cavity and Hydration*

The partial specific volume of a protein at infinite dilution, $v_2{}^{\mathrm{o}}$ (or $\phi_2{}^{\mathrm{o}}$), is expressed as the sum of three contributions (Kauzmann 1959): (i) the constitutive volume estimated as the sum of van der Waals volumes of the constitutive atoms (v_{c}), (ii) the volume of the cavities in the molecule due to imperfect atomic packing (v_{cav}), and (iii) the volume change due to solvation or hydration (Δv_{sol}) on the protein surface. Hereafter, a subscript "2" attached to $v_2{}^{\mathrm{o}}$ is abbreviated as v^{o} for simplifying the protein notation:

$$v^{\mathrm{o}} = v_{\mathrm{c}} + v_{\mathrm{cav}} + \Delta v_{\mathrm{sol}} \tag{5.10}$$

where the intrinsic volume of a protein is defined as $v_{\mathrm{c}} + v_{\mathrm{cav}}$. Figure 5.2 depicts the cross section of a protein molecule that illustrates the internal cavities and the solvent-accessible surface. This surface is defined as a surface traced by the center of a spherical probe of appropriate radius (this equals 1.4 Å for most purposes, which is that of a water molecule) rolling on the molecular surface (Lee and Richards 1971; Richards 1977), and has often been approximated as being proportional to the amount of hydration. Each cavity is also defined by rolling a probe of smaller radius ($\sim$ 1.2 Å) inside the protein molecule, although the obtained cavity volume is dependent on the size of the probe. The thermal volume, which is defined as an empty domain around the solute molecule resulting from the mutual thermal motions of the solute and solvent molecules in a scaled particle theory (Reiss 1965), is not depicted here, and is included in Δv_{sol} for simplicity (the thermal volume of a protein is described in more detail in Chalikian et al. 1996).

The Δv_{sol} value is negative at standard temperature and pressure because the hydration around the ionic, polar, and nonpolar groups accompanies the volume contraction of the solvent water as a whole. The contributions of cavities and

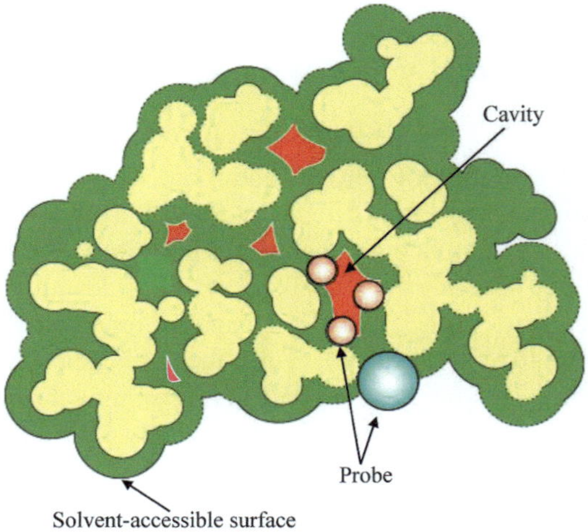

Fig. 5.2 Schematic cross section of a protein molecule illustrating the internal cavities and the solvent-accessible surface (Reproduced from Richards (1977) and Lee and Richards (1971))

hydration to v^o have been estimated based on certain assumptions. The v_{cav} values of some proteins were preliminarily estimated using the Δv_{sol} and v_c values calculated from the known amount of hydration and the van der Waals volumes of the constitutive atoms (Gekko and Noguchi 1979). Chalikian and colleagues evaluated the Δv_{sol} values using the intrinsic volumes ($v_c + v_{cav}$ values) calculated from X-ray crystal structures (Chalikian et al. 1996; Chalikian 2003). While precisely separating the two contributions would be difficult due to the complicated boundary between the protein and the solvent, it is apparent that v_{cav} is only a few percent of v^o and compensates the negative Δv_{sol} to result in only a small difference between v_c and v^o.

The volume change due to a conformational transition of the protein is defined as the difference in v^o between the two conformers before and after the transition. Since v_c does not change during a conformational transition, the observed volume change can be primarily attributed to the changes in v_{cav} and Δv_{sol}. In general, the volume decreases for processes that increase the solvent-accessible surface area, such as unfolding and dissociation into subunits, mainly due to the increased hydration. As discussed in Sect. 5.6, most experimental data meet this expectation, but some data show no change or even a slight volume increase, which are probably due to enlarged cavities. It should be noted that the volume change at a given protein concentration may differ—even in its sign—from that obtained at infinite dilution when significant protein–protein interaction is present.

5.3 Expansibility

The partial specific volume of a native globular protein in aqueous solution increases linearly with temperature in the range from 4 to 45 °C: positive values of (2.5–10) $\times 10^{-4}$ cm^3 g^{-1} K^{-1} have been reported for $\partial v^o/\partial T$, with most lying in the range of (3.5–5) $\times 10^{-4}$ cm^3 g^{-1} K^{-1} (Durchschlag 1986). Since the van der Waals volume is not expansible, the partial thermal expansion coefficient (α^o) is mainly attributable to the effects of temperature on cavities and hydration as follows:

$$\alpha^o = (1/v^o)\ (\partial v^o/\partial T) = (1/v^o)\ [(\partial v_{cav}/\partial T) + (\partial \Delta v_{sol}/\partial T)] \qquad (5.11)$$

Since the cavity enlarges and the hydration decreases with increasing temperature, both terms on the right-hand side of Eq. 5.11 would contribute positively to α^o. Although it is difficult to separate the two terms precisely, the temperature effects on the X-ray crystal structure may aid the estimation of the two terms (Gekko and Hasegawa 1989; Chalikian 2003). For myoglobin, volume expansions of 5 % in the crystal lattice and 3 % in the protein molecule occur as the temperature increases from 80 to 300 K: the latter corresponds to a thermal expansion coefficient of 1.15×10^{-4} K^{-1} (Frauenfelder et al. 1987). Ribonuclease A (RNase A) exhibits a smaller expansion of 0.9 % as the temperature increases from 98 to 320 K (Tilton et al. 1992). These values are clearly smaller than the thermal expansion coefficients of these proteins in solution. Considering that the thermal expansion of a crystal structure mainly comes from the cavities, the larger thermal expansion of a protein in solution would be dominantly attributed to a volume increase due to dehydration upon increasing temperature. In fact, the temperature effect on v^o is more pronounced for amino acids and small peptides with negligibly small cavities (e.g., 1.3×10^{-3} cm^3 g^{-1} K^{-1} for glycine). In this sense the denatured protein is expected to have a larger $\partial v^o/\partial T$ value than the folded one because unfolding is associated with an increase in the solvent-accessible surface or hydration. The $\partial v^o/\partial T$ value of RNase A (pH 1.6) increases from 5.0×10^{-4} to 7.8×10^{-4} cm^3 g^{-1} K^{-1} upon thermal denaturation (Tamura and Gekko 1995). Hawley reported that the $\partial v^o/\partial T$ values of RNase A (pH 2.0) and chymotrypsinogen (pH 2.07) increase by 0.18×10^{-4} and 0.51×10^{-4} cm^3 g^{-1} K^{-1}, respectively, upon pressure denaturation (Hawley 1971). Pressure-perturbation calorimetry, which has recently been developed by MicroCal (Northampton, MA, USA), is available for determining the change in partial molar expansibility due to thermal denaturation of a protein. In this technique, the apparent thermal expansion coefficient of a protein is calculated from the heat changes induced by small periodic changes in pressure above a protein solution. The technique and its application in biological systems are described elsewhere (Schweiker and Makhatadze 2009; Zhai et al. 2011).

5.4 Compressibility

The coefficient of partial specific compressibility of a protein in solution, β^o, is defined as the change in the partial specific volume (v^o) with increasing pressure (P): $\beta^o = -(1/v^o)(\partial v^o/\partial P)$. This coefficient is more useful than the partial specific compressibility ($\kappa^o = v^o \beta^o$) when comparing the pressure sensitivities and flexibilities of different proteins. Since the van der Waals volume of atoms is assumed to be incompressible, the experimentally observed β^o value of a protein can be mainly attributed to the effects of pressure on cavities and hydration (Gekko and Noguchi 1979):

$$\beta^o = -(1/v^o)\ (\partial v^o/\partial P) = -(1/v^o)\ [(\partial v_{cav}/\partial P) + (\partial \Delta v_{sol}/\partial P)] \tag{5.12}$$

The first and second terms on the right-hand side contribute positively and negatively, respectively, to β^o, and so β^o can be either positive or negative according to the relative magnitude of these two terms. This is prominently different from the partial expansion coefficient (α^o), for which temperature contributes positively to both the cavity and hydration terms (Eq. 5.11). Consequently, compressibility very sensitively reflects the structure of a protein. Two types of compressibilities are defined according to the experimental conditions: adiabatic ($\beta_s{}^o$) and isothermal ($\beta_T{}^o$). Most compressibility studies of proteins have been of $\beta_s{}^o$ because directly measuring $\beta_T{}^o$ is technically difficult for proteins.

5.4.1 Adiabatic Compressibility

The coefficient of partial specific adiabatic compressibility, $\beta_s{}^o$, can be determined by measuring the sound velocity (u) and the density (ρ) of a protein solution based on the Newton-Laplace equation ($\beta_s = 1/\rho u^2$) (Gekko 1991; Sarvazyan 1991; Taulier and Chalikian 2002). The sound velocity has been measured mainly by two types of techniques using ultrasound at 3–8 MHz: sing-around pulse method and resonance-ultrasound spectroscopy. In the sing-around pulse method, which was developed by Greenspan and Tschiegg (1956), sound pulses repeatedly transmitted by a transmitter into the protein solution are received by a receiver positioned opposite the transmitter. Each pulse detected at the receiver triggers a new sound pulse at the transmitter repeatedly, and the repetition time of the sequence of these pulses is transformed into the sound velocity with an accuracy of 1 cm sec^{-1} based on the electrical circuitry delay time and the distance between the transmitter and receiver. In contrast, resonance-ultrasound spectroscopy determines the sound velocity based on the resonance frequency produced when sweeping a continuous sound between two resonators. It has an accuracy of 0.1 cm sec^{-1} by means of differential measurements of the sample solution and solvent (Sarvazyan 1991). This spectroscopy method is currently commonly used with the aid of various types of commercially available spectrometer, including the ResoScan device (TF-Instruments, Heidelberg, Germany) and the HR-US 102 device (Ultrasonic Scientific, Dublin, Ireland).

The $\beta_s{}^o$ value at infinite dilution can be calculated using the following equation (Gekko and Noguchi 1979; Gekko 1991):

$$\beta_s^o = -(1/v^o)\ (\partial v^o/\partial P)_s = (1/v^o)\, lim_{c\to 0}\,(1/c)\ [\beta_s/\beta_{s,o} - (\rho - c)/\rho_o] \tag{5.13}$$

where β_s and $\beta_{s,o}$ are the adiabatic compressibility coefficients of the protein solution and solvent, respectively. Chalikian's group uses a somewhat different equation based on the relative sound-velocity increment of the protein (Taulier and Chalikian 2002), but the $\beta_s{}^o$ value obtained at infinite dilution is the same for both calculation methods. The $\beta_s{}^o$ values of more than 50 proteins have been measured in water or dilute buffer solutions by many groups (Millero et al. 1976; Gavish et al. 1983; Sarvazyan and Kharakoz 1977; Gekko and Hasegawa 1986; Chalikian et al. 1996), with the largest set of $\beta_s{}^o$ data of proteins (including mutants) being reported by Gekko and coworkers (Gekko 1991; Gekko et al. 2000, 2004). Table 5.2 lists the $\beta_s{}^o$ and v^o values of various proteins (including mutants) at 25 °C. $\beta_s{}^o$ is negative for fibrous proteins due to the large contribution of hydration overcoming the cavity effect, whereas most globular proteins show positive $\beta_s{}^o$ values due to the large cavity effect overcoming the hydration effect. Interestingly, the $\beta_s{}^o$ value of globular proteins varies over a wide range and is affected by amino acid substitution. The value of $\beta_s{}^o$ is greatly affected by the protein structure mainly due to the variation of the cavity volume among different proteins, because the amount of hydration does

Table 5.2 Partial specific volumes (v^o) and adiabatic compressibility coefficients ($\beta_s{}^o$) of native proteins in water at 25 °C[a, b]

Protein	v^o ($cm^3\ g^{-1}$)	$\beta_s{}^o$ ($Mbar^{-1}$)
Globular proteins		
Soybean trypsin inhibitor	0.713	0.17
Trypsin	0.717 (0.718)	0.92 (2.4)
Ribonuclease A	0.704 (0.702)	1.12 (1.3)
Peroxidase	0.702	2.36
Cytochrome c[c]	0.738 (0.735)	3.6 (4.2)
Chymotrypsinogen A	0.734 (0.727)	4.05 (4.4)
α-Chymotrypsin	0.717 (0.717)	4.15 (3.9)
Lysozyme (hen)	0.712 (0.699)	4.67 (1.9)
(turkey)	0.709	1.97
Conalbumin	0.728 (0.729)	4.89 (6.3)
Carbonic anhydrase	0.742	6.37
α-Lactalbumin	0.736 (0.711)	8.27 (3.8)
β-Lactoglobulin	0.751 (0.731)	8.45 (5.7)
Pepsin	0.743 (0.730)	8.60 (7.1)
Myoglobin	0.747 (0.742)	8.98 (9.6)
Ovalbumin	0.746 (0.735)	9.18 (8.2)
Bovine serum albumin	0.735 (0.735)	10.5 (9.0)
Dihydrofolate reductase (wild, 15 °C)[d]	0.723	1.7
mutant (G67D)	0.724	3.0
mutant (G121S)	0.721	0.7
mutant (A145G)	0.726	3.1
cAMP-Receptor protein (wild)[e]	0.750	7.98
mutant (S62F)	0.749	6.89
mutant (T127L)	0.748	9.68
mutant (L148R)	0.753	9.43
Fibrous proteins		
Gelatin	0.689	−2.5
F-actin (20 °C)[f]	0.720	−6.3
Myosin (20 °C)[f]	0.724	−18
Tropomyosin (20 °C)[f]	0.733	−41

[a]Taken from Gekko and Noguchi (1979), Gekko and Hasegawa (1986), and Gekko and Yamagami (1991). [b]The values in parentheses are taken from Chalikian et al. (1996). [c]Kamiyama et al. (1999). [d]Gekko et al. (2000). [e]Gekko et al. (2004). [f]Sarvazyan and Kharakoz (1977)

not differ greatly among globular proteins (being about 0.3 g per gram of protein). The $\beta_s{}^o$ values of various proteins are plotted against v^o in Fig. 5.3, in which it is evident that $\beta_s{}^o$ increases linearly with v^o (correlation coefficient: $r = 0.85$). This proportionality is reasonable because the cavities contribute positively and hydration contributes negatively to both parameters, confirming the dominant participation of cavities in the variation of $\beta_s{}^o$. The apparent adiabatic compressibility of the cavities is roughly estimated to be a few hundred per megabar (e.g., 550 $Mbar^{-1}$ for

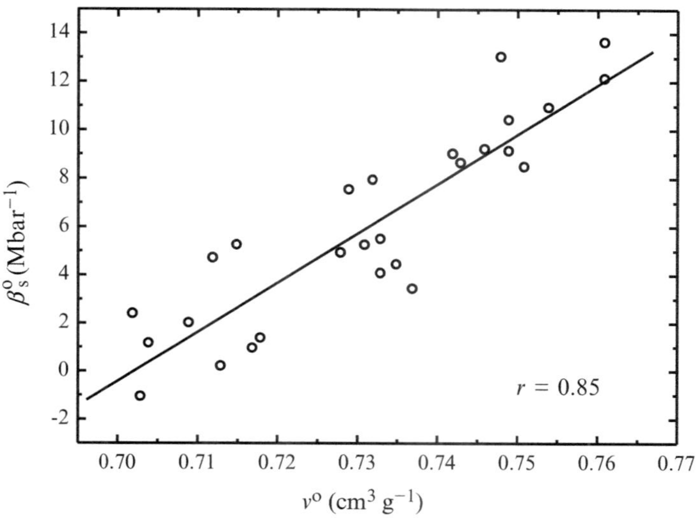

Fig. 5.3 Plot of $\beta_s{}^o$ against v^o for proteins (Reproduced from Gekko and Hasegawa (1986), with the addition of some unpublished data)

α-chymotrypsinogen A) (Gekko and Noguchi 1979), and hence only a small change in the cavity characteristics induces large changes in $\beta_s{}^o$. The intrinsic adiabatic compressibility of globular proteins free from hydration has been evaluated to be 10–30 Mbar^{-1} using various methods (Gekko and Noguchi 1979; Kharakoz and Sarvazyan 1993; Kharakoz 2000; Taulier and Chalikian 2002). These values are comparable to or higher than the adiabatic compressibility of ice ($\sim$ 13 Mbar^{-1}) and smaller than that of water ($\sim$ 45 Mbar^{-1}). This suggests that the protein interior exhibits considerable flexibility, which is necessary for its biological function. The compressibility–structure–function relationships of globular proteins are discussed in Sect. 5.5.

5.4.2 *Isothermal Compressibility*

The pressure-dependent properties of proteins are directly related to the coefficient of partial specific isothermal compressibility ($\beta_T{}^o$) rather than to $\beta_s{}^o$. Determination of $\beta_T{}^o$ requires the measurement of v^o as a function of pressure under hydrostatic pressure or a centrifugal force. However, such experiments are generally difficult for proteins because v^o varies little with pressure and applying a high pressure may induce a conformational change in a protein outside the range of its native fold and modify the preferential solvent interaction; these difficulties have resulted in only a few $\beta_T{}^o$ data being reported for proteins. A notable technique for measuring $\beta_T{}^o$ is oscillatory densitometry, which yields reliable solution density values under high pressures up to 80 MPa. A device for implementing this type of densitometry measurement is commercially available (DMA 512P, Anton Paar, Graz, Austria). Seemann et al. (2001) applied this technique to the pressure-induced unfolding of staphylococcal nuclease. They found that $\beta_T{}^o$ decreases with increasing pressure in both the native and unfolded states, which suggests that a protein molecule becomes less compressible due to enhanced atomic packing and/or water penetration when it is subjected to high pressure. We have recently measured the $\beta_T{}^o$ value of bovine serum albumin (BSA) at various temperatures, and confirmed that $\beta_T{}^o$ increases with temperature (as does $\beta_s{}^o$) due to the expanded cavities and the decreased amount of hydration at higher temperatures (Gekko et al. 2009).

Other methods are available for evaluating the isothermal compressibility coefficient of a protein molecule itself ($\beta_{T,M}$), although it is not a partial quantity in solution. Kundrot and Richards (1987) evaluated the $\beta_{T,M}$ value of tetragonal hen egg-white lysozyme to be 4.7 $Mbar^{-1}$ based on X-ray crystallographic data obtained at 0.1 and 100 MPa. Fourme et al. (2001) reported about twofold value of $\beta_{T,M}$ for this protein based on crystal structures at pressures up to 1 GPa. They confirmed that contraction of the molecule is anisotropic and decreases with increasing pressure. $\beta_{T,M}$ can also be estimated by computational methods such as molecular dynamics and normal-mode analysis (Yamato et al. 1993; Paci and Marchi 1996; Dadarlat and Post 2001; Mori et al. 2006).

5.4.3 Difference Between Adiabatic and Isothermal Compressibilities

The coefficients of isothermal (β_T) and adiabatic (β_s) compressibilities of a system (or substance) are related according to the following equation:

$$\beta_T = \beta_s + T\alpha^2/\rho c_p \tag{5.14}$$

where ρ is the density, α is the thermal expansion coefficient, and c_p is the specific heat capacity at constant pressure. Previous studies of protein compressibility have used this equation to estimate $\beta_T{}^o$ from experimentally obtained $\beta_s{}^o$ values based on the analogy of the relationship between $\beta_T{}^o$ and $\beta_s{}^o$ (Millero et al. 1976; Gekko and Noguchi 1979; Gekko and Hasegawa 1989). However, the underlying assumption is oversimplified because this equation does not take the thermal expansion of a bulk solvent into consideration. The relationship between the two compressibilities for a solute in solution is expressed rigorously as follows (Blandamer et al. 2001; Taulier and Chalikian 2002):

$$\beta_T^o = \beta_s^o + T\left(\alpha_o^2/\rho_o c_{p,o}\right)\left(2\alpha^o/\alpha_o - c_p^o/v^o \rho_o c_{p,o}\right) \tag{5.15}$$

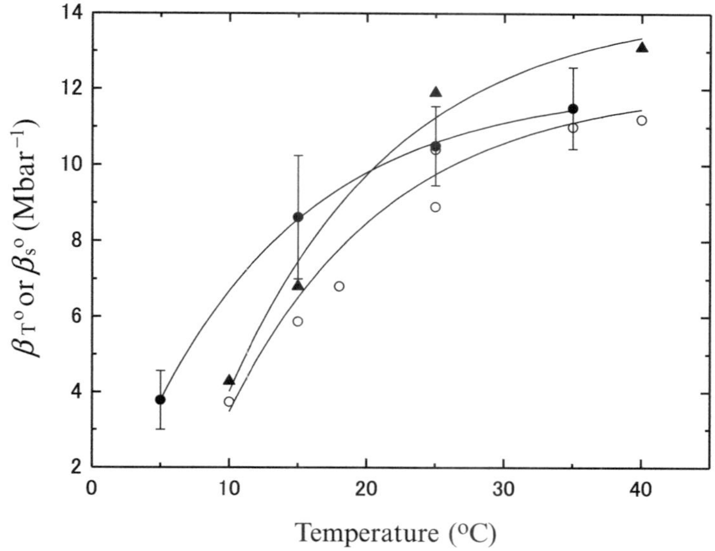

Fig. 5.4 Temperature dependence of $\beta_s{}^o$ and $\beta_T{}^o$ of BSA in water. ○ $\beta_s{}^o$ determined experimentally, ● $\beta_T{}^o$ determined experimentally, ▲ $\beta_T{}^o$ calculated using Eq. 5.15 based on the experimentally observed $\beta_s{}^o$ values (Reproduced from Gekko et al. (2009))

where ρ_o , α_o, and $c_{p,o}$ are the density, thermal expansion coefficient, and specific heat capacity of the solvent, respectively, and $c_p{}^o$ is the partial specific heat capacity of the protein at constant pressure. This equation applies at all pressures and temperatures. However, since the pressure dependences of $\beta_T{}^o$, $\beta_s{}^o$, α^o, and $c_p{}^o$ are unknown for most proteins, the difference between $\beta_T{}^o$ and $\beta_s{}^o$ of a protein may be evaluated within a small range of pressures and temperatures where the constancy of these parameters is guaranteed. Figure 5.4 shows plots of the experimentally observed $\beta_s{}^o$ and $\beta_T{}^o$ values of BSA against temperature at atmospheric pressure (Gekko et al. 2009). It is apparent that $\beta_T{}^o$ is slightly larger than $\beta_s{}^o$ at all temperatures examined, as expected theoretically. However, the experimentally observed $\beta_T{}^o$ values are in agreement with those calculated using Eq. 5.15 based on the experimental $\beta_s{}^o$ values only around room temperature. The difference between the two compressibilities increases with decreasing temperature, which contradicts the theoretical expectation that it should decrease to zero around 4 °C, where the solvent expansion is negligible ($\alpha_o = 0$). The origin of these discrepancies cannot be explained at present because the experimentally observed $\beta_T{}^o$ and $\beta_s{}^o$ values may involve some unknown contributions in addition to large experimental errors (greater than 5 %).

There are some theoretical arguments about the interpretation of the $\beta_s{}^o$ value of a protein determined by sound velocity measurements. Nölting (1995) suggested that the interior of a protein molecule is close to the isothermal condition in a sound velocity measurement, so that the experimentally obtained $\beta_s{}^o$ values are between the adiabatic and isothermal compressibilities (pseudoadiabatic theory). Pinfield and Povey (1997) proposed that propagation of the sound remains adiabatic under the normal experimental conditions, but the effects of thermal scattering of the sound must be taken into account when calculating adiabatic compressibility (thermal scattering theory). Since most cavities are essentially empty, they would be adiabatic for any thermal exchange that occurred during sound scattering. However, compressing a protein under an isothermal condition may squeeze out cavities and force water into them. A high-pressure NMR study provided evidence for such penetration of water molecules into cavities under high pressure (Akasaka 2006). Since the cavities are the most compressible part of a protein molecule, even a small perturbation of these cavities will cause large changes in $\beta_T{}^o$ and $\beta_s{}^o$. The difference in thermal conductivity between hydrating water and bulk water may induce a temperature difference between the protein and the solvent, resulting in sound velocity measurements not being performed under pure adiabatic conditions. Further, the thermal volume of a protein may induce additional modes in the mutual thermal motions of the protein and hydrating water. Thus, the interpretation of the $\beta_T{}^o$ and $\beta_s{}^o$ values of a protein molecule involving cavities is still theoretically and experimentally controversial, and thereby is different from those of small solute molecules (Gekko et al. 2009). However, the $\beta_s{}^o$ data determined from sound velocity measurements are of practical value in investigations of the hydration, conformation, and dynamics of a protein molecule, because the difference between $\beta_T{}^o$ and $\beta_s{}^o$ is not significant under normal experimental conditions. Accumulation of more experimental data on both $\beta_T{}^o$ and $\beta_s{}^o$ is required for a deeper understanding of the pressure effects on protein molecules.

5.4.4 Volume Fluctuations

According to statistical thermodynamics, the volume fluctuation $< (\delta V_M)^2 >$ of a molecule with volume V_M is related to its isothermal compressibility coefficient $\beta_{T,M}$ (Cooper 1976) according to

$$< (\delta V_M)^2 >= k_B T V_M \beta_{T,M} \tag{5.16}$$

where k_B is the Boltzmann constant and T is the absolute temperature. The volume fluctuation of a protein molecule may be estimated by assuming an analogous equation in which V_M and $\beta_{T,M}$ are replaced by the corresponding partial quantities V^o and $\beta_T{}^o$ (or $\beta_s{}^o$). The estimated volume fluctuations of proteins are within the range of 30–200 cm^3 mol^{-1}; this range is equivalent to a volume range of 2–11 mol of water, being dependent on the size of the protein molecule (Gekko and Hasegawa 1986). Although these volume fluctuations correspond to only about 0.3 % of the overall dimensions of a protein, if concentrated within one area at a particular moment they could produce sufficient cavities or channels to allow the entry of a solvent or probe molecules. This interpretation is consistent with a mobile defect model (Lumry and Rosenberg 1975; Pain 1987) and a local unfolding or solvent penetration model (Karplus and McCammon 1981; Woodward et al. 2004) for protein fluctuation, which have been used to account for hydrogen-exchange and fluorescence-quenching phenomena of folded proteins. The timescale of volume fluctuations and the underlying mechanics at the microscopic level have been discussed through a high-pressure NMR approach (Hattori et al. 2004; Williamson in this book).

5.5 Compressibility–Structural–Function Relationships

While β_s^o sensitively depends on the type of proteins (Table 5.2), its relationships to the structures and functions of proteins remain unknown. Knowledge of the structural factors affecting the β_s^o value is important because it gives new insights into the origin of protein dynamics that is essential for the expression of biological function. The compressibility–structure–function relationships of proteins have been investigated by analyses of the correlations of β_s^o values with some structural factors and biological functions of various globular proteins, including mutants.

5.5.1 *Structural Factors Determining Compressibility*

The correlation between the β_s^o values and various molecular parameters of 25 globular proteins was statistically analyzed by Gekko and Hasegawa (1986). Since the ratio of the accessible surface area to the volume of a protein molecule is lower for proteins with higher molecular weights, the hydration effect would be small relative to the cavity effect in large proteins, and then β_s^o may increase with the molecular weight of the protein. However, no definite correlation was found between β_s^o and the molecular weight, probably because large proteins often consist of some subunit and/or domain structures different from small proteins. The helix-rich proteins appear to be highly compressible even though the α-helix itself is rigid: typical helix proteins (myoglobin and BSA) show very large β_s^o values compared with essentially nonhelix proteins (soybean trypsin inhibitor and trypsin). It is unlikely that such a bulky structural unit could be packed in a protein molecule without producing some cavities around it, and hence α-helices could be a dynamic domain causing the thermal fluctuations of a protein. This is consistent with the results of a computational analysis that the α-helices in myoglobin have a very small internal compressibility, but that the spaces between the helices are greatly compressed by pressure (Yamato et al. 1993). On the other hand, β-strands did not show distinct correlations with β_s^o values, probably because most β-strands are connected with each other to form β-sheets over long distances, unlike α-helices. Proteins with a higher hydrophobicity were more compressible ($r = 0.70$); this is reasonable because the cavities would be mainly generated by imperfect packing of nonpolar amino acid residues localized in the interior of the protein molecule, and the nonpolar surface would act to decrease the volume change due to hydration. There was a linear correlation ($r = -0.70$) between β_s and the number of disulfide (S–S) bonds normalized by the total number of residues, implying that a larger fraction of S–S bonds will result in making a protein molecule more compact (Gekko et al. 2003). The plots of β_s^o against the average number of residues enclosed by the S–S linkage showed a weakly negative correlation ($r = -0.60$), suggesting that cross-linked cysteine residues being farther apart in the primary structure will make a greater contribution to the compactness of a protein. Thus, the S–S bonds of a protein play an active role in reducing structural fluctuations, as expected from the reduced conformational entropy of the polypeptide chain.

It is not possible to predict the β_s^o value of a protein from its primary structure, but the contribution of each amino acid residue to β_s^o can be estimated by analyzing the correlation between β_s^o and the amino acid composition of various globular proteins (Gekko and Hasegawa 1986). Four amino acid residues (Leu, Glu, Phe, and His) can strongly increase β_s^o, while another four (Asn, Gly, Ser, and Thr) decrease β_s^o. All four residues in the former group are strong helix-formers and two residues (Asn and Gly) in the latter group are strong helix-breakers, which is consistent with the positive correlation between β_s^o and the α-helix content. Multiple correlation analysis using both groups of amino acids as independent variables revealed a strong correlation ($r = 0.74$), meaning that it is possible to predict the β_s^o value of an unknown protein from its amino acid composition with a fair degree of certainty. Interestingly, a single amino acid substitution brought about a noticeable change in β_s^o (Table 5.2) (Gekko et al. 1996, 2000, 2004), indicating that changes in the local structure cooperatively affect the overall protein dynamics, as discussed below.

β_s^o may be influenced by many other factors such as flexible loops, domain structures, and prosthetic groups. At present it is difficult to estimate their contributions because all of them cooperatively affect β_s^o via changes in cavities and hydration. However, a statistical analysis of the correlations between β_s^o and various molecular parameters would be a suitable approach for deducing the structural factors regulating the fluctuations of a protein and for predicting the β_s^o values of unknown proteins. A better understanding of the structural factors affecting β_s^o at the microscopic level requires theoretical calculations of compressibility (Yamato et al. 1993; Paci and Marchi 1996; Mori et al. 2006).

5.5.2 *Protein–Ligand and Protein–Protein Interactions*

The formation of a protein–ligand complex generally proceeds via three stages: (i) formation of chemical bonds between the protein and ligand, (ii) changes in hydration of the interacting species, and (iii) conformational changes of the protein. In general, the solute–solute interaction in water accompanies dehydration of each solute molecule to increase β_s^o and v^o, but both parameters do not necessarily increase with the ligand binding of a protein. A typical example is the formation of a lysozyme–inhibitor complex. The v^{o} and $\beta_s{}^{o}$ values of lysozyme decrease by the addition of *N*-acetyl-D-glucosamine oligomers, in the order of monomer > dimer > trimer (Gekko and Yamagami 1998). Assuming that the decrease in $\beta_s{}^{o}$ is attributed to the protein moiety only, the volume fluctuations are estimated to decrease by about 3 %, 7 %, and 14 % on binding the monomer, dimer, and trimer, respectively. The effects of these inhibitors, which are bound in the cleft of the active site ("induced-fit" interaction), would propagate throughout the protein molecule to tighten the internal atomic packing and reduce the structural fluctuation of the protein molecule. A similar conclusion was drawn by Son et al. (2012) for the binding of trimer to lysozyme, although there was a less-significant reduction in the volume fluctuation. Binding of cytidine 2′-monophosphate or cytidine 3′-monophosphate to RNase A produces small increases in $\beta_s{}^{o}$ and v^{o} because the dehydration effect overcomes the atomic packing effect (Dubins et al. 2000).

More pronounced changes in $\beta_s{}^{o}$ and v^{o} by ligand binding were found for *Escherichia coli* dihydrofolate reductase (DHFR) (Kamiyama and Gekko 2000). DHFR catalyzes the reduction of dihydrofolate (DHF) to tetrahydrofolate (THF) with the aid of coenzyme NADPH (the reduced form of nicotinamide adenine dinucleotide phosphate) in the kinetic reactions via five intermediates: DHFR·NADPH, DHFR·NADPH·DHF, DHFR·$NADP^{+}$·THF, DHFR·THF, and DHFR·NADPH·THF ($NADP^{+}$ is the oxidized form of nicotinamide adenine dinucleotide phosphate). This enzyme reaction is strongly inhibited by methotrexate (MTX). Figure 5.5 shows plots of the $\beta_s{}^{o}$ values of these kinetic intermediates in the reaction pathway, in which the $\beta_s{}^{o}$ value of the transient state DHFR·NADPH·DHF is assumed to be the same as that of DHFR·NADPH·MTX due to their structural similarity. $\beta_s{}^{o}$ changes alternately by binding or releasing the coenzyme and substrate, the transient state is the most flexible, and DHFR·$NADP^{+}$·THF is the most rigid of the intermediates. X-ray crystal structures indicate that the solvent-accessible surface area decreases only slightly (at most 6 %), but the total cavity volume is greatly influenced (±40 %) by the binding of these ligands: the number, size, and distribution of cavities also change in parallel with the changes in compressibility. These results demonstrate that the cavity effect overcomes the hydration effect in ligand-induced compressibility changes in DHFR and that the enzyme function is closely related to the structural fluctuation (Gekko 2002).

Direct measurements of the v^{o} and $\beta_s{}^{o}$ changes due to protein–protein interactions are technically difficult, although they may be determined based on the pressure dependence of the equilibrium constant. Taulier and Chalikian reported the changes in v^{o} and κ^{o} associated with dimer-to-monomer dissociation of β-lactoglobulin to be $-0.008\ cm^3\ g^{-1}$ and $-0.7 \times 10^{-6}\ cm^3\ g^{-1}\ Mbar^{-1}$, respectively, and suggested that a 7 % increase in hydration was associated with the increased surface area of monomer (Taulier and Chalikian 2001). The $\beta_s{}^{o}$ and v^{o} changes due to protein–DNA interactions might also provide important information on molecular recognition in genes, but there have been no direct measurements (Chalikian 2003).

The digestibility of proteins by protease has been used as a measure of protein flexibility in food science, even though digestion itself is a chemical reaction

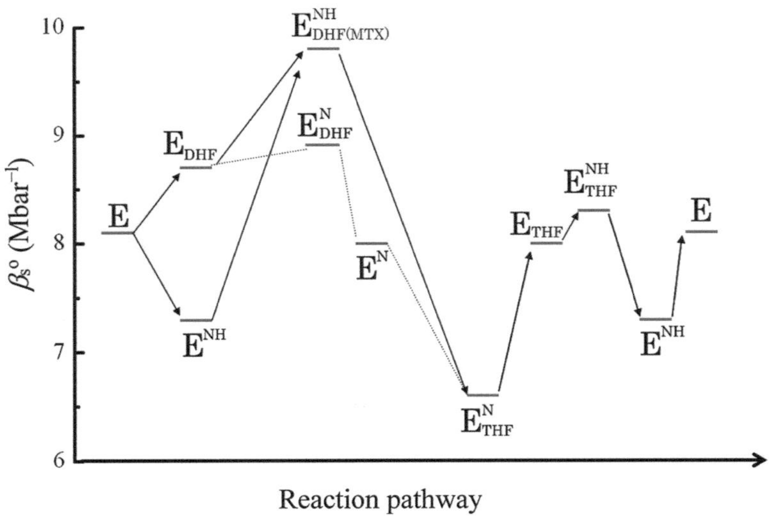

Fig. 5.5 Changes in $\beta_s{}^{o}$ of the kinetic intermediates of DHFR in the reaction pathway. Capital letters, E, NH, and N, in the intermediates refer to enzyme (DHFR), NADPH, and NAGP^{+}, respectively (Reproduced from Kamiyama and Gekko (2000))

involving limited peptide bonds. This is based on the idea that the peptide bonds in flexible proteins would be more frequently exposed to the solvent and hence be easily attacked by protease. The digestion velocities of proteins do tend to increase with their $\beta_s{}^{o}$ values (Gekko and Yamagami 1991). Most proteases and protease inhibitors appear to be less compressible than nutrient (storage) proteins such as egg and milk proteins (Table 5.2). The hydrogen–deuterium exchange rate of a protein was found to increase with the $\beta_s{}^{o}$ value (Gekko et al. 2004). Thus, the compressibility data complement digestibility and hydrogen–deuterium exchange experiments related to protein fluctuation.

5.5.3 Amino Acid Substitution

Only small visible changes have been observed in the X-ray crystal structures of mutants in many cases where amino acid substitution clearly induces large changes in function and stability. However, the $\beta_s{}^{o}$ and v^{o} values of six mutants of *E. coli* aspartate aminotransferase (AspAT) at Val39, which is located at the gate of the substrate-binding site, are different from those of the wild-type protein (Gekko et al. 1996). Since the amount of hydration should be hardly affected by substituting 1 of 396 amino acid residues, these changes in $\beta_s{}^{o}$ and v^{o} could be mainly due to modifications of internal cavities. The observed linear correlation between $\beta_s{}^{o}$ and the van der Waals volumes of the introduced amino acids suggests that the bulkiness of the side chain at this position makes the protein structure more flexible due to enlargement of the cavities. Interestingly, the enzyme activity (k_{cat}/K_m) of AspAT mutants increases with their $\beta_s{}^o$ values, indicating that the increased fluctuation of the structure enhances the enzyme function.

The enzyme activity and stability of DHFR (comprising 159 amino acid residues) are influenced by mutations at Gly67, Gly121, and Ala145, which are located at the center of 3 different flexible loops far from the catalytic site Asp27 (Gekko et al. 1994; Ohmae et al. 1996, 1998). Since these mutation sites do not participate directly in ligand binding, the modified enzyme activity would be mainly due to the change in k_{cat}, and not that in K_m. These loop mutations induce large changes in v^o (0.710–0.733 $cm^3\ g^{-1}$) and $\beta_s{}^o$ (−1.8–5.5 $Mbar^{-1}$) compared to the corresponding values of the wild-type enzyme ($v^o = 0.723\ cm^3\ g^{-1}$ and $\beta_s{}^o = 1.7\ Mbar^{-1}$) (Gekko et al. 2000). The $\beta_s{}^o$ values increase with the enzyme activity, indicating that the structural fluctuation also positively contributes to enzyme function. A high-pressure NMR study of folate-bound DHFR revealed that these loop regions are greatly affected by pressure as well as hinge motion of the active-site Met20 loop, supporting that the loop regions play important roles in protein dynamics and function (Kitahara et al. 2000). A hyperactive DHFR mutant with a 3.7-fold higher activity than the wild type, in which all five methionine and two cysteine residues were replaced by other amino acid residues using a molecular evolutional technique, showed a large $\beta_s{}^o$ value that was associated with increases in v^o and v_{cav}, as predicted from the plots of $\beta_s{}^o$ against the enzyme activity of loop mutants (Iwakura et al. 2006). It is noteworthy that the local structure changes induced by mutation affect the overall protein dynamics so as to regulate the enzyme function (Gekko 2002).

E. coli cyclic AMP receptor protein (CRP) is composed of two chemically identical subunits, each of which contains two domains: (i) the larger N-terminal domain (residues 1–133) contains the binding site of cyclic AMP (cAMP) and (ii) the smaller C-terminal domain (residues 139–209) binds to DNA through a helix–turn–helix motif. Binding of cAMP to CRP exhibits positive cooperativity to significantly enhance the DNA binding, and hence CRP is a typical allosteric protein. Amino acid substitutions at distal residues (K52N, D53H, S62F, G141Q, L148R, H159L, and K52N/H159L) from DNA- and cAMP-binding sites induce large changes in v^o (0.747–0.756 $cm^3\ g^{-1}$) and $\beta_s{}^o$ (6.89–9.68 $Mbar^{-1}$), increasing or decreasing from the corresponding values of the wild-type protein ($v^o = 0.750\ cm^3\ g^{-1}$ and $\beta_s{}^o = 7.98\ Mbar^{-1}$) (Gekko et al. 2004). $\beta_s{}^o$ exhibits strong linear correlations with both the free energy of DNA binding and the difference in the free energy of binding between two cAMP molecules, as shown in Fig. 5.6. These results demonstrate that the structural fluctuation plays an essential role in modulating the DNA-binding function and the allosteric properties of this protein. Together with the effects of mutation on the β_s values of proteins examined so far, a 1-unit increase in $\beta_s{}^o$ is assumed to enhance the function of a protein by about tenfold. This may be useful as a rough criterion when designing protein function based on dynamic aspects in protein engineering.

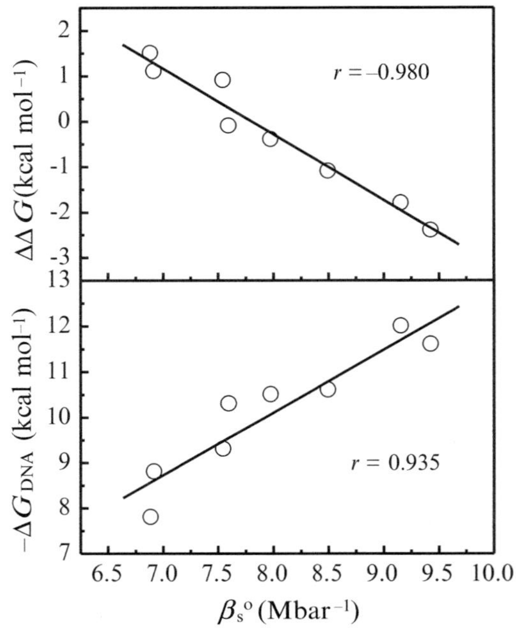

Fig. 5.6 Plots of $\beta_s{}^o$ of CRP mutants against the difference in the free energy of binding between two cAMP molecules (**a**) and the free energy of DNA binding (**b**) (Reproduced from Gekko et al. (2004))

In relation to amino acid substitution, it is of interest how the structural fluctuation of ancestral proteins has changed during molecular evolution as part of the process of environmental adaptation. Although the neutral theory of molecular evolution has been widely accepted (Kimura 1983), there is considerable controversy about whether protein evolution follows a neutral or nonneutral pathway at the molecular level, since few proteins have a series of evolutionary variants. As an approach to answer this problem, we calculated the $\beta_{T,M}$ values of chicken lysozyme and its six evolutionary mutants at Thr40, Ile55, and Ser91—in which a ternary mutant at these residues corresponds to bobwhite lysozyme—by applying normal-mode analysis to their X-ray crystal structures (Mimura et al. 2012). These evolutionary intermediates showed larger $\beta_{T,M}$ values than the two extant lysozymes from chicken and bobwhite. Based on this result it can be predicted that ancestral lysozymes had larger volume fluctuations than extant ones for adapting to diverse environments, and that the molecular evolution of lysozymes followed a nonneutral evolutionary pathway, since according to the neutral theory of molecular evolution, the overall properties of evolutionary intermediates should fall within the range exhibited by the extant species. The linear correlation found between $\beta_{T,M}$ and the total cavity volume also suggests that the presence of cavities or the atomic packing is an important factor for regulating volume fluctuations during the molecular evolution of this protein.

5.6 Volume and Compressibility Changes due to Conformational Transitions

Despite the huge number of thermodynamic and spectroscopic investigations of protein denaturation, the structures or compactness of unfolded states remain unclear. Since any conformational changes accompany modifications of cavities and hydration, the v^o and $\beta_s{}^o$ values are reliable for characterizing unfolded structures under various denaturing conditions. The volume and compressibility changes (Δv^o and $\Delta \kappa_s{}^o$) induced by certain typical conformational transitions are listed in Table 5.3. These results are briefly explained according to the changes in cavities and hydration below; more-detailed discussions are available elsewhere (Taulier and Chalikian 2002; Chalikian 2003).

Table 5.3 Changes in partial specific volume (Δv^o) and adiabatic compressibility ($\Delta \kappa_s{}^o$) due to conformational transitions of proteins[a]

Protein	Δv^o ($cm^3\ g^{-1}$)	$\Delta \kappa_s{}^o$ ($cm^3\ g^{-1}\ Mbar^{-1}$)
GuHCl-induced unfolding		
Ribonuclease A (pH 2, 15 °C)[b]	−0.06 (0.630)	−16 (−29.1)
Lysozyme (pH 4, 25 °C)[c]	−0.08 (0.579)	−9 (−19.9)
Reduction of S–S bond		
Lysozyme (pH 7, 25 °C)[d]	−0.008 (0.704)	−5.7 (−3.4)
Bovine serum albumin (pH 7, 25 °C)[d]	−0.009 (0.726)	−5.2 (3.4)
Acid-denaturation (Molten globule)		
Staphylococcal nuclease (20 m*M* NaCl, 25 °C)[e]	−0.013	−6.3
Cytochrome c (water, 25 °C)[f]	0.010	−2.0
CsCl-induced MG (pH 2, 25 °C)[f]	0.006	6.8
sorbitol-induced MG (pH 2, 20 °C)[g]	−0.003 (0.748)	11 (15)
Thermal denaturation		
Ribonuclease A (pH 1.9, 25.8 °C)[b]	−0.03 (0.675)	1.7 (4.7)
Chymotrypsinogen A (pH 2, 25 °C)[h]	0.003	4.3
Pressure denaturation		
Chymotrypsinogen (pH 2.07, 0 °C)[i]	−0.00056	1.2 in $\Delta \kappa_T{}^o$
Ribonuclease A (pH 2.0, 0 °C)[i]	−0.0035	1.4 in $\Delta \kappa_T{}^o$
Staphylococcal nuclease (pH 5.3, 24 °C)[j]	−0.0025	1.1 in $\Delta \kappa_T{}^o$
(pH 7, 45 °C)[k]	+0.005	2.1 in $\Delta \kappa_T{}^o$
(pH 7, 45 °C, 100 MPa)[k]	−0.0033	1.9 in $\Delta \kappa_T{}^o$

[a]The v^o and $\beta_s{}^o$ values of unfolded states are in parentheses, [b]Tamura and Gekko (1995), [c]Kamiyama and Gekko (1997), [d]Gekko et al. (2003), [e]Filfil and Chalikian (2000), [f]Chalikian et al. (1995), [g]Kamiyama et al. (1999), [h]Chalikian et al. (1997), [i]Hawley (1971), [j]Lassalle et al. (2000), [k]Seemann et al. (2001)

5.6.1 Unfolding by a Strong Denaturant

RNase A (pH 2, 15 °C) and lysozyme (pH 4, 25 °C) fully unfolded by GuHCl show large negative β_s^o values of −29.1 and −19.9 $Mbar^{-1}$, respectively, which are much smaller than the corresponding values for their native states (−4.0 and 4.6 $Mbar^{-1}$) (Tamura and Gekko 1995; Kamiyama and Gekko 1997). These β_s^o values are close to the average for unfolded proteins estimated from those of amino acid residues (Lee et al. 2008). The v^o values also decrease markedly in the fully unfolded states. These results indicate that most of the internal residues are exposed to the solvent upon GuHCl denaturation. However, the extent of their exposure would be influenced by the S–S bonds and the residual secondary structures being present even in the fully unfolded states. Interestingly, the v^o and β_s^o values deviate from the two-state transition curves obtained by circular dichroism (CD) spectroscopy (Tamura and Gekko 1995; Kamiyama and Gekko 1997). Although these values in the transition region involve the contributions of preferential binding of GuHCl in addition to the conformational change of the protein, this result suggests that the compactness of a protein molecule does not necessarily change in an all-or-none mechanism and that some intermediates are involved in the unfolding process. The free-energy changes of protein unfolding (ΔG_u) induced by GuHCl or urea tend to decrease as the β_s^o value of the native state increases, at room temperature (Gekko and Yamagami 1991), suggesting that a hard protein is more stable than a soft protein and that the flexibility of the native state significantly influences protein stability.

5.6.2 Reduction of S–S Bonds

Reduction of S–S bonds induces the unfolding of a protein even when a denaturant is not present. The v^o and β_s^o values of five globular proteins decreased when their S–S bonds were completely reduced by carboxyamidomethylation (Gekko et al. 2003). However, these values of the reduced state are considerably larger than those of the fully unfolded state induced by GuHCl (Table 5.3), indicating that the reduced state is still somewhat compact. The κ_s^o values of all the oxidized and reduced proteins decrease with increasing CD intensity at 222 nm ($r = -0.83$), suggesting that the protein structure becomes more compressible as the remaining secondary structures increase, and that the changes in volumetric properties induced by reduction are dominantly attributable to unfolding of the secondary structures. The effects of this reduction on β_s^o becomes more significant as the number of S–S bonds increases and as they are formed over larger distances in the primary structure, as expected from the effects of S–S bonds in reducing the fluctuation of the native state.

5.6.3 *Acid Denaturation (Molten Globule)*

Acid denaturation generally decreases β_s^o but does not necessarily decrease v^o (Table 5.3). For a pH-induced transition, the situation is complex since the observed v^o and β_s^o values include the contributions of protonation/deprotonation of ionizable groups in addition to the conformational change of the protein. Any change in pH induces a large change in hydration (especially around the carboxyl group), resulting in significant changes in v^o and β_s^o. Chalikian and colleagues estimated the contributions of ionizable groups to v^o and β_s^o, and found that the overall effect of a pH change seems to depend on the characteristics of the proteins investigated (Chalikian et al. 1995; Filfil and Chalikian 2000; Taulier and Chalikian 2001).

A similar degree of complexity is present for the molten globule (MG) states induced by salts and organic solvents. The MG state is generally formed by a delicate balance between electrostatic repulsion and the opposing stabilizing forces such as hydrophobic interaction. The MG states of cytochrome c induced by 200 mM NaCl and CsCl (pH 2) increase β_s^o without a large change in v^o (Nölting and Sligar 1993; Chalikian et al. 1995). This result indicates that the MG state is a highly fluctuating ensemble in which the cavity effect is greater than the hydration effect. However, it should be noted that the v^o and β_s^o values when cosolvents are present at high concentrations might be influenced by preferential interaction between the solvent and the protein.

Polyols such as sorbitol also induce the MG state of cytochrome c, with the effect being greater as their concentration and chain length increase (the latter corresponds to a larger number of OH groups) (Kamiyama et al. 1999). In this case the MG state is stabilized by the enhanced hydrophobic interaction overcoming the electrostatic repulsion between the ionizable groups, because polyols are preferentially excluded from the nonpolar surface of a protein without any significant change in the dielectric constant of the solvent (Gekko and Morikawa 1981). The v^o and β_s^o values in the absence of preferential solvent interaction were estimated by linearly extrapolating these values obtained at high concentrations of sorbitol to an infinite dilution. The obtained β_s^o value was much larger than those of the native and salt-induced MG states, while the v^o value did not change within the experimental error. These results suggest that the sorbitol-induced MG is more flexible than the salt-induced one and that the MG structure is sensitive to the solvent conditions.

5.6.4 *Thermal Denaturation*

The β_s^o and v^o values of α-chymotrypsinogen A increase upon thermal denaturation under acidic conditions following the two-state transition mechanism (Chalikain et al. 1997). On the other hand, β_s^o increases and v^o decreases upon thermal denaturation of RNase A, and both values deviate from the two-state transition curves obtained by CD spectroscopy (Tamura and Gekko 1995). The thermal denaturation of cytochrome c results in v^o decreasing but no change in $\beta_s{}^o$ (Dubins et al. 2003). Thus, the thermally denatured state is dependent on the type of proteins, probably due to differences in the secondary structures, S–S bonds, and the extent of exposure of internal residues. The generally observed increase in $\beta_s{}^o$ suggests that the thermally denatured state is a highly fluctuating ensemble in which the cavity effect is larger than the hydration effect as well as the MG states.

The thermal denaturation temperature (T_m) of a protein appears to be higher when its $\beta_s{}^o$ value in the native state is larger, although the correlation is weak ($r = 0.61$) (Gekko and Hasegawa 1986). This suggests that a rigid protein cannot necessarily tolerate heat, even though it is resistant to strong denaturants (GuHCl and urea). A flexible protein may be affected less by large thermal fluctuations due to the buffer action of its internal cavities. Since the temperature dependence of ΔG_u is approximated by a quadratic function with a convex upward curvature, and this curvature broadens when the difference in the molar heat capacity (ΔC_p) between the native and denatured states is smaller, it is possible that a flexible protein has a more gentle curvature, resulting in a higher T_m (where $\Delta G_u = 0$) than a rigid protein even if the flexible protein has a smaller ΔG_u value than the rigid one at room temperature.

5.6.5 Pressure Denaturation

In contrast to other types of protein denaturation, pressure denaturation is directly induced by the difference in partial volumes between the native and denature states. The pressure denaturation of a protein is complicated by the nonadditive effects of pressure and temperature and the variable water structure, as typically shown by the *P–T* diagram in Fig. 5.7a (Brandts et al. 1970; Hawley 1971; Zipp and Kauzmann 1973). This elliptic *P–T* diagram can be analyzed by the following equation:

$$\begin{aligned}\Delta G_u = {} & \Delta G_o - \Delta K_T^o (P - P_o)^2/2 + \Delta E^o (P - P_o)\ (T - T_o) \\ & - \Delta C_p [T\ (\ln T/T_o - 1) + T_o] + \Delta V_o (P - P_o) - \Delta S_o (T - T_o)\end{aligned} \tag{5.17}$$

where the subscripts in T_o and P_o indicate the arbitrarily chosen references to the T and P states (generally room temperature and 0.1 MPa), $K_T{}^o = \beta_T{}^o V^o = -(\partial V^o/\partial P)_T$, and $E^o = \alpha^o V^o = (\partial V^o/\partial T)_P$. Δ always means the change in the corresponding parameter in molar units during denaturation at T_o and P_o; for example, $\Delta V_o = V^o$ (denatured) – V^o (native). The thermodynamic parameters ($\Delta K_T{}^o$, ΔE^o, ΔC_p, ΔV_o, ΔS_o, and ΔG_o) that determine the actual shape, size, and orientation of the elliptic boundary can be estimated by fitting the curve of the *P–T* diagram to Eq. 5.17. However, this type of detailed analysis has been performed only for a small number of proteins (Brandts et al. 1970; Hawley 1971; Prehoda et al. 1998; Smeller 2002).

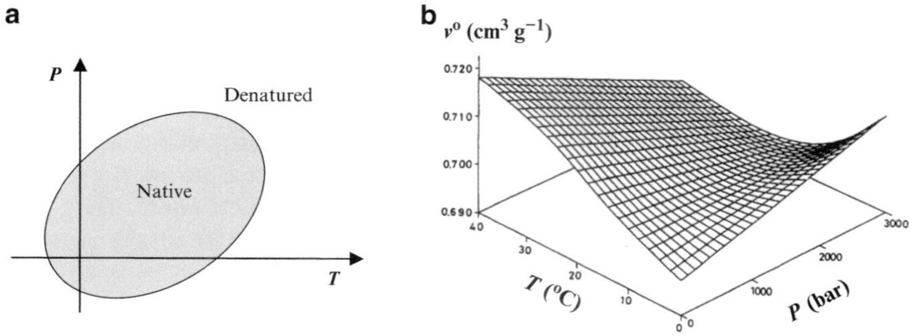

Fig. 5.7 Schematic of the elliptic *P*–*T* phase diagram of a protein (**a**) and the *P*–v^o–*T* diagram of native lysozyme (**b**) (Reproduced from Gekko and Hasegawa (1989))

An important finding in these studies is that v^o decreases but $\beta_s{}^o$ increases upon pressure denaturation in most cases (Table 5.3). This result cannot be explained by the simple unfolding model in which the atomic groups (mainly nonpolar groups) buried in the interior of a protein molecule are exposed to the solvent accompanying the increase in hydration and the decreased volume of the cavities. This contradiction might be explained by assuming that the internal nonpolar groups are not fully exposed to the solvent, with instead some of them remaining loosely packed and the local concentration of nonpolar groups being quite high in the denatured protein (Gekko and Noguchi 1979). Although the presence of loosely packed nonpolar groups might result in the production of a cavity, its volume increase would be offset by the volume decrease due to the increased hydration, resulting in the decrease in v^o. On the other hand, even a small increase in v_{cav} could make $\partial v_{cav}/\partial P$ sufficiently negative to overcome the increase in $\partial \Delta v_{sol}/\partial P$ of the denatured state, leading to the increase in $\kappa_s{}^o$, because the apparent compressibility of the cavities is very large. Such an enlargement of a cavity would be likely even if water molecules penetrate into it because the partial molar volume of water in nonpolar organic solvents (e.g., 22 cm^3 mol^{-1} in benzene) is larger than that of pure water (18 cm^3 mol^{-1}) (Masterton and Seiler 1968). The contributions of hydration and cavities to pressure denaturation of a protein are discussed more quantitatively in Chaps. 3 and 4 of this book.

The *P*–v^o–*T* diagram, which corresponds to the equation of state of a protein molecule, is indispensable for understanding the elliptic *P*–*T* diagram of pressure denaturation. However, while such analyses have been applied to many small compounds, they are very rare for proteins. The *P*–v^o–*T* diagrams of native lysozyme and BSA were predicted from the temperature dependence of $\beta_s{}^o$ (Gekko and Hasegawa 1989). The result for lysozyme is depicted in Fig. 5.7b, which shows curvature in the low-temperature and high-pressure regions due to the effective cross-factor of temperature and pressure: $\partial v^o/\partial T$ is positive over the entire temperature region at low pressure, but changes to negative in the low-temperature and high-pressure regions, and the critical temperature for $\partial v^o/\partial T = 0$ shifts to a higher temperature as the pressure increases. Similar inversion effects of temperature and pressure have been observed for the critical micelle concentration of surfactants and in the solubility of hydrocarbons. These results might be mainly attributable to the water-structure changes in the bulk and hydration phases under high pressure. $\partial v_{cav}/\partial T$ should always be positive, but it would become smaller under high pressure when the cavities are compressed. $\partial \Delta v_{sol}/\partial T$ is also positive at low temperature since the amount of hydration decreases with increasing temperature. Thus, both the cavities and hydration would contribute positively to $\partial v^{o}/\partial T$ at low pressure. In contrast, under high pressure the bulk water is compressed so as to decrease its molar volume, with the extent being greater at lower temperature due to the increased compressibility of water. For example, the molar volume of water, which is 18.0 $cm^3\ mol^{-1}$ at 25 °C and 0.1 MPa, is reduced to 17.0 $cm^3\ mol^{-1}$ at 25 °C and 150 MPa and to 16.2 $cm^3\ mol^{-1}$ at 0 °C and 300 MPa. These values are comparable with the molar volume of water of hydration around the solute molecule at standard temperature and pressure. Since the water of hydration is less compressible than bulk water, it is probable that Δv_{sol}—which is negative at low pressure—becomes positive under the high-pressure and low-temperature conditions where the molar volume of bulk water is greatly reduced. If this is the case, $\partial \Delta v_{sol}/\partial T$ could be negative and compensate for the positive $\partial v_{cav}/\partial T$, making $\partial v^{o}/\partial T$ negative in the high-pressure and low-temperature regions. A similar P–v^{o}–T diagram would also be expected for the denatured state, although its curvature would differ from that of native state, leading to an inversion of the volume change of denaturation at a given temperature and pressure in the P–T diagrams of pressure denaturation. Although these speculations should be confirmed by more-detailed analyses, it is indisputable that the water structure in bulk solvent and the hydration phases play a crucial role in the pressure denaturation of a protein.

5.6.6 Polymerization

Flagellin from *Salmonella typhimurium* is composed of a central portion with multiple compact domains and disordered N- and C-terminal regions (Ala1–Arg65 and Ser451–Arg494). Flagellin undergoes large conformational changes upon polymerization into a filament, and the disordered terminal regions become stabilized in the filament. Tamura et al. (1997) found that the v^{o} and $\beta_s{}^{o}$ values decrease in the order of the central portion (0.745 $cm^3\ g^{-1}$ and 6.7 $Mbar^{-1}$, respectively) > filament (0.734 $cm^3\ g^{-1}$ and 4.7 $Mbar^{-1}$) > flagellin (0.728 $cm^3\ g^{-1}$ and 4.0 $Mbar^{-1}$). The smaller v^{o} and $\beta_s{}^{o}$ values for flagellin than for the central portion are reasonably explained by the presence of disordered terminal regions that are thought to be highly hydrated. The volume increase upon polymerization of the flagellin into a filament is also consistent with the finding that the filament is depolymerized under high pressure. The smaller v^{o} and $\beta_s{}^{o}$ values of the filament compared with the central portion suggest the presence of extensive hydration of the filament on its complex surface, which surpasses the effects of cavities that are possibly generated at the subunit interfaces upon polymerization.

There are very few v^o and $\beta_s{}^o$ data available for amyloid fibrils, although they are important for a deep understanding of the compactness and fluctuation of the fibrils, which are crucial for amyloidosis. The genetically engineered disulfide-deficient variant of hen lysozyme, in which the eight Cys residues are replaced by Ala or Ser, constitutes an ideal system for such a volumetric study. This variant, which has no tertiary structure and only residual secondary structures in monomeric form, spontaneously forms protofibrils through a linear polymerization mechanism in 20 mM sodium acetate buffer (pH 4.0) containing 30 mM NaCl. Akasaka et al. (2007) found that v^o and $\beta_s{}^o$ increased greatly, from 0.684 to 0.724 $cm^3\ g^{-1}$ and from −7.48 to 1.35 $Mbar^{-1}$, respectively, as the monomer transformed into a protofibril in two phases, with the faster phase preceding the main development of the β-structure in the protofibrils. These results are consistent with the protofibrils dissociating under high pressure in a fully reversible manner, and demonstrate that protofibrils are highly voluminous and compressible, disclosing a cavity-rich fluctuating nature for amyloid protofibrils. However, caution is needed when extending these results to other amyloid systems since high pressure does not necessarily lead to the dissociation of amyloid fibrils.

5.7 Concluding Remarks

As reviewed in this chapter, the volumetric properties (partial specific volume, expansibility, and compressibility) of a protein provide important information on its structure and dynamics that cannot be obtained by other spectroscopic techniques. It should be kept in mind that the partial specific volume involves the hydration effect in addition to the intrinsic volume, and that it differs under isomolal and isopotential conditions of the cosolvent due to the preferential solvent interaction. The partial specific adiabatic compressibility is very sensitive to the structure of a protein due to the opposite effects of pressure on internal cavities and hydration. The compressibility–structure–function relationships reveal a decisive role of cavities in the structural fluctuation and function of a protein, which represents important new information that could be utilized for molecular designs in protein engineering. The partial volume and compressibility data also are reliable for detecting the compactness of nonnative proteins and the mechanism of conformational transitions (especially pressure denaturation). Although the interpretation of volume and compressibility data is still tentative because of the complicated contributions of cavities and hydration, these data are now being theoretically reconsidered based on molecular dynamics and solution chemistry (Smolin and Winter 2006; Imai et al. 2005, 2007). The temperature and pressure dependence of adiabatic and isothermal compressibilities should be experimentally and theoretically cross-examined in order to further advance the understanding of pressure effects on protein structure. These volumetric studies are complementary to recently developed high-pressure NMR and X-ray crystallography techniques (Akasaka 2006; Fourme et al. 2001, 2012), and will open up exciting new areas in high-pressure protein science.

Acknowledgement We thank Dr. Tadashi Kamiyama of Kinki University for his helpful comments and careful reading of the manuscript.

References

Akasaka K (2006) Probing conformational fluctuation of proteins by pressure perturbation. Chem Rev 106:1814–1835

Akasaka K, Latif AR, Nakamura A, Matsuo K, Tachibana H, Gekko K (2007) Amyloid protofibril is highly voluminous and compressible. Biochemistry 46:10444–10450

Arakawa T, Timasheff SN (1982) Preferential interactions of proteins with salts in concentrated solutions. Biochemistry 21:6545–6552

Arakawa T, Timasheff SN (1984) Protein stabilization and destabilization by guanidinium salts. Biochemistry 23:5924–5929

Blandamer MJ, Davis MI, Douhéret G, Reis JCR (2001) Apparent molar isentropic compressions and expansions of solutions. Chem Soc Rev 30:8–15

Brandts JF, Oliveira RJ, Westort C (1970) Thermodynamics of protein denaturation: effect of pressure on the denaturation of ribonuclease A. Biochemistry 9:1038–1047

Casassa EF, Eisenberg H (1964) Thermodynamic analysis of multicomponent solutions. Adv Protein Chem 19:287–395

Chalikian TV (2003) Volumetric properties of proteins. Annu Rev Biophys Biomol Struct 32:207–235

Chalikian TV, Gindikin VS, Breslauer KJ (1995) Volumetric characterizations of the native, molten globule and unfolded states of cytochrome c at acidic pH. J Mol Biol 250:291–306

Chalikian TV, Totrov M, Abagyan R, Breslauer KJ (1996) The hydration of globular proteins as derived from the volume and compressibility measurements: cross correlating thermodynamic and structural data. J Mol Biol 260:588–603

Chalikain TV, Völker J, Anafi D, Breslauer KJ (1997) The native and the heat-induced denatured states of α-chymotrypsinogen A: thermodynamic and spectroscopic studies. J Mol Biol 274:237–252

Cohn EJ, Edsall JT (1943) Density and apparent specific volume of proteins. In: Cohn EJ, Edsall JT (eds) Proteins, amino acids and peptides. Reinhold, New York, pp 370–381

Cooper A (1976) Thermodynamic fluctuations in protein molecules. Proc Natl Acad Sci U S A 73:2740–2741

Dadarlat VM, Post CB (2001) Insights into protein compressibility from molecular dynamics simulations. J Phys Chem B 105:715–724

Dubins DN, Filfil R, Macgregor RB Jr, Chalikian TV (2000) Role of water in protein–ligand interactions: volumetric characterization of the binding of 2'-CMP and 3'-CMP to ribonuclease A. J Phys Chem B 104:390–401

Dubins DN, Filfil R, Macgregor RB Jr, Chalikian TV (2003) Volume and compressibility changes accompanying thermally-induced native-to-unfolded and molten globule-to-unfolded transitions of cytochrome c: a high pressure study. Biochemistry 42:8671–8678

Durchschlag H (1986) Specific volumes of biological macromolecules and some other molecules of biological interest. In: Hinz HJ (ed) Thermodynamic data for biochemistry and biotechnology. Springer, Berlin, pp 45–128

Fasman GD (ed) (1976) Handbook of biochemistry and molecular biology, vol 2, 3rd edn. CRC Press, Cleveland

Filfil R, Chalikian TV (2000) Volume and spectroscopic characterizations of the native and acid-induced denatured states of staphylococcal nuclease. J Mol Biol 299:827–842

Fourme R, Kahn R, Mezouar M, Girard E, Hoerentrup C, Prangé T, Ascone I (2001) High-pressure protein crystallography (HPPX): instrumentation, methodology and results on lysozyme crystals. J Synchrotron Radiat 8:1149–1156

Fourme R, Girard E, Akasaka K (2012) High-pressure macromolecular crystallography and NMR: status, achievements and prospects. Curr Opin Struct Biol 22:636–642

Frauenfelder H, Hartmann H, Karplus M, Kuntz ID Jr, Kuriyan J, Parak F, Petsko GA, Ringe D, Titon RF Jr, Connolly ML, Max N (1987) Thermal expansion of a protein. Biochemistry 26:254–261

Gavish B, Gratton E, Hardy CJ (1983) Adiabatic compressibility of globular proteins. Proc Natl Acad Sci U S A 80:750–754

Gekko K (1991) Flexibility of globular proteins in water as revealed by compressibility. In: Levine H, Slade L (eds) Water relationship in food. Plenum Press, New York, pp 753–771

Gekko K (2002) Compressibility gives new insight into protein dynamics and enzyme function. Biochim Biophys Acta 1595:382–386

Gekko K, Hasegawa Y (1986) Compressibility-structure relationship of globular proteins. Biochemistry 25:6563–6571

Gekko K, Hasegawa Y (1989) Effect of temperature on the compressibility of native globular proteins. J Phys Chem 93:426–429

Gekko K, Koga S (1984) The stability of protein structure in aqueous propylene glycol: amino acid solubility and preferential solvation of protein. Biochim Biophys Acta 786:151–160

Gekko K, Morikawa T (1981) Preferential hydration of bovine serum albumin in polyhydric alcohol-water mixtures. J Biochem 90:39–50

Gekko K, Noguchi H (1979) Compressibility of globular proteins in water at 25 °C. J Phys Chem 83:2706–2714

Gekko K, Timasheff SN (1981) Mechanism of protein stabilization by glycerol: preferential hydration in glycerol-water mixtures. Biochemistry 20:4667–4676

Gekko K, Yamagami K (1991) Flexibility of food proteins as revealed by compressibility. J Agric Food Chem 39:57–62

Gekko K, Yamagami K (1998) Compressibility and volume changes of lysozyme due to inhibitor binding. Chem Lett 27:839–840

Gekko K, Kunori Y, Takeuchi H, Ichihara S, Kodama M (1994) Point mutations at glycine-121 of *Escherichia coli* dihydrofolate reductase: important roles of a flexible loop in the stability and function. J Biochem 116:34–41

Gekko K, Tamura Y, Ohmae E, Hayashi H, Kagamiyama H, Ueno H (1996) A large compressibility change of protein induced by a single amino acid substitution. Protein Sci 5:542–545

Gekko K, Kamiyama T, Ohmae E, Katayanagi K (2000) Single amino acid substitutions in flexible loops can induce large compressibility changes in dihydrofolate reductase. J Biochem 128:21–27

Gekko K, Kimoto A, Kamiyama T (2003) Effects of disulfide bonds on compactness of protein molecules revealed by volume, compressibility, and expansibility changes during reduction. Biochemistry 42:13746–13753

Gekko K, Obu M, Li J, Lee JC (2004) A linear correlation between the energetics of allosteric communication and protein flexibility in the *Escherichia coli* cyclic AMP receptor protein revealed by mutation-induced changes in compressibility and amide hydrogen-deuterium exchange. Biochemistry 43:3844–3852

Gekko K, Araga M, Kamiyama T, Ohmae E, Akasaka K (2009) Pressure dependence of the apparent specific volume of bovine serum albumin: insight into the difference between isothermal and adiabatic compressibilities. Biophys Chem 144:67–71

Greenspan MG, Tschiegg CE (1956) Sing-around ultrasonic velocimeter for liquids. Rev Sci Instrum 28:897–901

Hattori M, Li H, Yamada H, Akasaka K, Hengstenberg W, Gronwald W, Kalbitzer HR (2004) Infrequent cavity-forming fluctuations in HPr from *Staphylococcus carnosus* revealed by pressure- and temperature-dependent tyrosine ring flips. Protein Sci 13:3104–3114

Hawley SA (1971) Reversible pressure–temperature denaturation of chymotrypsinogen. Biochemistry 10:2436–2442

Hinz HJ (ed) (1986) Thermodynamic data for biochemistry and biotechnology. Springer, Berlin

Imai T, Kovalenko A, Hirata F (2005) Partial molar volume of proteins studied by the three-dimensional reference interaction site model theory. J Phys Chem B 109:6658–6665
Imai T, Ohyama S, Kovalenko A, Hirata F (2007) Theoretical study of the partial molar volume change associated with the pressure-induced structural transition of ubiquitin. Protein Sci 16:1927–1933
Iwakura M, Maki K, Takahashi H, Takenawa T, Yokota A, Katayanagi K, Kamiyama T, Gekko K (2006) Evolutional design of a hyperactive cysteine- and methionine-free mutant of *Escherichia coli* dihydrofolate reductase. J Biol Chem 281:13234–13246
Kamiyama T, Gekko K (1997) Compressibility and volume changes of lysozyme due to guanidine hydrochloride denaturation. Chem Lett 26:1063–1064
Kamiyama T, Gekko K (2000) Effects of ligand binding on the flexibility of dihydrofolate reductase as revealed by compressibility. Biochim Biophys Acta 1478:257–266
Kamiyama T, Sadahide Y, Nogusa Y, Gekko K (1999) Polyol-induced molten globule of cytochrome c: an evidence for stabilization by hydrophobic interaction. Biochim Biophys Acta 1434:44–57
Karplus M, McCammon JA (1981) The internal dynamics of globular proteins. CRC Crit Rev Biochem 9:293–349
Kauzmann W (1959) Some factors in the interpretation of protein denaturation. Adv Protein Chem 14:1–63
Kharakoz DP (2000) Protein compressibility, dynamics, and pressure. Biophys J 79:511–525
Kharakoz DP, Sarvazyan AP (1993) Hydration and intrinsic compressibilities of globular proteins. Biopolymers 33:11–26
Kimura M (1983) The neutral theory of molecular evolution. Cambridge University Press, Cambridge
Kitahara R, Sareth S, Yamada H, Ohmae E, Gekko K, Akasaka K (2000) High pressure NMR reveals active-site hinge motion of folate-bound *Escherichia coli* dihydrofolate reductase. Biochemistry 39:12789–12795
Kratky O, Leopold H, Stabinger H (1973) The determination of the partial specific volume of proteins by the mechanical oscillator technique. Methods Enzymol 27:98–110
Kundrot CM, Richards FM (1987) Crystal structure of hen egg-white lysozyme at a hydrostatic pressure of 1000 atmospheres. J Mol Biol 193:157–170
Kupke DW (1973) Density and volume change measurements. In: Leach SJ (ed) Physical principles and techniques of protein chemistry. Academic, New York, pp 1–75, part C
Lassalle MW, Yamada H, Akasaka K (2000) The pressure-temperature free energy-landscape of staphylococcal nuclease monitored by (1)H NMR. J Mol Biol 298:293–302
Lee BK, Richards FM (1971) The interpretation of protein structures: estimation of static accessibility. J Mol Biol 55:379–400
Lee JC, Timasheff SN (1974) Partial specific volumes and interactions with solvent components of proteins in guanidine hydrochloride. Biochemistry 13:257–265
Lee JC, Timasheff SN (1981) The stabilization of proteins by sucrose. J Biol Chem 256:7193–7201
Lee JC, Gekko K, Timasheff SN (1979) Measurements of preferential solvent interactions by densimetric techniques. Methods Enzymol 61:26–49
Lee S, Tikhomirova A, Shalvardjian N, Chalikian TV (2008) Partial molar volumes and adiabatic compressibilities of unfolded protein states. Biophys Chem 134:185–199
Lumry R, Rosenberg A (1975) The mobile defect hypothesis of protein function. Coll Int CNRS L'Eau Syst Biol 246:55–63
Masterton WL, Seiler HK (1968) Apparent and partial molal volumes of water in organic solvents. J Phys Chem 72:4257–4262
Millero FJ, Ward GK, Chetirkin P (1976) Partial specific volume, expansibility, compressibility, and heat capacity of aqueous lysozyme solutions. J Biol Chem 251:4001–4004
Mimura S, Yamato T, Kamiyama T, Gekko K (2012) Nonneutral evolution of volume fluctuations in lysozymes revealed by normal-mode analysis of compressibility. Biophys Chem 161:39–45
Mori K, Seki Y, Yamada Y, Matsumoto H, Soda K (2006) Evaluation of intrinsic compressibility of proteins by molecular dynamics simulation. J Chem Phys 125:054903

Na GC, Timasheff SN (1981) Interaction of calf brain tubulin with glycerol. J Mol Biol 151:165–178

Nölting B (1995) Relation between adiabatic and pseudoadiabatic compressibility in ultrasonic velocimetry. J Theor Biol 175:191–196

Nölting B, Sligar SG (1993) Adiabatic compressibility of molten globules. Biochemistry 32:12319–12323

Ohmae E, Iriyama K, Ichihara S, Gekko K (1996) Effects of point mutation at a flexible loop glycine-67 of *Escherichia coli* dihydrofolate reductase on its stability and function. J Biochem 119:703–710

Ohmae E, Ishimura K, Iwakura M, Gekko K (1998) Effects of point mutation at a flexible loop alanine-145 of *Escherichia coli* dihydrofolate reductase on its stability and function. J Biochem 123:839–846

Paci E, Marchi M (1996) Intrinsic compressibility and volume compression in solvated proteins by molecular dynamic simulation at high pressure. Proc Natl Acad Sci U S A 93:11609–11614

Pain RH (1987) Protein structure. New light on old defects. Nature 326:247

Pinfield VJ, Povey MJW (1997) Thermal scattering must be accounted for in the determination of adiabatic compressibility. J Phys Chem B 101:1110–1112

Prehoda KE, Mooberry ES, Markley JL (1998) Pressure denaturation of proteins: evaluation of compressibility effects. Biochemistry 37:5785–5790

Reiss H (1965) Scaled particle methods in the statistical thermodynamics of fluids. Adv Chem Phys 9:1–84

Richards FM (1977) Area, volumes, packing and protein structure. Annu Rev Biophys Bioeng 6:151–176

Sarvazyan AP (1991) Ultrasonic velocity of biological compounds. Annu Rev Biophys Biophys Chem 20:321–342

Sarvazyan AP, Kharakoz DP (1977) Acoustical studies of the conformation states of proteins in aqueous solutions. In: Frank GM (ed) Molecular and cell biophysics. Nauka, Moscow, pp 93–106

Schweiker KL, Makhatadze GI (2009) Use of pressure perturbation calorimetry to characterize the volumetric properties of proteins. Methods Enzymol 466:527–547

Seemann H, Winter R, Royer CA (2001) Volume, expansivity and isothermal compressibility changes associated with temperature and pressure unfolding of *Staphylococcal* nuclease. J Mol Biol 307:1091–1102

Smeller L (2002) Pressure-temperature phase diagrams of biomolecules. Biochim Biophys Acta 1595:11–29

Smolin N, Winter R (2006) A molecular dynamics simulation of SNase and its hydration shell at high temperature and high pressure. Biochim Biophys Acta 1764:522–534

Son I, Shek YS, Dubins DN, Chalikian TV (2012) Volumetric characterization of tri-N-acetylglucosamine binding to lysozyme. Biochemistry 51:5784–5790

Tamura Y, Gekko K (1995) Compactness of thermally and chemically denatured ribonuclease A as revealed by volume and compressibility. Biochemistry 34:1878–1884

Tamura Y, Gekko K, Yoshioka K, Vonderviszt F, Namba K (1997) Adiabatic compressibility of flagellin and flagellar filament of *Salmonella typhimurium*. Biochim Biophys Acta 1335:120–126

Taulier N, Chalikian TV (2001) Characterization of pH-induced transitions of β-lactoglobulin: ultrasonic, densimetric, and spectroscopic studies. J Mol Biol 314:873–889

Taulier N, Chalikian TV (2002) Compressibility of protein transitions. Biophim Biophys Acta 1595:48–70

Tilton RF Jr, Dewan JC, Petsko GA (1992) Effects of temperature on protein structure and dynamics: X-ray crystallographic studies of the protein ribonuclease-A at nine different temperatures from 98 to 320 K. Biochemistry 31:2469–2481

Timasheff SN (1995) Preferential interactions of water and cosolvents with proteins. In: Gregory RB (ed) Protein-solvent interactions. Marcel Dekker, New York, pp 445–482

Woodward CK, Carulla N, Barany G (2004) Native state hydrogen-exchange analysis of protein folding and protein motional domains. Methods Enzymol 380:379–400

Wyman J, Gill J (eds) (1990) Binding and linkage: functional chemistry of biological macromolecules. University Science, Mill Valley

Yamato T, Higo J, Seno Y, Go N (1993) Conformational deformation in deoxymyoglobin by hydrostatic pressure. Proteins 16:327–340

Zamyatnin AA (1984) Amino acid, peptide, and protein volume in solution. Annu Rev Biophys Bioeng 13:145–165

Zhai Y, Okoro L, Cooper A, Winter R (2011) Applications of pressure perturbation calorimetry in biophysical studies. Biophys Chem 156:13–23

Zipp A, Kauzmann W (1973) Pressure denaturation of metmyoglobin. Biochemistry 12:4217–4228

Chapter 6
Pressure-Dependent Conformation and Fluctuation in Folded Protein Molecules

Mike P. Williamson

Abstract Hydrostatic pressure leads to nonuniform compression of proteins. The structural change is on average only about 0.1 Å $kbar^{-1}$, and is therefore within the range of fluctuations at ambient pressure. The largest changes are around cavities and buried water molecules. Sheets distort much more than helices. Hydrogen bonds compress about 0.012 Å $kbar^{-1}$, although there is a wide range, including some hydrogen bonds that lengthen. In the presence of ligands and inhibitors, structural changes are smaller. Pressure has little effect on rapid fluctuations, but with larger scale slower motions, pressure increases the population of excited states (if they have smaller overall volume), and slows the fluctuations. In barnase, pressure is shown to be a useful way to characterise fluctuations on the timescale of microseconds, and helps to show that fluctuations in barnase are hierarchical, with the faster fluctuations providing a platform for the slower ones. The excited states populated at high pressure are probably functionally important.

Keywords Compression • Buried water • Fluctuation • Dynamics • Hierarchical motion

6.1 Overview

Proteins are large and complicated molecules, and can access a very large number of conformational states. The lowest in energy (under standard conditions) is the ground state, and is very close to the conformation observed in crystal and NMR structures. Proteins undergo fluctuations in solution, which move them away from the ground state. Evidence is accumulating that the most populated (i.e., the lowest energy) of these fluctuations are often functionally important: they may move an enzyme away from the ground state to a catalytically more active state, or a state better able to recognise or bind ligands. For example, ubiquitin needs to recognise

M.P. Williamson (✉)
Department of Molecular Biology and Biotechnology, University of Sheffield, Sheffield S10 2TN, UK
e-mail: m.williamson@sheffield.ac.uk

K. Akasaka, H. Matsuki (eds.), *High Pressure Bioscience*,
Subcellular Biochemistry 72, DOI 10.1007/978-94-017-9918-8_6

and bind to a large number of partner proteins. Recognition is achieved by low-energy fluctuations, which distort the ground state structure slightly and match the structure better to recognise its ligands (Lange et al. 2008). The identification of these fluctuations is therefore important, but is difficult because the higher energy states are only populated to a few percent under ambient conditions (Williamson 2011). Considerable progress in characterising such states has been made by using relaxation dispersion measurements, which can provide details of the 'invisible' higher energy states (Baldwin and Kay 2009), but these measurements are difficult both to obtain and to analyse. Pressure has been shown to be a good way to selectively populate and study such states (Kitazawa et al. 2013, 2014).

The application of hydrostatic pressure changes the energy landscape of proteins, by stabilising states with smaller partial molar volumes. It is a useful technique because the energy added to the system by pressure is very small: for a typical small protein it is $p\Delta V \approx 2$ kJ mol^{-1} (Wilton et al. 2008b), compared to thermal energy of $RT = 2.4$ kJ mol^{-1}. The ground state of a globular protein always contains cavities, some of which are partially occupied by water molecules, and some of which are empty. The ground state therefore always has a larger partial molar volume than the unfolded state, which is fully hydrated and contains no internal cavities. Therefore, pressure always destabilises the ground state compared to partially unfolded alternative states. Thus, pressure provides a way of stabilising and characterising alternative states relatively close in energy to the ground state, many of which are likely to be functionally important. A high enough pressure leads to denaturation of the protein, and thus protein unfolding. Pressure can therefore be used to study protein folding and unfolding, discussed elsewhere in this book. However, pressures lower than those needed to completely unfold proteins will destabilise the ground state relative to locally unfolded states, without leading to global unfolding. The focus of this chapter is the effects of pressure on protein structure close to the ground state. As discussed by Akasaka in this book and elsewhere (Akasaka 2006), pressure has two main effects on proteins: it leads to a general compression, and it also leads to a stabilisation of locally partially unfolded (or hydrated) states, many of which are likely to be important either for function or unfolding or both (Williamson 2003). In this chapter, we look first at what is known about pressure-induced compression of proteins in solution, and then discuss some of the conformational states stabilised by pressure and the fluctuations that produce them.

6.2 Effects of Pressure on Protein Structure

6.2.1 Effects on Water and Covalent Bonds

The structural effects of pressure on proteins as analysed by X-ray crystallography are discussed elsewhere in this book and have been reviewed (Collins et al. 2011), and are in good agreement with the results reported here. The other technique able to provide information at atomic resolution is NMR. This has been done in two different ways. One is the 'classic' NMR structure determination carried out at high pressure. The other is to use changes in chemical shifts, induced by pressure, to obtain structural information on the change in structure. The former method has been used by Kitahara and Akasaka on ubiquitin (Kitahara et al. 2005; Kitazawa et al. 2014), and the latter by Williamson on melittin (Iwadate et al. 2001), lysozyme (Refaee et al. 2003), BPTI (Williamson et al. 2003), protein G (Wilton et al. 2008b) and barnase (Wilton et al. 2009a). The two measurements are largely consistent, and also consistent with information obtained by other techniques such as crystallography (Collins et al. 2011), ultrasound (Gekko and Hasegawa 1986) and spectroscopy (Heremans and Smeller 1998; Shimizu et al. 1988).

Hydrostatic pressure acts to reduce the volume of the total system, consisting of the protein itself and the water in which it sits. Proteins are surrounded by a layer of water molecules 1–2 molecules thick, which is relatively incompressible. Most studies on the compressibility of the hydration layer have relied on ultrasound measurements on solutions, for which it is difficult to separate surface layer effects from bulk solution effects. A study of a B-DNA fragment at high pressure (Wilton et al. 2008a) showed that the DNA molecule itself is almost incompressible, because it has no significant internal cavities. The surface layer was found to have a compressibility approximately 75 % of the compressibility of the bulk water: less compressible than bulk water but not a vast difference. There was compression of the interstrand hydrogen bonds, with the AT base pairs being compressed to a greater extent than the GC base pairs, as one might expect. However this is not enough to produce much overall compression. The biggest structural change was a widening of the minor groove, which probably leads to a greater degree of hydration of the minor groove. Thus, the overall volume of the system is smaller because some water molecules are able to move from the bulk solution into the minor groove.

Covalent bonds are also almost incompressible. Measurements of crystals under high pressure show that covalent bonds are expected to compress by only about 0.01 % at a pressure of 1 kbar (100 MPa: 1 bar = 0.1 MPa = 0.987 atm) (Lynch and Drickamer 1966). The changes in ^{13}C chemical shift observed in proteins under pressure are consistent with length changes of this magnitude (Wilton et al. 2009b), and lead to changes in ^{13}C chemical shift of about −0.1 ppm kbar^{-1}. The compression is thus easily measurable, but leads to very small structural change. A structural change of this magnitude leads to very little change in electronic properties.

Hydrostatic pressure can also lead to changes in bond angle. This is observed most clearly in methyl groups. If the methyl group is compressed 'face-on', then there is some splaying out of the methyl protons, and thus a change in hybridisation of the molecular orbitals, in shielding of the nucleus and thus in ^{13}C chemical shift. Changes in shift of the order of +0.2 ppm kbar^{-1} are observed, indicating a small but measurable effect (Wilton et al. 2009b). Again, this change in electron distribution is too small to produce effects on the chemistry of the protein.

Table 6.1 Structural changes in proteins at 1 kbar pressure

Protein	Backbone rmsd change (Å)	Radius of gyration (%)	Volume (%)	Cavity volume (%)	H-bond length (Å)
Melittin	0.11	–	−0.6	–	−0.011
Lysozyme	0.11	−0.13	−0.46	−2.4	−0.006
BPTI	0.12	+0.23	+0.08	−5	−0.015
Protein G B1	0.09	−0.15	−0.37	−1	−0.011
Barnase	0.08	−0.1	−0.35	−6	−0.013

Data taken from (Iwadate et al. 2001), (Refaee et al. 2003), (Williamson et al. 2003), (Wilton et al. 2008b, 2009a)

6.2.2 Non-Uniform Compression Within the Protein

Hydrostatic pressure causes structural changes to proteins of about 0.1 Å $kbar^{-1}$ rmsd, shown in more detail in Table 6.1. The overall structural change is thus very small, though it is far from uniform.

Easily the most compressible part of a protein is the cavities, followed by hydrogen bonds. Other non-covalent bonds such as van der Waals interactions rapidly become strongly repulsive on compression, and are thus resistant to compression. Therefore the most obvious consequence of pressure is compression of cavities and shortening of hydrogen bonds.

All folded proteins contain cavities, because the packing of sidechains is unable to fill the space completely. Some of these cavities are large enough to be able to accommodate a water molecule. The number of such cavities is roughly linearly correlated with the size of the protein, with at least one cavity in most proteins of greater than 100 residues (Hubbard et al. 1994). It is not easy to tell whether a cavity actually contains a water molecule or not. The water will only be seen by X-ray crystallography if the water molecule occupies a well-defined location within the cavity. It is thought that most large cavities (large enough to hold several water molecules) are hydrated, but that many smaller ones are not (Rashin et al. 1986). Moving a water molecule from bulk solvent into a cavity is accompanied by only a small change in enthalpy, because a water molecule can often form roughly the same number and strength of hydrogen bonds in the cavity as it can in bulk water. A water molecule in a cavity inevitably has more restrictions on translational movement than a water in bulk solvent. Thus, it was long thought that putting a water molecule into a cavity would necessarily lead to a large reduction in entropy, and would therefore be energetically unfavourable (Dunitz 1994). More recent results have shown that the reduction in entropy is less severe than expected (Denisov et al. 1997; Fischer and Verma 1999). This is partly because water molecules in cavities often retain considerable rotational entropy as well as some translational entropy; and also because a water molecule moving around in a cavity is able to drag parts of the protein with it, and therefore increases the dynamical mobility and thus the entropy of the protein. The loss of entropy of the water is therefore counterbalanced to some extent by an increase in entropy of the protein. As a result, the average energy change in moving a water molecule from bulk solvent into a cavity is reasonably close to zero.

Structural changes in a number of proteins are summarised in Table 6.1. At a pressure of 1 kbar, the overall volume (i.e. the volume enclosed by the outer hydrated surface) decreases by about 0.4 % (a compressibility coefficient of 4 $Mbar^{-1}$), and the radius of gyration in consequence by roughly one third of this, around 0.13 %. However, the changes in cavity volumes are much larger than this, at around 3–4 % (though with a large standard deviation, because in detail cavities vary widely from one protein to another). A significant change is brought about by a shortening of hydrogen bonds, by about 0.012 Å $kbar^{-1}$, though again with a very large standard deviation. A comparable effect has been seen by Raman spectroscopy (Shimizu et al. 1988) and other spectroscopic techniques (Heremans and Smeller 1998). This results in a small but noticeable shortening of α-helices. There is also (on average) some straightening of hydrogen bonds, making hydrogen bonds more 'ideal' in geometry at higher pressure.

The shortening of hydrogen bonds has been examined in detail for protein G (Li et al. 2000) and (with greater sensitivity) for ubiquitin, using the scalar coupling across a hydrogen bond as a measure of the hydrogen bond strength (Nisius and Grzesiek 2012). Although most hydrogen bonds show the sort of shortening described above, with considerable spread, a few have much greater reduction in length (and one has a significant increase in length). These are all hydrogen bonds with large sequence separations, i.e. high contact order. Nisius and Grzesiek (2012) suggest that these hydrogen bonds are important for correct folding of the protein, and are therefore selected by evolution to be stabilised within the protein fold.

A key result to come out of these studies is that compression is not uniform, and varies widely across proteins, with some regions actually showing a local expansion (probably caused in some cases by insertion of water molecules into cavities). In all the proteins studied so far, helices and helical domains compress reasonably uniformly. However, β-sheets have lower compressibility (as seen also by Gekko and colleagues using ultrasound (Gekko and Hasegawa 1986)), and tend to get deformed rather than compressed. A typical example is the GB1 domain of protein G (Fig. 6.1), which consists of an α-helix sitting over the top of a four-stranded β-sheet. At high pressure, the helix is slightly shorter, is closer to the sheet, and is slightly more bent, curving in toward the protein (Wilton et al. 2008b). The β-sheet is curled up around the helix, with the maximum change being seen at one of the outside strands (strand 2).

A similar result has also been seen for lysozyme (Refaee et al. 2003). Lysozyme consists of two domains, with one composed of α-helices and the other of β-sheets. The helical domain is compressed fairly uniformly, by about 0.5 % $kbar^{-1}$, but the β-sheet domain is compressed less (around 0.3 % $kbar^{-1}$) and is markedly distorted, with a twist to the main β -hairpin.

The structural changes are a result of small changes to dihedral angles across the proteins. In lysozyme, the backbone dihedral angles φ and ψ altered by $2.0 \pm 2.0°$ and $2.1 \pm 1.8°$ respectively over 1 kbar. In BPTI, backbone dihedral angles changed

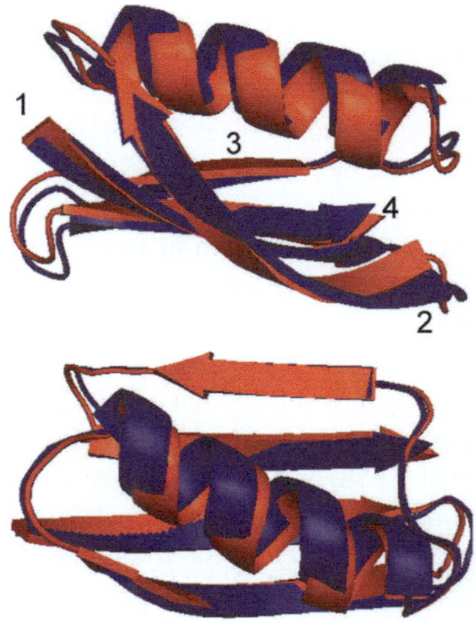

Fig. 6.1 Structure of the B1 domain of protein G, compared at low pressure (*blue*) and high pressure (*red*). The four β-strands are numbered in the *top panel*, which is rotated approximately 90° compared to the *lower panel*. The structural change has been magnified by a factor of 5 for clarity (Adapted from (Wilton et al. 2008b), with permission)

to a very similar extent ($2.3 \pm 2.5°$ and $2.6 \pm 2.3°$ respectively), while sidechain angles showed a larger variability ($-0.1 \pm 3.6°$ in χ_1) (Williamson et al. 2003). Similar structural changes have been seen in other proteins (Urayama et al. 2002) and in lysozyme by crystallography (Kundrot and Richards 1987), which also showed a similar pattern of structural change overall. It is interesting to note that in the NMR study of lysozyme (Refaee et al. 2003), there is a good negative correlation between the change in ψ of one residue and the change in φ of the next, implying that many of the angle changes are a result of reorientation of the peptide plane.

6.2.3 Buried Water Molecules and Active Sites

A striking result that is common to almost all the proteins studied by NMR is that the structural changes are largest close to buried water molecules. Buried water molecules are common in proteins, and tend to occupy the same positions in homologous structures, suggesting that they are functionally important (Kamatari et al. 2011). The function probably involves the increased local flexibility afforded to structures by the presence of mobile water molecules. The structural changes seen in BPTI and lysozyme were greatest close to buried water molecules. In lysozyme, buried water molecules are found at two locations (Fig. 6.2a): in the interface between the two domains, and at the end of the β-hairpin, and can be thought of as the lubrication that allows the β-hairpin to twist with pressure. This also locates the buried waters close to the active site. The largest changes in amino acid volume were observed close to buried water molecules (Fig. 6.2b). Interestingly, many of these volume changes were expansions. Although the overall effect of pressure is a compression of the structure, there is considerable variability on the local effects.

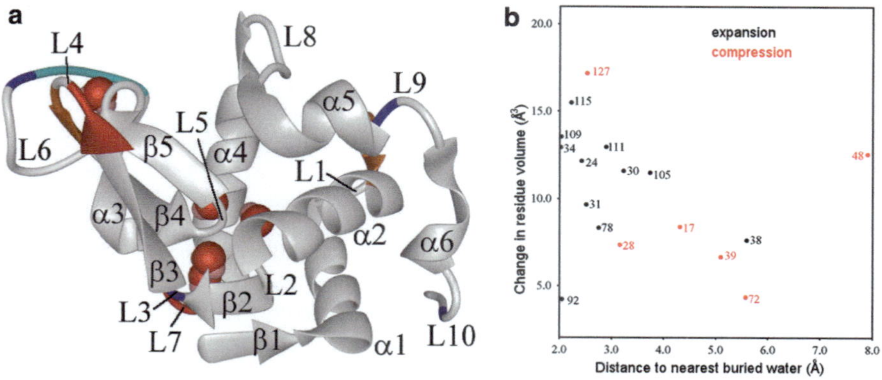

Fig. 6.2 Structural changes in lysozyme with pressure are close to buried water molecules. (**a**) The crystal structure of lysozyme, showing the α-domain (*right*) and β-domain (*left*), with locations of buried water molecules (*red/pink* spheres), and regions of the protein that have moved closer to the rest of the protein (*orange* 0.16 Å < d < 0.22 Å; *red* d > 0.22 Å) and further away (*cyan* −0.12 Å < d < −0.06 Å; *blue* d < −0.12 Å). (**b**) Absolute change in amino acid Voronoi volume plotted against distance from C_α atom to nearest buried water molecule, for all completely buried atoms. Only residues with volume changes > 2 $Å^3$ are shown (Adapted from (Refaee et al. 2003), with permission)

Similarly, in BPTI the structural changes localise close to buried water molecules, and thus also close to the active site (Williamson et al. 2003). The same is also true in barnase (Wilton et al. 2009a). Measurements of dynamics in azurin based on phosphorescence also suggest that buried waters enhance protein mobility (Cioni and Gabellieri 2011). These four proteins do not form a large enough sample to be able to generalise, but they do suggest that there may be a connection between local flexibility (observed using high pressure and related to the presence of buried water molecules) and function.

Of particular interest is the observation that for BPTI, lysozyme, protein G and barnase, the region that shows the biggest structural change with pressure is close to the active site. Protein G binds to immunoglobulins by forming an extended β-sheet with the immunoglobulin using strand 2 (Fig. 6.1). This is the strand that undergoes the greatest structural change with pressure. For the other three proteins the active site is well characterised and is where the pressure-induced structural changes are greatest. There is nothing 'special' about pressure in this respect – none of these proteins has had any evolutionary pressure to adapt to high pressure. Therefore, the observation that they all show pressure-induced structural change at the active site implies that the active site is more compressible than the rest of the protein, and in particular that the alternative conformation(s) have a smaller volume. In other words, pressure is a useful tool to characterise an interesting property that seems to be a common feature of active sites (possibly because of the presence of buried water molecules nearby), namely a large compressibility.

6.2.4 Compressibility and Volume Fluctuations

Compressibility and volume fluctuations are related by the thermodynamic relationship (Cooper 1976)

$$< (\delta V)^2 >= kTV\beta_T \tag{6.1}$$

where $< (\delta V)^2 >$ is the mean squared volume fluctuation, k is the Boltzmann constant, T the absolute temperature, V the volume, and β_T the isothermal compressibility coefficient. Because this is a thermodynamic relationship, it is true for the system as a whole but is not necessarily true for regions of the protein. However, it has been shown to apply to local regions at high frequencies and small volume changes, where the structural change is very localised (Yu and Leitner 2003), while crystallography (Tilton et al. 1992), optical spectroscopy (Pfeiffer et al. 2008) and our NMR results (Refaee et al. 2003; Wilton et al. 2008b, 2009a) suggest that it applies reasonably well across a wide range of fluctuations. This means that measurements of compressibility at the amino acid level can be translated into measures of local volume fluctuation. Therefore, the results discussed in the previous section imply that proteins have the largest volume fluctuations around their active sites.

Based on Eq. 6.1, it has been estimated that the root mean square volume fluctuation is approximately 60 $Å^3$ in barnase and 70 $Å^3$ in lysozyme, much of which is localised close to buried water molecules. It is interesting to compare these values to the activation volume required for flipping the conformationally restricted aromatic rings of Phe45 and Tyr35 in BPTI, estimated as 50 ± 10 and 60 ± 20 $Å^3$ respectively (Wagner 1980), and more recently re-measured as 46 ± 9 and 85 ± 20 $Å^3$ respectively (Li et al. 1999). These activation volumes are roughly half the total volume swept out by a rotating ring, and provide a measure for how tightly packed the protein is in the vicinity of these rings. In other words, the volume fluctuation in barnase and lysozyme at room temperature is significant, and is localised close to buried waters/active site.

The study on barnase measured the compressibility both in the absence and presence of an inhibitor, d(CGAC). The compressibility was shown to be markedly reduced on addition of inhibitor, indicating a reduction in volume fluctuations by 50 %: again, largely localised to the region around the active site. This observation is consistent with the conformational selection model for enzyme function (Csermely et al. 2010), which hypothesises that the free protein is in equilibrium between open and closed states, while the bound protein is largely in a closed state. Similar observations have been made on other proteins. Thus, for dihydrofolate reductase, the most compressible state was the Michaelis complex, while the least compressible was the reaction product complex (Kamiyama and Gekko 2000). Similarly for lysozyme, the least compressible complex was the one formed with the strongest inhibitor (Gekko and Yamagami 1998; Gekko 2002). The catalytic power of enzymes is generally explained by their binding most strongly to the transition state; and the best competitive inhibitors are normally compounds with structures closest to the transition state (Williamson 2011). Thus, the limited data so far suggest that the transition state has much smaller volume fluctuations than the free enzyme – a result widely predicted by transition state theory.

6.3 Timescales of Structure Fluctuations

In this section, we consider what is known about the timescales of different fluctuations, and how these are affected by pressure.

6.3.1 *Rapid Dynamics on the ps/ns Timescale*

Proteins are subject to rapid rotations about single bonds, and local disorder ('wobbling'). This occurs on a timescale of picoseconds to nanoseconds, i.e. faster than overall tumbling. Such motions can be characterised well by NMR relaxation methods, which can give measures of the extent and timescales of the motions. Measurements of ^{15}N T_1, T_2 and NOE were carried out for a fragment of bacteriorhodopsin in chloroform/methanol (Orekhov et al. 2000), together with measurements of translational diffusion. It was observed that high pressure decreases both the translational and rotational diffusion rate, by a factor of 0.325 at 2 kbar. This was shown to be due to the increased viscosity of the solvent at this pressure. Analysis using the standard model-free approach showed that motions on the nanosecond timescale were unchanged at 2 kbar, but showed a drop in the order parameter for picosecond motion, implying an unexpected increase in motion at this timescale. However, the authors showed that the decrease in the order parameter could be explained by a shortening of hydrogen bonds along the helix and thus a slight increase in H-N bond lengths, and possibly a change in ^{15}N chemical shift anisotropy. They therefore concluded that there is no evidence for any change in mobility on the ps/ns timescale.

Relaxation studies of ubiquitin (Kitahara et al. 2005) and BPTI (Sareth et al. 2000) in water reached similar conclusions (though modified because the viscosity of water is less affected than that of chloroform/methanol by pressure): the overall tumbling is slowed slightly at higher pressure, but there is no change in internal motions on the ps/ns timescales. There was however a slowing down of a slower motion, discussed in more detail below. The authors note that most pressure-dependent chemical shift changes in BPTI are linear with pressure (a behaviour observed in many other proteins since). Because the change in chemical shift is caused by a change in average structure, this result implies that the structure changes linearly with pressure, i.e. that compressibility is independent of pressure, at least for most residues. Most other studies of pressure effects on proteins have reached similar conclusions. Most authors agree that for a protein in water at ambient

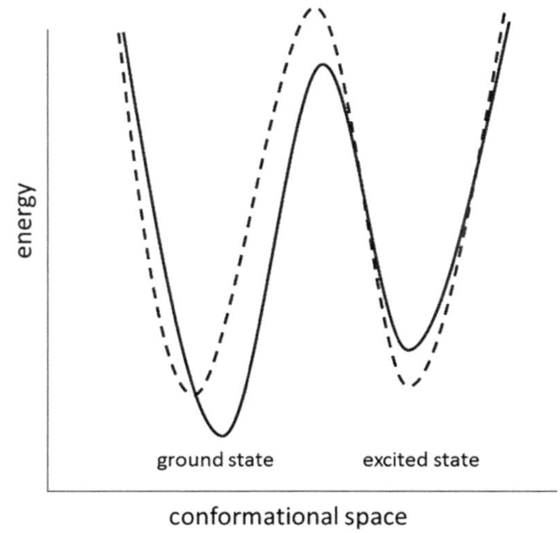

Fig. 6.3 Diagrammatic representation of the effect of pressure on proteins. *Solid line*: low pressure, *dashed line*: high pressure. An increase in pressure slightly shifts the distribution of conformers within the ground state ensemble towards a more compact structure, and narrows the distribution. It also lowers the energy of excited states relative to the ground state, and increases the energy barrier between them, leading to an increase in the population of the excited state and a slowing down of the exchange rate between them

temperature and pressure, the rms fluctuation in backbone coordinates is of the order of 0.6 Å (Lindorff-Larsen et al. 2005), with larger excursions for loops. The rms coordinate change per kbar is less than 0.1 Å for most residues, averaging to about 0.1 Å for all residues (Table 6.1). Thus the total structural change caused by pressure is less than 20 % of the normal structural fluctuation, and remains within it (Fig. 6.3). It is therefore not surprising if local fluctuations are not affected by pressure. [See also Kuwata et al. (2004), which reached a similar conclusion on the basis of ^{15}N chemical shifts.]

These results should be compared to results from quasielastic neutron scattering and molecular dynamics simulations (Calandrini et al. 2008), which saw a slowing down and a 20 % reduction in the amplitudes of motions on all timescales, although the effect was greater at slower timescales, and for larger length scales. Another molecular dynamics study (Meinhold et al. 2007) concluded that vibrations within conformational substates stiffen with pressure, although the barriers between one substate and another are not affected by pressure. This study concurs with the previous one in concluding that pressure selectively damps large-scale low-frequency collective modes, while having little effect on the more local rapid modes.

Relaxation measurements of sidechain methyls report on somewhat slower motions than backbone amides. With pressure, methyl motions become somewhat more restricted, though with a wide spread in changes, and the amplitude of motion is more restricted than that of backbone amides (Fu et al. 2012).

6.3.2 Intermediate Dynamics on the μs-ms Timescale

Characterisation of fluctuations on a timescale between ns and ms remains challenging, even at ambient pressure. Where it occurs on this timescale, the flipping of aromatic rings in proteins can be characterised by measuring lineshapes as a function of temperature. Most aromatic rings show a single sharp averaged signal for the two chemically equivalent ring protons on opposite edges of the ring, indicating a rapid ring flip, but a few have slower rates. Ring flips with pressure have been measured in BPTI (Li et al. 1999; Wagner 1980): the flip rates for Tyr35 and Phe45 are respectively 630 s^{-1} and 10^4 s^{-1} at ambient temperature and pressure, and slow down by factors of approximately 7 and 3 respectively per kbar. Tyr6 of *Staphylococcus carnosus* HPr slows by a factor of 4 $kbar^{-1}$ from an initial rate of about 2.5×10^4 s^{-1} (Hattori et al. 2004).

Motions on the millisecond timescale can also be examined using relaxation dispersion. There have been some relaxation dispersion experiments described at variable pressure to study protein unfolding, as described below. To our knowledge, the only published use of relaxation dispersion to look at internal motions within a folded protein concluded that motions in ubiquitin were slowed down by pressure but were still too fast to characterise, even at high pressure (Kitazawa et al. 2013).

6.3.3 Slow Dynamics: Amide Exchange and Protein Unfolding

As implied above, fluctuations on long timescales, which inherently are the largest scale motions (Wilton et al. 2009a), are slowed selectively at higher pressure. This is seen most clearly in the fact that NMR signals tend to broaden and disappear at higher pressure. The classic case of line broadening is the molten globule state of proteins, typically observed under destabilising conditions such as low pH, reduction of disulphides, intermediate concentrations of denaturants, etc. This state is also characterised by broad NMR signals, the reason for which is generally agreed to be exchange between different conformations on a slow timescale of the order of milliseconds or longer (comparable to their differences in chemical shift) (Alexandrescu et al. 1993). The same is true for proteins at high pressure: the rate of exchange between one conformer and another slows down sufficiently to lead to line broadening. Typically only a subset of residues shows dramatic line broadening with pressure, and these are assumed to represent the site of local unfolding (Inoue et al. 2000). An interesting example is provided by the hamster prion protein (Kuwata et al. 2004), which consists of a disordered N-terminus and a folded C-terminus. With pressure, signals from the N-terminus have almost unaltered intensity, but signals from the folded C-terminus become broad and are almost gone by 2 kbar. Such results have been reviewed (Akasaka 2006) and are discussed elsewhere in this book. The populations of globally and locally unfolded states increase with pressure, and the exchange rates slow down. A study using NMR and fluorescence (Roche et al. 2013) observed a several-fold slowing down of the global unfolding rate of staphylococcal nuclease with pressure. The effects were complex and location-dependent, with the rates becoming faster again at pressures in the range 1–1.5 kbar.

An interesting example is provided by a study on the mitogen-activated protein kinase p38α (Nielsen et al. 2013). Like many kinases, several of the most interesting residues are not visible in HSQC spectra, presumably being too broadened by exchange to be seen. These include residues in the catalytic loop, also known as the DFG loop. This loop is known to alternate between 'in' and 'out' conformations, the 'out' conformation being unable to bind ATP. Application of pressure to the apo state selectively populated the out conformation; it also altered the dynamics of the in/out exchange process (presumably slowing it down, though the authors do not explicitly comment on this), so as to make the resonance of the central Phe169 visible.

The exchange of amide protons provides an interesting example of the application of high pressure. At neutral or slightly acidic pH and under normal solution conditions, amide proton exchange is base-catalysed and usually follows the EX2 mechanism, under which the observed exchange rate is equal to the intrinsic exchange rate (i.e. the rate for a fully exposed peptide with the same sequence) multiplied by the equilibrium constant for opening of the hydrogen bond, K_{op}. Intrinsic exchange rates can be readily calculated (and vary over a relatively limited range), meaning that the exchange rate can be converted to a measure of K_{op}. The variation in exchange rate with pressure therefore directly provides a measure of the activation volume for local opening of the amide hydrogen bond.

Measurement of amide exchange rates in T4 lysozyme produced confusing results (Hitchens and Bryant 1998). Exchange rates were faster at higher pressure, implying a negative activation volume. This seems reasonable, given that protein unfolding corresponds to a reduction in overall volume: the local unfolding necessary to expose the amide proton to solvent might therefore be expected to correspond to a negative activation volume. However, activation volumes had no clear correlation with structural features. Neighbouring or structurally coupled residues could have very different activation volumes, with no clear dependence on secondary structure. Nor is there any correlation between activation volume and protection factor (the ratio between exchange rates in the folded protein and unfolded peptides). There is also no correlation with activation enthalpy or entropy (Dixon et al. 2000). An independent study on apocytochrome b_{562} (a fairly unstable protein, because of the absence of the haem cofactor) also saw little clearly explicable effect (Fuentes and Wand 1998). A subsequent theoretical study (Kharakoz 2000) concluded that the observed results are compatible with the rate of permeation of hydroxide ions into the protein interior, although no detailed calculations were presented.

6.3.4 Summary of Dynamic Fluctuations Probed by Pressure

The picture presented above is far from clear. Rapid motions in the ps/ns range are minimally affected by pressure. For all other fluctuations, pressure stabilises states with smaller partial molar volumes, which in general means partially unfolded states. Pressure therefore increases the population of such states: it also slows down the exchange rates between states, although in general not by large amounts: ratios of 4–10 kbar^{-1} seem the general rule. A detailed understanding of the features governing the rates of fluctuations is still some way off, particularly in comparison with the structural changes with pressure discussed in Sect. 6.2, which are much more interpretable.

6.4 An Example: Hierarchical Motion in Barnase

Fluctuations in the ribonuclease barnase over a wide range of timescales have been investigated using a number of methods (Fig. 6.4). Backbone ^{15}N relaxation (Sahu et al. 2000) has shown that rapid fluctuations are of similar magnitude throughout the protein, although with slightly greater magnitude at the termini and in a loop around residue 35. Molecular dynamics and related calculations (Hilser et al. 1998; Nolde et al. 2002; Giraldo et al. 2004; Zhuravleva et al. 2007) have described hinge-bending fluctuations in which the central β-sheet flexes, to bring the two ends of the molecule together and thus start to close over the active site. This hinge-bending motion starts to occur on a timescale of 10^9 s^{-1}. The fluctuation is a consequence of the architecture of the protein, which has a bulky hydrophobic cluster at each end, with a thinner and more flexible β-sheet in the middle (Fig. 6.4b).

The chemical shift-based methodology described in Sect. 6.2 was applied to barnase, to use changes in ^{1}H chemical shift as structural restraints to calculate the change in structure of barnase at high pressure. The structure change allowed the authors to read out the compressibility, which can then be related to the volume fluctuations at ambient pressure, using Eq. 6.1 (Wilton et al. 2009a). These are shown in Fig. 6.5a. Fluctuations in barnase bound to a competitive inhibitor, d(CGAC), are shown in Fig. 6.5b, and the difference between these two is shown in Fig. 6.5c. Ligand-bound barnase is considerably less flexible, the changes mapping again to the hinge, as well as to residues extending out from the hinge including Lys27, a key residue that binds the cleaved phosphate in the transition state and acts as an electrophilic catalyst (Mossakowska et al. 1989). Figure 6.5c shows that the residues whose fluctuations are quenched in the bound state are close both to the active site and to most of the buried water molecules. The timescale of these fluctuations was estimated based on the activation volume, lack of correlation with backbone ^{15}N relaxation, lack of exchange broadening, and lack of relaxation dispersion effects, to be between ns and μs, i.e. in the range 10^7–10^8 s^{-1}. It is thus

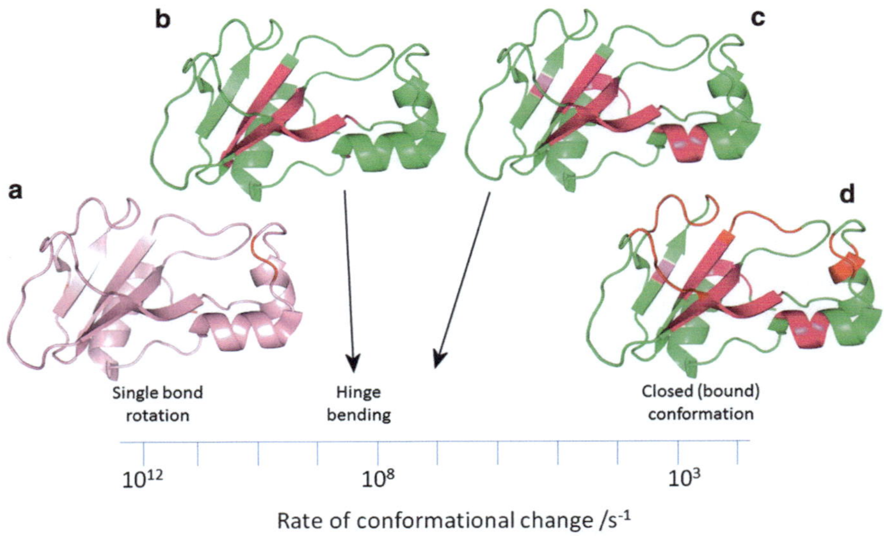

Fig. 6.4 Fluctuations in barnase over a range of timescales and length scales. (**a**) Picosecond fluctuations probed by ^{15}N backbone relaxation (Sahu et al. 2000) are almost uniform across the protein and confined to very local regions. (**b**) Molecular dynamics calculations reveal a hinge-bending motion occurring at about 10^9 s^{-1}, whose location is determined by the architecture of the protein (Zhuravleva et al. 2007). (**c**) These fluctuations extend across a larger part of the protein at longer timescales, characterised by pressure-dependent chemical shift changes (Wilton et al. 2009a), which describe compressibility and thus volume fluctuations, and occur at around 10^8 s^{-1}. (**d**) At a rate of about 1,000 s^{-1} the protein undergoes a fluctuation to a closed state, closely similar to the catalytically active (ligand bound) conformation. This fluctuation has been characterised by relaxation dispersion NMR (Pandya et al. 2015). Longer timescales correspond to greater numbers of residues and thus less frequent fluctuations

somewhat slower than the hinge-bending seen by molecular dynamics, as expected based on the larger number of residues involved (Fig. 6.4).

These fluctuations should be compared to the structural changes seen on comparison of free and bound enzyme, which are summarised in Figure 6.4d, and show bending of the hinge, as discussed above, but in addition with a closure of the lip around the active site to tighten the grip of the enzyme on its substrate mimic. Relaxation dispersion measurements have been used to characterise these motions (Pandya et al. 2015), and have shown that they occur at around 1,000 s^{-1}, which is a similar timescale to the catalytic turnover.

Several fluctuations have thus been characterised, on a range of timescales ranging from 10^{10} s^{-1} to 10^3 s^{-1}. The fastest motions are random and uncorrelated, and are directly linked to thermal motion of the solvent. The slowest motions are likely to be related to the catalytic turnover of the enzyme. The linkage between these is the hinge bending, which occurs in a timescale difficult to characterise by most techniques, between the ranges of backbone relaxation and relaxation dispersion, and which in this case was probed by pressure dependence. Analysis of

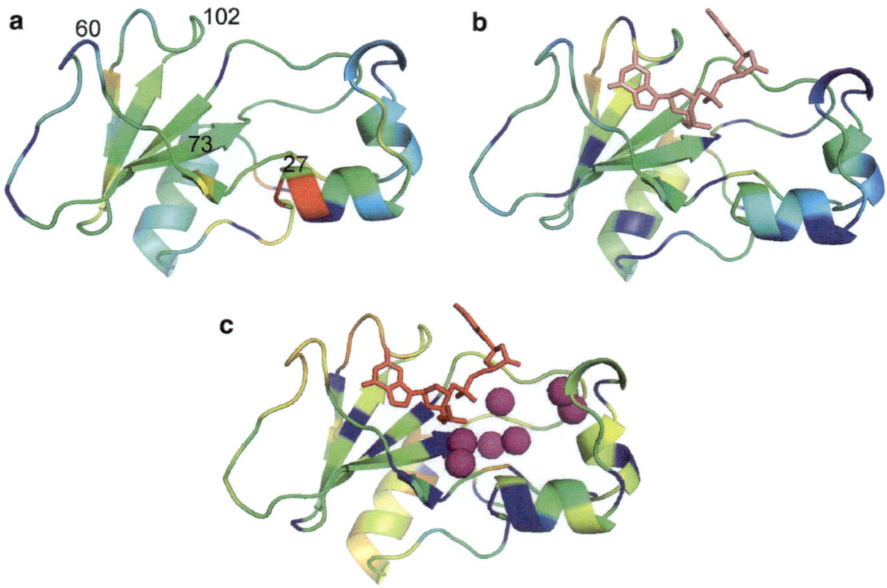

Fig. 6.5 Compressibility (volume fluctuations) in barnase. (**a**) Difference in C_{α} coordinates for free barnase between low and high pressure (2 kbar). *Red* indicates a region of the protein undergoing a pressure-dependent change that is much larger than any possible structural variability. *Orange*, *yellow* and *green* show progressively less significant changes, to *blue* which shows changes less than normal variation in solution, and thus unchanged with pressure. Lys27 forms hydrogen bonds to the phosphate that is cleaved in the reaction. (**b**) Structural differences for barnase bound to the competitive inhibitor d(CGAC). The central two residues of d(CGAC) are shown. (**c**) Difference between (**a**) and (**b**). *Blue* indicates residues that alter less in bound barnase than in free barnase. *Magenta* spheres show buried water molecules (Reproduced from (Wilton et al. 2009a) with permission)

the structural changes involved shows that the hinge-bending motion is a necessary precursor of the catalytically important lip closure. The timescales and structural requirements imply that these two different motions are hierarchical, in the sense that lip closure can only occur from a state in which hinge bending has taken place. In other words, there is an evolved dynamic pathway leading from random thermal motions to highly correlated lip closure. The information derived from high-pressure experiments is crucial in characterising this pathway, because it provides a window into an otherwise obscure region.

6.5 Conclusions

Proteins are remarkably complex molecules. They have evolved to adopt a range of conformations, both to recognise, bind and release partners, ligands and substrates, and to do this at rates appropriate to the function. Most analytical methods focus on the ground states of proteins, but these only tell a limited part of the story. A full description of a protein necessitates a characterisation of functionally important alternative states. Proteins probably have rather few excited states close in energy to the ground state that are local minima, because such states would act as kinetic traps in the folding pathway and would therefore lead to a risk of aggregation. Techniques for identifying excited states typically see a very small number for each protein, and such states are often functionally important (Williamson 2003; Sekhar and Kay 2013; Tzeng and Kalodimos 2013). This includes the excited states characterised using high pressure (Akasaka 2003; Kitazawa et al. 2013, 2014). High pressure is thus emerging as a powerful technique to stabilise these states, because it does not inject large amounts of energy into the system, but is a controllable way to alter the relative energies of different states. It may also be able to slow down the exchange between different states, particularly on larger distance scales, but the extent to which this may be possible is not yet clear.

References

Akasaka K (2003) Highly fluctuating protein structures revealed by variable-pressure nuclear magnetic resonance. Biochemistry 42:10875–10885

Akasaka K (2006) Probing conformational fluctuation of proteins by pressure perturbation. Chem Rev 106:1814–1835

Alexandrescu AT, Evans PA, Pitkeathly M, Baum J, Dobson CM (1993) Structure and dynamics of the acid-denatured molten globule state of α-lactalbumin: a two-dimensional NMR study. Biochemistry 32:1707–1718

Baldwin AJ, Kay LE (2009) NMR spectroscopy brings invisible protein states into focus. Nature Chem Biol 5:808–814

Calandrini V, Hamon V, Hinsen K, Calligari P, Bellissent-Funel MC, Kneller GR (2008) Relaxation dynamics of lysozyme in solution under pressure: combining molecular dynamics simulations and quasielastic neutron scattering. Chem Phys 345:289–297

Cioni P, Gabellieri E (2011) Protein dynamics and pressure: what can high pressure tell us about protein structural flexibility? Biochim Biophys Acta 1814:934–941

Collins MD, Kim CU, Gruner SM (2011) High-pressure protein crystallography and NMR to explore protein conformations. Ann Rev Biophys 40:81–98

Cooper A (1976) Thermodynamic fluctuations in protein molecules. Proc Natl Acad Sci U S A 73:2740–2741

Csermely P, Palotai R, Nussinov R (2010) Induced fit, conformational selection and independent dynamic segments: an extended view of binding events. Trends Biochem Sci 35:539–546

Denisov VP, Venu K, Peters J, Hörlein HD, Halle B (1997) Orientational disorder and entropy of water in protein cavities. J Phys Chem B 101:9380–9389

Dixon ME, Hitchens TK, Bryant RG (2000) Comparisons of pressure and temperature activation parameters for amide hydrogen exchange in T4 lysozyme. Biochemistry 39:248–254

Dunitz JD (1994) The entropic cost of bound water in crystals and biomolecules. Science 264:670

Fischer S, Verma CS (1999) Binding of buried structural water increases the flexibility of proteins. Proc Natl Acad Sci U S A 96:9613–9615

Fu Y, Kasinath V, Moorman VR, Nucci NV, Hilser VJ, Wand AJ (2012) Coupled motion in proteins revealed by pressure perturbation. J Am Chem Soc 134:8543–8550

Fuentes EJ, Wand AJ (1998) Local stability and dynamics of apocytochrome b_{562} examined by the dependence of hydrogen exchange on hydrostatic pressure. Biochemistry 37:9877–9883

Gekko K (2002) Compressibility gives new insight into protein dynamics and enzyme function. Biochim Biophys Acta 1595:382–386

Gekko K, Hasegawa Y (1986) Compressibility-structure relationship of globular proteins. Biochemistry 25:6563–6571

Gekko K, Yamagami K (1998) Compressibility and volume changes of lysozyme due to inhibitor binding. Chem Lett 27:839–840

Giraldo J, De Maria L, Wodak SJ (2004) Shift in nucleotide conformational equilibrium contributes to increased rate of catalysis of GpAp versus GpA in barnase. Proteins: Struct Funct Bioinf 56:261–276

Hattori M, Li H, Yamada H, Akasaka K, Hengstenberg W, Gronwald W et al (2004) Infrequent cavity-forming fluctuations in HPr from *Staphylococcus carnosus* revealed by pressure- and temperature-dependent tyrosine ring flips. Protein Sci 13:3104–3114

Heremans K, Smeller L (1998) Protein structure and dynamics at high pressure. Biochim Biophys Acta 1386:353–370

Hilser VJ, Dowdy D, Oas TG, Freire E (1998) The structural distribution of cooperative interactions in proteins: analysis of the native state ensemble. Proc Natl Acad Sci U S A 95:9903–9908

Hitchens TK, Bryant RG (1998) Pressure dependence of amide hydrogen-deuterium exchange rates for individual sites in T4 lysozyme. Biochemistry 37:5878–5887

Hubbard SJ, Gross KH, Argos P (1994) Intramolecular cavities in globular proteins. Protein Eng 7:613–626

Inoue K, Yamada H, Akasaka K, Hermann C, Kremer W, Maurer T et al (2000) Pressure-induced local unfolding of the Ras binding domain of RalGDS. Nature Struct Biol 7:547–550

Iwadate M, Asakura T, Dubovskii PV, Yamada H, Akasaka K, Williamson MP (2001) Pressure-dependent changes in the structure of the melittin α-helix determined by NMR. J Biomol NMR 19:115–124

Kamatari YO, Smith LJ, Dobson CM, Akasaka K (2011) Cavity hydration as a gateway to unfolding: an NMR study of hen lysozyme at high pressure and low temperature. Biophys Chem 156:24–30

Kamiyama T, Gekko K (2000) Effect of ligand binding on the flexibility of dihydrofolate reductase as revealed by compressibility. Biochim Biophys Acta 1478:257–266

Kharakoz DP (2000) Protein compressibility, dynamics, and pressure. Biophys J 79:511–525

Kitahara R, Yokoyama S, Akasaka K (2005) NMR snapshots of a fluctuating protein structure: ubiquitin at 30 bar–3 kbar. J Mol Biol 347:277–285

Kitazawa S, Kameda T, Yagi-Utsumi M, Sugase K, Baxter NJ, Kato K et al (2013) Solution structure of the Q41N variant of ubiquitin as a model for the alternatively folded N2 state of ubiquitin. Biochemistry 52:1874–1885

Kitazawa S, Kameda T, Kumo A, Yagi-Utsumi M, Baxter NJ, Kato K et al (2014) Close identity between alternatively folded state N2 of ubiquitin and the conformation of the protein bound to the ubiquitin-activating enzyme. Biochemistry 53:447–449

Kundrot CE, Richards FM (1987) Crystal structure of hen egg-white lysozyme at a hydrostatic pressure of 1000 atmospheres. J Mol Biol 193:157–170

Kuwata K, Kamatari YO, Akasaka K, James TL (2004) Slow conformational dynamics in the hamster prion protein. Biochemistry 43:4439–4446

Lange OF, Lakomek NA, Farès C, Schröder GF, Walter KFA, Becker S et al (2008) Recognition dynamics up to microseconds revealed from an RDC-derived ubiquitin ensemble in solution. Science 320:1471–1475

Li H, Yamada H, Akasaka K (1999) Effect of pressure on the tertiary structure and dynamics of folded basic pancreatic trypsin inhibitor. Biophys J 77:2801–2812

Li H, Yamada H, Akasaka K, Gronenborn AM (2000) Pressure alters electronic orbital overlap in hydrogen bonds. J Biomol NMR 18:207–216

Lindorff-Larsen K, Best RB, DePristo MA, Dobson CM, Vendruscolo M (2005) Simultaneous determination of protein structure and dynamics. Nature 433:128–132

Lynch RW, Drickamer HG (1966) Effect of high pressure on the lattice parameters of diamond, graphite, and hexagonal boron nitride. J Chem Phys 44:181–184
Meinhold L, Smith JC, Kitao A, Zewail AH (2007) Picosecond fluctuating protein energy landscape mapped by pressure-temperature molecular dynamics simulation. Proc Natl Acad Sci U S A 104:17261–17265
Mossakowska DE, Nyberg K, Fersht AR (1989) Kinetic characterization of the recombinant ribonuclease from *Bacillus amyloliquefaciens* (barnase) and investigation of key residues in catalysis by site-directed mutagenesis. Biochemistry 28:3843–3850
Nielsen G, Jonker HRA, Vajpai N, Grzesiek S, Schwalbe H (2013) Kinase in motion: insights into the dynamic nature of p38α by high-pressure NMR spectroscopic studies. Chem Bio Chem 14:1799–1806
Nisius L, Grzesiek S (2012) Key stabilizing elements of protein structure identified through pressure and temperature perturbation of its hydrogen bond network. Nature Chem 4:711–717
Nolde SB, Arseniev AS, Orekhov VY, Billeter M (2002) Essential domain motions in barnase revealed by MD simulations. Proteins: Struct Funct Genet 46:250–258
Orekhov VY, Dubovskii PV, Yamada H, Akasaka K, Arseniev AS (2000) Pressure effect on the dynamics of an isolated alpha-helix studied by ^{15}N-^{1}H NMR relaxation. J Biomol NMR 17:257–263
Pandya MJ, Schiffers S, Hounslow AM, Williamson MP (2015) The catalytically active conformation of barnase is generated by a hierarchy of promoting motions. (submitted)
Pfeiffer H, Heremans K, Wevers M (2008) The influence of correlated protein-water volume fluctuations on the apparent compressibility of proteins determined by ultrasonic velocimetry. Biochim Biophys Acta 1784:1546–1551
Rashin AA, Iofin M, Honig B (1986) Internal cavities and buried waters in globular proteins. Biochemistry 25:3619–3625
Refaee M, Tezuka T, Akasaka K, Williamson MP (2003) Pressure-dependent changes in the solution structure of hen egg-white lysozyme. J Mol Biol 327:857–865
Roche J, Dellarole M, Caro JA, Norberto DR, Garcia AE, Garcia-Moreno B et al (2013) Effect of internal cavities on folding rates and routes revealed by real-time pressure-jump NMR spectroscopy. J Am Chem Soc 135:14610–14618
Sahu SC, Bhuyan AK, Udgaonkar JB, Hosur RV (2000) Backbone dynamics of free barnase and its complex with barstar determined by ^{15}N NMR relaxation study. J Biomol NMR 18:107–118
Sareth S, Li H, Yamada H, Woodward CK, Akasaka K (2000) Rapid internal dynamics of BPTI is insensitive to pressure: ^{15}N spin relaxation at 2 kbar. FEBS Lett 470:11–14
Sekhar A, Kay LE (2013) NMR paves the way for atomic level descriptions of sparsely populated, transiently formed biomolecular conformers. Proc Natl Acad Sci U S A 110:12867–12874
Shimizu H, Nagata K, Sasaki S (1988) High-pressure Raman study of the hydrogen-bonded crystalline formamide. J Chem Phys 89:2743–2747
Tilton RF, Dewan JC, Petsko GA (1992) Effects of temperature on protein structure and dynamics: X-Ray crystallographic studies of the protein ribonuclease A at 9 different temperatures from 98 K to 320 K. Biochemistry 31:2469–2481
Tzeng S-R, Kalodimos CG (2013) Allosteric inhibition through suppression of transient conformational states. Nature Chem Biol 9:462–465
Urayama P, Phillips GN, Gruner SM (2002) Probing substates in sperm whale myoglobin using high-pressure crystallography. Structure 10:51–60
Wagner G (1980) Activation volumes for the rotational motion of interior aromatic rings in globular proteins determined by high-resolution ^{1}H NMR at variable pressure. FEBS Lett 112:280–284
Williamson MP (2003) Many residues in cytochrome c populate alternative states under equilibrium conditions. Proteins 53:731–739
Williamson MP (2011) How proteins work. Garland Science, New York
Williamson MP, Akasaka K, Refaee M (2003) The solution structure of bovine pancreatic trypsin inhibitor at high pressure. Protein Sci 12:1971–1979
Wilton DJ, Ghosh M, Chary KVA, Akasaka K, Williamson MP (2008a) Structural change in a B-DNA helix with hydrostatic pressure. Nucleic Acids Res 36:4032–4037

Wilton DJ, Tunnicliffe RB, Kamatari YO, Akasaka K, Williamson MP (2008b) Pressure-induced changes in the solution structure of the GB1 domain of protein G. Proteins: Struct Funct Bioinf 71:1432–1440

Wilton DJ, Kitahara R, Akasaka K, Pandya MJ, Williamson MP (2009a) Pressure-dependent structure changes in barnase reveal changes in dynamics on ligand binding. Biophys J 97:1482–1490

Wilton DJ, Kitahara R, Akasaka K, Williamson MP (2009b) Pressure-dependent ^{13}C chemical shifts in proteins: origins and applications. J Biomol NMR 44:25–33

Yu X, Leitner DM (2003) Vibrational energy transfer and heat conduction in a protein. J Phys Chem B 107:1698–1707

Zhuravleva A, Korzhnev DM, Nolde SB, Kay LE, Arseniev AS, Billeter M et al (2007) Propagation of dynamic changes in barnase upon binding of barstar: an NMR and computational study. J Mol Biol 367:1079–1092

Chapter 7
Water Turns the "Non-biological" Fluctuation of Protein into "Biological" One

Fumio Hirata

Abstract Structural fluctuation of protein is not just an mechanical "oscillation," but an event induced by an interplay of mechanical and thermodynamic processes in which *water* plays crucial role. The chapter is devoted to provide a theoretical description concerning the concept of structural fluctuation of protein, based on methods of the statistical mechanics.

Keywords Structural fluctuation • Generalized Langevin theory • 3D-RISM/RISM • Linear response theory

7.1 Introduction

Structural fluctuation of protein around its native state plays essential roles in a variety of process in which a biomolecule performs its intrinsic function. For example, so called "gating" mechanisms of ion channels are often regulated by the structural fluctuation of amino-acid residues consisting the gate region of the channel. Molecular recognition such as the formation of an enzyme-substrate complex in an enzymatic reaction is controlled often by structural fluctuation of protein. Few typical examples of structural fluctuations around a native conformation of protein, related to biological functions, are "breathing,"(Makowski et al. 2008) "hinge-bending," (Mchaourab et al. 1997) and "arm-rotating" motions (Burghardt et al. 2011). Those motions are *collective* in nature involving many atoms moving in the same direction. Those structural fluctuation associated with protein functions, whether it's large or small, stays within its native conformation, and does not induce global conformational change, such as native to denatured, unless the thermodynamic condition of solution, such as denaturant concentration, changes.

It is not surprising that considerable effort have been devoted to clarify the conformational fluctuation of protein theoretically, which has started at the end of the last century based on the molecular mechanics or dynamics. One of earliest

F. Hirata (✉)
College of Life Sciences, Ritsumeikan University, Kusatsu, Shiga 525-8577, Japan
e-mail: hirataf@fc.ritsumei.ac.jp

K. Akasaka, H. Matsuki (eds.), *High Pressure Bioscience*,
Subcellular Biochemistry 72, DOI 10.1007/978-94-017-9918-8_7

attempts was to relate the structural fluctuation to the normal mode of protein (Go et al. 1983; Brooks and Karplus 1983; Horiuchi and Go 1991). Those works have demonstrated the importance of the collective mode in the fluctuation. However, those efforts could not have provided a realistic physical insight into the dynamics of actual biological processes, since they are concerned with a protein in "vacuum," which of course cannot describe the fluctuation correlated with that of solvent.

In actual biological processes, solvent plays vital roles both in the equilibrium and in fluctuation of protein (Kuwajima et al. 2009; Bihan and Fukuyama 2010). It may not be necessary to spend many words for emphasizing the crucial role played by solvent for stabilizing or destabilizing native structure of protein, such as the hydrophobic interaction and hydrogen bonds. Here, let us consider roles played by water in fluctuation of protein around its native conformation, associated with recognition of a ligand by protein. The process is primarily a thermodynamic process, governed by the free energy difference between the two states before and after the recognition. It is obvious that water plays crucial role in the thermodynamics, since the equilibrium structures are determined by the free energies including the excess chemical potential or the solvation free energy of water. However, it's not a mere role of water in the process. Water actually regulates the kinetic pathway of the process as well by controlling the structural fluctuation of amino-acid residues consisting the active site. An example of such processes is a *mouth-like* motion of amino-acid residues. The open-and-close motion of the *mouth* is driven not only by the direct force acting among atoms in protein, but by that originated from the solvent induced force which in turn caused by the fluctuation in the solvation free energy, or the *non-equilibrium* free energy.

An earlier attempt to consider the effect of solvent on structural fluctuation of protein has been made by Kitao et al. using the molecular dynamics simulation (Kitao et al. 1991; Hayward et al. 1993). By diagonalizing the variance-covariance matrix of the fluctuation, obtained from the trajectory, they have characterized the fluctuation in terms of the frequency spectrum. Unfortunately, such analyses are largely limited by the sampling of the configuration space.

Recently, two novel methods to characterize the structural fluctuation of protein have been proposed, one due to an experiment, the other by a theory. Using the high pressure NMR, Akasaka and his coworkers have developed a unique method to realize the fluctuation of protein in solution (Akasaka 2003, 2006; Fourme et al. 2012). The idea is based on their finding that the NMR signals such as the chemical shift of amino acids change upon increasing pressure, and that the changes become more significant as the pressure enhanced. Based on the finding, Akasaka has proposed a hypothesis that an *equilibrium* structure at higher pressure should correspond to one of *fluctuated* structures at the ambient pressure. The idea suggests that the fluctuated states of protein can be realized by changing pressure as an *order parameter*.

The theoretical characterization concerning the structural fluctuation of protein in solution has been proposed by Kim and Hirata, combining two methods in the statistical mechanics, the 3D-RISM/RISM theory and the generalized Langevin theory (Kim and Hirata 2013; Yoshida et al. 2009; Hirata 2003). The theory has clarified that the structural fluctuation can be treated in a fashion analogous to the *normal mode analysis*, but on the thermodynamic potential surface including the solvation free energy, not on the mechanical potential-energy surface.

The present manuscript is devoted to review the topics related to the structural fluctuation of protein in solution. The general concept to describe the fluctuation phenomena is also outlined to help readers understand.

7.2 Mathematical Foundation of Fluctuation (Gaussian Fluctuation)

It has been well documented that randomly fluctuating variables takes the Gaussian distribution, or the normal distribution, around its average value, when deviations (fluctuation) of the variables from the average are not extraordinary large. The rigorously proved natural law has been known as the "central limiting theorem" (Chandrasekhar 1943; Kubo et al. 1991).

$$w(\Delta x) = \sqrt{\frac{A}{2\pi}} \exp\left[-\frac{1}{2} A \Delta x^2\right] \tag{7.1}$$

The half-width of the distribution is a variance that measures the magnitude of a fluctuation.

$$A = \frac{1}{\langle \Delta x^2 \rangle} \tag{7.2}$$

where $\Delta x \, (= x - \langle x \rangle)$ is the fluctuation of a random variable x from its average, $\langle x \rangle$, and

$$\langle \Delta x^2 \rangle = \int_{-\infty}^{\infty} \Delta x^2 w(\Delta x)\, dx \tag{7.3}$$

So, $(1/A)^{1/2}$ is the standard deviation or the mean square fluctuation.

The equation can be readily generalized to the case where several fluctuating variables are concerned.

$$w(\Delta x_1, \Delta x_2, \ldots, \Delta x_n) = \sqrt{\frac{A}{(2\pi)^n}} \exp\left[-\frac{1}{2}\sum_i \sum_j A_{ij} \Delta x_i \Delta x_j\right] \tag{7.4}$$

where $\Delta x_i = x_i - \langle x_i \rangle$ is the fluctuation of the i-th random variable, A_{ij} is related to the variance-covariance matrix $\langle \Delta x_i \Delta x_j \rangle$ of the fluctuation by

$$A_{ij} = \frac{1}{\left\langle \Delta x_i \Delta x_j \right\rangle} \tag{7.5}$$

$$\left\langle \Delta x_i \Delta x_j \right\rangle = \int_{-\infty}^{\infty} \int_{-\infty}^{\infty} \Delta x_i \Delta x_j w \left(\Delta x_1, \Delta x_2, \ldots, \Delta x_n \right) dx_i dx_j$$

and A is the determinant of the matrix $\{A_{\mathrm{ij}}\}$.

7.3 Earlier Treatment of the Structural Fluctuation of Protein in Vacuum

Suppose that the interactions among atoms in protein are infinitesimally small, or harmonic. Then, the deviation (or "fluctuation") of atoms in protein from their equilibrium positions follows the Hook equation of motion

$$M_\alpha \frac{d^2 \Delta \mathbf{R}_\alpha(t)}{dt^2} = -\sum_\beta H_{\alpha\beta} \cdot \Delta \mathbf{R}_\beta(t) \tag{7.6}$$

$$\Delta \mathbf{R}_\alpha(t) = \mathbf{R}_\alpha(t) - \langle \mathbf{R}_\alpha \rangle \tag{7.7}$$

where $\mathbf{R}_\alpha(t)$, $\langle \mathbf{R}_\alpha \rangle$, and M_a are the position at time t, the equilibrium position, and the mass, of atom α in protein. $H_{\alpha\beta}$ is the force constant or the Hessian matrix defined by,

$$H_{\alpha\beta} = \frac{\partial^2 U}{\partial \mathbf{R}_\alpha \partial \mathbf{R}_\beta} \tag{7.8}$$

where $U(\mathbf{R}_1, \mathbf{R}_2, \ldots, \mathbf{R}_\mathrm{N})$ is a harmonic potential, and it has a quadratic form with respect to $\Delta\mathbf{R}$, that is,

$$U\left(\Delta \mathbf{R}_1, \Delta \mathbf{R}_2, \ldots, \Delta \mathbf{R}_N\right) = \frac{1}{2} \sum_\alpha \sum_\beta H_{\alpha\beta} \Delta \mathbf{R}_\alpha \Delta \mathbf{R}_\beta \tag{7.9}$$

If one considers that the protein atoms are in thermal motion, the distribution of $\Delta\mathbf{R}_\alpha$ becomes a "Gaussian,"

$$p\left(\Delta \mathbf{R}_1, \Delta \mathbf{R}_2, \ldots, \Delta \mathbf{R}_N\right) = \sqrt{\frac{A}{(2\pi)^{3N}}} \exp\left[-\frac{1}{2} \sum_\alpha \sum_\beta A_{\alpha\beta} \Delta \mathbf{R}_\alpha \Delta \mathbf{R}_\beta\right] \tag{7.10}$$

where $A_{\alpha\beta}$ is defined by,

$$A_{\alpha\beta} = \frac{H_{\alpha\beta}}{kT} \tag{7.11}$$

and A is the determinant of the matrix $\{A_{\alpha\beta}\}$. The variance-covariance matrix of the "fluctuation" is related to the Hessian through

$$\langle \Delta\mathbf{R}_\alpha \Delta\mathbf{R}_\beta \rangle = \frac{kT}{H_{\alpha\beta}} \tag{7.12}$$

The equation implies that the structural "fluctuation" of a protein can be characterized theoretically by calculating the second derivative of the interaction potential with respect to atomic coordinates. The idea has been employed in the early 1990s to characterize the structural fluctuation of protein by means of the computational science. However, the "fluctuation" was essentially a mechanical oscillation as is clarified in this section. The physical origin of the apparent "Gaussian" distribution of "fluctuation" in this system lies in the harmonicity of the mechanical potential energy. Such a structural fluctuation readily causes a non-linear or plastic deformation of protein, when the fluctuation becomes larger, in the following two reasons. Firstly, the protein structure is not in the equilibrium state, since the system does not include solvent, thereby does not satisfy the proper thermodynamic condition of equilibrium. In another word, the native structures of protein, found in PDB, are extremely unstable in "vacuum," which should quickly relax to their equilibrium structure. Secondly, a larger fluctuation readily violates the condition of "harmonicity."

Such a fluctuation described by the normal mode analysis of protein in vacuum should not have biological significance by any means.

7.4 Thermodynamic Fluctuation

Here, we consider a thermodynamic system of constant volume (V) and temperature (T). The equilibrium state of the system is defined by the point where the Helmholtz free energy becomes minimum, or

$$\left(\frac{\partial F}{\partial T}\right)_V = 0, \left(\frac{\partial F}{\partial V}\right)_T = 0. \tag{7.13}$$

Suppose the thermodynamic variables are fluctuating around the equilibrium state. The deviation of the free energy can be expanded around the equilibrium state up to the second order as

$$\Delta F = \frac{1}{2}\left(\frac{\partial^2 F}{\partial V^2}\right)_T \Delta V^2 + \frac{1}{2}\left(\frac{\partial^2 F}{\partial V \partial T}\right)\Delta V \Delta T + \frac{1}{2}\left(\frac{\partial^2 F}{\partial T^2}\right)_{V,} \Delta T^2 \tag{7.14}$$

where $\Delta F = F - \langle F \rangle$, $\Delta T = T - \langle T \rangle$, $\Delta V = V - \langle V \rangle$ are the fluctuation of the Helmholtz free energy, volume, and temperature, respectively. The first order term in the expansion disappears due to Eq. 7.13. The equation can be transformed into the following equation by the standard manipulation in thermodynamics,

$$\Delta F = \frac{1}{2V\kappa_T}\Delta V^2 - \frac{1}{2}\beta \Delta V \Delta T - \frac{1}{2}\frac{C_V}{T}\Delta T^2 \tag{7.15}$$

where κ_T, β, and C_v are defined, respectively, by

$$\kappa_T = -\frac{1}{V}\left(\frac{\partial V}{\partial P}\right)_T, \ \beta = \left(\frac{\partial P}{\partial T}\right)_V, \ C_V = T\left(\frac{\partial S}{\partial T}\right)_V \tag{7.16}$$

The probability distribution of the free energy can be written as

$$w(\Delta F) \propto \exp\left[-\frac{\Delta F}{k_B T}\right] \tag{7.17}$$

or

$$w(\Delta F) \propto \exp\left[-\frac{1}{2Vk_B T\kappa_T}\Delta V^2 + \frac{\beta}{2k_B T}\Delta V \Delta T + \frac{1}{2}\frac{C_V}{k_B T^2}\Delta T^2\right] \tag{7.18}$$

Applying the theorems concerning the Gaussian distribution, described in the previous section, one can get the relation between the thermodynamic variables and the variance-covariance matrix of the fluctuation as,

$$\langle \Delta V^2 \rangle = Vk_B T\kappa_T, \ \langle \Delta V \Delta T \rangle = \frac{k_B T}{\beta}, \ \langle \Delta T^2 \rangle = \frac{k_B T^2}{C_V}. \tag{7.19}$$

Similar relations for the other statistical ensembles such as constant pressure and temperature can be derived. The relations originated by Landau scored a great landmark in the history of thermodynamics, and any phenomenological fluctuation should follow the general theorem (Landau and Lifshitz 1964). However, the theorem does not tell anything about the molecular process behind the fluctuation, concerning protein conformation and solution structure. It is a nontrivial problem for experiments and theories to characterize fluctuating states in terms of the molecular interaction or structure, since such states are high (free) energy states by definition, and their occurrence is rare in comparison with the equilibrium state.

One way of characterizing fluctuating states is to use a *reaction coordinate* or an *order parameter*. The next two sections are devoted to introduce methods to realizing fluctuated states using order parameters.

7.5 Realization of Solvent Fluctuation Around a Solute by Marcus

The credit as the first scientist to realize theoretically the solvent fluctuation conjugated with that in solute should go to R. Marcus (Marcus 1956). The theory is concerned with the rate of electron transfer reaction between a donor-acceptor pair in polar solvent, in which the reaction rate is controlled by a barrier crossing process. The free energy barrier is created, in this simple reaction, by fluctuation of the solvent polarizations around the donor-accepter pair. Thus, the problem is reduced to how to characterize the free energy barrier from the solvent property. The idea by Marcus is illustrated in Fig. 7.1.

In the figure, the state labeled (A) illustrates an equilibrium state in which the solute molecule is electrically neutral, and the solvent orientations are distributed more or less randomly according to their own energetic cause. On the other hand, the state (B) illustrates an equilibrium state after the solute having accepted an electron to become an anion, in which solvent dipoles are oriented along the electric field of the ion. The state (A*) illustrates a fluctuating state of solvent around the neutral solute, in which solvent is polarized as if the solute has a negative charge. (See the solvent polarization in the state B.) The state (B*) illustrates a fluctuating state of solvent around the anion, in which solvent is distributed as if the solute molecules are neutral electrostatically. (See the solvent polarization in the state A)

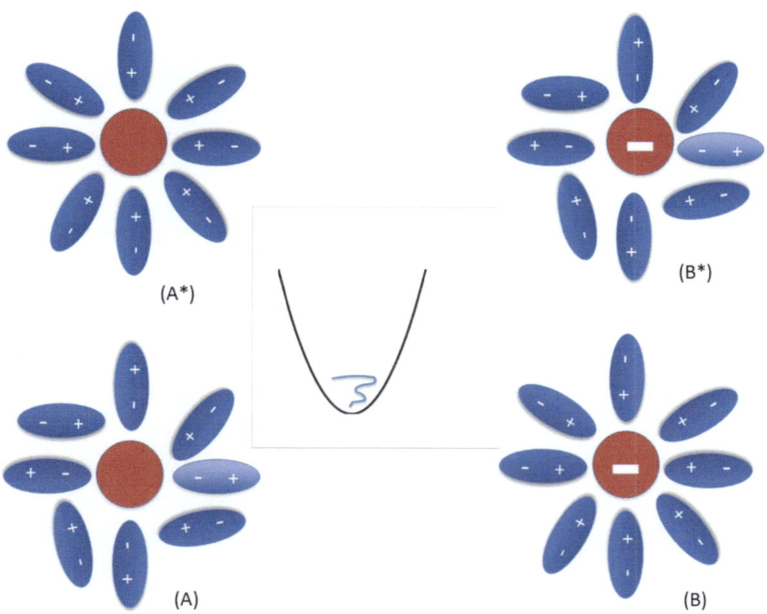

Fig. 7.1 Illustrative picture of solvent fluctuation around a solute, induced by a charge as an order parameter

The figure indicates that a fluctuated state of solvent can be realized by changing the hypothetical charge on the solute as an "order parameter," or a reaction coordinate.

Combining the idea with the Born model concerning the solvation free energy of an ion, Marcus has derived the famous equation, which rewarded him a Nobel prize, concerning the rate constant of the electron transfer reaction,

$$k_e = \nu \exp\left[\frac{-\Delta G^{\neq}}{k_B T}\right] \tag{7.20}$$

$$\Delta G^{\neq} = \frac{(\lambda + \Delta G_0)}{4\lambda} \tag{7.21}$$

where $\Delta G^{\neq}$, ΔG_0 are the activation barrier and the free energy difference between reactant and product states, respectively. λ is so called "reorganization energy" which represents the fluctuation of the free energy from the equilibrium state of reactants, expressed in its simplest form as

$$\lambda = \frac{z^2}{2}\left(1 - \frac{1}{\varepsilon}\right)\left(\frac{1}{R_D} + \frac{1}{R_A} - \frac{2}{L}\right) \tag{7.22}$$

where ε is the dielectric constant of solvent, R_D and R_A are the radius of donor and acceptor, respectively, and L is the separation between the donor and acceptor. It is interesting to note that the reorganization energy, or the fluctuation, depends quadratically on the "hypothetical charge" z as the reaction coordinate. This is another manifestation of the central limiting theorem concerning the fluctuation. Since the activation barrier is dominated by the reorganization energy, the reaction rate is essentially determined by the thermodynamic fluctuation of solvent.

Although the Marcus theory succeeded to characterize the solvent fluctuation along the reaction coordinates of the electron-transfer reaction, the theory was highly phenomenological based on the continuum solvent representation. An attempt to place the theory on the molecular-science foundation has been made by Chong, Hirata and their coworkers based on the statistical mechanics of molecular liquids, or the RISM theory (Chong et al. 1995; Chong and Hirata 1997).

7.6 Realization of Structural Fluctuation of Protein by Akasaka

An interesting characterization of the structural fluctuation of protein has been proposed experimentally by Akasaka about 10 years ago based on the high pressure NMR (Akasaka 2003, 2006; Fourme et al. 2012). The idea is conceptually akin to that proposed by Marcus to realize the reorganization energy in the electron-transfer reaction which was described in the preceding section. Instead of the hypothetical charges on a solute, Akasaka and coworkers employed the hydrostatic pressure as an order parameter, or the reaction coordinates, to realize the fluctuating states of protein empirically. When they applied the pressure on aqueous solutions of protein, they found a variety of different conformations of protein, from partially denatured to fully denatured, depending on the strength of pressure, which can be well characterized by the difference in the NMR chemical shift of each residue. They identified those different conformations under pressure as fluctuating states of protein when the solution is in the native condition, or ambient pressure. There is no doubt that their contribution is the first to have extracted experimentally the conformational fluctuation of protein in the ambient condition. The soundness of the method is verified theoretically in the following section, based on a new theory concerning the structural fluctuation of protein.

In the mean time, if the hypothesis is correct, the fluctuating state of protein in ambient condition should have the partial molar volume which is less compared to the native state, due to Le Chatelier's principle (Remember, a fluctuating conformation was produced at a higher pressure). The hypothesis cannot be readily verified just by an experiment, because the partial molar volume of the fluctuating state is not easy to be measured. (A fluctuating state is a rare event.)

The hypothesis has been verified theoretically by Imai and coworkers based on the 3D-RISM/RISM theory combined with the Kirkwood-Buff theory (or the site-site Kirkwood-Buff theory) (Imai et al. 2000, 2004; Kirkwood and Buff 1951; Imai et al. 2007).

$$\Delta\overline{V}\left(\{\mathbf{R}\}\right) = k_B T \chi_T^0 \left(1 - \rho \sum_{\gamma} \int_V c_{\gamma}\left(\mathbf{r}; \{\mathbf{R}\}\right) d\mathbf{r}\right) \tag{7.23}$$

In the equation, $\{\mathbf{R}\}$ represents the structure of protein, χ_T^0 and $c_\gamma(\mathbf{r}; \{\mathbf{R}\})$ are, respectively the isothermal compressibility of solvent, and so called the direct correlation function which is closely related to the density correlation function of solvent around protein. They have calculated the partial molar volume of ubiquitin at the ambient condition, using the two protein conformations determined based on NMR by Kitahara and Akasaka at two different pressures, 3 and 300 MPa (Kitahara et al. 2005). The conformation at higher pressure is supposed to be the one which is identified as a fluctuating state in the ambient condition. The results for the partial molar volume $\Delta\overline{V}$ are analyzed according to the following equation, which decomposes the volume into the contributions with different physical origins, as is illustrated in Fig. 7.2.

$$\Delta\overline{V} = V_{id} + V_W + V_V + V_T + V_I \tag{7.24}$$

In the equations, $V_{id}\ \left(= k_B T \chi_T^0\right)$ is the ideal contribution to the volume, when the protein is viewed as ideal gas; V_{w} is the volume of atoms enclosed by the van der Waals surface of protein as is illustrated in the figure; V_{c} is the volume of a cavity including solvent molecules; V_v is the volume of voids which do not include water molecules: V_0 is the partial molar volume of protein without charges;

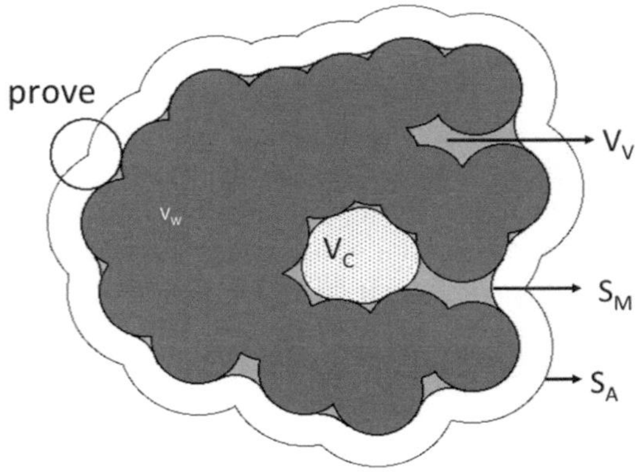

Fig. 7.2 Illustration of each contribution to the partial molar volume

Table 7.1 Partial molar volume ($\Delta\overline{V}$, cm^3 mol^{-1}) of low-pressure (3 MPa) and high-pressure (300 MPa) structure (LPS and HPS, respectively) of ubiquitin in aqueous solutions at 298.15 K, and its components

	$\Delta\overline{V}$	V_W	V_V	V_T	V_I
LPS	5,788.4	3,768.3	1,546.9	645.5	−173.6
HPS	5,741.2	3,764.7	1,445.6	680.5	−150.8
difference	−47.2	−3.6	−101.3	34.9	22.8

$V_T\left(=\overline{V}_0 - V_{id} - (V_W + V_V)\ \right)$ is the "thermal volume": $V_I\left(=\Delta\overline{V} - \overline{V}_0\right)$ is the contribution from the molecular interaction between protein and water. The results are shown in Table 7.1. As can be readily seen, the partial molar volume of the high pressure state (HPS) is less than that of the low pressure structure by about 47 cm^3 mol^{-1}, thereby the hypothesis is proved. Then, what is the physical cause of the reduction in the partial molar volume at the fluctuated conformation? It is obvious that the major contribution to the reduction is coming from V_v, the "void volume," which is essentially the contribution from internal cavities of protein. In order to inspect this point more closely, let's look at the distribution of water molecules in the two states of protein, which can be also calculated by 3D-RISM/RISM (Fig. 7.3). Apparently, HPS has well developed distribution of water molecules inside the cavity, while LPS has much less amount of water molecules. Therefore, it can be concluded that the lower partial molar volume of the fluctuated conformation is caused essentially by those water molecules penetrated into the protein cavity from the bulk.

The finding by Imai et al. has a great implication concerning the physical origin of the structural fluctuation of protein. That is, the fluctuation takes place as a result of interplay between those of the naked protein and of solvent. Just think of the difference in the partial molar volume between the native and fluctuated states of protein, which amounts to 47 cm^3 mol^{-1}. This difference contributes to the thermodynamic fluctuation, or the compressibility, since $\chi_T \sim \langle \Delta V^2 \rangle$. Therefore, the earlier characterization of the structural fluctuation of protein, based on the normal mode analysis of naked protein, or protein without solvent, does not make any sense.

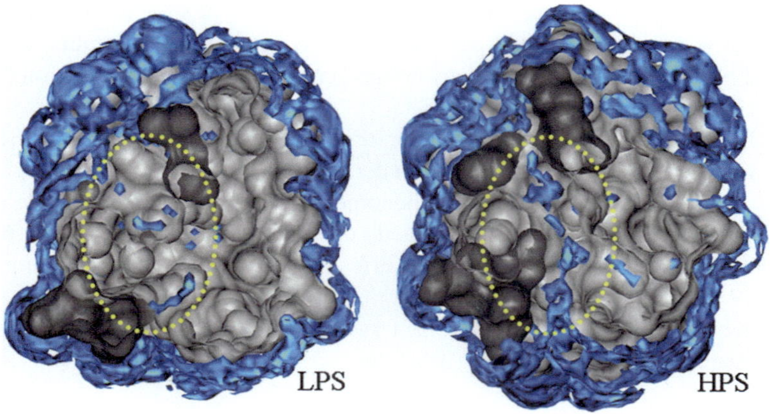

Fig. 7.3 Distribution of water inside cavities of ubiquitin. *LPS* the low pressure structure, *HPS* the high pressure structure

The results are also suggestive of the physical origin of the pressure denaturation. There are many cavities inside protein, which can be roughly categorized into two groups depending on whether the cavity includes water molecules or not. An example of the former is the active-site of an enzyme in apo-state, which includes one or few water molecules in general. The cavities are insensitive to pressure, because the cavities are already filled by water molecules, thereby the volume does not change much by applying pressure. An example of the cavities, which do not include water molecules, is the "void" examined in the present chapter. Such cavities are created due to imperfect packing of amino-acid residues in protein, which are hydrophobic in many cases. The cavities are sensitive to pressure, because water molecules can penetrate into the cavity from bulk, reducing the partial molar volume. The situation can be viewed in another way. The cavity with water molecules is created as a fluctuation of the native structure in the ambient condition, which has smaller partial molar volume, and higher free energy, compared to the native state, or to the equilibrium state. By applying pressure, the structure with lower partial molar volume is stabilized.

Then, how such a process involving the conjugated structural and thermodynamic fluctuations of protein in solution can be treated theoretically? The question is answered in the following section.

7.7 Theoretical Characterization of Structural Fluctuation of Protein in Solution

7.7.1 Phenomenological Langevin Theory

The structural fluctuation of protein is not just a mechanical process represented by the "normal mode analysis," but conjugated mechanical and thermodynamic processes, in which structure and thermodynamics of water play a crucial role.

Therefore, the theory should include both the structure of protein and thermodynamics of water as independent variables in the equation. Such a treatment can be realized by the theoretical framework of the generalized Langevin equation (GLE) (Mori 1962).

In this section, we introduce a theoretical treatment of the fluctuation based on GLE. In order to help the reader understand the theoretical treatment, let us begin with a brief explanation of the phenomenological Langevin theory, on which most of the conventional "coarse-grained" dynamics of protein are based (Wang and Uhlenbeck 1945; McQuarrie 1976; Ermak and McCammon 1978; Lamm and Szabo 1986).

As everybody knows, a particle in a system including N particle follows the Newtonian equation in good approximation,

$$m_i \frac{d^2\mathbf{r}_i}{dt^2} = \mathbf{f}_i(t) \quad (\mathrm{i} = 1,\ N) \tag{7.25}$$

where $\mathbf{r}_\mathrm{i}$ and m_i are position and mass of the i-th particle in the system, and $\mathbf{f}_\mathrm{i}$ the force acting on the particle. This is the equation employed in the all atom MD simulation. The MD simulation tries essentially to solve the equation directly by means of a numerical integration.

Then, let us imagine a system in which one big molecule is moving in an environment of solvent molecules, and that the motion of the big particle is accelerated and decelerated by a random force $\mathbf{W}(t)$ and a friction ζ, respectively, due to solvent's thermal motion. Such a motion can be described by the following Langevin equation

$$\mathbf{V} = \frac{d\mathbf{R}}{dt} \tag{7.26}$$

$$M\frac{d\mathbf{V}}{dt} = \mathbf{F}(t) - \zeta\mathbf{V} + \mathbf{W}(t) \tag{7.27}$$

where **R**, **V** and M are position, velocity, and mass of the particle, and **F** the force acting on the particle. The equation can be applied to the structural dynamics of protein in solvent by putting a suffix on **R**, **V**, M, **F** and **W** to specify atoms, and by making the friction ζ into a matrix indicating that the frictions onto atoms are not entirely independent but correlated. The correlation is called *hydrodynamic interaction*.

$$\mathbf{V}_a = \frac{d\mathbf{R}_a}{dt} \tag{7.28}$$

$$M_a\frac{d\mathbf{V}_a}{dt} = \mathbf{F}_a(t) - \sum_b \zeta_{ab}\mathbf{V}_b + \mathbf{W}_a(t) \tag{7.29}$$

This is essentially the equation proposed by Ermak and McCammon to describe a structural dynamics of protein, and employed by Szabo for the normal mode analysis of protein in solvent (Wang and Uhlenbeck 1945; McQuarrie 1976; Ermak and McCammon 1978; Lamm and Szabo 1986). Unfortunately, the success of the theory largely depends on the phenomenological parameter ζ. It is almost impossible to set up uniquely such parameters for each atom and atom pair, because the environment around each atom is extremely inhomogeneous both in space and in time. It is the generalized Langevin theory (GLT) that has great perspective to overcome the difficulty (Mori 1962; Kim and Hirata 2013).

7.7.2 Generalized Lagenvin Theory of Protein in Solution

The time evolution of a mechanical variable in the phase space can be described by the Liouville equation.

$$\frac{d\mathbf{B}(t)}{dt} = i\mathcal{L}\mathbf{B}(t) \tag{7.30}$$

where $\mathscr{L}$ is the Liouville operator defined in terms of the Poisson bracket,

$$\mathcal{L} \equiv i\sum_i \left[\left(\frac{\partial H}{\partial \mathbf{r}_i} \right) \cdot \left(\frac{\partial}{\partial \mathbf{p}_i} \right) - \left(\frac{\partial H}{\partial \mathbf{p}_i} \right) \cdot \left(\frac{\partial}{\partial \mathbf{r}_i} \right) \right] \tag{7.31}$$

The basic idea of GLE is to reduce the number of variables to describe the time evolution, by *projecting* all other degrees of freedom onto few variables which are essential to describe the dynamics under concern. Such variables are referred to as “dynamic variables.” So, let us briefly outline the concept of the projection.

The concept is a generalization of the projection of a vector onto another vector in the Euclidean space to that of the functional space or the Hilbert space. When one tries to project a vector **B** in the Euclidean space to another vector **C**, the projection is defined by the following equation (see the illustration in Fig. 7.4)

$$\mathcal{P}\mathbf{B} = \frac{\mathbf{C}}{C}\frac{\mathbf{C}}{C}\cdot \mathbf{B} \tag{7.32}$$

Now, we generalize the idea to the projection in the phase space or the Hilbert space. The scalar product is defined as,

$$(\mathbf{C}, \mathbf{B}) = \left\langle \mathbf{C}\tilde{\mathbf{B}} \right\rangle = \int d\Gamma f^{eq}(\Gamma)\, \mathbf{C}(\Gamma)\, \tilde{\mathbf{B}}(\Gamma) \tag{7.33}$$

where $< \ldots >$ indicates the canonical ensemble average. The projection of a variable **B** onto **C** in the phase space can be defined as

$$\mathcal{P}\mathbf{B}(t) = \frac{\mathbf{C}\cdot(\mathbf{C}, \mathbf{B}(t))}{(\mathbf{C}, \mathbf{C})} \tag{7.34}$$

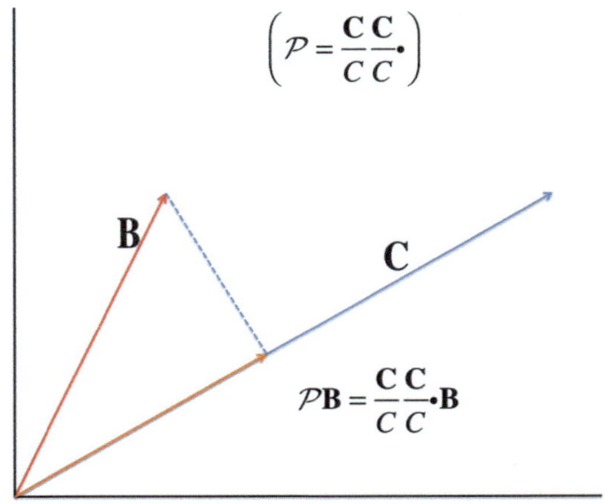

Fig. 7.4 Projection of a vector B onto a vector C

where $\mathcal{P}$ is the projection operator,

$$\mathcal{P} = \frac{\mathbf{C} \cdot (\mathbf{C},\ \ )}{(\mathbf{C}, \mathbf{C})} \tag{7.35}$$

which satisfies the idempotence.

$$\mathcal{P}\mathcal{P} = \mathcal{P} \tag{7.36}$$

Now, we apply the method to a system which consists of a protein molecule and solvent molecules. The Hamiltonian of the system is defined by,

$$H = H_0 + H_1 + H_2, \tag{7.37}$$

where H_0, H_1, H_2 are, respectively, the Hamiltonian concerning solvent, protein, and solute-solvent interaction, defined by the following equations. (Kim and Hirata 2013)

$$H_0 = \sum_{i=1}^{N} \sum_{a=1}^{n} \left[\frac{\mathbf{p}_i^a \cdot \mathbf{p}_i^a}{2m_a} + \sum_{j \neq i} \sum_{b \neq a} U_0 \left(\left| \mathbf{r}_i^a - \mathbf{r}_j^b \right| \right) \right] \tag{7.38}$$

$$H_1 = \sum_{\alpha=1}^{N_u} \left[\frac{\mathbf{P}_\alpha \cdot \mathbf{P}_\alpha}{2M_\alpha} + \sum_{\beta \neq \alpha} U_1 \left(\left| \mathbf{R}_\alpha - \mathbf{R}_\beta \right| \right) \right] \tag{7.39}$$

$$H_2 = \sum_{\alpha=1}^{N_u} \sum_{i=1}^{N} \sum_{a=1}^{n} U_{\text{int}} \left(\left| \mathbf{R}_\alpha - \mathbf{r}_i^a \right| \right) \tag{7.40}$$

In the equations, $\mathbf{p}_i^{a}$ and $\mathbf{r}_i^{a}$ denote the momentum and position of the atom a in solvent, U_0 the interaction energy between a pair of atoms in solvent, $\mathbf{P}_\alpha$ and $\mathbf{R}_\alpha$ the momentum and position of the atom α of protein, U_1 the interaction energy between a pair of atoms in the protein, U_{int} the interaction energy between an atom of solvent and an atom of the protein.

A crucial step of developing the theory is how to chose the dynamic variables, which should be able to describe the structural and thermodynamic fluctuations of both protein and water. We chose the deviation or fluctuation of atomic positions from its equilibrium states, $\Delta\mathbf{R}_\alpha$, and their conjugated momentum $\mathbf{P}_\alpha$ as the dynamic variables for protein, defined as

$$\Delta\mathbf{R}_\alpha \equiv \mathbf{R}_\alpha(t) - \langle\mathbf{R}_\alpha\rangle\,, \tag{7.41}$$

$$\mathbf{P}_\alpha = M_\alpha\dot{\mathbf{R}}_\alpha. \tag{7.42}$$

On the other hand, the collective density fluctuation of atom, $\delta\rho_a$, and the momentum density $\mathbf{J}$ are chosen for solvent, defined as follows.

$$\delta\rho_a = \sum_i \delta\left(\mathbf{r} - \mathbf{r}_i^a(t)\right) - \langle\rho_a\rangle \tag{7.43}$$

$$J_a = \sum_i \mathbf{p}_i^a\delta\left(\mathbf{r} - \mathbf{r}_i^a(t)\right) \tag{7.44}$$

Applying the projection operator to the Liouville equation gives rise to the following equations which describe the spatial and temporal fluctuation of protein structure as well as the solvent density around their equilibrium states. (Kim and Hirata 2013)

$$M_\alpha\frac{d^2\Delta\mathbf{R}_\alpha(t)}{dt^2} = -k_B T\sum_\beta\left(\mathbf{L}^{-1}\right)_{\alpha\beta}\cdot\Delta\mathbf{R}_\beta(t) - \int_0^t ds\sum_\beta\Gamma_{\alpha\beta}(t-s)\cdot\frac{\mathbf{P}_\beta(s)}{M_\beta} + W_\alpha(t) \tag{7.45}$$

$$\begin{aligned}\frac{d^2\delta\rho_{\mathbf{k}}^a(t)}{dt^2} &= -k^2\sum_{b,c} J_{ac}(\mathbf{k})\,\chi_{cb}^{-1}(\mathbf{k})\delta\rho_{\mathbf{k}}^b(t)\\ &\quad - \frac{i\mathbf{k}}{N}\sum J_{ac}^{-1}(k)\int_0^t ds M_{\mathbf{k}}^{bc}(t-s)\cdot J_k^b(s) + i\mathbf{k}\cdot\Xi_{\mathbf{k}}^a(t)\end{aligned} \tag{7.46}$$

In the equations, $\Gamma_{\alpha\beta}(t-s)$, $W_\alpha(t)$, $J_{ac}(k)$, $M_{\mathbf{k}}^{bc}(t-s)$, and $\Xi_{\mathbf{k}}^a(t)$ have the following physical meaning.

$\Gamma_{\alpha\beta}(t-s)$: frictional force due to solvent, acting on protein atoms
$W_\alpha(t)$: random force acting on protein atoms
$J_{ac}(k)$: correlation of momentum density between a pair of atoms in solvent

$\chi_{cb}(k)$: density correlation function between a pair of solvent atoms under influence of protein
$M_{\mathbf{k}}^{bc}(t-s)$: force acting on solvent molecules
$\Xi_{\mathbf{k}}^{a}(t)$: random force acting on solvent atoms

The second equation describes the time evolution of the density fluctuation of water (solution). The equation looks quite similar to the one derived by one of the authors for dynamics of *pure* water (Hirata 1992). The main difference is that the new equation describes the dynamics of solvent in inhomogeneous field produced by protein, while the old equation is valid just for dynamics in a homogeneous environment.

The first equation concerns the dynamics of protein atoms in the field exerted by the protein. The equation has a general form of the Langevin equation, including the friction and random-force terms. The matrix **L** included in Eq. 7.45 is defined by

$$L_{\alpha\beta} = \langle \Delta\mathbf{R}\Delta\mathbf{R} \rangle_{\alpha\beta}, \tag{7.47}$$

where $\Delta\mathbf{R}$ is the displacement, or fluctuation, of atoms in protein from their equilibrium position, defined by Eq. 7.41. Therefore, the matrix represents a correlation function of structural fluctuation, or the variance-covariance matrix explained in Sect. 7.2. The structural fluctuation and dynamics of protein in solution can be explored by solving the Eqs. 7.45 and 7.46 coupled together. However, Eq. 7.45 provides a deep physical insight into the structural fluctuation of protein in solution, already, before actually solving the equation.

If one ignores the friction and random force terms, Eq. 7.45 leads

$$\frac{d^2\Delta\mathbf{R}(t)}{dt^2} = -k_B T \mathbf{L}^{-1} \cdot \Delta\mathbf{R}(t) \tag{7.48}$$

The equation is *formally* identical with a "harmonic oscillator," described in Sect. 7.3, the "Hessian" of which is $k_B T\mathbf{L}^{-1}$. With Eq. 7.47, the Hessian is related to the variance-covariance matrix of the structural fluctuation through

$$k_B T \mathbf{L}^{-1} \equiv k_B T \langle \Delta\mathbf{R}\Delta\mathbf{R} \rangle^{-1} \tag{7.49}$$

The observation strongly suggests that the dynamics described by Eq. 7.48 is the Gaussian fluctuation around a *minimum* of the *free energy surface* consisting not only of the interactions among atoms in the protein, but of the solvation free energy. In this respect, the configuration corresponding to the free energy minimum is not just one but an ensemble of distinguishable configurations concerning protein and solvent, which can be converted among each other due to the thermal noise. Then, the Hessian matrix in Eq. 7.48 should be the second derivative of the *free energy*, not of the *potential energy*, with respect to the position of protein atoms. The physical insight further provides an *ansatz* to evaluate the variance-covariance matrix based on the 3D-RISM theory (Yoshida et al. 2009).

The free energy surface of protein can be given by

$$F(\{\mathbf{R}\}) = U(\{\mathbf{R}\}) + \Delta\mu(\{\mathbf{R}\}), \tag{7.50}$$

where $U(\{\mathbf{R}\})$ is the interaction among atoms in a protein, $\Delta\mu(\{\mathbf{R}\})$ is the solvation free energy of protein, the conformation of which is represented by $\{\mathbf{R}\}$. The variance-covariance matrix or the Hessian matrix is the second derivative of the free energy surface, namely,

$$k_B T\left(\mathbf{L}^{-1}\right)_{\alpha\gamma} = \frac{\partial^2 F}{\partial \mathbf{R}_\alpha \partial \mathbf{R}_\gamma} \tag{7.51}$$

Since the free energy $F(\{\mathbf{R}\})$ can be obtained by solving the 3D-RISM/RISM equation, Eq. 7.51 provides a practical way to evaluate the variance-covariance matrix.

The variance-covariance matrix by itself is quite informative for characterizing the structural fluctuation of protein around its native state in atomic detail. As an example, let us consider a hinge-bending motion of protein. In this case, the variance-covariance matrix is expected to have a structure in which a block of elements $<\Delta\mathbf{R}_\alpha \Delta\mathbf{R}_\beta>$ for atom pairs, α, β, belonging to the two sides of the hinge-axis have the negative sign, because the direction of the displacements $\Delta\mathbf{R}_\alpha$ and $\Delta\mathbf{R}_\beta$ is opposite.

It will be worthwhile to make a further remark on Eq. 7.45 in order to clarify the essential physics involved in the equation. The equation looks similar to the Langevin harmonic-oscillator treated by Wang and Uhlenbeck (Wang and Uhlenbeck 1945) and by Lamm and Szabo (Lamm and Szabo 1986). However, the two equations are essentially different in their physics. The earlier treatments assume that the oscillation is infinitesimally small, or equivalently that the *interaction potential* among atoms is harmonic. Such an assumption allows to represent the Hessian in terms of the second derivative of the *potential energy* with respect to atom positions. On the other hand, no such assumption concerning the interaction potential was made to derive Eq. 7.45. The *harmonic-oscillator-like* equation has been derived from the Liouville equation through a projection of solvent degrees of freedom on to those of protein in the phase-space. The apparent linearity of the dynamics in Eq. 7.45, therefore, is not due to making the assumption on the potential energy of protein, but originates from the projection that makes the process stochastic. The linearity of the stochastic process is a manifestation of the central limiting theorem.

7.7.3 Characterizing a Fluctuating Conformation with an Order Parameter

Equation 7.48 implies that the structural fluctuation of protein around its equilibrium state is the Gaussian fluctuation in the free energy surface described by

$$F\left(\{\Delta\mathbf{R}\}\right) = \frac{1}{2}k_B T \sum_{\alpha,\beta} \Delta\mathbf{R}_\alpha \cdot \left(\mathbf{L}^{-1}\right)_{\alpha,\beta} \cdot \Delta\mathbf{R}_\beta. \tag{7.52}$$

Let us consider a structural change of protein induced by some force, or a perturbation. The situation can be described by the following equation

$$F\left(\{\Delta\mathbf{R}\}\right) = \frac{1}{2}k_B T \sum_{\alpha,\beta} \Delta\mathbf{R}_\alpha \cdot \left(L^{-1}\right)_{\alpha,\beta} \cdot \Delta\mathbf{R}_\beta - \sum_{\alpha} \Delta\mathbf{R}_\alpha \cdot \mathbf{f}_\alpha. \tag{7.53}$$

In the equation, the first term describes the free energy surface of an unperturbed system, and the second term the perturbation. By applying the variational principle, or

$$\frac{\partial F\left(\{\Delta\mathbf{R}\}\right)}{\partial \Delta\mathbf{R}_\alpha} = 0, \tag{7.54}$$

one can get the following equation for the structural change due to the perturbation.

$$\langle \Delta\mathbf{R}_\alpha \rangle_1 = (k_B T_0)^{-1} \sum_{\beta} \langle \Delta\mathbf{R}_\alpha \Delta\mathbf{R}_\beta \rangle_0 \cdot \mathbf{f}_\beta \tag{7.55}$$

So, if one thinks $\mathbf{f}_\beta$ as an order parameter, this represents a fluctuated state produced by the parameter. The equation first derived by Ikeguchi et al. based on the linear response theory (Ikeguchi et al. 2005), and reformulated based on the variational principle (Kim and Hirata 2013), can be employed to put the method by Akasaka, described in Sect. 7.6, on a solid theoretical foundation (Hirata and Akasaka 2014).

Let us consider that the structural change of protein is induced by pressure. The force causing the structural change can be written in the following equality (Hirata and Akasaka 2014).

$$\mathbf{f}_\alpha = -\left(\frac{\partial F_s}{\partial \mathbf{R}_\alpha}\right) = -\left(\frac{\partial F_s}{\partial V}\right)_T \left(\frac{\partial V}{\partial \mathbf{R}_\alpha}\right) = P \left(\frac{\partial \Delta \overline{V}}{\partial \mathbf{R}_\alpha}\right) \tag{7.56}$$

Combining Eqs. 7.55 and 7.56, one get

$$\langle \Delta\mathbf{R}_\alpha \rangle = (k_B T_0)^{-1} \sum_{\beta} \langle \Delta\mathbf{R}_\alpha \Delta\mathbf{R}_\beta \rangle_0 \left(\frac{\partial \Delta \overline{V}}{\partial \mathbf{R}_\beta}\right)_{P,T} P \tag{7.57}$$

In the equality, F_s, V and, P are the Helmholtz free energy, volume, and pressure of the *entire system*, while $\Delta\overline{V}$ is the partial molar volume of a protein in the solution. (Note that the notation F_s is used to distinguish it from the *free energy surface* of the protein defined by $\mathbf{F}(\{\Delta\mathbf{R}\})$. The last equality is resulted from the heuristic consideration that the change in the volume due to the structural change is only concerned with the partial molar volume. Equation 7.55 provides a method to produce a fluctuated structure of the protein by changing pressure as an order parameter (Hirata and Akasaka 2014).

The equation has a great implication that the structural fluctuations produced by pressure correspond to those occurring naturally in the ambient condition. According to the equations, the structural response to pressure is realized through the successive response functions. The first one is the first derivative of the partial molar volume with respect to the atomic coordinates of protein. The second is the variance-covariance matrix of the fluctuation in the unperturbed system. So, when pressure is applied, the deformation first occurs so that the partial molar volume decreases. Such a process is described in the preceding section for ubiquitin as an example. Then, the deformation is propagated to the entire protein through the variance-covariance matrix.

It is worthwhile to make remarks on the new equation. The first point to be made is that the both response functions are *intrinsic* properties of a protein solution, meaning that the properties are determined just by the interactions among atoms in the protein and water. The pressure plays just a role of *magnifying* the intrinsic properties, as anticipated in Akasaka's experiment. Secondly, the equation indicates that the entropy of the system does not change due to the perturbation. It in turn implies that the dispersion of the structural fluctuation is not enhanced by applying pressure. The theoretical finding is in accord with the experimental observation by Akasaka, which shows that conformational variation produced by applying pressure is limited within its original ensemble (Akasaka et al. 1999).

Another important point to be made is that the fluctuating states of protein can be reproduced theoretically, which are identical in its physics to those found by the experiments due to Akasaka et al. (Akasaka 2003, 2006; Fourme et al. 2012). It has been already shown in this section by Eq. 7.51 that the variance-covariance matrix can be calculated from the 3D-RISM/RISM theory. The first derivative of the partial molar volume can be also calculated based on the 3D-RISM/RISM theory combined with the site-site Kirkwood-Buff theory, or Eq. 7.23. In the equation, $c_\gamma(\mathbf{r};\{\mathbf{R}\})$ is the direct correlation function of the water molecules around protein, which is a functional of the protein structure $\{\mathbf{R}\}$.

7.8 Concluding Remarks

In the present article, we have reviewed the latest development in theories concerning the structural fluctuation of protein. The development has been made based on two findings, one by the pressure NMR experiment, the other by the statistical mechanics of solution, both indicating strongly that the structural fluctuation of protein around equilibrium conformation is linear, and that the distribution of the conformation is Gaussian. Apparently, the statement sounds similar to that made earlier in the study of structural fluctuation of protein, based on the "normal mode analysis." However, the physical origins of the linearity and Gaussian behavior in the two processes are entirely different.

The linearity and the Gaussian behavior derived from the normal mode analysis is sensible only for a system in which the interactions among atoms are strictly "harmonic." Such a condition is satisfied only for the case where the displacement of atoms from their equilibrium positions are extremely small. Any finite displacement can readily cause non-linearity or the plastic deformation. If the analysis is made for a protein in vacuum as in the earlier studies, the results will be a disaster. It is because an equilibrium conformation in native conditions is nowhere near the equilibrium structure in vacuum. Only a small perturbation will results in entire destruction. Such a fluctuation can never be *biological.*

On the other hand, the linearity and the Gaussian behavior derived from the statistical mechanics of solution, or combined generalized Langevin and 3D-RISM/RISM theories, is a natural result without making any assumption for the interactions, such as "harmonic." Since the theory takes all the inter- and intra-molecular interactions between protein and water, the equilibrium conformation is that of the free energy minimum. The fluctuation around the equilibrium structure would therefore be the real *biological* fluctuation.

The following question might be raised. Why does the structural fluctuation against a thermodynamic perturbation stay in the linear regime, or in another word, why is the free energy surface quadratic with respect to the structural fluctuation, or equivalently why is the distribution of structure Gaussian? First of all, we have excluded any perturbation which might cause *plastic deformation*, for examples, aggregation among proteins, breaking of chemical bonds, complete evaporation of solvent, and so on. In another word, we are considering the thermodynamic condition which is not far from that in which a living body survives. In such a relatively mild condition, the central limiting theorem governs the stochastic processes. As is well regarded, the Gaussian distribution is a manifestation of the central limiting theorem. After all, protein structure is designed to be consistent with the general law of nature. Of course, the general law does not exclude the existence of specific sub-states which play key roles in the function of protein. Rather, the fact that the structural fluctuation follows the general law assures the robustness of a protein to maintain our life.

Characterizing the structural fluctuation of protein is important not only in the analyses of bio-functions, but also in the computer-aided drug discovery, because the affinity of a drug compound to a target protein is significantly affected by the structural fluctuation. Currently, the molecular dynamic simulation has been exclusively employed to realize the structural fluctuation. However, the fluctuation produced by the simulation is largely limited within a small dispersion, since a large fluctuation is a rare event. The method proposed in the present article will provide the best alternative to the molecular simulation.

References

Akasaka K (2003) Highly fluctuating protein structures revealed by variable-pressure nuclear magnetic resonance. Biochemistry 42:10877–10885
Akasaka K (2006) Probing conformational fluctuation of proteins by pressure perturbation. Chem Rev 106:1814–1835
Akasaka K, Li H, Yamada H, Li R, Thoresen T, Woodward CK (1999) Pressure response of protein backbone structure. Pressure-induced amide ^{15}N chemical shifts in BPTI. Protein Sci 8:1946–1953
Bihan DL, Fukuyama H (eds) (2010) Water: The forgotten biological molecule. Pan Stanford Publishing, Singapore
Brooks B, Karplus M (1983) Harmonic dynamics of protein: normal modes and fluctuations in bovine pancreatic trypsin inhibitor. Proc Natl Acad Sci U S A 80:6571–6575
Burghardt TP, Josephson MP, Ajtai K (2011) Single myosin cross-bridge orientation in cardiac papillary muscle detects lever-arm shear strain in transduction. Biochemistry 50:7809–7821
Chandrasekhar S (1943) Stochastic problems in physics and astronomy. Rev Mod Phys 15:1–89
Chong SH, Hirata F (1997) Nonlinear electrical potential fluctuations of solvent around solutes: an integral equation study. J Chem Phys 106:5225–5238
Chong SH, Miura S, Bsasu G, Hirata F (1995) Molecular theory for the non-equilibrium free energy profile in electron transfer reaction. J Phys Chem 99:10526–10529
Ermak DL, McCammon JA (1978) Brownian dynamics with hydrodynamic interactions. J Chem Phys 69:1352–1360
Fourme R, Girard E, Akasaka K (2012) High-pressure macromolecular crystallography and NMR: status, achievements and prospects. Current Opinion in Struct Biol 22:1–7
Go N, Noguchi T, Nishikawa T (1983) Dynamics of a small globular protein in terms of low-frequency vibrational modes. Proc Natl Acad Sci U S A 80:3696–3700
Hayward S, Kitao A, Hirata F, Go N (1993) Effect of solvent on collective motions in BPTI. J Mol Biol 234:1207–1217
Hirata F (1992) Interaction-site representation of the Smoluchowski-Vlasov equation – the space-time correlation-functions in a molecular-liquid. J Chem Phys 96:4619–4624
Hirata F (ed) (2003) Molecular theory of solvation. Kluwer Academic Publishers, Dordrecht
Hirata F, Akasaka K (2015) Structural fluctuation of protein induced by thermodynamic perturbation. J Chem Phys 142:044110–044118
Horiuchi T, Go N (1991) Investigating protein dynamics in collective coordinate space. Proteins 10:106–116
Ikeguchi M, Ueno J, Sato M, Kidera A (2005) Protein structural change upon ligand binding: linear response theory. Phys Rev Lett 94:078102–078104
Imai T, Kinoshita M, Hirata F (2000) Theoretical study for partial molar volume of amino acids in aqueous solution: implication of ideal fluctuation volume. J Chem Phys 112:9469–9478
Imai T, Kovalenko A, Hirata F (2004) Solvation thermodynamics of protein studied by the 3D-RISM theory. Chem Phys Lett 395:1–6
Imai T, Ohyama S, Kovalenko A, Hirata F (2007) Theoretical study of the partial molar volume change associated with the pressure-induced structural transition of ubiquitin. Protein Sci 16:1927–1933
Kim B, Hirata F (2013) Structural fluctuation of protein in water around its native state: a new statistical mechanics formulation. J Chem Phys 138:054108–054119
Kirkwood JG, Buff FP (1951) The statistical mechanical theory of solutions. J Chem Phys 19:774–777
Kitahara R, Yokoyama S, Akasaka K (2005) NMR snapshots of a fluctuating protein structure: ubiquitin at 30 bar-3 kbar. J Mol Biol 347:277–285
Kitao A, Hirata F, Go N (1991) The effect of solvent on the conformation and the collective motions of protein: normal mode analysis and molecular dynamics simulations of melittin in water and in vacuum. Chem Phys 158:447–472

Kubo R, Toda M, Hashitsume N (1991) Statistical physics II, nonequilibrium statistical mechanics. Springer, Berlin
Kuwajima K, Goto Y, Hirata F, Terazima M, Kataoka M (eds) (2009) Water and biomolecules. Springer, Berlin/Heidelberg
Lamm G, Szabo A (1986) Langevin modes of macromolecules. J Chem Phys 85:7334–7348
Landau LD, Lifshitz EM (1964) Statistical physics (Japanese translation). Iwanami, Tokyo
Makowski L, Rodi DJ, Mandava S, Minh D, Gore DB, Fischetti RF (2008) Molecular crowding inhibits intramolecular breathing motions in proteins. J Mol Biol 375:529–546
Marcus R (1956) On the theory of oxidation–reduction reactions involving electron transfer. I. J Chem Phys 24:966–979
Mchaourab HS, Oh KJ, Fang CJ, Hubbell WL (1997) T4 lysozyme and mutants thereof crystallize in different conformations that are related to each other by a bend about a hinge in the molecule. Biochemistry 36:307–316
McQuarrie DA (1976) Statistical mechanics. Harper and Row Publishers, New York
Mori H (1962) Collective motion of particles at finite temperatures. Prog Theor Phys 28:763–783
Wang MC, Uhlenbeck GE (1945) On the theory of the Brownian motion II. Rev Mod Phys 17:323–342
Yoshida N, Imai T, Phongphanphanee S, Kovalenko A, Hirata F (2009) Molecular recognition in biomolecules studied by statistical-mechanical integral-equation theory of liquids. J Phys Chem B 113:873–886

Chapter 8
Pressure Effects on the Intermolecular Interaction Potential of Condensed Protein Solutions

Roland Winter

Abstract Knowledge of the intermolecular interaction potential of proteins as a function of their solution conditions is essential for understanding protein aggregation, crystallization, and the phase behavior of proteins in general. Here, we report on a combined small-angle X-ray scattering and liquid-state theoretical approach to study dense lysozyme solutions as a function of temperature and pressure, but also in the presence of salts and osmolytes of different nature. We show that the pressure-dependent interaction potential of lysozyme changes in a nonlinear fashion over a wide range of temperatures, salt and protein concentrations, indicating that changes of the bulk water structure mediate the pressure dependence of the intermolecular forces. We present also results on the effect of high hydrostatic pressure on the phase behavior of dense lysozyme solutions in the liquid-liquid phase-coexistence region. As also shown in this study, the application of pressure can be used to fine-tune the second virial coefficient of protein solutions, which can be used to control nucleation rates and hence protein crystallization, or to prevent protein aggregation. Moreover, these results are also important for understanding the hydration behavior of biological matter under extreme environmental conditions, and the high stability of dense protein solutions (as they occur intracellularly) in organisms thriving under hydrostatic pressure conditions such as in the deep sea, where pressures up to the 100 MPa-level are reached.

Keywords Protein-protein interactions • High pressure • Liquid-liquid phase separation • Protein crystallization • Salt effects • Hofmeister series • Cosolvents • TMAO • Lysozyme • Interaction potential • Small-angle X-ray scattering

---

R. Winter (✉)
Physical Chemistry I – Biophysical Chemistry, TU Dortmund University, Otto-Hahn Str. 6, D-44227 Dortmund, Germany
e-mail: roland.winter@tu-dortmund.de

K. Akasaka, H. Matsuki (eds.), *High Pressure Bioscience*,
Subcellular Biochemistry 72, DOI 10.1007/978-94-017-9918-8_8

8.1 Introduction

The study of protein-protein interactions and protein condensation phenomena is not only an intellectually challenging subject in soft condensed matter physics, physical chemistry and liquid-state theory, a detailed understanding of the formation and properties of such interactions and of protein phases is also prerequisite for a number of applications (Gunton et al. 2007; Stradner et al. 2004, 2006, 2008; Poon et al. 2000; Dumetz et al. 2008; Tardieu et al. 2002; Zhang et al. 2011). For example, from a protein crystallization point of view, the main challenge is to determine the initial solution conditions so that optimal crystal nucleation takes place. In this respect, the existence of the metastable liquid-liquid phase separation (demixing) curve found for proteins, such as lysozyme, at high concentrations and low temperatures is quite important, as it has a dramatic impact on protein crystallization. The crystal nucleation rate of aqueous solutions of lysozyme is much larger in the liquid-liquid phase separation region compared to homogeneous protein solutions. One factor that favors the enhanced nucleation is that the supersaturation in the high-concentration phase is much larger than in the initial solution. In addition, the surface energy difference between the high-concentration liquid droplet and the crystal is smaller.

Next to exploring the full intermolecular interaction potential, $V(r)$, the second virial coefficient, B_2, is often used as a measure of the net attractive or repulsive interaction between protein molecules. The sign of B_2 is positive/negative for net repulsive/attractive protein-protein interactions, and the window for optimal crystallization corresponds to a weak net attractive interaction. Thus, knowing the second virial coefficient and its dependence on the physical-chemical parameters of the solution provides, in principle, a powerful tool in determining the conditions for optimal crystallization (Tardieu et al. 2002). Given its importance, there have been extensive experimental studies of B_2 and its dependence on protein and precipitant concentration. These include static and dynamic light scattering, small-angle X-ray scattering (SAXS), small-angle neutron scattering (SANS), membrane osmometry, sedimentation equilibrium, self-interaction chromatography (SIC), and size exclusion chromatography (SEC).

A second important subject of protein condensation phenomena involves diseases that occur due to undesired protein nucleation and, eventually, fibril formation. Classic examples are the formation of polymer fibers of sickle hemoglobin molecules within the red blood cells leading to sickle cell anemia, or age-related cataracts produced by the undesired aggregation of γ-crystalline within the vitreous fluid of the eye. Moreover, many examples involve the role of amyloid fibres via cross-β-sheet formation in a series of debilitating diseases such as Alzheimer's, Parkinson's, Creutzfeld-Jakob disease or diabetes mellitus type 2 (Grudzielanek et al. 2006; Seeliger et al. 2013; Silva et al. 2014). This list is likely to grow as scientists become more aware of the molecular origins of different diseases. To combat such diseases, strategies for preventing nucleation of protein fibers must be developed. In recent years, also functional amyloids have been identified in all kingdoms of life, demonstrating a ubiquitous role of amyloids in living systems.

The phase diagram of globular proteins, including the solubility line, depends on many control parameters, such as temperature, pressure, pH, and precipitant (e.g. salt) concentration. The phase behavior is quite complex and is still not well understood theoretically up to now. The use of the pressure variable is least studied so far and is hence in the focus of this review. A detailed knowledge of the intermolecular interactions and the complete temperature-pressure-concentration phase diagram of proteins would be needed to control and fine-tune proteins' physical-chemical properties, which is, for example, demanded for the challenging task of protein crystallization (Poon et al. 2000; Cheng et al. 2006; Asherie 2004; Curtis and Lue 2006; Rosenberger et al. 1993; Muschol and Rosenberger 1997; Thomson et al. 1987; Russo et al. 2013). Moreover, pressure is a rather mild perturbating agent and pressure-dependent studies on protein solutions have generally been found to be fully reversible (Smeller et al. 2006; Mishra and Winter 2008; Möller et al. 2012; Winter et al. 2007; Akasaka 2006; Silva et al. 2001, 2014; Krywka et al. 2008; Heremans and Smeller 1998). Furthermore, they have been shown to lead to a better understanding of the volumetric and hydrational properties and hence the stability and function of biomolecular systems under extreme environmental conditions, such as in the deep sea where pressures up to the 100 MPa-level and beyond are encountered (Meersman et al. 2013). Finally, in a technological context, knowledge of the phase behavior of proteins is essential in areas such as protein purification, storage and high pressure food processing (Silva et al. 2001; Winter et al. 2007).

We (and others) essentially focused on the model protein lysozyme in our studies of protein-protein interactions and protein condensation phenomena. Lysozyme is a small enzyme that protects humans from bacterial infection. It attacks the cell walls of bacteria by breaking the carbohydrate chains, thereby destroying the structural integrity of the bacterial cell walls. It consists of 129 amino acid residues, with a molecular weight of 14.3 kDa. The net charge on the protein varies with pH. Its isoelectric point (i.e., the pH value at which the sum of the positive and negative charges on the protein is zero) is $pI = 11.1$. At pH 7.0, titration experiments on lysozyme yield a titrable positive charge of about +8 electronic charges. Chicken-egg-white lysozyme has been the paradigm for protein crystallization studies. Thus this protein is also an ideal candidate for theoretical and experimental investigations of its intermolecular interactions and phase behavior under high-pressure conditions.

8.2 From Scattering Data to Interaction Potentials and the Second Virial Coefficient

In this chapter we describe how quantitative information about intermolecular interactions can be obtained from small-angle scattering experiments. The theoretical approaches needed to calculate the measured structure factor of the protein solution from effective pair potentials, and vice versa, to extract $V(r)$ from scattering experiments, are extensions of developments made in the framework of the theory of simple liquids and colloidal suspensions (Hansen and McDonald 2013; de With 2013; Israelachvili 2011; Zemb and Lindner 2002; Zhang et al. 2007).

The SAXS intensity, $I(q)$, from diluted protein solutions can be described by the form factor $P(q)$, with $q = (4\pi/\lambda)\sin\theta$ being the wave vector transfer and 2θ the scattering angle. For lysozyme in solution, the form factor can be described by that of an ellipsoid of revolution with the semi-axes a and b and molecular volume $(\pi/6) \times 4.5 \times 3.0 \times 3.0\ \text{nm}^3$.

For solutions of higher protein concentration as discussed in this review, an additional scattering contribution has to be taken into account, which originates from the intermolecular interactions of the particles. The resulting SAXS signal can be described, within the so-called decoupling approximation, as product of the form factor $P(q)$ and an effective structure factor (Schroer et al. 2011a; Möller et al. 2012)

$$S_{\text{eff}}(q) = 1 + \frac{\langle F(q) \rangle_\Omega^2}{P(q)} \left(S(q) - 1\right). \tag{8.1}$$

Here, $\langle F(q) \rangle_\Omega$ is the spherical average of the Fourier-transform of the protein electron density, and $S(q)$ is the intermolecular structure factor of interest. Hence, by measuring the scattering pattern at both dense and highly diluted concentrations, the structure factor $S(q)$ can be determined. In practice, $S(q)$ is obtained by dividing the experimental scattering pattern of a concentrated protein solution by that of a very dilute solution scaled to the same concentration, or, eventually, a scattering curve calculated from a crystallographic model of the single particle (Semenyuk and Svergun 1991).

The structure of an isotropic liquid (consisting of spherically symmetric particles) can be described by distribution functions, such as the pair distribution (correlation) function, $g(r)$, which is the probability to a find a particle at distance r relative to one located at the origin ($r = 0$). The function $g(r)$ thus represents the ratio between the local density and the average or bulk number density, ρ, and thus describes the deviations from a complete random distribution, which, in turn, depends on the intermolecular interaction potential of the particles. The properties of Fourier-transforms imply that $g(r)$ is the inverse Fourier-transform of the measured structure factor $S(q)$:

$$S(q) = 1 + 4\pi\rho \int_0^\infty r^2 \left[g(r) - 1\right] \frac{\sin(q\,r)}{q\,r} \mathrm{d}r \tag{8.2}$$

and hence experimentally accessible through $S(q)$.

In order to relate the measured solution structure, i.e. $g(r)$, or its Fourier-transform $S(q)$, to the underlying interaction potential $V(r)$, the so-called Ornstein-Zernike (OZ) relation can be used, which describes the fact that the total correlation

$h(r_{12}) = g(r_{12}) - 1$ between two particles 1 and 2 results from a direct correlation $c(r_{12})$ between these particles 1 and 2 and from indirect correlations of particle 1 and particle 2 propagated through the influence of particle 1 on a particle 3 and the influence of particle 3 on particle 2, summed over all such third particles:

$$h(r_{12}) = c(r_{12}) + \rho \int c(r_{13})\, h(r_{32})\, \mathrm{d}r_3, \tag{8.3}$$

which leads to a multicenter integral representing a convolution (Hansen and McDonald 2013; de With 2013; Svergun et al. 2013). As the Ornstein-Zernike relationship contains two unknowns, a so-called closure relationship is required to solve the equation. The usefulness and accuracy of such closure relation depends, among others, on the importance of long-range electrostatic interactions in the potential. Here, we used the mean spherical approximation (MSA), i.e. $c(r > \sigma) = -V(r)/(\mathrm{k_B}T)$, $g(r \leq \sigma) = 0$ (σ, hard-core diameter of the particle), for linking $S(q)$ to $V(r)$.

To calculate measured structure factors from pair potentials, and vice versa, a number of simplifications are usually made. For globular macromolecules such as proteins, one generally assumes that they are spherical and that their interactions are independent of their relative orientations, and hence can be described by a central potential $V(r)$. The forces between protein molecules in solution usually include screened repulsive Coulomb interactions and attractive van der Waals interactions. In certain cases, protein-ion dispersion forces, hydrophobic and hydration forces may play a role as well. Please note that subtle mutations on the surface of the protein molecules (such as in the case of sickle hemoglobin or γ-crystalline) can have profound effects on their short-range intermolecular interactions as well, thereby significantly increasing their propensity to aggregate.

Most quantitative studies of protein-protein interactions have been carried out with potentials derived from a particularly successful model in colloid science, the DLVO (Derjaguin, Landau, Verwey, Overbeek) potential consisting of a hard core, repulsive Coulomb interactions and an attractive Van der Waals interaction (Verwey and Overbeek 1948; Hiemenz 1997; Israelachvili 2011). In that sense, proteins can also be viewed as model colloids, as they are monodisperse and of the order of a few nanometers. The DLVO theory is based on a continuum model in which both forces (the attractive van der Waals and repulsive Coulomb forces between the particles) are described in terms of the bulk properties of the intervening solvent (water or buffer/salt solution), i.e., its dielectric permittivity, ionic strength and number density (Fig. 8.1). Van der Waals interactions (and hydrogen bonding) are short-range, extending at most over a few tenths of nm, whereas electrostatic Coulombic interactions – dependent on the ionic strength of the solution - extend over up to a few nm. Generally, when the long-range repulsive interactions dominate (e.g., for highly charged proteins in low-salt solutions), the positions of the macromolecules become correlated and the solution may become structured, whereas when the weaker short-range attractive forces dominate, solutions containing oligomers or clusters may even develop.

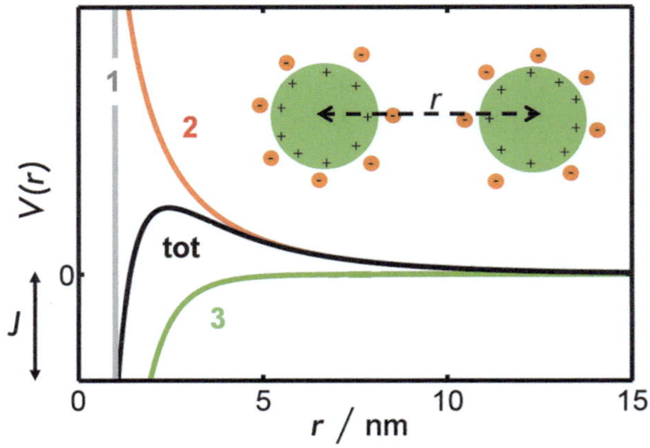

Fig. 8.1 Protein-protein interaction potential, $V(r)$, as a function of separating distance r. The depicted DLVO potential (tot) is given by the sum of a screened Coulomb (*2*) and a van der Waals potential (*3*). The impenetrable protein surface is modeled as an infinite high hard-sphere potential (*1*) at $r \leq \sigma$, with σ being the diameter of the particles. J is the depth of the attractive part of the interaction potential

Although the DLVO potential very successfully describes many of the commonly observed protein condensation phenomena, due, for example, to changes in protein concentration, temperature, pH and ionic strength, it fails to explain special effects such as the Hofmeister series (see below) or the effects of addition of polymers such as polyethylene glycol (PEG), the latter resulting in attractive depletion forces. In that case, the theoretical explanation requires the introduction of additional forces (e.g., dispersion forces related to polarisability), which can become dominant at high salt concentrations when Coulomb interactions are strongly reduced by charge screening.

The intermolecular structure factor can be calculated theoretically in the mean spherical approximation (MSA) using a modified DLVO potential $V(r)$ to model the interaction of the proteins (Liu et al. 2005; Javid et al. 2007a, b; Schroer et al. 2011a). Here, the potential $V(r)$ is given by the sum over a hard sphere potential, $V_{\mathrm{HS}}(r)$, a repulsive screened Coulomb potential, $V_{\mathrm{SC}}(r)$, and an attractive part modeled as a Yukawa potential, $V_{\mathrm{Y}}(r)$:

$$V(r) = V_{\mathrm{HS}}(r) + V_{\mathrm{SC}}(r) + V_{\mathrm{Y}}(r). \tag{8.4}$$

The single contributions can be calculated as follows:

$$V_{\mathrm{HS}}(r) = \begin{cases} \infty \ , & r \leq \sigma \\ 0 \ , & r > \sigma \end{cases}$$

$$V_{\mathrm{SC}}(r) = \begin{cases} 0 & , \ r \leq \sigma \\ \frac{Z^2 e^2}{4\pi\varepsilon_0\varepsilon_r (1+0.5\kappa\sigma)^2} \frac{\mathrm{e}^{-\kappa(r-\sigma)}}{r} & , \ r > \sigma \end{cases}$$

$$V_{\mathrm{Y}}(r) = \begin{cases} 0 & , r \leq \sigma \\ -J\sigma \frac{\mathrm{e}^{-(r-\sigma)/d}}{r} & , \ r > \sigma \end{cases} \tag{8.5}$$

Here, e is the elementary charge, ε_0 the vacuum dielectric constant, ε_r the dielectric permittivity of the solution, and κ the reciprocal Debye-Hückel screening length. The pressure and temperature dependence of the dielectric permittivity of the solution has to be taken into account as well: ε_r shows a slight monotonic increase with pressure and a decrease with temperature. At pH 7, the effective protein charge in the model was kept constant at a value of $Z = 8$ (Kuehner et al. 1999). Additionally, an effective protein hard-sphere diameter of $\sigma = 2.99$ nm and a width of the attractive part of $V_Y(r)$ of $d = 0.3$ nm was used. With this, only the strength J of the attractive potential (see Fig. 8.1) is unknown, which can now be determined from fits to the experimental structure factor data.

Knowledge of $V(r)$ also allows us to determine the second virial coefficient:

$$B_2 = 2\pi \int_0^\infty \left(1 - e^{-V(r)/k_B T}\right) r^2 dr. \tag{8.6}$$

B_2 is thus an integral characteristics for the two-body intermolecular interaction, $V(r)$, revealing – by its sign – overall attractive or repulsive interactions of the particles. In order to compare the values for proteins of different sizes, the dimensionless normalized second virial coefficient, b_2, is calculated by factoring out the hard sphere part, ${B_2}^{HS}$, of the integral,

$$b_2 = \frac{B_2}{B_2^{HS}}. \tag{8.7}$$

Positive b_2-values correspond to repulsive interactions whereas negative values reveal dominating attractive forces.

8.3 Nonlinear Pressure Dependence of the Interaction Potential of Dense Protein Solutions

Biological macromolecules such as proteins function within intracellular environments that are highly crowded. Such crowding may result in large effects on the stability, phase behavior, dynamics and folding rate of proteins (Minton 2006; Ellis 2001, 2007; Ravindra et al. 2004; Zhao et al. 2007; Erlkamp et al. 2014). Quantitative data are largely unknown, however, in particular under high pressure conditions (Winter et al. 2007; Ortore et al. 2009; Russo et al. 2013; Schroer et al. 2011a). The properties of water upon compression may also be important for the interactions in dense protein solutions, where only a few water layers separate neighboring protein molecules. On these grounds, we performed pressure dependent SAXS measurements on dense lysozyme solutions in bis-Tris buffer solution (in order to keep the pH value constant upon compression). To generate high pressure conditions, a custom-built high pressure cell employing two flat diamond windows was used (Krywka et al. 2007, 2008). The scattering pattern at the lowest concentration (0.45 wt.%) exhibits negligible interference effects and represents the form factor $P(q)$ of the particle. It is adequately described by the computed scattering form factor of a prolate ellipsoid of revolution with radius of gyration $R_g = 14.8$ Å. Comparing the $P(q)$ values obtained for all pressures and temperatures studied, no differences due to changes of the tertiary structure of the protein were detected, indicating that the protein remains in its native, folded conformation at all conditions studied, which is also in agreement with high-pressure FTIR spectroscopic data.

As depicted in Fig. 8.2a, the SAXS intensity curves $I(q)$ of the concentrated protein solution (here $c_p = 100$ mg mL^{-1}) show a pronounced correlation peak due to the presence of a structure factor reflecting strong protein-protein interactions. The structure factors $S(q)$, obtained by refinement of the scattering data $I(q)$, are shown in Fig. 8.2b. Increasing pressurization from 0.1 up to 150 MPa leads, as expected, to a slight shift of the correlation peak to larger q-values, i.e., smaller

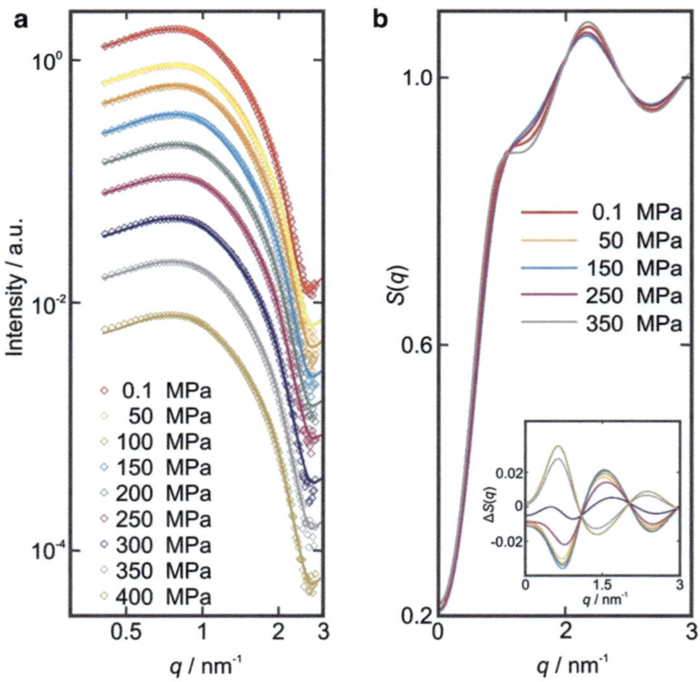

Fig. 8.2 (**a**) Experimental SAXS data $I(q)$ of a lysozyme solution ($c_p = 100$ mg mL^{-1}) at $T = 25$ °C for different pressures with corresponding fitting curves using the DLVO potential and MSA model superimposed on the experimental data. (**b**) Structure factors $S(q)$ obtained for selected pressures. To emphasize the pressure-induced changes, the inset shows the differences $\Delta S(q)$ between $S(q)$ at high pressures and at 1 bar: $\Delta S(q) = \Delta S(q)_p - \Delta S(q)_{0.1\ \mathrm{MPa}}$ (Modified from Schroer et al. 2011a)

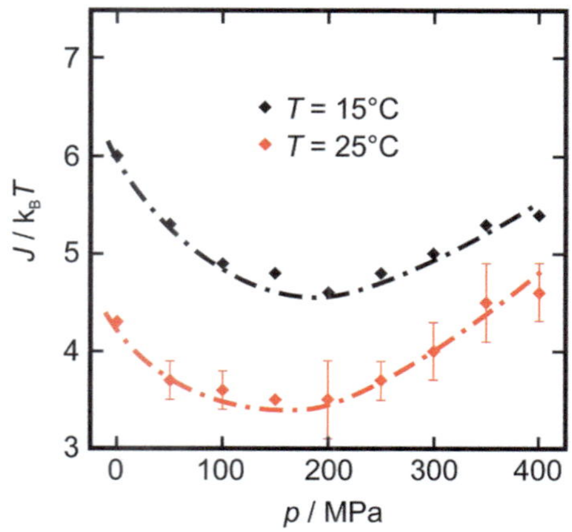

Fig. 8.3 Experimentally determined strength J of the attractive part of the interaction potential $V(r)$ of a 100 mg mL^{-1} lysozyme solution as a function of pressure for two different temperatures (pH 7). Lines have been included to aid in visualization (Modified from Schroer et al. 2011a)

distances. Interestingly, as the pressure is further increased, this shift is reversed. This effect is better visible in the $\Delta S(q)$ data, which change sign in the intermediate pressure range (inset of Fig. 8.2b). From the refinement of the data employing the liquid-state-physics approach described above, the pressure dependence of the strength of the attractive part of the interaction potential, J, could be determined, which is depicted in Fig. 8.3. For all temperatures measured, $J(p)$ first decreases with pressure, displays a minimum at about 150–200 MPa, and then increases again. As expected, decreasing the temperature results in an increase of J, i.e., an increase in intermolecular attraction of the proteins.

In the absence of protein unfolding and conformational changes, $V(r)$ can still be influenced by pressure through changes in the hydration properties, i.e., through solvent-mediated effects. In fact, multi-hundred MPa pressures modify the structure of liquid water, which is also reflected in changes of the transport properties, such as the viscosity and diffusion coefficient, which exhibit changes in their pressure coefficients around 200 MPa (Ludwig 2001; Weingärtner et al. 1996). The pressure-induced volume reduction is accompanied by a marked change in local molecular arrangement below 500 MPa: the volume reduction is caused by a rapid and monotonous increase in the coordination number N_1, from ~4 to 6.5 at 500 MPa, which can be attributed to the penetration of nonhydrogen-bonded water molecules into the first shell (Soper and Ricci 2000; Weck et al. 2009; Katayama et al. 2010). These structural changes in the hundreds MPa pressure range seem to coincide with the observed nonlinear pressure response of $V(r)$, i.e., beyond ~200 MPa, changes in the water structure may lead to the effective small increase of J.

With increasing protein concentration, the minimum in $J(p)$ becomes shallower and disappears for concentrations above about 200 mg mL^{-1} (Grobelny et al. 2014). For these high protein concentrations, $J(p)$ remains essentially pressure-insensitive above ~150 MPa, indicating that no pressure-induced minimum in $J(p)$ is observed anymore, and the interaction potential turns rather insensitive to compression. This should not be too surprising, as in this concentration range intermolecular

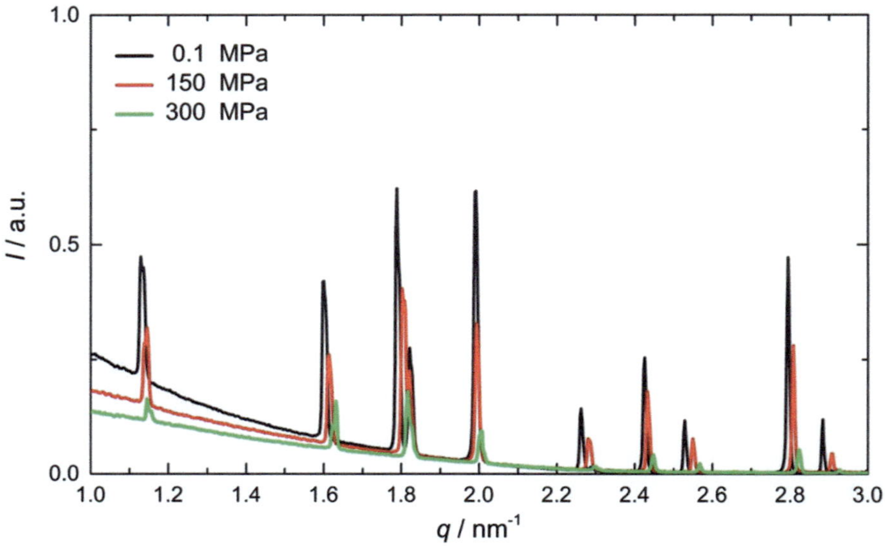

Fig. 8.4 Pressure dependent SAXS curves of lysozyme solutions at a concentration of 366 mg mL^{-1} ($T = 20$ °C), showing the intermediate q-range where Bragg reflections are observed. With increasing pressure, protein crystal formation is suppressed (Modified from Grobelny 2014)

distances have decreased to a few monolayers of water only, and no bulk-like solvent properties can be expected anymore.

Increasing the concentration to 300–400 mg mL^{-1}, a protein concentration range is reached, where, within the duration of the experiment (a few hours), nucleation and partial crystallization of the supersaturated protein solution takes place at 20 °C in the absence of salt. Figure 8.4 displays the q-range where Bragg reflections are observed. Indexing of the Bragg reflections reveals a primitive tetragonal unit cell (space group $P4_32_12$) with axis $a = b = 7.9$ nm and $c = 3.8$ nm, which is in good agreement with literature data (Fourme et al. 2001; Cheng et al. 2006; Kundrot and Richards 1986; Sauter et al. 2001). The application of pressure leads to an anisotropic compression along the axis of the unit cell, i.e., a compression of the a axis, whereas the c-axis-length stays approximately constant. These data are in line with pressure dependent measurements on grown lysozyme crystals (Fourme et al. 2001). Of note, the intensity of the Bragg peaks decreases upon pressurization, which indicates suppression of protein crystallization upon hydrostatic compression of the protein solution. This observation implies – following Le Châtelier's principle – that formation of lysozyme crystals is connected to an overall volume increase of the system.

Our results show that pressure suppresses protein nucleation, aggregation and finally crystallization in supersaturated condensed lysozyme solutions. These findings may thus also be of importance for understanding the high stability of dense protein solutions (as they occur intracellularly) in biosystems thriving under hydrostatic pressure conditions such as in the deep sea, where pressures up to the 100 MPa-level and beyond are reached (Daniel et al. 2006; Lauro and Bartlett 2008). In the crowded cell, where highly concentrated protein solutions (200–300 mg mL^{-1}) are encountered and macromolecular crowding reaches levels of 30 vol.% and more, pressure is expected to effectively suppress protein aggregation.

8.4 Re-entrant Liquid-Liquid Phase Separation in Protein Solutions at Elevated Hydrostatic Pressures

The intermolecular interaction of proteins in solution combines short-range van der Waals attraction with long-range electrostatic double layer repulsion. It is believed that the existence of many phases of protein solutions, including the liquid-liquid phase separation (LLPS) region, can be explained by this interaction. In the LLPS region, the protein solution gets divided into two phases of high and low protein concentration, respectively (Fig. 8.5). The liquid-liquid phase coexistence region is metastable and is also the phase region where protein crystallization can take place by reducing the nucleation barrier. Generally, to obtain optimal crystallization conditions, gel and amorphous aggregate states must be avoided. Instead, the formation of protein-rich liquid droplets via this metastable protein-poor, protein-rich liquid-liquid phase separation can be beneficial. Although the dense liquid

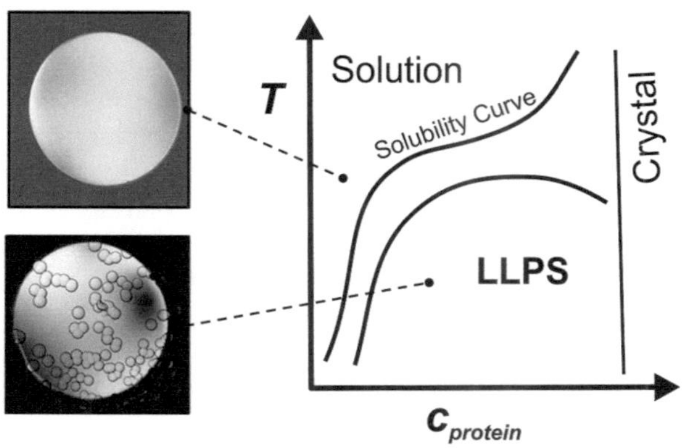

Fig. 8.5 Schematic temperature-concentration phase diagram of protein solutions. A transition to a (metastable) liquid-liquid phase separation (LLPS) region occurs at lower temperatures than the separation between the solution and solid (crystal) phase. Studies exploring nucleation and crystallization cover the supersaturated regime beyond the solubility curve, where crystals and solution are in equilibrium. In this metastable region, supersaturation is usually too low for spontaneous nucleation to occur, but liquid-liquid phase separation may occur at high protein concentrations and low temperatures. In the crystallization zone, supersaturation is sufficient for spontaneous nucleation and crystallization

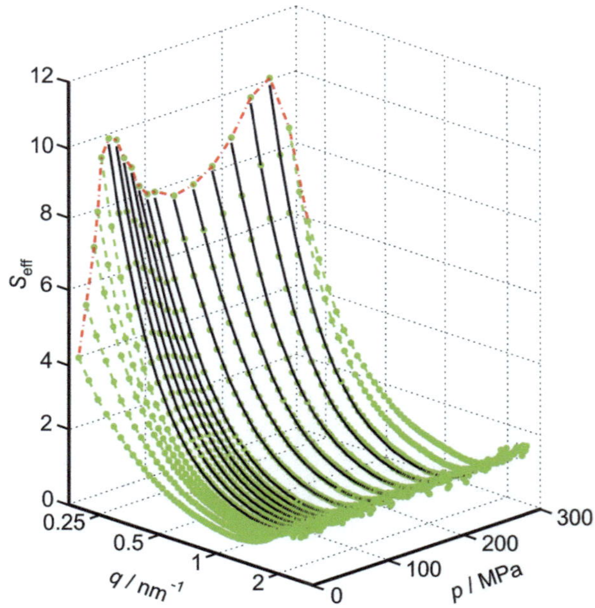

Fig. 8.6 Measured effective structure factor $S_{eff}(q)$ of a 20 wt.% lysozyme solution as a function of pressure at $T = 16$ °C. *Black lines* display the refinement of the data using a sticky hard sphere model (Modified from Möller et al. 2014a)

phase is metastable and decays over time into a crystal phase, this crystal formation in lysozyme solutions is still slow so that the liquid-liquid phases can be studied in detail. The occurrence of such a metastable liquid-liquid phase separation (LLPS) region in the phase diagram of proteins has been reported, e.g., for lysozyme, γ-crystalline, and hemoglobin (Gunton et al. 2007). Formation of a LLPS in protein solutions has been studied as a function of pH, salt concentration, type of salt, and temperature (Tardieu et al. 2002; Gunton et al. 2007; Zhang et al. 2011; Annunziata et al. 2008). Recently, we set out to experimentally explore the effect of high pressure on lysozyme solutions in the vicinity of the LLPS region using SAXS.

SAXS experiments were carried out on lysozyme solutions at various concentrations in the presence of 500 mM NaCl (3 wt.%) and the pressure-stable buffer bis-Tris at pH 7. For solution conditions close to the liquid-liquid phase boundary, $V(r)$ can be modeled by a sticky hard sphere potential (Möller et al. 2014a). Here, the potential is described by a hard sphere part, which presents the impenetrable protein surface, and an attractive potential at the protein surface, which is modeled by an infinitesimal narrow, infinitely deep square-well potential.

A typical pressure dependence of the measured effective structure factor $S_{eff}(q)$ is shown in Fig. 8.6, together with the refinement of the data. The $S_{eff}(q)$ data for the 20 wt.% lysozyme solution at 16 °C reveal a strong increase of the scattering intensity at low q-values upon compression. The sample is in the liquid-liquid phase-separated state at 0.1 MPa at these salt concentrations. Highest scattering intensities at small q are observed at pressures of 40 and 250 MPa, respectively, where the sample crosses the phase boundary out of the LLPS region into the homogenous one-phase region and vice versa at the higher pressure.

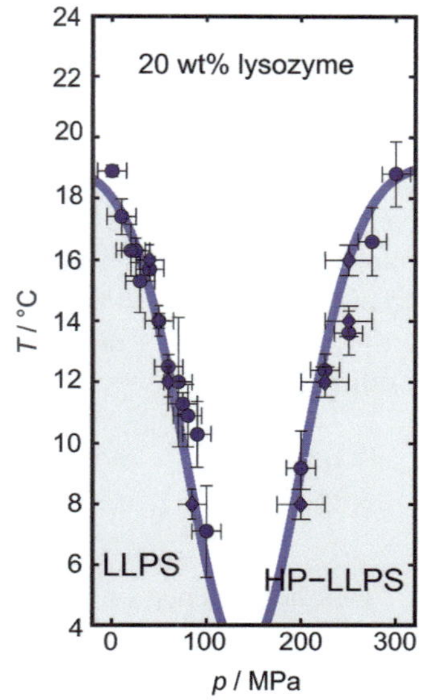

Fig. 8.7 *p-T* phase diagram of the LLPS region of lysozyme for a concentration of 20 wt.%. Low and high pressure (HP) areas of the LLPS are marked in dark (Modified from Möller et al. 2014a)

The reduced second virial coefficient b_2, obtained from refinement of the sticky sphere potential to the SAXS data, has been determined as a function of temperature and pressure as well. $b_2(p)$ nicely reflects the non-linear pressure dependence of $S_{\mathrm{eff}}(q)$, exhibiting a maximum between 0.1 and 150 MPa. Interestingly, when the protein interactions start to become more attractive again at higher pressures, the system is able to reenter the LLPS regime, which is indicated by the second maximum in $S_{\mathrm{eff}}(q = 0.2\ \mathrm{nm}^{-1})$ at 250 MPa for $T = 16\ ^{\circ}\mathrm{C}$. The values obtained for b_2 are in good agreement with theoretical predictions for the LLPS to occur, i.e. $b_2 \approx -1.5$ (Vliegenthart and Lekkerkerker 2000; Noro and Frenkel 2000).

From the combined temperature and pressure dependent b_2 data, a *p-T* phase diagram can be constructed for the LLPS region of lysozyme, which is displayed in Fig. 8.7. Solution conditions where the protein samples undergo LLPS are marked in dark. As can be clearly seen, a re-entrant LLPS is found at high pressures (denoted HP-LLPS). The observed re-entrant liquid-liquid phase transition is consistent with the increasing attractivity of $V(r)$ of the dense protein solution at 100–200 MPa pressures, which, as discussed above, may be explained by a change of the solvent structure in that pressure range (Schroer et al. 2011a).

8.5 Salt Effects on Intermolecular Interactions of Proteins at High Pressures

Salt-induced precipitation of proteins from solution is one of the common methods used in the crystallization of proteins. Hence, a theoretical understanding of the role of ionic strength on the intermolecular interaction potential of proteins is desirable as well. The use of salts (e.g., NaCl, Na_2SO_4) as precipitation agent is largely due to electrostatic screening effects, which decreases the effective repulsive electrostatic interaction of charged proteins (Tardieu et al. 2002; Zhang et al. 2007, 2011; Javid et al. 2007b; Möller et al. 2014b). The original experimental studies date back to the late 1800s, which showed that this precipitation depends not only on the salt concentration, but also on the specific salt used. Hofmeister and his co-workers found that salts with a common cation but with different anions have different precipitant abilities. In the so-called direct lyotropic series of Hofmeister, the effect on protein solubility due to anions are in the following order (Hofmeister 1888; Collins 2004):

$$SO_4^{2-} > HPO_4^{2-} > CH_3COO^- > HCO_3^- > Cl^- > Br^- > NO_3^- > ClO_3^- > I^- > SCN^-$$

This dependence on the background salt is also of crucial importance in cell biology, affecting, for example, the thermal stability of proteins and nucleic acids and intermolecular interactions, such as ligand binding. Many ideas have been proposed to explain the Hofmeister effect, including salt-induced changes in the water structure, hydration forces, and specific co-ion and counter-ion exclusion and adsorption. The ability of ions to influence the local water structure is often discussed in terms of their water-structure-making (kosmotropic) or water-structure-breaking (chaotropic) propensity.

Such salt-specific effects cannot be explained by the classical DLVO theory since the only ionic property that enters this theory is the ionic charge. What is clear, however, is that for any macromolecule, the addition of salt screens the electrostatic interactions between the protein molecules, but in addition changes the short-range attractive potential that may be ion-specific. The latter interaction increases with decreasing temperature and increasing salt concentration. Moreover, a direct or reverse order of the Hofmeister series may be observed, depending on whether the pH of the solution is lower or higher than the pI value of the protein. Salt effects are generally less significant at the pI of the protein.

Combined effects of the concentration and type of salt added, temperature and hydrostatic pressure are little studied up to now. The latter parameter was explored in a few studies only, such as on subtilisin, glucose isomerase, thaumatin, SNase, RNase A, and lysozyme (Webb et al. 1999; Visuri et al. 1990; Waghmare et al. 2000; Crisman and Randolph 2010; Kadri et al. 2003, 2005; Gross and Jaenicke 1991, 1993; Schall et al. 1994; Saikumar et al. 1995; Lorber et al. 1996; Takano et al. 1997; Sazaki et al. 1999; Suzuki et al. 2002a,b, 2005; Nagatoshi et al. 2003; Möller et al. 2012, 2014b; Javid et al. 2007a, b). Some of those studies investigated the solubility, nucleation, and growth rates of protein crystals under pressure, often yielding diverse results for different proteins, however. Less attention has been paid to the intermolecular interaction potential controlling protein crystallization and phase behavior in general. Recently, we studied the combined effects of temperature, pressure, and ionic strength (NaCl concentration) on the protein-protein interaction

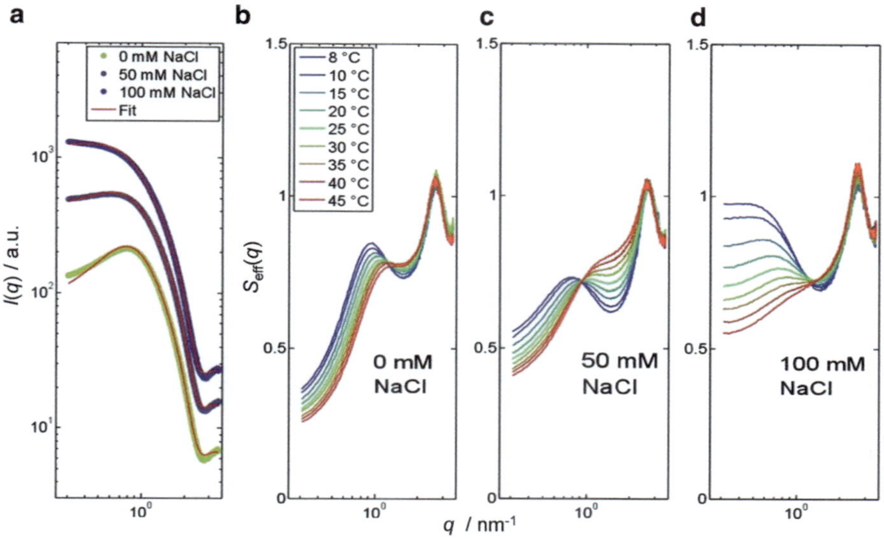

Fig. 8.8 (**a**) SAXS intensities $I(q)$ of a 10 wt.% lysozyme solution at 25 °C for NaCl concentrations of 0, 50, and 100 mM. The SAXS curves are shifted vertically. Also the refinement of the data is shown. (**b–d**) Corresponding effective structure factors, $S_{\mathrm{eff}}(q)$, at various salt concentrations and temperatures (Modified from Möller et al. 2012)

potential of dense lysozyme solutions and the use of pressure to modulate protein crystallization in situ (Möller et al. 2012).

As an example, we show the measured X-ray scattering data of a 10 wt.% lysozyme solution upon addition of 0, 50, and 100 mM NaCl in Fig. 8.8. Clearly, a drastic increase of the attractive interaction of the protein is seen with increasing ionic strength, indicated by a drastic increase of the scattering intensity at small q-values. Furthermore, as expected, an increasing attractivity is observed upon a decrease of temperature from 45 to 8 °C.

$V(r)$ and J have been obtained by fitting the DLVO/MSA model to the experimental data. Figure 8.9 also depicts the temperature dependence of J at various NaCl concentrations for three protein concentrations. A marked temperature dependence of the attractive interaction is observed for all salt concentrations: J decreases by a factor of 2–3 upon a temperature increase from about 5 to 45 °C. Moreover, with increasing ionic strength of the protein solution, an increase of the protein concentration reduces J more effectively, i.e., leads to an increase of repulsive interaction.

The effective structure factors for a 10 wt.% lysozyme 100 mM NaCl solution at pressures from 0.1 up to 300 MPa are depicted in Fig. 8.10a. The nonlinear behavior of the structure factor with increasing pressure is still visible in the salt solution. The minimum in $J(p)$ is shifted to smaller J-values, but also to slightly lower pressures with increasing protein concentration (Fig. 8.10b). The results for $J(p)$ of the 10 wt.% lysozyme solution are shown in Fig. 8.10c, d for various temperatures

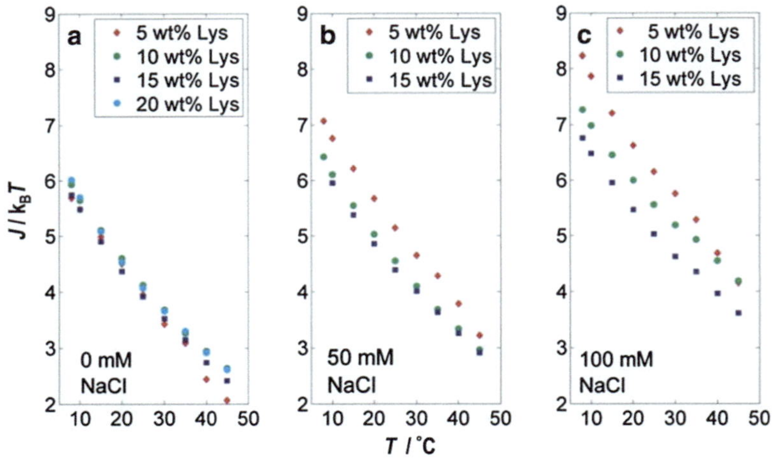

Fig. 8.9 Temperature dependence of the strength of the inter-protein attractive interaction, J, as a function of NaCl and protein concentration at $p = 0.1$ MPa (Modified from Möller et al. 2012)

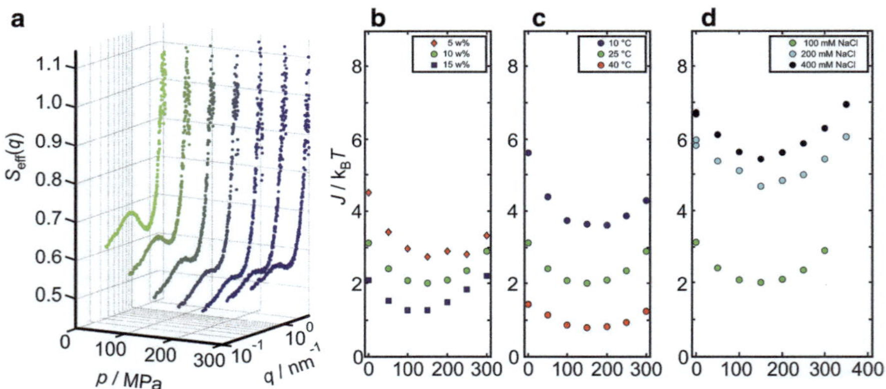

Fig. 8.10 (**a**) Effective structure factor, $S_{\mathrm{eff}}(q)$, of a 10 wt.% lysozyme solution in 100 mM NaCl at 25 °C as a function of pressure. (**b**, **c**) Results of the refinement for $J(p)$ as a function of protein concentration (25 °C, 100 mM NaCl) and temperature (10 wt.% lysozyme, 100 mM NaCl), respectively. (**d**) Strength of the attractive interaction, J, for different NaCl concentrations (10 wt.% lysozyme, 25 °C) (Modified from Möller et al. 2012)

and NaCl concentrations. The minimum in $J(p)$ around 150 MPa remains up to 400 mM NaCl, and the J-value increases with increasing NaCl concentration, i.e., with increased screening of the positively charged lysozyme (Fig. 8.10d). As shown in Fig. 8.10c, the location of the minimum in $J(p)$ is also independent of temperature, and J-values decrease by about 30 % upon an increase of temperature from 10 to 40 °C.

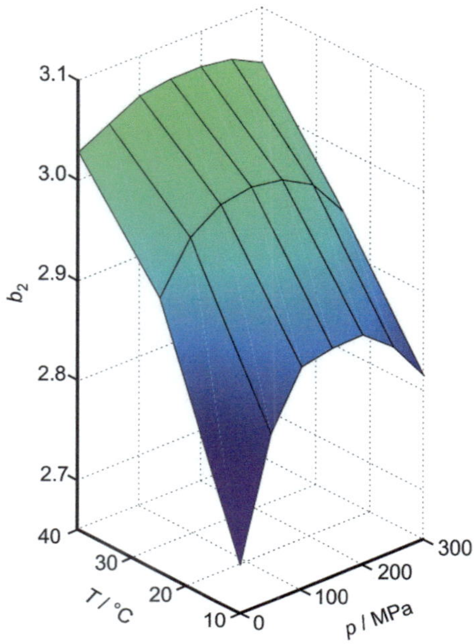

Fig. 8.11 Normalized second virial coefficient, b_2, for 10 wt.% lysozyme in the presence of 100 mM NaCl solution as a function temperature and pressure (Modified from Möller et al. 2012)

To reveal optimal crystallization conditions under the influence of all thermodynamic potentials, the pressure dependence of the second virial coefficient should be implemented as well for salty protein solutions. We calculated the pressure dependence of the normalized second virial coefficient from the measured interaction potentials. A suitable crystallization slot is generally found for b_2-values lying in a narrow range (George and Wilson 1994), typically between about −0.85 and −3.2 (Poon et al. 2000). As an example, Fig. 8.11 shows measured values for b_2 as a function of temperature and pressure for a salt concentration of 100 mM NaCl: b_2 increases with increasing temperature and displays a maximum at a pressure of ~200 MPa. As expected, the nonlinear pressure dependence of $V(r)$ is also reflected in $b_2(p)$. At pressures up to 200 MPa, an increase of pressure results in an increase of b_2. For pressures higher than about 200 MPa, b_2 decreases again. The increase of b_2 with increasing pressure is much steeper at low temperatures, resulting in a more pronounced maximum of $b_2(p)$. Favorable crystallization conditions, corresponding to b_2 values lower than about −0.85, are reached for higher NaCl concentrations. For example, a salt concentration of 500 mM NaCl shifts b_2 to values around −1 at ambient pressure (Sedgwick et al. 2007). The increase in pressure has a nonlinear effect on the inter-protein interaction potential even under such high salt concentrations. The observed pressure dependence of the second virial coefficient is in good agreement with observations by Crisman and Randolph (2010) who studied the crystallization behavior of a human growth hormone and found an increase of the virial coefficient with increasing pressure from 0.1 to 250 MPa.

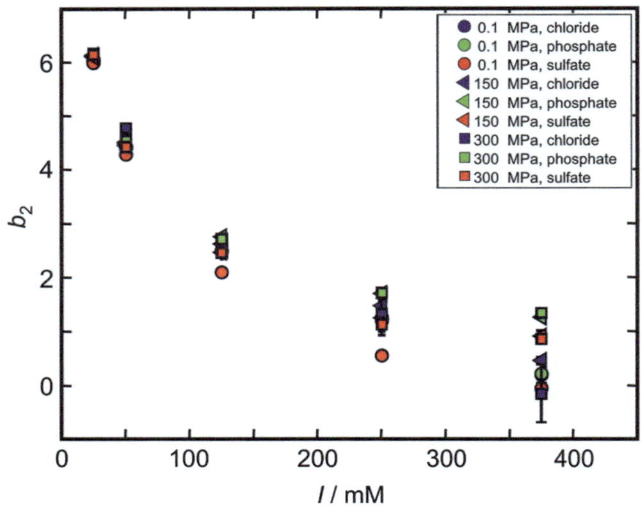

Fig. 8.12 Normalized second virial coefficient, b_2, of a 5 wt.% lysozyme solution as a function of ionic strength for sodium chloride, sulfate, and phosphate at pressures of 0.1 (*circles*), 150 (*triangles*), and 300 MPa (*squares*) (Modified from Möller et al. 2014b)

We also explored the effect of pressure on the intermolecular interaction potential of lysozyme in the presence of salts anions of different water structure breaking/making properties (Cl^-, SO_4^{2-}, PO_4^{3-}), the counterion being Na^+ for all salts (Möller et al. 2014b). The normalized second virial coefficient, b_2, is plotted as a function of ionic strength in Fig. 8.12 for lysozyme in these salt solutions. Generally, with increasing salt concentration, b_2 and hence the strength of the repulsive interaction between the protein molecules decreases. At low ionic strengths, i.e. in the dilute Debye-Hückel region, variations of the salt type and pressure have only a minor effect on the interaction potential. Conversely, with increasing ionic strength, differences in the b_2 values increase, owing to an increased influence of the anion type on the protein-protein interaction potential.

In order to reveal the sole pressure dependence in more detail, J is presented in Fig. 8.13 relative to the ambient pressure data. The following observations could be made: First, a nonlinear pressure dependence of $J(p)$ with a broad minimum at about 150–200 MPa is again visible at low to medium ionic strengths. Second, the course of the interaction strength as a function of pressure is markedly influenced by the type of anion. Compared to the "salt free" (i.e., pure buffer) solution, small differences in $J(p)$ can be found for the NaCl solution only at very high ionic strengths, where the pressure dependence tends to be less pronounced. The position of the minimum of $J(p)$ stays essentially constant over the entire salt concentration range, however. Obviously, the chloride anions exert only a modest effect on the water structure and hence the pressure dependence of the interaction potential. Only at the highest salt concentration studied ($I = 375$ mM), the addition of NaCl

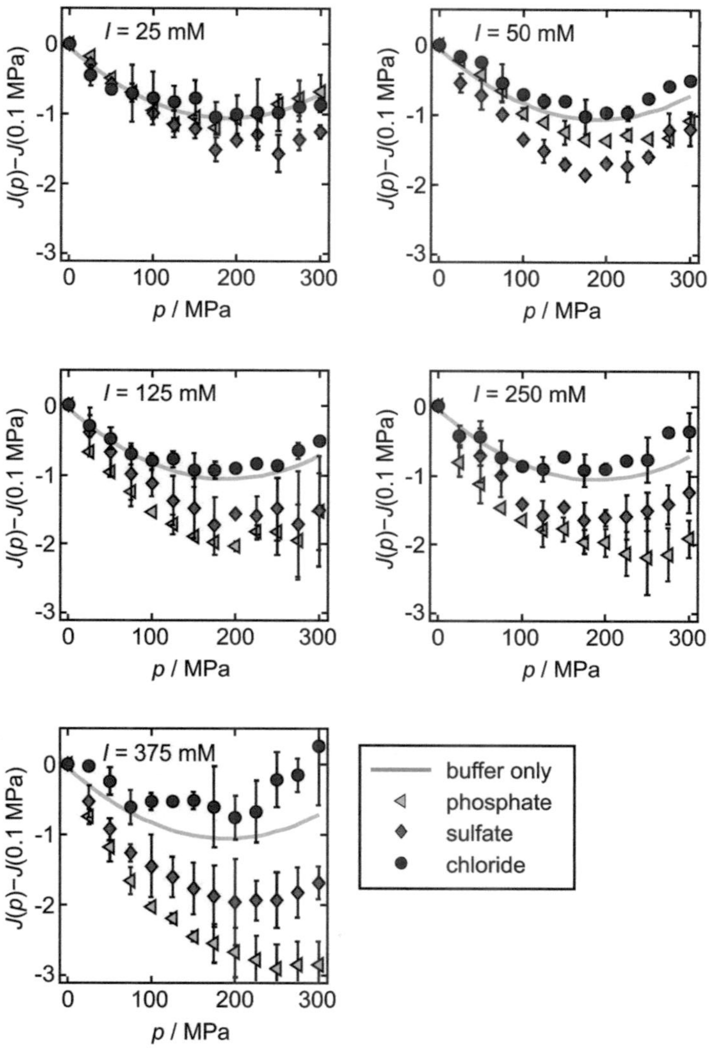

Fig. 8.13 Pressure dependence of the attractive strength, J, of the protein-protein interaction potential for various salt solutions. The pressure dependence of $J(p)$ of the corresponding salt-free lysozyme solution is shown as *grey line* (Modified from Möller et al. 2014b)

slightly reduces the pressure-induced increase in repulsion between the protein molecules. In contrast, an increasing influence on $V(r)$ is observed with increasing concentration of the multivalent anions - the effect being more pronounced for the phosphate. They tend to increase the repulsion with increasing pressure. Moreover, the minimum in $J(p)$ is shifted to higher pressures for both kosmotropic anions or even disappears for $I = 325$ mM in the case of phosphate. This effect might be due to a counteracting effect of these kosmotropic anions on the pressure perturbation of the water structure. Hence, the particular structural properties of the salt solutions seem to be able to influence the spatial organization and the intermolecular interaction of the protein, in particular in the compressed state at high pressure conditions.

8.6 The Effect of Cosolvents on Intermolecular Interactions of Proteins at High Pressures

Cells are able to compensate for extreme environmental conditions and hence rescue proteins from denaturation and subsequent aggregation by using particular osmolytes (Javid et al. 2007b). Organic osmolytes are accumulated under anhydrobiotic, thermal and pressure stresses. Among those are amino acids, polyols and sugars (e.g., glycerol and trehalose), methylamines such as trimethylamine-*N*-oxide (TMAO), and urea (Yancey 2005). Osmolytes are able to destabilize proteins (such as urea) or stabilize them (such as TMAO, glycerol or sugars) against partial or full unfolding. A significant decrease of the unfolding temperature, T_m, is observed upon addition of urea, reflecting its destabilizing effect. Conversely, a marked increase of T_m is observed upon addition of TMAO. For the 1:1 and 2:1 urea: TMAO mixture, a counteracting behavior has been observed. Interestingly, TMAO has been found to counteract perturbations imposed by the waste product urea and hydrostatic pressure in deep-sea animals most effectively (Yancey 2005). High hydrostatic pressure generally destabilizes the protein structure, inhibits polymerization of proteins and ligand binding (Silva et al. 2001; Heremans and Smeller 1998; Mishra and Winter 2008). TMAO has also been shown to largely offset these pressure effects. In fact, it was found that the amount of TMAO in the cells of a series of marine organisms increases linearly with the ocean's depth. For that reason, TMAO is thought to serve even as pressure counteractant. The term "piezolyte" has been coined for this cosolute (Yancey 2005).

Regarding the underlying mechanism of stabilization by TMAO at ambient pressure conditions, several experimental and theoretical (MD simulations) works have been published (Panuszko et al. 2009; Paul and Patey 2007; Bennion and Daggett 2004; Wei et al. 2010; Kuffel and Zielkiewicz 2010). It is generally agreed that TMAO is largely excluded from the protein surface and enhances the water structure by increased hydrogen bonding. However, the mechanism of this "chemical chaperon" at high hydrostatic pressure conditions is still unclear. To yield a deeper understanding of this phenomenon, we determined the intermolecular interaction of dense protein solutions in the absence and presence of cosolvent mixtures of TMAO and urea also under HHP conditions.

Figure 8.14 depicts J as a function of pressure for several TMAO concentrations, which has been obtained from the fit of the DLVO potential to the $I(q)$ data within the MSA approximation. For comparison, the data for the protein in pure buffer solution are included (dashed line). At atmospheric pressure, the presence of TMAO

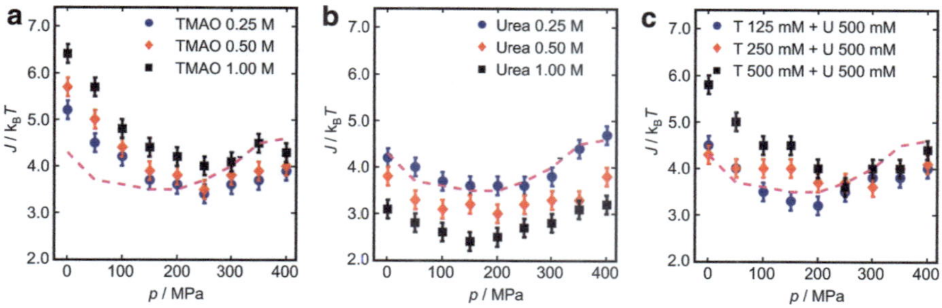

Fig. 8.14 Pressure dependence of the attractive interaction (J) of lysozyme molecules in different osmolyte solutions. (**a**) TMAO, (**b**) urea, (**c**) different mixtures of TMAO and urea. The data for the pure buffer solution are shown as *dashed line* (Modified from Schroer et al. 2011b)

in the protein solution leads to a marked increase of J with increasing osmolyte concentration ($+2\ k_BT$ for 1 M TMAO), in accord with literature data (Niebuhr and Koch 2005). Upon pressurization, a drastic decrease of J is observed up to a pressure of ~250 MPa. At higher pressures, no significant change of J is observed anymore. For the protein in pure buffer solution, the decrease of J upon pressurization is less pronounced, whereas, as discussed above, J increases again at higher pressures. Furthermore, the broad minimum in $J(p)$ is shifted to higher pressures when TMAO is added to the buffer solution, i.e., the collapse of the second hydration shell seems to be suppressed by the addition of TMAO.

Figure 8.14 exhibits also the pressure dependence of J for different urea concentrations. With increasing urea concentration, the interaction becomes more repulsive, in accord with previous ambient-pressure data (Niebuhr and Koch 2005; Ortore et al. 2008). The pressure dependence of J is similar to that in the pure buffer solution. An increase in the amount of urea results in a systematic shift of J towards smaller values, only. This might be due to the fact that the water structure is not significantly perturbed in the presence of urea, also upon compression. The data also show that even in the presence of TMAO concentrations as low as 0.125 M, the tendency of urea to foster the repulsive nature of the intermolecular protein interaction is largely compensated by the TMAO at ambient pressure. Up to a pressure of 250 MPa, the 0.5 M urea/0.125 M TMAO mixture displays a $J(p)$ dependency which is similar to that of the pure buffer solution. At higher pressures, J is slightly below the value of the pure buffer solution, which is also observed for the pure TMAO and urea cosolvent solutions. An increase of the TMAO concentration in the urea-TMAO cosolvent mixture to 0.25 M causes a shift of the minimum of $J(p)$ from about 100 to 200 MPa. For the equimolar mixture 0.5 M TMAO + 0.5 M urea, the $J(p)$ data look similar to those of the pure 0.5 M TMAO solution. These data thus reveal a strong and counteracting influence of TMAO on the intermolecular protein-protein interaction potential also under HHP conditions (Schroer et al. 2011b).

According to experimental and molecular dynamics simulation studies, TMAO seems to enhance the number of strong hydrogen bonds in the water solution, i.e., serves as "water structure maker" (Meersman et al. 2009). In the presence of proteins, a direct interaction between the protein and the osmolyte is disfavored, and it is the depletion of TMAO from the protein's surface that gives rise to the increased protein stability as the cosolvent is added. Hence, TMAO increases the amount of strong H-bonds of water, leads to a strengthening of the H-bond network structure, and the second hydration shell of water seems to move slightly outwards – contrary to the effect of HHP (Meersman et al. 2009). Addition of urea increases the repulsive interaction between lysozyme molecules ($\Delta J = -1\ k_B T$/M urea). Conversely, the addition of TMAO increases the attractive interaction ($\Delta J = +2\ k_B T$/M TMAO). For urea-TMAO mixtures, a large counteracting effect on the intermolecular interaction, qualitatively similar to their effect on protein stability (and also T_m), is observed. Hence, indirect, i.e., solvent-mediated, effects seem to play a major role in protein stabilisation also under HHP conditions, where TMAO and pressure have counteracting effects on the water structural properties. These findings may thus be of importance for understanding the upregulation of TMAO under HHP stress conditions and the compensatory effect of urea-TMAO mixtures in deep-sea organisms (Yancey 2005).

Acknowledgments Financial support from the DFG Research Unit FOR 1979 and in part of the Cluster of Excellence RESOLV (EXC 1069) funded by the Deutsche Forschungsgemeinschaft (DFG) is gratefully acknowledged.

References

Akasaka K (2006) Probing conformational fluctuation of proteins by pressure perturbation. Chem Rev 106:1814–1835

Annunziata O, Payne A, Wang Y (2008) Solubility of lysozyme in the presence of aqueous chloride salts: common-ion effect and its role on solubility and crystal thermodynamics. J Am Chem Soc 130:13347–13352

Asherie N (2004) Protein crystallization and phase diagrams. Methods 34:266–272

Bennion BJ, Daggett V (2004) Counteraction of urea-induced protein denaturation by trimethylamine N-oxide: a chemical chaperone at atomic resolution. Proc Natl Acad Sci U S A 101:6433–6438

Cheng YC, Lobo RF, Sandler SI, Lenhoff AM (2006) Kinetics and equilibria of lysozyme precipitation and crystallization in concentrated ammonium sulfate solutions. Biotechnol Bioeng 94:177–188

Collins KD (2004) Ions from the Hofmeister series and osmolytes: effects on proteins in solution and in the crystallization process. Methods 34:300–311

Crisman RL, Randolph TW (2010) Crystallization of recombinant human growth hormone at elevated pressures: pressure effects on PEG-induced volume exclusion interactions. Biotechnol Bioeng 107:663–672

Curtis R, Lue L (2006) A molecular approach to bioseparations: protein-protein and protein-salt interactions. Chem Eng Sci 61:907–923

Daniel I, Oger P, Winter R (2006) Origins of life and biochemistry under high-pressure conditions. Chem Soc Rev 35:858–875

De With G (2013) Liquid-state physical chemistry. Wiley-VCH, Weinheim
Dumetz AC, Chockla AM, Kaler EW, Lenhoff AM (2008) Protein phase behavior in aqueous solutions: crystallization, liquid-liquid phase separation, gels, and aggregates. Biophys J 94:570–583
Ellis RJ (2001) Macromolecular crowding: obvious but underappreciated. Trends Biochem Sci 26:597–604
Ellis RJ (2007) Protein misassembly: macromolecular crowding and molecular chaperones. Adv Exp Med Biol 594:1–13
Erlkamp M, Grobelny S, Winter R (2014) Crowding effects on the temperature and pressure dependent structure, stability and folding kinetics of staphylococcal nuclease. Phys Chem Chem Phys 16:5965–5976
Fourme R, Kahn R, Mezouar M, Girard E, Hoerentrup C, Prangé T, Ascone I (2001) High-pressure protein crystallography (HPPX): instrumentation, methodology and results on lysozyme crystals. J Synchrotron Rad 8:1149–1156
George A, Wilson WW (1994) Predicting protein crystallization from a dilute solution property. Acta Cryst D50:361–365
Grobelny S (2014) Strukturuntersuchungen von Biomolekülen und Stimuli-sensitiven Polymeren unter hohen hydrostatischen Drücken. PhD thesis, Department of Chemistry and Chemical Biology, TU Dortmund University
Grobelny S, Erlkamp M, Möller J, Tolan M, Winter R (2014) Intermolecular interactions in highly concentrated protein solutions upon compression and the role of the solvent. J Chem Phys 141:22D506
Gross M, Jaenicke R (1991) Growth inhibition of lysozyme crystals at high hydrostatic pressure. FEBS Lett 284:87–90
Gross M, Jaenicke R (1993) A kinetic model explaining the effect of hydrostatic pressure on nucleation and growth of lysozyme crystals. Biophys Chem 45:245–252
Grudzielanek S, Smirnovas V, Winter R (2006) Solvation-assisted pressure tuning of insulin fibrillation: from novel aggregation pathways to biotechnological applications. J Mol Biol 356:497–509
Gunton JD, Shiryayev A, Pagan DL (2007) Protein condensation. Cambridge University Press, Cambridge
Hansen J-P, McDonald IR (2013) Theory of simple liquids. Elsevier, Amsterdam
Heremans K, Smeller L (1998) Protein structure and dynamics at high pressure. Biochim Biophys Acta 1386:353–370
Hiemenz PC (1997) Principles of colloid and surface chemistry. CRC Press, Boca Raton
Hofmeister F (1888) Zur Lehre der Wirkung der Salze. Arch Exp Pathol Pharmakol 24:247–260
Israelachvili JN (2011) Intermolecular and surface forces. Elsevier, Amsterdam
Javid N, Vogtt K, Krywka C, Tolan M, Winter R (2007a) Capturing the interaction potential of amyloidogenic proteins. Phys Rev Lett 99:028101
Javid N, Vogtt K, Krywka C, Tolan M, Winter R (2007b) Protein-protein interactions in complex cosolvent solutions. Chem Phys Chem 8:679–689
Kadri A, Damak M, Jenner G, Lorber B, Giegé R (2003) Investigating the nucleation of protein crystals with hydrostatic pressure. J Phys Condens Matter 15:8253–8262
Kadri A, Lorber B, Charron C, Robert MC, Capelle B, Damak M, Jenner G, Giegé R (2005) Crystal quality and differential crystal-growth behaviour of three proteins crystallized in gel at high hydrostatic pressure. Acta Cryst D61:784–788
Katayama Y, Hattori T, Saitoh H, Ikeda T, Aoki K, Fukui H, Funakoshi K (2010) Structure of liquid water under high pressure up to 17 GPa. Phys Rev B 81:014109
Krywka C, Sternemann C, Paulus M, Javid N, Winter R, Al-Sawalmih A, Yi S, Raabe D, Tolan M (2007) The small-angle and wide-angle x-ray scattering set-up at beamline BL9 of DELTA. J Synchrotron Rad 14:244–251
Krywka C, Sternemann C, Paulus M, Tolan M, Royer C, Winter R (2008) Effect of osmolytes on pressure-induced unfolding of proteins: a high-pressure SAXS study. Chem Phys Chem 9:2809–2815

Kuehner DE, Engmann J, Fergg F, Wernick M, Blanch HW, Prausnitz JM (1999) Lysozyme net charge and ion binding in concentrated aqueous electrolyte solutions. J Phys Chem B 103:1368–1374
Kuffel A, Zielkiewicz J (2010) The hydrogen bond network structure within the hydration shell around simple osmolytes: urea, tetramethylurea, and trimethylamine-N-oxide, investigated using both a fixed charge and a polarizable water model. J Chem Phys 133:035102
Kundrot CE, Richards FM (1986) Collection and processing of x-ray diffraction data from protein crystals at high pressure. J Appl Cryst 19:208–213
Lauro FM, Bartlett DH (2008) Prokaryotic lifestyles in deep sea habitats. Extremophiles 12:15–25
Liu Y, Chen W-R, Chen S-H (2005) Cluster formation in two-Yukawa fluids. J Chem Phys 122:44507
Lorber B, Jenner G, Giege R (1996) Effect of high hydrostatic pressure on nucleation and growth of protein crystals. J Cryst Growth 158:103–117
Ludwig R (2001) Water: from clusters to the bulk. Angew Chem Int Ed 40:1808–1827
Meersman F, Bowron D, Soper AK, Koch MHJ (2009) Counteraction of urea by trimethylamine N-oxide is due to direct interaction. Biophys J 97:2559–2566
Meersman F, Daniel I, Bartlett D, Winter R, Hazael R, McMillan PF (2013) High-pressure biochemistry and biophysics. Rev Mineral Geochem 75:607–648
Minton AP (2006) How can biochemical reactions within cells differ from those in test tubes? J Cell Sci 119:2863–2869
Mishra R, Winter R (2008) Cold- and pressure-induced dissociation of protein aggregates and amyloid fibrils. Angew Chem Int Ed 47:6518–6521
Möller J, Schroer MA, Erlkamp M, Grobelny S, Paulus M, Tiemeyer S, Wirkert FJ, Tolan M, Winter R (2012) The effect of ionic strength, temperature, and pressure on the interaction potential of dense protein solutions: from nonlinear pressure response to protein crystallization. Biophys J 102:2641–2648
Möller J, Grobelny S, Schulze J, Bieder S, Steffen A, Erlkamp M, Paulus M, Tolan M, Winter R (2014a) Reentrant liquid-liquid phase separation in protein solutions at elevated hydrostatic pressures. Phys Rev Lett 112:028101
Möller J, Grobelny S, Schulze J, Steffen A, Bieder S, Paulus M, Tolan M, Winter R (2014b) Specific anions effects on the pressure dependence of the protein-protein interaction potential. Phys Chem Chem Phys 16:7423–7429
Muschol M, Rosenberger F (1997) Liquid–liquid phase separation in supersaturated lysozyme solutions and associated precipitate formation/crystallization. J Chem Phys 107:1953–1962
Nagatoshi Y, Sazaki G, Suzuki Y, Miyashita S, Matsui T, Ujihara T, Fujiwara K, Usami N, Nakajima K (2003) Effects of high pressure on the growth kinetics of orthorhombic lysozyme crystals. J Cryst Growth 254:188–195
Niebuhr M, Koch MHJ (2005) Effects of urea and trimethylamine-N-oxide (TMAO) on the interactions of lysozyme in solution. Biophys J 89:1978–1983
Noro MG, Frenkel D (2000) Extended corresponding-states behavior for particles with variable range attractions. J Chem Phys 113:2941–2944
Ortore MG, Sinibaldi R, Spinozzi F, Carsughi F, Clemens D, Bonincontro A, Mariani P (2008) New insights into urea action on proteins: a SANS study of the lysozyme case. J Phys Chem B 112:12881–12887
Ortore MG, Spinozzi F, Mariani P, Paciaroni A, Barbosa LR, Amenitsch H, Steinhart M, Ollivier J, Russo D (2009) Combining structure and dynamics: non-denaturing high-pressure effect on lysozyme in solution. J R Soc Interface 6:S619–S634
Panuszko A, Bruździak P, Zielkiewicz J, Wyrzykowski D, Stangret J (2009) Effects of urea and trimethylamine-N-oxide on the properties of water and the secondary structure of hen egg white lysozyme. J Phys Chem B 113:14797–14809
Paul S, Patey GN (2007) Structure and interaction in aqueous urea-trimethylamine-N-oxide solutions. J Am Chem Soc 129:4476–4482
Poon WCK, Egelhaaf SU, Beales PA, Salonen A, Sawyer L (2000) Protein crystallization: scaling of charge and salt concentration in lysozyme solutions. J Phys Condens Matter 12:L569–L574

Ravindra R, Zhao S, Gies H, Winter R (2004) Protein encapsulation in mesoporous silicate: the effects of confinement on protein stability, hydration, and volumetric properties. J Am Chem Soc 126:12224–12225

Rosenberger F, Howard SB, Sowers JW, Nyce TA (1993) Temperature dependence of protein solubility – determination and application to crystallization in x-ray capillaries. J Cryst Growth 129:1–12

Russo D, Ortore MG, Spinozzi F, Mariani P, Loupiac C, Annighofer B, Paciaroni A (2013) The impact of high hydrostatic pressure on structure and dynamics of β-lactoglobulin. Biochim Biophys Acta 1830:4974–4980

Saikumar MV, Glatz CE, Larson MA (1995) Crystallization of lysozyme at high pressures. J Cryst Growth 151:173–179

Sauter C, Otálora F, Gavira JA, Vidal O, Giegé R, García-Ruiz JM (2001) Structure of tetragonal hen egg-white lysozyme at 0.94 Å from crystals grown by the counter-diffusion method. Acta Cryst D57:1119–1126

Sazaki G, Nagatoshi Y, Suzuki J, Durbin SD, Miyashita S, Nakada T, Komatsu H (1999) Solubility of tetragonal and orthorhombic lysozyme crystals under high pressure. J Cryst Growth 196:204–209

Schall CA, Wiencek JM, Yarmush M, Arnold E (1994) Lysozyme crystal growth reduced at high pressure. J Cryst Growth 135:548–554

Schroer MA, Markgraf J, Wieland DC, Sahle CJ, Möller J, Paulus M, Tolan M, Winter R (2011a) Nonlinear pressure dependence of the interaction potential of dense protein solutions. Phys Rev Lett 106:178102

Schroer MA, Zhai Y, Wieland DC, Sahle CJ, Nase J, Paulus M, Tolan M, Winter R (2011b) Exploring the piezophilic behavior of natural cosolvent mixtures. Angew Chem Int Ed 123:11613–11616

Sedgwick H, Cameron JE, Poon WC, Egelhaaf SU (2007) Protein phase behavior and crystallization: effect of glycerol. J Chem Phys 127:125102

Seeliger J, Werkmüller A, Winter R (2013) Macromolecular crowding as a suppressor of human IAPP fibril formation and cytotoxicity. PLoS One 8:e69652

Semenyuk AV, Svergun DI (1991) GNOM – a program package for small-angle scattering data processing. J Appl Cryst 24:537–540

Silva JL, Foguel D, Royer CA (2001) Pressure provides new insights into protein folding, dynamics and structure. Trends Biochem Sci 26:612–618

Silva JL, Oliveira AC, Vieira TCRG, de Oliveira GAP, Suarez MC, Foguel D (2014) High-pressure chemical biology and biotechnology. Chem Rev 114:7239–7267

Smeller L, Meersman F, Heremans K (2006) Refolding studies using pressure: the folding landscape of lysozyme in the pressure-temperature plane. Biochim Biophys Acta 1764:497–505

Soper AK, Ricci MA (2000) Structures of high-density and low-density water. Phys Rev Lett 84:2881–2884

Stradner A, Sedgwick H, Cardinaux F, Poon WC, Egelhaaf SU, Schurtenberger P (2004) Equilibrium cluster formation in concentrated protein solutions and colloids. Nature 432:492–495

Stradner A, Cardinaux F, Schurtenberger P (2006) Comment on "Effective long-range attraction between protein molecules in solution studied by small angle neutron scattering". Phys Rev Lett 96:219801

Stradner A, Cardinaux F, Egelhaaf SU, Schurtenberger P (2008) Do equilibrium clusters exist in concentrated lysozyme solutions? Proc Natl Acad Sci U S A 105:E75

Suzuki Y, Sazaki G, Miyashita S, Sawada T, Tamura K, Komatsu H (2002a) Protein crystallization under high pressure. Biochim Biophys Acta 1595:345–356

Suzuki Y, Sazaki G, Visuri K, Tamura K, Nakajima K, Yanagiya S (2002b) Significant decrease in the solubility of glucose isomerase crystals under high pressure. Cryst Growth Des 2:321–324

Suzuki Y, Sazaki G, Matsui T, Nakajima K, Tamura K (2005) High-pressure acceleration of the growth kinetics of glucose isomerase crystals. J Phys Chem B 109:3222–3226

Svergun DI, Koch MHJ, Timmins PA, May RP (2013) Small angle X-ray and neutron scattering from solutions of biological macromolecules. Oxford University Press, Oxford

Takano KJ, Harigae H, Kawamura Y, Ataka M (1997) Effect of hydrostatic pressure on the crystallization of lysozyme based on in situ observations. J Cryst Growth 171:554–558

Tardieu A, Bonneté F, Finet S, Vivarès D (2002) Understanding salt or PEG induced attractive interactions to crystallize biological macromolecules. Acta Cryst D58:1549–1553

Thomson JA, Schurtenberger P, Thurston GM, Benedek GB (1987) Binary liquid phase separation and critical phenomena in a protein/water solution. Proc Natl Acad Sci U S A 84:7079–7083

Verwey EJW, Overbeek JTG (1948) Theory of the stability of lyophobic colloids. Elsevier, New York

Visuri K, Kaipainen E, Kivimäki J, Niemi H, Leisola M, Palosaari S (1990) A new method for protein crystallization using high pressure. Nat Biotechnol 8:547–549

Vliegenthart GA, Lekkerkerker HNW (2000) Predicting the gas-liquid critical point from the second virial coefficient. J Chem Phys 112:5364–5369

Waghmare RY, Webb JN, Randolph TW, Larson MA, Glatz CE (2000) Pressure dependence of subtilisin crystallization kinetics. J Cryst Growth 208:678–686

Webb JN, Waghmare RY, Carpenter FJ, Glatz CE, Randolph TW (1999) Pressure effect on subtilisin crystallization and solubility. J Cryst Growth 205:563–574

Weck G, Eggert J, Loubeyre P, Desbiens N, Bourasseau E, Maillet J-B, Mezouar M, Hanfland M (2009) Phase diagram and isotopic effects of normal and deuterated water studied via x-ray diffraction up to 4.5 GPa and 500 K. Phys Rev B 80:180202

Wei H, Fan Y, Gao YQ (2010) Effects of urea, tetramethyl urea, and trimethylamine N-oxide on aqueous solution structure and solvation of protein backbones: a molecular dynamics simulation study. J Phys Chem B 114:557–568

Weingärtner H, Franck EU, Wiegand G, Dahmen N, Frimmel FH, Gordalla BC, Johannsen K, Summers RS, Höll W, Jekel M, Gimbel R, Rautenbach R, Glaze WH (1996) Water. In: Ullmann's encyclopedia of industrial chemistry. Wiley-VCH, Weinheim

Winter R, Lopes D, Grudzielanek S, Vogtt K (2007) Towards an understanding of the temperature/pressure configurational and free-energy landscape of biomolecules. J Non-Equilib Thermodyn 32:41–97

Yancey PH (2005) Organic osmolytes as compatible, metabolic and counteracting cytoprotectants in high osmolarity and other stresses. J Exp Biol 208:2819–2830

Zemb T, Lindner P (2002) Neutron, X-rays and light scattering methods applied to soft condensed matter. Elsevier, Amsterdam

Zhang F, Skoda MW, Jacobs RM, Martin RA, Martin CM, Schreiber F (2007) Protein interactions studied by SAXS: effect of ionic strength and protein concentration. J Phys Chem B 111:251–259

Zhang F, Roosen-Runge F, Skoda MW, Jacobs RM, Wolf M, Callow P, Frielinghaus H, Pipich V, Prévost S, Schreiber F (2011) Hydration and interactions in protein solutions containing concentrated electrolytes studied by small-angle scattering. Phys Chem Chem Phys 14:2483–2493

Zhao S, Gies H, Winter R (2007) Stability of proteins confined in MCM-48 mesoporous molecular sieves. Z Phys Chem 221(139):154

Part III
Pressure and Functional Sub-states in Proteins

Editors' Foreword of Part III

Chapter 3 focuses on sub-states of proteins located higher in free energy than the most stable folded state N. High-energy sub-states have become the explicit target of NMR spectroscopic measurement, and partly of X-ray crystallography, under variable pressure, thanks to the technical developments in recent years (cf. Chaps. 31 and 33). These two techniques give direct spectroscopic information as to structures and, in case of high pressure NMR spectroscopy, thermodynamic stabilities as well, of sub-states. In Chap. 9, Kalbitzer describes the general principle for detecting and analyzing the high-energy sub-states in proteins by high pressure NMR, with actual examples of application to Ras proteins. In particular, he discusses the general utility of linear and non-linear pressure shift coefficients for analyzing sub-states. In Chap. 10, Kitahara shows that high-energy sub-states are commonly found in ubiquitin and ubiquitin-like proteins, for which they could express the high-energy structures in atomic coordinates by trapping them stably under pressure. In Chap. 11, Dhaussy and Girard give a survey of their pioneering works on high pressure macromolecular crystallography performed on proteins and nucleic acids. Chapter 12 discusses widely on cavities in proteins, whether existing transiently or in equilibrium, and show they are related to protein dynamics, high-energy sub-states and folding intermediates.

Chapter 9	High Pressure NMR Methods for Characterizing Functional Substates of Proteins	Hans Robert Kalbitzer
Chapter 10	High-Pressure NMR Spectroscopy Reveals Functional Sub-states of Ubiquitin and Ubiquitin-Like Proteins	Ryo Kitahara
Chapter 11	Functional Sub-states of Proteins by High-pressure Macromolecular Crystallography	Anne-Claire Dhaussy and Eric Girard
Chapter 12	Cavities and Excited States in Proteins	Hua Li and Yuji O. Kamatari

Chapter 9
High Pressure NMR Methods for Characterizing Functional Substates of Proteins

Hans Robert Kalbitzer

Abstract Proteins usually exist in multiple conformational states in solution. High pressure NMR spectroscopy is a well-suited method to identify these states. In addition, these states can be characterized by their thermodynamic parameters, the free enthalpies at ambient pressure, the partial molar volumes, and the partial molar compressibility that can be obtained from the analysis of the high pressure NMR data. Two main types of states of proteins exist, functional states and folding states. There is a strong link between these two types, the functional states represent essential folding states (intermediates), other folding states may have no functional meaning (optional folding states). In this chapter, this concept is tested on the Ras protein, an important proto-oncogen in humans where all substates required by theory can be identified experimentally by high pressure NMR spectroscopy. Finally, we show how these data can be used to develop allosteric inhibitors of proteins.

Keywords Conformational states • Drug design • Essential folding intermediates • High pressure NMR spectroscopy • Ras

9.1 Overview

In solution, proteins are not rigid but exist in a multitude of different conformations. This conformational ensemble (typically $3 \cdot 10^{17}$ molecules in solution NMR spectroscopy) is usually visible in the NMR spectra. Therefore, a single lowest energy or crystal structure usually cannot describe the NMR spectra satisfactorily as we have shown for the chemical shifts of a protein data set. Here, the agreement between experimentally observed chemical shifts and chemical shifts recalculated from the structures increases with the number of structures used (Baskaran et al. 2010). Figure 9.1 shows the error ε of the calculated shifts for HPr (histidine

H.R. Kalbitzer (✉)
Institute of Biophysics and Physical Biochemistry, University of Regensburg,
D-93040 Regensburg, Universitätsstr. 31, Germany
e-mail: Hans-Robert.Kalbitzer@biologie.uni-regensburg.de

K. Akasaka, H. Matsuki (eds.), *High Pressure Bioscience*,
Subcellular Biochemistry 72, DOI 10.1007/978-94-017-9918-8_9

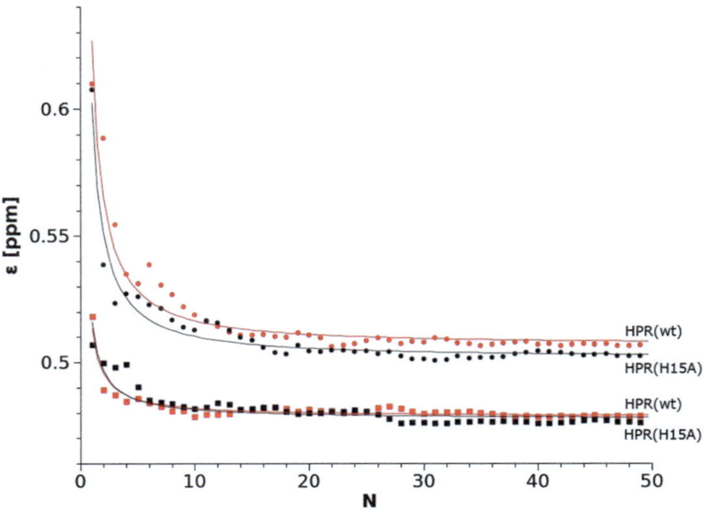

Fig. 9.1 Minimum ensemble size required for a prediction of the backbone chemical shifts. The mean chemical shift error ε of the backbone resonances is plotted as a function of the number N of structures used for the calculation of the backbone chemical shifts by the program SHIFTX (Neal et al. 2003). HPr-wild type (red) and HPr(H15A) (black) before (circle) and after (square) water refinement. The data were fitted with the lognormal function. $\varepsilon = \frac{1}{N\sigma\sqrt{2\pi}} e^{-\frac{(\ln N)^2}{2\sigma^2}} + C$ (Modified from Baskaran et al. (2010))

containing protein), a small protein from *Staphyloccocus aureus* as a function of the number of structures used for the calculations. The chemical shifts of different nuclei were weighted according to Schumann et al. (2007). HPr is a protein that probably occurs in a dominant main conformation that means in first approximation it can be represented by an ensemble of rapidly interconverting conformations with a single minimum of free energy. It turns out that for optimally representing the chemical shifts of the backbone atoms H, N, C^{α}, H^{α}, and C' 10–20 different structures are required. Refinement of the structures in explicit water results in some improvement, but does not remove the necessity of taking more than one structure (Fig. 9.1). Note that chemical shift calculations are not very accurate; there is still a remarkable systematic error in the calculation. Presently it is approximately 0.5 ppm, it seems possible that a higher accuracy of chemical shift calculation would increase substantially the number of structures required to reach an optimal prediction.

As we will see later, most probably for the majority of proteins more than one functional state exists; these (sub)states also have to be represented in the ensemble of structures by additional minima in the free energy landscape. The transition between these substates can be much slower than inside an ensemble that is part of a given substate. In this chapter we will focus on the application of high pressure

NMR spectroscopy for the elucidation of conformational equilibria of proteins that include a number of lowly occupied substates. In contrast to other methods it is possible by high pressure NMR spectroscopy to handle systems with multiple conformational states since it is possible to separate them on the basis of their pressure dependence and their specific thermodynamic parameters as their partial molar volumes and partial molar compressibilities.

9.2 Description of Pressure Effects in Multistate Systems

Most high pressure NMR experiments and also the experiments we discuss in the following are performed under equilibrium conditions and therefore can be described by classical equilibrium thermodynamics. If a protein occurs in N states i ($i = 1, \ldots,$ N), the populations p_i of the different states in thermodynamic equilibrium are determined by their differences in the free Gibbs energies ΔG_{ij} (Fig. 9.2) with

$$\frac{p_j}{p_i} = K_{ij} = \exp\left(-\frac{\Delta G_{ij}}{\mathrm{R}T}\right) \tag{9.1}$$

The equilibrium constant K_{ij} is dependent on pressure P and the absolute temperature T (R, gas constant). The pressure dependence of ΔG_{ij} at constant temperature T and hence of K_{ij} can be approximated by (Heremans and Smeller 1998)

$$\Delta G_{ij}(P) = \Delta G_{ij}^{0}(P_0) + \Delta V_{ij}^{0}(P - P_0) + \frac{1}{2}\frac{\partial \Delta V_{ij}^{0}}{\partial P}(P - P_0)^2 \tag{9.2}$$

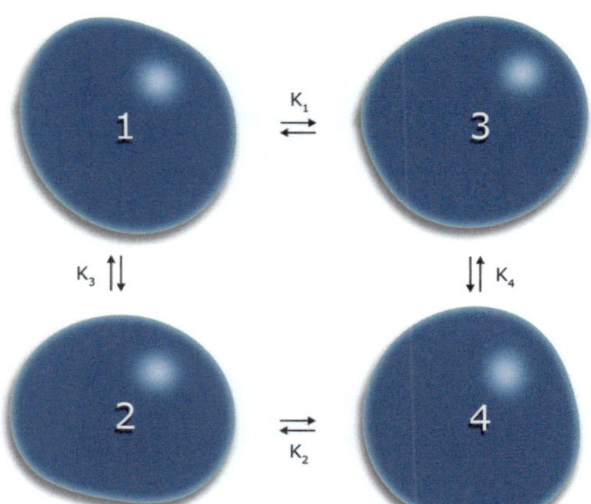

Fig. 9.2 Equilibrium between different conformational states in a four state model. $K_1 = [3]/[1]$, $K_2 = [4]/[2]$, $K_3 = [2]/[1]$, $K_4 = [4]/[3]$

The derivative in Eq. 9.2 is often abbreviated by

$$\Delta\beta_{ij}^{\prime 0} = -\frac{\partial \Delta V_{ij}^{0}}{\partial P} \tag{9.3}$$

$\Delta G_{ij}{}^{0}$, and $\Delta V_{ij}{}^{0}$ are the differences of free energies and partial molar volumes, respectively, at temperature T_0 and pressure P_0.

The actual response observed in an NMR spectrum, strongly depends on the timescale of the exchange between the different states characterized by the exchange correlation times τ_{ij} and the differences of the respective parameters in the two states measured in frequency units. The parameter commonly evaluated in high pressure NMR spectroscopy is the chemical shift δ. Here the chemical shift difference $\Delta\omega_{ij}$ (measured in angular frequency units) between states *i* and *j* determines the NMR time scale. When the fast exchange condition $|\Delta\omega_{ij}\ \tau_{ij}| << 1$ holds for transitions between M^* states, then the correspondent chemical shifts are population averaged.

The chemical shift dependence of nucleus *k* on pressure is then given by (Baskaran et al. 2010)

$$\delta^k = \sum_{j=1}^{M*} p_j^* \delta_j^k = \frac{1}{Z*}\sum_{j=1}^{M*} \delta_j^k e^{\frac{-G_j}{RT}} = \frac{\sum_{i=1}^{M*} \delta_j^k e^{-\frac{\Delta G_{1j}}{RT}}}{\sum_{i=1}^{M*} e^{-\frac{\Delta G_{1j}}{RT}}} \tag{9.4}$$

For separating unspecific effects of pressure from the conformational effects of interest (Kremer et al. 2007) the chemical shifts can be corrected by subtracting the known pressure dependence of chemical shifts of the random-coil peptides such as Gly-Gly-Xxx-Ala or Ac-Gly-Gly-Xxx-Ala-NH_2 from the experimental data (Arnold et al. 2002; Koehler et al. 2012).

The size of the chemical shift changes is correlated to the size of the local conformational changes. Often the pressure dependence can be represented phenomenologically by the first and second order pressure coefficients B_1^* and B_2^* obtained from a Taylor expansion at the pressure P_0 and temperature T_0. The corrected chemical shifts δ^* are described by

$$\delta^*(P, T_0) = \delta_0^*(P_0, T_0) + B_1^*(P - P_0) + B_2^*(P - P_0)^2 \tag{9.5}$$

The pressure coefficients B_1^* and B_2^* can be interpreted in thermodynamic terms in a two-state system when the condition holds.

$$\left|\frac{\Delta G^0}{2RT}\right| << 1 \tag{9.6}$$

In this case the ratio of B_2^*/B_1^* is (Beck Erlach et al. 2014)

$$\frac{-2B_2^*}{B_1^*} = \frac{\Delta\beta'(P_0)}{\Delta V(P_0)} =< \beta^0 > + < V^0 > \frac{\Delta\beta^0}{\Delta V^0} \tag{9.7}$$

with the mean partial molar compressibility $< \beta >=< \beta'/V >$, the difference of the partial molar volumes and compressibilities ΔV and $\Delta\beta$, and the mean partial molar volumes $<V>$ of the two states. Note that the definition of B_2^* given by Beck Erlach et al. (2014) differs to that given in Eq. 9.5 by a factor of 2 taken into account in Eq. 9.7. For

$$\left|\frac{\Delta G^0}{2RT}\right| >> 1 \tag{9.8}$$

the following approximation holds

$$\frac{-2B_2^*}{B_1^*} = \frac{\Delta\beta'(P_0)}{\Delta V(P_0)} + \frac{\Delta V(P_0)}{RT_0} =< \beta^0 > + < V^0 > \frac{\Delta\beta^0}{\Delta V^0} + \frac{\Delta V^0}{RT_0} \tag{9.9}$$

When the conformational exchange between the states i and j is slow relative to the time scale of the NMR experiment ($|\Delta\omega_{ij}| >> 1/\tau_{ex}$), the signal volume V_i of state i is proportional to its concentration c_i. The equilibrium constant K_{ij} can be calculated from the cross peak volumes V of the HSQC spectra by

$$K_{ij} = \frac{V_j}{V_i} = e^{-\frac{\Delta G_{ij}}{RT}} \tag{9.10}$$

If only one peak i can be observed then V_i is given by

$$V_i(P) = \frac{V_T(P)}{\sum_{j=1}^{N} e^{-\frac{\Delta G_{ij}}{RT}}} \tag{9.11}$$

With $V_T(P)$ the total cross peak volume corresponding to the sum of all states at pressure P.

The relative population p_j of a state j in an N-state system can be calculated from the equilibrium constants $K_{ij} = [j]/[i]$ as

$$p_j(P, T_0) = \frac{K_{ij}(P, T_0)}{\sum_{j=1}^{N} K_{ij}(P, T_0)} \tag{9.12}$$

with i an arbitrarily selected state.

9.3 Ras-Dependent Signal Transduction

Proteins that interact with different partners are typical candidates for a multistate equilibrium. An excellent example of such a protein is the rat sarcoma (Ras) protein, a small guanine nucleotide binding protein that is involved in multiple signal transduction pathways that control cellular proliferation, differentiation, and apoptosis. It represents a kind of molecular switch, in its "off" state GDP is bound in its active center, in its "on" state GDP is replaced by GTP and the signal transmission is activated. Ras is activated by guanine nucleotide exchange factors (GEFs) and is deactivated by GTPase activating proteins (GAPs) (Fig. 9.3). Only in the GTP-bound state effector proteins bind to Ras with high affinity (Wittinghofer and Waldmann 2000; Herrmann 2003; Wennerberg et al. 2005; Rajalingam et al. 2007). Mutations of Ras that lead to a permanent activation of the Ras pathway are found in more than 30 % of all human tumors (Bos 1989; Karnoub and Weinberg 2008; Baines et al. 2011). Inhibiting the activity of Ras thus represents an interesting strategy in cancer therapy (for reviews see e.g. Adjei 2001; Friday and Adjei 2005; Baines et al. 2011).

When the Ras activation cycle (Fig. 9.3) is described by a multistate equilibrium, activated Ras with GTP bound has at least three different states, the GEF-interacting state 1(T), the effector binding state 2(T), and the GAP interacting state 3(T). Corresponding states for GDP can be defined when GDP is bound, however an interacting partner has not known for state 2(D). In addition, two states 1(0) and 3(0) are required where in state 1(0) the active centre is not occupied by a nucleotide and where in state 3(0) P_i is not yet released after GTP-hydrolysis (Fig. 9.3).

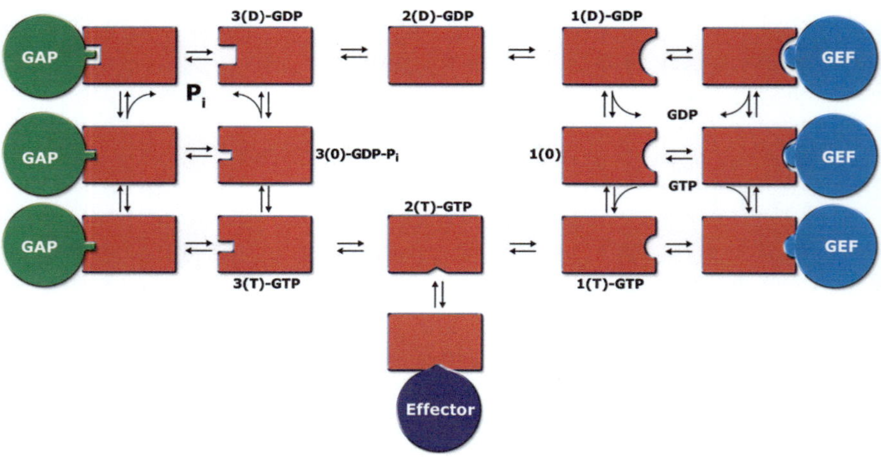

Fig. 9.3 Generalized Ras activation cycle. Ras, rat sarcoma protein, GAP, GTPase activating protein, GEF, guanine nucleotide exchange factor. 0, D, and T designate the nucleotide-free or GDP-Pi containing, the GDP bound, and the GTP bound state of the Ras protein (Modified from Kalbitzer et al. (2013a))

9.4 A Fundamental Link Between Functional States and Folding Intermediates

An important application of high pressure NMR spectroscopy is the investigation of folding/unfolding of proteins and the detection of folding intermediates. Classically, folding studies and functional studies are two separate areas that are dealt with by different groups of scientists. However, when considering the two processes in the light of the free energy landscape model it becomes evident that there is a fundamental link between folding and function (Kalbitzer et al. 2009). In thermodynamic equilibrium all energetically possible states become populated and are accessible; the same free energy landscape describes folding and function. Only the transient populations of the different states (local free energy minima) after a perturbation depend on the selected initial conditions. To show the connection between the two domains we performed high pressure studies as well as classical denaturation studies with the Ras protein.

By ^{31}P NMR spectroscopy of Ras protein complexed with the non-hydrolysable GTP analog GppNHp two structural states of the protein can be identified, since the resonance lines of the γ- as well as the α-phosphate groups are split (Fig. 9.4). The integration of the peak areas gives the corresponding populations of the states. At low temperature the transitions between the two states are slow on the NMR time scale ($\tau_{ex} = 7.7$ ms) (Geyer et al. 1996). The functional importance of the two states was determined by adding GTPase activating proteins, effectors or exchange factors to a solution of Ras. Addition of effectors leads to the increase of one set of upfield shifted resonance lines, addition of the exchange factor SOS (son

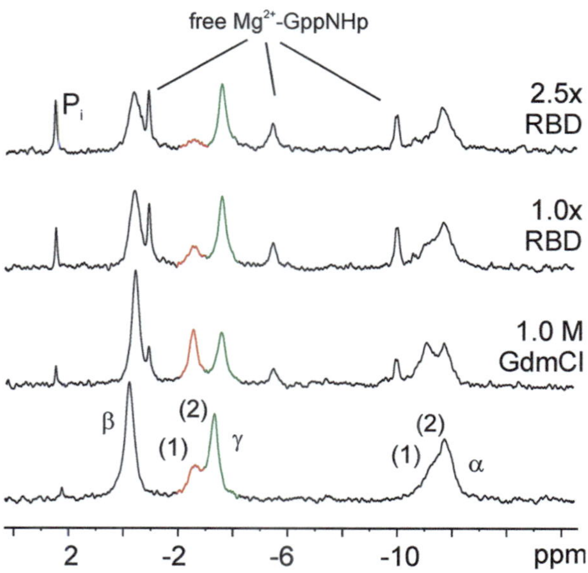

Fig. 9.4 Denaturation of Ras complexed with Mg^{2+}·GppNHp studied by ^{31}P NMR spectroscopy. ^{31}P NMR spectrum of Ras·Mg^{2+}·GppNHp changes after successive addition of Guanidinium hydrochloride (GdmCl) and the Ras binding domain (RBD) of Raf kinase. α, β, and γ, signals of the α-, β-, and γ-phosphate group of the nucleotide. (1) and (2), Ras in state 1(T) and 2(T) (modified from Kalbitzer et al. (2009)). Temperature 278 K, ^{31}P resonance frequency 202.4 MHz

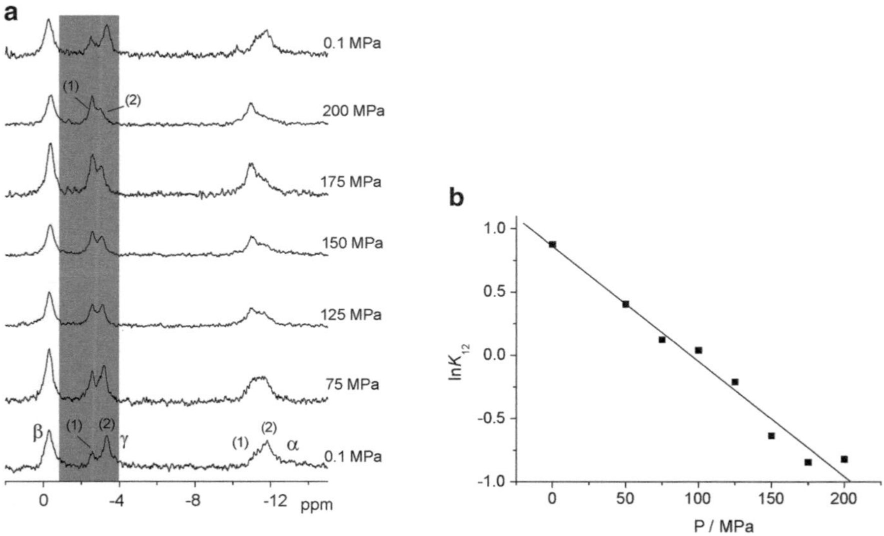

Fig. 9.5 ^{31}P NMR spectroscopy of Ras complexed with Mg^{2+}·GppNHp at different pressures. (**a**) ^{31}P NMR spectra of Ras·Mg^{2+}·GppNHp at different pressures. α, β, and γ, signals of the α-, β-, and γ-phosphate group of the nucleotide. (1) and (2), Ras in state 1(T) and 2(T). (**b**) Plot of ln $K = [2(T)]/[1(T)]$ as a function of pressure P. Temperature 278 K, ^{31}P resonance frequency 202.4 MHz. Difference of the free energy $\Delta G_{21}{}^0$ and the partial molar volume $\Delta V_{21}{}^0$ 1.42 kJ mol^{-1} and -17.2 mL mol^{-1} (Modified from Kalbitzer et al. (2009))

of sevenless) to an increase of the other set of downfield shifted resonance lines (Geyer et al. 1996; Spoerner et al. 2001; Kalbitzer et al. 2009). This means that after binding the population of the corresponding states 1(T) or 2(T) increases (Fig. 9.4). The equilibrium constant between the two states K_{12} is 1.9 at 278 K for the wild type protein (Spoerner et al. 2004) that corresponds to a difference of free energies $\Delta G_{12}{}^0$ of -1.5 kJ mol^{-1}. Applying pressure to the sample (Fig. 9.5a) leads to a shift of the equilibrium to the state with the smaller partial volume, in our case to state 1(T). A fit of ln K_{12} as function of the pressure P leads to a difference of the partial molar volumes ΔV_{12} of -17.2 mL mol^{-1}. According to the above suppositions, in classical denaturation experiments with a denaturant such as guanidininum hydrochloride (GdmCl) a shift of the equilibrium of the functional states should also be observed during the denaturation process. Figure 9.4 shows that this prediction is true, at 1 M GdmCl, the relative occupancy of state 1(T) is strongly increased and simultaneously some GppNHp is set free indicating beginning denaturation of the protein. As required from theory the population of state 1(T) is shifted again to the effector binding state 2(T) when the effector-binding domain (RBD) of the Raf kinase is added.

Thus the following paradigm can be formulated: the minimum number of folding intermediates is determined by the number of all required functional states of a protein ("essential" folding intermediates). Additional "optional folding" intermediates may be observable (Kalbitzer et al. 2009). The main difference between the two perturbations of the equilibrium applied here (and also other perturbations such as temperature changes) is the observation, that the pressure perturbation usually separates the different states much better than the other perturbations. In addition, protein aggregation and virtually irreversible protein aggregation is suppressed at high pressures.

9.5 Conformational States of the Ras Protein Identified by High Pressure NMR Spectroscopy

When we describe protein-protein and protein-ligand interaction in the Monod-Wyman-Changeux model (Monod et al. 1965), different conformational substates of a protein with different affinities for the interaction partners are required. For the Ras protein, such a minimal scheme has been already introduced in Fig. 9.3. A possibility to detect these substates is the analysis of the high pressure NMR data that we recorded for the Ras protein with the GTP analog GppNHp bound. Figure 9.6 shows a superposition of HSQC-spectra of uniformly ^{15}N-enriched Ras protein recorded at different pressures in the range from 3 to 180 MPa. The signals of the backbone amide groups shift continuously with pressure, most of them in downfield direction. The pressure dependence of the chemical shift varies from amide group to amide group. In addition, some cross peak intensities are influenced by pressure and a few new cross peaks appear in the spectra at high pressures.

A first analysis of the size of the chemical shift changes was performed using the Taylor expansion described in Eq. 9.5. Before calculating the pressure coefficients,

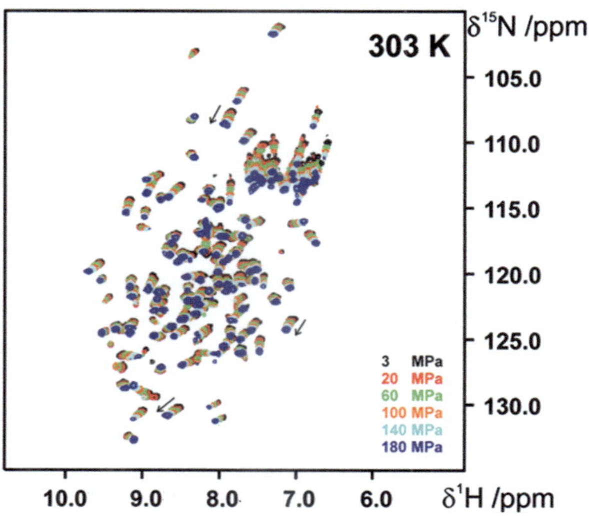

Fig. 9.6 [^1H,^{15}N]-HSQC spectra of Ras·Mg^{2+}·GppNHp at different pressures. Temperature 278 K. Proton resonance frequency 800.2 MHz (Modified from Kalbitzer et al. (2013a))

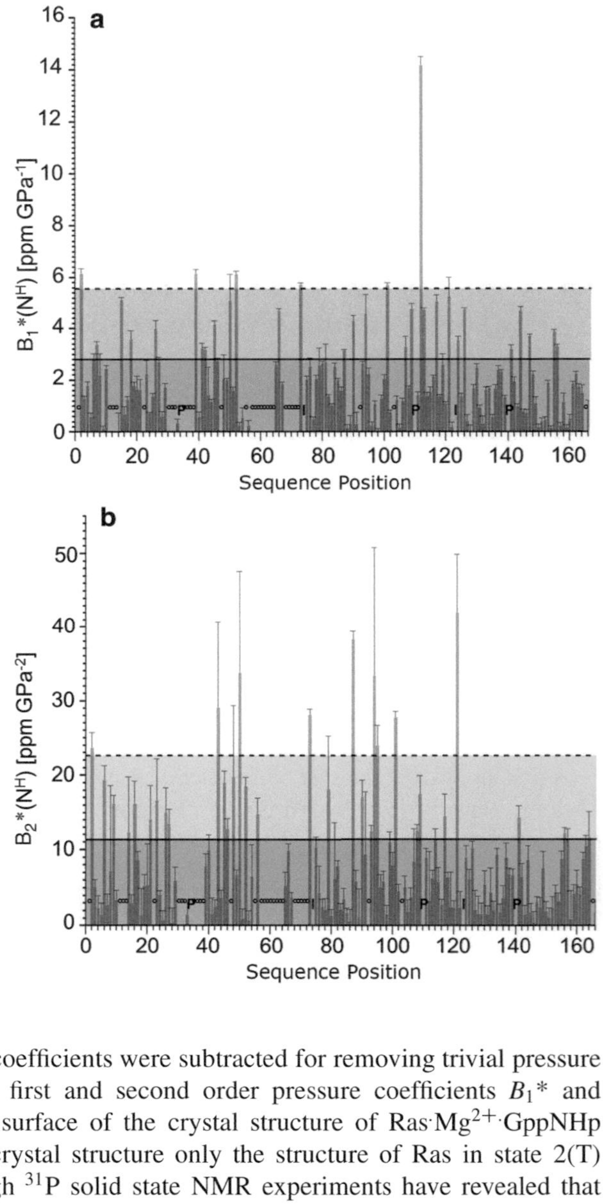

Fig. 9.7 Pressure induced chemical shift changes in wild type Ras·Mg^{2+} · GppNHp at 303 K. The absolute values of the first order and second order pressure coefficients B_1*(N^H) and B_2*(N^H) of the amide nitrogen shifts are plotted as a function of the position in amino acid sequence. Data were corrected for random coil effects before calculation of the pressure coefficients $B_{1,2}$*. Solid line, standard deviation σ_0 to zero, dashed line 2 σ_0. P, prolines; o, not assigned residues; I, residues for which the coefficients could not be determined due to the loss of signal intensity within the pressure series (Modified from Kalbitzer et al. (2013))

the random-coil pressure coefficients were subtracted for removing trivial pressure effects. The ^{1}H and ^{15}N first and second order pressure coefficients B_1* and B_2* were plotted on the surface of the crystal structure of Ras·Mg^{2+}·GppNHp (Pai et al. 1990). In this crystal structure only the structure of Ras in state 2(T) could be detected although ^{31}P solid state NMR experiments have revealed that the single crystals used for the X-ray structure determination also simultaneously contain molecules in state 1(T) (Stumber et al. 2002; Iuga et al. 2004). In Fig. 9.7 the absolute values of the nitrogen first and second order pressure coefficients B_1* and B_2* are shown. For a number of residues values larger than the standard deviation to zero σ_0 are visible representing residues that are strongly influenced by pressure. The pressure effects are not restricted to the switch I and switch II regions but are also visible at the backside of the protein. The largest effects are observed

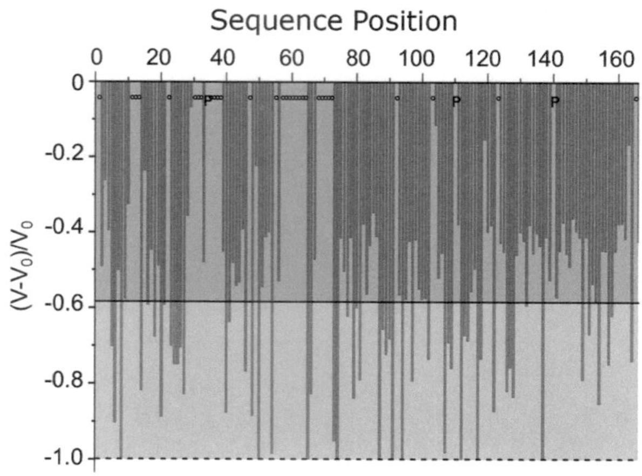

Fig. 9.8 Pressure-induced cross peak volume changes. Plot of the relative change of the [^{1}H, ^{15}N]-HSQC cross peak volumes V_0 measured at 3 MPa and V measured at 180 MPa. Temperature 303 K, solid line, standard deviation to zero σ_0

for Thr2, Ser39, Gln43, Thr50, Leu53, Ser65, Arg73, Thr87, His84, His94, Glu95, Ala121, Tyr141, and Lys147. Unfortunately, most of the residues in the switch regions are not visible in the [^{1}H,^{15}N]-HSQC spectra, meaning that pressure effects of these residues are not observable. However, Ser39 is located in switch I, Ser65 and Arg73 in the switch II region, indicating that these regions involved in protein-protein interactions probably are strongly influenced by pressure. Lys147 is part of the G3 motif that recognizes the guanine base of GTP.

In addition to chemical shift changes also position specific changes of the cross peak volumes can be observed. Strong cross peak volume changes usually are observed when slow exchange conditions apply. Typically, a cross peak assignable to the dominant state at ambient pressure loses intensity when at higher pressures the population of this state is reduced. A new cross peak corresponding to the high pressure state is to be expected. Such a peak often cannot be identified since it overlaps with other peaks or simply because it has not been assigned. Figure 9.8 shows the cross peak reductions at 180 MPa, the highest pressure used at 303 K. Again significant pressure effects are observable. Most of these residues influenced by pressure are located close to regions where large pressure dependent chemical shift changes occur.

The above analysis mainly describes the magnitude of the pressure-induced changes and gives only partial information on the mechanisms behind the observed changes. However, as a rule large second order terms B_2^* often are associated with larger conformational changes in the neighbourhood of the corresponding residue. In contrast, a thermodynamic analysis of the pressure-induced changes of NMR parameters such as chemical shifts or cross peak volumes gives information on the free energies and differences in the partial molar volumes of different states. Although the analysis is primarily restricted to the local probes such as the amide protons, identical structural transitions should also be characterized by a set of free energy differences ΔG^0, partial molar volume differences ΔV^0, and differences of the compressibility coefficients $\Delta\beta'^0$ measured at a temperature T_0 and a pressure

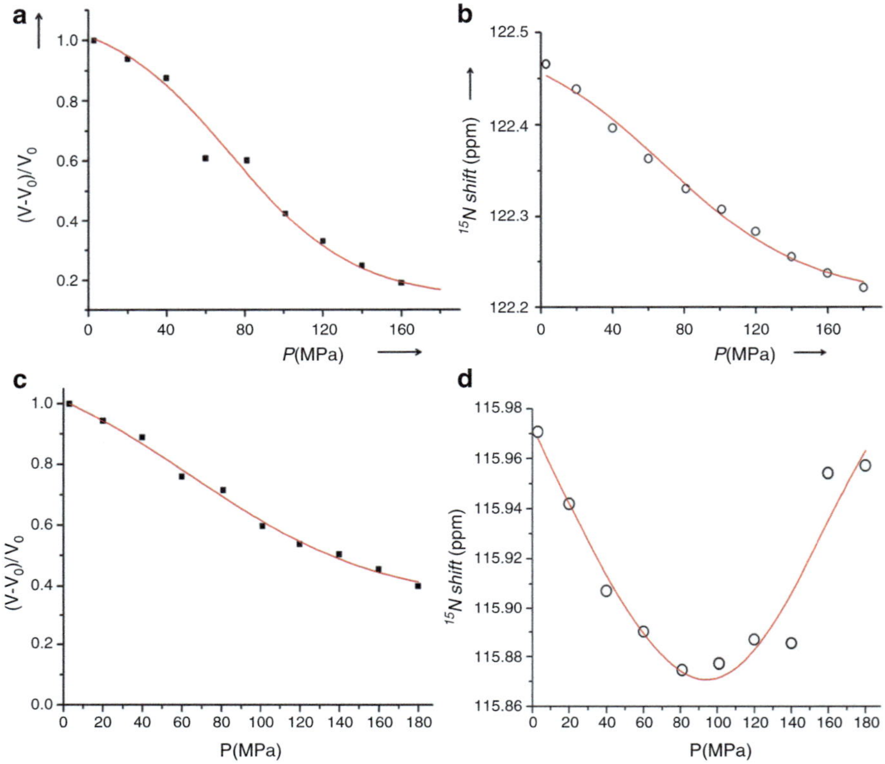

Fig. 9.9 Pressure dependence of ^{15}N chemical shifts and amide cross peak volumes in [^{1}H, ^{15}N]-HSQC spectra of Ras(wt) $\cdot$ Mg^{2+} $\cdot$ GppNHp. Temperature 303 K. Fit of the chemical shift changes (Eq. 9.5) and the normalized pressure dependent volume changes (Eq. 9.7). (**a**) Fit of the volume changes of Ala18, (**b**) fit of the chemical shift changes of Glu49, (**c**) fit of the volumes changes of Glu49. Data can be fitted with the same values for ΔG_{24} (12.4 kJ mol^{-1}), and ΔV_{24} (-115 mL mol^{-1}). (**d**) Fit of the chemical shift changes of Glu99, fit with a three-state model, ΔG_{21} 1.48 kJ mol^{-1}, ΔV_{21} -18 mL mol^{-1}, ΔG_{23} 5.2 kJ mol^{-1}, ΔV_{23} -81 mL mol^{-1} (from Kalbitzer et al. (2013))

P_0. Conversely, if similar thermodynamic parameters (identical within limits of error) are found for different residues, it is likely (but not mandatory) that these residues are involved in the same transition. A point making the analysis more difficult is the effect that in principle all residues are influenced by all transitions but to a different extent.

We have evaluated the high pressure induced chemical shift changes and cross peak volume changes of Ras$\cdot$$Mg^{2+}$$\cdot$GppNHp at two different temperatures, 278 and 303 K. For a satisfactory fit of the data (at least) four different states were required. In Fig. 9.9 some examples of such a fit are depicted. Biphasic curves such as shown in Fig. 9.9d can only be fitted when assuming at least three different states. Although the individual fit of different resonance lines does not give identical fit parameters because of the limited precision of the data, the obtained ΔG^0 and ΔV^0 values can be

Table 9.1 Conformational transitions and corresponding molar free energies ΔG^0 and molar volumes ΔV^0 of Ras(wt) $\cdot Mg^{2+} \cdot$ GppNHp at 303 K[a]

Transition	K^0 at 0.1 MPa	ΔG^0[kJ mol^{-1}]	ΔV^0[mL mol^{-1}]
2(T) → 1(T)	0.55	1.5 ± 0.2	−18 ± 1
2(T) → 3(T)	0.13	5.2 ± 0.3	−81 ± 3
2(T) → 1(0)	0.007	12.4 ± 0.4	−115 ± 2
1(T) → 1(0)	0.013	10.9 ± 0.3	−97 ± 2
1(T) → 1(0)[b]	0.03[b]	8.3[b]	
2(T) → 1(0)[b]	0.01[b]	9.8[b]	

[a]Values for $T = 303$ K. Within the limits of error the differences in free energies ΔG^0 and partial molar volumes ΔV^0 for the transitions 2(T) → 1(T) and 2(T) → 3(T) are independent of temperature in the range between 278 K and 303 K. ΔG^0 and ΔV^0 were determined from the pressure dependence of chemical shifts and cross peak volumes (data from Kalbitzer et al. 2013)
[b]Values from Kalbitzer et al. (2009) obtained at 278 K from a denaturation study with GdmCl. ΔG^0 was determined from the release of the nucleotide and interpreted as denaturation

grouped in the assumed three groups with overlapping error ranges. The obtained mean values of ΔG^0 and ΔV^0 then can be used for a satisfactory fit of all data corresponding to a given transition. That is shown in Fig. 9.9a, b where the same parameter set can be used to fit the pressure dependent ^{15}N chemical shift changes of Ala18 and the cross peak volume changes of Glu49.

The obtained parameters are summarized in Table 9.1. In this Table, the probable assignment of the transitions is already indicated, although that is by no means trivial. Only the assignment of the transition between state 2(T) to 1(T) is straightforward since this transition is already functionally defined by the ^{31}P NMR experiments that showed that one of the states (2(T)) interacts with effector proteins, the other state (1(T)) with the exchange factor SOS. The corresponding difference of the free energy $\Delta G_{21}{}^0$ and the partial molar volume $\Delta V_{21}{}^0$ obtained from the fit of the ^{31}P NMR data (Kalbitzer et al. 2009) were 1.42 kJ mol^{-1} and −17.2 mL mol^{-1} (see Fig. 9.5). Within the limits of error these values are identical to the corresponding values obtained from the analysis of the HSQC spectra for the 2(T) to 1(T) transition (Table 9.1).

Actually, the assignment of the other transitions in the proposed model (Fig. 9.3) can only be made on the basis of functional and structural properties. It is expected (and experimentally observed for Ras-mutants) that at very high pressures the nucleotide is released as it should occur in state 1(0). In fact, such a release of the nucleotide is also observed in denaturation studies of Ras with GdmCl (Kalbitzer et al. 2009). From this study also two ΔG^0 values are obtained for the transitions of 1(T) and 2(T) to 1(0) (Table 9.1). Taken into account that this study was performed at a different temperature, there is a good agreement between these values and the ΔG^0 values obtained from the analysis of the two-dimensional HSQC spectra. By exclusion, the third transition then can be associated with the 2(T) to 3(T) transition (Table 9.2). These assignments to the different transitions can be confirmed also by a structural analysis (see below).

Table 9.2 Populations of the different conformational states in Ras·Mg^{2+}·GppNHp.[a]

State i	p_i at 0.1 MPa
1(T)	0.326
2(T)	0.593
3(T)	0.077
1(0)	0.004

[a]The populations p_i were calculated with Eq. 9.12 from the equilibrium constants given in Table 9.1. Temperature 303 K, pressure 0.1 MPa

From the obtained free energies and partial molar volume differences it is possible to calculate the relative populations of the different states at any pressure. For practical purposes, the populations at ambient pressure are the most important (Table 9.2). The relative populations of state 1(T) and 2(T) are with 33 % and 59 % rather high and therefore they can be easily detected by ^{31}P NMR spectroscopy at ambient pressure. However, since the populations of states 3(T) and 1(0) are only 8 % and 0.4 %, they are almost undetectable by ^{31}P NMR. However, addition of GAP to a sample of Ras·Mg^{2+}·GppNHp leads to the growth of a new γ-phosphate peak (Geyer et al. 1996) at −3.0 ppm located just between the γ-phosphate resonances of state 1(T) and 2(T) at −2.6 and −3.3 ppm. If this would be the position of the γ-phosphate resonance of state 3(T), it could explain that the resonances of state 1(T) and 2(T) are not completely separated but show a kind of weak shoulder. A peak with a relative intensity smaller than 1 % as estimated for state 1(0) would disappear in the noise of the phosphorous resonance spectra.

It seems natural to assume that state 3(T), the interaction state with the GTPase activating protein GAP, has the highest intrinsic hydrolysis rate. Since in the reaction scheme (Fig. 9.3) state 2(T) is the direct neighbour of 3(T), it should also have a higher intrinsic hydrolysis activity than state 1(T). Experimental results support this assertion: the intrinsic hydrolysis rate increases when the population of state 2(T) is increased by site specific mutations or by effector binding stabilizing state 2(T) (Spoerner et al. 2010). An important point to have in mind when discussing the populations of the different states quantitatively is the dependence of the occupancy of the different states on the nucleotide analog used. Replacing the nucleotide analog GppNHp by the analog GppCH$_2$p leads only to a small shift of the ratio of the populations K_{12} of state 2(T) to state 1(T) from 1.9 to 2.0 (Spoerner et al. 2005a). However, binding of the GTP analog GTPγS or the natural nucleotide GTP causes a dramatic decrease of the occupancy of state 1(T) with $K_{12} > 10$ (Spoerner et al. 2007) and $K_{12} > 20$ (Spoerner et al. 2010), respectively. Also C-terminal truncation after amino acid 166 influences the equilibrium between the two states and leads to a decrease of K_{12} for GTP to 11.3.

A powerful NMR method that provides complementary dynamical information in a multiple state system is the chemical exchange saturation transfer (CEST) spectroscopy. Essentially, a resonance line of a nucleus i in a conformational state a of a protein with a resonance position δ^i_a (that can be too weak to be observable in the spectrum) is saturated by a long, frequency selective weak pulse and the saturation is transferred to the same resonance in conformational state *b* with a chemical shift δ^i_b. When the frequency of the saturation pulse is scanned through the spectral range, a decrease of the intensity of the resonance at position δ^i_b is observed when the saturation pulse hits the frequency position δ^i_a. This method has been introduced originally by Forsén and Hoffman (1963) in 1D-spectroscopy and has later been applied to identify the very weak ^{31}P resonance of GTP bound to Ras (Spoerner et al. 2010). An analysis of the data also gives the exchange rate between the two states as 7 s^{-1}. It can also be used in 2D spectroscopy e.g. [^{1}H, ^{15}N]-HSQC spectroscopy by introducing a long frequency selective ^{15}N pulse in the standard pulse sequence. Again, in the 2D-spectra cross peaks can be identified that are involved in chemical exchange. Surprisingly, its application to *H*-Ras(1–171)·Mg^{2+}·GTP led to an exchange rate that is with 72.3 s^{-1} almost one order of magnitude larger (Long et al. 2013). One reason could be that the CEST-experiment did not identify the correct transition since detailed state information is not available from CEST experiments.

An alternative method that is also able to obtain information about exchange processes and thus the existence of more than one conformational state are CPMG relaxation dispersion measurements (Loria et al. 1999) that where applied to *H*-Ras(1–171)·Mg^{2+}·GppNHp by O'Connor and Kovrigin (2008). However, it is not usable when the population of one state is extremely low as in the case of the GTP complex.

9.6 Drug-Design and High Pressure NMR Spectroscopy

Permanent activation of the Ras pathway leads to cell proliferation, a hallmark of tumor development. Looking at the scheme presented in Fig. 9.3 it is obvious that a new type of allosteric inhibition we call intrinsic allosteric inhibition is feasible and is in principle applicable to all proteins involved in signal transduction pathways (Kalbitzer et al. 2013; Kalbitzer and Spoerner 2013). It is obvious that the exclusive stabilization of one of the many structural states by selective binding of a small compound to that state will also modify the populations of all other states and thus the activity linked to them. For this mechanism it is not required that the small ligand binds on the interaction site of another protein such as the effector protein but it has to bind state selective. We have already shown that this mechanism is working since Zn^{2+}-cyclen binds selectively to Ras·Mg^{2+}·GppNHp in state 1(T), the weak effector binding state (Spoerner et al. 2005b). As to be expected the effector Raf has a substantially reduced affinity to activated Ras in the presence of this compound (Rosnizeck et al. 2010). Chemical shift perturbation studies reveal that this compound binds to the γ-phosphate of the bound nucleotide inside the nucleotide-binding pocket. Shortly after that we found a different compound, Zn^{2+}-BPA (bis(2-piclolyl)amine), that binds at a different position outside the active center but nevertheless mainly recognizes state 1(T). Again it suppresses the effector interaction (Rosnizeck et al. 2012).

a

MTEYKLVVVG AGGVGKSALT IQLIQNHFVD EYDPTIEDSY
RKQVVIDGET CLLDILDTAG QEEYSAMRDQ YMRTGEGFLC
VFAINNTKSF EDIHQYREQI KRVKDSDDVP MVLVGNKCDL
AARTVESRQA QDLARSYGIP YIETSAKTRQ GVEDAFYTLV
REIRQH

b

MTEYKLVVVG AGGVGKSALT IQLIQNHFVD EYDPTIEDSY
RKQVVIDGET CLLDILDTAG QEEYSAMRDQ YMRTGEGFLC
VFAINNTKSF EDIHQYREQI KRVKDSDDVP MVLVGNKCDL
AARTVESRQA QDLARSYGIP YIETSAKTRQ GVEDAFYTLV
REIRQH

c

MTEYKLVVVG AGGVGKSALT IQLIQNHFVD EYDPTIEDSY
RKQVVIDGET CLLDILDTAG QEEYSAMRDQ YMRTGEGFLC
VFAINNTKSF EDIHQYREQI KRVKDSDDVP MVLVGNKCDL
AARTVESRQA QDLARSYGIP YIETSAKTRQ GVEDAFYTLV
REIRQH

Fig. 9.10 Residues involved in different transitions in Ras. Grey areas correspond to known motifs, P-loop, PM1 motif (G10xxxxGKS17), G1-motif (F28), switch I (30–40), PM2 (T35), PM3 (D57xxGQ), switch II (60–76), G2-motif (N116KxD), G3 motif, (S145AK147). Highlighted are the residues involved in (**a**) the 2(T)-1(T) transition, in (**b**) the 2(T)-3(T) transition, and (**c**) in the 2(T)-1(0) transition (see Kalbitzer et al. 2013)

The question arises how to identify possible state specific binding sites. Here, an analysis of the high pressure response of the protein may help since residues can be identified that are predominantly influenced by a given transition. We have plotted the residues that are mainly influenced by a given transition as function of the position in the amino acid sequence (Fig. 9.10). In addition, the known structural and functional motifs are also indicated in Fig. 9.10. The residues sensible to pressure are located in known functional regions such as switch I and switch II as well as outside these regions. Another example is the 2(T) to 1(0) transition that is the transition to the nucleotide release state. It is also visible in the G3-motif (S145, A146, K147) known to be involved in an interaction of the base of the nucleotide.

Meanwhile, a number of small compounds have been published that either inhibit the effector binding or they influence the GEF-assisted nucleotide exchange. For Zn^{2+}-cyclen and Zn^{2+}-BPA we have shown that they stabilize state 1(T) but also lead to an increase of the population of state 1(0). Zn^{2+}-cyclen interacts with Gly12, Asp33, Thr35, and Ala59; unfortunately all these residues except Asp33 are not visible in the NMR-spectra of wild type Ras·Mg^{2+}·GppNHp. From the residues binding Zn^{2+}-BPA (Asp38, Ser39, Tyr40), only Ser39 is primarily influenced by the 2(T) to 1(0) transition.

Small molecules that inhibit the GDP to GTP exchange by the guanidine nucleotide exchange factor SOS were reported by several groups. One of them was DCAI (4,6-dichloro-2-methyl-3-aminoethyl-indole) interacting with Lys5, Leu6,

Val7, Ile55, Leu56, and Thr74 in Ras·Mg^{2+}·GDP (Maurer et al. 2012). None of these residues could be identified in Ras·Mg^{2+}·GppNHp as selectively influenced by a 2(T) to 1(T) or 2(T) to 1(0) transition. However, with Asp3, Cys51, Leu52, and Ile53 residues close to the interacting residues Lys5, Leu6, Val7, Ile55, and Leu56 are identified by high pressure NMR spectroscopy that are selective for the 1(0) and 1(T) state in Ras·Mg^{2+}·GppNHp, respectively. However, primarily a very close agreement is not to be expected since for the compound only the binding to the GDP state has been studied. Benzimidazole was reported to have similar effects by Maurer et al. (2012). Here, we could show experimentally by ^{31}P NMR spectroscopy that it induces a shift to state 1(T) when the GTP analog GppNHp is bound and thus stabilizes also a SOS binding state in triphospho nucleoside bound state (Kalbitzer et al. 2013; Kalbitzer and Spoerner 2013).

Sun et al. (2012) reported the SOS-inhibitor 2-((1*H*-Indol-3-yl)methyl)-3*H*-imidazo[4,5-*c*]pyridine that interacts with Lys5, Val7, Ser39, Asp54, Leu56, Tyr71, Thr74. Ser39 is characteristic for state 1(0), the nucleotide free state (Fig. 9.1) that could be permanently stabilized by SOS binding and thus could inhibit the nucleotide exchange. Also Lys5 and Val7 are close to Asp3 that selectively reacts in the transition to the nucleotide-free state 1(0) of Ras. Bisphenol introduced by Schöpel et al. (2013) as SOS inhibitor recognizes essentially the same binding pocket as the inhibitors described above.

9.7 Summary

To summarize, high pressure NMR spectroscopy provides the possibility to observe different conformational states of a protein that are involved in function or folding. Both processes are intimately linked together as derived from the free energy landscape model. For an analysis of the high pressure data it is necessary to carefully observe chemical shift and cross peak volume changes and to evaluate them with an appropriate model. The obtained data can be used to devise state specific inhibitors of functional proteins.

References

Adjei AA (2001) Blocking oncogenic Ras signaling for cancer therapy. J Natl Cancer Inst 93:1062–1074

Arnold MR, Kremer W, Luedemann H-D, Kalbitzer HR (2002) ^{1}H NMR parameters of common amino acid residues measured in aqueous solutions of the linear tetrapeptides Gly-Gly-X-Ala at pressures between 0.1 and 200 MPa. Biophys Chem 96:129–140

Baines AT, Xu D, Der CJ (2011) Inhibition of Ras for cancer treatment: the search continues. Future Med Chem 3:1787–1808

Baskaran K, Brunner K, Munte CE, Kalbitzer HR (2010) Mapping of protein structural ensembles by chemical shifts. J Biomol NMR 48:71–83

Beck Erlach M, Koehler J, Moeser B, Horinek D, Kremer W, Kalbitzer HR (2014) Relationship between non-linear pressure induced chemical shift changes and thermodynamic parameters. J Chem Phys B 118:5681–5690

Bos JL (1989) Ras oncogenes in human cancer: a review. Cancer Res 49:4682–4689

Forsén S, Hoffman RA (1963) Study of moderately rapid chemical exchange reactions by means of nuclear magnetic double resonance. J Phys Chem 39:2892–2901

Friday BB, Adjei AA (2005) K-ras as a target for cancer therapy. Biochim Biophys Acta 1756:127–144

Geyer M, Schweins T, Herrmann C, Prisner T, Wittinghofer A, Kalbitzer HR (1996) Conformational transitions in p21ras and in its complexes with the effector protein raf-RBD and the GTPase activating protein GAP. Biochemistry 35:10302–10308

Heremans K, Smeller L (1998) Protein structure and dynamics at high pressure. Biochim Biophys Acta 1386:353–370

Herrmann C (2003) Ras-effector interactions: after one decade. Curr Opin Struct Biol 13:122–129

Iuga A, Spoerner M, Kalbitzer HR, Brunner E (2004) Solid-state 31P NMR spectroscopy of microcrystals of the Ras protein and its effector loop mutants: comparison of solution and crystal structures. J Mol Biol 342:1033–1040

Kalbitzer HR, Spoerner M (2013) State 1(T) inhibitors of activated Ras. In: Tamanoi F (ed) The enzymes, vol 33. Academic, Burlington, pp 69–94

Kalbitzer HR, Spoerner M, Ganser P, Hosza C, Kremer W (2009) Fundamental link between folding states and functional states of proteins. J Am Chem Soc 131:16714–16719

Kalbitzer HR, Rosnizeck IC, Munte CE, Puthenpurackal Narayanan S, Kropf V, Spoerner M (2013) Intrinsic allosteric inhibition of signaling proteins by targeting rare interaction states detected by high pressure NMR spectroscopy. Angew Chem Int Ed 52:14242–14246

Karnoub AE, Weinberg RA (2008) Ras oncogenes: split personalities. Mol Cell Biol 9:517–531

Koehler J, Beck Erlach M, Crusca E Jr, Kremer W, Munte CE, Kalbitzer HR (2012) On the pressure dependence of 15N chemical shifts in the model peptides Ac-Gly-Gly-X-Ala-NH2. Mater 5:1774–1786

Kremer W, Kachel N, Kuwata K, Akasaka K, Kalbitzer HR (2007) Species specific differences in the intermediate states of human and Syrian hamster prion protein detected by high pressure NMR spectroscopy. J Biol Chem 282:22689–22698

Long D, Marshall CB, Bouvignies G, Mazhab-Jafari MT, Smith MJ, Ikura M, Kay LE (2013) A comparative CEST NMR study of slow conformational dynamics of small GTPases complexed with GTP and GTP analogues. Angew Chem Int Ed 52:10107–10771

Loria JP, Rance M, Palmer AG (1999) A relaxation-compensated Carr-Purcell-Meiboom-Gill sequence for characterizing chemical exchange by NMR spectroscopy. J Am Chem Soc 121:2331–2332

Maurer T, Garrenton LS, Oh A, Keith P, Anderson DJ, Skelton NJ, Fauber BP, Pan B, Malek S, Stokoe D, Ludlam MJ, Bowman KK, Wu J, Giannetti AM, Starovasnik MA, Mellman I, Jackson PK, Rudolph J, Wang W, Fang G (2012) Small-molecule ligands bind to a distinct pocket in Ras and inhibit SOS-mediated nucleotide exchange activity. Proc Natl Acad Sci U S A 109:5299–5304

Monod J, Wyman J, Changeux JP (1965) On the nature of allosteric transitions: a plausible model. J Mol Biol 12:88–118

Neal S, Nip AM, Zhang H, Wishart DS (2003) Rapid and accurate calculation of protein ^{1}H, ^{13}C and ^{15}N chemical shifts. J Biomol NMR 2:215–240

O'Connor C, Kovrigin EL (2008) Global conformational dynamics in Ras. Biochemistry 47:10244–10246

Pai EF, Krengel U, Petsko GA, Goody RS, Kabsch W, Wittinghofer A (1990) Refined crystal structure of the triphosphate conformation of H-ras p21 at 1.35 a resolution: implications for the mechanism of GTP hydrolysis. EMBO J 9:2351–2359

Rajalingam K, Schreck R, Rapp UR, Albert S (2007) Ras oncogenes and their downstream targets. Biochim Biophys Acta 1773:1177–1195

Rosnizeck IC, Graf T, Spoerner M, Tränkle J, Filchtinski D, Herrmann C, Gremer L, Vetter IR, Wittinghofer A, König B, Kalbitzer HR (2010) Stabilizing a weak binding state for effectors in the human Ras-protein by cyclen complexes. Angew Chem Int Ed 49:3830–3833

Rosnizeck IC, Spoerner M, Harsch T, Kreitner S, Filchtinski D, Herrmann C, Engel D, König B, Kalbitzer HR (2012) Metal-bis(2-picolyl)amine complexes as state 1(T) inhibitors of activated Ras protein. Angew Chem Int Ed 51:10647–10651

Schöpel M, Jockers KFG, Düppe PM, Autzen J, Potheraveedu VN, Semra I, Yip KT, Heumann R, Herrmann C, Scherkenbeck J, Stoll R (2013) Bisphenol A binds to Ras proteins and competes with guanine nucleotide exchange: implications for GTPase-selective antagonists. J Med Chem 2013(56):9664–9672

Schumann FH, Riepl H, Maurer T, Gronwald W, Neidig K-P, Kalbitzer HR (2007) Combined chemical shift changes and amino acid specific chemical shift mapping of protein-protein interactions. J Biomol NMR 39:275–289

Spoerner M, Herrmann C, Vetter IR, Kalbitzer HR, Wittinghofer A (2001) Dynamic properties of the Ras switch I region and its importance for binding to effectors. Proc Natl Acad Sci U S A 98:4944–4949

Spoerner M, Wittinghofer A, Kalbitzer HR (2004) Perturbation of the conformational equilibria in Ras by selective mutations as studied by 31P NMR spectroscopy. FEBS Lett 578:305–310

Spoerner M, Nuehs A, Ganser P, Herrmann C, Wittinghofer A, Kalbitzer HR (2005a) Conformational states of Ras complexed with the GTP-analogs GppNHp or GppCH$_2$p: implications for the interaction with effector proteins. Biochemistry 44:2225–2236

Spoerner M, Graf T, König B, Kalbitzer HR (2005b) A novel mechanism for the modulation of the Ras-effector interaction by small molecules. Biochem Biophys Res Comm 334:709–713

Spoerner M, Nuehs A, Herrmann C, Steiner G, Kalbitzer HR (2007) Slow conformational dynamics of the guanine nucleotide-binding protein Ras complexed with the GTP analogue GTPγS. FEBS J 274:1419–1433

Spoerner M, Hozsa C, Poetzl JA, Reiss K, Ganser P, Geyer M, Kalbitzer HR (2010) Conformational states of rat sarcoma (Ras) protein complexed with its natural ligand GTP and their role for effector binding and GTP hydrolysis. J Biol Chem 285:39768–39778

Stumber M, Geyer M, Graf R, Kalbitzer HR, Scheffzek K, Haeberlen U (2002) Observation of slow dynamic exchange processes in Ras protein crystals by ^{31}P solid state NMR spectroscopy. J Mol Biol 323:899–907

Sun Q, Burke JP, Phan J, Burns MC, Olejniczak ET, Waterson AG, Lee T, Rossanese OW, Fesik SW (2012) Discovery of small molecules that bind to K-Ras and inhibit Sos-mediated activation. Angew Chem Int Ed 51:6140–6143

Wennerberg K, Rossman KL, Der CJ (2005) The Ras superfamily at a glance. J Cell Sci 118:843–846

Wittinghofer A, Waldmann H (2000) Ras-a molecular switch involved in tumor formation. Angew Chem Int Ed 39:4192–4214

Chapter 10
High-Pressure NMR Spectroscopy Reveals Functional Sub-states of Ubiquitin and Ubiquitin-Like Proteins

Ryo Kitahara

Abstract High-pressure nuclear magnetic resonance (NMR) spectroscopy has revealed that ubiquitin has at least two high-energy states – an alternatively folded state N_2 and a locally disordered state I – between the basic folded state N_1 and totally unfolded U state. The high-energy states are conserved among ubiquitin-like post-translational modifiers, ubiquitin, NEDD8, and SUMO-2, showing the E1-E2-E3 cascade reaction. It is quite intriguing that structurally similar high-energy states are evolutionally conserved in the ubiquitin-like modifiers, and the thermodynamic stabilities vary among the proteins. To investigate atomic details of the high-energy states, a Q41N mutant of ubiquitin was created as a structural model of N_2, which is 71 % populated even at atmospheric pressure. The convergent structure of the "pure" N_2 state was obtained by nuclear Overhauser effect (NOE)-based structural analysis of the Q41N mutant at 2.5 kbar, where the N_2 state is 97 % populated. The N_2 state of ubiquitin is closely similar to the conformation of the protein bound to the ubiquitin-activating enzyme E1. The recognition of E1 by ubiquitin is best explained by conformational selection rather than by induced-fit motion.

Keywords High-pressure NMR spectroscopy • High-energy states • Locally disordered state • Ubiquitin • NEDD8 • SUMO-2

10.1 Overview

Despite significant progress in our knowledge about basic folded conformations of proteins, high-Gibbs free energy states of proteins are poorly understood. We believe that high-pressure NMR spectroscopy has the potential to significantly improve our

R. Kitahara (✉)
College of Pharmaceutical Sciences, Ritsumeikan University, 1-1-1 Noji-Higashi, Kusatsu, Shiga 525-8577, Japan
e-mail: ryo@ph.ritsumei.ac.jp

K. Akasaka, H. Matsuki (eds.), *High Pressure Bioscience*,
Subcellular Biochemistry 72, DOI 10.1007/978-94-017-9918-8_10

understanding of rarely populated high-energy states of proteins. In this chapter, I will discuss the structure, dynamics, and thermodynamics of high-energy states of ubiquitin and ubiquitin-like proteins.

10.2 Introduction to High-Energy States in Proteins

A protein in solution is a thermodynamic entity existing as multiple conformers ranging from folded to unfolded states, and the Gibbs free energy difference between the thermodynamic states is a function of pressure and temperature when the amount of each component in solution is constant. In recent years, the presence of high-Gibbs free energy states of proteins, the so-called high-energy states, including the open conformation of enzymes, partially unfolded states, and molten globule states has been reported by using various NMR approaches, including hydrogen-deuterium exchange (Li and Woodward 1999), relaxation dispersion (Mulder et al. 2001; Boehr et al. 2006a; Boehr et al. 2006b; Bouvignies et al. 2011), and variable pressure-temperature NMR experiments (Jonas and Jonas 1994; Zhang et al. 1995; Inoue et al. 2000; Akasaka and Yamada 2001; Akasaka 2006; Li and Akasaka 2006; Kalbitzer et al. 2009; Roche et al. 2012; Kitahara et al. 2013). The high-energy states can be involved in protein functions, in particular in the process of molecular recognition, and may lead to protein aggregation such as amyloid fibril formation of proteins.

Although NMR spectroscopy, in particular relaxation dispersion techniques, is a powerful tool for the characterization of high-energy states of proteins in terms of kinetics and thermodynamics, the conventional NOE-based NMR approach for structure determination so far reveals atomic coordinates of the most stable conformation of a protein, the so-called "native state" or "ground state". Since the approach can be achieved by searching for the most provable single conformation satisfying distance and angle constraints obtained by NMR experiments, the realized structure is dominated in population by the basic folded conformer, which in most cases is at ambient pressure.

There is a practical rule for a globular protein in solution that *a partial molar volume of a protein decreases in parallel with the loss of conformational order* (Kitahara et al. 2002). Therefore, application of pressure leads to an increase in the population of high-energy states, which have fewer ordered (lower volume) conformations. NMR spectroscopy combined with pressure perturbation, or "high-pressure NMR spectroscopy", is consequently a novel and straightforward method for investigating the structure, dynamics, and thermodynamics of high-energy states of proteins.

10.3 Alternatively Folded State N_2 of Ubiquitin

Ubiquitin consists of 76 amino acids, and numerous studies have been conducted to determine its physical properties. Global unfolding of ubiquitin was investigated along temperature and pressure axes (Herberhold and Winter 2002). Although atomic details of folding and backbone dynamics of ubiquitin have also been examined by solution and solid-state NMR spectroscopy (Briggs and Roder 1992; Fu et al. 2012; Salvi et al. 2012; Tollinger et al. 2012; Fu and Wand 2013), high-energy states of the protein are still largely unknown. I have performed high-pressure NMR spectroscopy to explore the large conformational space of ubiquitin using atomic details, and have found two high-energy states, the alternatively folded state N_2 and the locally disordered state I, which lie between the basic folded state N_1 and totally unfolded U state (Kitahara et al. 2001; Kitahara and Akasaka 2003; Kitahara et al. 2005; Kitahara et al. 2006a). In this section, the structure and dynamic aspects of N_2 will be described.

We measured $^{1}H/^{15}N$ HSQC (heteronuclear single-quantum correlation) spectra of ubiquitin at different pressures. All HSQC cross-peaks showed continuous and reversible changes in chemical shifts. Several backbone ^{1}H and ^{15}N nuclei of the protein in the folded state show downfield linear shifts with increasing pressure. This general downfield tendency with pressure was also observed in ^{1}H and ^{15}N chemical shifts of various globular proteins (Kitahara et al. 2013). When an amide proton is involved in a hydrogen bond to a carbonyl oxygen or oxygen atom of solvent water, the downfield shifts originating from the deshielding of the proton by repulsion of the bonding electrons and bond magnetic anisotropy of the C=O double bond indicate a contraction of the hydrogen bond with increasing pressure (Asakura et al. 1995; Williamson et al. 1995). In the case of the backbone amide nitrogen, interpretation of the pressure-induced chemical shifts is more complicated since the ^{15}N chemical shifts are dependent on several factors including backbone torsion angles, side-chain conformation, and hydrogen bonding. However, a previous study has shown that the pressure-induced downfield shift is correlated with a decrease in the hydrogen bond N–O distance, and additionally is also strongly dependent on local variations in the ϕ and ψ torsion angles (Akasaka et al. 1999). According to theoretical analysis, the pressure-induced linear changes in the ^{1}H and ^{15}N shifts of BPTI (bovine pancreatic trypsin inhibitor) and lysozyme by 2 kbar correspond to a 0.01 Å contraction of the hydrogen bond and to a 3–4° change in the backbone dihedral angles, which are within the range of conformational fluctuations observed for the folded state ensemble of proteins in solution (Refaee et al. 2003; Williamson et al. 2003). Therefore, we think that the pressure-induced linear shifts observed for ubiquitin, which are the same order as those of BPTI and lysozyme, indicate a small increase in the population of a conformer with a smaller partial molar volume which is within the conformational ensemble of the basic folded state N_1.

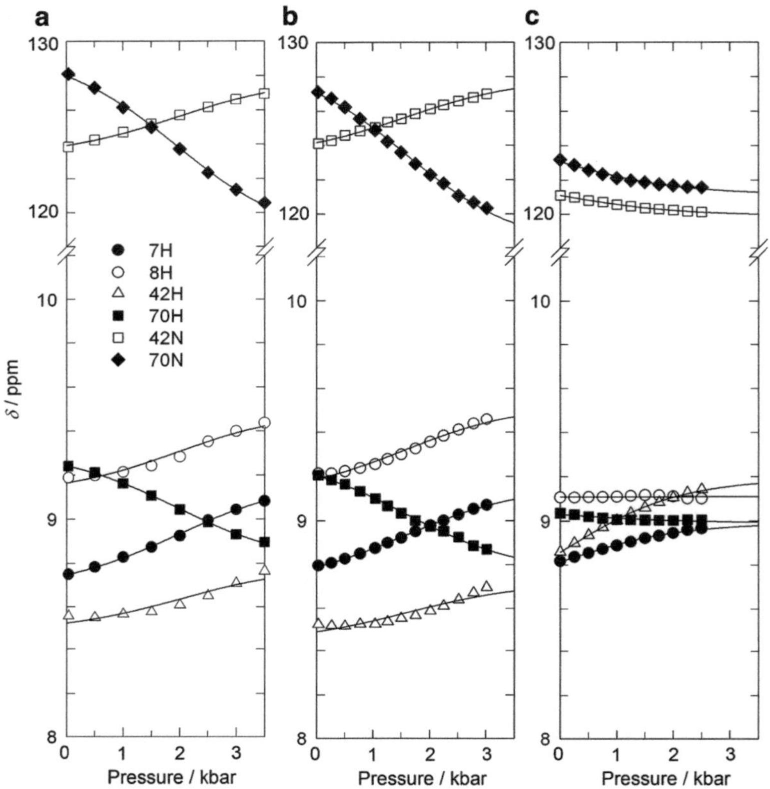

Fig. 10.1 Plots of ^{1}H and ^{15}N chemical shifts of amide groups of selected residues (7, 8, 42, and 70) as a function of pressure. (**a**) Wild-type ubiquitin at pH 4.7 and 293 K. (**b**) Wild-type ubiquitin at pH 7.2 and 298 K. (**c**) Q41N variant of ubiquitin at pH 7.2 and 298 K. The curves are obtained by global fitting of the chemical shifts (Kitazawa et al. 2013)

Several peaks showed nonlinear changes in the ^{1}H and ^{15}N chemical shifts as a function of pressure. In particular, sigmoid-shape changes were observed at several residues, namely 7, 8, 42, and 70, as shown in Fig. 10.1a and b (Kitahara et al. 2001; Kitazawa et al. 2013). Since the sigmoid-shape change is a representative of a two-state transition, we may assume that ubiquitin is in equilibrium with the basic folded state N_1 and alternatively folded state N_2. In general, non-linear shifts with pressure could be attributable to alternative conformers that have different chemical shifts and/or compressibility compared to the basic conformer. Assuming a two-state exchange between N_1 and N_2, a global fitting was performed for the protein using seven chemical shift data sets (^{1}H shifts of residues 7, 8, 33, 42, and 70 and ^{15}N shifts of residues 42 and 70), with four variables for each data set: chemical shifts for N_1 and N_2, the Gibbs free energy difference $\Delta G°$ between N_1 and N_2, and the partial molar volume difference $\Delta V°$ between N_1 and N_2. The global parameters $\Delta G°$ and $\Delta V°$ were estimated both at pH 4.7 (293 K) and pH 7.2 (298 K) (Table 10.1) and gave very similar values under the two conditions studied.

Table 10.1 Thermodynamic parameters for the transition from N_1 to N_2

	WT (pH4.7, 293 K)	Error	WT (pH 7.2, 298 K)	Error	Q41N (pH 7.2, 298 K)	Error
$\Delta G°$ (kJ/mol)	3.9	0.16	3.4	0.13	−2.2	0.3
$\Delta V°$ (mL/mol)	−22.7	0.9	−24.7	0.8	−24.7	–

For the analysis of Q41N, $\Delta V°$ value of WT at pH 7.2 (−24.7 mL/mol) is used, as a constant. Limiting chemical shifts for ^{1}H and ^{15}N of selected residues are listed in the literature (Kitazawa et al. 2013)

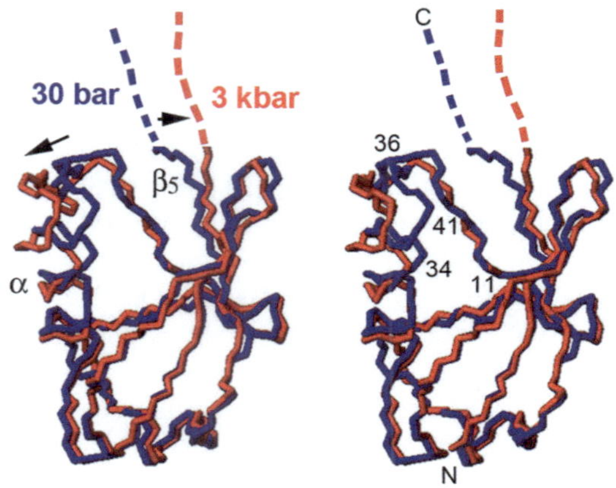

Fig. 10.2 A superposition of two NMR structures of wild-type ubiquitin (pH 4.6 and 293 K) at 30 bar (*blue*) and 3 kbar (*red*). Each structure is represented as an average of ten convergent structures (Kitahara et al. 2005). Displacement of the α-helix and the β_5-strand is indicated by arrows. The program MOLMOL was used to visualize NMR structures (Koradi et al. 1996)

At 30 bar and 3 kbar, where the N_2 conformer is 15 % and 77 % populated, respectively, ^{1}H/^{1}H two-dimensional NOESY experiments were performed at 293 K (Kitahara et al. 2005). Structure determination was performed by using the program, CYANA (Herrmann et al. 2002), with >1,000 distance constraints and 80 torsion angle constraints. Convergent structures were obtained at both pressures. Figure 10.2 shows a superposition of the representative structures at 30 bar and 3 kbar. The resultant structure at 3 kbar reveals a large displacement of the α-helix together with the following loop, as well as the C-terminal β_5-strand following residue 70, creating an opening of the core at the C-terminal face of the protein.

To investigate the backbone dynamics in more detail, ^{15}N-spin relaxation parameters, *i.e.* the ^{15}N-longitudinal relaxation rate R_1, the ^{15}N-transverse relaxation rate R_2, and heteronuclear steady-state ^{15}N-{^{1}H} NOE were measured for individual amide groups at 30 bar and 3 kbar (Kitahara et al. 2005). Model-free analyses of Lipari & Szabo (Lipari and Szabo 1982a; Lipari and Szabo 1982b) based on the spin relaxation parameters were also performed for the protein at both pressures. The ^{15}N-R_1 and NOE values report on the rapid segmental motions of the backbone in the picosecond to nanosecond range, and the order parameter S^2 represents a degree of spatial restriction of individual NH vectors against the ps-ns motions. The ^{15}N-R_1 and NOE values as well as the calculated S^2 values are similar at both pressures, indicating that the amplitude of the ps-ns motions for the individual amide groups of the protein is closely similar at both pressures. In contrast, the ^{15}N-R_2 values are more strongly related to the microsecond-to-millisecond time scale motions. We found increases in the exchange contributions to R_2, with R_{ex} values in the range 1–10 s^{-1} for residues 21–44 and 70 at 3 kbar, which nearly coincides with those residues showing large backbone displacements at 3 kbar. The coincidence of the results indicates that the increase in R_2 originates from conformational exchange between the N_1 and N_2 states.

10.4 The N_2 Model Mutant, Q41N Ubiquitin

Comparing the structures at 30 bar and 3 kbar, we found two key interactions for controlling the N_1 to N_2 conformational fluctuation; a hydrogen bond between the I36 carboxyl group and the Q41 side chain amide group and a salt bridge between the K11 amino group and the E34 carboxyl group (Fig. 10.2) (Kitazawa et al. 2013). To stabilize the N_2 state of ubiquitin by the point mutagenesis approach, we created mutants, K11A, E34A, Q41A, and Q41N and compared the ^1H and ^{15}N chemical shifts of the mutants with those of the wild-type (WT) protein. Chemical-shift changes by mutation were correlated among the variants, and the magnitude of the changes were consistently in the following order, K11A $\sim$ E34A < Q41A < Q41N. In addition, the chemical-shift changes by mutation were significantly correlated with those obtained by pressure perturbation of WT ubiquitin. This result strongly implies that an increase in the population of N_2 can be achieved independently by either progressing through the mutation series or via the application of pressure to WT ubiquitin.

To investigate the combined effect of these observations on the population of N_2, high-pressure NMR spectroscopy was conducted for the Q41N mutant of ubiquitin at pH 7.2 and 298 K. Figure 10.1c shows the ^1H and ^{15}N chemical shifts of the amide groups of selected residues as a function of pressure. The pressure-dependent chemical shifts in Q41N are simply non-linear, in contrast to those of the WT protein that are sigmoid-shaped. Since the transition curves of the residues in Q41N are highly similar to the latter half of those in WT, we considered that N_2 is predominantly populated in Q41N even at 1 bar. Assuming again a two-state exchange between N_1 and N_2, thermodynamic parameters for Q41N were obtained by using global fitting with three variables for each data set: chemical shifts for N_1 and N_2, and $\Delta G°$. Here, $\Delta V°$ between N_1 and N_2 for Q41N was assumed to be same as that for WT (*i.e.*, −24.7 mL mol^{-1}), because much less of the transition is experimentally accessible. The estimated $\Delta G°$ was -2.2 ± 0.3 kJ mol^{-1}, showing that the N_2 state is 71 % populated at 1 bar (Table 10.1). These results show that the Q41N mutant preferentially adopts the N_2 conformation of ubiquitin.

We conducted an NOE-based structural determination of the Q41N mutant at 2.5 kbar, where the N_2 state is 97 % populated (Kitazawa et al. 2014). This allowed us to characterize the structure of the "pure" N_2 state of ubiquitin. To obtain high-quality structural constraints, we used a ceramic pressure-resistant cell (Daedalus Innovations), which has a tenfold higher signal-to-noise ratio than that

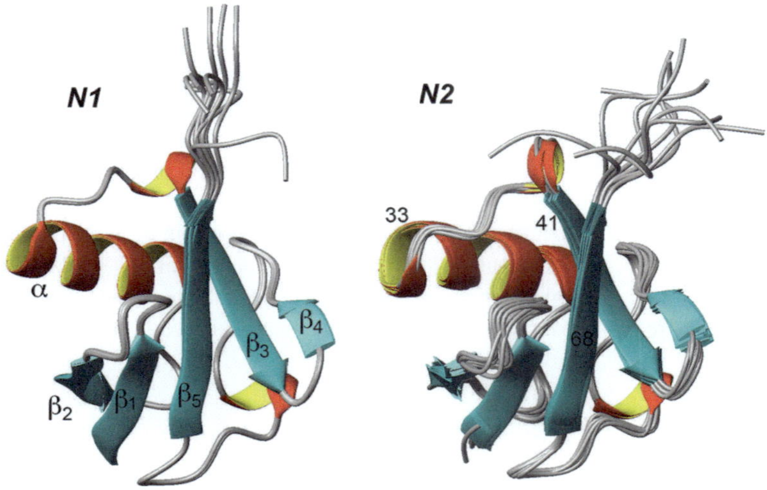

Fig. 10.3 Solution structures of wild-type ubiquitin at 1 bar (*left*, PDB entry 1D3Z), which is a model for N_1, and Q41N variant at 2.5 kbar (*right*, PDB entry 2RU6), which is a model for N_2 (Kitazawa et al. 2014). The program MOLMOL was used to visualize NMR structures (Koradi et al. 1996)

of our previous quartz capillary NMR cell. Structural calculation for $^{13}C/^{15}N$-double labeled Q41N ubiquitin was performed using the program CYANA3.93 (Herrmann et al. 2002) with 1,245 distance and 72 torsion angle constraints. The geometrical statistics showed that the quality of the structural models by high-pressure NMR spectroscopy is almost the same as that of the conventional 1 bar structure. Figure 10.3 shows the solution structures of WT ubiquitin at 1 bar, which is a model for the N_1 state, and Q41N at 2.5 kbar, which is a model for the N_2 state. Comparison of the structures reveals a large displacement of the C-terminal β_5-strand after residue 68, as well as a displacement of the α-helix and the following loop regions. These structural characteristics of Q41N under high-pressure conditions are consistent with the results obtained by residual dipolar couplings (Fu and Wand 2013) and hydrogen-bond scalar couplings (Nisius and Grzesiek 2012) for the WT protein at 2.5 kbar, as well as the NOE-based structure of the WT protein at 3 kbar (Kitahara et al. 2005).

10.5 Locally Disordered Conformation of Ubiquitin

To explore a much wider conformational space, we applied pressure on WT ubiquitin at a low temperature. Proteins generally undergo destabilization at low temperatures, which is also known as cold-denaturation (Privalov 1990; Nishii et al. 1994; Nash and Jonas 1997a; Nash and Jonas 1997b; Kunugi and Tanaka 2002;

Babu et al. 2004; Kitahara et al. 2006b). The partial molar volume difference between the folded and unfolded states is negatively larger at lower temperatures because the expansivity (temperature dependence of the partial molar volume) of the unfolded state is larger than that of the folded state (Mitra et al. 2008). On considering these advantages at low temperature, we have observed the locally disordered state I and totally unfolded U state of ubiquitin along the pressure-axis experiment.

Figure 10.4 shows $^{1}H/^{15}N$ HSQC spectra of WT ubiquitin at various pressures between 30 bar and 3.7 kbar at 273 K (Kitahara and Akasaka 2003). All of the cross-peaks show reversible changes in chemical shifts and intensities with pressure. Similar to the results observed at 293 K, non-linear pressure-induced changes in

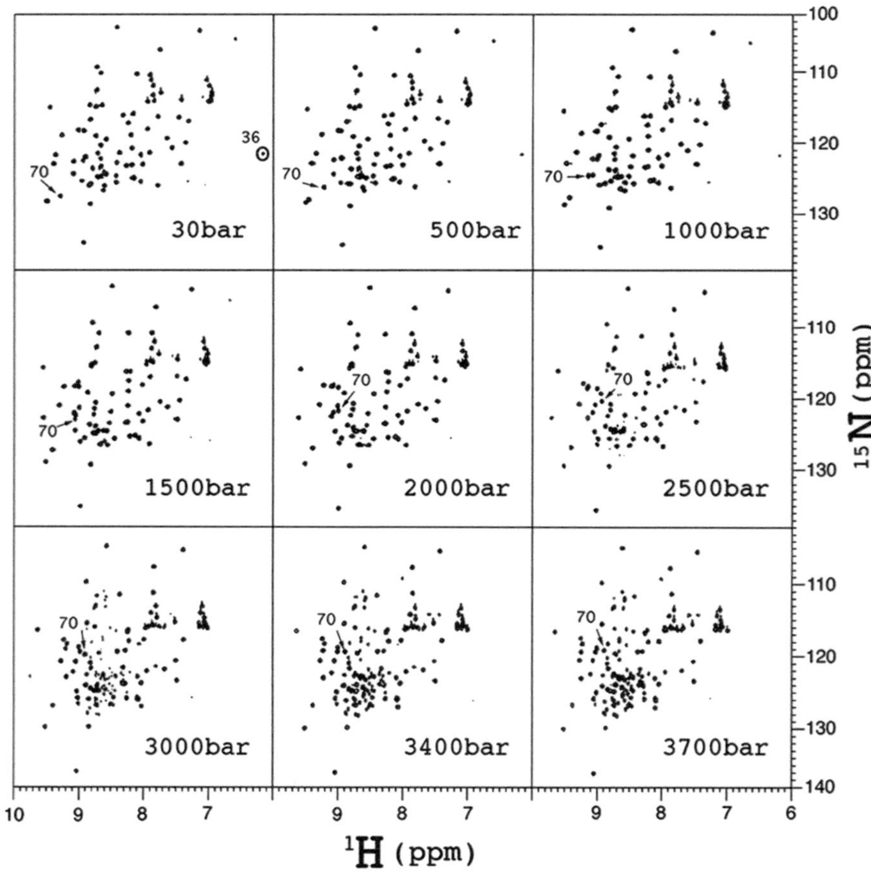

Fig. 10.4 $^{1}H/^{15}N$ HSQC spectra of ^{15}N uniformly labeled ubiquitin at various pressures from 30 bar to 3.7 kbar at 273 K. The cross-peak of 36 showing preferential signal broadening below 1.5 kbar is encircled. The cross-peak of 70 is marked with arrows (Kitahara and Akasaka 2003) (Copyright 2005 National Academy of Sciences)

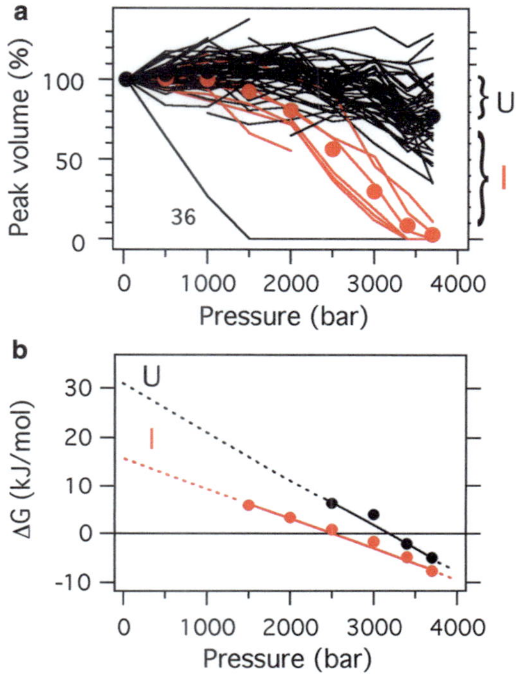

Fig. 10.5 (**a**) Plots of cross-peak volumes of all observable signals for the backbone amide group in ^{1}H/^{15}N HSQC spectra at various pressures. (**b**) Plots of ΔG_{NI} and ΔG_{NU} as a function of pressure (Kitahara and Akasaka 2003). The unfolding transition curves are classified into two groups: one is the rapidly decaying group (*red*), and the other is a slowly decaying group (*black*). Average-values for each group are represented by closed circles. The averaged values were used for the thermodynamic analysis (Kitahara and Akasaka 2003) (Copyright 2005 National Academy of Sciences)

the ^{1}H and ^{15}N chemical shifts were detected for the cross-peaks, indicating an increase in the N_2 conformation with pressure. In addition, changes in cross-peak intensities were observed with further increases in pressure. Figure 10.5a shows the changes in intensities (peak-volumes) of all observable original signals for backbone amide groups in the HSQC spectra. Intensities of residues 8, 33, 35, 36, and 39–42 of the folded conformations were preferentially decreased above 2 kbar and became null at 3.7 kbar (red profiles), although other peaks remained at 80 % of their original values (black profiles). Concomitantly with the decrease in the cross-peak intensities, new peaks appeared above 2 kbar in a narrow spectral region typical for disordered polypeptide chains. These results indicate that local disordering at residues 33–42 takes place above 2 kbar, and a total disordering of the entire polypeptide chain occurs with further increases in pressure. Since two sets of separate signals were observed in the HSQC spectra for these residues, the conformational transition between the folded state N (N_1 and N_2) and locally disordered state I takes place at a slower time scale than the chemical shift difference between the states ($\tau >>$ milliseconds; slow-exchange regime). At 1 bar, the locally disordered conformer I was undetectable in HSQC spectra. However at elevated pressures, increases in the population were observed driven by an apparent smaller partial molar volume in solution compared with that of the basic folded conformer.

For thermodynamic analysis, the unfolding transition curves were classified into two groups: the rapidly decaying group (Fig. 10.5a, red profiles), and the

Table 10.2 Thermodynamic parameters for the transitions from N to I and from N to U

	Ubiquitin (pH 4.7, 273 K)	Error	NEDD8 (pH 6.0, 303 K)	Error	SUMO-2 (pH 7.0, 298 K)	Error
$\Delta G°_{NI}$ (kJ/mol)	15	1	<5	3	9.1	0.7
$\Delta V°_{NI}$ (mL/mol)	−58	4	<−70	20	−44	3
$\Delta G°_{NU}$ (kJ/mol)	31	5	11	2	23	2
$\Delta V°_{NU}$ (mL/mol)	−85	7	−70	9	−85	8

slowly decaying group (Fig. 10.5a, black profiles, excluding data for 36). The rapidly and slowly decaying groups showed a decrease in the population of folded conformers (N_1 and N_2) and an increase in the population of the totally unfolded U state, respectively. The difference in behavior between the rapidly and slowly decaying curves is indicative of a conformer I population. We can estimate the proportion of N, I, and U conformers based on the average intensities of each group, and calculate changes in the Gibbs energy difference (ΔG) from N to I and from N to U using the function of pressure. Linear extrapolation of ΔG to 1 bar results in $\Delta G°_{NI} = 15 \pm 1$ kJ mol^{-1} and $\Delta G°_{NU} = 31 \pm 5$ kJ mol^{-1}, assuming a zero change in compressibility, and $\Delta V°_{NI} = -58 \pm 4$ mL mol^{-1} and $\Delta V°_{NU} = -85 \pm 7$ mL mol^{-1} (Fig. 10.5b, Table 10.2). The populations of the locally disordered state I and the totally unfolded U state are predicted to be 0.1 % and 10^{-4} %, respectively, at 1 bar and 273 K.

10.6 Evolutionally Conserved High-Energy States Among Ubiquitin-Like Proteins

I extended the high-pressure NMR study to NEDD8, SUMO-2, and ubiquitin-like domain (UBL) of parkin (Hochstrasser 2000). Despite the fairly modest sequence identity of NEDD8, SUMO-2, and parkin-UBL with ubiquitin (57 %, 18 %, and 32 %, respectively), these proteins and domain have identical folding topologies to ubiquitin. NEDD8 and SUMO-2 act as post-translational modifiers through the E1-E2-E3 cascade reaction, which is similar to ubiquitin, whereas parkin-UBL inhibits the autoubiquitination of parkin. Here, we discuss characteristics of structure, dynamics, and thermodynamics of ubiquitin-like proteins.

Firstly, we discuss the N_1 to N_2 conformational fluctuation of NEDD8 and SUMO-2. We investigated the backbone dynamics of ^{15}N-labeled NEDD8 (Kitahara et al. 2006b) and SUMO-2 (Kitahara et al. 2008) using ^{15}N-spin relaxation measurements of individual amide groups and obtained values for the ^{15}N-longitudinal relaxation rate R_1, the ^{15}N-transverse relaxation rate R_2, and the heteronuclear steady-state ^{15}N-{^{1}H} NOE. In the case of NEDD8, the model-free analysis was performed (Lipari and Szabo 1982a; Lipari and Szabo 1982b). The order parameter S^2 values that show the amplitude of ps-ns motions at 1 bar were almost the same as those of ubiquitin at 30 bar and 3 kbar. Interestingly, the R_2 values were larger (10–20 s^{-1}) than the average (7 s^{-1}) for residues 33–42, 68, 70, and 72, even

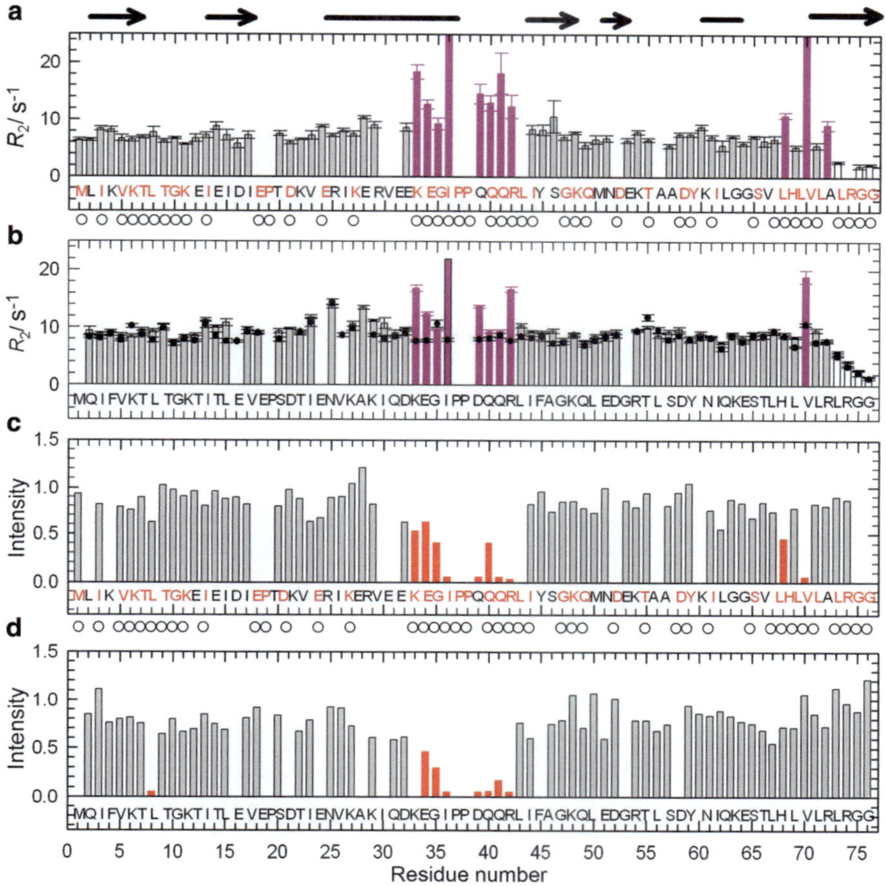

Fig. 10.6 Comparison of conformational fluctuations between NEDD8 and ubiquitin. (**a**) ^{15}N-R_2 values for NEDD8 at 1 bar and 303 K. (**b**) ^{15}N-R_2 values for ubiquitin at 30 bar (*circles*) and 3 kbar and 293 K. Quantitative evaluation of R_2 values for residues 36 and 70 of NEDD8 and for residue 36 of ubiquitin at 3 kbar were prohibited because of severe line broadening, but they are predicted to be more than 22 s^{-1}. The residues showing larger R_2 values are colored purple (residues 33–42, 68, 70, and 72 for NEDD8, residues 33–42 and 70 for ubiquitin). Types of amino acids are shown at the bottom of each panel. The residues indicated by open circles (and *red color*) are conserved amino acid sequences between human NEDD8 and ubiquitin. (**c and d**) Cross-peak intensities for individual residues in the ^{1}H/^{15}N HSQC spectra of NEDD8 at 1 kbar and 303 K (**c**) and of ubiquitin at 3.7 kbar and 273 K (**d**), each normalized to the intensity at 30 bar. The residues showing rapid decrease of the cross-peak intensities with increasing pressure are colored red. The positions of α-helix (*lines*) and β-strand (*arrows*) are shown at the top of the figure (Kitahara et al. 2006b)

at 30 bar. The larger R_2 values can be explained by larger R_{ex} contributions to R_2. Figure 10.6a and b show the comparison of R_2 values between NEDD8 and ubiquitin. Most of the residues showing larger R_2 values in NEDD8 at 1 bar match those of ubiquitin at 3 kbar. The consistency of the residues strongly indicates that the N_1 to N_2 conformational fluctuation involving the displacement of segments with residues 33–42, 68, 70, and 72 occurs in NEDD8, as well as in ubiquitin. Because the exchange contribution is significant in NEDD8 even at 1 bar, it is likely that the alternatively folded state N_2 is more populated in NEDD8 than it is in ubiquitin at 30 bar.

In the case of SUMO-2, residue-specific increases in R_2 values (*i.e.* R_{ex} contributions) were not observed at 30 bar and 1 kbar. Unfortunately, relaxation measurements could not be extended beyond 1 kbar for SUMO-2 as a result of substantial decreases in cross-peak intensities. Within the limited pressure range sampled, SUMO-2 does not significantly populate alternative high-energy conformations. However, alternative conformations have been observed in temperature-axis experiments for SUMO-1, which has a 47 % sequence identity and similar topology to SUMO-2 (Kumar et al. 2007). Further examinations are required to detect the occurrence of an alternatively folded state N_2, similar to those observed for ubiquitin and NEDD8.

Secondly, we discuss the locally disordered conformers I of NEDD8 and SUMO-2. For these proteins, Fig. 10.7a and b show the changes in intensities of all observable cross-peaks in $^1H/^{15}N$ HSQC spectra recorded at different pressures. With increasing pressure, significant decreases in cross-peak intensities were observed for several residues in both proteins. Concomitantly, several new cross-peaks appeared in the spectral region corresponding to a typical disordered polypeptide chain. These results indicate that a population of a locally disordered conformation increases with the application of pressure. For instance, Fig. 10.6c and d show cross-peak volumes for individual residues in the HSQC spectra of NEDD8 at 1 kbar and 303 K and of ubiquitin at 3.7 kbar and 273 K, with each normalized to the volume at low pressure (1 bar or 30 bar). The residues showing the formation of a disordered conformation in NEDD8 under high pressure are highly similar to those observed for ubiquitin. Furthermore, these residues involved in a transition to a locally disordered state I map to almost identical regions of the basic folded N_1 state for ubiquitin, NEDD8, and SUMO-2 (Fig. 10.7c). Since the locally disordered conformer I was not detected in parkin-UBL, this high-energy intermediate is only common to the ubiquitin-like post translation modifiers.

Finally, we discuss the stabilities of the locally disordered state I of ubiquitin, NEDD8, and SUMO-2. Based on the pressure-induced changes in cross-peak intensities, as in the case for ubiquitin, the unfolding transition curves were classified into two groups. We calculated the changes in the Gibbs energy difference (ΔG) from N to I and from N to U as a function of pressure. By linear extrapolation of ΔG to 1 bar, ΔG°_{NI} and ΔG°_{NU} were estimated together with ΔV°_{NI} and ΔV°_{NU} for each transition. The data for ubiquitin, NEDD8, and SUMO-2 are summarized in Table 10.2. The ΔV°_{NI} values for the three proteins were comparable, indicating that the structural changes for the $N_{(1+2)}$ to I transition are fairly similar among these proteins. Because local disordering is mainly focused at regions proximate to the C-terminal face, a decrease in the partial molar volume can be readily achieved by using a common opening mechanism together with increased solvation of these segments. In contrast, the ΔG°_{NI} values vary among the proteins, which represents

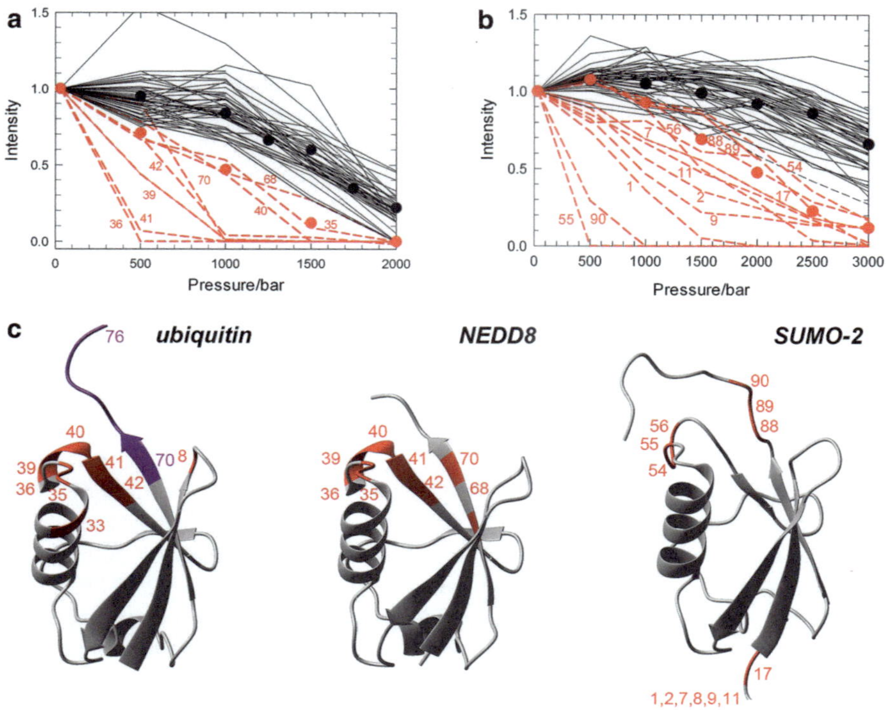

Fig. 10.7 (**a** and **b**) Plots of intensities (volumes normalized at 30 bar) of all observable cross-peaks for the backbone amide groups in the $^{1}H/^{15}N$ HSQC spectra of NEDD8 (**a**) and SUMO-2 (**b**). The rapidly decaying group is indicated by red broken lines. The locally unfolded conformer is apparently heterogeneous so that we can only evaluate some averages of stabilities. Average intensities corresponding to the rapidly decaying group (*red circles*) and showing a fairly cooperative transition are obtained by using values for residues 35, 40, and 68 in NEDD8 and residues 54, 56, 88, and 89 in SUMO-2. (**c**) Common conformational fluctuations among ubiquitin (*left*), NEDD8 (*center*), and SUMO-2 (*right*). The regions showing characteristic disorder are indicated by red and purple residue numbers (Kitahara, et al. 2006b; Kitahara et al. 2008). The program MOLMOL was used to visualize NMR structures (Koradi et al. 1996)

large differences in the equilibrium population of the locally disordered state I. These results indicate that these proteins are structurally similar, whereas their thermodynamic stabilities are different.

10.7 Functional Significance of High-Energy States

We have presented high-pressure NMR data which shows that ubiquitin has at least two high-energy states, an alternatively folded state N_2 and a locally disordered state I, between the basic folded state N_1 and totally unfolded U state. The high-energy states that are evolutionary conserved among the ubiquitin-like post-translational modifiers and which engage in the E1-E2-E3 cascade reaction are structurally similar but different in thermodynamic stabilities (Kitahara et al. 2006b, Kitahara et al. 2008). Here we discuss the functional significance of the high-energy states.

The N_2 conformer was observed in ubiquitin and NEDD8. Surprisingly, the N_2 structure of ubiquitin adopted the same conformation as that observed during the binding of the protein to E1, but was not seen in complexes with E2 or other ubiquitin-interacting proteins and motifs. In addition, similar structural changes were observed during the binding of NEDD8 to the NEDD8-activating enzyme E1. Therefore, we strongly suggest that N_2 would be involved in the E1-E2-E3 cascade reaction commonly conserved in the ubiquitin-like post-translational modifiers. N_2 in the structural ensembles of these proteins would thus serve as a prerequisite for the binding of the proteins to their E1, and that the recognition of E1 by the proteins is best explained by conformational selection rather than by induced-fit motion (Kitazawa et al. 2014).

Finally, We briefly refer to the functional significance of conformer I. The atomic coordinates of conformer I are unknown; so far we have identified segments showing local disordering in conformer I (Kitahara and Akasaka 2003). The residues undergoing local disordering, namely residues 8 and 33–42 in ubiquitin, were almost coincident with the residues involved in the N_1 to N_2 conformational fluctuation, although the time scales taking place in the fluctuations were very different, namely 1–10 μs for the N_1 to N_2 fluctuation and slower time scales greater than milliseconds for the $N_{(1+2)}$ to I fluctuations. The consistency in the regions for the conformational fluctuations implies that the larger amplitude motion, namely to the transition into conformer I, may follow the N_1 to N_2 transition. We also revealed that conformer I is conserved in the post-translational modifiers, ubiquitin, NEDD8, and SUMO-2. Since the C-terminal surface of the protein is involved in interactions with E1 and E2 proteins, the high-energy conformation having largely flexible C-terminal face could favor the binding reaction kinetically, compared with the basic folded state N_1, if it imposed a smaller activation energy to the binding transition.

Acknowledgement The author would like to thank all the collaborators in the study of ubiquitin and ubiquitin like proteins. This work was supported by a Grant-in-Aid for Young Scientists(B) (25840025) and a Grant-in-Aid for Scientific Research on Innovative Areas (23107729) from MEXT/JSPS of Japan to R.K.

References

Akasaka K (2006) Probing conformational fluctuation of proteins by pressure perturbation. Chem Rev 106:1814–1835

Akasaka K, Yamada H (2001) On-line cell high-pressure nuclear magnetic resonance technique: application to protein studies. Methods Enzymol 338:134–158

Akasaka K, Li H, Yamada H, Li R, Thoresen T, Woodward CK (1999) Pressure response of protein backbone structure. Pressure-induced amide ^{15}N chemical shifts in BPTI. Protein Sci 8:1946–1953

Asakura T, Taoka K, Demura M, Williamson MP (1995) The relationship between amide proton chemical shifts and secondary structure in proteins. J Biomol NMR 6:227–236

Babu CR, Hilser VJ, Wand AJ (2004) Direct access to the cooperative substructure of proteins and the protein ensemble via cold denaturation. Nat Struct Mol Biol 11:352–357

Boehr DD, Dyson HJ, Wright PE (2006a) An NMR perspective on enzyme dynamics. Chem Rev 106:3055–3079

Boehr DD, McElheny D, Dyson HJ, Wright PE (2006b) The dynamic energy landscape of dihydrofolate reductase catalysis. Science 313:1638–1642

Bouvignies G, Vallurupalli P, Hansen DF, Correia BE, Lange O, Bah A et al (2011) Solution structure of a minor and transiently formed state of a T4 lysozyme mutant. Nature 477:111–114

Briggs MS, Roder H (1992) Early hydrogen-bonding events in the folding reaction of ubiquitin. Proc Natl Acad Sci U S A 89:2017–2021

Fu Y, Wand AJ (2013) Partial alignment and measurement of residual dipolar couplings of proteins under high hydrostatic pressure. J Biomol NMR 56:353–357

Fu Y, Kasinath V, Moorman VR, Nucci NV, Hilser VJ, Wand AJ (2012) Coupled motion in proteins revealed by pressure perturbation. J Am Chem Soc 134:8543–8580

Herberhold H, Winter R (2002) Temperature- and pressure-induced unfolding and refolding of ubiquitin: a static and kinetic Fourier transform infrared spectroscopy study. Biochemistry 41:2396–2401

Herrmann T, Guntert P, Wuthrich K (2002) Protein NMR structure determination with automated NOE assignment using the new software CANDID and the torsion angle dynamics algorithm DYANA. J Mol Biol 319:209–227

Hochstrasser M (2000) Evolution and function of ubiquitin-like protein-conjugation systems. Nat Cell Biol 2:E153–E157

Inoue K, Yamada H, Akasaka K, Herrmann C, Kremer W, Maurer T et al (2000) Pressure-induced local unfolding of the Ras binding domain of RalGDS. Nat Struct Biol 7:547–550

Jonas J, Jonas A (1994) High-pressure NMR spectroscopy of proteins and membranes. Annu Rev Biophys Biomol Struct 23:287–318

Kalbitzer HR, Spoerner M, Ganser P, Hozsa C, Kremer W (2009) Fundamental link between folding states and functional states of proteins. J Am Chem Soc 131:16714–16719

Kitahara R, Akasaka K (2003) Close identity of a pressure-stabilized intermediate with a kinetic intermediate in protein folding. Proc Natl Acad Sci U S A 100:3167–3172

Kitahara R, Yamada H, Akasaka K (2001) Two folded conformers of ubiquitin revealed by high-pressure NMR. Biochemistry 40:13556–13563

Kitahara R, Yamada H, Akasaka K, Wright PE (2002) High pressure NMR reveals that apomyoglobin is an equilibrium mixture from the native to the unfolded. J Mol Biol 320:311–319

Kitahara R, Yokoyama S, Akasaka K (2005) NMR snapshots of a fluctuating protein structure: ubiquitin at 30 bar-3 kbar. J Mol Biol 347:277–285

Kitahara R, Okuno A, Kato M, Taniguchi Y, Yokoyama S, Akasaka K (2006a) Cold denaturation of ubiquitin at high pressure. Magn Reson Chem 44:S108–S113

Kitahara R, Yamaguchi Y, Sakata E, Kasuya T, Tanaka K, Kato K et al (2006b) Evolutionally conserved intermediates between ubiquitin and NEDD8. J Mol Biol 363:395–404

Kitahara R, Zhao C, Saito K, Koshiba S, Ioune M, Kigawa T et al (2008) Basic folded and low-populated locally disordered conformers of SUMO-2 characterized by NMR spectroscopy at varying pressures. Biochemistry 47:30–39

Kitahara R, Hata K, Li H, Williamson MP, Akasaka K (2013) Pressure-induced chemical shifts as probes for conformational fluctuations in proteins. Prog Nucl Magn Reson Spectrosc 71:35–58

Kitazawa S, Kameda T, Yagi-Utsumi M, Sugase K, Baxter NJ, Kato K et al (2013) Solution structure of the Q41N variant of ubiquitin as a model for the alternatively folded N_2 state of ubiquitin. Biochemistry 52:1874–1885

Kitazawa S, Kameda T, Kumo A, Yagi-Utsumi M, Baxter NJ, Kato K et al (2014) Close identity between alternatively folded state N_2 of ubiquitin and the conformation of the protein bound to the ubiquitin-activating enzyme. Biochemistry 53:447–449

Koradi R, Billeter M, Wuthrich K (1996) A program for display and analysis of macromolecular structures. J Mol Graph 14:51–55
Kumar A, Srivastava S, Hosur RV (2007) NMR characterization of the energy landscape of SUMO-1 in the native-state ensemble. J Mol Biol 367:1480–1493
Kunugi S, Tanaka N (2002) Cold denaturation of proteins under high pressure. Biochim Biophys Acta 1595:329–344
Li H, Akasaka K (2006) Conformational fluctuations of proteins revealed by variable pressure NMR. Biochim Biophys Acta 1764:331–345
Li R, Woodward C (1999) The hydrogen exchange core and protein folding. Protein Sci 8:1571–1590
Lipari G, Szabo A (1982a) Model-free approach to the interpretation of nuclear magnetic-resonance relaxation in macromolecules. 1. Theory and range of validity. J Am Chem Soc 104:4546–4559
Lipari G, Szabo A (1982b) Model-free approach to the interpretation of nuclear magnetic-resonance relaxation in macromolecules. 2. Analysis of experimental results. J Am Chem Soc 104:4559–4570
Mitra L, Rouget JB, Garcia-Moreno B, Royer CA, Winter R (2008) Towards a quantitative understanding of protein hydration and volumetric properties. Chemphyschem 9:2715–2721
Mulder FA, Mittermaier A, Hon B, Dahlquist FW, Kay LE (2001) Studying excited states of proteins by NMR spectroscopy. Nat Struct Biol 8:932–935
Nash DP, Jonas J (1997a) Structure of pressure-assisted cold denatured lysozyme and comparison with lysozyme folding intermediates. Biochemistry 36:14375–14383
Nash DP, Jonas J (1997b) Structure of the pressure-assisted cold denatured state of ubiquitin. Biochem Biophys Res Commun 238:289–291
Nishii I, Kataoka M, Tokunaga F, Goto Y (1994) Cold denaturation of the molten globule states of apomyoglobin and a profile for protein folding. Biochemistry 33:4903–4909
Nisius L, Grzesiek S (2012) Key stabilizing elements of protein structure identified through pressure and temperature perturbation of its hydrogen bond network. Nat Chem 4:711–717
Privalov PL (1990) Cold denaturation of proteins. Crit Rev Biochem Mol Biol 25:281–305
Refaee M, Tezuka T, Akasaka K, Williamson MP (2003) Pressure-dependent changes in the solution structure of hen egg-white lysozyme. J Mol Biol 327:857–865
Roche J, Dellarole M, Caro JA, Guca E, Norberto DR, Yang Y, Garcia AE, Roumestand C, Garcia-Moreno B, Royer CA (2012) Remodeling of the folding free energy landscape of staphylococcal nuclease by cavity-creating mutations. Biochemistry 51:9535–9546
Salvi N, Ulzega S, Ferrage F, Bodenhausen G (2012) Time scales of slow motions in ubiquitin explored by heteronuclear double resonance. J Am Chem Soc 134:2481–2484
Tollinger M, Sivertsen AC, Meier BH, Ernst M, Schanda P (2012) Site-resolved measurement of microsecond-to-millisecond conformational-exchange processes in proteins by solid-state NMR spectroscopy. J Am Chem Soc 134:14800–14807
Williamson MP, Kikuchi J, Asakura T (1995) Application of ^{1}H NMR chemical shifts to measure the quality of protein structures. J Mol Biol 247:541–546
Williamson MP, Akasaka K, Refaee M (2003) The solution structure of bovine pancreatic trypsin inhibitor at high pressure. Protein Sci 12:1971–1979
Zhang J, Peng X, Jonas A, Jonas J (1995) NMR study of the cold, heat, and pressure unfolding of ribonuclease A. Biochemistry 34:8631–8641

Chapter 11
Functional Sub-states by High-pressure Macromolecular Crystallography

Anne-Claire Dhaussy and Eric Girard

Abstract At the molecular level, high-pressure perturbation is of particular interest for biological studies as it allows trapping conformational substates. Moreover, within the context of high-pressure adaptation of deep-sea organisms, it allows to decipher the molecular determinants of piezophily. To provide an accurate description of structural changes produced by pressure in a macromolecular system, developments have been made to adapt macromolecular crystallography to high-pressure studies. The present chapter is an overview of results obtained so far using high-pressure macromolecular techniques, from nucleic acids to virus capsid through monomeric as well as multimeric proteins.

Keywords Diamond-anvil cell • High-pressure macromolecular crystallography • Pressure-induced conformers • Protein cavity

11.1 A Tribute

The present chapter is a tribute to Roger Fourme (1942–2012), Emeritus Professor at Paris Sud University.

Roger was behind the revival of high-pressure macromolecular crystallography (HPMX). At the beginning of 2000, together with Isabella Ascone, he launched the project with Richard Kahn (1946–2011) from Institut de Biologie Structurale (Grenoble) and the support of Mohamed Mezouar from ESRF (Grenoble). It is important to remind that the interest of Roger Fourme for high-pressure studies had existed well before. Indeed, he was one of the first to develop a diamond anvil cell in France for single crystals crystallography at variable temperature and variable

A.-C. Dhaussy
CRISMAT, ENSICAEN, 6 Boulevard du Marechal Juin, F-14000 Caen, France

E. Girard (✉)
University of Grenoble Alpes, IBS, F-38044 Grenoble, France

CNRS, IBS, F-38044 Grenoble, France

CEA, DSV, IBS, F-38044 Grenoble, France
e-mail: eric.girard@ibs.fr

K. Akasaka, H. Matsuki (eds.), *High Pressure Bioscience*,
Subcellular Biochemistry 72, DOI 10.1007/978-94-017-9918-8_11

Fig. 11.1 From *left* to *right*: E. Girard, R. Fourme and R. Kahn during the first HPMX experiments on CRISTAL beamline at SOLEIL in September 2011

pressure, with application on the polymorphism of molecular crystals, leading to a pioneer article in Journal of Applied Crystallography in 1968 (Fourme 1968). Together with Richard Kahn, he also developed the use of synchrotron radiation for macromolecular crystallography, first at LURE (Orsay, France), then at synchrotron SOLEIL where he was the first Scientific Director for Life Science. We had the great chance to work with the scientist but, more importantly, the great chance to know the man (Fig. 11.1).

11.2 Interest of High-Pressure in Biology and the Potential of High-Resolution Techniques

The field of high-pressure biology was launched in 1914 when Percy Bridgman reported that egg white coagulates under pressure (Bridgman 1914). The interest of pressure in biology was, for a long time, related to marine biology. Indeed, more than 60 % of the total volume of the biosphere is submitted to pressures higher than 10 MPa (Somero 1992). The question of life adaptation to extreme hydrostatic pressure conditions is still up-to-date as exemplified by the observation of fishes at 6,000 m depth (Jamieson et al. 2009) or the discovery of the first obligate piezophilic hyperthermophilic archaeon (Zeng et al. 2009; Xu et al. 2011). The latter, named *Pyrococcus yayanosii* CH1, is an organism that presents optimal growth at 50 MPa and 98 °C. More interestingly, it can withstand pressure up to 150 MPa and is unable to grow at atmospheric pressure. Even if we know that pressure influences living cells, the mechanisms by which an organism can counterbalance the effects of pressure are far from being understood (Oger and Jebbar 2010). Related to life adaptation to extreme conditions, the question of the origin of life and the potential role of pressure within the emergence of life on Earth is still open (Daniel et al. 2006). Consequently, it is interesting to study the proteins isolated from theses deep-sea organisms to understand piezophilic adaptation and the molecular determinants associated with it. Such studies may have important biotechnological implications (Rivalain et al. 2010).

The second main interest of pressure in biology, especially at the molecular level, is the exploration of the conformational landscape. In a modern view, a protein in solution is described using a multi-conformational free-energy landscape (Frauenfelder and McMahon 2000; Akasaka 2006). Proteins equilibrate among multiple conformational states including their basic folded state, *i.e.* "the native state", which is dominantly populated, as well as their unfolded states and different substates that include functionally and biologically important ones. These various substates differ substantially in their partial molar volume (Royer 2002). It should be noted that protein should be here understood as protein system including the protein molecule, but also solvent, ions and ligands that surround it. Pressure is a fundamental thermodynamic variable for defining and exploring the conformational landscape of a protein. Indeed, while entropy is the conjugate variable of temperature, volume is the conjugate variable for pressure. As volume is more intimately related with the structure of bio-macromolecules than entropy, pressure is an ideal tool to modify the partial molar volume of a protein system, allowing investigating the various conformational substates.

Getting details at atomic resolution of these pressure-induced substates as well as of proteins isolated from piezophilic organisms is thus of great importance and justifies the adaptation of high-resolution techniques such as solution NMR and macromolecular crystallography to high-pressure studies (Fourme et al. 2012a).

11.3 From the Early Days to Today: An Experimental Overview

To understand the potential of HPMX, it is important to have an overview of the experimental methods. Indeed, three methods have been used to perform high-pressure studies on proteins by X-ray crystallography. These methods can be divided in two categories: techniques performed at cryogenic temperature and those used at room temperature. This distinction has implications on the definition of the physico-chemical state observed for the macromolecules, as discussed in Sect. 11.3.3.

11.3.1 High-Pressure Cryocooling Method

Gruner's group at Cornell developed a high-pressure cryocooling device (Kim et al. 2005) based on the pioneering work of Thomanek et al. (1973). The crystal is compressed in helium gas and then dropped, still at high pressure, in a zone at liquid nitrogen temperature. Helium pressure is then released and the crystal is kept at cryogenic temperature. This method has been successfully applied to study the effects of hydrostatic pressure (Sect. 11.7.3) on the yellow fluorescent protein, citrine (Barstow et al. 2008, 2009), but it remains mainly exploited in the framework of cryopreservation of macromolecular crystals (Kim et al. 2005, 2013) for conventional protein crystallography.

11.3.2 Ambient Temperature High-Pressure Methods

Two techniques have been used to perform high-pressure crystallography on biological macromolecules. Both techniques run at room temperature.

The first one is based on the use of a Beryllium pressure cell. In the mid-1980s, Kundrot and Richards specifically designed a beryllium specimen chamber for X-ray crystallography (Kundrot and Richards 1986). They used it to determine the structure of hen egg-white lysozyme (HEWL) up to 100 MPa (Kundrot and Richards 1987). This study allowed to identify the most compressible regions within the protein and was completed by the analysis of pressure effects on solvent within the crystal (Kundrot and Richards 1988). Despite launching this new field of research, the Kundrot and Richards work had several unexpected consequences. First, based on their observations, the magnitude of the effects of pressure were found to be very small, generally less than an Angstrom, leading to the conclusion that such effects were not biologically relevant. However, we now know that such displacements around active site of proteins, for example, may greatly modify protein activity. In addition, their experience of crystals cracking during pressurization gave the impression that protein crystals are fragile. However, with the advance in the field, various experiments with dozen of crystals have shown recently that protein crystals are robust while pressurized (Collins et al. 2010; Nagae et al. 2012b).

The second method is the use of a diamond anvil cell (DAC). The first application of diamond anvil cell to HPMX was the determination of unit cell compressibility of orthorhombic as well as tetragonal crystal forms of HEWL, without subsequent data collection for structural study (Katrusiak and Dauter 1996). Since 2001, following Roger Fourme's leadership, using DAC specifically modified, as well as ultra-short wavelength X-rays, diffraction data on a variety of compressed macromolecular crystals has been routinely collected (Fourme et al. 2009, 2012a), showing that HPMX is a full-fledged technique (Girard et al. 2007a). Together with an increasing interest of pressure perturbation for molecular biophysics, these successful achievements led other groups to adopt a similar approach (Nagae et al. 2012b).

11.3.3 Comparison of HP-Cooling Method and DAC-Based Method

The high-pressure cooling method presents the major advantage of allowing conventional biocrystallography. Indeed, after cryo-cooling, the pressurized sample is mounted on a classical loop and can thus be manipulated by usual techniques. Consequently, data collection can be performed on conventional bio-crystallography beamlines with no limitations in terms of crystal symmetry and available rotation range. However, it should be kept in mind that the physico-chemical state of the sample reflects the high-pressure treatment combined to the cryocooling process.

In the case of DAC-based data collection, the sample is kept in equilibrium with pressure that is maintained constant during data collection. The physico-chemical state of the sample is therefore precisely defined as data collection is performed at room temperature. The latter can be considered as a disadvantage in terms of radiation damage. However, recent studies have shown the interest of considering data collection at room temperature in the trapping of conformational substates in X-ray crystallography (Fraser et al. 2009, 2011) and illustrated in the development of X-ray free-electron laser (Boutet et al. 2012; Liu et al. 2013).

The drawback of the DAC lies in its geometry that imposes constraints on data collection. The aperture of the pressure cell limits the available rotation range and may limit data resolution. However it is important to notice that this default is largely compensated by the use of a beam with energy between 30 and 40 keV, as proposed by Fourme et al. (2003, 2012b). Hence, the use of high-energy X-rays has a double advantage: first it reduces the absorption of the two diamond windows, but more importantly, it naturally increases the accessible reciprocal space allowing to reach very high resolution in the structural determination using DACs.

11.4 Protein Crystals Are Well Suited for High-Pressure Studies

Proteins crystals should not be considered as rigid static systems. Indeed, protein crystals are biphasic systems in which a solid phase, the protein, coexists with a solvent one made of channels that cross the crystals, which may represent 30–80 % of the total crystal volume (Fig. 11.2). These solvent channels communicate with the surrounding liquid in which the crystal is grown. Consequently, this situation ensures that a true hydrostatic pressure is applied to the protein.

The majority of protein crystals grow in low-symmetry space groups, mainly in monoclinic or orthorhombic lattices (Fig. 11.3). Consequently, getting a high-completeness data set on a single crystal, by the rotation method, is difficult with the limited aperture of DAC. To overcome this difficulty, a complete data set can be obtained by merging the data from several crystals. With the aim of limiting the number of crystals to be used, X-ray transparent tiny splinters (Girard et al. 2007a),

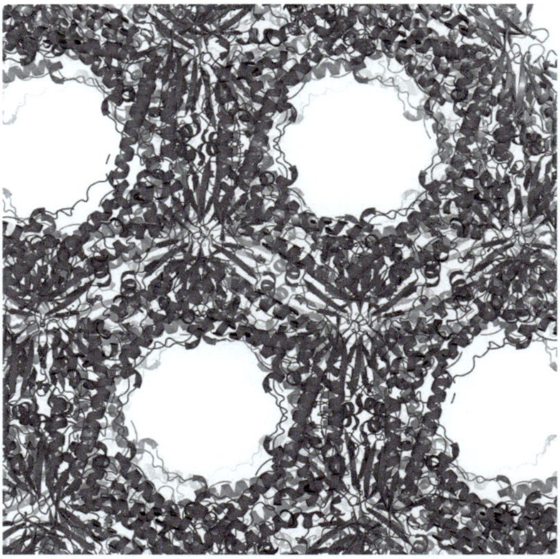

Fig. 11.2 Crystal packing in protein crystals. The protein region is surrounded by a solvent region that may include large channel that communicates with the exterior. The present figure is made using structure corresponding to PDB ID code: 2FM6

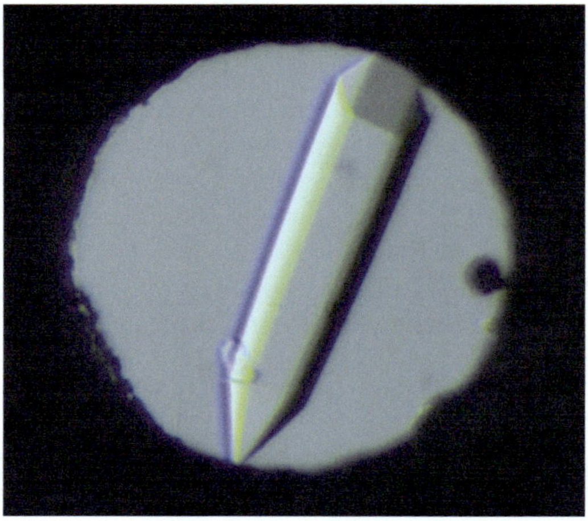

Fig. 11.3 Crystal of a protein isolated from the piezophilic organism *Pyrococcus yayanosii* CH1 loaded in an optimized pneumatic DAC specifically designed for HPMX with a geometrical aperture of 100° (Girard et al. 2010a). The sample cavity is 400 μm in diameter according to the 800 μm diamond culet diameter and 200 μm in depth, allowing to accommodate large crystals or several smaller ones. The cavity is filled with the mother liquor, which acts as the pressure-transmitting medium, prior to sample loading

such as small pieces of a broken diamond anvil can be introduced within the sample cavity to change the orientation of the crystal with respect to the diamond culets. Based on the same idea, Nagae et al. proposed to use cigarette-filter fibres to modify the sample orientation (Nagae et al. 2012b).

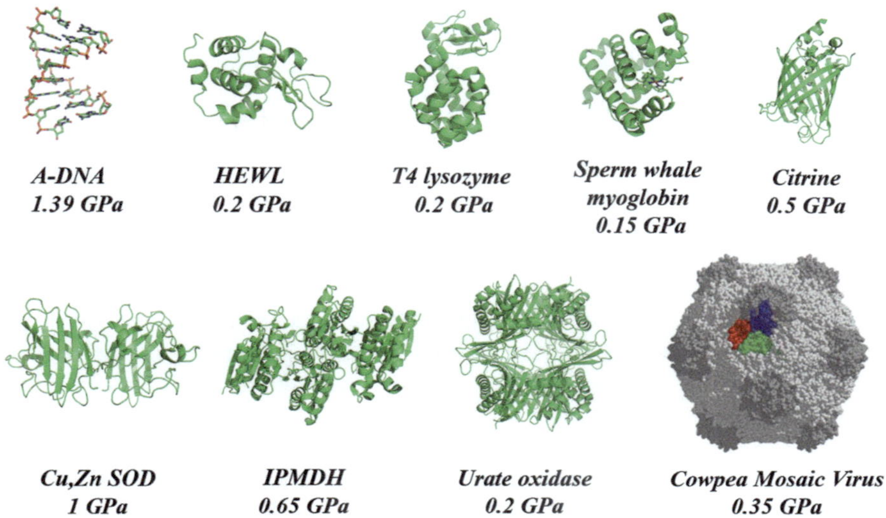

Fig. 11.4 Macromolecular structures investigated by HPMX

11.5 Macromolecular Systems Studied by High-Pressure Macromolecular Crystallography

Figure 11.4 summarizes the various biomolecules studied by HPMX. It includes nucleic acids, virus capsid, monomeric as well as multimeric proteins in a pressure range from 150 MPa to 1.39 GPa.

In the following paragraphs, the different high-pressure X-ray studies have been related to three main effects of pressure: elastic compression of biomolecules, trapping of conformational substates and, related with the previous one, the role of internal cavities in pressure response.

11.6 Elastic Compression Within a Single Conformational Substate

Pressure-induced modifications of the protein volume may arise from a global elastic compression within closely related conformational substates. These modifications originate from changes in the various interactions between amino acids (hydrogen bonds, ionic, van der Waals as well as hydrophobic interactions), from changes in the internal voids and cavities found in proteins, and from changes in the hydration properties (Li et al. 1998; Marchi and Akasaka 2001; Boonyaratanakornkit et al. 2002; Refaee et al. 2003; Girard et al. 2005; Nisius and Grzesiek 2012). The next three examples are related to the order of magnitude of such elastic compression on proteins as observed by HPMX.

11.6.1 Effects of Pressure on a Hen-Egg White Lysozyme: A Seminal Work

As already mentioned, the work by Kundrot and Richard on HEWL was the first structural study of pressure effects on a protein by X-ray crystallography. Besides setting the required instrumentation (Kundrot and Richards 1986), the authors analyzed the pressure-induced modifications of the protein as well as of the hydration shell (Kundrot and Richards 1987, 1988). First, the whole molecule volume decreased (estimated isothermal compressibility of 4.7×10^{-3} kbar^{-1}) but the compression depends on the domain. While domain 2 (residues 40–88) is incompressible, domain 1 (residues 1–39 and 89–129) together with the interdomain region are compressible. Kundrot and Richards analyzed the evolution of Debye-Waller factors, also called B-factors. B-factors can be taken as indicating the relative vibrational motion of different parts of the structure. Consequently, atoms with low B-factors can be considered to belong to a part of the structure that is well ordered, while atoms with higher B-factors generally belong to parts of the structure that are more flexible. In the case of HEWL, the B-factors decreased on average even if there was no clear individual trend. While the beta-sheet was only slightly deformed with increasing pressure, the three helices responded differently to pressure, as the volume of one increased, whilst another decreased, the third one remaining unchanged. Finally, contrary to what was expected, the two domains of HEWL did not move about a hypothetic hinge axis that would have closed the active site cleft. The Kundrot and Richards analysis of the effect of hydrostatic pressure on the solvent in HEWL crystals (Kundrot and Richards 1988) led to observe more water sites in the 100 MPa structure. They showed a tendency of these water sites to increase the number of hydrogen bonds made with other water molecules at the expense of hydrogen bonds made with the protein.

11.6.2 Effects of Pressure on a Viral Capsid: Elastic Compression on a Large System

The high-pressure study on the plant virus CpMV (Cowpea Mosaic Virus) is an example of elastic compression within the same conformational substate of a large assembly (Girard et al. 2005). Such macromolecular system was an incentive to get the order of magnitude of the effects of pressure on the different component of a protein assembly, as observed by HPMX. The CpMV structure at 330 MPa, obtained using the DAC method, was compared with the ambient structure.

This study showed that multimeric proteins can withstand large pressure values, as diffraction was observed up to 400 MPa for CpMV. CpMV is not an isolated case as exemplified by dimeric superoxide dismutase that can withstand pressure up to 1 GPa (see Sect. 11.7.2) or by the large dodecameric TET aminopeptidase from the archaeon *Pyrococcus horikoshii* that conserves its quaternary structure up to 350 MPa and 90 °C (Rosenbaum et al. 2012). Globally, the average B-factors for the 330 MPa structure were lower than those of the ambient pressure by about 8–9 $Å^2$, as previously observed for HEWL (Kundrot and Richards 1987) and twice as many water molecules were modelled in the high-pressure structure (Girard et al. 2005).

The CpMV structure comparison led to observe a shortening of H-bond lengths by about 0.1 Å GPa^{-1} on average (Girard et al. 2005). Even if this shortening can be considered as small, it is statistically relevant as it was calculated on a large set of measurements (about 600 including hydrogen bonds both between amino-acids and between amino-acids and water molecules). The value obtained is in agreement with the values determined by molecular dynamics (Paci and Marchi 1996; Marchi and Akasaka 2001; Paci 2002) or observed by high-pressure NMR (Li et al. 1998; Refaee et al. 2003). Moreover, the compression of the CpMV virus capsid is associated with a large reduction of the volume of the internal cavities. The total volume of internal cavities is reduced by 40 % at 330 MPa, corresponding to a mean value of 10 % per 100 MPa. However, this mean value does not reflect the individual behaviour of each cavity, as some shows a volume diminution greater than 40 % whereas others are only slightly affected by pressure.

The high-pressure study on CpMV led also to an unexpected observation that may be of interest in conventional macromolecular crystallography. Data resolution that can be obtained in a crystallographic experiment depends on the quality of crystal order. The quality of pre-existing crystals may be improved by pressure application within a DAC. Indeed, a spectacular effect was observed for CpMV crystals (Fourme et al. 2002; Girard et al. 2005). Low-diffracting crystals (diffraction limit of 4 Å at best) became good diffracting ones (resolution limit around 2.8 Å) when a pressure higher than 250 MPa was applied. The effects observed for CpMV are maybe a particular case, as pressure improves the packing of nearly spherical particles featuring some orientation disorders at atmospheric pressure (Lin et al. 2005). However, application of pressure might become a standard way of improving the long-range order of crystals, even if such application would require systematic studies to confirm the diffraction improvement observed for CpMV.

11.6.3 Effects of Pressure on Nucleic Acids: Prebiotic Chemistry

Single crystals of d(GGTATACC) octanucleotide were gradually compressed from ambient pressure to 2 GPa (Girard et al. 2007b). Four complete data sets at 1.60–1.65 Å resolution were collected at ambient pressure and at three pressures, up to 1.39 GPa, and the corresponding structures were fully refined. The main observation is a large axial compressibility of the double helix. The base-stacking shrinkage is about 10 %. The average base-pair step varies from 2.92 to 2.73 Å at ambient pressure and 1.39 GPa respectively (Fig. 11.5). Surprisingly, the transversal

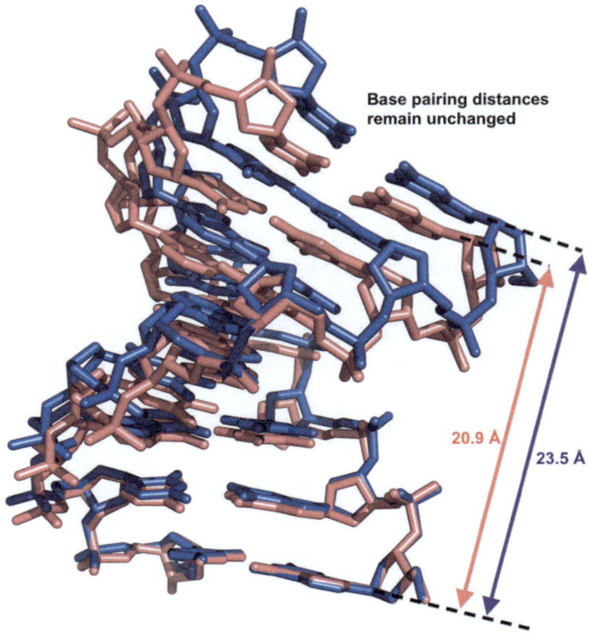

Fig. 11.5 Comparison of the A-DNA structure at 1.39 GPa (in salmon) with the ambient pressure one (in *blue*) showing a large axial compression of the double helix (Girard et al. 2007b)

compressibility is negligible. Under high pressure, the double-helix reacts as a molecular spring and during compression, the geometry of Watson-Crick, base-pair is preserved. Based on these observations, Girard et al. proposed that the double helix topology is adapted to high pressure, and that the adaptation of this singular architecture to harsh conditions may have played an important role at the prebiotic stage and in the first steps of the emergence of life (Girard et al. 2007b). However, the behaviour observed with A-DNA has to be confirmed as study of B-DNA by high-pressure NMR led to different results (Wilton et al. 2008).

11.7 Pressure Trapping of Conformational Substates

Pressure is a way to promote conformational substates that present lower specific volumes. It is thus a convenient way to trap and study such conformational substates as well as folding/unfolding processes in proteins (Akasaka 2006; Fourme et al. 2012a; Akasaka et al. 2013). However, one could consider that the trapping of such substates in crystals is difficult, as the crystal packing would prevent their promotion. As showed by the three examples below, as well as the ones in Sect. 11.8, it is not the case.

11.7.1 *Effects of Pressure on Sperm-Whale Myoglobin: Mapping Conformational Substates*

In 2002, Urayama et al. probed the possibility of mapping conformational substates in sperm whale myoglobin using HPMX (Urayama et al. 2002). Using the beryllium cell method, they solved multiple structures from which they identified various spatial displacements in 6 regions of the protein. The order of magnitude of these displacements, even if they were small, was sufficient to relate them to potential substates of myoglobin. The comparison of the high-pressure structures with the ones obtained by varying the pH confirmed that pressure, in combination with crystallography, could be used to trap and characterize molecular modifications in biologically-relevant conformational substates of a given protein. Indeed, it was known that lowering pH and increasing pressure had similar effects on the populations of two spectroscopically defined conformational states of myoglobin (Frauenfelder et al. 1990). Based on the displacements of the peptide backbone, Urayama et al. showed that the change due to pH (as several structures of myoglobin determined at various pH were available) and the ones from pressure are correlated, confirming that high-pressure crystallography is a convenient way to identify structural changes associated with conformational substates.

11.7.2 *Effects of Pressure on a Metalloprotein: Cu-Zn Superoxide Dismutase*

Superoxide dismutases (SODs) catalyze the dismutation of superoxide radicals into dioxygen and hydrogen peroxide. They are proteins functionally active as homodimer and ubiquitous to all oxygen-utilising organisms. As metalloproteins, SODs can bind various types of metals in their active-site center, such as copper, zinc, manganese or iron. Ascone et al. have determined the behaviour of bovine erythrocyte Cu-Zn superoxide dismutase (CuZnSOD) with pressure. The first observation was that crystals stability extended beyond 1 GPa, indicating that the dimeric assembly remains stable on a large pressure range, at the opposite of what was generally observed for multimeric proteins (Ascone et al. 2010). Ascone et al. performed a detailed analysis of the structure of CuZnSOD determined at 570 MPa. Based on atomic displacement and variation of B-factors, they determined a region of particular flexibility corresponding to a loop, situated at the rim of the active site, named the electrostatic loop. Such flexibility may have an important role in facilitating interactions between Cu-carriers and CuZnSOD in the copper uptake mechanism (see references in Ascone et al. 2010). The high-pressure structure also revealed slightly modified metal-binding sites. In particular, one of the two monomers showed an elongated electron density at the Cu site, which was interpreted as an alternate Cu position. One position would correspond to a characteristic Cu^{II} coordination, while the second one has a typical Cu^{I} coordination. These observations are in agreement with high-pressure X-ray absorption experiments (Ascone et al. 2000) that indicated no local modification around the Zn site whereas the Cu atom presented structural modifications.

11.7.3 Effects of Pressure on a Yellow Fluorescent Protein: Citrine

The yellow fluorescent protein citrine has been pressure-characterized by Barstow et al. (2008, 2009). Citrine differs from green fluorescent proteins (GFPs) by the presence of a tyrosine at position 203 instead of a threonine. As a consequence, the phenol ring of Tyr203 interacts with the main chromophore of citrine through an overlap with a characteristic distance of 3.4 Å (Fig. 11.6 panel a), leading to a characteristic fluorescence peak at 530 nm. Barstow et al. performed a complete high-pressure structure-function study of citrine. First, they followed the pressure-induced shift of the fluorescence peak, *i.e.* of the activity of citrine. Indeed, the peak moves from 530 nm to 510 nm when the pressure is increased from 50 MPa to 300 MPa, with a decrease of about 2 of the fluorescence yield. It should be noticed that 510 nm corresponds to the GFP fluorescence peak. Second, they determined a series of citrine structures at pressures ranging from 50 MPa to 500 MPa using the high-pressure cooling methods (Kim et al. 2005). The analysis of the different structures (Barstow et al. 2008) reveals a minor but progressive reorientation of the two stacked aromatic rings that form the citrine active site, *i.e.* the phenol ring of Tyr203 and the imidazolinone ring of the main chromophore. Relative to Tyr203, the main chromophore shows a sliding motion essentially confined to a plane parallel to the Tyr203 phenol ring and, above 300 MPa, the main chromophore slightly rotates, thus changing the relative orientation of the two rings (Fig. 11.6 panel b). These small deformations, less than 1 Å, explain the pressure-induced shift toward green of citrine's fluorescence peak. As the relative positions of the main chromophore and of Tyr203, acting as a perturbation of the green fluorescence of the main chromophore, changes with increasing pressure, the perturbing influence of Tyr203 is removed and the main chromophore returns to its unperturbed green fluorescent state. In 2009, Barstow et al. pushed further the analysis by linking the changes observed in the active site with subtle modifications of some regions of the citrine β barrel (Barstow et al. 2009). For that, they identified groups of atoms that move together with increasing pressure and determined two distinct regions. One region contains Tyr203, whereas the second one supports the main chromophore, explaining why the two parts of the active site move relative to each other. The interesting conclusion demonstrated by high-pressure is that small displacements of the two regions (estimated to a couple tenths of angstrom) led to important

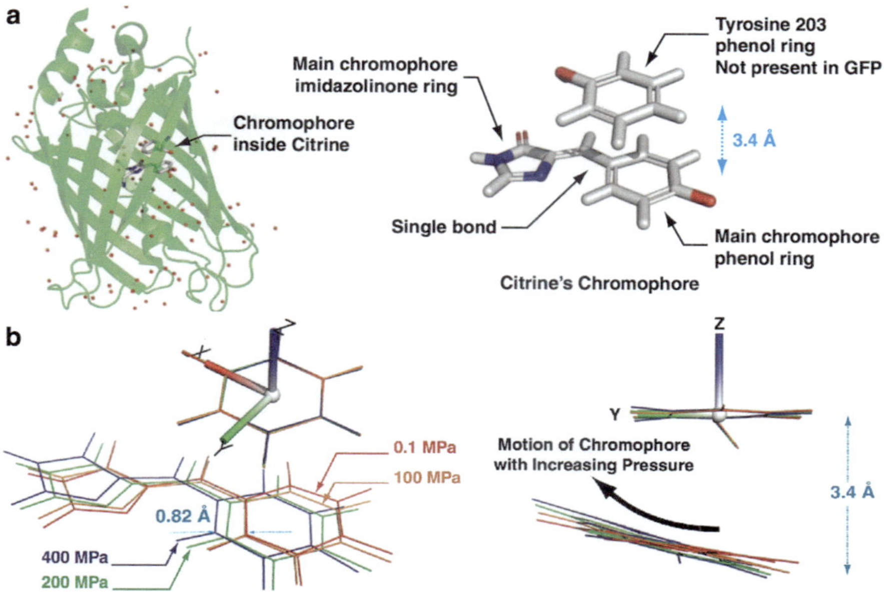

Fig. 11.6 (**a**) Overall structure of the fluorescent protein citrine (PDB ID code: 3DPW). The insert shows the active site of citrine with Tyr203 and the main chromophore. (**b**) Relative motions induced by pressure of the two components of the citrine active site (Reprinted with permission from (Barstow et al. 2008) (Copyright (2008) National Academy of Sciences, U.S.A))

modifications of the fluorescence properties as a consequence of lever arm effect. This probes the link between minute overall displacements, of some part of the β barrel in the case of citrine, and the active site centre.

11.8 On the Importance of Cavities in Pressure Molecular Response of Macromolecules

As previously mentioned, pressure affects the volume of internal cavities. These effects have been proposed as the origin of the pressure-induced unfolding of proteins. Indeed, water penetration to the protein interior, through the internal cavities, is the principal explanation of pressure unfolding of protein, as exemplified recently ((Roche et al. 2012) and references within). The next two examples show what HPMX can bring within this research field, and point the potential role of cavities in molecular pressure adaptation. Finally, the third example on urate oxidase shows how pressure combined to different biophysical methods, including crystallography, shed light on the biological role of a large hydrophobic cavity.

11.8.1 T4 Lysozyme Mutant: Water Penetration of a Large Internal Cavity

The L99A mutant of T4 lyzozyme was originally designed to study the stabilizing interactions between buried non-polar residues (Eriksson et al. 1992, 1993). Indeed, the leucine to alanine mutations creates a larger cavity of about 150 $Å^3$ that induces a destabilizating effect of the folded protein. Thus, this mutant provides an excellent system to explore the effects of pressure on a large cavity. Collins et al. performed a pressure study of the L99A T4 lysozyme mutant (Collins et al. 2005, 2007). The crystal structure was determined at four pressures up to 200 MPa using the Be-cell system (Kundrot and Richards 1986). Unexpectedly, one of the main results was that the cavity remained unchanged as its volume only decreased by less than 3 % at 200 MPa. A careful analysis of the electron density maps allowed determining an increase of the electron density within the cavity with increasing pressure. Collins et al. interpreted this increase by water filling of the cavity with, on average, half a water molecule at 100 MPa to approximately two water molecules at 200 MPa (Collins et al. 2005). Molecular dynamics simulations were also performed to get insights into the dynamic of the water inside the cavity. Apart from being in agreement with the experiments in terms of the presence of water in the cavity, two of these simulations have shown that water can escape from the cavity through a transient opening between the side-chain of four amino-acids. Collins et al. concluded that it was possible that non-polar cavities could be transiently populated with polar molecules, even if there is no clear direct access to the cavity, as revealed by the crystallographic structure. This led the authors to further exploit the data on L99A T4 lysozyme structures, by taking into account the pressure behaviour of the wild-type protein (Collins et al. 2007). They observed that N-terminal domain moves relative to the C-terminal one. Associated with this domain displacement, three helices displace. Helix C not only displaces but also deforms with increasing pressure, as helix C kinks near Leu66. Helix D is associated to these changes and helix H rotates toward the cavity (Fig. 11.7). These observations led the authors to an interesting discussion on the role of side-chain steric interactions on the rigidity of the large cavity (Collins et al. 2007).

11.8.2 Effects of Pressure on 3-Isopropylmalate Dehydrogenase: High-Pressure Adaptation

Nagae et al. used 3-isopropylmalate dehydrogenase as a model enzyme to investigate molecular determinants of pressure adaptation of macromolecules (Nagae et al. 2012a, b). Indeed, the enzyme from the obligate piezophile *Shewanella benthica* DB21MT-2 (SbIPMDH) isolated from the Mariana Trench remains active at high pressure, whereas the one isolated from the land bacterium *S. oneidensis*

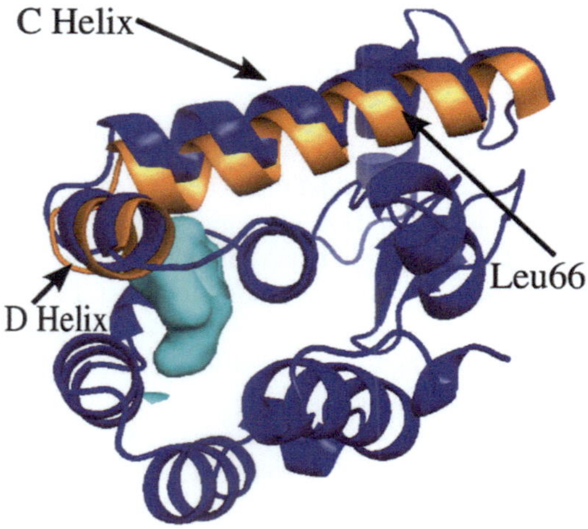

Fig. 11.7 Pressure-induced displacements of helices C and D in L99A T4 lysozyme. A kink in helix C is visible near Leu66. The cavity is shown in *light blue*. The ambient structure is shown in cartoon mode *coloured blue* (Reprinted with permission from (Collins et al. 2007) (Copyright (2007) Elsevier))

MR-1 (SoIPMDH) becomes inactivated (Kasahara et al. 2009). IPMDH catalyzes the reduction of isopropylmalate to 2-isopropyl-3-oxosuccinate in the presence of divalent cations and nicotinamide adenine dinucleotide, and is biologically active as a dimer. A HPMX experiment was performed on the surface protein SoIPMDH using the DAC method (Nagae et al. 2012b). A total of six structures were determined from ambient to 650 MPa. The structures revealed no characteristic structural changes, in particular around the active site, which can explain a reduction in enzymatic activity of SoIPMDH. Nevertheless, they observed global displacements induced by pressure as illustrated by the closure of the entrance to the active site and the simultaneous opening of a groove, situated at the opposite side of the active site. Moreover, changes in volume of internal cavities were also observed. Whereas most cavities are compressed under pressure, the volume of a cavity situated at the dimer interface increases and is gradually occupied by water molecules. At pressures higher than 500 MPa, pressure induces the generation of an additional cleft at the bottom of the groove associated with the presence of three water molecules, two being observed directly in the cleft and the third one near the groove. These observations are in agreement with the study of the T4 lysozyme (Collins et al. 2005, 2007), and may represent conformational changes associated with the reduction of the partial molar volume induced by pressure.

In a second study, Nagae et al. used conventional crystallography to compare the structures of SoIPMDH and SbIPMDH (Nagae et al. 2012a). In particular, at atmospheric pressure, SbIPMDH has a larger dimer cavity volume than SoIPMDH. This loosely packed structure of SbIPMDH may limit pressure-induced modification of the native structure keeping it active at higher pressures compared to SoIPMDH. This study points out the potential role of internal cavity in high-pressure adaptation.

11.8.3 Effects of Pressure on Urate Oxidase: Probing the Role of a Large Internal Cavity in Enzymatic Mechanism

Urate oxidase (UOx) is an enzyme that belongs to the purine degradation pathway. It catalyses the oxidation of uric acid to a primary reaction intermediate, 5-hydroxyisourate. This enzyme, which is not produced in humans, can be used medically to manage hyperuricemic disorders that can occur in chemotherapy. UOx is functionally active as a homotetramer. Its crystallographic structure (Colloc'h et al. 1997) has revealed the presence of a large hydrophobic cavity, nearby the active site pocket which is situated at the largest interface between two monomers that form the tetrameric assembly.

A multi-faceted high-pressure study was performed by combining high-pressure small-angle x-ray scattering (SAXS), functional assays at high pressure as well as on high-pressure incubated UOx, HP fluorescence spectroscopy, and HPMX (Girard et al. 2010b). Based on both fluorescence and SAXS measurements, a complete description of the pressure-induced destabilisation of the tetrameric assembly, occurring at around 200 MPa, was provided (Girard et al. 2010b). Combined with activity measurements performed both under pressure and on pressure-incubated protein, it allowed to propose a model that highlighted the transient presence of monomers prior to aggregation. More importantly, the study highlighted the presence of a less-active tetrameric conformational state of UOx.

To characterize this particular conformational state of the tetramer, the structure of UOx was determined at 150 MPa and compared to a reference one obtained at ambient pressure. The structure modifications were linked to the activity measurements and to the UOx tetramer stability. Indeed, the UOx enzymatic activity is lost while the protein remains soluble, *i.e.* tetrameric, as illustrated in Fig. 11.8 panel a. The major result of the structures comparison is related to the hydrophobic cavity (Fig. 11.8 panel b). Indeed, the 150 MPa structure shows that this large cavity collapses with pressure whereas the active site pocket expands. This observation was related to the pressure induced loss of activity of UOx. Girard et al. proposed that the hydrophobic cavity acts as a ballast to allow different active site geometries, in particular an enlarged active site, required for the accommodation of the different enzymatic intermediates as well as of the reaction product (Girard et al. 2010b) (Fig. 11.8 panel b).

The proposed role for the hydrophobic cavity derived from the high-pressure study was then confirmed by Marassio et al. (2011). Indeed, inert gases, such as xenon, bind to the hydrophobic cavity leading to its enlargement. Enzymatic measurements in presence of such inert gases concluded to an inactivation of UOx, confirming the role of the hydrophobic cavity in the enzymatic process (Colloc'h and Prangé 2014).

As mentioned, the active site lies at the largest interface between two monomers. In UOx, the plasticity of the active site has to be sufficient to ensure enzymatic activity, but too large modifications may induce destabilisation of the whole edifice.

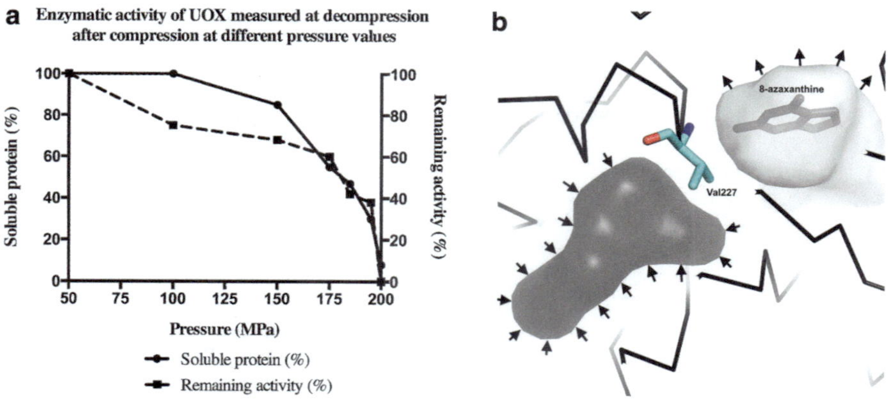

Fig. 11.8 (**a**) Evolution of loss of urate oxidase activity and of the remaining protein in solution with increasing pressure (based on Table 1 in (Girard et al. 2010b)). (**b**) With increasing pressure, the hydrophobic cavity volume decreases (as highlighted by the *small arrows*) while the active site pocket (shown in presence of the substrate-like inhibitor, 8-azaxanthine) expands

Consequently, this study also highlights the subtle balance between activity and stability, and the power of pressure to reveal such a relation.

11.9 Summary

HPMX can now be considered as a fully-fledged technique. Technical problems arising from macromolecular crystals have been overcome as illustrated by the wide scientific results obtained, ranging from nucleic acids to virus capsid through proteins of various sizes. The reader will have noticed that initial studies were mainly focused on the elastic compression of macromolecules and to the molecular effects of pressure, in particular on the stability as well as on the unfolding process. Then, combination of activity measurements and crystallography allowed describing precisely conformational substates promoted by pressure, as in the case of citrine. Finally, the work on urate oxidase shows the power of a multi-technique approach combined to pressure perturbation to decipher the close link between stability and activity in proteins, and to propose new clues of the biological role of macromolecules.

Note Original figures were prepared using Pymol (http://www.pymol.org) and Hollow (Ho and Gruswitz 2008) for Fig. 11.8.

References

Akasaka K (2006) Probing conformational fluctuation of proteins by pressure perturbation. Chem Rev 106:1814–1835. doi:10.1021/cr040440z

Akasaka K, Kitahara R, Kamatari YO (2013) Exploring the folding energy landscape with pressure. Arch Biochem Biophys 531:110–115. doi:10.1016/j.abb.2012.11.016

Ascone I, Cognigni A, Le Godec Y, Itié JP (2000) X-ray absorption study of Cu, Zn SOD under high pressure. High Pressure Res 19:277–283. doi:10.1080/08957950008202565

Ascone I, Savino C, Kahn R, Fourme R (2010) Flexibility of the Cu, Zn superoxide dismutase structure investigated at 0.57 GPa. Acta Crystallogr D Biol Crystallogr 66:654–663. doi:10.1107/S0907444910012321

Barstow B, Ando N, Kim CU, Gruner SM (2008) Alteration of citrine structure by hydrostatic pressure explains the accompanying spectral shift. Proc Natl Acad Sci U S A 105:13362–13366. doi:10.1073/pnas.0802252105

Barstow B, Ando N, Kim C, Gruner SM (2009) Coupling of pressure-induced structural shifts to spectral changes in a yellow fluorescent protein. Biophys J 97:1719–1727

Boonyaratanakornkit BB, Park CB, Clark DS (2002) Pressure effects on intra- and intermolecular interactions within proteins. Biochim Biophys Acta 1595:235–249

Boutet S, Lomb L, Williams GJ et al (2012) High-resolution protein structure determination by serial femtosecond crystallography. Science 337:362–364. doi:10.1126/science.1217737

Bridgman PW (1914) The coagulation of albumen by pressure. J Biol Chem 19:511–512

Collins MD, Hummer G, Quillin ML, Matthews BW, Gruner SM (2005) Cooperative water filling of a nonpolar protein cavity observed by high-pressure crystallography and simulation. Proc Natl Acad Sci U S A 102:16668–16671. doi:10.1073/pnas.0508224102

Collins MD, Quillin ML, Hummer G, Matthews BW, Gruner SM (2007) Structural rigidity of a large cavity-containing protein revealed by high-pressure crystallography. J Mol Biol 367:752–763. doi:10.1016/j.jmb.2006.12.021

Collins MD, Kim CU, Gruner SM (2010) High-pressure protein crystallography and NMR to explore protein conformations. Annu Rev Biophys. doi:10.1146/annurev-biophys-042910-155304

Colloc'h N, Prangé T (2014) Functional relevance of the internal hydrophobic cavity of urate oxidase. FEBS Lett 588:1715–1719. doi:10.1016/j.febslet.2014.03.017

Colloc'h N, El Hajji M, Bachet B, L'Hermite G, Schiltz M, Prangé T, Castro B, Mornon J-P (1997) Crystal structure of the protein drug urate oxidase-inhibitor complex at 2.05 Å resolution. Nat Struct Mol Biol 4:947–952

Daniel I, Oger P, Winter R (2006) Origins of life and biochemistry under high-pressure conditions. Chem Soc Rev 35:858–875. doi:10.1039/b517766a

Eriksson AE, Baase WA, Zhang XJ et al (1992) Response of a protein structure to cavity-creating mutations and its relation to the hydrophobic effect. Science 255:178–183

Eriksson AE, Baase WA, Matthews BW (1993) Similar hydrophobic replacements of Leu99 and Phe153 within the core of T4 lysozyme have different structural and thermodynamic consequences. J Mol Biol 229:747–769. doi:10.1006/jmbi.1993.1077

Fourme R (1968) Appareillage pour etudes radiocristallographiques sous pression et a temperature variable. J Appl Crystallogr 1:23–30. doi:10.1107/S0021889868004954

Fourme R, Ascone I, Kahn R, Mezouar M, Bouvier P, Girard E, Lin T, Johnson JE (2002) Opening the high-pressure domain beyond 2 kbar to protein and virus crystallography–technical advance. Structure 10:1409–1414

Fourme R, Girard E, Kahn R, Ascone I, Mezouar M, Dhaussy A-C, Lin T, Johnson JE (2003) Using a quasi-parallel X-ray beam of ultrashort wavelength for high-pressure virus crystallography: implications for standard macromolecular crystallography. Acta Crystallogr D Biol Crystallogr 59:1767–1772

Fourme R, Girard E, Kahn R, Dhaussy A-C, Ascone I (2009) Advances in high-pressure biophysics: status and prospects of macromolecular crystallography. Annu Rev Biophys 38:153–171. doi:10.1146/annurev.biophys.050708.133700

Fourme R, Girard E, Akasaka K (2012a) High-pressure macromolecular crystallography and NMR: status, achievements and prospects. Curr Opin Struct Biol 22:636–642. doi:10.1016/j.sbi.2012.07.007

Fourme R, Honkimaki V, Girard E, Medjoubi K, Dhaussy A-C, Kahn R (2012b) Reduction of radiation damage and other benefits of short wavelengths for macromolecular crystallography data collection. J Appl Crystallogr 45:652–661. doi:10.1107/S0021889812019164

Fraser JS, Clarkson MW, Degnan SC, Erion R, Kern D, Alber T (2009) Hidden alternative structures of proline isomerase essential for catalysis. Nature 462:669–673. doi:10.1038/nature08615

Fraser JS, van den Bedem H, Samelson AJ, Langa PT, Holtonc JM, Echolsd N, Alber T (2011) Accessing protein conformational ensembles using room-temperature X-ray crystallography. Proc Natl Acad Sci U S A 108:16247–16252

Frauenfelder H, McMahon BH (2000) Energy landscape and fluctuations in proteins. Annalen Der Physik 9:655–667

Frauenfelder H, Alberding NA, Ansari A, Braunstein D, Cowen BR, Honk MK, Iben IET, Johnson JB, Luck S, Marden MC, Mourant JR, Ormos P, Reinisch L, Scholl R, Schulte A, Shyamsunder E, Sorensen LB, Steinbach PJ, Xie A, Young RD, Yue KT (1990) Proteins and pressure. J Phys Chem 94:1024–1037

Girard E, Kahn R, Mezouar M, Dhaussy A-C, Lin T, Johnson JE, Fourme R (2005) The first crystal structure of a macromolecular assembly under high pressure: CpMV at 330 MPa. Biophys J 88:3562–3571. doi:10.1529/biophysj.104.058636

Girard E, Dhaussy A-C, Couzinet B, Chervin J-C, Mezouar M, Kahn R, Ascone I, Fourme R (2007a) Toward fully fledged high-pressure macromolecular crystallography. J Appl Crystallogr 40:912–918. doi:10.1107/S0021889807033833

Girard E, Prangé T, Dhaussy A-C et al (2007b) Adaptation of the base-paired double-helix molecular architecture to extreme pressure. Nucleic Acids Res 35:4800–4808. doi:10.1093/nar/gkm511

Girard E, Fourme R, Ciurko R, Joly J, Bouis F, Legrand P, Jacobs J, Dhaussy A-C, Ferrer J-L, Mezouar M, Kahn R (2010a) Macromolecular crystallography at high pressure with pneumatic diamond anvil cells handled by a six-axis robotic arm. J Appl Crystallogr 43:762–768. doi:10.1107/S0021889810016146

Girard E, Marchal S, Perez J et al (2010b) Structure-function perturbation and dissociation of tetrameric urate oxidase by high hydrostatic pressure. Biophys J 98:2365–2373. doi:10.1016/j.bpj.2010.01.058

Ho BK, Gruswitz F (2008) HOLLOW: generating accurate representations of channel and interior surfaces in molecular structures. BMC Struct Biol 8:49. doi:10.1186/1472-6807-8-49

Jamieson AJ, Fujii T, Solan M, Matsumoto AK, Bagley PM, Priede IG (2009) Liparid and macrourid fishes of the hadal zone: in situ observations of activity and feeding behaviour. Proc Biol Sci 276:1037–1045. doi:10.1098/rspb.2008.1670

Kasahara R, Sato T, Tamegai H, Kato C (2009) Piezo-adapted 3-isopropylmalate dehydrogenase of the obligate piezophile *Shewanella benthica* DB21MT-2 isolated from the 11,000-m depth of the mariana trench. Biosci Biotechnol Biochem 73:2541–2543. doi:10.1271/bbb.90448

Katrusiak A, Dauter Z (1996) Compressibility of lysozyme protein crystals by X-ray diffraction. Acta Crystallogr D Biol Crystallogr 52:607–608. doi:10.1107/S0907444996000431

Kim CU, Kapfer R, Gruner SM (2005) High-pressure cooling of protein crystals without cryoprotectants. Acta Crystallogr D Biol Crystallogr 61:881–890. doi:10.1107/S090744490500836X

Kim CU, Wierman JL, Gillilan R, Lima E, Gruner SM (2013) A high-pressure cryocooling method for protein crystals and biological samples with reduced background X-ray scatter. J Appl Crystallogr 46:234–241. doi:10.1107/S0021889812045013

Kundrot C, Richards F (1986) Collection and processing of X-ray diffraction data from protein crystals at high pressure. J Appl Crystallogr 19:208–213

Kundrot CE, Richards FM (1987) Crystal structure of hen egg-white lysozyme at a hydrostatic pressure of 1000 atmospheres. J Mol Biol 193:157–170

Kundrot CE, Richards FM (1988) Effect of hydrostatic pressure on the solvent in crystals of hen egg-white lysozyme. J Mol Biol 200:401–410

Li H, Yamada H, Akasaka K (1998) Effect of pressure on individual hydrogen bonds in proteins. Basic pancreatic trypsin inhibitor Biochemistry 37:1167–1173. doi:10.1021/bi972288j

Lin T, Schildkamp W, Brister K, Doerschuk PC, Somayazulu M, Mao H, Johnson JE (2005) The mechanism of high-pressure-induced ordering in a macromolecular crystal. Acta Crystallogr D Biol Crystallogr 61:737–743. doi:10.1107/S0907444905000053

Liu W, Wacker D, Gati C et al (2013) Serial femtosecond crystallography of G protein-coupled receptors. Science 342:1521–1524. doi:10.1126/science.1244142

Marassio G, Prangé T, David HN et al (2011) Pressure-response analysis of anesthetic gases xenon and nitrous oxide on urate oxidase: a crystallographic study. FASEB J 25:2266–2275. doi:10.1096/fj.11-183046

Marchi M, Akasaka K (2001) Simulation of hydrated BPTI at high pressure: changes in hydrogen bonding and its relation with NMR experiments. J Phys Chem B 105:711–714. doi:10.1021/jp002539p

Nagae T, Kato C, Watanabe N (2012a) Structural analysis of 3-isopropylmalate dehydrogenase from the obligate piezophile Shewanella benthica DB21MT-2 and the nonpiezophile Shewanella oneidensis MR-1. Acta Crystallogr F Struct Biol Cryst Commun 68:265–268. doi:10.1107/S1744309112001443

Nagae T, Kawamura T, Chavas LMG, Niwa K, Hasegawa M, Kato C, Watanabe N (2012b) High-pressure-induced water penetration into 3-isopropylmalate dehydrogenase. Acta Crystallogr D Biol Crystallogr 68:300–309. doi:10.1107/S0907444912001862

Nisius L, Grzesiek S (2012) Key stabilizing elements of protein structure identified through pressure and temperature perturbation of its hydrogen bond network. Nature Chem 4:711–717. doi:10.1038/nchem.1396

Oger PM, Jebbar M (2010) The many ways of coping with pressure. Res Microbiol 161:799–809. doi:10.1016/j.resmic.2010.09.017

Paci E (2002) High pressure simulations of biomolecules. Biochim Biophys Acta 1595:185–200

Paci E, Marchi M (1996) Intrinsic compressibility and volume compression in solvated proteins by molecular dynamics simulation at high pressure. Proc Natl Acad Sci U S A 93:11609–11614

Refaee M, Tezuka T, Akasaka K, Williamson MP (2003) Pressure-dependent changes in the solution structure of hen egg-white lysozyme. J Mol Biol 327:857–865

Rivalain N, Roquain J, Demazeau G (2010) Development of high hydrostatic pressure in biosciences: pressure effect on biological structures and potential applications in biotechnologies. Biotechnol Advan 28:659–672. doi:10.1016/j.biotechadv.2010.04.001

Roche J, Caro JA, Norberto DR et al (2012) Cavities determine the pressure unfolding of proteins. Proc Natl Acad Sci U S A 109:6945–6950. doi:10.1073/pnas.1200915109

Rosenbaum E, Gabel F, Durá MA, Finetd S, Cléry-Barraude C, Massona P, Franzetti B (2012) Effects of hydrostatic pressure on the quaternary structure and enzymatic activity of a large peptidase complex from Pyrococcus horikoshii. Arch Biochem Biophys 517:104–110. doi:10.1016/j.abb.2011.07.017

Royer CA (2002) Revisiting volume changes in pressure-induced protein unfolding. Biochim Biophys Acta 1595:201–209

Somero GN (1992) Adaptations to high hydrostatic pressure. Annu Rev Physiol 54:557–577. doi:10.1146/annurev.ph.54.030192.003013

Thomanek UF, Parak F, Mossbauer RL, Formanek H, Schwager P, Hoppe W (1973) Freezing of myoglobin crystals at high pressure. Acta Cryst A 29:263–265

Urayama P, Phillips GN, Gruner SM (2002) Probing substates in sperm whale myoglobin using high-pressure crystallography. Structure 10:51–60

Wilton DJ, Ghosh M, Chary KVA, Akasaka K, Williamson MP (2008) Structural change in a B-DNA helix with hydrostatic pressure. Nucleic Acids Res 36:4032–4037. doi:10.1093/nar/gkn350

Xu J, Liu L, Xu M, Oger P, Wang F, Jebbar M, Xiao X (2011) Genome announcement: complete genome sequence of the first obligate piezophilic hyperthermophilic archaeon *Pyrococcus yayanosii* CH1. J Bacteriol. doi:10.1128/JB.05345-11

Zeng X, Birrien J-L, Fouquet Y, Cherkashov G, Jebbar M, Querellou J, Oger P, Cambon-Bonavita M-A, Xiao X, Prieur D (2009) Pyrococcus CH1, an obligate piezophilic hyperthermophile: extending the upper pressure-temperature limits for life. ISME J 3:873–876. doi:10.1038/ismej.2009.21

Chapter 12
Cavities and Excited States in Proteins

Hua Li and Yuji O. Kamatari

Abstract Protein cavities or voids are observed as defects in atomic packing. Cavities have long been suggested to play important roles in protein dynamics and function, but the underlying origin and mechanism remains elusive. Here, recent studies about the cavities characterized by high-pressure NMR spectroscopy have been reviewed. Analysis of the pressure-dependent chemical shifts showed both linear and nonlinear response of proteins to pressure. The linear response corresponded to compression within the native ensemble, while the nonlinear response indicated the involvement of low-lying excited states that were different from the native state. The finding of non-linear pressure shifts in various proteins suggested that the existence of the low-lying excited states was common for globular proteins. However, the absolute nonlinear coefficient values varied significantly from protein to protein, and showed a good correlation with the density of cavities. Extensive studies on hen lysozyme as a model system showed that the cavity hydration and water penetration into the interior of proteins was an origin of the conformational transition to the excited states. The importance of cavities for protein function and evolution has also been explained. In addition to these "equilibrium" cavities, there are also "transient" cavities formed in the interior of the protein structure, as manifested by the ring flip motions of aromatic rings. The significance of transient cavities, reflecting an intrinsic dynamic nature within the native state, has also been discussed.

Keywords Cavity • Conformational fluctuation • High-energy sub-states • High-pressure NMR spectroscopy • Hydration • Low-lying excited states

---

H. Li
National Center for Protein Science Shanghai, State Key Laboratory of Molecular Biology, Institute of Biochemistry and Cell Biology, Shanghai Institutes for Biological Sciences, Chinese Academy of Sciences, 320 Yueyang Road, Shanghai 200031, China

Y.O. Kamatari (✉)
Life Science Research Center, Gifu University, 1-1 Yanagido, Gifu 501-1193, Japan
e-mail: kamatari@gifu-u.ac.jp

K. Akasaka, H. Matsuki (eds.), *High Pressure Bioscience*,
Subcellular Biochemistry 72, DOI 10.1007/978-94-017-9918-8_12

12.1 Introduction

Protein cavities or voids are observed stably as "static" defects in atomic packing in crystal structures, but their occurrence and size vary from protein to protein. In addition to those "static" cavities, there are "transient" cavities formed also in the interior of the protein structure that are rather rarely detected in solution. The latter demonstrates directly the intrinsically dynamic nature of the folded protein structure in solution. On the other hand, the "static" cavities are not really static, but have been considered the source of dynamics for a long time based on the recognition of their contribution to the compressibility of protein molecules (Gekko et al. 1986). Later, high-pressure fluorescence study revealed their roles in pressure denaturation of proteins (Pin et al. 1990). In more recent years, high pressure NMR spectroscopy is beginning to reveal their roles in dynamics and high-energy sub-states in proteins. The manner how cavities provide sources for internal hydration, dynamics and high-energy sub-states is becoming increasingly clear, together with their roles in function. In this chapter, by reviewing some representative examples using pressure as a perturbation, we will discuss the effect of both "transient" and "static" cavities on protein dynamics, high-energy sub-states and function.

12.2 Cavities and Excited States in Proteins

Under physiological conditions, a protein molecule folds into a stable three-dimensional structure with close atomic packing. In the crystal structure, the folded structure of a protein is often encountered with a void or atomic defect where the density of atoms of the protein is extremely low. This is called a "static" cavity or, may be more correctly, an "equilibrium" cavity (See Fig. 5.2 in Chap. 5), as they are considered to exist in thermodynamic equilibrium in solution. There are two categories of the equilibrium cavities. One is a nonpolar (or hydrophobic) cavity that is surrounded by non-polar side chains. The other is a polar (or hydrophilic) cavity that is surrounded, at least in part, by polar side chains and/or main chain of peptide groups. In the latter case, water molecules inside the cavity are hydrogen bonded to polar groups of the main chain or side chains (Liu et al. 2008). In the former case, the question still remains whether nonpolar cavities in proteins are truly empty (Matthews and Liu 2009). In this Sect. 12.2, we describe the definition of the equilibrium cavities, high pressure NMR spectroscopy, and linearity analysis of the high pressure NMR data. This analysis clearly shows that the existence of low-lying excited states is usually hidden inside the large population of the native state.

12.2.1 Equilibrium Cavities Detected in Protein Structures

The intrinsic volume of a protein, which can be determined from its X-ray crystal structure, may be defined as the sum of van der Waals volumes of the constitutive atoms (v_c), the volume of the cavities (v_{cav}) and the volume change due to solvation or hydration (Δv_{sol}) on protein surface:

$$v^o = v_c + v_{cav} + \Delta v_{sol} \tag{12.1}$$

(cf. Gekko, Chap. 5). The cavity is defined by rolling a probe of smaller radius (~1.2 Å) inside the protein molecule. The internal cavity and solvent-accessible surface (SAS) are illustrated in Fig. 5.2. SAS is defined as a surface traced by the center of a spherical probe of appropriate radius (water molecule of 1.4 Å for most purposes) rolling on the molecular surface.

Several softwares and algorithms have been published for the calculation of cavities and/or SAS, such as VOIDOO (Kleywegt and Jones 1994), GRASP (Nicholls et al. 1991), VOLBL (Liang et al. 1998a, b), POCKET (Levitt and Banaszak 1992), and a Monte Carlo (MC) procedure (Chakravarty et al. 2002), etc. Recently, several web servers have also been developed to calculate protein cavities, such as 3 V server (Voss and Gerstein 2010) located at http://3vee.molmovdb.org, and CASTp (Computed Atlas of Surface Topography of proteins) server (Binkowski et al. 2003) located at http://sts.bioengr.uic.edu/castp/.

12.2.2 Pressure as a Tool for Investigating Cavity and Hydration of Proteins

Pressure is a fundamental thermodynamic variable for defining protein conformational states, and volume is the conjugate variable for pressure. The partial molar volume is composed of the volume of the protein constituent atoms, the cavity, and the hydration layer. Therefore, pressure is a unique and appropriate perturbation to study the cavity and its hydration of the protein, since the volume of the atoms does not change with pressure.

In the past, pressure and volume effects on the protein structure, dynamics, and function have been studied macroscopically by measuring compressibility (Gekko and Hasegawa 1986). More recently, with the advancement of high-pressure NMR technique (Akasaka and Yamada 2001), high-pressure macromolecular crystallography (HPMX) (Girard et al. 2005; Fourme et al. 2011), and molecular dynamics simulation at high pressure (Imai et al. 2007; Sgourakis et al. 2008; Imai and Sugita 2010), pressure is increasingly used to study structural fluctuations and transitions into alternate conformations in atomic level details. Among these, high-pressure NMR has been the most versatile technique to probe a wide range of conformational fluctuations of protein molecules in solution (Kamatari et al. 2004; Akasaka 2006; Li and Akasaka 2006).

12.2.3 Excited States in Proteins Revealed from Nonlinear Chemical Shifts

The chemical shift is a sensitive probe to study the pressure response of protein structure. Both amide ^{1}H and ^{15}N and ^{13}C shifts reflect site-specific structural changes in the polypeptide backbone. The amide ^{1}H pressure shift is particularly well correlated with the H…O hydrogen-bond (H-bond) distance (Li et al. 1998), whereas ^{15}N and ^{13}C pressure shifts are more sensitive to variations in the main chain torsion angles ϕ and ψ (Akasaka et al. 1999; Kalbitzer et al. 2000; Wilton et al. 2009). Therefore, the dependence of the ^{1}H, ^{15}N, and ^{13}C chemical shifts on pressure would indicate the response of the H-bond distance and torsion angles to pressure.

12.2.3.1 Analysis of Pressure Response of Chemical Shifts

When performing high-pressure NMR measurements of proteins, we encountered cases showing linear and distinct nonlinear pressure shifts of amide ^{1}H and ^{15}N signals in many proteins (Fig. 12.1a–c), and even sigmoidal curves (Fig. 12.1d–e). Then we performed the estimation of linearity and nonlinearity of chemical shift changes with pressure (Akasaka and Li 2001; Kitahara et al. 2013), which is derived from least-square fits of experimental data for individual ^{1}H and ^{15}N signals to the following equation:

$$\delta_i = a_i + b_i P + c_i P^2 \tag{12.2}$$

where P is the pressure (bars), δ_i is the chemical shift (parts per million) for the ith residue, a_i (parts per million) is the chemical shift at 1 bar, and b_i (parts per million per bar) and c_i (parts per million per squared bar) are the linear (first-order) and nonlinear (second-order) coefficients, respectively.

In Fig. 12.2, the correlation plots of the linear (vertical axis) and nonlinear (horizontal axis) coefficients of amide ^{15}N are shown for 16 proteins with different structures. As clearly seen in Fig. 12.2, the distribution of the nonlinear coefficients varies from protein to protein, while that of the linear coefficients is similar for all proteins. Therefore, the quantitative analysis of the linear and nonlinear coefficients was performed for each protein.

As shown in Fig. 12.3a (dotted columns), the average value of linear coefficients varies little for the proteins investigated. On the other hand, the degree of nonlinearity varies greatly from protein to protein and appears to be characteristic for each protein (Fig. 12.3a, filled columns).

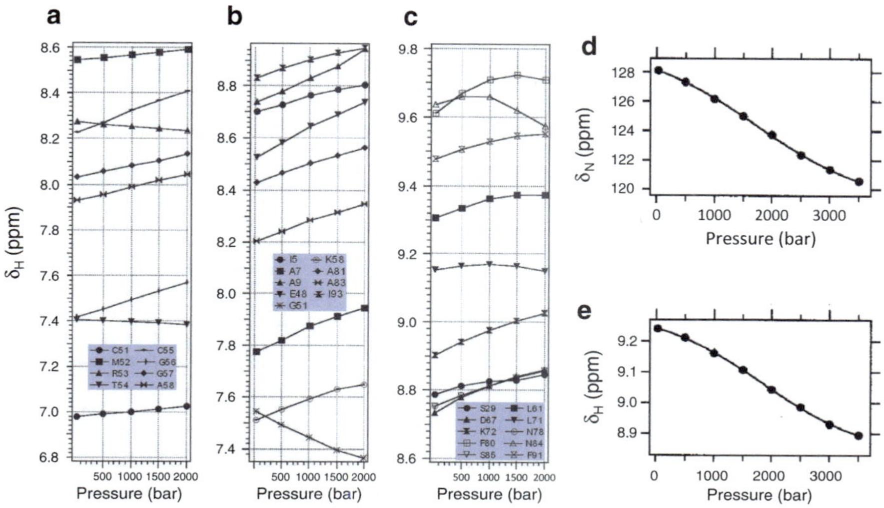

Fig. 12.1 Nature of the linear and nonlinear pressure response. Plot of chemical shifts of selected amide protons of (**a**) BPTI, (**b**) DHFR, and (**c**) RalGDS-RBD as a function of pressure. Pressure dependence of (**d**) ^{15}N and (**e**) ^{1}H chemical shifts of Val70 of ubiquitin at 20 °C (Adapted from Fig. 1 in Akasaka and Li (2001) and Fig. 6 in Kitahara et al. (2001), respectively, by permission of American Chemical Society)

12.2.3.2 Nature of the Linear and Nonlinear Pressure Response

The linear chemical-shift change with pressure results from a linear change in the averaged inter-nuclear distances and torsion angles, the combination of which determines the compressibility coefficient (β_T) of the protein. The observation of a linear pressure shift, therefore, indicates that β_T is independent of pressure within the pressure range studied (approximately 0.1–200 MPa) and arises from a small shift of population within the basic folded ensemble. The similarity of linear coefficients among proteins would mean that the amplitude of structural fluctuation (inter-nuclear distances and torsion angles of the main chain) is of a similar magnitude for all these proteins at a pressure of 0.1 MPa, namely in the ground state of the protein conformation.

On the other hand, the remarkable nonlinearity of pressure shifts is explained by the involvement of an ensemble N′ different from the basic folded ensemble N. N′ not only has a compressibility different from N, giving different slopes of pressure shifts, but also has a partial molar volume smaller than that of N so that its relative population ([N′]/[N]) increases with increasing pressure. The distinct nonlinear pressure shifts over a relatively low-pressure range (approximately 50–100 MPa) indicate that the population of N′ increases significantly with pressure, even at relatively low pressure. This means that the Gibbs-free energy difference between N and N′ (ΔG_0) is not very large (of the order of a few kJ mol^{-1}), or in other words, that N′ is a low-lying excited state of a protein. Furthermore, in general, it is likely that N′ involves a family of different conformers (such as N′, N″,

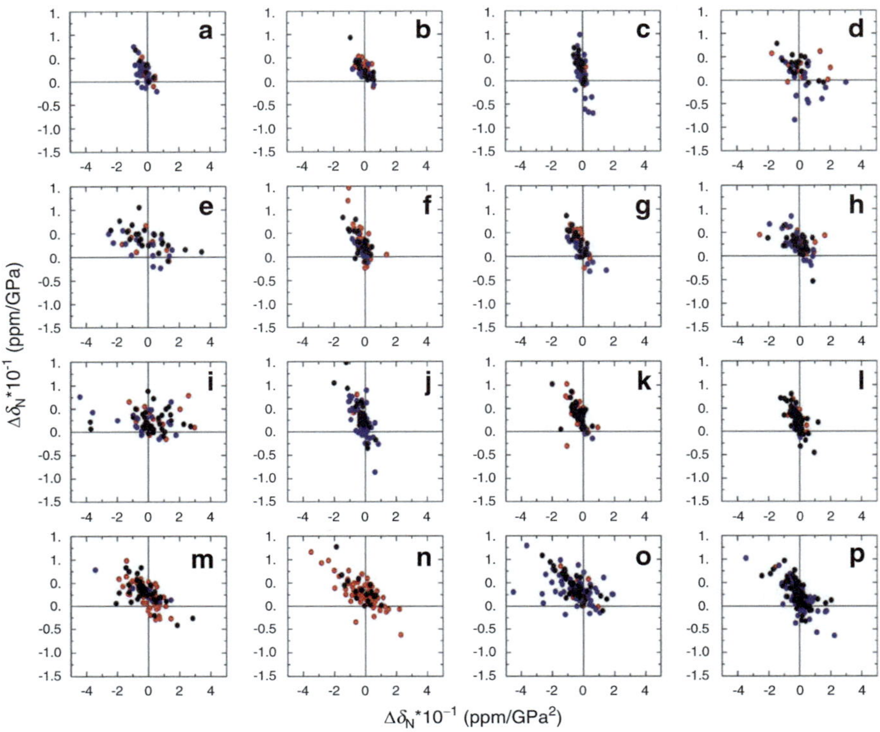

Fig. 12.2 Correlations between the linearity and nonlinearity coefficients of pressure-induced ^{15}N shifts. In each figure, the linear coefficient is plotted against the non-linear coefficient for individual residues of each protein: (**a**) protein G, (**b**) BPTI, (**c**) ubiquitin, (**d**) NEDD8, (**e**) parkin-UBL, (**f**) HPr (WT), (**g**) HPr (H15A), (**h**) RasRBD, (**i**) RasRBD complex, (**j**) p13, (**k**) α-lactalbumin (MG), (**l**) lysozyme, (**m**) prion protein, (**n**) apomyoglobin, (**o**) β-lactoglobulin, (**p**) OspA. α-helix, β-sheet, and other regions are colored by *red*, *blue*, and *black*, respectively (Adapted from Fig. 17 in Kitahara et al. (2013) by permission of Elsevier Inc)

N‴, etc., evidenced in Sect. 12.3.2 later), or subensembles, with different ΔG_0 and ΔV_0 values. The finding of non-linear pressure shifts in various proteins suggests that low-lying excited states are common in many globular proteins, but that their fractional populations vary significantly from protein to protein.

12.2.3.3 Structure of the Excited States

As an ideal nonlinear response, sigmoidal changes in pressure-induced chemical shifts have been observed in ubiquitin in the pressure range from 3 to 350 MPa at 20 °C (Kitahara et al. 2001). This is a clear evidence for a two-state like transition between the basic folded state (N) and a low-lying excited state (N′). In this case, the detailed structural analysis of an excited-state conformer is even possible using

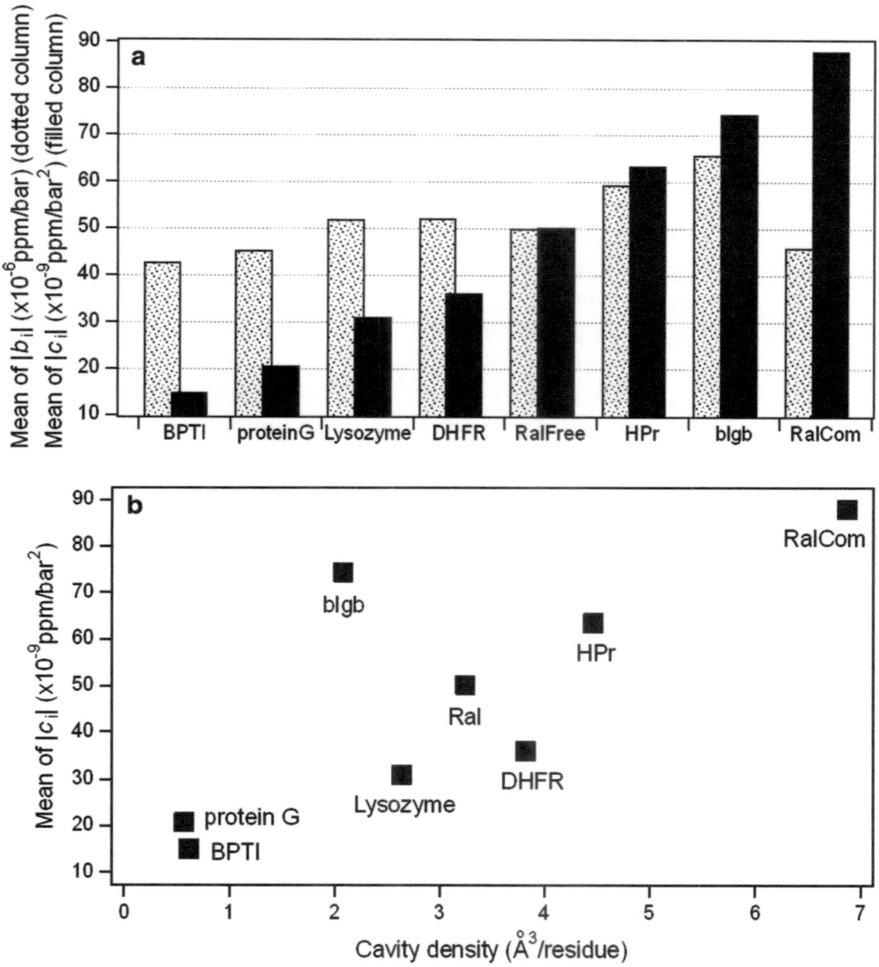

Fig. 12.3 (**a**) Histograms of the mean of the absolute values of the first-order coefficient b_i (*dotted columns*) and the second-order coefficient c_i (*filled columns*) for ^{1}HN in eight globular proteins. (**b**) Plot of the mean of the absolute values of the second-order coefficient c_i of ^{1}HN pressure shifts vs the density of cavities (the total cavity volume divided by the number of amino acid residues) for the eight proteins. Cavity volumes of >20 Å^3 are employed (calculated by GRASP using PDB coordinates) (Adapted from Fig. 3 in Akasaka and Li (2001) by permission of American Chemical Society)

NMR spectroscopy, which was manifested by the first excited-state structure of ubiquitin at 300 MPa (Kitahara et al. 2005). Kitahara et al. found that the major difference between N and N′ of ubiquitin is that N′ has an open structure in its central cleft.

12.2.3.4 Origin of Nonlinearity and Excited States

The population of N′ increases with pressure implies that the N′ state has a lower partial volume than the N state, suggesting that the N′ state is likely to be more hydrated than N. In support of this view, a nonlinear shift is observed for residues close to water-accessible cavities within the folded protein structure in BPTI (Williamson et al. 2003), hen lysozyme (Kamatari et al. 2001), RalGDS-RBD (Inoue et al. 2000), and β-lactoglobulin (Kuwata et al. 2001). Furthermore, a good correlation is found between the mean of the absolute nonlinear coefficient values for amide protons and the density of cavities (the total volume of cavities divided by the number of amino acid residues) in the native ensemble of these proteins (Fig. 12.3b). These findings suggest that the nonlinear shift is related to the hydration of water-accessible cavities. Indeed, an X-ray crystallography study on the L99A cavity mutant of T4 lysozyme showed an increase in the number of internal water molecules within the created cavity at higher pressures (Collins et al. 2005).

An intriguing finding from an exceptionally long (μs) molecular dynamics simulation carried out for ubiquitin at 1 bar and 6,000 bar by Imai and Sugita is that the larger amplitude transition occurs simultaneously with the penetration of water molecules into the central cleft of the ubiquitin (Imai and Sugita 2010). This MD result is consistent with an experimental observation about the volume loss from N to N′ ($\Delta V = -24$ mL mol^{-1}) (Kitahara et al. 2005). In the conformer N′, most of the residues involved in the enzyme-binding are exposed. Thus, it is likely that the low-lying excited state does not correspond to a single structure but spans a continuous range of structures with different hydration numbers and locations (Kamatari et al. 2011). When the hydration becomes substantial, it will collapse into local or total unfolding, making partial or the entire polypeptide segment hydrated and exposed, as is the case for RalGDS-RBD (Inoue et al. 2000), ubiquitin (Kitahara et al. 2005), and β-lactoglobulin (Kuwata et al. 2001).

12.2.3.5 Timescale of the Equilibrium of Excited States

N and N′ states exist in equilibrium and undergo conformational exchange. The continuous chemical shift changes upon pressure suggest that the equilibrium between N and N′ is fast within the NMR timescale, i.e. faster than μs (or 10^6 s^{-1}). At even higher pressure, an ideal nonlinear response, i.e. sigmoidal changes, in pressure-induced chemical shifts would be observed for some proteins. This is a clear evidence for a two-state like transition between the basic folded state (N) and a low-lying excited state (N′), as illustrated above for ubiquitin in the pressure range from 3 to 350 MPa at 20 °C (Kitahara et al. 2001). Kitahara et al. further found that N and N′ are in dynamic equilibrium, and the exchange rate between the two states are on the timescale of ∼10 μs (or ∼10^5 s^{-1}). Furthermore, depending on the protein, when the equilibrium between N and N′ is slow with respect to chemical shift differences (on the timescale of 10^4 s^{-1} or even slower), instead of the continuous chemical shift changes, peak intensity decrease of N will be observed, implying the existence of an excited state with higher energy, or an intermediate state.

12.2.3.6 Comparison with Temperature Response of Chemical Shifts

Williamson group had analyzed the curvature of temperature dependences of amide proton NMR chemical shifts of proteins before (Baxter et al. 1998) and demonstrated the existence of the low free-energy excited states in folded proteins. They found that the source of conformational heterogeneity includes many independent causes, such as the existence of a minor conformer, an alternative hydrogen bond network of the amide proton involved, etc. Although the factor of cavity is not included in their analysis, the importance of buried waters is also demonstrated from the temperature response.

12.3 Water Penetration into Cavities: Hen Lysozyme

In this section, we use hen lysozyme as a model system to show the significant role of the cavities played in protein dynamics, structural transition, protein function, and evolution.

12.3.1 Compression and Mobility in the Folded Hen Lysozyme

Kamatari et al. used high-pressure NMR techniques to characterize the conformational fluctuations of hen lysozyme, in its native state and when denatured in 8 M urea, over the pressure range 30–2,000 bar (Kamatari et al. 2001). All the NMR signals showed continuous and fully reversible changes in chemical shifts with no sign of pressure denaturation in this pressure range (Fig. 12.4a). Significant variations in pressure shifts from their average values are observed for both ^{1}H and ^{15}N in native lysozyme (Fig. 12.4a, c and d). Especially, some residues in the β domain or at the α/β domain interface have anomalously large ^{15}N and ^{1}H chemical-shift changes. All these residues lie close to water-containing cavities (Fig. 12.5), suggesting that the cavities are the source of conformational changes. In contrast, the pressure-induced ^{1}H and ^{15}N shifts for lysozyme denatured in 8 M urea, which is unfolded and has no cavity, are much more uniform than those for the folded lysozyme (Fig. 12.4b, e and f). All these results suggest that the anomalous pressure shifts of amide ^{1}H and ^{15}N signals in the state arise mainly from structural rearrangements of these cavities, including the buried water molecules inside, with increasing pressure. Therefore, cavities provide a source of compression and mobility in the folded lysozyme.

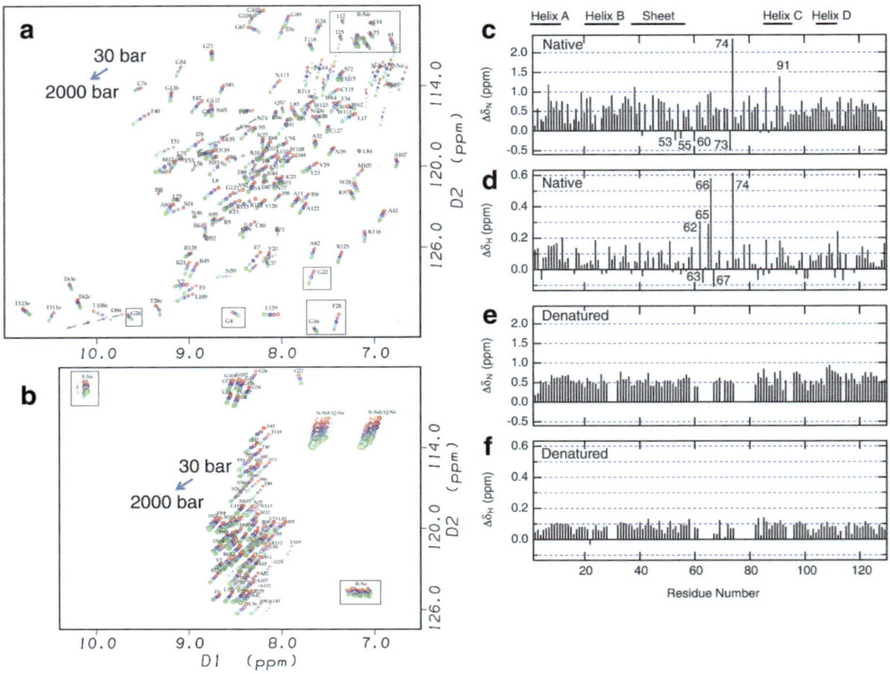

Fig. 12.4 Overlay of ^{15}N-^{1}H HSQC spectra of (**a**) native hen lysozyme and (**b**) hen lysozyme denatured in 8 M urea at pH 2 recorded at high pressure. Cross-peaks from spectra recorded at 30, 500, 1,000, 1,500 and 2,000 bar. The folded cross-peaks are shown enclosed in boxes in the spectra. ^{15}N and ^{1}H pressure shifts ($\Delta\delta_N$ and $\Delta\delta_H$, respectively) for the main-chain amide groups of native (**c** and **d**) and denatured (**e** and **f**) hen lysozyme [$\Delta\delta = \delta$ (2,000 bar) − δ (30 bar)]. The positions in the protein sequence of the α helices and the triple-stranded antiparallel β sheet in native lysozyme are indicated above the diagram. (Adapted from Fig. 1 and 3 in Kamatari et al. (2001) by permission of John Wiley & Sons, Inc)

Williamson and co-workers have calculated the protein structure under high-pressure using the pressure-induced ^{1}H chemical shift changes as restraints (Refaee et al. 2003). The structural changes of hen lysozyme with pressure calculated by them are shown in Fig. 6.2 in Chap. 6. An interesting finding is that, consistent with the above pressure response of chemical shifts, the largest volume changes tend to occur close to water-containing cavities.

12.3.2 Cavity Hydration as a Gateway to Unfolding

We have been focusing on conformational fluctuations localized around water contained in cavities within the ensemble of native conformers of hen lysozyme. The next question is how the initial step of the conformational transition would be caused from the native state to excited states of this protein. For this purpose,

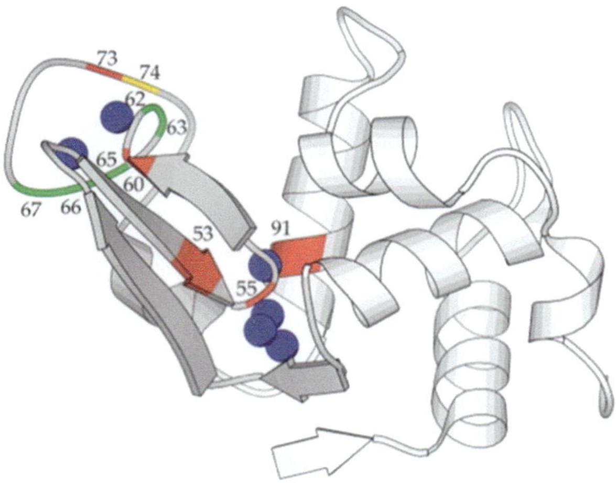

Fig. 12.5 Compression and mobility in the native hen lysozyme. Main-chain trace of native hen lysozyme (structure 2lzt) showing the positions of the six buried water molecules (indicated by *blue spheres*) in cavities within the structure. The α domain of the protein (residues 1–35, 85–129) is shown in *light grey* and the β domain (residues 36–94) in *dark grey*. Residues with large main chain ^{15}N pressure shifts (>1.3 or < 0.20 ppm) are shown in *red* (Tyr53, Ile55, Ser60, Arg73, Ser91), those with large ^{1}H pressure shifts (>0.25 or < −0.08 ppm) in *green* (Trp62, Trp63, Asn65, Asp66, Gly67), and that with large ^{15}N and ^{1}H pressure shifts in *yellow* (Asn74) (Adapted from Fig. 5 in Kamatari et al. (2001) by permission of John Wiley & Sons, Inc)

Kamatari et al. used low temperatures (down to −20 °C) and high pressures (up to 2,000 bar) to populate low-lying excited state conformers of hen lysozyme, and analyzed their structures site-specifically using ^{1}H one-dimensional and ^{15}N/^{1}H two-dimensional HSQC NMR spectroscopy (Fig. 12.6). The experiment condition is taking the advantage of the fact that the freezing point of water is decreased to −21 °C at 2,000 bar (Jonas 1982). As shown in the ^{1}H one-dimensional spectra (Fig. 12.6a), the signals gradually broaden as the temperature is lowered, although the chemical shifts are largely unchanged even at the lowest temperature, −20 °C, reached at 2,000 bar, revealing that the overall structure of the native state is retained down to −20 °C at 2,000 bar. However, the features of the ^{15}N/^{1}H HSQC spectra recorded over the corresponding temperature range show very different behavior (Fig. 12.6b), the cross-peaks being selectively broadened above −10 °C and almost completely disappearing below −15 °C. The resulting disappearance of cross-peaks includes those of residues in the α-domain of the protein and the cleft between the α- and β-domains, both located close to water-containing cavities (Fig. 12.7a). The disappearance of some cross-peaks is often accompanied by the appearance of new peaks, indicating that a segment of the protein is denatured. However, this is not the case. Instead, the data shown in Fig. 12.6 revealed that the α-domain is not yet fully denatured but is likely to be partially disordered and heterogeneous in conformation around the original natively folded structure N. The high degree of broadening can be attributed to exchange effect that simply requires an increased dispersion of ^{1}H

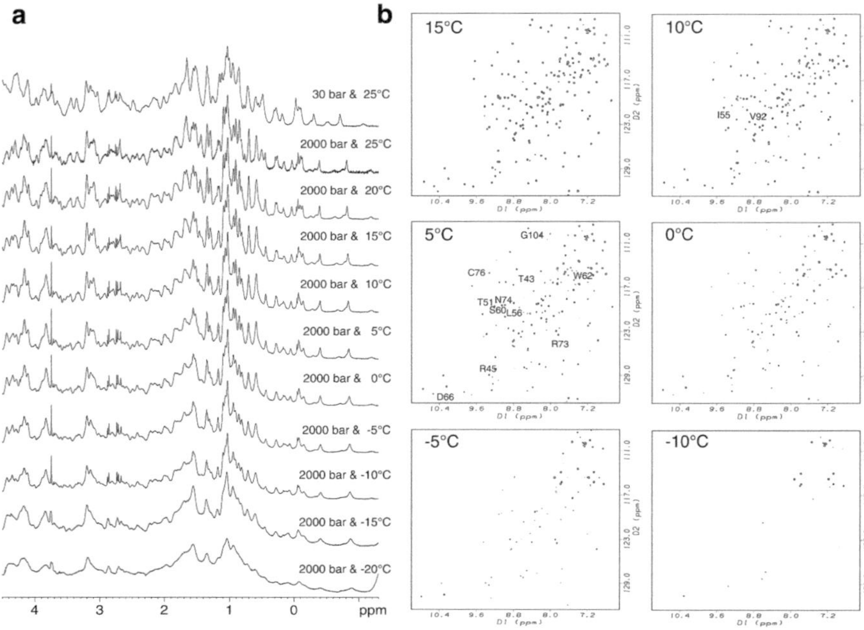

Fig. 12.6 Changes in NMR spectra of hen lysozyme at high pressure and low temperature. (**a**) ^{1}H one-dimensional NMR spectra of ^{15}N-uniformly labeled hen lysozyme (1.7 mM, pH 3.8) at 2,000 bar and various temperatures indicated. (**b**) $^{15}N/^{1}H$ two-dimensional HSQC spectra at 2,000 bar and various temperatures. Cross-peaks that disappear between 15 °C and 10 °C are labeled on the spectrum recorded at 10 °C and that disappear between 10 °C and 5 °C are labeled on the spectrum at 5 °C (Adapted from Fig. 1 in Kamatari et al. (2011) by permission of Elsevier Inc)

and ^{15}N chemical shifts. To cause the broadening of the NMR signals, the rate of exchange among these heterogeneous conformations must be rather slow on the NMR time scale of $\sim\mu s\sim ms$ (or $\sim 10^3 \sim 10^6\ s^{-1}$). The ensemble of conformers with substantial heterogeneity in the β-domain can be attributed to an ensemble of low-lying excited state conformers (N′, N″, N‴, etc.) stabilized under the condition of low temperature and high pressure relative to the stable folded conformer N (Fig. 12.7b). The fact that their populations are increased with increasing pressure indicates that they have slightly smaller partial molar volumes than the conformer N, consistent with a view that they are more hydrated than the N conformer.

In our previous $^{15}N/^{1}H$ HSQC NMR study of hen lysozyme carried out at the conditions under which the folded conformer is quite stable, we showed from pressure-induced chemical shifts that the fluctuations of the folded structure are most evident in the vicinity of the water-containing cavities (Fig. 12.5). If we compare Figs. 12.5 and 12.7a, residues in the vicinity of the water-containing cavities are commonly perturbed. Then is there any correlation between the two dynamic processes, i.e., conformational fluctuations within the ensemble of folded

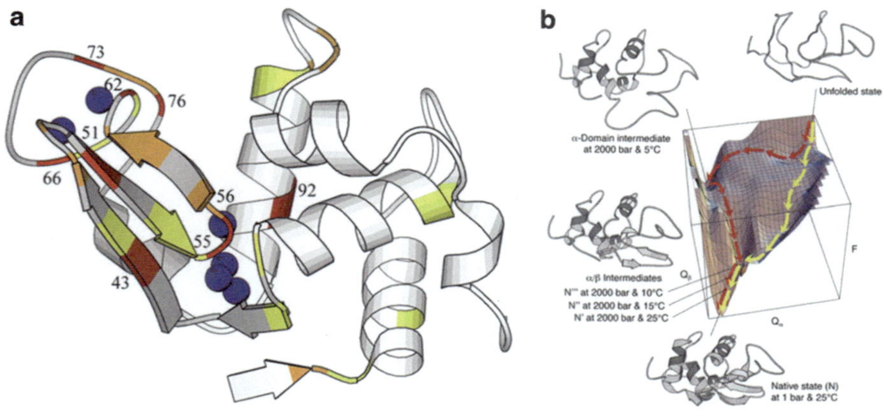

Fig. 12.7 NMR images of native hen lysozyme. (**a**) From $^{15}N/^{1}H$ HSQC cross-peak intensities at low temperature and 2,000 bar. The residues whose main-chain amide resonances disappear from the HSQC spectrum at temperatures above 7.5 °C, 2.5 °C, and 0 °C are shown in *red*, *orange*, and *yellow*, respectively. The main-chain trace of native hen lysozyme (structure 2lzt) showing the positions of the six buried water molecules (indicated by *blue spheres*) in cavities within the structure. (**b**) Probable locations of low-lying excited states (N′, N″, N‴, . . .) and α-domain intermediate depicted in this study on the schematic free-energy (F) surface representing features of the folding of hen lysozyme. Q_α and Q_β are the numbers of contacts in the α-and β-domains. The right trajectory represents a "fast track" in which α/β intermediates are populated only transiently. The left trajectory represents a "slow track" in which both α/β and α intermediates are populated transiently, depending on the time-scale (Adapted from Fig. 3 and 4 in Kamatari et al. (2011) by permission of Elsevier Inc)

conformations and conformational transitions to low-lying excited states? The finding suggests that the conformational fluctuations within the folded manifold are localized to the vicinity of the cavities but give rise to significant heterogeneity and disorder in the conformational ensemble involving a large number of residues in the α-domain.

Furthermore, $^{15}N/^{1}H$ HSQC NMR study of hen lysozyme carried out at higher pressure and lower temperature shown here provides an example to link the equilibrium intermediate stabilized by pressure and the kinetic intermediate in folding, the two conformers being generally coincident (Kitahara et al. 2003). Therefore, high-pressure NMR can provide a unique opportunity for studying structural details of kinetic intermediates involved in protein folding. In lysozyme, the preferential disappearance of signals from the entire β-domain seems to represent the hydration event just prior to the complete solvation or denaturation of the β-domain, leading to the formation of the "α-domain intermediate", found in the kinetic experiments (Radford et al. 1992; Miranker et al. 1993). Probable locations of these low-lying excited states and the α-domain intermediate on the free-energy surface are shown in Fig. 12.7b. Thus the study supports the view that cavities are the source of water penetration into the protein interior and lead to the formation of low-lying excited states.

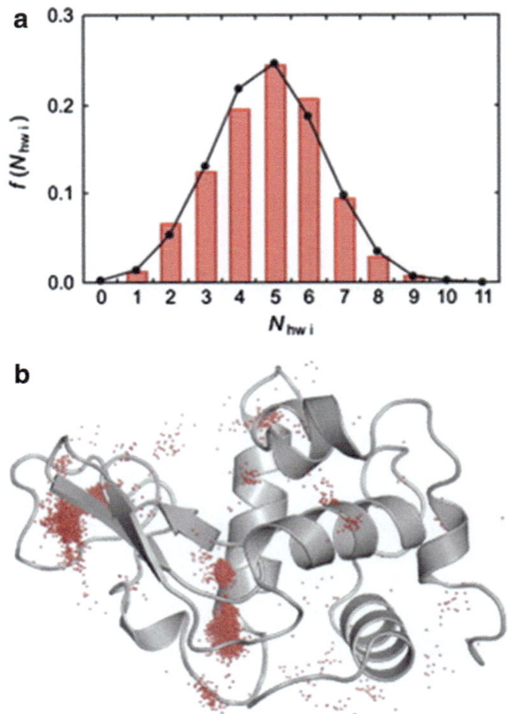

Fig. 12.8 (**a**) Probability distribution of the total number of internal hydration waters for the whole hydration sites, with its mean of 4.81. (**b**) Distribution of internal hydrated waters in the interior of hen lysozyme. The internal hydrated waters distribute fairly widely over the whole protein as well as in the major internal hydration sites of the hinge region (*center*) and the loop region in the β-domain (*left side*) (Adapted from Fig. 2 in Soda et al. (2011) by permission of Elsevier Inc)

12.3.3 Water Penetration into Hen Lysozyme Interior: MD Simulation Study

Recently, Soda et al. presented a new method for determining the hydration site of proteins using MD simulation, where the effect of structural fluctuations and hydration water was explicitly considered (Soda et al. 2011). Probability distribution of the total number of internal hydration waters and the spatial distribution of the internal hydration waters in the interior of hen lysozyme is shown in Fig. 12.8. It is interesting to find that internal hydration waters distribute fairly widely over the whole protein as well as in the major internal hydration sites of the hinge region (center) and the loop region in the β-domain (left side). These regions are just the places where the non-linear pressure shifts are distributed from the high-pressure NMR study (Kamatari et al. 2001).

12.3.4 Cavity is Evolutionarily Conserved for Protein Function

As mentioned in the previous section, cavity and hydration play important roles for protein dynamics. Therefore, cavity and hydration also expected to play important roles for protein function. Here, we show an example related to cavity and function.

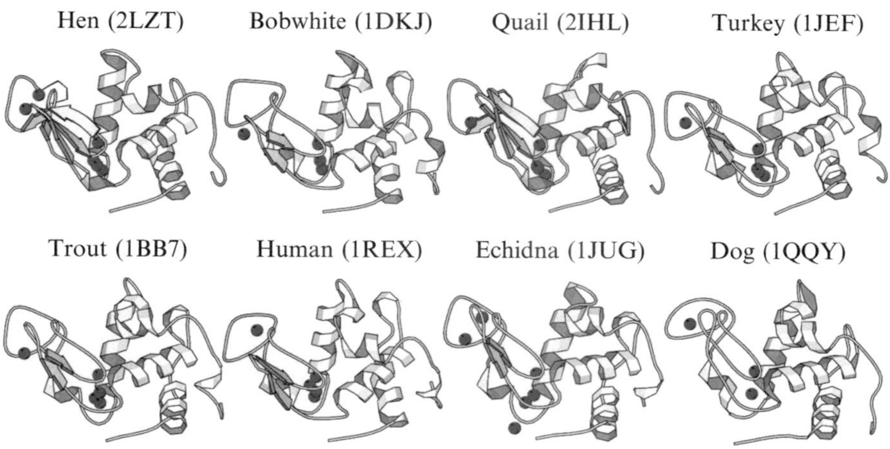

Fig. 12.9 Locations of cavities and buried water molecules (indicated by *blue spheres*) in a variety of structures of lysozyme from different biological species: hen, bobwhite, quail, turkey, trout, human, echidna, and dog (Adapted from Fig. 5 in Kamatari et al. (2011) by permission of Elsevier Inc)

Locations of water-containing cavities in the crystal structures of lysozymes from a variety of species are shown in Fig. 12.9 (Kamatari et al. 2011). The figure shows that the locations of the water-containing cavities are remarkably similar among the lysozyme molecules from different biological origin, despite the fact that the amino acid sequences are not highly conserved (for example, the sequence identity of hen and dog lysozyme is less than 50 %). This observation suggests that the cavities could be an important feature of the lysozyme function, and is consistent with the view that the cavities play a crucial role in the catalytic cycle of lysozymes by giving a degree of mobility in the active site, always with a well-defined population of water molecules that could contribute to the hydrolysis of the substrate molecules. This result also suggests cavities are evolutionarily conserved for protein function.

12.4 Transient Cavities and Protein Dynamics

All cavities described above in this chapter exist in equilibrium. However, cavity can also appear transiently by the conformational fluctuations of proteins. One example is flip-flop motions of aromatic rings (Fig. 12.10a). In spite of close packing of amino acid residues in globular proteins, aromatic side chains buried in the protein interior undergo rotational motion. This phenomenon indicates that protein flexibility provides the necessary space for a ring to flip. The flip-flop motions reflect the dynamic feature of a protein, and the flip rates of the side chains serve as probes reporting the fluctuation of the interior of the folded structure.

Temperature and pressure dependence of the flip rate is expected to bring crucial information on energy and volume fluctuations in the interior of the folded protein.

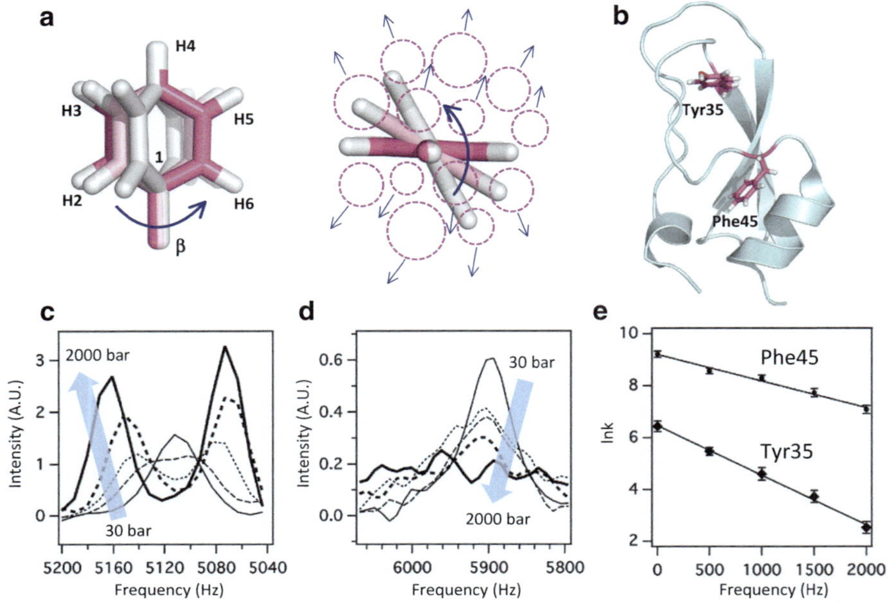

Fig. 12.10 Transient void detected through ring flip motions. (**a**) Illustration of the aromatic ring flip motion taking Phe as an example. The aromatic ring undergoes 180° flip motion along its C_β–C_1 axis. *Right panel* is the view from along C_β–C_1 axis. The *dotted circles* represent the atoms around Phe schematically. Flip-flop motion of Phe would result in the formation of transient cavities. (**b**) A ribbon model for BPTI with locations of Tyr35 and Phe45. (**c**) Superposition of the 3H and 5H region of the TOCSY slice spectra taken at 2H resonance position of Tyr35 measured at 57 °C at five different pressures. (**d**) Superposition of the 3H and 5H region of the TOCSY slice spectra taken at 2H resonance position of Phe45 measured at 57 °C at five different pressures. In (**c**) and (**d**), the arrow shows the direction of pressure increase: 1 bar (*line*), 500 bar (*dotted line*), 1,000 bar (*dashed line*), 1,500 bar (*thick dotted line*), and 2,000 bar (*thick line*). (**e**) Logarithmic plot of the flip-flop rate of the ring, k (in s^{-1}), for Tyr35 and Phe45 against pressure (Adapted from Fig. 5 in Li et al. (1999) by permission of Elsevier Inc)

The temperature dependence of the flip rate derives the activation enthalpy ($\Delta H^{\ddagger}$), while the pressure dependence of the flip rate derives the activation volume ($\Delta V^{\ddagger}$). Li et al. (1999) investigated the pressure-dependent ring flip motions of BPTI at 57 °C over the pressure range 30–2,000 bar using the 2D ^{1}H NMR spectroscopy. The flip rate of each residue is estimated by the line shape simulation, and the activation volume is calculated from the pressure-dependent flip rate. As a result, the flip rate of each residue is estimated to be on the ms timescale, ranging from 10 s^{-1} to 10^4 s^{-1}, and is decreased with the increase of pressure. Furthermore, the slope of the plot of the logarithms of the flip rates obtained from the simulation against pressure gives the $\Delta V^{\ddagger}$ value through the relation:

$$\Delta V^{\ddagger} = -k_{\rm B}T\left(\frac{\partial \ln k}{\partial P}\right)_T \qquad (12.3)$$

Table 12.1 Thermodynamic parameters for aromatic residues in BPTI, HPr, and cytochrome *c*

Protein (experimental condition)	Residue	Proton (or carbon)	$\Delta H^{\ddagger}$ (kJ mol^{-1})	$\Delta S^{\ddagger}$ (J K^{-1} mol^{-1})	$\Delta V^{\ddagger}$ (mL mol^{-1})
BPTI					
(0.1 MPa/329 K)[a,b,c]	Tyr23	Hδ	109	146	–
(0.1 MPa/338 K)[d]		Cε	86	81	–
(0.1 MPa/329 K)[a,b,c]	Tyr35	Hδ	155	284	36
(0.1 MPa/330 K)[e]		Hδ	139	223	–
(0.1 MPa/330 K)[f]		Hδ	–	–	51
(0.1 MPa/328 K)[d]		Cε	83	41	–
(0.1 MPa/329 K)[a,b,c]	Phe45	Hδ	71	46	30
(0.1 MPa/330 K)[f]		Hε	–	–	28
(0.1 MPa/338 K)[d]		Cδ	86	86	–
HPr					
(3 MPa/275 K)[g]	Tyr6	Hδ	87	16	26
		Hε	87	16	28
(200 MPa/298 K)[g]	Tyr6	Hδ	89	15	25
		Hε	92	16	27
Cytochrome *c*					
(0.1 MPa)[h]	Tyr	Hδ	97	96	–

[a]Data from Wagner et al. (1976)
[b]Data from Wagner (1980)
[c]Data from Wagner (1983)
[d]Data from Weininger et al. (2014)
[e]Data from Otting et al. (1993)
[f]Data from Li et al. (1999)
[g]Data from Hattori et al. (2004)
[h]Data from Campbell et al. (1976)

where k_B is the Boltzmann constant, T is the absolute temperature, and k is the flip rate. The experimental results and data analysis of BPTI are shown in Fig. 12.10c–e. The $\Delta V^{\ddagger}$ value is defined as a difference between the volume occupied by the ring in the ground state and the volume in the activated state. The positive $\Delta V^{\ddagger}$ value shown in Table 12.1 indicates that the flip requires opening of the space surrounding the aromatic ring at the activation state. It is noteworthy that, despite the variety of microenvironments and types of the individual aromatic rings, the $\Delta V^{\ddagger}$ value is comparable to each other, ranging from 25 to 51 mL mol^{-1}. This observation fits well with the notion that the flip requires a minimum common space or cavity to be created around the ring. This extra space must be created as a result of cooperative thermal fluctuations of atoms surrounding the aromatic ring, occurring infrequently (10^3–10^5 s^{-1}) around it at ambient pressure. This means that a cavity is infrequently created in the most buried part of the protein (the core).

The first and the best-characterized case for the slowly exchanging aromatic residues is found in BPTI. Temperature and pressure dependence of the ring flip motions were investigated using 1D ^{1}H NMR spectroscopy (Wagner et al. 1976, 1980, 1983). Li et al. renewed the pressure dependence the ring flip motions using 2D ^{1}H NMR spectroscopy. Besides BPTI, temperature dependence of the rate of ring flip has been reported for phenylalanine and tyrosine residues of cytochrome *c* (Campbell et al. 1976) and hen lysozyme (Campbell et al. 1975). Later, pressure dependence of the rate of ring flip has been reported for HPr using high-pressure NMR techniques (Hattori et al. 2004). Recently, ^{13}C transverse relaxation dispersion methods have been developed to characterize the ring-flip motion and renewed the rank order of ring-flip rates in BPTI (Weininger et al. 2014). In Table 12.1, the thermodynamic parameters for aromatic residues in BPTI, HPr, and cytochrome *c* are summarized.

It would be interesting to compare the activation volume with the equilibrium volume fluctuation of the protein. The root-mean-square fluctuation would be 0.3 ± 0.1 % of the total volume of the protein (Gekko and Hasegawa 1986). In the case of BPTI, the root-mean-square fluctuation is estimated to be only $\sim$24 Å^3. Thus the local volume fluctuation, reflected by the activation volume, is far exceeding the fluctuation of the entire protein molecule. This result suggests that the formation of a large local cavity may be a consequence of accumulation of smaller cavities, and appears to be consistent with the dynamic nature of cavity formation (Kocher et al. 1996) or a mobile defect model such that water occasionally penetrates into the core of the protein (Lumry and Rosenberg 1975).

It is also interesting to notice that the activation volume of the ring-flip motions gives a positive value, while the volume change induced by pressure is negative. The positive value of the activation volume indicates transient opening of the space surrounding the aromatic ring at the activation state, reflecting the intrinsic dynamic nature of the N state. On the other hand, the negative volume change induced by pressure is caused by decrease of the cavity size or even water penetration into the cavities. The critical fluctuations of protein molecules, as manifested by ring-flips occurring in ms range, take place in a range far from the range of equilibrium fluctuation. As discussed in Sect. 12.2.3.5, however, the equilibrium of the low-lying excited states is in μs-ms range.

12.5 Summary

In this chapter, we have highlighted the importance of cavities for protein dynamics, stability, and function. The equilibrium cavities, caused by atomic packing defects, are commonly found inside protein molecules, and sometimes filled with internal waters molecules. Using MD simulation and high-pressure NMR spectroscopy, it has been shown that the equilibrium cavities provide the structural fluctuation, serve as sources of water penetration and hence the creation of low-lying excited states and even the gateway to unfolding. The analysis of the nonlinear chemical shifts induced by pressure has revealed the existence of the low-lying excited states in many globular proteins, while the equilibrium between the ground state and the excited states may be fast or slow. Furthermore, in the N state, besides the equilibrium cavities, the transient cavities have also been detected, as manifested by the ring-flip motions. The transient cavities accompanied by transient space opening, are intrinsic dynamic nature of the N state. Therefore, cavities, both equilibrium and transient, serve as the source of protein dynamics, and the evolutionarily conserved features of cavities and low-lying excited states of proteins are expected to shed new light on the molecular dynamism of proteins.

References

Akasaka K (2006) Probing conformational fluctuation of proteins by pressure perturbation. Chem Rev 106:1814–1835

Akasaka K, Li H (2001) Low-lying excited states of proteins revealed from nonlinear pressure shifts in ^{1}H and ^{15}N NMR. Biochemistry 40:8665–8671

Akasaka K, Yamada H (2001) On-line cell high-pressure nuclear magnetic resonance technique: application to protein studies. Methods Enzymol 338:134–158

Akasaka K, Li H, Yamada H, Li R, Thoresen T, Woodward CK (1999) Pressure response of protein backbone structure. Pressure-induced amide ^{15}N chemical shifts in BPTI. Protein Sci 8:1946–1953

Baxter NJ, Hosszu LL, Waltho JP, Williamson MP (1998) Characterisation of low free-energy excited states of folded proteins. J Mol Biol 284:1625–1639

Binkowski TA, Naghibzadeh S, Liang J (2003) CASTp: computed atlas of surface topography of proteins. Nucleic Acids Res 31:3352–3355

Campbell ID, Dobson CM, Williams RJ (1975) Proton magnetic resonance studies of the tyrosine residues of hen lysozyme-assignment and detection of conformational mobility. Proc R Soc Lond B Biol Sci 189:503–509

Campbell ID, Dobson CM, Moore GR, Perkins SJ, Williams RJP (1976) Temperature dependent molecular motion of a tyrosine residue of ferrocytochrome C. FEBS Lett 70:96–100

Chakravarty S, Bhinge A, Varadarajan R (2002) A procedure for detection and quantitation of cavity volumes proteins. Application to measure the strength of the hydrophobic driving force in protein folding. J Biol Chem 277:31345–31353

Collins MD, Hummer G, Quillin ML, Matthews BW, Gruner SM (2005) Cooperative water filling of a nonpolar protein cavity observed by high-pressure crystallography and simulation. Proc Natl Acad Sci U S A 102:16668–16671

Fourme R, Girard E, Dhaussy AC, Medjoubi K, Prangé T, Ascone I, Mezouar M, Kahn R (2011) A new paradigm for macromolecular crystallography beamlines derived from high-pressure methodology and results. J Synchrotron Radiat 18:31–36

Gekko K, Hasegawa Y (1986) Compressibility-structure relationship of globular proteins. Biochemistry 25:6563–6571

Girard E, Kahn R, Mezouar M, Dhaussy AC, Lin T, Johnson JE, Fourme R (2005) The first crystal structure of a macromolecular assembly under high pressure: CpMV at 330 MPa. Biophys J 88:3562–3571

Hattori M, Li H, Yamada H, Akasaka K, Hengstenberg W, Gronwald W, Kalbitzer HR (2004) Infrequent cavity-forming fluctuations in HPr from *Staphylococcus carnosus* revealed by pressure- and temperature-dependent tyrosine ring flips. Protein Sci 13:3104–3114

Imai T, Sugita Y (2010) Dynamic correlation between pressure-induced protein structural transition and water penetration. J Phys Chem B 114:2281–2286

Imai T, Hiraoka R, Kovalenko A, Hirata F (2007) Locating missing water molecules in protein cavities by the three-dimensional reference interaction site model theory of molecular solvation. Proteins 66:804–813

Inoue K, Yamada H, Akasaka K, Herrmann C, Kremer W, Maurer T, Döker R, Kalbitzer HR (2000) Pressure-induced local unfolding of the Ras binding domain of RalGDS. Nat Struct Biol 7:547–550

Jonas J (1982) Nuclear magnetic resonance at high pressure. Science 216:1179–1184

Kalbitzer HR, Görler A, Li H, Dubovskii PV, Hengstenberg W, Kowolik C, Yamada H, Akasaka K (2000) ^{15}N and ^{1}H NMR study of histidine containing protein (HPr) from *Staphylococcus carnosus* at high pressure. Protein Sci 9:693–703

Kamatari YO, Yamada H, Akasaka K, Jones JA, Dobson CM, Smith LJ (2001) Response of native and denatured hen lysozyme to high pressure studied by ^{15}N/^{1}H NMR spectroscopy. Eur J Biochem 268:1782–1793

Kamatari YO, Kitahara R, Yamada H, Yokoyama S, Akasaka K (2004) High-pressure NMR spectroscopy for characterizing folding intermediates and denatured states of proteins. Methods 34:133–143

Kamatari YO, Smith LJ, Dobson CM, Akasaka K (2011) Cavity hydration as a gateway to unfolding: an NMR study of hen lysozyme at high pressure and low temperature. Biophys Chem 156:24–30

Kitahara R, Akasaka K (2003) Close identity of a pressure-stabilized intermediate with a kinetic intermediate in protein folding. Proc Natl Acad Sci U S A 100:3167–3172

Kitahara R, Yamada H, Akasaka K (2001) Two folded conformers of ubiquitin revealed by high-pressure NMR. Biochemistry 40:13556–13563

Kitahara R, Yokoyama S, Akasaka K (2005) NMR snapshots of a fluctuating protein structure: ubiquitin at 30 bar–3 kbar. J Mol Biol 347:277–285

Kitahara R, Hata K, Li H, Williamson MP, Akasaka K (2013) Pressure-induced chemical shifts as probes for conformational fluctuations in proteins. Prog Nucl Magn Reson Spectrosc 71:35–58

Kleywegt GJ, Jones TA (1994) Detection, delineation, measurement and display of cavities in macromolecular structures. Acta Crystallogr D Biol Crystallogr 50:178–185

Kocher JP, Prevost M, Wodak SJ, Lee B (1996) Properties of the protein matrix revealed by the free energy of cavity formation. Structure 4:1517–1529

Kuwata K, Li H, Yamada H, Batt CA, Goto Y, Akasaka K (2001) High pressure NMR reveals a variety of fluctuating conformers in β-lactoglobulin. J Mol Biol 305:1073–1083

Levitt DG, Banaszak LJ (1992) POCKET: a computer graphics method for identifying and displaying protein cavities and their surrounding amino acids. J Mol Graph 10:229–234

Li H, Akasaka K (2006) Conformational fluctuations of proteins revealed by variable pressure NMR. Biochim Biophys Acta 1764:331–345

Li H, Yamada H, Akasaka K (1998) Effect of pressure on individual hydrogen bonds in proteins. Basic pancreatic trypsin inhibitor. Biochemistry 37:1167–1173

Li H, Yamada H, Akasaka K (1999) Effect of pressure on the tertiary structure and dynamics of folded basic pancreatic trypsin inhibitor. Biophys J 77:2801–2812

Liang J, Edelsbrunner H, Fu P, Sudhakar PV, Subramaniam S (1998a) Analytical shape computation of macromolecules: I. Molecular area and volume through alpha shape. Proteins 33:1–17

Liang J, Edelsbrunner H, Woodward C (1998b) Anatomy of protein pockets and cavities: measurement of binding site geometry and implications for ligand design. Protein Sci 7:1884–1897

Liu L, Quillin ML, Matthews BW (2008) Use of experimental crystallographic phases to examine the hydration of polar and nonpolar cavities in T4 lysozyme. Proc Natl Acad Sci U S A 105:14406–14411

Lumry R, Rosenberg A (1975) The mobile defect hypothesis of protein function. Coll Int CNRS L'Eau Syst Biol 246:55–63

Matthews BW, Liu L (2009) A review about nothing: are apolar cavities in proteins really empty? Protein Sci 18:494–502

Miranker A, Robinson CV, Radford SE, Aplin RT, Dobson CM (1993) Detection of transient protein folding populations by mass spectrometry. Science 262:896–900

Nicholls A, Sharp KA, Honig B (1991) Protein folding and association: insights from the interfacial and thermodynamic properties of hydrocarbons. Proteins 11:281–296
Otting G, Liepinsh E, Wüthrich K (1993) Disulfide bond isomerization in BPTI and BPTI(G36S): an NMR study of correlated mobility in proteins. Biochemistry 32:3571–3582
Pin S, Royer CA, Gratton E, Alpert B, Weber G (1990) Subunit interactions in hemoglobin probed by fluorescence and high-pressure techniques. Biochemistry 29:9194–9202
Radford SE, Dobson CM, Evans PA (1992) The folding of hen lysozyme involves partially structured intermediates and multiple pathways. Nature 358:302–307
Refaee M, Tezuka T, Akasaka K, Williamson MP (2003) Pressure-dependent changes in the solution structure of hen egg-white lysozyme. J Mol Biol 327:857–865
Sgourakis NG, Day R, McCallum SA, Garcia AE (2008) Pressure effects on the ensemble dynamics of ubiquitin inspected with molecular dynamics simulations and isotropic reorientational eigenmode dynamics. Biophys J 95:3943–3955
Soda K, Shimbo Y, Seki Y, Taiji M (2011) Structural characteristics of hydration sites in lysozyme. Biophys Chem 156:31–42
Voss NR, Gerstein M (2010) 3V: cavity, channel and cleft volume calculator and extractor. Nucleic Acids Res 38:W555–W562
Wagner G (1980) Activation volumes for the rotational motion of interior aromatic rings in globular proteins determined by high resolution ^{1}H NMR at variable pressure. FEBS Lett 112:280–284
Wagner G (1983) Characterization of the distribution of internal motions in the basic pancreatic trypsin inhibitor using a large number of internal NMR probes. Q Rev Biophys 16:1–57
Wagner G, DeMarco A, Wüthrich K (1976) Dynamics of the aromatic amino acid residues in the globular conformation of the basic pancreatic trypsin inhibitor (BPTI). I. ^{1}H NMR studies. Biophys Struct Mech 2:139–158
Weininger U, Modig K, Akke M (2014) Ring flips revisited: ^{13}C relaxation dispersion measurements of aromatic side chain dynamics and activation barriers in basic pancreatic trypsin inhibitor. Biochemistry 53:4519–4525
Williamson MP, Akasaka K, Refaee M (2003) The solution structure of bovine pancreatic trypsin inhibitor at high pressure. Protein Sci 12:1971–1979
Wilton DJ, Kitahara R, Akasaka K, Williamson MP (2009) Pressure-dependent ^{13}C chemical shifts in proteins: origins and applications. J Biomol NMR 44:25–33

Part IV
Pressure and Protein Folding and Assembly

Editors' Foreword of Part IV

This part deals with protein unfolding and assembly. Chapter 13 gives the real-time and steady-state kinetics of protein unfolding using multi-dimensional high pressure NMR spectroscopy. It shows detailed folding pathway that consists of both a structural description of the existence of intermediate states and a dynamic description on the rates of local rearrangements. Then, a fairly complete structural and energetic description of the folding landscape for a model protein and its mutants is given. Chapter 14 deals with the association and dissociation processes of amyloid fibrils performed on hen lysozyme mutants whose disulfide bonds are strategically removed. It gives the first detailed description on the molecular mechanism of amyloid fibril formation and dissociation. Chapter 15 is a topic dealing with the pressure-induced viral inactivation of many viruses of clinical importance and the production of inactivated viral vaccines for animals and humans. Here we can understand that pressure can bring about virus inactivation while preserving immunogenic properties.

Chapter 13
Exploring the Protein Folding Pathway with High-Pressure NMR: Steady-State and Kinetics Studies

Julien Roche, Mariano Dellarole, Catherine A. Royer, and Christian Roumestand

Abstract Defining the physical-chemical determinants of protein folding and stability, under normal and pathological conditions has constituted a major subfield in biophysical chemistry for over 50 years. Although a great deal of progress has been made in recent years towards this goal, a number of important questions remain. These include characterizing the structural, thermodynamic and dynamic properties of the barriers between conformational states on the protein energy landscape, understanding the sequence dependence of folding cooperativity, defining more clearly the role of solvation in controlling protein stability and dynamics and probing the high energy thermodynamic states in the native state basin and their role in misfolding and aggregation. Fundamental to the elucidation of these questions is a complete thermodynamic parameterization of protein folding determinants. In this chapter, we describe the use of high-pressure coupled to Nuclear Magnetic Resonance (NMR) spectroscopy to reveal unprecedented details on the folding energy landscape of proteins.

Keywords High pressure • NMR spectroscopy • Protein folding

13.1 Introduction

The phenomenon of spontaneous protein folding underlies nearly all key biological processes, yet despite decades of intense research and significant progress in both experiment and theory (Wolynes et al. 2012; Dill and McCallum 2012) the

---

J. Roche • M. Dellarole • C. Roumestand (✉)
Centre de Biochimie Structurale, UMR UM1&UM2/5048 CNRS/1054 INSERM, 29 rue de Navacelles, 34090 Montpellier, France
e-mail: christian.roumestand@cbs.cnrs.fr

C.A. Royer
Department of Biology, Rensselaer Polytechnic Institute, Troy, NY 12180, USA
e-mail: royerc@rpi.edu

K. Akasaka, H. Matsuki (eds.), *High Pressure Bioscience*,
Subcellular Biochemistry 72, DOI 10.1007/978-94-017-9918-8_13

mechanisms and determinants of folding remain incompletely understood. A protein is a hetero-polymer, the basic motif for which is one of the 20 natural amino acids. The chaining of these motives constitutes the sequence of the protein. Individual protein sequences have evolved in order to adopt a tridimensional (3D) structure that confers the level of stability, cooperativity and conformational dynamics required for their optimal function under the conditions in which the organism must survive. However, we have yet to reveal the subtle rules by which sequence determines these properties. Protein folding energy landscapes remain to be experimentally mapped. A general description of folding transition states and routes cannot be predicted for arbitrary amino acid sequences. Likewise, a protein's propensity to aggregate cannot be deduced from its sequence. Finally, we do not know how sequence encodes the conformational fluctuations and heterogeneity required for function in specific contexts. Nevertheless such a high level of understanding will be required for further progress in the conception and modulation of proteins, enzymes and small molecules for therapeutic or biotechnological applications. Importantly, numerous "conformational" diseases (Alzheimer, Parkinson, Prion diseases) are associated with a deregulation of the folding/unfolding equilibrium for a specific protein (Dobson 2005), which justifies – if needed – the efforts dedicated to the study of this phenomenon.

Most of the information gathered on this phenomenon comes from folding/unfolding experiments performed in vitro. Several methods are possible to unfold a protein: adding chaotropic reagents (urea, guanidinium chloride...), modifying the physicochemical parameters of the sample (P, T, pH)... In spite of difficulties for implementation, pressure is a method of choice to unfold a protein: it is a "soft" method, generally reversible, that give access to a large panel of thermodynamic parameters, specific to the folding/unfolding reaction (Kamatari et al. 2004; Akasaka et al. 2013). It is generally used in conjunction with spectroscopic methods, like Circular Dichroism, fluorescence or IR spectroscopy (Dellarole and Royer 2013), which bring global information on the evolution of the tertiary or secondary structure during protein unfolding. Associated to a method which provided access to local information, such as Nuclear Magnetic Resonance (NMR), pressure can yield extremely detailed, residue specific information on protein folding. The combination of high pressure and NMR constitutes a powerful tool that can lead to a new knowledge about the role of residue packing in protein stability, of conformational fluctuations in water penetration, or that can be used to describe folding intermediates or other details of the energy landscape, otherwise invisible when using other approaches.

13.2 Unfolding Protein with High Hydrostatic Pressure

As noted in other contributions to this volume, the physical basis for the effects of pressure on protein structure and stability remains controversial, in contrast to a relatively clear physical understanding of temperature effects on protein structure (Kauzmann 1959; Privalov and Gill 1988; Murphy et al. 1990). The fundamental observation of pressure effects has been that over most of the accessible temperature range, the application of pressure leads to the unfolding of proteins, indicating that the volume change upon unfolding is negative, i.e., *the specific molar volume of the unfolded state is smaller that that of the folded state* (reviewed in Royer 2002). This observation is rather counter-intuitive, as one might expect that pressure would simply compress the folded state of the protein. While this compression does occur, pressure eventually unfolds most proteins, due to this negative volume change upon unfolding and Le Chatelier's principle, which states that that application of any perturbation will shift a system in equilibrium toward the state that alleviates the perturbation. In the case of pressure perturbation, this is the state that occupies the smallest volume, i.e., the unfolded state.

The relation establishing the difference of free energy between the native and the unfolded states at a given pressure ($\triangle G(p)$) depends on the volume difference ($\triangle V$, first order term) and the difference in compressibility ($\triangle\beta$, second order term):

$$\Delta G(p) = \Delta G^0 + \Delta V\,(p - p_0) - \frac{\Delta\beta}{2}(p - p_0)^2 \qquad (13.1)$$

where $\triangle G^0$ stands for the difference in free energy between native and unfolded states at atmospheric pressure (p_0). The difference in compressibility between the native and unfolded states is typically found to be negligible compared to the experimental error in $\triangle G$(p) measurement, such that Eq. 13.1 is usually simplified to:

$$\Delta G(p) = \Delta G^0 + \Delta V\,(p - p_0) \qquad (13.2)$$

As noted previously (Royer 2002), and in other contributions to this volume, the origin of $\triangle V$ remains matter of debate. Several effects have been put forward: (i) a difference in compressibility between the native and the unfolded states of proteins (Chalikian and Macgregor 2009), (ii) pressure induced changes in the structure of bulk water (Grigera and McCarthy 2010), (iii) density difference for water molecules close to polar or apolar amino acids (Mitra et al. 2006), (iv) lost of dehydrated cavities inside the native structure of a protein, or a combination of these different effects. Recently, using a model protein (the Staphyloccocal Nuclease hyper-stable variant: $\triangle$ + PHS SNase) and several of its mutants, we have shown that the main contribution to $\triangle V$ was the presence of residual cavities inside the native 3D structure (Fig. 13.1) (Roche et al. 2012a; Rouget et al. 2011).

The $\triangle V$ values accessible through techniques like fluorescence spectroscopy represent model dependent global values and as such are not thermodynamic quantities, per se. Techniques allowing atomic resolution, such as NMR, provide not one spectroscopic probe, but dozens, which are site-specific, allowing observation of pressure effects over the entire sequence and structure of the protein. Comparing the $\triangle V$ values from all of these observables reveals deviations from the two state model. Indeed, if unfolding is truly 2-state, then the pressure unfolding NMR profiles from all of the site-specific resonances should be identical within experimental error. Resonances with significantly different profiles can reveal the existence of intermediates and what is more, provide information about their structure.

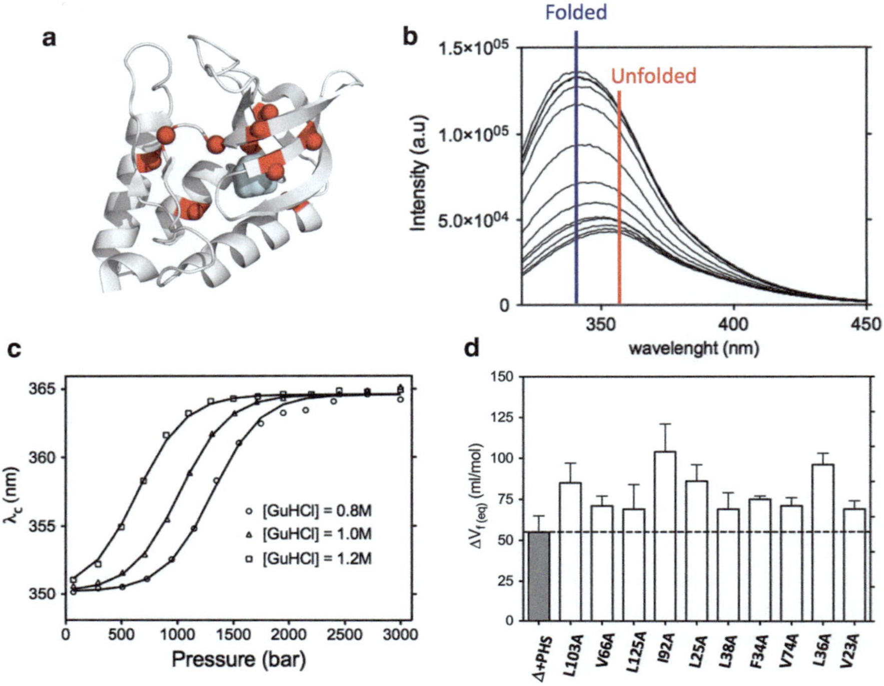

Fig. 13.1 Unfolding of SNase and of its cavity mutants: a fluorescence study. (**a**) SNase 3D structure (hyper-stable variant $\Delta + $PHS): the native cavity is colored in light blue, the mutated residues in *red*. (**b**) Pressure dependence of the SNase tryptophan (Trp140) emission fluorescence. At low pressure, the fluorescence emission is intense, with a frequency located in the «blue» region of the visible light spectrum, indicative of a tryptophan residue buried in the hydrophobic core of a folded protein. At high pressure, the emission drop and shifts to «red», suggesting a solvent (water) exposed tryptophan residue, as expected in an unfolded protein. (**c**) The center of mass of the fluorescence spectra reported in (**b**) are plotted as a function of the pressure. The 3 curves stand for 3 different samples with 3 different guanidinium chloride concentrations. Addition of this chaotropic reagent is necessary to unfold SNase in the pressure range experimentally available ($\leq$300 MPa). The sigmoidal behavior of these curves suggests a two-states equilibrium between the native and the unfolded protein. Note the similar slope within the three curves, suggesting that addition of guanidinium chloride has little effect on ΔV. (**d**) Measured ΔV on $\Delta + $PHS SNase and 10 of its cavity mutants: introducing an additional cavity (or enlarging the original native cavity) brings a concomitant increase in ΔV. Modified from results published (Roche et al. 2012a)

13.3 NMR Spectroscopy: A Method of Investigation at a "Local" Scale

With X-Ray crystallography, NMR is the only technique available for solving the three-dimensional structure of a molecule at atomic resolution. An NMR structure is obtained "indirectly", by building a 3D model from inter-atomic distances deduced from the quantification of magnetic interactions between the magnetically active nuclei of the molecule. Thus, the classical strategy used to solve the 3D structure of a protein can be divided into two steps: the resonance assignment of the active nuclei (^{1}H, ^{13}C and ^{15}N), then the measurement of the interactions between these nuclei. These interactions, essentially the dipolar through-space interactions, are translated into distance restraints used within molecular dynamics algorithms for the calculation of the 3D structure. Resonance assignment is an important step, requiring the exploitation of state-of-the-art spectroscopy techniques that can be grouped together under the generic term of "correlation spectroscopy". Given the number of nuclei concerned, protein spectra are eminently complex. Thus, correlation spectroscopy is generally coupled to multidimensional spectroscopy, allowing distributing the information initially contained in a 1D spectrum along two or three (sometimes more!) dimensions, yielding a considerable gain in resolution and facilitating their interpretation.

For a biophysicist, the interest of NMR to study protein folding lies in the close relationship between the resonance frequencies and the chemical, but also tridimensional, structure of the protein: a proton covalently bound to a nitrogen atom has a different resonance frequency than a proton covalently bound to a carbon atom, and protons borne by the same chemical group (methyl groups, for instance . . .) will have a different resonance frequency if similar chemical groups are involved in different structures (β-sheet, α-helix . . .), if they "see" a different environment. Thus, the $\Delta +$ PHS SNase 2D correlation spectrum reported in Fig. 13.2 concerns nuclei (^{1}H and ^{15}N) belonging to similar chemical groups: the amide group (HN) that contributes to the peptide bond along the protein backbone. From a residue to another, their resonance frequencies differ, since they are involved in different environments inside the protein 3D structure.

Each amino acid bears an HN group, with the exception of proline, generally a minority residue in the composition of soluble proteins (<3 %), and will give rise to a specific correlation in the 2D [^{1}H-^{15}N] HSQC spectrum of the protein. This HSQC spectrum (Bodenhausen and Ruben 1980) can be seen as a protein's "finger-print": any "chemical" (mutation) or structural modification will bring about a change in the amide resonance frequencies in the region concerned by the modification. Thus, these resonances constitute local probes, allowing detection of any perturbation in the 3D structure of the protein.

While protein unfolding does not bring any chemical modification of its sequence, it considerably modifies its 3D structure. This perturbation can be measured locally, residue-per-residue, by following the evolution of the correlation peaks on HSQC experiments recorded with increasing pressures (Fig. 13.3). In a two-state model of the unfolding/refolding equilibrium, the exchange between the native and unfolded forms occurs on a time scale slower than the time scale of the NMR measurement. Thus, we do not observe chemical shift variations (a displacement of the cross-peaks on the HSQC spectra, indicative of fast exchange) as a function of pressure, but the disappearance of correlation peaks belonging to the native form, with the concomitant appearance of new peaks (centered at 8.5 ppm on the proton chemical-shift axis) that correspond to the spectrum of the unfolded species. The weak spectral dispersion of these cross-peaks corresponds to an

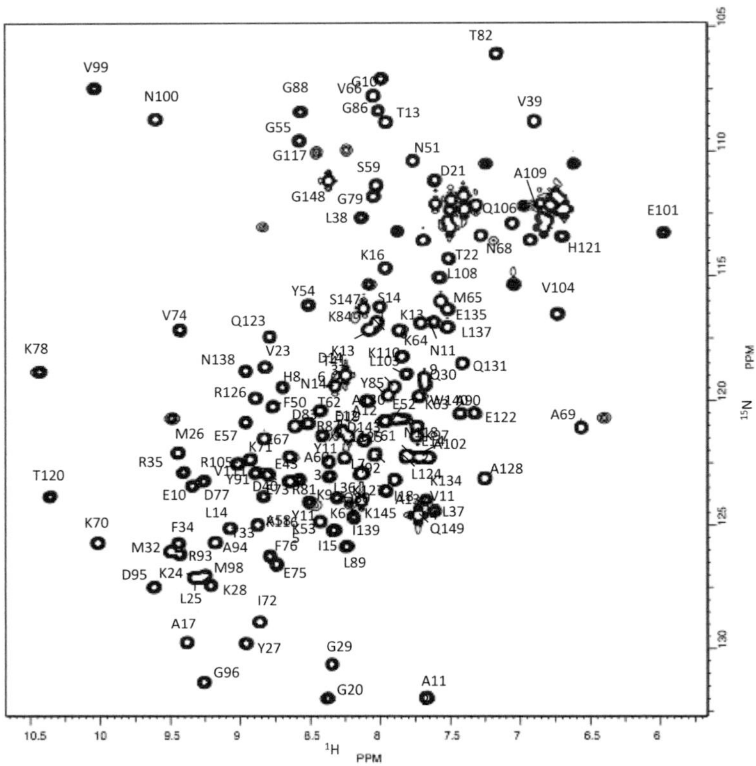

Fig. 13.2 2D [^{1}H-^{15}N] HSQC spectrum recorded on a Δ + PHS SNase sample. This spectrum reveals heteronuclear scalar (through-bond) correlations between the proton and the nitrogen nuclei from a given amide group (HN). The cross-peaks are labeled with the «one-letter» code for amino acids. ^{1}H and ^{15}N resonance frequencies are displayed on the horizontal and the vertical axes, respectively. Instead of the absolute frequency (in Hz), the nuclei resonances are reported in ppm, a relative measurement that allows ignoring the magnetic induction delivered by the magnet used for the measurement

unfolded protein: the native 3D structure turns into a "random coil-like" structure, yielding a similar environment for all residues, hence for all amide groups. The residual dispersion comes from a weak "chemical" effect due to the different nature of the side chain between different the 20 natural amino acids used in a protein.

Coupling high-pressure and NMR raises complicated technological issues. The experimental device should of course support high pressure, but must be non-magnetic and permeable to radiofrequency. Among all the alternatives proposed (reviewed in Fourme et al. 2012), a satisfying commercial solution has emerged, based on the use of ceramic tubes (zirconium oxide). These tubes can support up to 300 MPa (3 kbar), usually enough to unfold protein. If proteins are too stable to be unfolded by pressure over this range, one can adjust the protein stability, generally by adding low concentration of chaotropic reagents. The ceramic high

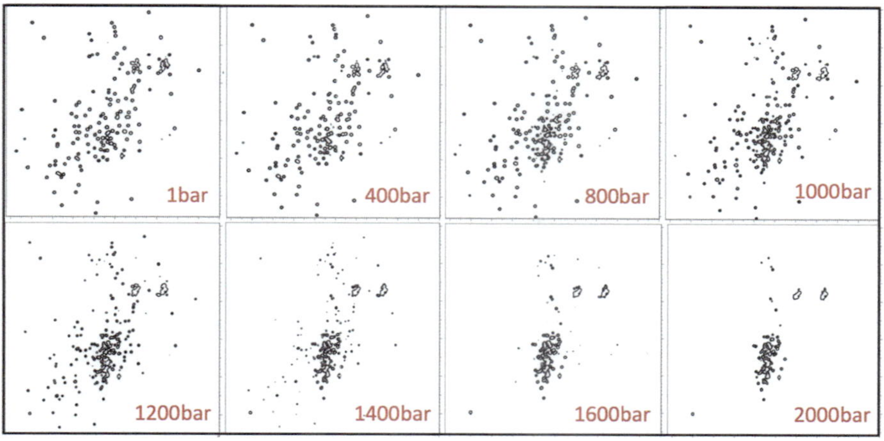

Fig. 13.3 Pressure dependence of the Δ + PHS SNase 2D [^{1}H-^{15}N] HSQC spectrum

Fig. 13.4 Experimental device installed at the Centre de Biochimie Structurale (Montpellier, France) for High Pressure NMR. From right to left. – Avance III 600 MHz NMR Bruker spectrometer used for this study. This spectrometer is equipped with a classical «hot» double-resonance (BBI) probe, with Z gradients. – Deadalus Innovation™ High Pressure automatic pump. – Deadalus Innovation™ ceramic tube with its titan manifold. This system is connected to the pump by a stainless steel pipe (ø = 0.25′), and pressurized via a pushing fluid: a light paraffin wax (Sigma™), immiscible with the aqueous buffer used to dissolve the protein sample

pressure NMR tubes can be used in combination with any commercial NMR probe, and maintain a sensitivity of ~50 % relative to that obtained with classical 3 mm Borosilicate or Pyrex glass tubes (Fig. 13.4).

13.4 High-Pressure NMR and Protein Unfolding

In combination with high pressure, NMR is a technique able to bring unprecedented details, residue-per-residue, on protein unfolding (Roche et al. 2012b, 2013a). Equilibrium measurements give access to thermodynamic parameters such as the free energy difference between the folded and unfolded states of the protein, and

ΔV. As noted, comparison of the values from the different residues informs on the cooperativity of the unfolding/folding reaction. Moreover, given the volumetric properties of protein folding transition states, kinetic measurements using multi-dimensional NMR are also possible (See also Chap. 4 by Royer in this volume).

13.4.1 High-Pressure NMR and Protein Unfolding: Steady-State Study

13.4.1.1 Measurement of "Local" Thermodynamic Parameters

Steady-state measurements are recorded when the equilibrium between native population/unfolded population has been reached after a pressure jump. Reaching equilibrium after a change in pressure requires a variable period of time (the relaxation time). Because the activation volume for protein folding is invariably positive, relaxation times can be quite long, ranging from seconds to a few minutes (Δ + PHS SNase L125A variant or WT SNase for example), up to several hours (Δ + PHS SNase). Classically, steady-state measurements consist in recording 2D [^{1}H-^{15}N] HSQC – type NMR experiments as a function of pressure, as reported in Fig. 13.3. One follows the evolution of the residue specific cross-peak intensities (or volumes), as a function of the applied pressure. As the native population decreases compared to the unfolded one, the intensity of the cross-peaks corresponding to the native state decreases with pressure, following a sigmoidal curve that can be fitted with the characteristic equation for a two-state equilibrium:

$$I = \frac{I_F + I_U e^{-(\Delta G^0 + p\Delta V)/RT}}{1 + e^{-(\Delta G^0 + p\Delta V)/RT}} \tag{13.3}$$

where I is the cross-peak intensity at a given pressure, I_F the cross-peak intensity at atmospheric pressure (folded protein, I_F), and I_U the residual cross-peak intensity when the protein is fully unfolded (I_U). The four parameters I_F, I_U, ΔG^0 and ΔV have to be fitted to the experimental cross-peak intensities measured at variable pressure. This equation has been obtained from the following relations:

$$\Delta G(p) = \Delta G^0 + p\Delta V = -RT\ln(K(p)) \quad \text{where} \quad K(p) = \frac{[U]}{[F]} = \frac{I_F - I}{I - I_U} \tag{13.4}$$

where [U] and [F] stands for the protein unfolded and folded populations, respectively.

Contrary to fluorescence spectroscopy, for instance, which gives a "global" value for the parameters ΔG^0 and ΔV, NMR yields "local" residue specific values of ΔG^0 and ΔV (Fig. 13.5). For our model reference protein (Δ + PHS SNase), unfolding is a nearly cooperative phenomenon: most of the residues "see" a similar ΔV

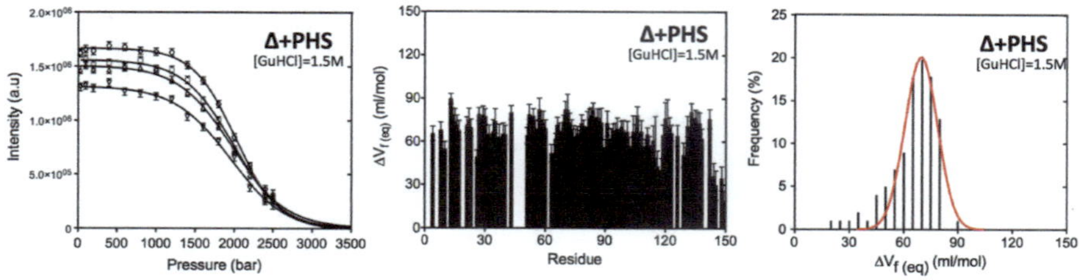

Fig. 13.5 From left to right. – Sigmoidal decrease of the intensity measured for 4 cross-peaks in the native 2D [^{1}H-^{15}N] HSQC spectrum of $\Delta + $PHS SNase (see Figs. 13.2 and 13.3) as a function of pressure. The lines come from the fit of the experimental intensities with the characteristic equation for a 2-states equilibrium (Eq. 13.3). – ΔV values obtained through the fit of the intensity decrease of the 2D [^{1}H-^{15}N] HSQC cross-peaks with pressure, plotted versus the protein sequence. – Histogram of the ΔV values measured for $\Delta + $PHS SNase. This histogram has been fitted with a Gaussian function (in *red*) that gives an average value for ΔV close to the one measured with fluorescence. Modified from results published (Roche et al. 2012a)

of ≈ 80 ml mol^{-1}, a value close to the "global" value measured with fluorescence spectroscopy. Nevertheless, in some areas of the protein, the measured residue-specific ΔV values fall below this average value (<30 ml mol^{-1}), skewing the distribution in Fig. 13.5, and suggesting the presence of folding intermediates, i.e. partially folded conformers having some degree of stability, in the protein energy landscape.

13.4.1.2 Structural Characterization of Folding Intermediates

High pressure NMR measurements highlight differences in local stability for $\Delta + $PHS SNase: some elements of structure unfold before others. That means that partially unfolded structures exist along the coordinate axis of the folding/unfolding reaction. These partially unfolded structures are mandatory steps in the folding pathway of a protein.

We used the following procedure to extract structural and energetic information about the folding intermediates from the pressure-dependent multi-dimensional NMR data. After normalizing the residue-specific denaturation (Fig. 13.6a) curves obtained from the amide cross-peak intensity decays measured on the HSQC experiments recorded at variable pressure, the value of one for a given cross-peak ($I = I_F = 1$) can be associated with a probability of one (100 %) for the corresponding residue to be in the native state, with all the native contacts present. Similarly, a residue for which, at the same pressure, the corresponding cross-peak has disappeared ($I = I_U = 0$) from the HSQC spectrum has a probability equal to zero to be in a native state: it belongs to an unfolded state where all the native contacts are lost.

Now, we consider two residues *i* and *j*, in an intermediate situation where the probability to be in a folded states are $p(i) = 0.99$ and $p(j) = 0.68$, respectively, at a

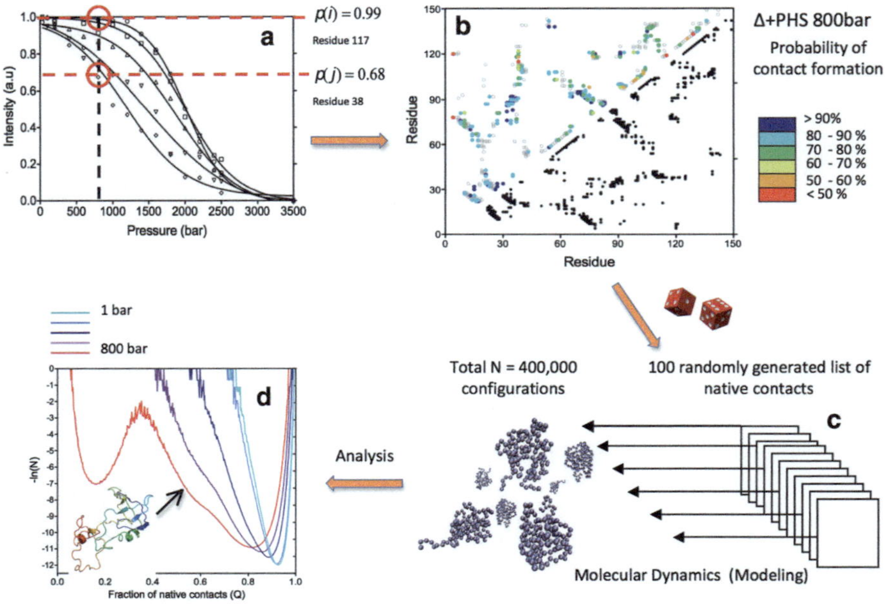

Fig. 13.6 Tracking folding intermediates with high pressure NMR. (**a**) Residue-specific normalized denaturation curves can be used to calculate contact probabilities between two residues (here between residue 38 and 117) at a given pressure (80 MPa). (**b**) Contact probabilities are reported with a color code (above the diagonal) in a contact map. This map will be used to define the lists of native contacts at a given pressure (80 MPa). (**c**) These contact lists will be translated into restraint lists for use by molecular dynamics program allowing the modeling of the corresponding conformation. (**d**) Conformation analysis: on the graph, the populations (Ln(P)) are plotted as a function of the residual native contacts (Q), at five different pressures. The red curve shows the result for a pressure of 80 MPa. It presents three minima: the most populated ($Q = 0.85$) corresponds to the population of native conformers at this specific pressure; the minimum at $Q = 0.1$ corresponds to the population of unfolded states; the minimum (shoulder) at $Q = 0.55$ corresponds to the population of an intermediate states which structure is represented in the insert. Modified from results published (Roche et al. 2012a)

given pressure (80 MPa in the example displayed in Fig. 13.6a). For two residues that are in contact in the native state, we assume the probability *p*(*i*,*j*) to be in contact at pressure of 80 MPa in this example to be given by the product of the individual probabilities $p(i,j) = p(i) \text{ x } p(j) = 0.67$. Pressure dependent contact maps can be constructed based on the contact map of the folded protein. These folded state contact maps are constructed from the 3D crystal or NMR structure of the protein by measuring all contacts between different atoms: usually, only the distances between C_α atoms of the different residues are used, and a "contact" between two residues *i* and *j* is defined by a distance $C_\alpha{}^i - C_\alpha{}^j \leq 4$ Å. These contacts are then plotted in a diagonal diagram with the protein sequence numbering on the X and Y axes (Fig. 13.6b). It is now possible to use a color code (for example) in order to report in this contact map the pressure dependent probability of contact, as defined above (Fig. 13.6b).

Given the folded state contact map and our contact probabilities for each contact pair at each pressure we work backwards to obtain structural models and population probabilities for the folding intermediates. Indeed, if it is easy to calculate a contact map from a known 3D structure, it is also possible to calculate a 3D structure from a contact map, using the contacts as restraints in a molecular modeling program. This is the classical method used by NMR spectroscopists to build a protein 3D structure from inter-nuclear distances deduced from magnetic interactions between nuclei. Similarly, a restraint list can be obtained from a contact map and used for molecular modeling. It is also possible to "weight" the restraint list with contact probabilities calculated at a given pressure. In this manner, several (usually 100) restraint lists are generated, instead of one. A contact between two residues associated to a probability of 0.8 will be randomly added to 80 lists over 100, if the probability is only 0.4, the corresponding restraints will be in 40 lists over 100, etc.

Then we model, class and analyze the 3D structures obtained from coarse grained simulations using the 100 different restraint lists. In the case of our model protein $\Delta + \mathrm{PHS}$ SNase, this analysis allowed us to identify a population of conformers corresponding to a folding intermediate where the C-terminal α-helix is unfolded, whereas the N-terminal β-barrel maintains its native structure (Fig. 13.6d). It is worth noting that this folding intermediate has been also identified through other high-pressure NMR experiments, concerning the measurement of amide proton/deuteron exchange (Roche et al. 2012b). These experiments allow probing the evolution of the H-bond network involved in the stabilization of the protein 3D structure. This approach has been developed in detail elsewhere (Fuentes and Wand 1998) and will not be discussed here further.

13.4.2 High-Pressure NMR and Protein Unfolding: Kinetics Study

A complete understanding of the protein folding/unfolding phenomenon requires, in addition to the spatial, structural description of the energy landscape, a temporal description of the sequence of events along the folding pathway followed by the protein. Such a description relies on the measurement of kinetic parameters, after perturbation of the thermodynamic equilibrium between the folded and unfolded conformers of the protein. This perturbation can be achieved by a fast mixing with chaotropic reagent, a pH jump, a temperature jump… and of course a pressure jump (P-jump)!

In addition to kinetic parameters, these measurements allow characterizing the transition state associated to a first order kinetics, commonly used to describe an equilibrium reaction between two states. Thus, using pressure to unfold the protein will give access to folding (k_f) and unfolding (k_u) rates, as well as to the volume of

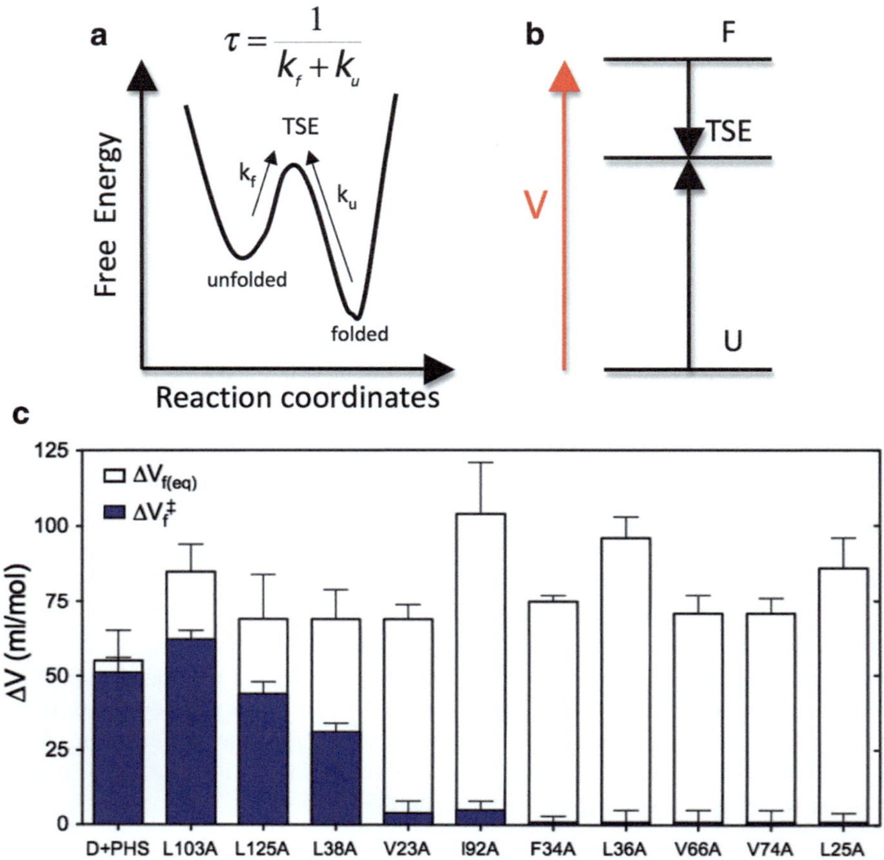

Fig. 13.7 Protein folding and kinetics measurements. (**a**) Unfolding reaction coordinates as a function of the free energy. The constants k_f and k_u stand for the folding and unfolding rates, respectively, and τ is the experimentally measured relaxation time as depicted in Fig. 13.8. (**b**) Scheme of the volumetric diagram of the folding/unfolding reaction. F, TSE and U represents the folded, transition, and unfolded state, respectively. (**c**) Experimental volumetric diagrams obtained for $\Delta +$ PHS SNase and 10 of its "cavity" variants. ΔV_{et} stands for the volume difference between the folded and unfolded states as measured at thermodynamic equilibrium (steady state measurements), $\Delta V_f^{\ddagger}$ is the activation volume between the folded and transition states obtained from kinetics measurements. Modified from results published (Roche et al. 2013b)

the transition state ensemble (TSE): the transition state of a protein being relatively heterogeneous, it is better described as an ensemble of (close) states rather than an unique conformer (Fig. 13.7).

The return to a new equilibrium after perturbation can be monitored by different spectroscopy techniques that give access to a "global" measurement of the kinetic parameters for the folding/unfolding reaction (Fluorescence, IR…). A "local" description of the kinetic parameters and of the transition state ensemble implies the use of a technique combining spatial resolution, allowing a precise local description of the time evolution of the structure of the protein, and a sufficient time resolution. At atmospheric pressure the return to thermodynamic equilibrium after a perturbation can be relatively fast (few ms seconds to a few minutes, generally at most). NMR has high spatial resolution, but its time resolution is limited: the recording time of the 2D [^{1}H-^{15}N] HSQC spectra described above ranges from 10 to 40 min, depending on the sample concentration and the expected spectral resolution. Thus, such experiments can be used only in the case of proteins having extremely slow relaxation times.

Pressure, because of the positive activation volume for folding, has the unique feature to slow down the folding reaction: a reaction completed in few seconds at atmospheric pressure will take several minutes to few hours at higher pressure, making P-jump NMR monitoring feasible. For example, $\Delta +$PHS SNase at pressure above 1 kbar, exhibits residue-specific relaxation times greater than 10 h! (see for instance Fig. 13.9b). This is all the more true since methodological advances have been realized during the last decade in the field of "real-times" measurement of NMR multidimensional experiments (Frydman et al. 2002; Gal et al. 2007). Now, 2D correlation experiments can be acquired in few tens of seconds (sometimes even in less than one second!) instead of few tens of minutes, with a good sensitivity and enough spectral resolution. For example for following faster P-jump kinetics we have used the so-called 2D [1H-15 N]- "SOFAST-HMQC" experiments (Schanda and Brutscher 2005), recorded in only 25 s (Fig. 13.8), to monitor the unfolding reaction kinetics of the L125A variant of $\Delta +$PHS SNase, exhibiting relaxation times shorter than 10 min after P-jump of 20 MPa (Roche et al. 2013b). In the case of proteins exhibiting even faster relaxation times, it is possible to combine NMR measurements with stop-flow techniques where the sample evolution is "frozen" at different times along the exponential decay of the cross-peak intensities (Balbach et al. 1996). This approach is easy to implement when using addition of denaturing reagents, or pH-jump to unfold the protein, but has severe technological limitations when combined to high pressure.

Practically, P-jump kinetics measurements consist in recording a series of 2D HSQC (or SOFAST-HMQC) after a pressure jump, in order to correctly sample the exponential decay of the cross-peak intensity during unfolding (in the case of a "positive" P-jump, where pressure is increased) (Fig. 13.9), but can be done as well for negative pressure jumps. Several P-jumps are realized in the pressure range where unfolding appears (typically between 40 and 200 MPa for $\Delta +$PHS SNase and its variants). P-jump amplitude should be enough to get a measurable intensity decay for the amide cross peaks, but should remain moderate to avoid any imbalance between the folding and the unfolding reactions: an excessive positive P-jump, for instance, will favor the unfolding reaction at the expense of the folding reaction. In the present study, we have used pressure jumps of 20 MPa, which is about 10 % of the pressure range needed to fully unfold the protein $\Delta +$PHS SNase and its variants (200 MPa).

The fit of the amide cross-peak decays with exponential functions gives a residue-specific exponential time t for each P-jump, equal to the inverse of the sum of the

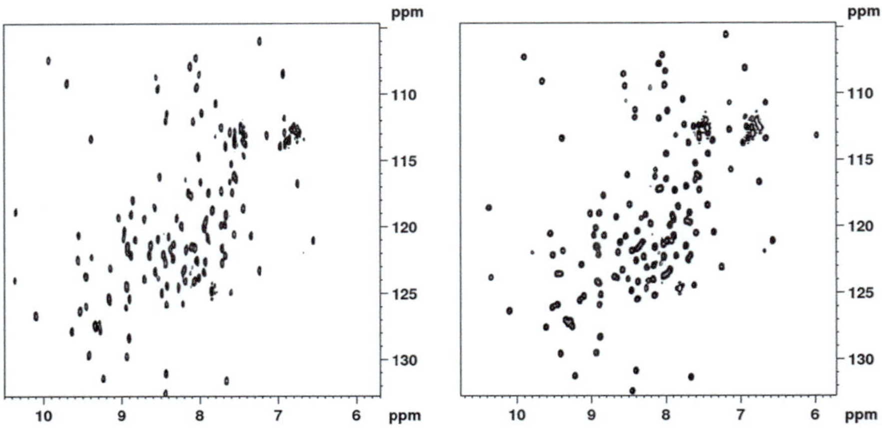

Fig. 13.8 2D [^{1}H-^{15}N] correlation experiments recorded on the L125A variant of $\Delta + $PHS SNase. Left: classical HSQC experiment recorded at equilibrium in 40 min. Right: SOFAST-HMQC used for kinetic measurements, recorded in 25 s after a 20 MPa P-jump (60 to 80 MPa)

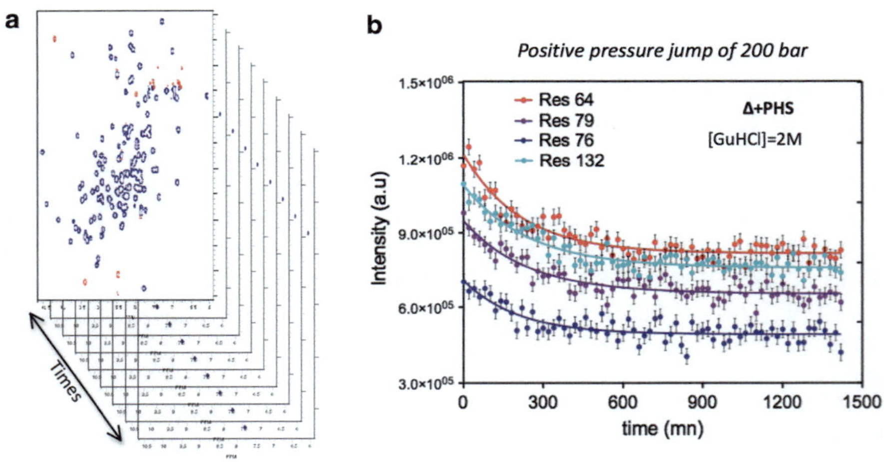

Fig. 13.9 High-pressure NMR and kinetics measurements. (**a**) Series of 2D [^{1}H-^{15}N] HSQC spectra used to sample a positive P-jump of 20 MPa (100 to 120 MPa) on a sample of $\Delta + $PHS SNase. Individual measuring time for each HSQC experiment was 20 min, for a protein concentration of 1 mM. (**b**) Times evolution of the amide cross-peak intensity for 4 residues. The curves were obtained by exponential fitting of the intensity values, giving the value of $\tau = 1/(k_u + k_f)$

folding and unfolding rates ($t = 1/(k_u + k_f)$). It is thus possible to obtain the value of the activation volume (TSE volume) for the unfolding reaction at atmospheric pressure by the fit of the evolution of t values at different pressure with the following equation:

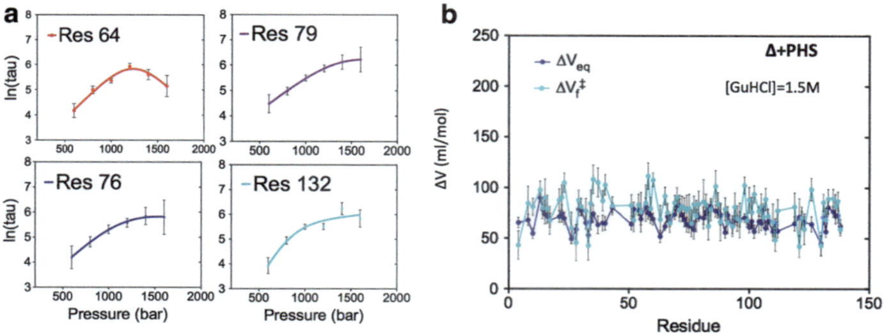

Fig. 13.10 High-pressure NMR kinetics measurements. (**a**) Examples for the evolution of the residue-specific relaxation time τ values as a function of the final pressure for the pressure jump. Curves are obtained with the fit of the experimental values with Eq. 13.5 and allow determining the value of the activation volume for the folding reaction "see" by each residue. (**b**) Comparison of ΔV at equilibrium (dark blue) and $\Delta V_f^{\ddagger}$ activation volumes (light blue) as a function of the protein sequence. Modified from results published (Roche et al. 2013b)

$$\tau(p) = \left[k_{u0} e^{\left(\frac{-p\Delta V_{u0}^{\ddagger}}{RT}\right)} + k_{u0} K_{eq} e^{\left(\frac{-p\left(\Delta V + \Delta V_{u0}^{\ddagger}\right)}{RT}\right)} \right]^{-1} \tag{13.5}$$

where $\Delta V_{et} = \Delta V_f - \Delta V_u$ and $K_{eq} = k_f/k_u$ stand for the volume difference between the folded and unfolded states and the equilibrium constant measured at thermodynamic equilibrium. Only two variables need to be fitted: k_{u0} (the unfolded rate at atmospheric pressure) and $\Delta V^{\ddagger}{}_{u0}$ (the activation volume for unfolding at atmospheric pressure) because of the equilibrium constraints (Fig. 13.10).

Analogous to the steady state measurements, combining high pressure and NMR spectroscopy in relaxation kinetics allows obtaining a local dynamic view of the folding pathway being taken by each residue, through the measurement of the kinetic parameter τ and the extraction of the activation volume for each residue. It is worth noting that when looking to the activation volumes, average values obtained for the reference protein Δ + PHS SNase and its mutants, very different results are obtained: the reference protein shows an activation volume $\Delta V^{\ddagger}{}_f$ close to the equilibrium ΔV value, indicating that the molar volume of the transition state is quite close to that of the folded state, suggesting a "dehydrated" TSE where most of the native cavities are still present. In contrast, some mutants (I92A, V66A . . .) show very different values for these two parameters, with a TSE volume close to that of the unfolded state, reflecting a more "hydrated" TSE where most of the native cavities have disappeared and are hydrated (Fig. 13.7) (Roche et al. 2013b).

13.5 Conclusions

Combining modern NMR techniques with high pressure gives unprecedented details on protein folding. The spatial and temporal resolution of NMR spectroscopy is necessary to describe the folding pathway at a residue level, giving both a structural description of the pathway, bringing to light the existence of intermediate states, and a dynamic description informing on the rates of local rearrangements involved in the phenomenon. Thus, we have obtained a fairly complete structural and energetic description of the folding landscape for our model protein (Fig. 13.11) and several of its mutants. Even more information can be obtained on the protein folding/unfolding mechanism by combining high-pressure NMR with other perturbations: temperature for instance, or addition of denaturing agents, allowing in-depth mapping of folding free energy landscapes and how they can be modulated.

Beyond the fundamental interest of studying the protein folding/unfolding mechanism, a better understanding of this phenomenon will find applications in many fields. Is it necessary to recall that the so-called "conformational" neurodegenerative diseases (Alzheimer, Parkinson, prion diseases . . .) are due to misfolding of proteins having originally a physiological function? A better understanding of the folding/unfolding mechanism for each of these specific proteins may allow exploring new avenues for drug rational design. Similarly, understanding phe-

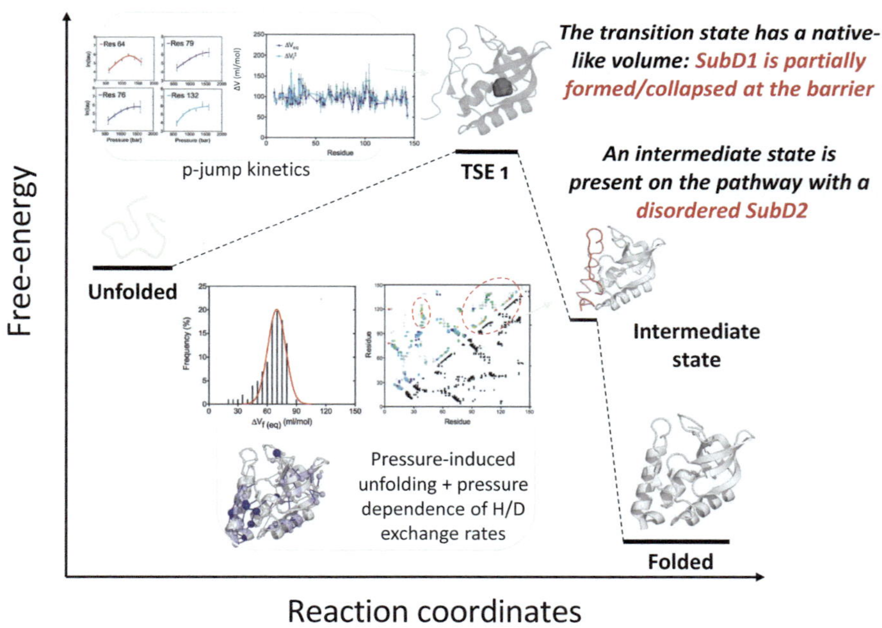

Fig. 13.11 Reconstruction of the folding pathway for Δ + PHS SNase through steady state and kinetics NMR measurements at high pressure

nomenon underlying protein stability is a major issue for the design of industrial enzymes or other biotechnological products, able to function at high pressure or high temperature. Thus, understanding how a protein can accommodate mutations to gain stability, keeping intact its function, could have a major economic impact. The challenge is huge, but combining NMR with high pressure may prove to be extremely useful.

Acknowledgements Authors are indebted to our long and ongoing collaborations with J.A. Caro, B. Garcia-Moreno (*John Hopkins University, Baltimore, USA*) and Angel E. Garcia *(Rensselaer Polytechnic Institute, Troy, U.S.A.)*, in the work that led to the publications reviewed here. High-pressure NMR equipment (except the spectrometers) used in the work of the original manuscripts cited here was funded by the French National Agency for Research (ANR, PiriBio 09–455024; Project Coordinator: C.A. Royer). This work was also supported by the French Infrastructure for Integrated Structural Biology (FRISBI) ANR-10-INSB-05-01.

References

Akasaka K, Kitahara R, Kamatari YO (2013) Exploring the folding energy landscape with pressure. Arch Biochem Biophys 531:110–115

Balbach J, Forge V, Lau WS, van Nuland NA, Brew K, Dobson CM (1996) Protein folding monitored at individual residues during a two-dimensional NMR experiment. Science 274:1161–1163

Bodenhausen G, Ruben DJ (1980) Natural abundance nitrogen-15 NMR by enhanced heteronuclear spectroscopy. Chem Phys Lett 69:185–189

Chalikian TV, Macgregor RB Jr (2009) Origins of pressure-induced protein transitions. J Mol Biol 394:834–842

Dellarole M, Royer C (2013) High-pressure fluorescence applications. In: Engelborghs Y, Visser T (eds) Methods in molecular biology, Springer, pp 53–74

Dill KA, MacCallum JL (2012) The protein-folding problem, 50 years on. Science 338:1042–1046

Dobson CM (2005) An overview of protein misfolding diseases. In: Buchner J, Kiefhaber T (eds) Protein folding handbook. Wiley-VCH Verlag GMBH & Co. KgaA, Weinheim, pp 1093–1113

Fourme R, Hamel G, Loupiac C, Perrier-Cornet J-M, Pin S, Roche J (2012) La biologie sous pression: de la molecule au procédé. Hautes Pressions: les Nouveaux Enjeux. Ed.: Publications Mission Ressources et Compétences Technologiques 2012.

Frydman L, Scherf T, Lupulescu A (2002) The acquisition of multidimensional NMR spectra within a single scan. Proc Natl Acad Sci U S A 99:15858–15862

Fuentes EJ, Wand AJ (1998) Local stability and dynamics of apocytochrome b562 examined by the dependence of hydrogen exchange on hydrostatic pressure. Biochemistry 37:9877–9883

Gal M, Schanda P, Brutscher B, Frydman L (2007) UtraSOFAST HMQC NMR and repetitive acquisition of 2D protein spectra at Hz rates. J Am Chem Soc 129:1372–1377

Grigera JR, McCarthy AN (2010) The behavior of the hydrophobic effect under pressure and protein denaturation. Biophys J 98:1626–1631

Kamatari YO, Kitahara R, Yamada H, Yokoyama S, Akasaka K (2004) High-pressure NMR spectroscopy for characterizing folding intermediates and denatured states of proteins. Methods 34:133–143

Kauzmann W (1959) Some factors in the interpretation of protein denaturation. Adv Protein Chem 14:1–63

Mitra L, Smolin N, Ravindra R, Royer CA, Winter R (2006) Pressure perturbation calorimetric studies of the solvation properties and the thermal unfolding of proteins in solution: experiments and theoretical interpretation. Phys Chem Chem Phys 8:1249–1265
Murphy KP, Privalov PL, Gill SJ (1990) Common features of protein unfolding and dissolution of hydrophobic compounds. Science 247:559–561
Privalov PL, Gill SJ (1988) Stability of protein structure and hydrophobic interaction. Adv Protein Chem 39:191–234
Roche J, Caro JA, Norberto DR, Barthe P, Roumestand C, Schlessman JL, Garcia AE, Garcia-Moreno B, Royer CA (2012a) Cavities determine the pressure unfolding of proteins. Proc Natl Acad Sci U S A 109:6945–6950
Roche J, Dellarole M, Caro JA, Guca E, Norberto DR, Yang YS, Garcia AE, Roumestand C, García-Moreno B, Royer CA (2012b) Remodeling of the folding free-energy landscape of staphylococcal nuclease by cavity-creating mutations. Biochemistry 51:9535–9546
Roche J, Caro JA, Dellarole M, Guca E, Royer CA, García-Moreno B, Garcia AE, Roumestand C (2013a) Structural, energetic and dynamic responses of the native state ensemble of staphylococcal nuclease to cavity-creating mutations. Proteins 81:1069–1080
Roche J, Dellarole M, Caro JA, Norberto DR, Garcia AE, García-Moreno B, Roumestand C, Royer CA (2013b) J Am Chem Soc 135:14610–14618
Rouget JB, Aksel T, Roche J, Saldana JL, Garcia AE, Barrick D, Royer CA (2011) Size and sequence and the volume change of protein folding. J Am Chem Soc 133:6020–6027
Royer CA (2002) Revisiting volume changes in pressure-induced protein unfolding. Biochim Biophys Acta 1595:201–209
Schanda P, Brutscher B (2005) Very fast two-dimensional NMR spectroscopy for real-time investigation of dynamic events in proteins on the time scale of seconds. J Am Chem Soc 127:8014–8015
Wolynes PG, Eaton WA, Fersht AR (2012) Chemical physics of protein folding. Proc Natl Acad Sci U S A 109:17770–17771

Chapter 14
Basic Equations in Statics and Kinetics of Protein Polymerization and the Mechanism of the Formation and Dissociation of Amyloid Fibrils Revealed by Pressure Perturbation

Hideki Tachibana

Abstract Studies of the pressure-dissociation of several amyloid or amyloid-like fibrils have shown that the fibril state is considerably voluminous. Quantitative characterization of the protein fibrillation reaction with respect to volumetric parameters is necessary to elucidate mechanisms of amyloid fibrillation in molecular terms such as protein cavity and hydration. Here we discuss, firstly, basic equations in statics and kinetics of protein polymerization as employed to obtain thermodynamic, volumetric, and kinetic parameters. Equilibrium treatment of the reactions with the scheme such as one-step polymerization, linear-association polymerization, or nucleation-dependent polymerization, and kinetic treatment of seeded linear-polymerization or spontaneous nucleation-elongation polymerization are described. In particular we will detail kinetics of the dissociation of fibrils which have been produced under the linear-association mechanism and therefore the length-distribution of which conforms to a geometric sequence in the degree of polymerization with a common ratio r, which is less than, and usually very close to, unity. In this case, an observed macroscopic rate of dissociation is shown to be a product of the microscopic elementary dissociation rate constant and a factor $(1-r)$, extremely reduced compared with the intrinsic elementary rate. Secondly, we discuss protein conformational states in fibrillogenesis with molecular and volumetric observations reported, such as the unfolded state responsible for the association with seeds and the extension of amyloid fibrils, the transition state in which protein cavity formation and dehydration occur to intermediate levels, and the fibril state in which they occur to final respective levels which, in some cases, depend on the maturity of the fibril.

Keywords Amyloid fibril • β2-microglobulin • Cavity and hydration • Lysozyme • Partial specific volume • Unfolded state

---

H. Tachibana (✉)
Department of Biotechnological Science, School of Biology-Oriented Science and Technology, and High Pressure Protein Research Center, Institute of Advanced Technology, Kinki University, 930 Nishimitani, Kinokawa 649-6493, Japan
e-mail: tachi887@waka.kindai.ac.jp

K. Akasaka, H. Matsuki (eds.), *High Pressure Bioscience*,
Subcellular Biochemistry 72, DOI 10.1007/978-94-017-9918-8_14

14.1 Overview

Natively folded state of proteins, which is not a completely well-packed state, generally contains cavities inside and has a partial specific volume which measures slightly higher than that of the unfolded and fully hydrated state (Royer 2002). High-pressure NMR studies have shown that the presence of the cavities plays a pivotal role in the dynamics and functioning of proteins (Akasaka 2006), as described in chapters of this monograph. The folding of protein into a native state involves intricate association and packing processes of constitutive peptide segments. Likewise, supramolecular protein assemblies such as multi-subunit enzymes, virus capsids, and subcellular structural and functional filaments, constructed through the association and packing of component subunit proteins, often have higher partial specific volumes compared with the dissociated monomeric state, which makes them dissociate with high hydrostatic pressure (Ikkai et al. 1965; Josephs and Harrington 1967; Silva and Weber 1993; Royer 1995). In addition to these assemblies, the stability, structure, and conformational transitions of the proteins and their aggregates involved in amyloidogenic and neurodegenerative diseases have been studied in recent decades with high-pressure methods, leading to elucidation of amyloidogenic state of the transthyretin oligomer, pressure-populated conformational intermediates of prion protein, difference in pressure-sensitivity between early and mature fibrillar species of prion protein, etc. (Silva et al. 2001; Foguel and Silva 2004; Silva et al. 2006). Pressure-studies in all these research fields have been comprehensively reviewed (Silva et al. 2014).

As mentioned above, pressure can tweeze association-dissociation equilibrium of proteins on the one hand, but it also induces protein denaturation on the other hand. Unless the two reactions occur in respective separate pressure range this causes difficulties in interpretation of volumetric parameters obtained. We have constructed a series of disulfide-deficient variants of hen lysozyme to dissect the folding reaction of the protein into various levels such as embryonic, core-nucleated, and domain-docking (Tachibana 2000; Tachibana and Segawa 2004; Matsuo et al. 2009; Noda et al. 2012). Among the variants the one, named 0SS, which contains no disulfide bonds, has only a residual amount of secondary structure and no tertiary structure in the absence of denaturant. This intrinsically unfolded variant protein served as a good model for a denatured state of lysozyme that can be studied without the complications of added denaturants or extremes of temperature and pH, and was useful as a starting reactant of amyloid fibrillation. The discussion in this chapter is primarily concerned with the reported results with this variant protein. Firstly, we describe basic conventional equilibrium analyses of polymerization and discuss volumetric parameter values obtained. Secondly, kinetic analyses of polymerization are described. Thirdly, kinetic analyses of the pressure-dissociation of fibrils, and volumetric properties of the transition state are described. Lastly, molecular volumetric aspects of amyloid fibrillation are discussed.

14.2 Statics of Fibrillation Viewed on Pressure Axis

Amyloid fibril formation of proteins under denaturing solution conditions has been reported for a number of proteins. However, detailed thermodynamic analyses have rarely been reported because of complexity of experimental conditions for denaturation. By contrast, a genetically engineered disulfide-deficient variant of hen lysozyme, 0SS, which has all eight cysteinyl residues replaced by alanine or serine, represents a chemically-stable, simple model for a denatured state and a reaction system for spontaneous amyloid fibrillation (Niraula et al. 2004). The pressure-dissociation of the assemblage of lysozyme 0SS variant, which was formed early in the fibrillation reaction, as monitored with high-pressure NMR spectroscopy (Fig. 14.1) showed a reversible dissociation-association reaction as a function of applied pressure.

The normalized volume of cross-peaks in HSQC spectra represents the fraction of monomeric state at each pressure, and thermodynamic analyses described in the following subsections give the stability of the assembly, ΔG_0, and the change in partial molar volume, ΔV, on dissociation.

14.2.1 *One-Step Polymerization*

Firstly, we consider the case when the shape of the assembly is not known to be fibrous, as may be in the case of the precursor formation of fibrils. When n monomers associate, in one step, to form a polymer, $nM \rightleftarrows P$, the dissociation constant K_d is expressed in terms of the equilibrium monomer concentration C_1 and the total protein concentration C_T (in monomer unit) as follows:

$$K_\mathrm{d} = C_1^n / [P] = nC_1^n / (C_\mathrm{T} - C_1) \tag{14.1}$$

where $[P]$ is the number concentration of polymers. Therefore,

$$\begin{aligned} \Delta G &= \Delta G_0 + p\Delta V \\ &= -RT \ln K_\mathrm{d} = -RT\,(\ln n + n \ln C_1 - \ln (C_\mathrm{T} - C_1)) \end{aligned} \tag{14.2}$$

where p and ΔV are pressure and the volume change (per mol of polymer) on dissociation, and ΔG and ΔG_0 represent thermodynamic stability of the polymer at (high) pressure p and at ambient pressure, respectively.

The plot of ΔG, calculated with observed C_1 values at each pressure (Eq. 14.2), against p is expected to give a straight line, the slope and ordinate-intercept of which give ΔV and ΔG_0, respectively. In the case of lysozyme 0SS, n was estimated to be around 30, and ΔV was calculated to be – 52.7 ± 11.3 ml mol^{-1} (per monomer unit), and ΔG_0 was 23.3 ± 0.8 kJ mol^{-1} (per monomer unit) (Niraula et al. 2004).

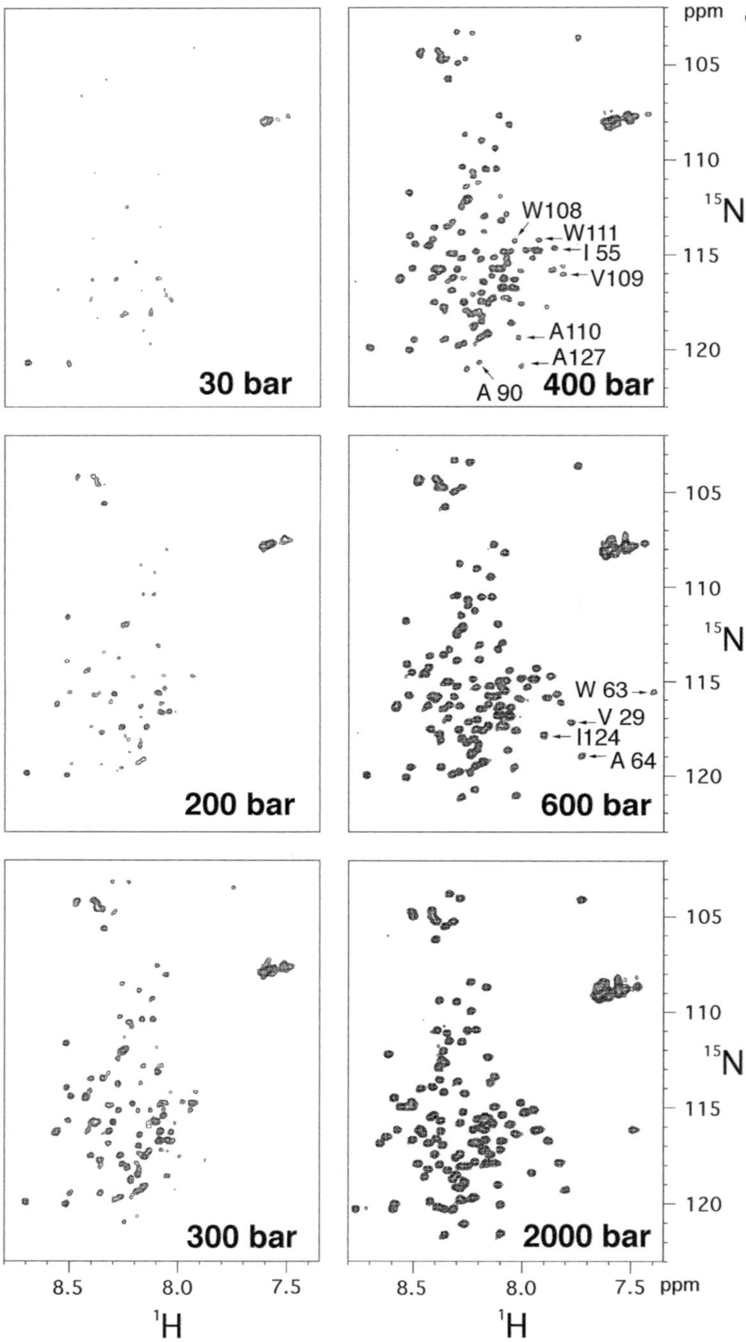

Fig. 14.1 ^{15}N /^{1}H-heteronuclear single-quantum coherence (HSQC) spectra of the uniformly ^{15}N-labeled lysozyme 0SS variant in 50 mM deuterated sodium maleate buffer, pH 2.0, at various pressures ranging from 30 to 2,000 bar. With increasing pressure, cross peaks appear and progressively increase in their intensity at positions expected for an unfolded and hydrated state of protein, indicating that pressure dissociates the early assembly into monomers. A total of 55 cross peaks were assigned (Niraula et al. 2004) Copyright (2004) National Academy of Science, USA

14.2.2 Linear-Association Polymerization

The assembly treated above, which was formed early in the association reaction, was later been shown to be a short fibril (Abdul Latif et al. 2007). Therefore a scheme of linear association, $C_i + M \rightleftarrows C_{i+1}$, is more suited, in which a polymer grows through successive addition of a monomeric unit to its end with an association equilibrium constant K which is independent of the polymer length (Oosawa and Asakura 1975; Cantor and Schimmel 1980). In this case, the number concentration of i-mer, C_i, equals to $K^{i-1}C_1^i = C_1(KC_1)^{i-1} = C_1 r^{i-1}$, where C_1 is the equilibrium concentration of monomer, and $r = KC_1$ (<1) is a common ratio in the geometric sequence. With increasing polymer length i, the number concentration C_i gradually decreases like an exponential-type decay function, and the parameter r characterizes the degree of decay. The number-averaged degree of polymerization, $<\mathrm{DP}>_{\mathrm{N}}$, is given by:

$$\begin{aligned} <\mathrm{DP}>_{\mathrm{N}} &= \Sigma i C_i / \Sigma C_i = \left(C_1/(1 - KC_1)^2\right) / \left(C_1/(1 - KC_1)\right) \\ &= 1/(1 - KC_1) = 1/(1 - r)\,. \end{aligned} \tag{14.3}$$

Likewise, the weight-averaged degree of polymerization, $<\mathrm{DP}>_{\mathrm{W}}$, is given by:

$$<\mathrm{DP}>_{\mathrm{W}} = \Sigma i^2 C_i / \Sigma i C_i = (1 + KC_1)/(1 - KC_1) = (1 + r)/(1 - r)\,. \tag{14.4}$$

These quantities become high when r approaches unity, or C_1 approaches K^{-1}. This limit concentration is called a "critical monomer concentration" $C_{\mathrm{crit}}\left(= K^{-1}\right)$.

The equilibrium monomer concentration C_1 is related with the total (weight) concentration of protein C_{T} and the association equilibrium constant K through the conservation relation:

$$C_{\mathrm{T}} = \Sigma i C_i = C_1/(1 - KC_1)^2 \tag{14.5}$$

The dissociation equilibrium constant K_{d} (=1/K) under the linear-association scheme can be calculated from Eq. 14.5 in terms of C_1 and C_{T} :

$$K_{\mathrm{d}} = C_1 / \left(1 - \sqrt{\frac{C_1}{C_T}}\right) \tag{14.6}$$

$$\therefore \Delta G = -RT \ln K_d = -RT\left(\ln C_1 - \ln\left(1 - \sqrt{\frac{C_1}{C_T}}\right)\right) = \Delta G_0 + p\Delta V \tag{14.7}$$

The slope of the plot of ΔG against p gives ΔV, the partial volume change per mol of monomeric unit upon dissociation, and extrapolation of ΔG to 1 bar gives ΔG_0

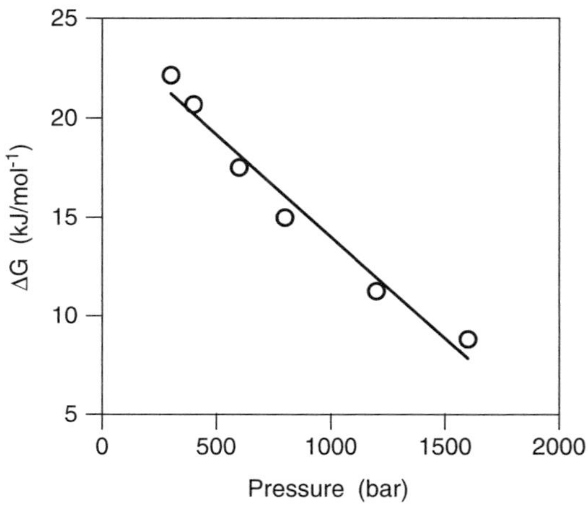

Fig. 14.2 Free energy change for the dissociation of the lysozyme 0SS fibrils as a function of pressure, calculated under the scheme of linear-association polymerization. The line was obtained by regression, to Eq. 14.7, of the averaged normalized peak volume, in HSQC spectra of the sort shown in Fig. 14.1 at pressures ranging from 300 to 1,600 bar, which represents the concentration of dissociated monomer

per mol of monomeric unit (Fig. 14.2). When recalculated with this scheme, ΔG_0 for the 0SS assembly was shown to be 24.3 ± 0.8 kJ mol^{-1} (per monomer unit), and ΔV on dissociation was shown to be -100 ± 9 ml mol^{-1}.

14.2.3 Nucleation-Dependent Polymerization

If the equilibrium length-distribution of polymers obeys an exponential type, then, the value of r can be estimated. In an ideal linear-association mechanism, if, for instance, the number-averaged degree of polymerization $\langle DP \rangle_N = 1/(1-r)$ is around 1000, then $r = 0.999$. From Eq. 14.5, the fractional equilibrium monomer amount is given by:

$$C_1/C_T = (1 - KC_1)^2 = (1 - r)^2 \tag{14.8}$$

which, in this case, is as small as $0.001^2 = 10^{-6}$. If, however, the experimentally determined fractional equilibrium monomer amount is significantly higher than the value predicted by Eq. 14.8, it indicates that polymers are formed through a nucleation-dependent reaction mechanism as described in the following. The existence of nucleation reaction imposes polymers to exist in lower number concentrations because the association equilibrium constant in the pre-nucleation steps is low compared with that in the elongation steps beyond the nucleus.

Let n and σ be, respectively, the size of nucleus in monomeric unit (note that n here is used differently from that in Eq. 14.1) and the ratio of association equilibrium constant in the pre-nucleation step to that beyond the nucleation step. The factor σ represents the degree of difficulty of nucleation in an equilibrium thermodynamic term. In usual cases, n is larger than three and $\sigma \ll 1$. Then (Oosawa and Asakura 1975),

$$\begin{aligned} C_T &\approx C_1 + \sum_{i=n}^{\infty} iC_i = C_1 + \sum_{i=n}^{\infty} i\sigma^{n-2}C_1(KC_1)^{i-1} \\ &= C_1 + \sigma^{n-2}\sum_{i=n}^{\infty} iC_1(KC_1)^{i-1} \end{aligned} \tag{14.9}$$

Here, K represents the association equilibrium constant beyond the nucleation step, and the second term, $\sigma^{n-2}\sum_{i=n}^{\infty} iC_1(KC_1)^{i-1}$ represents the total amount of polymer. Note that the polymer amount is apparently reduced by the factor, σ^{n-2} compared with that in the linear-association case (Eq. 14.5). When $\sigma \ll 1$ and $C_1 \ll \sum_{i=n}^{\infty} iC_1(KC_1)^{i-1}$ (the second condition is satisfied when $KC_1 = r \to 1$),

$$\begin{aligned} C_T &\approx C_1 + \sigma^{n-2}\sum_{i=1}^{\infty} iC_1(KC_1)^{i-1} \\ &= C_1 + \sigma^{n-2}C_1/(1-KC_1)^2 = C_1 + \sigma^{n-2}C_1/(1-r)^2 \end{aligned} \tag{14.10}$$

Thus, the formation of polymer in the nucleation-elongation reaction is delicately achieved through the product of two mutually opposing factors, i.e. very small σ^{n-2} and very large $1/(1-r)^2$.

When $r \to 1$, the fractional equilibrium monomer amount $C_1/C_T\left(= 1/\left(1 + \sigma^{n-2}/(1-r)^2\right)\right)$ approximates to:

$$C_1/C_T = (1-r)^2/\sigma^{n-2} \tag{14.11}$$

In a linear-association mechanism without nucleation ($\sigma^{n-2} = 1$), C_1/C_T is as small as 10^{-6}, as mentioned above, if r value is so close to the limiting value 1.00 as to be 0.999. In a nucleation-dependent polymerization mechanism, however, the denominator in Eq. 14.11, σ^{n-2}, takes a small value such as 10^{-4}, which makes C_1/C_T moderately low as to be 10^{-2}. Supplementary calculations show that, for a single common value of K, although C_1 approaches the limiting value K^{-1}, and therefore $r = KC_1$ approaches 1.0 in both mechanisms, there exists a very slight but significant increment in r value in the latter mechanism (where r is 0.9999, for instance) compared with the former mechanism ($r = 0.9900$), which makes $\langle DP \rangle_N = 1/(1-r)$ far much higher in the latter, i.e. nucleation-dependent, mechanism as compensation for the low number concentrations of polymers.

From Eq. 14.10 it follows that:

$$K_d = C_1/\left(1-\sqrt{\sigma^{n-2}C_1/(C_T-C_1)}\right) \tag{14.12}$$

$$\therefore \Delta G = -RT\left(\ln C_1 - \ln\left(1-\sqrt{\sigma^{n-2}C_1/(C_T-C_1)}\right)\right) = \Delta G_0 + p\Delta V \tag{14.13}$$

Our experiences on the same data set described in previous subsections show that the plot of ΔG vs. p with a selected σ^{n-2} value indeed fit to a linear regression line (not shown), from which ΔG_0 of 25.9 ± 0.2 kJ mol^{-1} (per monomer unit), and

ΔV of -102 ± 3 ml mol^{-1} were obtained. As for the value of σ^{n-2}, however, much more numbers of experimental data points should be acquired to obtain a reliable value.

The analyses described in this section show that ΔG_0 value does not highly depend on the reaction scheme chosen. The value of ΔV neither vary significantly between the linear-association and the nucleation-dependent association schemes. The kinetics of fibrillation reaction of 0SS shown in the next section, however, clearly indicates that it proceeds in a nucleation-dependent way. In any way, the result that the 0SS assemblage is reversibly dissociated with a decrease in partial molar volume of about 100 ml·(mol monomer)$^{-1}$ indicates that the 0SS amyloid fibril is voluminous to the extent similar to most of the natively folded proteins.

14.3 Kinetics of Fibrillation

14.3.1 Seeded Linear-Polymerization

Spontaneous amyloid fibrillation from folded state of protein is generally difficult to occur, and addition of seeds (prepared mostly by means of fragmentation of preformed amyloid fibrils) is frequently employed ("seeded-fibrillation"). In this case, spontaneous nucleation is neglected and only the elongation reaction at the ends of preformed polymers takes place. Therefore, the total number concentration of polymer, N_p, which is equal to that of added seeds, is fixed throughout the polymerization reaction. The kinetics of polymerization, described in terms of the change in the monomer concentration C_1, is:

$$\mathrm{d}C_1/\mathrm{dt} = -k_+ C_1 N_\mathrm{p} + k_- N_\mathrm{p} \tag{14.14}$$

where k_+ and k_- are the rate constants of association and dissociation of a monomer at a polymer end, respectively. Integration of this equation gives

$$\left(C_1\,(\mathrm{t}) - C_1\,(\infty)\right) / \left(C_1(0) - C_1\,(\infty)\right) = \exp\left(-k_+ N_\mathrm{p} t\right), \tag{14.15}$$

where $C_1(0)$ is an initial monomer concentration, and $C_1\,(\infty) = k_-/k_+ = K^{-1} = C_\mathrm{crit}$. Thus, the observed rate of the decrease in monomer concentration, and therefore that of the increase in the polymer weight concentration, is proportional to k_+ as well as N_p.

14.3.2 Spontaneous Nucleation and Elongation

In a spontaneous nucleation and elongation scheme, in which the nucleus, consisting of n monomers, spontaneously forms by multiple interaction of n monomers with a rate constant $k_+{}^*$, and decomposes with a rate constant $k_-{}^*$, the temporal change in the number concentration of the nucleus C_n as well as that of polymer C_i ($i > n$) are given by:

$$dC_n/dt = k_+^* C_1^n - k_-^* C_n - k_+ C_n C_1 + k_- C_{n+1} \tag{14.16}$$

$$dC_i/dt = k_+ C_{i-1} C_1 - k_- C_i - k_+ C_i C_1 + k_- C_{i+1} \tag{14.17}$$

where k_+ and k_- are the rate constants of attachment (elongation) and detachment, respectively, of a monomer at the end of a polymer. If $k_-{}^*$ and k_- are negligible and the overall rate of nucleation is low, the following approximate equation is obtained for the temporal increase in the total polymer weight concentration W_p (Oosawa and Asakura 1975):

$$W_p(t) = C_1(0)\left(1 - \left(\cosh\left(\left(\left(k_+^* k_+ n C_1(0)^n\right)/2\right)^{1/2} t\right)\right)^{-2/n}\right) \tag{14.18}$$

This equation indicates that the half-polymerization period, $t_{1/2}$, at which W_p reaches half the initial monomer concentration, $C_1(0)/2$, is inversely proportional to the $n/2$-th power of $C_1(0)$:

$$1/t_{1/2} \propto (C_1(0))^{n/2} \tag{14.19}$$

Therefore, the size of nucleus, n, can be estimated by doubling the slope of the plot of the logarithm of an apparent rate of polymerization vs. the logarithm of an initial monomer concentration. An example of such a plot in the case of the fibril formation reaction of lysozyme 0SS variant is shown in Fig. 14.3, which indicates that the size of the nucleus is five in monomeric unit under the solution conditions chosen.

Although the above-mentioned simple relationship between the overall rate of polymerization and the initial monomer concentration appears to work for the estimation of the size of nucleus, it is a relationship derived under simplifying assumptions and does not always hold. A numerical calculation study of the time course of the nucleation-dependent polymerization has shown that the power-dependence of the overall rate of polymerization on the initial monomer concentration varies its exponent when the monomer concentration range varies (Kodaka 2004): the exponent becomes significantly lower when the initial monomer concentration becomes higher.

14.3.3 Advanced Treatment of Polymerization Kinetics

Morris et al. (2009) have extensively reviewed various treatment in previous literatures in the research field of protein polymerization, starting from a classical

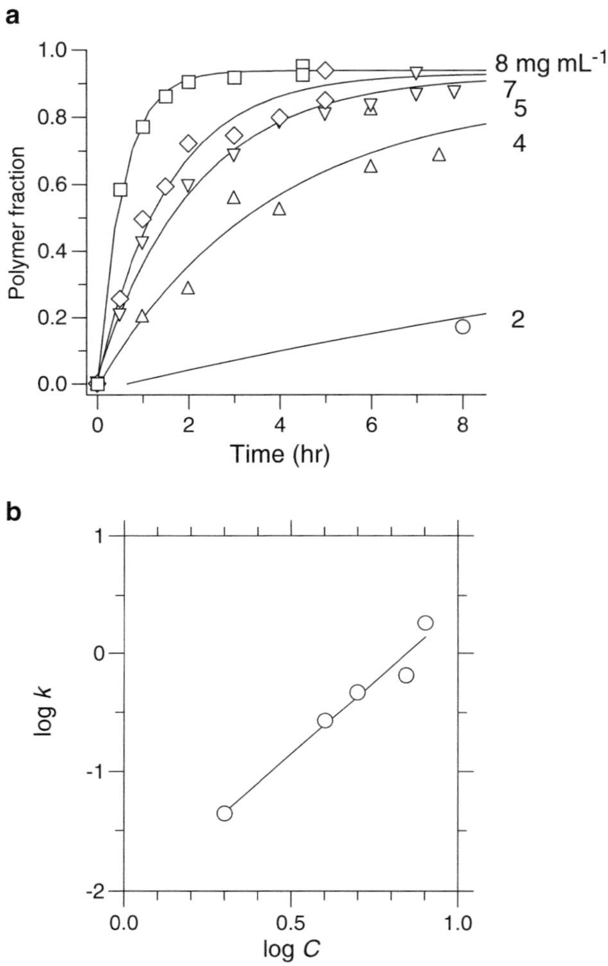

Fig. 14.3 (**a**) Time-course of the lysozyme 0SS fibrillation in 20 mM sodium acetate, pH 4.0, 30 mM sodium chloride at 25 °C and various protein concentrations from 2 to 8 mg ml^{-1} as labeled to each curve. Fractional polymer content was determined by size-exclusion-chromatography on a TSK G4000SW column (Tosoh). Data points which are not shown here but extend to the time range from 10 to 90 h were included in the regression to a simple single exponential. (**b**) A full logarithmic plot of the apparent association rate k against the protein concentration C. The slope of the regression line is 2.45 ± 0.27

work of Oosawa et al. (1959) on actin polymerization system, through contributions on sickle cell hemoglobin, tubulin, "living" polymers, multi-subunit enzymes, prion, amyloid-β protein, calcitonin, etc., leading to a minimalistic two-step model introduced by Watzky and Finke (1997) which employs two simple pseudo-elementary steps. Knowles et al. (2009), on the other hand, have presented an analytical solution of the kinetics of polymerization taking elementary microscopic processes, such as monomer addition at the ends of a filament, internal breaking of filaments and spontaneous nucleation, rigorously into account. They have demonstrated the dominance of secondary nucleation events which take place with the fragmentation of preformed filaments in the kinetics of amyloid growth, and showed that the lag time in this case scaled with an exponent of minus 0.5 with respect to the monomer concentration.

14.4 Kinetics of the Dissociation of Fibrils

A detailed treatment of the kinetics of the dissociation of fibrils by utilizing pressure as an accelerator of the dissociation has been reported (Abdul Latif et al. 2007), where an equilibrium exponential-type length distribution of fibrils, $C_{i+1}(0) = rC_i(0)$, is attainted as an initial state ($t = 0$). When dissociation occurs through detachment of a monomer from the end of a polymer, and repolymerization reaction can be neglected since the fibril solution is highly diluted before starting the pressure-dissociation reaction, the temporal change in C_i is given by:

$$dC_i(t)/dt = -k_{-}C_i(t) + k_{-}C_{i+1}(t) \tag{14.20}$$

Here, k_{-} is an intrinsic first-order dissociation rate constant. As will be described later, k_{-} should be considered to be a quantity quite distinct from, although closely related to, the apparent dissociation rate obtained from bulk (i.e., not single-molecular) measurement on a solution of the fibrils of heterogeneous lengths.

Let us intuitively assume that throughout the dissociation reaction, the above-described exponential-type length distribution, $C_{i+1} = rC_i$ is conserved. Then,

$$\begin{aligned} dC_i(t)/dt &= -k_{-}C_i(t) + k_{-}rC_i(t) = -k_{-}(1-r)\,C_i(t) \\ \therefore C_i(t) &= C_i(0)\exp(-k_{-}(1-r)\,t) = Ar^{i-1}\exp(-k_{-}(1-r)\,t) \end{aligned} \tag{14.21}$$

Here, the factor A equals to $C_1(0)$ and $\sigma^{n-2}C_1(0)$ for a linear-association and a nucleation-elongation schemes, respectively. It is easily shown that this function form indeed satisfies Eq. 14.20:

$$\begin{aligned} dC_i(t)/dt &= -Ak_{-}(1-r)\,r^{i-1}\exp(-k_{-}(1-r)\,t) \\ &= -Ak_{-}r^{i-1}\exp(-k_{-}(1-r)\,t) + Ak_{-}r^{i}\exp(-k_{-}(1-r)\,t) \\ &= -k_{-}C_i(t) + k_{-}rC_i(t) \\ &= -k_{-}C_i(t) + k_{-}C_{i+1}(t) \end{aligned} \tag{14.22}$$

Thus, during the progress of dissociation reaction, the number concentration of each polymer decreases exponentially with a rate constant $k_{-}(1 - r)$, while the length distribution of polymers remains to conform to a geometric sequence with a common ratio r. Figure 14.4 shows confirmation of this point with direct numerical integration of Eq. 14.20.

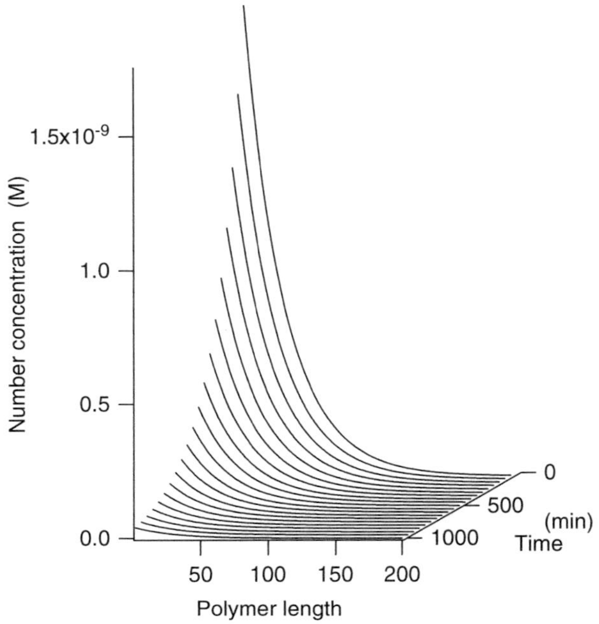

Fig. 14.4 The change in the distribution of fibril length during the progress of dissociation numerically integrated according to Eq. 14.20. Polymer lengths from 2 to 200 were taken into calculation. The boundary effect due to the imposed upper limit of the polymer length was not obvious since we employed r value of 0.9667, which was significantly lower than the value 0.995 that was actually observed for 0SS fibrils. The total protein concentration C_t and the elementary microscopic first-order dissociation rate constant were set to be 2 μM and 0.002 s^{-1}, respectively. Neither the initial monomer concentration in equilibrium with polymers, 2.22 nM, nor the subsequent increase in the monomer concentration is shown here

Now, the temporal change in the total (weight) amount of polymers, C_p, is given by summation of $iC_i(t)$ over polymeric species:

$$\begin{aligned} C_p(t) &= \Sigma i C_i(t) = \Sigma i C_i(0) \exp\left(-k_-\,(1-r)\,t\right) \\ &= C_p(0) \exp\left(-k_-\,(1-r)\,t\right) \end{aligned} \tag{14.23}$$

Thus, the dissociation rate which is experimentally obtained through the exponential decay in C_p equals to $k_-(1-r)$, a product of the intrinsic dissociation rate k_- and a factor $(1-r)$, and therefore is considerably reduced compared to the intrinsic rate k_-, since r usually takes a value very close to one. The reduction by a factor $(1-r)$ is understandable when we realize that the detachment of monomer occurs only at the polymer ends, whose number concentration is reduced compared to the total concentration of monomers within polymers by a factor '$1/ < DP>_N$', which equals to '$1-r$' in the case of the exponential-type length distribution.

The kinetics of the pressure-dissociation of the lysozyme 0SS fibril has been analyzed with high-pressure fluorescence measurement (Abdul Latif et al. 2007), utilizing a large difference in the quantum yield of intrinsic fluorescence from tryptophan residues between fibrillar and dissociated monomeric states. The intensity change followed a single-exponential function of time with an apparent rate constant k_{obs}, which can be simply related to an intrinsic first-order dissociation rate constant k_{-} by using Eq. 14.23. The observed fluorescence intensity I is the sum of the contributions from fibrillar and monomeric states:

$$\begin{aligned} I &= I_{\text{p}}C_{\text{p}} + I_1 C_1 = I_{\text{p}}C_{\text{p}} + I_1\left(C_{\text{T}} - C_{\text{p}}\right) = \left(I_{\text{p}} - I_1\right)C_{\text{p}} + I_1 C_{\text{T}} \\ &= (\Delta I)\,C_{\text{p}} + I_1 C_{\text{T}} \\ &= (\Delta I)\,C_{\text{p}}(0)\exp\left(-k_{-}(1-r)\,\text{t}\right) + I_1 C_{\text{T}} \end{aligned} \tag{14.24}$$

where I_{p} and I_1 are the fluorescence intensities of the polymeric and monomeric states, respectively, of the 0SS solution of unit concentration, and ΔI is the difference between them. Under the solution conditions favoring complete dissociation, the fluorescence intensity at an infinite time I^{∞} equals to $I_1 C_{\text{T}}$ and that at time zero I^0 equals to $(\Delta I)\,C_{\text{p}}(0) + I_1 C_{\text{T}}$. Then, a normalized fluorescence intensity, $(I - I^{\infty})/\left(I^0 - I^{\infty}\right)$, will decay exponentially,

$$(I - I^{\infty})/\left(I^0 - I^{\infty}\right) = \exp\left(-k_{-}(1-r)\,t\right) \tag{14.25}$$

with an apparent rate constant $k_{\text{obs}} = k_{-}(1\text{–}r)$.

The length distribution of the fibrils in the starting 0SS solution, as determined with AFM, conformed well to a geometric sequence with a common ratio r of 0.995. From the observed dissociation rate k_{obs} at a pressure of 200 MPa, 0.0157 min^{-1}, the intrinsic dissociation rate constant k_{-} was estimated:

$$k_{-} = k_{\text{obs}}/(1-r) = k_{\text{obs}} \times 200 = 3.1\ \text{min}^{-1}.$$

Also with an extrapolation to an ambient pressure (0.1 MPa) on the plot of k_{obs} vs. pressure, the intrinsic dissociation rate at 0.1 MPa can be estimated:

$$k_{-} \sim 0.00747\ \text{min}^{-1} \times 200 = 0.149\ \text{min}^{-1}.$$

Therefore, the dissociation of monomer 0SS from the fibril end is expected to take place once for every ~20 s and ~6.7 min at pressures of 200 and 0.1 MPa, respectively.

14.5 Volumetric Properties of the Transition State Revealed with Pressure- Dissociation Kinetics

The regression of the plot of the logarithm of the dissociation rate constant against pressure to a quadratic polynomial of pressure (Eq. 14.26) gives two important volumetric properties of the transition state: activation volume $\Delta V^{\circ\ddagger}$ at ambient pressure P° (=0.1 MPa) and activation compressibility $\Delta\kappa^{\ddagger}$:

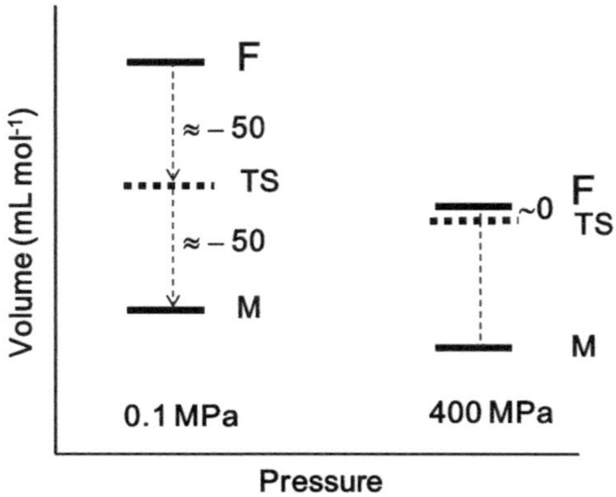

Fig. 14.5 The volume diagram for the dissociation reaction of 0SS fibrils. At 0.1 MPa, the transition state (TS) is about 50 ml mol monomer^{-1} below the fibril state (F) in volume. This difference becomes almost null at 400 MPa, because the activation compressibility is negative. At 0.1 MPa the dissociated monomer state (M) is further 50 ml below the TS in volume as described in Sect. 14.2.2. In the diagram, only the difference between the volume levels within each pressure condition is of concern (Adapted with permission from Abdul Latif et al. (2007). Copyright 2007 The Biophysical Society)

$$\ln k_{-} = \ln\left(k_B T/h\right) - \Delta G^{\circ\ddagger}/RT - \left(\Delta V^{\circ\ddagger}/RT\right)\left(P - P^{\circ}\right) + \left(\Delta\kappa^{\ddagger}/2RT\right)\left(P - P^{\circ}\right)^2 \quad (14.26)$$

Note that the activation volume $\Delta V^{\ddagger}$ at any pressure P is given by:

$$\Delta V^{\ddagger} = V^{\circ\ddagger} - \Delta\kappa^{\ddagger}\left(P - P^{\circ}\right). \quad (14.27)$$

In the case of lysozyme 0SS fibril (Abdul Latif et al. 2007), a negative $\Delta V^{\circ\ddagger}$ of -50.5 ± 1.6 ml mol monomer^{-1} and a negative $\Delta\kappa^{\ddagger}$ of -0.013 ± 0.0007 ml mol monomer^{-1} bar^{-1} have been reported, indicating that partial molar volume V° as well as compressibility κ of 0SS protein in the fibril state decreases in the transition state for dissociation; namely, the transition state is less voluminous and less compressible than the fibril state (Fig. 14.5). Due to the negative $\Delta\kappa^{\ddagger}$, the $\Delta V^{\ddagger}$ value, which is negative at $P = P^{\circ}$, becomes less negative with the increase in P, and will change its sign to positive at extremely high P, i.e. while in a relatively low pressure region the dissociation rate increases with increasing pressure, it stops to increase and turns to decrease with further increase in pressure.

14.6 Molecular Aspects of Amyloid Fibrillation

14.6.1 Protein Conformational State Responsible for Amyloid Fibrillation

The amyloid fibril formation of the lysozyme 0SS variant indicates that, despite the general notion that partial unfolding is a prerequisite for fibrillogenesis (Booth et al. 1997; Fink 1998; Kelly 1998; Lansbury Jr 1999), essentially unfolded conformer attains amyloid fibrillation (Niraula et al. 2004). This alternative notion has been reinforced by the temperature-dependence observed in the seeded-fibrillation reaction of WT lysozyme (Sasaki et al. 2008), which indicated that the amyloid fibrillation proceeds through the interaction of seeds with fully heat-denatured conformer of WT lysozyme. Recently, more elaborate evidence for this notion was reported for β2-microglobulin (β2m) fibrillation (Yanagi et al. 2012): the analysis of the NMR transverse relaxation rates and transferred cross-saturation experiments on acid-denatured β2m revealed that an extensively unfolded form, which is in equilibrium with a state that contains residual structures, is responsible for the association with seeds and the extension of amyloid fibrils. The hydrophobic residues which cluster and constitute residual structures in the acid-denatured state become exposed in the extensively unfolded state and are committed to the association with seeds.

In agreement with this mechanism, hydrophobic residues which belong, collectively, to five of the six hydrophobic clusters in the unfolded state of lysozyme (Klein-Seetharaman et al. 2002) have previously been shown to be preferentially involved in the intermolecular association of the intrinsically unfolded lysozyme 0SS variant fibrils (Niraula et al. 2004).

Introduction of a single disulfide bond to 0SS at each of the four residue positions, Cys6-Cys127, Cys30-Cys115, Cys64-Cys80, and Cys76-Cys94, found in WT lysozyme promotes structure formation only to the level of the pre-molten globule state and is coupled with the structure formation of submolecular size and low stability (Tachibana 2000). These four molecular species of single-disulfide lysozyme variants, however, show highly differential fibrillation kinetics, acceleration or deceleration as compared with that of 0SS, altogether by three orders of magnitude (Kono et al. 2009), indicating that the fibrillation process is highly dependent on the properties of submolecular residual structures in proteins.

14.6.2 Volumetric Properties of Amyloid Fibrils

In section two, it has been mentioned based on the reported value of partial volume change on pressure-dissociation that the lysozyme 0SS fibril is quite voluminous.

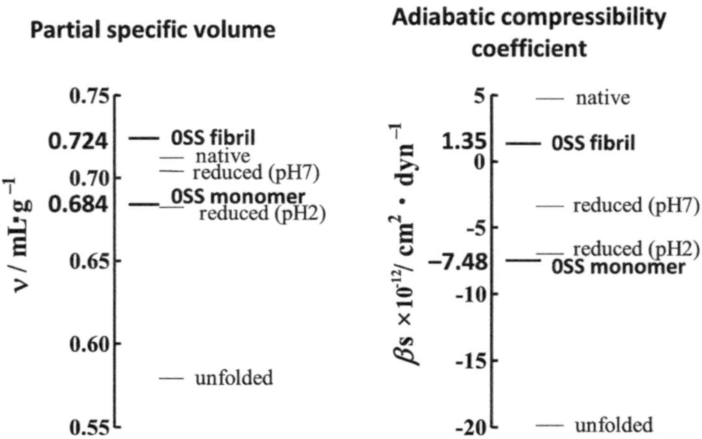

Fig. 14.6 Comparison of volume (ν) and adiabatic compressibility coefficient (β_S) of 0SS fibril and 0SS monomer states. For the purpose of reference, ν and β_S values of various conformational states of WT hen lysozyme are also shown (Adapted with permission from Akasaka et al. (2007). Copyright 2007 American Chemical Society)

Conclusive evidence for this has been obtained from direct measurement of the change in the absolute values of partial specific volume, ν, and the partial specific adiabatic compressibility coefficient, β_s, during the 0SS amyloid fibril formation by using a high precision density meter and an ultrasonic velocity meter (Akasaka et al. 2007). The value of ν increased from 0.684 ml g^{-1} in a monomeric state to 0.724 ml g^{-1} in a fibrillar state, and that of β_s increased too from -7.48×10^{-12} cm^2 dyn^{-1} in a monomeric state to $+1.35 \times 10^{-12}$ cm^2 dyn^{-1} in a fibrillar state (Fig. 14.6), demonstrating that the fibril is highly voluminous and compressible.

Volume and compressibility reflect atom packing and hydration state of protein or protein assembly. The presence of cavity due to imperfect atomic packing positively contributes to ν and β_s, while the hydration contributes negatively to both of them. The results described above that both volume and compressibility increase upon lysozyme 0SS fibril formation, therefore, indicates that an increase in cavity and a decrease in hydration take place on fibril formation. The decrease in hydration may simply be expected to occur when a monomer protein is associated with the end of a fibril, expelling originally hydrated waters in the interfacial regions.

Upon lysozyme 0SS fibril dissociation into monomers, therefore, both volume and compressibility decrease. There exists uncertainty, however, about the value of volume decrease: 100 ml mol^{-1} determined from an equilibrium analysis with variable pressure NMR spectroscopy (Sect. 14.2.2), and as much as 570 ml mol^{-1} from the measurement of partial specific volume by using a high precision density meter (this section). The amount of the decrease in the (adiabatic) compressibility coefficient corresponds to a decrease in (adiabatic) compressibility of 0.087 ml mol^{-1} bar^{-1}. These volume and compressibility decreases are considerably larger than the respective values of decrease observed as the activation volume and activation compressibility of the dissociation of the 0SS fibril, 50.5 ml mol^{-1} and 0.013 ml mol^{-1} bar^{-1} (Sect. 14.5). Thus, in the transition state from fibrillar to monomeric states, cavity decreases and hydration increases (i.e., the protein becomes partly "wet") by their respective amounts, and they further decreases and increases, respectively, by multiple amounts when going further to the monomeric state, which is nearly fully hydrated. The negativity of both the activation volume and activation compressibility for the dissociation reaction has also been reported for the amyloid fibril of WT hen lysozyme (Shah et al. 2012).

Unlike the above example of lysozyme 0SS fibrillation in which volume and compressibility increase with the progress of fibrillation and pressure efficiently dissociates fibrils, the cases when fibrils are insensitive to pressure have also been reported (Cordeiro et al. 2004; Dirix et al. 2005; Meersman and Dobson 2006). In particular, the volumetric results converse to those described here have been reported in the amyloid fibrillation of insulin (Smirnovas et al. 2006), in which the decrease in void volume upon fibrillation, i.e. well-packing of side-chains in the fibril, appears to have a dominant contribution to volumetric properties. It has also been reported, however, that the formation of the early nonfibrillar aggregate of insulin in the presence of salt accompanies increases in partial specific volume and compressibility (Smirnovas and Winter 2008). Altogether, these results can be taken to indicate the unique sensitiveness of volumetric measurements in distinguishing different types of aggregation reaction. We also note that pressure promotes circularly shaped amyloid fibrils of insulin (Jansen et al. 2004), which suggests the use of pressure not only for dissociation but also for alteration of the shape of supramolecular structures.

Lee et al. (2009) observed, by using a high precision densitometer, distinct volumetric changes for three different types of amyloid fibrils, derived from β2-m or its amyloidogenic core peptide K3, and have presented a comprehensive model which shows how apparently converse volumetric properties can be explained from their conformational state, including voids and hydration, as evidenced by their previous measurements such as calorimetry, high-pressure fluorescence, solid-state NMR, and H/D-exchange. The partial specific volume of a mature β2-m amyloid fibril is higher than that of acid-denatured monomeric β2-m, and moreover than that of native β2-m, indicating a voluminous nature of the mature amyloid fibril. By contrast, the partial specific volume of the mature K3 fibril is lower than that of acid-denatured monomeric β2-m, indicating a highly compact core structure. The partial specific volume of an immature β2-m fibril is intermediate between those of the mature β2-m amyloid fibril and K3 fibril. In their model, the mature K3 peptide fibrils assume a highly packed cross-β-structure representative of a core structure of amyloid fibrils. Although the immature β2-m fibril has the core peptide structure in the center too, peptide regions outside the core are loosely packed, not extensively hydrogen-bonded, and hydrated. Likewise, the mature β2-m fibril has the core peptide structure in the center as well, but the peptide regions outside the core are now extensively hydrogen-bonded, become solvent-inaccessible, still loosely-packed, and contain internal cavities to the extent more than the natively folded β2-m.

14.7 Summary and Concluding Remarks

The analyses of equilibrium pressure-dissociation data of lysozyme 0SS based on three different reaction schemes show a regression of similar degree among the three equations (Eqs. 14.2, 14.7, and 14.13), and while the estimated volumetric change does vary in magnitude depending in part on the choice of the scheme, the thermodynamic stability of the assembly varies little. Equilibrium measurements with much more dense acquisition of data than those presented here will distinguish the nucleation-dependent polymerization from the simple linear-association mechanism. In phenomenological kinetic analyses, application of the minimalistic two-step model (Watzky and Finke 1997) will be sufficient for general purposes. Fundamental questions about the nucleus, however, such as the number of component monomers, its structure, and the energetics of stabilization from prenucleus to the nucleus remain to be solved. For instance, the relationship for estimation of the nucleus size (Oosawa and Asakura 1975) from the apparently constant power-dependence between the overall rate of polymerization and the initial monomer concentration does not hold over an entire concentration range of monomer (Kodaka 2004). Further elucidation of this issue will require rigorous treatment of the reaction steps of successive formation and degradation of pre-nuclear oligomeric species, which have so far been often simplified. As for the kinetics of the dissociation of fibrils, free from the complexity of nucleation, a simple relationship between the macroscopic apparent rate of dissociation and the elementary microscopic dissociation rate constant was described. This relationship combined with extrapolation to an ambient pressure of the apparent dissociation rate experimentally obtained at high pressure enabled us to estimate the extremely low value of an elementary microscopic dissociation rate at an ambient pressure to be about once for every $\sim$6.7 min for the 0SS fibril (Abdul Latif et al. 2007). This kind of relationship will in principle be valuable in linking experimental results obtained by macroscopic methods with those by single-molecule techniques on amyloid fibrils (Ban et al. 2006). We also note that although in the case when the distribution, observed in weight concentration, of the fibril length appears to conform to a single peak, it actually could often be, in number concentrations, an exponential-type distribution to which the above relationship can be applied. Concerning the intermediately voluminous transition state of the 0SS fibril for dissociation, we note that our preliminary result indeed show that, due to the negative activation compressibility, the activation volume turns to positive at extremely high pressure. Namely, the transition state becomes more voluminous than the fibrillar state at extremes of high pressure. Although the notion that the partial specific volume increases with the fibrillation starting from the least-voluminous unfolded state of protein, and fibrils will dissociate upon pressurization, is simple and acceptable, in actual cases the pressure effect is not simple and often varies among the proteins, or even among the same protein under different solution conditions or different maturity of the fibril, partly because variously structured states of monomer and/or oligomers, and therefore with various volumetric properties, are the potentially dissociated species in equilibrium with the fibril, and also because a number of interactions in addition to hydrophobic interaction make up the fibrillar state. This apparent complexity, on the other hand, promises the power of pressure study to distinguish variously structured polymorphic forms of amyloid fibrils as well as diverse non-amyloidogenic assemblies. Lastly, we add that a method of pressure-accelerated proteolysis (Akasaka et al. 2008; Akasaka et al. 2014), which utilizes pressure-induced population shift of protein to the unfolded or dissociated state, is also a promising research field with industrial and hygienic application.

Acknowledgements The author thanks Dr K Sakurai for valuable discussions about β2-m studies, and Prof K Akasaka for inspiring general discussions.

References

Abdul Latif QR, Kono R, Tachibana H, Akasaka K (2007) Kinetic analysis of amyloid protofibril dissociation and volumetric properties of the transition state. Biophys J 92:323–329

Akasaka K (2006) Probing conformational fluctuation of proteins by pressure perturbation. Chem Rev 106:1814–1835

Akasaka K, Abdul Latif AR, Nakamura A, Matsuo K, Tachibana H, Gekko K (2007) Amyloid protofibril is highly voluminous and compressible. Biochemistry 46:10444–10450

Akasaka K, Nagahata H, Maeno A, Sasaki K (2008) Pressure acceleration of proteolysis: a general mechanism. Biophysics 4:29–32

Akasaka K, Maeno A, Murayama T, Tachibana H, Fujita Y, Yamanaka H, Nishida N, Atarashi R (2014) Pressure-assisted dissociation and degradation of "proteinase K-resistant" fibrils prepared by seeding with scrapie-infected hamster prion protein. Prion 8:314–318

Ban T, Yamaguchi K, Goto Y (2006) Direct observation of amyloid fibril growth, propagation, and adaptation. Acc Chem Res 39:663–670

Booth DR, Sunde M, Belloti V, Robinson CV, Hutchinson WL, Fraser PE, Hawkins PN, Dobson CM, Radford SE, Blake CC, Pepys MB (1997) Instability, unfolding and aggregation of human lysozyme variants underlying amyloid fibrillogenesis. Nature 385:787–793

Cantor CR, Schimmel PR (1980) Biophysical chemistry, Part I. WH Freeman & Company, New York

Cordeiro Y, Kraineva J, Ravindra R, Mauricio L, Lima TR, Gomes MPB, Foguel D, Winter R, Silva JL (2004) Hydration and packing effects on prion folding and beta-sheet conversion. J Biol Chem 279:32354–32359

Dirix C, Meersman F, MacPhee CE, Dobson CM, Heremans K (2005) High hydrostatic pressure dissociates early aggregates of TTR105-115, but not the mature amyloid fibrils. J Mol Biol 347:903–909

Fink AL (1998) Protein aggregation: folding aggregates, inclusion bodies and amyloid. Fold Des 3:R9–R23

Foguel D, Silva JL (2004) New insights into the mechanism of protein misfolding and aggregation in amyloidogenic disease derived from pressure studies. Biochemistry 43:11361–11370

Ikkai T, Ooi T, Noguchi H (1965) Actin: volume change on transformation of G-form to F-form. Science 152:1756–1757

Jansen R, Grudzielanek S, Dzwolak W, Winter R (2004) High pressure promotes circularly shaped insulin amyloid. J Mol Biol 338:203–206

Josephs R, Harrington WF (1967) An unusual pressure dependence for a reversibly associated protein system; sedimentation studies on myosin. Proc Natl Acad Sci U S A 58:1587–1594

Kelly JW (1998) The alternative conformations of amyloidogenic proteins and their multi-step assembly pathways. Curr Opin Struct Biol 8:101–106

Klein-Seetharaman J, Oikawa M, Grimshaw SB, Wirmer J, Duchardt E, Ueda T, Imoto T, Smith LJ, Dobson CM, Schwalbe H (2002) Long-range interactions within a non-native protein. Science 295:1719–1722

Knowles TJ, Waudby CA, Devlin GL, Cohen SA, Aguzzi A, Vendruscolo M, Terenjev EM, Welland ME, Dobson CM (2009) An analytical solution to the kinetics of breakable filament assembly. Science 326:1533–1537

Kodaka M (2004) Interpretation of concentration-dependence in aggregation kinetics. Biophys Chem 109:325–332

Kono R, Fujisawa T, Akasaka K, Tachibana H (2009) Amyloid fibrillation rate differs greatly among single-disulfide variants of hen lysozyme. J Biol Macromol 9:23–30

Lansbury PT Jr (1999) Evolution of amyloid: what normal protein folding may tell us about fibrillogenesis and disease. Proc Natl Acad Sci U S A 96:3342–3344

Lee YH, Chatani E, Sasahahara K, Naiki H, Goto Y (2009) A comprehensive model for packing and hydration for amyloid fibrils of beta2-microglobulin. J Biol Chem 284:2169–2175

Matsuo K, Watanabe H, Tate S, Tachibana H, Gekko K (2009) Comprehensive secondary-structure analysis of disulfide variants of lysozyme by synchrotron-radiation vacuum-ultraviolet circular dichroism. Proteins 77:191–201

Meersman F, Dobson CM (2006) Probing the pressure-temperature stability of amyloid fibrils provides new insights into their molecular properties. Biochim Biophys Acta 1764:452–460

Morris AM, Watzky MA, Finke RG (2009) Protein aggregation kinetics, mechanism, and curve-fitting: A review of the literature. Biochim Biophys Acta 1794:375–397

Niraula TN, Konno T, Li H, Yamada H, Akasaka K, Tachibana H (2004) Pressure-dissociable reversible assembly of intrinsically denatured lysozyme is a precursor for amyloid fibrils. Proc Natl Acad Sci U S A 101:4089–4093

Noda Y, Narama K, Kasai K, Tachibana H, Segawa S (2012) Glycerol-enhanced detection of a preferential structure latent in unstructured 1SS-variants of lysozyme. Biopolymers 97:539–549

Oosawa F, Asakura S (1975) Thermodynamics of the polymerization of protein. Academic, London

Oosawa F, Asakura S, Hotta K, Nobuhisa I, Ooi T (1959) G-F transformation of actin as a fibrous condensation. J Polym Sci 37:323–336

Royer CA (1995) Application of pressure to biochemical equilibria: the other thermodynamic variable. Methods Enzymol 259:357–377

Royer CA (2002) Revisiting volume changes in pressure-induced protein unfolding. Biochim Biophys Acta 1595:201–209

Sasaki K, Nakatsuka K, Hayashi I, Shah BR, Morimoto K, Akasaka K (2008) Efficient conversion of intact hen lysozyme into amyloid fibrils by seeding. J Biol Macromol 8:11–18

Shah BR, Maeno A, Matsuo H, Tachibana H, Akasaka K (2012) Pressure-dissociation of amyloid fibrils in wild-type hen lysozyme. Biophys J 102:121–126

Silva JL, Weber G (1993) Pressure stability of proteins. Annu Rev Phys Chem 44:89–113

Silva JL, Foguel D, Royer CA (2001) Pressure provides new insights into protein folding, dynamics and structure. Trends Biochem Sci 26:612–618

Silva JL, Cordeiro Y, Foguel D (2006) Protein folding and aggregation: two sides of the same coin in the condensation of proteins revealed by pressure studies. Biochim Biophys Acta 1764:443–451

Silva JL, de Oliveira AC, Vieira TCR, de Oliveira GAP, Suarez MC, Foguel D (2014) High-pressure chemical biology and biotechnology. Chem Rev 114:7239–7267

Smirnovas V, Winter R (2008) Revealing different aggregation pathways of amyloidogenic proteins by ultrasound velocimetry. Biophys J 94:3241–3246

Smirnovas V, Winter R, Funk T, Dzwolak W (2006) Protein amyloidogenesis in the context of volume fluctuations: a case study of insulin. Chem Phys Chem 7:1046–1049

Tachibana H (2000) Propensities for the formation of individual disulfide bonds in hen lysozyme and the size and stability of disulfide-associated submolecular structures. FEBS Lett 480:175–178

Tachibana H, Segawa S (2004) Disulfide-bond associated protein folding. In: Nalwa HS (ed) Encyclopedia of nanoscience and nanotechnology, vol 2. American Scientific Publishers, Los Angeles, pp 443–473

Watzky MA, Finke RG (1997) Transition metal nanocluster formation kinetic and mechanistic studies. A new mechanism when hydrogen is the reductant: slow, continuous nucleation and fast autocatalytic surface growth. J Am Chem Soc 119:10382–10400

Yanagi K, Sakurai K, Yoshimura Y, Konuma T, Lee YH, Sugase K, Ikegami T, Naiki H, Goto Y (2012) The monomer-seed interaction mechanism in the formation of the beta2-microglobulin amyloid fibril clarified by solution NMR techniques. J Mol Biol 422:390–402

Chapter 15
Pressure-Inactivated Virus: A Promising Alternative for Vaccine Production

Jerson L. Silva, Shana P.C. Barroso, Ygara S. Mendes, Carlos H. Dumard, Patricia S. Santos, Andre M.O. Gomes, and Andréa C. Oliveira

Abstract In recent years, many applications in diverse scientific fields with various purposes have examined pressure as a thermodynamic parameter. Pressure studies on viruses have direct biotechnological applications. Currently, most studies that involve viral inactivation by HHP are found in the area of food engineering and focus on the inactivation of foodborne viruses. Nevertheless, studies of viral inactivation for other purposes have also been conducted. HHP has been shown to be efficient in the inactivation of many viruses of clinical importance and the use of HHP approach has been proposed for the development of animal and human vaccines. Several studies have demonstrated that pressure can result in virus inactivation while preserving immunogenic properties. Viruses contain several components that can be susceptible to the effects of pressure. HHP has been a valuable tool for assessing viral structure function relationships because the viral structure is highly dependent on protein-protein interactions. In the case of small icosahedral viruses, incremental increases in pressure produce a progressive decrease in the folding structure when moving from assembled capsids to ribonucleoprotein intermediates (in RNA viruses), free dissociated units (dimers and/or monomers) and denatured monomers. High pressure inactivates enveloped viruses by trapping their particles in a fusion-like intermediate state. The fusogenic state, which is characterized by a smaller viral volume, is the final conformation promoted by HHP, in contrast with the metastable native state, which is characterized by a larger volume. The combined effects of high pressure with other factors, such as low or subzero temperature, pH and agents in sub-denaturing conditions (urea), have been a formidable tool in the assessment of the component's structure, as well as pathogen inactivation. HHP is

J.L. Silva (✉) • C.H. Dumard • P.S. Santos • A.M.O. Gomes • A.C. Oliveira
Instituto de Bioquímica Médica Leopoldo de Meis, Universidade Federal do Rio de Janeiro, Rio de Janeiro, RJ 21941-902, Brazil

Instituto Nacional de Ciência e Tecnologia de Biologia Estrutural e Bioimagem, Rio de Janeiro, Brazil
e-mail: jerson@bioqmed.ufrj.br

S.P.C. Barroso • Y.S. Mendes
Instituto de Bioquímica Médica Leopoldo de Meis, Universidade Federal do Rio de Janeiro, Rio de Janeiro, RJ 21941-902, Brazil

K. Akasaka, H. Matsuki (eds.), *High Pressure Bioscience*,
Subcellular Biochemistry 72, DOI 10.1007/978-94-017-9918-8_15

a technology for the production of inactivated vaccines that are free of chemicals, safe and capable of inducing strong humoral and cellular immune responses. Here we present a current overview about the pressure-induced viral inactivation and the production of inactivated viral vaccines.

Keywords Pressure disassembly • Pressure inactivation • Vaccines • Viruses

15.1 Introduction

Similar to temperature and pH, pressure is an important parameter that can influence living systems. However, compared with heat, the effects of high hydrostatic pressure (HHP) on living systems and biomolecules have not historically received similar attention. Approximately 70 % of the planet is covered by water, and the organisms that inhabit aquatic environments are subjected to hydrostatic pressure. The biosphere pressure varies from 0.1 MPa (atmospheric pressure) to 110 MPa at the deepest point in the ocean, Mariana Trench at $\sim$11 km deep (Somero 1992). As living beings are composed primarily of water, their tissues, cells and constitutive macromolecules may experience some degree of hydrostatic pressure (Rastogi et al. 2007; Aertsen et al. 2009).

The first studies on the effects of pressure on biological systems originated more than 100 years ago. The pioneers in this area were Regnard (1884), Royer (1895) and Hite (1899). However, until the twentieth century, the effects of pressure on living matter were underestimated or unknown because of the severe technical limitations imposed by the lack of materials and technology that could sustain pressures ranging from a few to 100 MPa (Silva et al. 2001; Mignaco et al. 2005).

For a long period of time, pressure was studied as a thermodynamic parameter primarily in the fields of engineering, physics and chemistry. Pressure reappeared in biological applications in 1980–1990, first in Japan and then in Europe and the US, in the food industry. The increase in the study of pressure applications in the previous three decades has resulted in a large number of studies in biological areas (Silva et al. 2001; Rivalain et al. 2010; Akasaka et al. 2013).

In recent years, many applications in diverse scientific fields with various purposes have examined pressure as a thermodynamic parameter. The effects of pressure are quite broad; pressure can act on simple molecular systems, such as proteins and macromolecular complexes, as well as complex systems, such as cells and food (Silva and Weber 1993; Silva and Foguel 2009; Winter and Jeworreck 2009; Aertsen et al. 2009; Akasaka et al. 2013). Several years ago, a very clear application for high-pressure technology was reported in the food industry with the aim of denaturing enzymes to modulate the functionality of food and eliminate pathogens, such as *Escherichia coli* and *Salmonella*, without the influence of other chemical or physical factors, such as temperature. Thus, HHP is a viable alternative to the thermal inactivation of food during processing (Balasubramaniam and Farkas 2008).

In proteins, HHP has been important in studies of structural dynamics, molecular stability, denaturation, disaggregation, and protein interactions with ligands, including other proteins, lipids, nucleic acids or drugs (Silva et al. 2001; Ishimaru et al. 2004; Schwarcz et al. 2004; Gonçalves et al. 2007; Gaspar et al. 2008; Aertsen et al. 2009; Silva and Foguel 2009; Rivalain et al. 2010; Cordeiro et al. 2013; Akasaka et al. 2013). HHP has been very effective in dissociating large assemblies, such as microtubules, amyloid aggregates, amorphous aggregates and inclusion bodies (Engelborghs et al. 1976; Silva et al. 1996, 2002; Foguel et al. 1999; Ferrão-Gonzales et al. 2000; Foguel and Silva 2004).

Pressure studies on viruses have direct biotechnological applications, especially concerning the production of new vaccines. The ability of pressure to inactivate viruses, prions and bacteria has been evaluated in relation to the applications of vaccine development and virus sterilization. Several studies have demonstrated that pressure can result in virus inactivation while preserving immunogenic properties (Silva et al. 1992; Oliveira et al. 1999; Tian et al. 2000; Barroso et al. 2012; Dumard et al. 2013).

15.2 Pressure Effects on Viral Particles

Viral particles present an excellent model of macromolecular assembly, which involves stringent packing constraints for the study of high-pressure effects. Viruses are composed of a membrane-enveloped or non-enveloped protein shell and nucleic acids (Johnson 1996). Therefore, viruses contain several components that can be susceptible to the effects of pressure. HHP has been a valuable tool for assessing viral structure-function relationships because the viral structure is highly dependent on protein-protein interactions (Silva and Weber 1988; Silva et al. 1996, 2002).

Due to compressibility, aqueous solutions occupy a smaller volume under pressure. Hydrophobic bonds are especially susceptible to destabilization by pressure because hydrophobic bond formation is accompanied by an increase in volume. In the interior of the protein, there are voids that can be destabilized by the penetration of water molecules. Hydrophobic regions are forced to occupy a smaller volume when pressure is applied, as pressure and volume are inversely proportional. The interaction with water is simultaneously increased. Hydrophobic bonds are important for the maintenance of a protein's tertiary structure and interactions. Therefore, the structure and interactions of proteins may be affected by the destabilization that results from interaction with a polar solvent (Weber and Drickamer 1983; Rivalain et al. 2010; Sarupria et al. 2010; Silva et al. 2010).

As the viral particle structure complexity increases, the sensitivity of the particle to pressure may increase or decrease depending on the protein-protein and protein-nucleic acid interactions. Studies have shown that monomeric or dimeric species of capsid coat proteins are much more sensitive to pressure compared with assembled icosahedral particles (Da Poian et al. 1993; Prevelige et al. 1994; Silva et al. 1996; Lima et al. 2006). In addition, empty capsids are usually much less stable compared with viral ribonucleoprotein particles (Silva et al. 1996; 2002). Viruses with a helical structure are also destabilized by high pressure (Lauffer and Dow 1941; Bonafe et al. 1998; Bispo et al. 2012).

Ribonucleoprotein intermediates have been detected in many icosahedral viruses after being subjected to pressure (Da Poian et al. 1993; Foguel et al. 1995; Silva et al. 1996, 2002). These intermediates may act as scaffolds for the regeneration of the particle (Gaspar et al. 1997; Silva et al. 2004). In the absence of RNA, the subunits do not renature when the perturbation is withdrawn, but instead tend toward a disorganized structure.

In the case of small icosahedral viruses, incremental increases in pressure produce a progressive decrease in the folding structure when moving from assembled capsids to ribonucleoprotein intermediates (in RNA viruses), free dissociated units (dimers and/or monomers) and denatured monomers (Silva et al. 1996, 2002, 2004; Gaspar et al. 1997).

We have found that high pressure inactivates enveloped viruses by trapping their particles in a fusion-like intermediate state. This effect on the viral particles prevents enveloped animal viruses from delivering their genomes into the appropriate sites in host cells (Silva et al. 1992; Skehel and Wiley 2000; Gaspar et al. 2002; Gomes et al. 2003). The fusogenic state, which is characterized by a smaller viral volume, is the final conformation promoted by a high hydrostatic pressure, in contrast with the metastable native state, which is characterized by a larger volume.

The conformational change promoted by pressure in the glycoproteins of membrane-enveloped viruses leads these proteins to a receptor-activated conformational state (Gaspar et al. 2002; Gomes et al. 2003; Freitas et al. 2006). In the influenza virus, this change is usually referred to as the "spring-loaded" model and is trigged by low pH, which indicates a mechanism in which the fusion peptide region is inserted into the target membrane in an early step of the fusion process (Skehel and Wiley 2000). We observed that subjecting the influenza virus to pressure exposes hydrophobic domains even in neutral pH (Gaspar et al. 2002; Barroso et al. 2012; Dumard et al. 2013) (Fig. 15.1). These data indicate a change in the labile native state of the envelope complex to a more stable state, thereby mimicking the fusion-active conformation.

15.3 Hydrostatic Pressure as a Viral Inactivating Agent

The use of HHP has resulted in full and partial viral inactivation (Lou et al. 2011; Dumard et al. 2013) because the pressure resistance of viruses varies greatly among different viruses and viral strains (Wouters et al. 1998; Shimasaki et al. 2009). Although a recent work described the lack of correlation between virus barosensitivity and the presence of a viral envelope (Lou et al. 2011), some authors have suggested that non-enveloped viruses are usually more pressure-resistant compared with membrane-enveloped viruses (Oliveira et al. 2008).

Full inactivation is usually demonstrated by the inoculation of inactivated virus samples onto permissive cell lines, hen eggs and animals (blind assay), and then

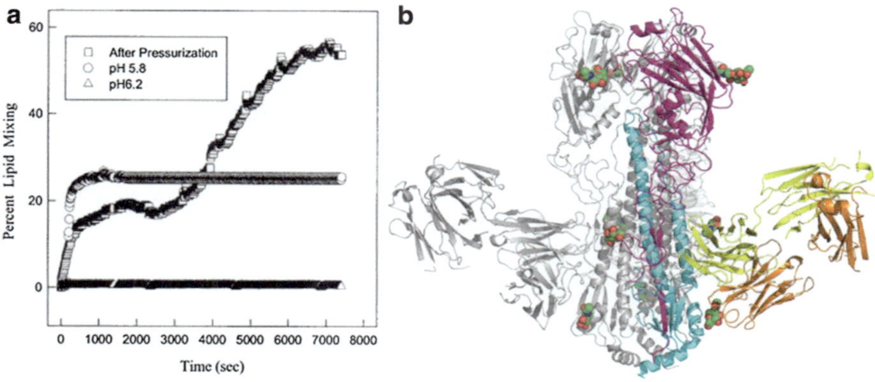

Fig. 15.1 (**a**) Pressurization for 2 h at 289.6 MPa of the human influenza virus at pH 7.4 leads to the fusogenic state. (*square*) Pressurized virus at pH 7.4, (*triangle*) non-pressurized virus at 7.4 and (*circle*) native virus at pH 5.8 (Reproduced with permission from (Gaspar et al. 2002). Copyright (2002) the American Society for Biochemistry and Molecular Biology) (**b**) Broadly neutralizing antibodies against the stem region of HA. Exposure of the fusion peptide by pressurization may be important to elicit antibodies against this region (Reproduced with permission from (Ekiert et al. 2009). Copyright (2009) American Association for the Advancement of Science)

further analyzed by infectivity assays and PCR (Ishimaru et al. 2004; Gaspar et al. 2008; Barroso et al. 2012; Dumard et al. 2013).

Currently, most studies that involve viral inactivation by HHP are found in the area of food engineering and focus on the inactivation of foodborne viruses (Kingsley 2013). Nevertheless, studies of viral inactivation for other purposes have also been conducted. HHP has been shown to be efficient in the inactivation of many viruses, such as the avian and human influenza viruses, yellow fever virus, human rhinovirus, foot-and-mouth disease virus (FMDV), and infectious bursal disease virus (IBDV) in chickens (See Table 15.1).

Many of these pressure-inactivated viruses have clinical importance. Thus, the use of the HHP approach has been proposed for the development of animal and human vaccines (Silva et al. 1992; Tian et al. 2000; Gomes et al. 2003; Dumard et al. 2013). To date, vaccines using inactivated viruses are produced by treating the virus with toxic chemicals, such as formaldehyde. HHP is a robust technology for enhancing the safety and reducing the production costs associated with viral vaccine production.

15.4 Influence of Temperature on the Effects of Pressure Inactivation

The application of high pressure based on the phase diagram of water has been the focus of many studies, especially in the food science. As pressure can alter the freezing point of water, a sample can be cooled to temperatures below 0 °C

Table 15.1 Viruses inactivated by pressure

Inactivated virus	Authors
Bovine rotavirus	Pontes et al. 1997
Simian rotavirus	Pontes et al. 2001
Infectious bursal disease virus	Tian et al. 2000
Human immunodeficiency virus (type1)	Otake et al. 2005*
Herpes simplex virus (type 1)	Kishida et al. 2013*, Nakagami et al. 1992
Encephalomyocarditis virus	Kishida et al. 2013*
Avian influenza viruses H3N8/H7N7	Barroso et al. 2012, Isbarn et al. 2007
Human influenza virus H3N2	Dumard et al. 2013
Yellow fever 17DD virus	Gaspar et al. 2008
Norovirus	Kovač et al. 2012
Vesicular stomatitis virus	Silva et al. 1992
Hepatitis A virus	Calci et al. 2005, Shimasaki et al. 2009
Human noroviruses (GI.1 Norwalk strain)	Kingsley et al. 2007
Murine norovirus	Leon et al. 2011, Tang et al. 2010
Simian immunodeficiency viruses (SIVmac251 and SIVagm)	Jurkiewicz et al. 1995
Bovine viral diarrhea virus type 1	Ceylan et al. 2012
Human cytomegalovirus	Nakagami et al. 1992
Aichivirus	Kingsley et al. 2004
Coxsackievirus A9	Kingsley et al. 2004
Human parechovirus-1	Kingsley et al. 2004
Avian metapneumovirus	Lou et al. 2011
Foot-and-mouth disease virus	Ishimaru et al. 2004
Human rotavirus	Lou et al. 2011

*Inactivated by pressure generation method (FPGM)

without freezing. In the early twentieth century, Bridgman showed that water goes through different phases when subjected to high pressure at different temperatures (Bridgman 1949). The freezing of water at atmospheric pressure allows the formation of ice I, which has a lower density compared with liquid water and results in an increased volume as the freezing temperature decreases (Kalichevsky et al. 1995). In contrast, pressurized water is able to form other types of ice (II – IX) that do not expand in volume and exhibit greater densities compared with liquid water. This minimizes tissue damage because the ice crystals are smaller, which is an advantage for the preservation of food characteristics (Li and Sun 2002). Moreover, in some cases, the high pressure-induced ice I-to-ice III phase transition might be the cause of microbial inactivation (Luscher et al. 2004).

The combined effects of high pressure with other factors, such as low or subzero temperature, pH and agents in sub-denaturing conditions (urea), have been a formidable tool in the assessment of the component's structure, as well as pathogen inactivation (Privalov 1990; Da Poian et al. 1995; Oliveira et al. 1999; Norberto et al. 2012). Many studies have sought to develop experimental vaccines and to understand the structure-function relationship of proteins through the processes of folding/unfolding and dissociation/aggregation (Silva et al. 2001; Dumay et al. 2006; Cordeiro et al. 2013). Low temperatures cause the exposure of nonpolar side chains to water. These side chains are more compressible, which explains the synergic effect of low temperature and pressure.

Poliovirus and rhinovirus exhibit high stability against HHP, and inactivation can only be attained through the association of HHP with low temperatures and sub-denaturing concentrations of urea (Oliveira et al. 1999). This association was also used to inactivate FMDV (Ishimaru et al. 2004). The freeze-pressure temperature method was also efficient in inactivating type 1 Human Immunodeficiency virus (HIV-1) (Otake et al. 2005). At room temperature, the reversible denaturation of Cowpea Mosaic Virus (CPMV) was achieved in the presence of 5 M urea. However, low temperatures (−20 °C) allow for denaturation of the viral particles in the absence of urea (Da Poian et al. 1995). Moreover, the combined effect of HHP and low temperatures markedly enhanced the dissociation of Tobacco Mosaic Virus (TMV). Other conditions, such as an increasing pH and sub-denaturing concentrations of urea, can also facilitate the pressure-induced dissociation of TMV (Santos et al. 2004, 2008; Bispo et al. 2012).

15.5 Freeze-Pressure Generation Method as a Viral Inactivating Agent

Freezing is a very old technology used for food preservation that was adopted in China in 1000 BC (Lund 2000). Although freezing is also an excellent way to preserve the viability of some pathogenic microorganisms, such as *Salmonella enteritidis, Giardia lamblia, Entamoebahistolytica, Vibrio cholera, Escherichia coli*, hepatitis A virus and norovirus, for long periods of time, freezing foods has historically represented a safe approach for preventing food-borne illnesses (Lund 2000).

The freeze-pressure generation method (FPGM) has been used since the 1990s for the sterilization and processing of food products, and it represents an alternative method for the inactivation of a number of pathogens, including bacteria, yeast, fungi and viruses. The pressure obtained by this method originates from the expansion of water during the liquid-crystalline phase transition. In a sealed vessel, water increases in volume and generates pressure in a temperature-dependent manner upon freezing. To estimate the pressure generated, a high-pressure chamber connected to a pressure gauge is usually filled with ethanol, sealed and then cooled with a temperature-controlled bath (Hayakawa et al. 1998). Ethanol is used because water would freeze and stop transmitting the pressure from the chamber to the pressure gauge. Under these conditions, it is possible to achieve approximately 200–250 MPa of pressure, depending on the temperature reached (0–30 °C, respectively) (Kishida et al. 2013).

Freezing kills microorganisms via physical and chemical effects and possibly through the induction of genetic changes. This death is mainly due to intra- and extracellular ice formation, changes in intra- and extracellular solute concentrations, and the low temperature (Archer 2004).

Microorganisms that survive high-pressure treatments can be evaluated for colony growth or infectivity titer. For example, using a pressure generator at 200 MPa at room temperature to inactivate *Saccharomyces cerevisiae* has the same effect as pressurization at 100 MPa and -10 °C (Hayakawa et al. 1998). Moreover, different strains of HIV-1, including clinical isolates, can be completely inactivated by FPGM, possibly due to a drastic reduction of the viral tropism for $CD4^+$ cells and a decrease in reverse transcriptase activity (Otake et al. 2005). The influence of FPGM was also evaluated with respect to the infection of other supposedly more resistant viruses, such as DNA and non-enveloped RNA viruses. Cells infected with FPGM-treated Encephalomyocarditis virus (EMCV) failed to induce an interferon response, and FPGM-treated human herpes virus type 1 (HSV-1) had showed a dramatically affected infectivity in addition to viral inactivation. The protection conferred by HSV-1 was confirmed by testing the inoculation in mice; all evaluated animals survived after a challenge with a lethal dose (Kishida et al. 2013).

Therefore, FPGM represents a simple, quick, easy and inexpensive technique that does not require large-scale generators or chemical compounds that can destroy the original taste and flavor of foods. Moreover, FPGM represents an interesting approach for the inactivation of a broad spectrum of viruses, and it can be used as either a single or complementary method to manufacture inactivated viral vaccines.

15.6 Immune Response to Pressurized Virus

Many viruses have been fully inactivated by HHP without drastic structural changes, as demonstrated by published reports of spectroscopic measurements and electron microscopy analysis (Dumard et al. 2013; Jurkiewicz et al. 1995; Silva et al. 1992). This interesting finding highlights the potential application of this tool for the preparation of whole viral vaccines in a simple, reproducible and inexpensive manner (Barroso et al. 2012; Silva et al. 2004). In addition, HHP is a practical approach for large-scale immunization (Gaspar et al. 2008).

The effects of HHP on the structure and biological activity of IBDV, a commercially important pathogen of chickens, have been investigated. Electron microscopy analysis revealed clear morphological changes in IBDV after pressure treatment, but pressure-inactivated IBDV retained its original immunogenic properties and elicited high titers of virus-neutralizing antibodies (Tian et al. 2000).

Rotavirus is an important viral agent that causes a form of gastroenteritis responsible for millions of cases of fatal diarrhea in children in developing countries. Studies have shown that pressure-inactivated rotavirus maintains its immunogenicity. Antibodies raised against pressurized material were as effective as antibodies raised against the intact virus, based on their neutralization titer in plaque reduction assays, enzyme-linked immunosorbent assays and direct interaction with the particle, as measured by gel-filtration chromatography (Pontes et al. 2001).

FMDV infectivity and immunogenic capacity have been evaluated after HHP treatment at 250 MPa and −15 °C with 1 M urea. This treatment completely abolished FMDV infectivity, while maintaining capsid structure integrity. Moreover, the ability of the virus to elicit neutralizing antibody production in rabbits was preserved (Ishimaru et al. 2004).

The effect of HHP has been evaluated in Yellow Fever Virus (YFV) using the live vaccine strain YFV 17DD as a study model. HHP treatment at 310 MPa for 3 h at 4 °C completely abolished YFV 17DD infectivity, as shown by a plaque assay on Vero cells and a blind assay conducted with C6/36 and Vero cells. Under these conditions, pressure-inactivated YFV 17DD failed to cause disease or kill mice (Fig. 15.2). Moreover, immunization with the inactivated virus was effective in eliciting a neutralizing antibody response and conferring protection after an intracerebral challenge with a lethal dose of live 17DD vaccine virus (Gaspar et al. 2008).

Recently, Barroso and colleagues assessed the immunogenic and protective capacity of pressure-inactivated avian influenza virus in a Balb/C mouse model.

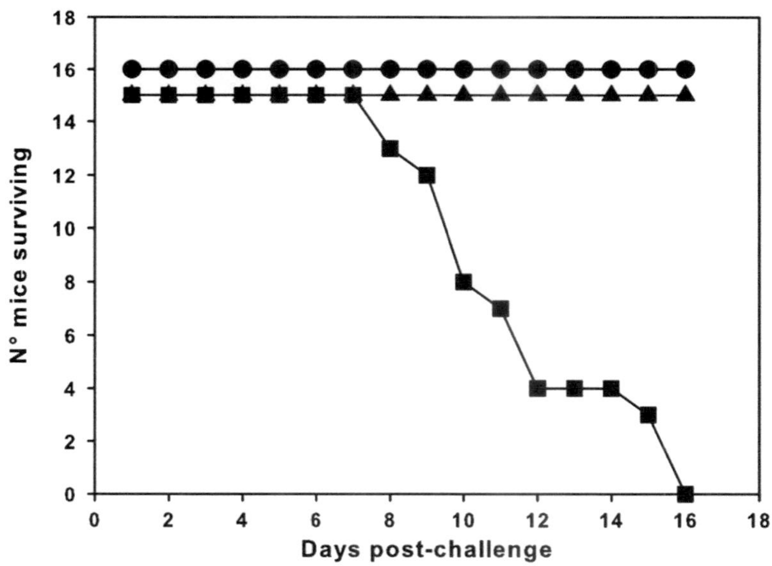

Fig. 15.2 Protection of mice immunized with 17DD virus inactivated by high pressure (310 MPa for 3 h at 4 °C). The immunized mice were challenged with the lethal dose of 5.0 log10 PFU inoculated by intracerebral route. The percentage of mice succumbing to infection was compared to that of mock-infected mice. *Circles* represent groups immunized with live YF 17DD vaccine control, *triangles* stands for mice immunized with 17DD inactivated virus and *squares* represent the mock-immunized group (Reproduced with permission from (Gaspar et al. 2008). Copyright (2008) Elsevier Ltd)

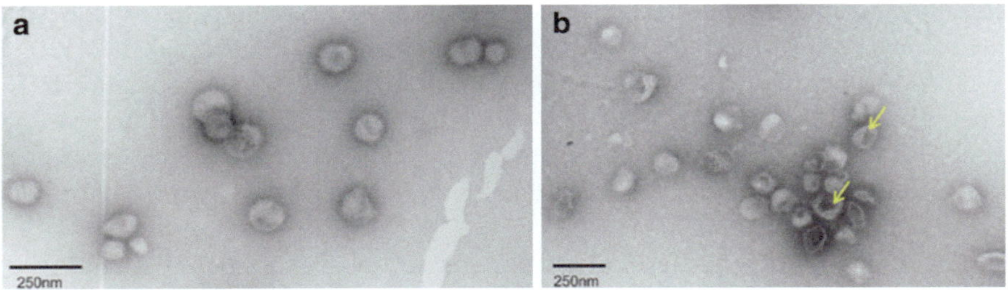

Fig. 15.3 Pressure effects (289.6 MPa) on viral particles. (**a**) Non pressurized virus. (**b**) Pressurized virus. *Arrows* indicates the formation of pores on pressurized viruses. Preservation of general morphology with exposure of hidden epitopes may be associated with a strong immune response (Dumard et al. 2013)

After immunization, there was an increase of IgG1 and IgG2a in the serum and IgA in the nasal lavage. A viral neutralization assay showed that the antibodies produced were neutralizing. After the challenge, none of the vaccinated animals developed clinical signs compared with the control group (Barroso et al. 2012). Vaccinated animals showed increased IgA in the feces and less Evans Blue dye leakage and lower cell counts in their bronchoalveolar lavage fluid compared with the challenged non-vaccinated group. After vaccination and challenge, Th1/Th2 cytokine secretion with a prevalence of IFN-γ was observed. At 3 months after vaccination, the animals remained protected after a challenge.

Recently, Dumard et al. (2013) described the inactivation of the human influenza virus H3N2 and analyzed the effects of HHP on the structure and immunogenic capacity of the virus. Electron microscopy data showed pore formation on the viral envelope, but the general morphology was preserved, and only small variations were observed in the particle structure (Fig. 15.3). Infectious activity ceased after 3 h of pressurization at 289 MPa, and the mice were protected from infection after vaccination. The results revealed full viral inactivation with an overall preservation of viral structure, hemmaglutinin and neuraminidase activities, as well as the maintenance of fusogenic capacity, thereby conferring protection against infection. A response that consisted of serum immunoglobulin IgG1, IgG2a, and serum and mucosal IgA was detected after vaccination (Dumard et al. 2013).

Structural information on human and avian influenza viruses, Newcastle disease virus, IBDV and other viruses inactivated by HHP is available in the literature. The small structural changes observed in these cases improve our understanding of the effective immunity obtained for the pressure-inactivated particles (Tian et al. 2000; Dumard et al. 2013). We believe that HHP inactivation preserves important epitopes in the virus structure, which are responsible for the production of neutralizing antibodies in vaccinated animals. Furthermore, previously hidden epitopes can be exposed, thereby increasing the immunogenicity of the pressure-inactivated viruses (Ishimaru et al. 2004; Gaspar et al. 2008; Oliveira et al. 2008).

15.7 Why Would a Pressurized Vaccine Result in a Better Vaccine?

The different commercially available viral vaccines can be classified as live attenuated, inactivated or killed and can further be divided into whole virus vaccines and split or subunit vaccines. Most vaccines are administered intramuscularly or subcutaneously. Additionally, all vaccines contain other substances (excipients) that are present because they improve the immune response (an adjuvant), are necessary for ensuring the stability of the product (stabilizers and preservatives), are the vehicle for delivering the vaccine (carrier) or are a residual product of the manufacturing process (e.g., antibiotics) (Baxter 2007).

Inactivated vaccines are usually safer than attenuated viruses because they do not allow the virus to reproduce in the host; however, they have disadvantages, such as the toxicity of the inactivating agent and changes in the immunogenic properties of the virus. Attenuated live viruses can revert after a certain time and may cause the disease that they are intended to prevent. Moreover, there are issues with vaccination with isolated subunits as well, most notably that the immune system recognizes the isolated antigen less effectively than the whole virus.

Using HHP to develop antiviral vaccines also has important advantages compared with other inactivation methods, as high pressure does not introduce exogenous substances into the vaccine and is often selective in its action on macromolecular structures. Therefore, HHP is a technology for the production of inactivated vaccines that are free of chemicals, safe and capable of inducing strong humoral and cellular immune responses (Silva et al. 1992; Tian et al. 1999; 2000; Barroso et al. 2012; Dumard et al. 2013). Pressurized viruses are able to maintain their immunogenic potential most likely because the structural changes are very subtle. The humoral and cellular responses, including the memory response, indicate that important epitopes are preserved.

The use of pressure as an inactivating agent has resulted in a whole inactivated vaccine (WIV) formulation. In the case of the influenza vaccine, many studies have demonstrated that whole inactivated influenza viruses are more immunogenic than split or subunit vaccines (Geeraedts et al 2008; Grebe et al. 2008). Recently, the influenza WIV was shown to promote cross-protection more efficiently if the fusogenic activity was preserved. The usual mechanism of chemical inactivation uses formalin and does not preserve the fusogenic activity. Thus, the pressure inactivation of the influenza virus, which maintains the fusogenic activity, may be more effective. The preservation of fusogenic activity allows the release of antigens in the cytoplasm, where they are processed and cross-presented to $CD8^{+}$ T cells (Huckriede et al. 2005; Rock and Shen 2005; Budimir et al. 2010; Budimir et al. 2013).

Partial conservation of the protein structure of viral particles is essential for the maintenance of immunogenic epitopes. The influenza virus, for example, was shown to be completely inactivated, but with hemagglutinating and neuraminidase activity, two major immunogenic regions were conserved (Dumard et al. 2013). Furthermore, the exposure of previously inaccessible epitopes could be an advantage of this mechanism of inactivation. The use of electron microscopy has demonstrated that HHP can induce pore formation on the membranes of enveloped viruses (Silva et al. 1992; Dumard et al. 2013). In influenza viruses, internal proteins, such as matrix protein (M1), nucleoprotein (NP) and RNA polymerase (PB1 and PB2), are associated with cytotoxic T lymphocyte (CTL) immune responses, which are critical for inhibiting viral clearance and spreading from infected cells (Brown and Kelso 2009). Although the pores indicate morphological changes to the viral particles, the changes at the level of protein structure are much more discrete, which is a crucial factor in the maintenance of immunogenic epitopes. The production of antibodies against the influenza hemmaglutinin binding site in mice vaccinated with pressure-inactivated viruses supports the idea that pressurization preserves this epitope, which enables antibody production (Dumard et al. 2013).

Despite reactogenicity being a common concern for WIV, a study in humans showed no significant differences in the side effects between WIV and split vaccines. WIV are available for influenza immunization, which shows that WIV can be well-tolerated. More studies on the humoral and cellular immunogenicities of inactivated viruses are needed. Nevertheless, these data indicate that HHP is a powerful tool for the development of animal and human vaccines that have been neglected.

15.8 Alternative Use for HHP in the Viral Vaccine Field

In general, viruses have variable intrinsic susceptibilities to high pressure. However, the mechanisms of resistance to pressure are not fully understood. In addition to promoting the sterilization of food and the inactivation of different microorganisms, HHP treatment has also shown potential for preventing the inactivation of viruses and viral-promoted thermal stabilization (Ferreira et al. 2009).

In particular, polioviruses show high resistance to different HHP treatments, such as pressurization at 600 MPa at 20 °C for 60 min (Oliveira et al. 1999; Wilkinson et al. 2001). In the absence of thermostabilizers, the three poliovirus strains were resistant to inactivation when subjected to prolonged and combined pressure and temperature treatments, including pressurization at 310 MPa for up to 65 h and incubation at 37 °C or −10 °C. Although these treatments were not able to inactivate the particles, viral stability tests showed that the pressurized viruses were able to resist exposure to 37 °C for 65 h in the absence of commercial stabilizers, such as $MgCl_2$ and L-arginine (Ferreira et al. 2009).

Live vaccines are usually very heat labile, and the Sabin oral poliovirus vaccine (OPV) is considered the most thermolabile vaccine according to the World Health Organization's Expanded Program on Immunization. For many developing countries, the proper implementation of a cold storage system for vaccines from the manufacturer to final vaccination sites is very poor (Wu et al. 1995; Shiomi et al.

2004). HHP could be used as a stabilizing method, which would circumvent the problems associated with the requirement for refrigeration during the storage and transport of the vaccines prior to use (Ferreira et al. 2009). Moreover, HHP treatment could be an interesting alternative to replace the thermostabilizers currently used, such as $MgCl_2$.

15.9 Perspectives and Conclusions

Many studies have been published on viral inactivation by HHP at room or cold temperatures. Most of these studies have applied HHP in food engineering for either the full or partial inactivation of foodborne viruses (Oliveira et al. 2008; Chawla et al. 2011; Kingsley 2013). Other uses involve the sterilization of biological products, such as blood and plasma derivatives (Bradley et al. 2000), the preparation of adeno-associated viral gene delivery systems (Leonard et al. 2007) and the development of animal and human vaccines (Silva et al. 1992; Oliveira et al. 1999; Tian et al. 2000; Dumard et al. 2013).

HHP vaccines appear to preserve fusogenic activity, which is not the case in vaccines produced using formalin inactivation (Budimir et al. 2010). These characteristics are great advantages of HHP vaccines. Although the reactogenicity of WIV commonly raises concerns, a study in humans showed that there were no significant differences in the side effects between the split vaccine and WIV groups (Hehme et al. 2002). This type of inactivation produces an antigen with many of the same chemical and physical properties of intact viral particles, and these properties are essential for inducing a satisfactory immune response. More investigations into HHP-based WIV are necessary to evaluate the reactogenicity, dose levels, and cellular and cytokine responses.

Live vaccines represent an effective strategy for controlling many diseases, especially because immunity is often long-term. Nevertheless, there is a risk of the emergence of collateral effects of varying severity, especially in immunocompromised individuals. Therefore, the complete substitution or addition of a sequential scheme involving a live vaccine and an inactivated vaccine has been proposed as a safer strategy for vaccination.

Clearly, OPV was essential for the eradication of poliomyelitis in the Americas, but the attenuated poliovirus used is genetically unstable, and vaccine-associated paralytic poliomyelitis cases have been increasing in recent years. However, the discontinuation of its use must be carefully programmed and planned. To address this issue, many countries have adopted the strategy of replacing the oral polio vaccine with the inactivated polio vaccine (IPV) or have performed sequential vaccinations to prevent the reintroduction of the wild virus and the resurgence of poliomyelitis (Heinsbroek and Ruitenberg 2010; Salas-Peraza et al. 2010). Thus, HHP is a promising tool for the development of vaccines, even in cases where a very effective vaccine is already available, such as the yellow fever vaccine.

Therefore, vaccines based on pressurized viruses represent a simple, clean, safe and low cost model, which makes use of industrially mature technology and provides an important alternative for the large-scale production of vaccines to protect against viral diseases.

Acknowledgements Our laboratories were supported by grants from the Conselho Nacional de Desenvolvimento Científico e Tecnológico (CNPq awards and INCT Program), Fundação Carlos Chagas Filho de Amparo à Pesquisa do Estado do Rio de Janeiro (FAPERJ), Ministério da Saúde (Decit Program) and Coordenação de Aperfeiçoamento de Pessoal de Nível Superior (CAPES).

References

Aertsen A, Meersman F, Hendrickx ME, Vogel RF, Michiels CW (2009) Biotechnology under high pressure: applications and implications. Trends Biotechnol 27:434–441

Akasaka K, Kitahara R, Kamatari YO (2013) Exploring the folding energy landscape with pressure. Arch Biochem Biophys 531:110–115

Archer DL (2004) Freezing: an underutilized food safety technology? Int J Food Microbiol 90:127–138

Balasubramaniam VM, Farkas D (2008) High-pressure food processing. Food Sci Technol Int 14:413–418

Barroso SP, Nico D, Gomes DC, Santos ACVD, Couceiro JNS, de Sousa CB, Silva JL, Oliveira AC (2012) Mice vaccination with high hydrostatic pressure-inactivated H3N8 virus protects against experimental avian flu. Procedia Vaccinol 6:98–105

Baxter D (2007) Active and passive immunity, vaccine types, excipients and licensing. Occup Med (Lond) 57:552–556

Bispo JA, Bonafe CF, Joekes I, Martinez EA, Carvalho GB, Norberto DR (2012) Entropy and volume change of dissociation in tobacco mosaic virus probed by high pressure. J Phys Chem B 116:14817–14828

Bonafe CF, Vital CM, Telles RC, Gonçalves MC, Matsuura MS, Pessine FB, Freitas DR, Vega J (1998) Tobacco mosaic virus disassembly by high hydrostatic pressure in combination with urea and low temperature. Biochemistry 37:11097–11105

Bradley DW, Hess RA, Tao F, Sciaba-Lentz L, Remaley AT, Laugharn JA Jr, Manak M (2000) Pressure cycling technology: a novel approach to virus inactivation in plasma. Transfusion 40:193–200

Bridgman PW (1949) The physics of high pressure. G. Bell, London

Brown LE, Kelso A (2009) Prospects for an influenza vaccine that induces cross-protective cytotoxic T lymphocytes. Immunol Cell Biol 87:300–308

Budimir N, Meijerhof T, Wilschut J, Huckriede A, de Haan A (2010) The role of membrane fusion activity of a whole inactivated influenza virus vaccine in (re)activation of influenza-specific cytotoxic T lymphocytes. Vaccine 28:8280–8287

Budimir N, de Haan A, Meijerhof T, Waijer S, Boon L, Gostick E, Price DA, Wilschut J, Huckried A (2013) Critical role of TLR7 signaling in the priming of cross-protective cytotoxic T lymphocyte responses by whole inactivated influenza virus vaccine. PLoS One 8:e63163

Calci KR, Meade GK, Tezloff RC, Kingsley DH (2005) High-pressure inactivation of hepatitis A virus within oysters. Appl Environ Microbiol 71:339–343

Ceylan C, Severcan F, Ozkul A, Severcan M, Bozoglu F, Taheri N (2012) Biophysical and microbiological study of high hydrostatic pressure inactivation of Bovine Viral Diarrheavirus type 1 on serum. Vet Microbiol 154:266–271

Chawla R, Patil GR, Singh AK (2011) High hydrostatic pressure technology in dairy processing: a review. J Food Sci Technol 48:260–268

Cordeiro Y, Foguel D, Silva JL (2013) Pressure-temperature folding landscape in proteins involved in neurodegenerative diseases and cancer. Biophys Chem 183:9–18
Da Poian AT, Oliveira AC, Gaspar LP, Silva JL, Weber G (1993) Reversible pressure dissociation of R17 bacteriophage. The physical individuality of virus particles. J Mol Biol 231:999–1008
Da Poian AT, Oliveira AC, Silva JL (1995) Cold denaturation of an icosahedral virus. The role of entropy in virus assembly. Biochemistry 34(8):2672–2677
Dumard CH, Barroso SPC, Oliveira GAP, Carvalho CAM, Gomes AMO, Couceiro JNSS, Ferreira DF, Nico D, Oliveira AC, Silva JL, Souza-Santos P (2013) Full inactivation of human influenza virus by high hydrostatic pressure preserves virus structure and membrane fusion while conferring protection to mice against infection. PLoS One 8:e80785
Dumay E, Picart L, Regnault S, Thiebaud M (2006) High pressure-low temperature processing of food proteins. Biochim Biophys Acta 1764(3):599–618
Ekiert DC, Bhabha G, Elsliger MA, Friesen RH, Jongeneelen M, Throsby M, Goudsmit J, Wilson IA (2009) Antibody recognition of a highly conserved influenza virus epitope. Science 324:246–251
Engelborghs Y, Heremans KA, De Maeyer LC, Hoebeke J (1976) Effect of temperature and pressure on polymerization equilibrium of neuronal microtubules. Nature 259:686–689
Ferrão-Gonzales AD, Souto SO, Silva JL, Foguel D (2000) The preaggregated state of an amyloidogenic protein: hydrostatic pressure converts native transthyretin into the amyloidogenic state. Proc Natl Acad Sci U S A 97:6445–6450
Ferreira E, Mendes YS, Silva JL, Galler R, Oliveira AC, Freire MS, Gaspar LP (2009) Effects of hydrostatic pressure on the stability and thermostability of poliovirus: a new method for vaccine preservation. Vaccine 27:5332–5337
Foguel D, Silva JL (2004) New insights into the mechanisms of protein misfolding and aggregation in amyloidogenic diseases derived from pressure studies. Biochemistry 43:11361–11370
Foguel D, Teschke CM, Prevelige PE Jr, Silva JL (1995) Role of entropic interactions in viral capsids: single amino acid substitutions in P22 bacteriophage coat protein resulting in loss of capsid stability. Biochemistry 34(4):1120–1126
Foguel D, Robinson CR, de Sousa PC, Jr SJL, Robinson AS (1999) Hydrostatic pressure rescues native protein from aggregates. Biotechnol Bioeng 63:552–558
Freitas M, Da Poian AT, Barth OM, Rebello MA, Silva JL (2006) The fusogenic state of mayaro virus induced by low pH and by hydrostatic pressure. Cell Biochem Biophys 44:325–335
Gaspar LP, Johnson JE, Silva JL, Da Poian AT (1997) Partially folded states of the capsid protein of cowpea severe mosaic virus in the disassembly pathway. J Mol Biol 273:456–466
Gaspar LP, Silva AC, Gomes AM, Freitas MS, Ano Bom AP, Schwarcz WD, Mestecky J, Novak MJ, Foguel D, Silva JL (2002) Hydrostatic pressure induces the fusion-active state of enveloped viruses. J Biol Chem 277:8433–8439
Gaspar LP, Mendes YS, Yamamura AM, Almeida LF, Caride E, Gonçalves RB, Silva JL, Oliveira AC, Galler R, Freire MS (2008) Pressure-inactivated yellow fever 17DD virus: implications for vaccine development. J Virol Methods 150:57–62
Geeraedts F, Bungener L, Pool J, ter Veer W, Wilschut J, Huckriede A (2008) Whole inactivated virus influenza vaccine is superior to subunit vaccine in inducing immune responses and secretion of proinflammatory cytokines by Dcs. Influenza Other Respir Viruses 2:41–51
Gomes AM, Pinheiro AS, Bonafe CF, Silva JL (2003) Pressure-induced fusogenic conformation of vesicular stomatitis virus glycoprotein. Biochemistry 42:5540–5546
Gonçalves RB, Mendes YS, Soares MR, Katpally U, Smith TJ, Silva JL, Oliveira AC (2007) VP4 protein from human rhinovirus 14 is released by pressure and locked in the capsid by the antiviral compound WIN. J Mol Biol 366(1):295–306
Grebe KM, Yewdell JW, Bennink JR (2008) Heterosubtypic immunity to influenza A virus: where do we stand? Microbes Infect 10:1024–1029
Hayakawa K, Ueno Y, Kawamura S, Kato T, Hayashi R (1998) Microorganism inactivationusing high-pressure generation in sealed vessels under sub-zero temperature. Appl Microbiol Biotechnol 50:415–418

Hehme N, Engelmann H, Kunzel W, Neumeier E, Sanger R (2002) Pandemic preparedness: lessons learnt from H2N2 and H9N2 candidate vaccines. Med Microbiol Immunol 191:203–208

Heinsbroek E, Ruitenberg EJ (2010) The global introduction of inactivated polio vaccine can circumvent the oral polio vaccine paradox. Vaccine 28:3778–3783

Hite B (1899) The effect of pressure in the preservation of milk. Bull W Virginia Univ Agric Exp Sta 58:15–35

Huckriede A, Bungener L, Stegmann T, Daemen T, Medema J, Palache AM, Wilschut J (2005) The virosome concept for influenza vaccines. Vaccine 23(Suppl 1):S26–S38

Isbarn S, Buckow R, Himmelreich A, Lehmacher A, Heinz V (2007) Inactivation of avian influenza virus by heat and high hydrostatic pressure. J Food Prot 70(3):667–673

Ishimaru D, Sá-Carvalho D, Silva JL (2004) Pressure-inactivated FMDV: a potential vaccine. Vaccine 22:2334–2339

Johnson JE (1996) Functional implications of protein-protein interactions in icosahedral viruses. Proc Natl Acad Sci U S A 93:27–33

Jurkiewicz E, Vilas-Boas M, Silva JL, Weber G, Gerhard H (1995) Inactivation of simian immunodeficiency virus by hydrostatic pressure. Proc Natl Acad Sci U S A 92:6935–6937

Kalichevsky MT, Knorr D, Lillford PJ (1995) Potential food applications of high-pressure effects on ice-water transitions. Trends Food Sci Technol 6:253–258

Kingsley DH (2013) High pressure processing and its application to the challenge of virus-contaminated foods. Food Environ Virol 5:1–12

Kingsley DH, Chen H, Hoover DG (2004) Inactivation of selected picornaviruses by high hydrostatic pressure. Virus Res 102:221–224

Kingsley DH, Holliman DR, Calci KR, Chen H, Flick GJ (2007) Inactivation of a norovirus by high-pressure processing. Appl Environ Microbiol 73:581–585

Kishida T, Cui FD, Ohgitani E, Gao F, Hayakawa K, Mazda O (2013) High pressure treatment under subfreezing temperature results in drastic inactivation of enveloped and non-enveloped viruses. Biotechnol Lett 35:1297–1301

Kovač K, Diez-Valcarce M, Raspor P, Hernández M, Rodríguez-Lázaro D (2012) Effect of high hydrostatic pressure processing on norovirus infectivity and genome stability in strawberry puree and mineral water. Int J Food Microbiol 152:35–39

Lauffer MA, Dow RB (1941) The denaturation of tobacco mosaic virus at high pressures. J Biol Chem 140:509–518

Leon JS, Kingsley DH, Montes JS, Richards GP, Lyon GM, Abdulhafid GM, Seitz SR, Fernandez ML, Teunis PF, Flick GJ, Moe CL (2011) Randomized, double-blinded clinical trial for human norovirus inactivation in oysters by high hydrostatic pressure processing. Appl Environ Microbiol 77(15):5476–5482

Leonard JN, Ferstl P, Delgado A, Schaffer DV (2007) Enhanced preparation of adeno-associated viral vectors by using high hydrostatic pressure to selectively inactivate helper adenovirus. Biotechnol Bioeng 97:1170–1179

Li B, Sun DW (2002) Novel methods for rapid freezing and thawing of foods – a review. J Food Eng 54:175–182

Lima SM, Vaz AC, Souza TL, Peabody DS, Silva JL, Oliveira AC (2006) Dissecting the role of protein-protein and protein-nucleic acid interactions in MS2 bacteriophage stability. FEBS J 273:1463–1475

Lou F, Neetoo H, Li J, Chen H, Li J (2011) Lack of correlation between vírus barosensitivity and the presence of a viral envelope during inactivation of human rotavirus, vesicular stomatitis virus, and avian metapneumovirus by high-pressure processing. Appl Environ Microbiol 77:8538–8547

Lund BM (2000) Freezing. In: Lund BM, Baird Parker TC, Gould GW (eds) The microbiological safety and quality of food, vol 1. Aspen Publishers, Gaithersburg, pp 122–145

Luscher C, Balasa A, Fröhling A, Ananta E, Knorr D (2004) Effect of high-pressure-induced ice I-to-ice III phase transitions on inactivation of Listeria innocua in frozen suspension. Appl Environ Microbiol 70:4021–4029

Mignaco JA, Lima LM, Rosenthal A, Foguel D, Silva JL (2005) Highlights of the 3rd international conference on high pressure bioscience and biotechnology. Braz J Med Biol Res 38(8):1147–1155

Nakagami T, Shigehisa T, Ohmori T, Taji S, Hase A, Kimura T, Yamanishi K (1992) Inactivation of herpes viruses by high hydrostatic pressure. J Virol Methods 38:255–261

Norberto DR, Vieira JM, Souza AR, Bispo JAC, Bonafe CFS (2012) Pressure- and urea-induced denaturation of bovine serum albumin: considerations about protein heterogeneity. Open J Biophys 2:4–14

Oliveira AC, Ishimaru D, Gonçalves RB, Smith TJ, Mason P, Sa-Carvalho D, Silva JL (1999) Low temperature and pressure stability of picornaviruses: implications for virus uncoating. Biophys J 76:1270–1279

Oliveira AC, Gomes AMO, Lima SMB, Gonçalves RB, Schwarcz WD, Silva ACB, Cortines JR, Silva JL (2008) Effects of hydrostatic pressure on viruses. In: Michiels C, Bartlett D, Aertsen A (eds) High-pressure microbiology. ASM press, Washington, DC, pp 19–34, Chapter 2

Otake T, Kawahata T, Mori H, Kojima Y, Hayakawa K (2005) Novel method of inactivation of human immunodeficiency virus type 1 by the freeze pressure generation method. Appl Microbiol Biotechnol 67:746–751

Pontes L, Fornells LA, Giongo V, Araujo JRV, Sepulveda A, Villas-Boas M, Bonafe CFS, Silva JL (1997) Pressure inactivation of animal viruses: potential biotechnological applications. In: Heremans K (ed) High pressure research in the bioscience and biotechnology. Leuven University Press, Leuven, pp 91–94

Pontes L, Cordeiro Y, Giongo V, Villas-Boas M, Barreto A, Araújo JR, Silva JL (2001) Pressure-induced formation of inactive triple-shelled rotavirus particles is associated with changes in the spike protein VP4. J Mol Biol 307:1171–1179

Prevelige PE Jr, King J, Silva JL (1994) Pressure denaturation of the bacteriophage P22 coat protein and its entropic stabilization in icosahedral shells. Biophys J 66:1631–1641

Privalov PL (1990) Cold denaturation of proteins. Crit Rev Biochem Mol Biol 25:281–305

Rastogi NK, Raghavarao KS, Balasubramaniam VM, Niranjan K, Knorr D (2007) Opportunities and challenges in high pressure processing of foods. Crit Rev Food Sci Nutr 47(1):69–112

Regnard P (1884) Recherches expérimentales sur l'influence des très hautes pressions sur les organismes vivants. CR Hebd Acad Sci 98:745–747

Rivalain N, Roquain J, Demazeau G (2010) Development of high hydrostatic pressure in biosciences: pressure effect on biological structures and potential applications in biotechnologies. Biotechnol Adv 28:656–672

Rock KL, Shen L (2005) Cross-presentation: underlying mechanisms and role in immune surveillance. Immunol Rev 207:166–183

Royer H (1895) Action des hautes pressions sur quelques bactéries. Arch Phys Norm Physiol 7:12–17

Salas-Peraza D, Avila-Agüero ML, Morice-Trejos A (2010) Switching from OPV to IPV: are we behind the schedule in Latin America? Expert Rev Vaccines 9:475–483

Santos JL, Bispo JA, Landini GF, Bonafe CF (2004) Proton dependence of tobacco mosaic virus dissociation by pressure. Biophys Chem 111(1):53–61

Santos JL, Aparicio R, Joekes I, Silva JL, Bispo JA, Bonafe CF (2008) Different urea stoichiometries between the dissociation and denaturation of tobacco mosaic virus as probed by hydrostatic pressure. Biophys Chem 134(3):214–224

Sarupria S, Ghosh T, García AE, Garde S (2010) Studying pressure denaturation of a protein by molecular dynamics simulations. Proteins 78(7):1641–1651

Schwarcz WD, Barroso SP, Gomes AM, Johnson JE, Schneemann A, Oliveira AC, Silva JL (2004) Virus stability and protein-nucleic acid interaction as studied by high-pressure effects on nodaviruses. Cell Mol Biol (Noisy-le-grand) 50(4):419–427

Shimasaki N, Kiyohara T, Totsuka A, Nojima K, Okada Y, Yamaguchi K, Kajioka J, Wakita T, Yoneyama T (2009) Inactivation of hepatitis A virus by heat and high hydrostatic pressure: variation among laboratory strains. Vox Sang 96:14–19

Shiomi H, Urasawa T, Urasawa S, Kobayashi N, Abe S, Taniguchi K (2004) Isolation and characterisation of poliovirus mutants resistant to heating at 50 °C for 30 min. J Med Virol 74:484–491
Silva JL, Foguel D, Suarez M, Gomes AMO, Oliveira AC (2004) High-pressure applications in medicine and pharmacology. J Phys Condens Matter 16:S929
Silva JL, Foguel D (2009) Hydration, cavities and volume in protein folding, aggregation and amyloid assembly. Phys Biol. 6(1):015002. doi:10.1088/1478-3975/6/1/015002
Silva JL, Weber G (1988) Pressure-induced dissociation of brome mosaic virus. J Mol Biol 199:149–159
Silva JL, Weber G (1993) Pressure stability of proteins. Annu Rev Phys Chem 44:89–113
Silva JL, Luan P, Glaser M, Voss E, Weber G (1992) Effects of hydrostatic pressure on a membrane-enveloped virus: high immunogenicity of the pressure-inactivated virus. J Virol 66:2111–2117
Silva JL, Foguel D, Da Poian AT, Prevelige PE (1996) The use of hydrostatic pressure as a tool to study viruses and other macromolecular assemblages. Curr Opin Struct Biol 6:166–175
Silva JL, Foguel D, Royer CA (2001) Pressure provides new insights into protein folding, dynamics and structure. Trends Biochem Sci 26:612–618
Silva JL, Oliveira AC, Gomes AM, Lima LM, Mohana-Borges R, Pacheco AB, Foguel D (2002) Pressure induces folding intermediates that are crucial for protein-DNA recognition and virus assembly. Biochim Biophys Acta 1595:250–265
Silva JL, Vieira TC, Gomes MP, Bom AP, Lima LM, Freitas MS, Ishimaru D, Cordeiro Y, Foguel D (2010) Ligand binding and hydration in protein misfolding: insights from studies of prion and p53 tumor suppressor proteins. Acc Chem Res 43:271–279
Skehel JJ, Wiley DC (2000) Receptor binding and membrane fusion in virus entry: the influenza hemagglutinin. Annu Rev Biochem 69:531–569
Somero GN (1992) Adaptations to high hidrostatic pressure. Annu Rev Physiol 54:557–577
Tang Q, Li D, Xu J, Wang J, Zhao Y, Li Z, Xue C (2010) Mechanism of inactivation of murine norovirus-1 by high pressure processing. Int J Food Microbiol 137(2–3):186–189
Tian SM, Qian JF, Shao GQ, Ruan KC (1999) High immunogenicity of the pressure-inactivated virus. Sheng Wu Hua Xue Yu Sheng Wu Wu Li Xue Bao (Shanghai) 31(3):334–336
Tian SM, Ruan KC, Qian JF, Shao GQ, Balny C (2000) Effects of hydrostatic pressure on the structure and biological activity of infectious bursal disease virus. Eur J Biochem 267(14):4486–4494
Weber G, Drickamer HG (1983) The effects of high pressure upon proteins and other biomolecules. Q Rev Biophys 116:89–112
Wilkinson N, Kurdziel AS, Langton S, Needs E, Cook N (2001) Resistance of poliovirus to inactivation by high hydrostatic pressure. Innovative Food Sci Emerg Technol 2:95–98
Winter R, Jeworreck C (2009) Effect of pressure on membranes. Soft Matter 5:3157–3173
Wouters PC, Glaasker E, Smelt JP (1998) Effects of high pressure on inactivation kinetics and events related to proton efflux in Lactobacillus plantarum. Appl Environ Microbiol 64:509–514
Wu R, Georgescu MM, Delpeyroux F, Guillot S, Balanant J, Simpson K, Crainicl R (1995) Thermostabilization of live virus vaccines by heavy water (D_2O). Vaccine 13:1058–1063

Part V
Pressure Effects on Biological Membranes

Editors' Foreword of Part V

One of the pressure-sensitive parts of a living cell is the membrane. This part is devoted to the behavior of membranes under pressure. Chapter 16 deals with the basic behavior of lipid bilayers under variable pressure and temperature. The thermotropic and barotropic bilayer phase behavior of various kinds of phospholipids with different molecular structures is systematically explained on the basis of their temperature–pressure phase diagrams. Chapter 17 discusses pressure-induced effects on artificial and cellular membranes. The pressure effects on the structure, dynamics and phase behavior of lyotropic lipid mesophases, model (homogeneous and hetrogeneous) membranes and natural (reconstituted and biological) membranes are comprehensively revealed by thermodynamic, spectroscopic, scattering and kinetic experiments. Effect of pressure on cell membrane dynamics and function is dealt with in Chap. 18. Here the chapter focuses on membrane properties of microbes (piezophiles and mesophiles) and describes membrane dynamics in deep-sea piezophiles obtained by time-resolved fluorescence anisotropy measurement. Finally, the environmental adaptation of biological membranes is discussed with archaea as example in Chap. 19. The possibility and limits of homeoviscous adaptation in archaea, which is membrane adaptation achieved by a modification of the membrane lipid composition as well as structure and properties of archaeal lipids, are discussed.

Chapter 16
How Do Membranes Respond to Pressure?

Hitoshi Matsuki

Abstract Bilayers formed by phospholipids are fundamental structures of biological membranes. The mechanical perturbation brought about by pressure significantly affects the membrane states of phospholipid bilayers. In this chapter, we focus our attention on the pressure responsivity for bilayers of some major phospholipids contained in biological membranes. At first, the membrane states and phase transitions of phospholipid bilayers depending on water content, temperature and pressure are explained by using the bilayer phase diagrams of dipalmitoylphosphatidylcholine (DPPC), which is the most familiar phospholipid in model membrane studies. Subsequently, the thermotropic and barotropic bilayer phase behavior of various kinds of phospholipids with different molecular structures is discussed from the comparison of their temperature – pressure phase diagrams to that of the DPPC bilayer. It turns out that a slight change in the molecular structure of the phospholipids produces a significant difference in the bilayer phase behavior. The systematic pressure studies on the phase behavior of the phospholipid bilayers reveal not only the pressure responsivity for the bilayers but also the role and meaning of several important phospholipids existing in real biological membranes.

Keywords Bilayer membrane • Dipalmitoylphosphatidylcholine • Phase stability • Phase transition • Phospholipid • Pressure

16.1 Introduction

A cell and organelles in the cell are shaped by biological membranes. Biological membranes consist of lipids and proteins, especially bilayer aggregates formed by the self-association of lipids are fundamental structures of the membranes. They are regularly arranged in organized but flexible sheets there. Although there are many kinds of lipids in biological membranes such as glycero-, sphingo-, and glycophospholipids and sterols, main constituent lipids distributed in the

---

H. Matsuki (✉)
Department of Life System, Institute of Technology and Science, Tokushima University,
2-1 Minamijosanjima-cho, Tokushima 770–8506, Japan
e-mail: matsuki@tokushima-u.ac.jp

K. Akasaka, H. Matsuki (eds.), *High Pressure Bioscience*,
Subcellular Biochemistry 72, DOI 10.1007/978-94-017-9918-8_16

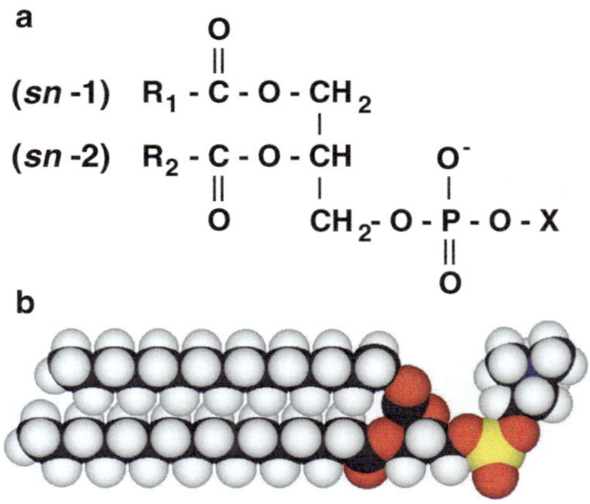

Fig. 16.1 (**a**) Molecular structure of glycerophospholipids. (**b**) Three-dimensional picture of DPPC. R_1, R_2 and X denote the hydrophobic chain in the *sn*-1 position, that in the *sn*-2 position and the hydrophilic group, respectively

membranes are glycerophospholipids. The molecular structure of glycerophospholipids is shown in Fig. 16.1a. They are amphiphilic molecules composed of two hydrophobic chains and a hydrophilic group that are respectively attached to a glycerol backbone by dehydration condensation, and there are many kinds of lipid species by combinations of hydrophobic and hydrophilic groups. Especially, phosphatidylcholines (PCs) phosphatidylethanolamines (PEs) distributed in great amounts in membranes of eukaryotes have been widely used as model membranes for life-science research (Koynova and Caffrey 1994; Koynova and Caffrey 1998).

High-pressure study of phospholipid bilayers has been done to elucidate action mechanisms of pressure in biological functions such as (1) pressure – anesthetic antagonism (Kaneshina et al. 1983; Wann and Macdonald 1988), (2) pressure adaptation in deep-sea organisms (Cossins and Macdonald 1989; Behan et al. 1992; Usui et al. 2012) and (3) high-pressure sterilization in food processing (Ludwig 1999; Hayashi 2002; Winter 2003). Pressure effect on various artificial and cellular membranes consisted by phospholipid bilayers has been described in recent reviews (Winter 2001; Winter and Jeworrek 2009) and also in Chap. 17.

The phospholipids targeted in the high-pressure study are almost all symmetric PCs with two equivalent hydrophobic chains in the *sn*-1 and *sn*-2 positions on the glycerol backbone, and most information regarding the pressure responsivity of the phospholipid bilayers has been concentrated on a certain PC, dipalmitoylphosphatidylcholine (DPPC), shown in Fig. 16.1b (Trudell et al. 1974; Srinivasan et al. 1974; Chong and Weber 1983; Lakowicz and Thompson 1983; Utoh and Takemura 1985; Wong and Mantsch 1985; Braganza and Worcester 1986; Winter and Pilgrim 1989; Driscoll et al. 1991). It is said that DPPC is one of main constituent lipids in biological membranes of eukaryotes and found in abundance in the membranes, hence, the bilayer is desirable as a model membrane. However, this fact is not rigorously correct. Biological membranes of prokaryotes like *Escherichia coli* do not contain PCs at all. Besides, major lipids in the membranes of eukaryotes are not symmetric phospholipids but asymmetric phospholipids, of which acyl chains attached to a glycerol backbone in the lipid molecule, are different. That is, the acyl chain in the *sn*-1 position on the glycerol backbone is exclusively saturated, whereas that in the *sn*-2 position is usually unsaturated and sometimes has a different carbon number compared with the *sn*-1 acyl chain. Actually, PCs contained in a membrane of a human erythrocyte are about 35 % of total lipids and DPPC is about 10 % of the PCs, therefore DPPC is only about 3.5 % of total lipids. In spite of these facts, DPPC plays the leading role in lipid membrane studies including high-pressure research. This seems to originate from facts that DPPC occupies the maximum percent in symmetric PCs of biological membranes and the membranes formed by DPPC have an advantage over those of other lipids experimentally: the gel-to-liquid-crystalline transition of the bilayer and the expanded-to-condensed transition of the insoluble monolayer can be easily observed in the vicinity of room temperatures under atmospheric pressure.

Our pressure study on biological membranes have begun with membrane states of the DPPC bilayer under high pressure (Kaneshina et al. 1992; Maruyama et al. 1997a), and we have investigated how the membrane state, in other words, phase behavior changes by various influencing factors such as lipid molecular structures, solvents, ionic strength, membrane targeting drugs under high pressure (Matsuki et al. 2013). In this chapter, we first explain the membrane states that phospholipid bilayers can take by using the DPPC bilayer as an example. Next, the characteristic bilayer phase behavior of various kinds of phospholipids with different molecular structures for hydrophobic, chain linkage and hydrophilic portions is overviewed in terms of their temperature – pressure phase diagram and by making a comparison between the diagrams and that of the DPPC bilayer. The functions and roles deduced from the systematic pressure studies on the phase behavior of the phospholipid bilayers are also described.

16.2 Phase Transitions of Phospholipid Bilayers

A phospholipid crystal takes lamellar structures, each of the lamellar is formed by a well-ordered bilayer of phospholipid molecules (Hauser et al. 1981). When water was added to the crystal, water molecules hydrate the polar head group of the lipid molecule and accumulated in the interlayer between bilayers. Then, the lamellar structures are swollen. Since the thickness of the interlayer in the lamellar structures have limitations and only definite water molecules are stored (in the case of neutral lipids, ca. 10 molecules for PC and 6 molecules for PE in the gel phase (Kodama et al. 2001)), excess water molecules detach from the lamellar structures and a closed bilayer microsome of unilamellar or multilamellar with a water core is formed. The bilayer microsome formed by the hydration has a diameter from about 20 nm to about 1,000 nm although the size of the microsome is dependent on the molecular structure of the phospholipid and the preparation method. This microsome is called a vesicle or a liposome.

The most notable feature of phospholipid bilayers is to undergo the structural change of the bilayer, namely a phase transition, in response to environmental changes of their surroundings. The hydration of phospholipid bilayers produces several membrane states depending on temperature and pressure and induces lyotropic phase transitions. The lyotropic phase transitions induced by temperature have been investigated for PC bilayers in detail (Kodama et al. 1982; Kodama 1986). Figure 16.2a shows the temperature-water content phase diagram of the DPPC bilayer together with the membrane state of each phase. The membrane states induced by hydration are, hydrated crystal or subgel (L_c) phase, gel (L_β) phase and liquid crystalline (L_α) phase. The three phases are the fundamental phases of hydrated phospholipid bilayers. The solid–liquid transition of the DPPC bilayer, which corresponds to the chain melting transition resulting from the *trans-gauche* conformational change of fatty-acid chains of DPPC molecules in the bilayer crystal, is observed at about 115 °C. The melting point of nonhydrated bilayer crystal of DPPC merely decreases steeply with increasing water content up to ca. 10 %. However, exceeding 10 %, another phase transition due to the hydration change in polar head groups appears at a low temperature in addition to the chain melting transition at a high temperature. Furthermore, at ca. 15 % of water content, the third phase transition due to the fluctuation of molecular packing in the bilayer appears at temperatures between the above two phase-transition temperatures. These phase transitions are referred to as the sub-, pre- and main transitions from the low temperature in turn and correspond to the transition between the L_c phase and the gel (lamellar gel ($L_{\beta'}$)) phase, the transition between the gel ($L_{\beta'}$ and ripple gel ($P_{\beta'}$)) phases, the transition between the $P_{\beta'}$ phase and the L_α phase, respectively. Here the superscript of prime implies that lipid molecules are oriented on the tilt from the bilayer plane. The tilted orientation of lipid molecules in bilayers is one of characteristics for the bilayers of phospholipids with a large and bulky head group like PCs. The three phase-transition temperatures become nearly constant independent of water content beyond ca. 25 %. Accordingly, although lipid concentrations are considerably different in respective lipid-membrane studies depending on the measurement methods such as X-ray diffraction (60–95 % of water content) and DSC (>99 % of water content), the observed phase states in these methods are equivalent to each other.

We can definitely specify the thermotropic phase states of phospholipid bilayers in the region where the phase-transition temperatures are not affected by water content. Moreover, by adopting pressure as an experimental variable, we can also specify the barotropic phase states of the bilayers in the region. Then, the thermotropic and barotropic phase behavior can be drawn by two-dimensional temperature (T)–pressure (p) phase diagram of the bilayer. We have measured the phase transitions of various phospholipid bilayers as a function of temperature and pressure by using DSC under atmospheric pressure and optical measurement methods under high pressure and constructed the T–p phase diagrams of the bilayers. Here we show the T–p phase diagrams of the various phospholipid bilayers and consider the thermotropic and barotropic phase behavior. The detailed experimental techniques for the phase-transition determination and the discussion for thermodynamic quantities associated with the phase transitions are described in the references given in the following sections.

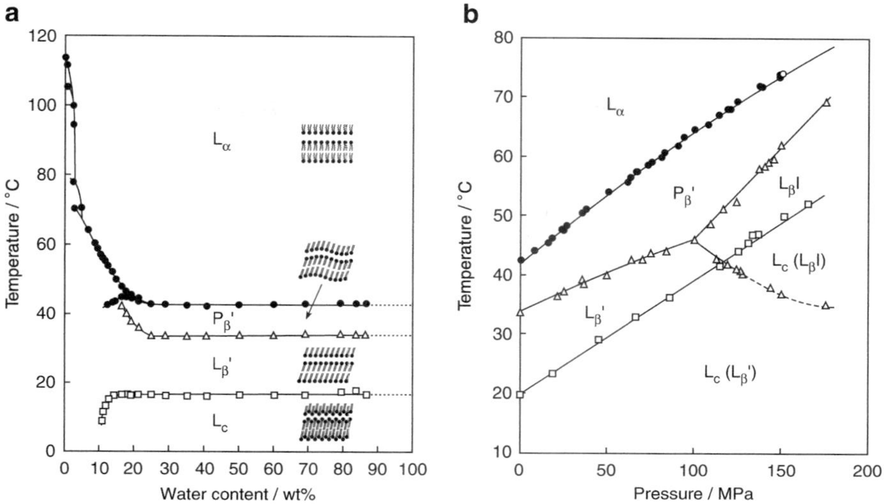

Fig. 16.2 (**a**) Effect of water content on phase-transition temperatures of DPPC bilayer. The membrane state of each phase is also depicted in the figure. (**b**) Temperature-pressure phase diagrams of DPPC bilayer. Phase transitions: (□) subtransition, (△) pretransition or transitions between gel phases, (●) main transition. *Dashed line* in figure (**b**) indicates the transition between metastable gel phases (Modified from (Kusube et al. 2005; Matsuki et al. 2013))

Figure 16.2b represents the *T–p* phase diagram of the DPPC bilayer (Kusube et al. 2005). All phase-transition temperatures increase linearly with increasing pressure. Interestingly, the DPPC bilayer forms another gel phase with nonbilayer structure called interdigitated gel (L_βI) phase at high pressures above about 100 MPa. In the L_βI phase, the acyl chains of the DPPC molecules in a monolayer constituting a bilayer extend beyond the region of the bilayer midplane and interpenetrate into the other opposing monolayer. Further pressurization extends the L_βI-phase region. The polymorphism of the gel phases such as the $L_{\beta'}$ and $P_{\beta'}$ phases with tilted molecular orientation and the nonbilayer L_βI phase is one of characteristics of bilayers of PCs with a bulky choline head group. This is caused by the large steric hindrance among adjacent choline groups.

16.3 Change in Phase Behavior by Differences in Phospholipid Molecular Structure

In biological membranes, there exist so many species of phospholipids with different molecular structures, symmetric and asymmetric lipids, saturated and unsaturated lipids, ester-linked and ether-linked lipids, PC and PE lipids, etc. We now systematically explain the thermotropic and barotropic phase behavior of bilayers formed by various phospholipids that can be obtained by changing the molecular structure of DPPC in terms of their *T–p* phase diagrams constructed.

16.3.1 Effect of Hydrophobic Acyl Chain

Figure 16.3a illustrates the overlays of the *T*–*p* phase diagrams for the bilayers of DPPC (C16PC) and two homologues of DPPC, dimyristoyl-PC (DMPC: C14PC) and dipalmitoyl-PC (DSPC: C18PC) (Ichimori et al. 1998). Here the subtransition curves are omitted from the figure for simplicity. The phase diagrams of three PC bilayers are similar in shape. We have shown (Ichimori et al. 1997, 1998; Kusube et al. 2005) that the slopes (d*T*/d*p*) of the phase boundaries depend on the kinds of the transitions and have similar values if the transition is the same. Therefore, the similarity of the phase boundaries among the diagrams suggests that the pressure responsivity of each phase transition is almost the same irrespective of the acyl chain length. On the other hand, with increasing acyl chain length, both the pre- and main-transition temperatures elevate and the diagram moves to the low-pressure and high-temperature region (Nagle and Wilkinson 1978; Blume 1983; Stümpel et al. 1983; Lewis et al. 1987; Ichimori et al. 1997, 1998). Then, the regions of the L_α and $P_{\beta'}$ phases shrink while those of the $L_{\beta'}$ and $L_\beta I$ phases extend in the order of DMPC, DPPC, and DSPC. We can understand that the enhancement of van der Walls interaction in the PC molecules makes the bilayer more rigid. It is noteworthy that the pressure of the triple point at which the three gel ($L_{\beta'}$, $P_{\beta'}$, and $L_\beta I$) phases coexist, we call the pressure the minimum interdigitation pressure (MIP), steeply decreases with the chain elongation. Therefore, the $L_\beta I$ phase extends the region markedly to lower pressures.

Recently, we have investigated the bilayer phase transitions of symmetric PCs with longer saturated acyl chains of carbon number of 19 and over (Goto et al. 2011a, 2015). The *T*–*p* phase diagrams constructed for diarachidoyl-PC (DAPC: C20PC) and dibehenoyl-PC (DBPC: C22PC) are shown in Fig. 16.3b. Further increases of the acyl chain length brings about the shrinkage of the L_α- and $P_{\beta'}$-phase regions while the extension of the $L_{\beta'}$- and $L_\beta I$-phase regions, which also produces a great decrease in the MIP value. As a result, in the diagram of the C22PC bilayer, the phase boundary between $L_{\beta'}$ and $L_\beta I$ phases disappears and no triple point is observed, indicating that the DBPC bilayer can induce the $L_\beta I$ phase under atmospheric pressure, not under high pressure. We confirmed that the DBPC bilayer forms the $L_\beta I$ phase by only hydration under atmospheric pressure by the neutron scattering and fluorometry under high pressure and the dihenearachidoyl-PC (C21PC) bilayer is the longest chain PC that requires pressure to induce the interdigitation (Goto et al. 2011a, 2015). On the other hand, the pressure-induced interdigitation occurs at ca. 300 MPa for the DMPC bilayer. We have shown (Ichimori et al. 1998) from volumetric consideration of pressure-induced interdigitation by using temperature and MIP at the triple point that the pressure-induced interdigitation substantially does not occur for bilayers of PCs with shorter saturated acyl chains of carbon number of 13 and under. Therefore, we can say that the pressure-induced interdigitation of the C*n*PC bilayers is a specific phenomenon at limited acyl chain lengths from 14 to 21, that is, only eight kinds of PC bilayers can induce the structure under high pressure. This finding for

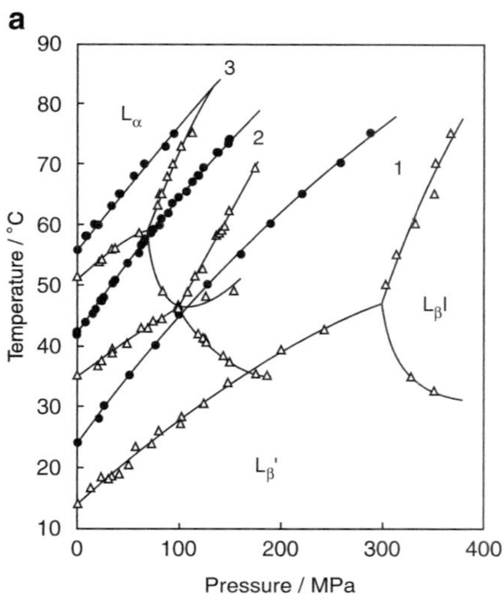

Fig. 16.3 (**a**) Temperature-pressure phase diagrams of bilayers of symmetric PCs: (1) DMPC (C14PC), (2) DPPC (C16PC), (3) DSPC (C18PC). (**b**) Temperature-pressure phase diagrams of bilayers of symmetric PCs: (1) DAPC (C20PC), (2) DBPC (C22PC). Phase transitions: (Δ) pretransition or transitions between gel phases, (●) main transition (Modified from (Ichimori et al. 1998; Matsuki et al. 2013))

the pressure-induced interdigitation suggests that the interdigitation is caused by a delicate balance between the repulsive head group interaction and the attractive acyl-chain interaction among lipid molecules in the bilayer.

Most phospholipids contained in biological membranes of eukaryotic cells are asymmetric phospholipids with two different acyl chains, not symmetric phospholipids with two equivalent acyl chains (Huang and Mason 1986) as mentioned above. Next we show the effect of the chain asymmetry on the phase behavior of PC bilayer membranes. We take up here asymmetric saturated PCs with difference of carbon number between the *sn*-1 and *sn*-2 chains being two, 1-palmitoyl-2-stearoyl-PC (PSPC) and its positional isomer 1-stearoyl-2-palmitoyl-PC (SPPC). In symmetric PCs like DPPC, two hydrophobic acyl chains in the *sn*-1 and *sn*-2 positions are not originally equivalent with each other although they have the chains of the same length. The *sn*-2 chain bends at C2-C3 position in the molecule, and this bend results in the *sn*-2 chain being 1.5 carbon – carbon bond lengths virtually shorter at the end of the acyl chains than the *sn*-1 acyl chain if they take *all-trans* conformation in the gel phase (Huang and Mason 1986; Bultmann et al. 1991). Accordingly, the terminal methyl ends between the *sn*-1 and *sn*-2 chains in the molecule are not aligned and the difference at the ends changes variously. In the case of PSPC and SPPC, the distance at the terminal methyl ends between the *sn*-1 and *sn*-2 chains are 0.5 and 3.5 carbon – carbon lengths, respectively.

The *T*–*p* phase diagrams of the PSPC and SPPC bilayers are shown in Fig. 16.4 (Matsuki et al. 2005; Goto et al. 2008, 2009). Both bilayers undergo three phase

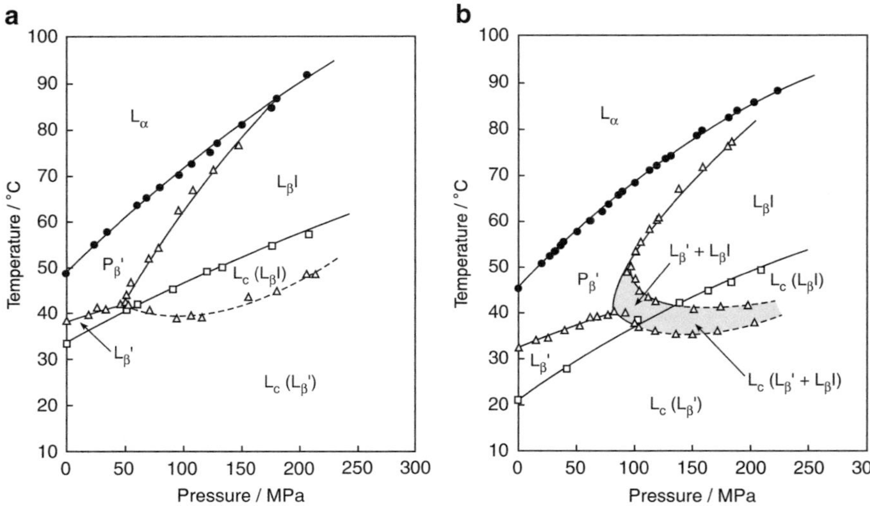

Fig. 16.4 Temperature-pressure phase diagrams of bilayers of asymmetric PCs with saturated acyl chains: (**a**) PSPC, (**b**) SPPC. Phase transitions: (□) subtransition, (Δ) pretransition or transitions between gel phases, (●) main transition. The *gray-colored* area in the SPPC bilayer is the coexistence region between the $L_{\beta'}$ and $L_\beta I$ phases. *Dashed line* in each figure indicates the transition between metastable gel phases (Modified from (Goto et al. 2008; Matsuki et al. 2013))

transitions (sub-, pre- and main) similar to the DPPC bilayer and induce the $L_\beta I$ phase at a certain pressure irrespective of the chain asymmetry. The three transition temperatures of the PSPC bilayer are higher than those of the SPPC bilayer. In the bilayer of SPPC with large chain asymmetry, the stabilities of the L_c and gel phases are reduced on account of the instability resulting from a portion of the chain difference between the *sn*-1 and *sn*-2 chains (Stümpel et al. 1981, 1983; Lin et al. 1991). By contrast, the effect of the alignment of terminal methyl groups in acyl chains appears in the subtransition and the interdigitation of the PSPC bilayer. The alignment causes the significant elevation of the subtransition temperature and the remarkable depression of the MIP value. It can be said that the degree of the alignment of terminal methyl groups in the acyl chains definitely influences the membrane states with dense packing like the L_c and $L_\beta I$ phases. Furthermore, there is the two-phase coexistence region (gray-colored area in the figure) of the $L_{\beta'}$ and $L_\beta I$ phases in the SPPC bilayer. We clarified by the high-pressure fluorometry (Goto et al. 2008, 2011b) that the region results from a less polar "pocket" in the membrane, which is formed from the chain asymmetry in the terminal parts of acyl chains of SPPC.

Phospholipids of most biological membranes contain a high percentage of unsaturated fatty acids. About a half of phospholipids in a human erythrocyte membrane has unsaturated fatty acids such as oleic, linoleic, arachidonic and docosahexaenoic acids (Heimburg 2007). We next explain the effect of unsaturated acyl chain on the phase behavior of PC bilayers. Here one symmetric unsaturated

PC, dioleoyl-PC (DOPC), which has two fatty acids with carbon number of 18 and a *cis* double bond at C9-C10 position in each chain and two mixed-chain PCs with an unsaturated acyl chain in the *sn*-1 or *sn*-2 position, 1-oleoyl-2-stearoyl-PC (OSPC) and 1-stearoyl-2-oleoyl-PC (SOPC), are taken up. The *T–p* phase diagrams of the DOPC, OSPC and SOPC bilayers (Kaneshina et al. 1998; Ichimori et al. 1999; Tada et al. 2009) are depicted in Fig. 16.5a together with that of the bilayer of DSPC that is a symmetric saturated homologue with acyl chains of the same carbon number. We can immediately notice that these unsaturated PC bilayers do not undergo the pretransition and interdigitation unlike the DSPC bilayer, namely they exhibit no polymorphism of the gel phases. The acyl chain having a double bond in unsaturated PCs hinders the close molecular orientation in the bilayer gel phase, and then the unsaturated PC bilayers undergo phase transitions among the L_c, L_β and L_α phases. The DOPC and OSPC bilayers undergo two transitions while the SOPC bilayer does only one transition. We identified the transitions of these bilayers from the thermal history under atmospheric pressure, the emergence patterns under high pressure and the dT/dp values in water and in 50 wt% ethylene glycol solutions (Kaneshina et al. 1998; Ichimori et al. 1999; Kusube et al. 2006; Tada et al. 2009). Consequently, the DOPC bilayer undergoes the L_c/L_α at low pressures or sub- (L_c/L_β) transition at high pressures and the main transition at all pressures and the OSPC does the subtransition and the main transition at all pressures while the SOPC bilayer does only the main transition at all pressures.

Comparing the main-transition temperatures of three kinds of unsaturated PC bilayers under atmospheric pressure with that of the DSPC bilayer, the temperatures are 55.6, 8.7, 6.7 and −40.3 °C for the DSPC, OSPC, SOPC and DOPC bilayers, respectively. The substitution of two *cis* unsaturated chains for two saturated chain of DSPC produces the great depression of the main-transition temperature by about 48 °C per a chain, and the introduction of a *cis* double bonds to the *sn*-2 position is slightly more effective than that to the *sn*-1 position (Tada et al. 2009). The *cis* unsaturated chain induces a bend in the chain. This leads to the reduction of van der Waals interaction between PC molecules in the gel phase, and then the gel phase is markedly destabilized (Heimburg 2007; Lewis and McElhaney 2005). To maintain high fluidity in biological membranes, the membrane state is required to be the L_α phase. The great stabilization of the L_α phase by introducing unsaturated fatty acids with a *cis* double bond to a lipid molecule is well correlated with the fact that membranes of marine and deep-sea organisms contain a high percentage of unsaturated phospholipids. A similar tendency is observed for the bilayer transitions related to the L_c phase under atmospheric pressure (Lewis et al. 1987; Goto et al. 2015), that is, the subtransition temperatures are 28.2 °C for the DSPC bilayer, 5.2 °C for the OSPC bilayer. It is evident that the structure of acyl chain also affects the stability of the L_c phase as well.

In the DOPC bilayer, the main-transition curve intersects the curve of the transition related to the L_c phase at about 234 MPa, and the stability of gel (L_β) phase changes at the intersection point: the L_β phase changes from metastable to stable at the pressure. This implies that the L_β phase at a low pressure becomes rather unstable, and only the L_c/L_α transition can be observed under atmospheric

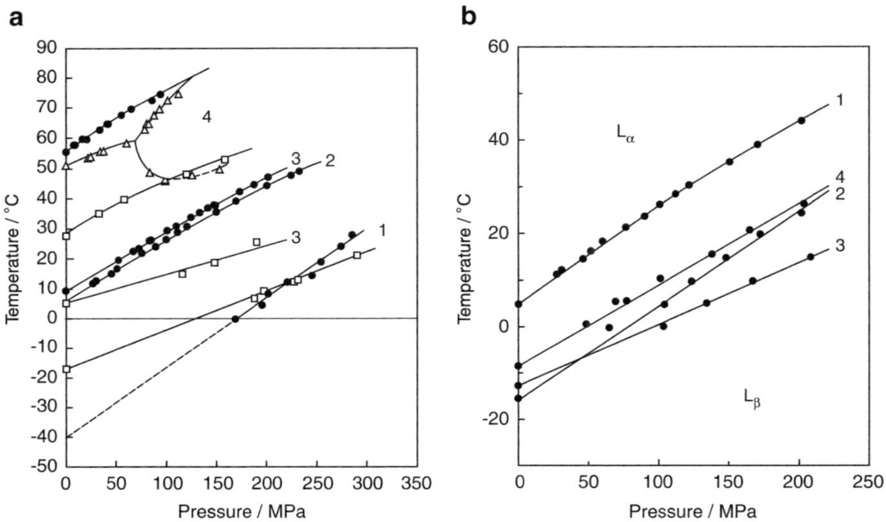

Fig. 16.5 (**a**) Temperature-pressure phase diagrams of unsaturated PC bilayers: (1) DOPC, (2) SOPC, (3) OSPC, (4) DSPC (for comparison). (**b**) Temperature-pressure phase diagrams of bilayers of asymmetric PCs with an unsaturated acyl chain in the *sn*-2 position: (1) SOPC, (2) SLPC, (3) SAPC, (4) SDPC. Phase transitions: (□) subtransition or L_c/L_α transition, (△) pretransition or transitions between gel phases, (●) main transition. *Dashed lines* in figure (**a**) indicates the unstable main transition (*curve* 1) and the transition between metastable gel phases (*curve* 4) (Modified from (Tada et al. 2010; Matsuki et al. 2013))

pressure (Lewis et al. 1988; Kusube et al. 2006). Here great caution should be taken in dealing with a phase transition of unsaturated phospholipids at a low temperature below the freezing point of water. The low-temperature phase transition of unsaturated phospholipid bilayers is generally regarded as the main (gel-liquid crystal) transition because there exists only one transition at the low temperature. However, at temperatures below the freezing point of water, the bilayers are prone to form the L_c phase since the bulk-water molecules freeze and simultaneously the molecules pull out the water molecules from the interlamellar region between the bilayers. Accordingly, the L_β phase becomes metastable or unstable, and the observed phase transition is the L_c/L_α transition, not the main transition as observed in the phase diagram of the DOPC bilayer. Recently, we have also investigated the phase transitions of a series of *N*-methylated dioleoylphosphatidylethanolamine (DOPE) bilayers (Kusube et al. 2006) and clarified that the low-temperature phase transition of these DOPE bilayers is the L_c/L_α transition, although the transitions have been generally considered to be the main transition (Gruner et al. 1988; Koynova and Caffrey 1994).

It is known that most of naturally occurring asymmetric phospholipids are phospholipids with a saturated chain at the *sn*-1 position and an unsaturated chain at the *sn*-2 position. We now consider the effect of the polyunsaturation in the *sn*-2 acyl chain on the bilayer phase behavior. Figure 16.5b shows the *T*–*p* phase diagrams for the bilayers of SOPC, 1-stearoyl-2-linoleoyl-PC (SLPC), 1-stearoyl-2-arachidonoyl-PC (SAPC) and 1-stearoyl-2-docosahexaenoyl-PC (SDPC), which are respectively containing a stearic acid chain in the *sn*-1 position and a mono- or poly-unsaturated acyl chain in the *sn*-2 position (Tada et al. 2010). These unsaturated PC bilayers undergo the main transition under a normal condition. The main-transition temperatures of the SOPC, SLPC, SAPC and SDPC bilayers under atmospheric pressure are 6.7, −15.0, −13.0 and −9.0 °C, respectively. The result indicates that the main-transition temperature of unsaturated PC bilayers does not decrease monotonously with increasing degree of unsaturation. The introduction of the first *cis* double bond into DSPC (yielding SOPC) lowers the main transition temperature by 48.9 °C and two double bonds (SLPC) lowers it by a further 21.7 °C, however, four and six double bonds (SAPC and SDPC) bring about no further decrease and unexpectedly cause a slight increase in the transition temperature. We have thermodynamically revealed from the smallest dT/dp value of the SAPC bilayer that the SAPC bilayer forms the gel phase with the loosest packing among these PC bilayers. Meanwhile, the relatively high temperature of the main transition of the SDPC bilayer suggests that the bilayer exists in a rather regular and relatively stable conformation in the gel phase as compared to the other bilayers (Holte et al. 1996; Koenig et al. 1997). In the gel state, since the acyl chain of SDPC must presumably be assumed to take a conformation approximately parallel to the bilayer normal and the motional freedom is severely limited by the lack of rotation at the six double bonds, allowed conformations are highly constrained. We speculate one of the possible conformations of SDPC in the gel state that the saturated *sn*-1 chain is well stretched, and the marked effect of the *cis* double bonds in the *sn*-2 chain, including the decrease in the effective chain length, yields an approximately helical conformation because the hydrophobic chain of docosahexaenoic acid can take the conformation (Applegate and Glomset 1986).

16.3.2 Effect of Linkage of Glycerol Backbone

Diacyl-PCs are phospholipids, of which each of the two hydrophobic chains binds to the glycerol backbone by an ester linkage. They are called ester-linked phospholipids. On the other hand, there exist other kinds of phospholipids, of which the hydrophobic chain binds to the glycerol backbone by an ether linkage in the cell membranes of some organisms (Paltauf 1994). They are called ether-linked phospholipids. Plasmalogens (1-alkenyl-2-acyl-PC or -PE) in brain and myelin membranes and macrocyclic lipids with many ether-linked isoprenoid chains in archaebacteria and deep-sea microorganisms are representative examples of ether-linked phospholipids (Han and Gross 1990; Lohner 1996; Broniec et al. 2009; Chong 2010). We next compare the phase behavior of a bilayer of dihexadecyl-PC (DHPC), which are PCs with two alkyl chains linking to the glycerol backbone by methylene groups ($-CH_2-$) instead of carbonyl groups ($>C{=}O$) in DPPC, with that of the DPPC bilayer. The substitution of the linkage affects the bilayer phase behavior significantly. For example, the DHPC bilayer undergoes the pre- and main transitions with increasing temperature under atmospheric pressure like the DPPC bilayer, where the main transition is the transition from the $P_{\beta'}$ phase to L_α the phase irrespective of the linkage difference, but the pretransition is different in both bilayers. The pretransition of the DHPC bilayer is the transition from the $L_\beta I$ phase to the $P_{\beta'}$ phase (Ruocco et al. 1985; Laggner et al. 1987; Kim et al. 1987), not the transition from the $L_{\beta'}$ phase as observed in the DPPC bilayer. Namely, the DHPC bilayer induces the interdigitation by only hydration under atmospheric pressure because the interaction between adjacent lipid molecules in the bilayer is weaker that that in the DPPC bilayer. Since the pre- and main-transition temperatures of the DHPC and DPPC bilayers under atmospheric pressure are very close to each other, and both bilayers have similar thermodynamic quantities of the transitions, it is difficult to distinguish the phase states by a conventional method as DSC. By contrast, pressurization clearly distinguishes the phase states of both bilayers.

The T–p phase diagram constructed for the DHPC bilayer is illustrated in the upper part of Fig. 16.6 (Maruyama et al. 1996; Matsuki et al. 2007). The difference in the pretransition between DHPC and DPPC appears in the slope of the pretransition curve on the phase diagrams: the slope of the pretransition ($L_\beta I/P_{\beta'}$ transition) for the DHPC bilayer is larger than that of the main-transition curve, while that of the pretransition ($L_{\beta'}/P_{\beta'}$ transition) for the DPPC bilayer is conversely smaller than it. Therefore, in the DHPC bilayer, the pretransition and main transition curves intersect each other at a certain pressure, then the $P_{\beta'}$ phase disappears at high pressures above the intersection point and only the $L_\beta I/L_\alpha$ transition occurs in the high-pressure region. It is understandable that pressure stabilizes the $L_\beta I$ phase but destabilizes the $P_{\beta'}$ and L_α phases and there is no pressure-induced phase. We have shown from the phase diagrams of a series of dialkyl-PC bilayers that the effect of the alkyl chain length on the phase behavior synergizes the pressure effect (Matsuki et al. 2007).

The phase diagram of the DHPC bilayer is appreciably different from that of the DPPC bilayer and becomes rather simpler (Ichimori et al. 1998; Kusube et al. 2005) because of the lack of a pressure-induced phase. The substitution of an ether linkage for an ester linkage brings about the appearance of the $L_\beta I$ phase under atmospheric pressure. Here, overlapping the phase diagram of the DHPC bilayer with that of the DPPC bilayer (the lower diagram in Fig. 16.6), we immediately notice that the phase behavior of the DPPC bilayer in the high-temperature and pressure region corresponds to that of the DHPC bilayer in the normal temperature and pressure region. When considering that membrane lipids of archaebacteria and deep-sea microorganisms are almost all ether-linked phospholipids and that an ether-linkage is chemically more stable than an ester linkage, it is so intriguing that the phase behavior of the ether-linked PC bilayers exhibits a striking similarity to that of the ester-linked PC bilayers under such extreme conditions as high temperature and pressure. The phase behavior of the ether-linked PC bilayers signifies that the bilayers become more rigid than the ester-linked PC bilayer, hence this fact seems to be inconsistent at a glance that organisms living in extreme conditions have softer membranes. We presume that the linkage-type changes

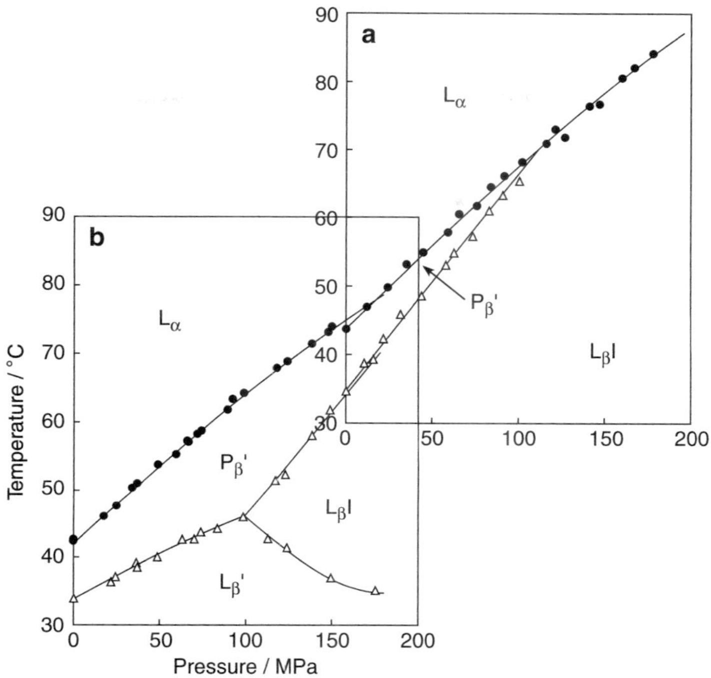

Fig. 16.6 Comparison of temperature-pressure phase diagram of (**a**) DHPC bilayer with that of (**b**) DPPC bilayer. Phase transitions: (Δ) pretransition or transitions between gel phases, (●) main transition (Modified from (Maruyama et al. 1996; Matsuki et al. 2013))

from the chemically stable ether-linkage at a high temperature and pressure to the chemically degradable ester-linkage at a normal temperature and pressure may be caused by the environmental adaptation in the organic evolutionary process. However, a recent study on biological membranes of deep-sea microorganisms under high pressure have revealed that the membranes are not necessarily softer than those of microorganisms under atmospheric pressure and actually do have a rather rigid structure (Usui et al. 2012; Abe 2013).

16.3.3 Effect of Polar Head Group

The bilayer phase behavior of various PCs with different molecular structures was shown in the preceding sections. On the other hand, there are also many phospholipids with different hydrophilic groups from PCs in biological membranes and this is the main factor of a diversity of membrane lipids. Here we describe the results of two kinds of phospholipids with the same acyl chains as DPPC and a different polar head group, dipalmitoylphosphatidylethanolamine (DPPE) and dipalmitoylphosphatidylglycerol (DPPG).

A polar head part of DPPC, the choline group has three methyl groups attached to a nitrogen atom ($-CH_2-CH_2-N^+(CH_3)_3$), whereas that of DPPE, the ethanolamine group has three hydrogen atoms instead ($-CH_2-CH_2-N^+H_3$). That is why the polar head group of DPPE is smaller in size than that of DPPC. The DPPE molecules in gel phase of the bilayer can be oriented perpendicular to the bilayer plane due to the smaller head group and form transient intermolecular hydrogen bonding between the protons of the quaternary nitrogen of the PE head group and the phosphate group on adjacent PE head group (Kodama et al. 1995). Figure 16.7a depicts the *T–p* phase diagram constructed for the DPPE bilayer (Kusube et al. 2005). The DPPE bilayer undergoes two-phase transitions with increasing temperature under atmospheric pressure depending on the thermal history of the sample. The transition in the first heating scan corresponds to the L_c/L_α transition while the transition in the second heating scan immediately after the first scan corresponds to the main (L_β/L_α) transition. Since it takes time to form the L_c phase from the L_α phase at a normal temperature and the L_c/L_α transition is observed at a higher temperature than the main-transition temperature, the L_β phase exists as the metastable phase under atmospheric pressure. Although both temperatures of the L_c/L_α and main transitions increase by applying pressure, the dT/dp value of the main transition is larger than that of the L_c/L_α transition in the whole range of pressure. Then the curves for the L_c/L_α and main transitions intersect each other on the phase diagram at ca. 22 MPa. Under high pressure above the pressure, the sub- (L_c/L_β) transition instead of the L_c/L_α transition and the main transition are consecutively observed with increasing temperature because the L_β phase converts to the stable phase. From the comparison of the *T–p* phase diagrams between the DPPE and DPPC bilayers (Fig. 16.2b), it turns out that the phase transition temperatures of the DPPE bilayer are much higher than those of the DPPC bilayer and the gel phase produces no polymorphism and exists stably only in the high-pressure region. These are originated from the stronger interaction of the DPPE molecule than the DPPC one in the bilayer.

Furthermore, DPPE having saturated acyl chains and a small head group is so different in molecular structure from DOPC having unsaturated acyl chains and a large head group and thus the interaction between lipid molecules in the DOPC bilayer is much weaker than that in the DPPE bilayer. But, comparing the *T–p* phase diagram of the DPPE bilayer with that of the DOPC bilayer given in Fig. 16.5a, we can find that the phase diagrams of both bilayers are qualitatively similar to each other (Kusube et al. 2005, 2006). They have a common feature that the gel phase of both bilayers exists as the stable phase only under high pressure: the stability of the gel phase in the DPPE bilayer is changed in the high-temperature and low-pressure region whereas that in the DOPC bilayer is changed in the low-temperature and high-pressure region. The similarity of membrane states between both bilayers suggests that bilayers of unsaturated phospholipids can take almost the same membrane states as those of saturated phospholipids depending on temperature and pressure. It seems that this is an example of environmental adaptations of lipid membranes.

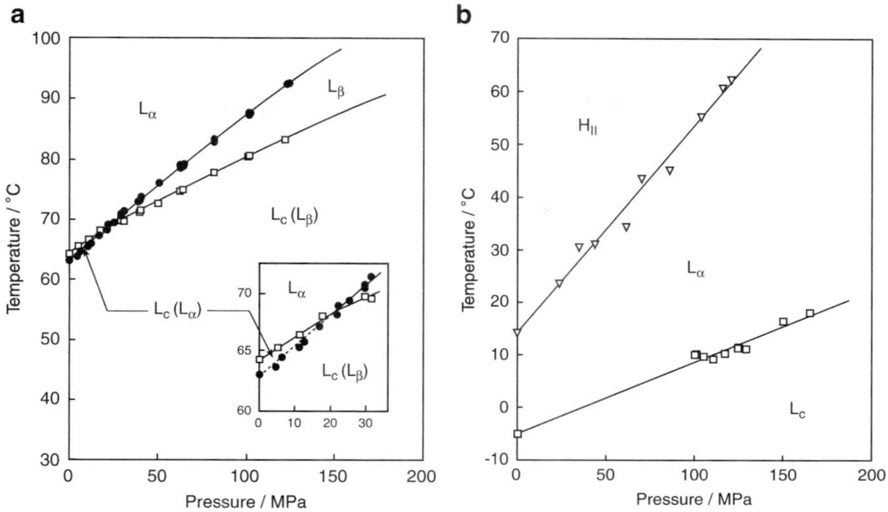

Fig. 16.7 (**a**) Temperature-pressure phase diagram of DPPE bilayer membrane (Modified from (Kusube et al. 2005)). (**b**) Temperature-pressure phase diagram of DOPE bilayer membrane. Phase transitions: (□) subtransition or L_c/L_α transition, (●) main transition, (∇) L_α/H_{II} transition. Here the results were obtained in water

The strong interaction between polar head groups of the DPPE molecule in the bilayer destabilizes the L_α phase and makes the membrane hard in comparison with the DPPC bilayer. Although PE rigidifies membranes greater than PC in this way, this strong interaction produces the inverted hexagonal (H_{II}) phase in the high-temperature region (Epand 1985). The H_{II} phase is one of nonbilayers like the $L_\beta I$ phase and takes a packing structure of the hexagonal lattices of a cylinder formed by lipid molecules surrounding a water-cored rod. Because the formation of the H_{II} phase is essential for phenomena relating to morphological changes of membranes such as cell fission and vesicle transport, it is considered that the formation ability of the H_{II} phase is an important roll of PE lipids. The formation of the H_{II} phase for the DPPE bilayer cannot be observed because it forms the phase at a temperature above the boiling point of water under atmospheric pressure (Caffrey 1993). However, we can observe the formation by decreasing the phase-transition temperature due to introducing unsaturated acyl chains into the PE molecules (Sueyoshi et al. 2006). The T–p phase diagram constructed for the dioleoyl-PE (DOPE) bilayer in water is shown in Fig. 16.7b. The DOPE bilayer undergoes the L_c/L_α transition, not the main transition at a temperature below the melting point of water under atmospheric pressure, and the L_α/H_{II} transition clearly appears at about 20 °C higher temperature. Both transition temperatures of the DOPC bilayer increase by applying pressure, but it is noteworthy that the L_α/H_{II} transition has the larger dT/dp value than any other kinds of transitions. The pronounced pressure responsivity of the L_α/H_{II} transition is attributable to the fact that the transition is the volume-controlled transition for the reconstruction of molecular packing (Sueyoshi et al. 2006).

In addition to neutral phospholipids as we explained above, biological membranes also include a pretty quantity of acidic phospholipids with a negative charge. And the membranes are charged negatively by the acidic phospholipids. Finally, we take up one of acidic phospholipids, DPPG, to show how the presence of a negative charge on a polar head group affects the barotropic bilayer phase behavior. The interaction among head groups of the anionic PG molecule in the bilayer is greatly different from that of the neutral PC and PE molecules since the electrostatic interaction among the head group acts in the bilayers (Watts et al. 1978, 1981; Riske et al. 2004; Pabst et al. 2007). The phase behavior of PC and PE that we described here is the results of the multilamellar vesicles obtained in water. Contrastingly, PG forms the unilamellar vesicles, not mutilamellar ones in water because the vesicle curvature increases and the multiplicity of membranes decreases on account of the strong repulsive interaction among the head group (Kodama et al. 1993; Kodama and Miyata 1995; Kodama et al. 1999). In this case, the cooperativity of phase transitions drastically decreases, and it is difficult to observe the transitions experimentally. Accordingly, the phase behavior of the DPPG bilayer is examined under conditions that electrostatic interaction among the head group is shield by adding sodium chloride (NaCl).

The DPPG bilayer undergoes two-phase transitions (pre- ($L_{\beta'}/P_{\beta'}$), and main ($P_{\beta'}/L_{\alpha}$) transitions) with increasing temperature at a normal temperature under atmospheric pressure like the DPPC bilayer. The dependence of NaCl concentration on the pre- and main-transition temperatures are drawn in Fig. 16.8a. Both transition temperatures increase with an increase in NaCl concentration. Since the slope of the pretransition curve is larger than that of the main-transition curve, both curves intersect each other at about 1.4 mol kg^{-1} NaCl concentration and the pretransition disappears the concentration. Sodium ions of added NaCl shield the negative charge of the DPPG molecules in the bilayer concentration-dependently, and then the electrostatic repulsion among the head group is relaxed and the membrane becomes rigid. Therefore, the addition of NaCl makes the $L_{\beta'}$ phase stabilize while it does the $P_{\beta'}$ and L_{α} phases destabilize. Further the effect of NaCl is enhanced in the pretransition as compared with the main transition, indicating that NaCl acts on the membrane surface greater than the membrane core.

Taking account of the shielding effect of NaCl on the DPPG bilayer, we fixed the NaCl concentration at 1.0 mol kg^{-1} and examined the barotropic phase behavior. Figure 16.8b demonstrates the *T–p* phase diagram of the DPPG bilayer. The DPPG bilayer exhibits the complicated phase behavior because the observed phase states are markedly affected by the thermal history of samples in a pretreatment called annealing before measurements (Zhang et al. 1997; Tanaka et al. 2012). The bilayer of DPPG treated with short-term annealing undergoes the subtransition in addition to the pre- and main transitions under atmospheric pressure as mentioned above. By applying pressure under this annealing condition, three kinds of the phase transition temperatures increase and the pressure-induced $L_{\beta}I$ phase is induced (Singh et al. 2008). This phase behavior of the DPPG bilayer is qualitatively consistent with that

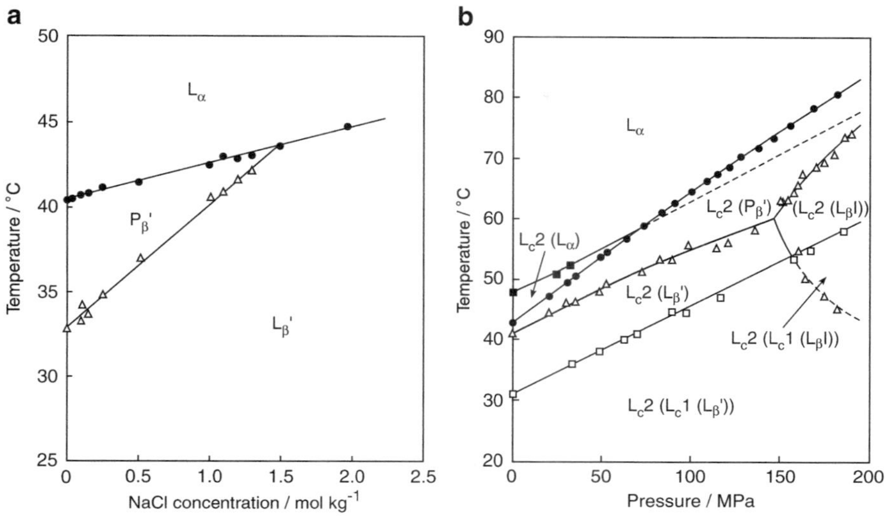

Fig. 16.8 (**a**) Effect of NaCl concentration on phase-transition temperatures of DPPG bilayer. (**b**) Temperature-pressure phase diagram of DPPG bilayer membrane at 1 mol kg^{-1} NaCl solution. Phase transitions: (□) subtransition, (■) L_c2/L_α transition, (△) pretransition or transitions between gel phases, (●) main transition. *Dashed line* in figure (**b**) indicates the transition between metastable gel phases

of the DPPC bilayer. On the other hand, treating the DPPG bilayer by long-term annealing, only one phase transition instead of three transitions appears at a higher temperature than the main-transition temperature under atmospheric pressure. We have investigated the phase transition in detail by DSC (Tanaka et al. 2012) that the transition corresponds to the transition between the stable phases, that is, the stable hydrated crystal (L_c2) phase, in which the DPPG molecules are oriented in rod-like shape, not in spherical shape, to the L_α phase. And we identified the phase observed by short-term annealing as the metastable hydrated crystal (L_c1) phase. The temperature of the L_c2/L_α transition also increases with an increase in pressure and the curve of the transition intersects the main-transition curve since the dT/dp value of the transition is smaller than that of the main transition. This phase behavior coincides with that of the DPPE bilayer as explained above. Therefore, we can say that the phase behavior of the DPPG bilayer has both characteristics of the phase behavior of the DPPC and DPPE bilayers, although the phase diagram seems to be complicated at a glance.

Then, why does the DPPG bilayer have both characteristics in the phase behavior? We speculate that the reason is closely related to the difference in lipid compositions of biological membranes between prokaryotes and eukaryotes. The lipids consisting the membranes of prokaryotes, e.g. in the membranes of *Escherichia coli*, are PE, PG and cardiolipin (dimer of PG), PC is not included, whereas those of eukaryotes, e.g. in the membranes of human erythrocyte, are PC, sphingomyelin, PE and phosphatidylserine, PG is not included. Taking into account that PC or PG with a large polar head group can form the bilayer structure more stably than PE with a small head group, prokaryotes that are acting at a certain salt concentration can maintain the stable membrane states and change the membrane shape by using PG and PE under the condition. In contrast to this, in eukaryotes that have inner membrane systems in the cell, the membrane systems existing under various conditions such as environments of inconstant salt concentrations are required to firmly form the stable bilayer structure irrespective of the salt concentrations. Then, it seems that eukaryotes acquired PC in the evolutionary process and began to construct the more stable bilayer structures.

16.4 Concluding Remarks

No one never knows DPPC if anyone treats a lipid membrane before. Although DPPC is the most standard and familiar phospholipid in model membrane studies, judging from the existing ratio in the membrane lipids of real biological membranes, it is rather classified as the special kind of phospholipids from the standpoint that it has symmetric saturated hydrophobic chains. It is rarely known in general that the greatest percentage of PC (ca. 20 %) in PCs of human erythrocyte membranes is an asymmetric mixed chain PC with a saturated chain in the *sn*-1 position and an unsaturated chain in the *sn*-2 position, palmitoyl-linoleyl-PC.

There is a criticism that the behavior of real biological membranes cannot be understood unless we investigate real biological membranes, not model membrane. However, we cannot appreciably obtain the useful information even if we measure the properties of the real membranes of a cell by physico-chemical methods. For example, performing DSC measurements on biological membranes of *Escherichia coli* and *Bacillus subtilis*, only two kinds of broad endothermic peaks appear in the thermogram: one is originated from the chain melting transition of membrane-constituent lipids in the low-temperature region and the other is from the thermal denaturation of membrane proteins in the high-temperature region (Heimburg and Jackson 2005; Heimberg 2007). Since lipids of model membranes are far less than those of real membranes, they oversimplify the membrane structure. But the lipid interaction in the bilayer can be clarified in detail because of the small number of constituent lipids.

In this chapter, by variously changing the molecular structure of DPPC, we described the effect of the lipid molecular structure on the bilayer phase behavior under high pressure by using the *T*–*p* phase diagram of each phospholipid bilayer. The effects of solvents and membrane targeting drugs such as anesthetics on the barotropic phase behavior are omitted for want of space. For the readers who are interested in the subject, please refer to the references (Maruyama et al. 1997a, b; Nishimoto et al. 2009). Although DPPC is regarded as a typical lipid in model membranes rather than a lipid in biological membranes, the systematic study on the lipid phase transitions by using the DPPC bilayer as the bilayer standard reveals that a slight change in the molecular structure of the phospholipids brings about a significant difference in the bilayer phase behavior. And simultaneously it enables us to consider the role and meaning of several important phospholipids existing in real biological membranes.

Acknowledgements Financial support by Grant-in-Aid for Scientific Research (No. 26410016) from Japan Society for the Promotion of Science is gratefully acknowledged.

References

Abe F (2013) Dynamic structural changes in microbial membranes in response to high hydrostatic pressure analyzed using time-resolved fluorescence anisotropy measurement. Biophys Chem 183:3–8

Applegate KR, Glomset TA (1986) Computer-based modeling of the conformation and packing properties of docosahexaenoic acid. J Lipid Res 27:658–680

Behan MK, Macdonald AG, Jones GR, Cossins AR (1992) Homeoviscous adaptation under pressure: the pressure dependence of membrane order in brain myelin membranes of deep-sea fish. Biochim Biophys Acta 1103:317–323

Blume A (1983) Apparent molar heat capacities of phospholipids in aqueous dispersion. Effects of chain length and head group structure. Biochemistry 22:5436–5442

Braganza LF, Worcester DL (1986) Hydrostatic pressure induces hydrocarbon chain interdigitation in single component phospholipid bilayers. Biochemistry 25:2591–2596

Broniec A, Goto M, Matsuki H (2009) A peculiar phase transition of plasmalogen bilayer membrane under high pressure. Langmuir 25:11265–11268

Bultmann T, Lin H, Wang Z, Huang C (1991) Thermotropic and mixing behavior of mixed-chain phosphatidylcholines with molecular weights identical with that of L-α-dipalmitoylphosphatidylcholine. Biochemistry 30:7194–7202

Caffrey M (1993) LIPIDAT, a database of thermodynamic data and associated information on lipid mesomorphic and polymorphic transitions. CRC Press, Boca Raton

Chong PLG (2010) Archaebacterial bipolar tetraether lipids; physico-chemical and membrane properties. Chem Phys Lipids 163:253–265

Chong PLG, Weber G (1983) Pressure dependence of 1,6-diphenyl-1,3,5-hexatrien fluorescence in single-component phosphatidylcholine liposomes. Biochemistry 22:5544–5550

Cossins AR, Macdonald AG (1989) The adaptation of biological membranes to temperature and pressure: fish from the deep and cold. J Bioenerg Biomembr 21:115–135

Driscoll DA, Samarasinghe J, Adamy S, Jonas J, Jonas A (1991) Pressure effects on dipalmitoylphosphatidylcholine bilayers measured by ^{2}H nuclear magnetic resonance. Biochemistry 30:3322–3327

Epand RM (1985) High sensitivity differential scanning calorimetry of the bilayer to hexagonal phase transitions of diacylphosphatidylethanolamines. Chem Phys Lipids 36:387–393

Goto M, Kusube M, Tamai N, Matsuki H, Kaneshina S (2008) Effect of hydrostatic pressure on the bilayer phase behavior of symmetric and asymmetric phospholipids with the same total chain length. Biochim Biophys Acta 1778:1067–1078

Goto M, Ishida S, Tamai N, Matsuki H, Kaneshina S (2009) Chain asymmetry alters thermotropic and barotropic properties of phospholipid bilayer membranes. Chem Phys Lipids 161:65–76

Goto M, Wilk A, Kazama A, Chodankar S, Kohlbrecher J, Matsuki H (2011a) Chain elongation of diacylphosphatidylcholine induces fully bilayer interdigitation under atmospheric pressure. Colloids Surf B: Biointerfaces 84:44–48

Goto M, Matsui T, Tamai N, Matsuki H, Kaneshina S (2011b) Prodan fluorescence detects the bilayer packing of asymmetric phospholipids. Colloids Surf B: Biointerfaces 84:55–62

Goto M, Endo T, Yano T, Tamai N, Kohlbrecher J, Matsuki H (2015) Comprehensive characterization of temperature- and pressure-induced phase transitions for saturated phosphatidylcholines containing longer chain homologs. Colloids Surf B Biointerfaces 128:389–397

Gruner SM, Tate MW, Kirk GL, So PTC, Turner DC, Keane DT, Tilcock CPS, Cullis PR (1988) X-ray diffraction study of the polymorphic behavior of N-methylated dioleoylphosphatidylethanolamine. Biochemistry 27:2853–2866

Han X, Gross RW (1990) Plasmenylcholine and phosphatidylcholine membrane bilayers possess distinct conformational motifs. Biochemistry 29:4992–4996

Hauser H, Pascher I, Pearson RH, Sundell S (1981) Preferred conformation and molecular packing of phosphatidylethanolamine and phosphatidylcholine. Biochim Biophys Acta 650:21–51

Hayashi R (ed) (2002) Trends in high pressure bioscience and biotechnology. Elsevier, Amsterdam

Heimburg T (2007) Thermal biophysics of membranes. Wiley-VCH, Weiheim

Heimburg T, Jackson AD (2005) On soliton propagation in biomembranes and nerves. Proc Natl Acad Sci U S A 102:9790–9795

Holte LL, Separovic F, Gawrisch K (1996) Nuclear magnetic resonance investigation of hydrocarbon chain packing in bilayers of polyunsaturated phospholipids. Lipids 31:199–203

Huang C, Mason JT (1986) Structure and properties of mixed-chain phospholipid assemblies. Biochim Biophys Acta 864:423–470

Ichimori H, Hata T, Yoshioka T, Matsuki H, Kaneshina S (1997) Thermotropic and barotropic phase transition on bilayer membranes of phospholipids with varying acyl chain-lengths. Chem Phys Lipids 89:97–105

Ichimori H, Hata T, Matsuki H, Kaneshina S (1998) Baroropic phase transitions and pressure-induced interdigitation on bilayer membranes of phospholipids with varying acyl chain-lengths. Biochim Biophys Acta 1414:165–174

Ichimori H, Hata T, Matsuki H, Kaneshina S (1999) Effect of unsaturated acyl chains on the thermotropic and barotropic phase transitions of phospholipid bilayer membranes. Chem Phys Lipids 100:151–154

Kaneshina S, Kamaya K, Ueda I (1983) Thermodynamics of pressure-anesthetic antagonism on the phase transition of lipid membranes. Displacement of anesthetic molecules. J Colloid Interface Sci 93:215–224

Kaneshina S, Tamura K, Kawakami H, Matsuki H (1992) Effects of pressure and ethanol on the phase behavior of dipalmitoylphosphatidylcholine multilamellar vesicles. Chem Lett 21:1963–1966

Kaneshina S, Ichimori H, Hata T, Matsuki H (1998) Barotropic phase transitions of dioleoylphosphatidylcholine and stearoyl-oleoylphosphatidylcholine bilayer membranes. Biochim Biophys Acta 1374:1–8

Kim JT, Mattai J, Shipley GG (1987) Gel phase polymorphism in ether-linked dihexadecylphosphatidylcholine bilayers. Biochemistry 26:6592–6598

Kodama M (1986) Phase transition phenomena induced by the successive appearances of new types of aggregation states of water molecules in the "L-dipalmitoylphosphatidylcholine-water" system. Thermochim Acta 109:81–89

Kodama M, Miyata T (1995) Effect of Na^+ concentrations on both size and multiplicity of multilamellar vesicles composed of negatively charged phospholipid as revealed by differential scanning calorimetry and electron microscopy. Thermochim Acta 267:365–372

Kodama M, Kuwabara M, Seki S (1982) Successive phase-transition phenomena and phase diagram of the phosphatidylcholine-water system as revealed by differential scanning calorimetry. Biochim Biophys Acta 689:567–570

Kodama M, Miyata T, Yokoyama T (1993) Crystalline cylindrical structures of Na^+-bound dimyristoylphosphatidylglycerol as revealed by microcalorimetry and electron microscopy. Biochim Biophys Acta 1168:243–248

Kodama M, Inoue H, Tsuchida Y (1995) The behavior of water molecules associated with structural changes in phosphatidylethanolamine assembly as studied by DSC. Thermochim Acta 266:373–384

Kodama M, Aoki H, Miyata T (1999) Effect of Na^+ concentration on the subgel phases of negatively charged phosphatidylglycerol. Biophys Chem 79:205–217

Kodama M, Kato H, Aoki H (2001) Comparison of differently bound molecules in the gel and subgel phases of a phospholipids bilyaer system. J Therm Anal Calorim 64:219–230

Koenig BW, Strey HH, Gawrisch K (1997) Membrane lateral compressibility determined by NMR and X-ray diffraction: effect of acyl chain polyunsaturation. Biophys J 73:1954–1966

Koynova R, Caffrey M (1994) Phases and phase transitions of the hydrated phosphatidylethanolamines. Chem Phys Lipids 69:1–34

Koynova R, Caffrey M (1998) Phases and phase transitions of the phosphatidylcholines. Biochim Biophys Acta 1376:91–145

Kusube M, Matsuki H, Kaneshina S (2005) Thermotropic and barotropic phase transitions of *N*-methylated dipalmitoylphosphatidylethanolamine bilayers. Biochim Biophys Acta 1668:25–32

Kusube M, Goto M, Tamai N, Matsuki H, Kaneshina S (2006) Bilayer phase transitions of *N*-methylated dioleoylphosphatidylethanolamines under high pressure. Chem Phys Lipids 142:94–102

Laggner P, Lohner K, Degovics G, Müller K, Schuster A (1987) Structure and thermodynamics of the dihexadecylphosphatidylcholine-water system. Chem Phys Lipids 44:31–60

Lakowicz JR, Thompson RB (1983) Differential polarized phase fluorometric studies of phospholipid bilayers under high hydrostatic pressure. Biochim Biophys Acta 732:359–371

Lewis RNAH, McElhaney RN (2005) The mesomorphic phase behavior of lipid bilayers. In: Yeagle PL (ed) The structure of biological membranes, 2nd edn. CRC Press, New York, pp 66–69

Lewis RNAH, Mak N, McElhaney RN (1987) A differential scanning calorimetric study of the thermotropic phase behavior of model membranes composed of phosphatidylcholines containing linear saturated fatty acyl chains. Biochemistry 26:6118–6126

Lewis RNAH, Sykes BD, McElhaney RN (1988) Thermotropic phase behavior of model membranes composed of phosphatidylcholines containing *cis*-monounsaturated acyl chain homologues of oleic acid: differential scanning calorimetric and ^{31}P NMR spectroscopic studies. Biochemistry 27:880–887

Lin H, Wang Z, Huang C (1991) The influence of acyl chain-length asymmetry on the phase transition parameters of phosphatidylcholine dispersions. Biochim Biophys Acta 1067:17–28

Lohner K (1996) Is the high propensity of ethanolamine plasmalogens to form non-lamellar lipid structures manifested in the properties of biomembranes? Chem Phys Lipids 81:167–184

Ludwig H (ed) (1999) Advances in high pressure bioscience and biotechnology. Springer, Heidelberg

Maruyama S, Matsuki H, Ichimori I, Kaneshina S (1996) Thermotropic and barotropic phase behavior of dihexadecylphosphatidylcholine bilayer membrane. Chem Phys Lipids 82:125–132

Maruyama S, Hata T, Matsuki H, Kaneshina S (1997a) Effects of pressure and local anesthetic tetracaine on dipalmitoylphosphatidylcholine bilayers. Biochim Biophys Acta 1325:272–280

Maruyama S, Hata T, Matsuki H, Kaneshina S (1997b) Effects of pressure and local anesthetic tetracaine on dihexadecylphosphatidylcholine bilayer membrane. Colloids Surf B: Biointerfaces 8:261–266

Matsuki H, Goto M, Kusube M, Tamai N, Kaneshina S (2005) Barotropic phase transitions of 1-palmitoyl-2-stearoylphosphatidylcholine bilayer membrane. Chem Lett 34:270–271

Matsuki H, Miyazaki E, Sakano F, Tamai N, Kaneshina S (2007) Thermotropic and barotropic phase transitions in bilayer membranes of ether-linked phospholipids with varying alkyl chain lengths. Biochim Biophys Acta 1768:479–489

Matsuki H, Goto M, Tada K, Tamai N (2013) Thermotropic and barotropic phase behavior of phosphatidylcholine bilayers. Int J Mol Sci 14:2282–2302

Nagle JF, Wilkinson DA (1978) Density measurements and molecular interactions. Biophys J 23:159–175

Nishimoto M, Hata T, Goto M, Tamai N, Kaneshina S, Matsuki H, Ueda I (2009) Interaction modes of long-chain fatty acids in dipalmitoylphosphatidylcholine bilayer membrane: contrast to mode of inhalation anesthetics. Chem Phys Lipids 158:71–80

Pabst G, Danner S, Karmakar S, Deutsch G, Raghunathany VA (2007) On the propensity of phosphatidylglycerols to form interdigitated phases. Biophys J 93:513–525

Paltauf F (1994) Ether lipids in biomembranes. Chem Phys Lipids 74:101–139

Riske KA, Amaral LQ, Döbereiner HG, Lamy MT (2004) Mesoscopic structure in the chain-melting regime of anionic phospholipid vesicles: DMPG. Biophys J 86:3722–3733

Ruocco MJ, Siminovitch DJ, Griffin RG (1985) Comparative study of the gel phases of ether- and ester-linked phosphatidylcholines. Biochemistry 24:2406–2411

Singh H, Emberley J, Morrow MR (2008) Pressure induces interdigitation differently in DPPC and DPPG. Eur Biophys J 37:783–792

Srinivasan KR, Kay RL, Nagle JF (1974) The pressure dependence of the lipid bilayer phase transition. Biochemistry 13:3494–3496

Stümpel J, Nichsch A, Eibl H (1981) Calorimetric studies on saturated mixed-chain lecithin-water systems. Nonequivalence of acyl chains in the thermotropic phase transition. Biochemistry 20:662–665

Stümpel J, Eibl H, Nichsch A (1983) X-ray analysis and calorimetry on phosphatidylcholine model membranes. The influence of length and position of acyl chains upon structure and phase behaviour. Biochim Biophys Acta 727:246–254

Sueyoshi R, Tada K, Goto M, Tamai N, Matsuki H, Kaneshina S (2006) Barotropic phase transition between the lamellar liquid crystal phase and the inverted hexagonal phase of dioleoylphosphatidylethanolamine. Colloids Surf B: Biointerfaces 50:85–88

Tada K, Miyazaki E, Goto M, Tamai N, Matsuki H, Kaneshina S (2009) Barotropic and thermotropic bilayer phase behavior of positional isomers of unsaturated mixed-chain phosphatidylcholines. Biochim Biophys Acta 1788:1056–1063

Tada K, Goto M, Tamai N, Matsuki H, Kaneshina S (2010) Pressure effect on the bilayer phase transition of asymmetric lipids with an unsaturated acyl chain. Ann N Y Acad Sci 1189:77–85

Tanaka S, Tamai N, Goto M, Kaneshina S, Matsuki H (2012) Morphological change of vesicle particles can produce a peculiar stepwise transition in dipalmitoylphosphatidylglycerol bilayer at high NaCl concentration. Chem Lett 41:304–306

Trudell JR, Payan DG, Chin JH, Cohen EN (1974) Pressure-induced elevation of phase transition temperature in dipalmitoylphosphatidylcholine bilayers. An electron spin resonance measurement of the enthalpy of phase transition. Biochim Biophys Acta 373:436–443

Usui K, Hiraki T, Kawamoto J, Kurihara T, Nogi Y, Abe F (2012) Eicosapentaenoic acid plays a role in stabilizing dynamic membrane structure in the deep-sea piezophile *Shewanella violacea*: a study employing high-pressure time-resolved fluorescence anisotropy measurement. Biochim Biophys Acta 1818:574–583

Utoh S, Takemura T (1985) Phase transition of lipid multilamellar aqueous suspension under high pressure I. Investigation of phase diagram of dipalmitoylphosphatidylcholine bimembrane by high pressure-DTA and -dilatometry. Jpn J Appl Phys 24:356–360

Wann KT, Macdonald AG (1988) Actions and interactions of high pressure and general anesthetic. Prog Neurobiol 30:271–307

Watts A, Harlos K, Maschke W, Marsh D (1978) Control of the structure and fluidity of phosphatidylglycerol bilayers by pH titration. Biochim Biophys Acta 510:63–74

Watts A, Harlos K, Marsh D (1981) Charge-induced tilt in ordered-phase phosphatidylglycerol bilayers evidence from x-ray diffraction. Biochim Biophys Acta 645:91–96

Winter R (2001) Effects of hydrostatic pressure on lipid and surfactant phases. Curr Opin Colloid Interface Sci 6:303–312

Winter R (ed) (2003) Advances in high pressure bioscience and biotechnology, vol II. Springer, Heidelberg

Winter R, Jeworrek C (2009) Effect of pressure on membranes. Soft Matter 5:3157–3173

Winter R, Pilgrim WC (1989) A SANS study of high pressure phase transitions in model biomembranes. Ber Bunsenges Phys Chem 93:708–717

Wong PTT, Mantsch HH (1985) Effects of hydrostatic pressure on the molecular structure and endothermic phase transitions of phosphatidylcholine bilayers: a Raman scattering study. Biochemistry 24:4091–4096

Zhang YP, Lewis RNAH, McElhaney RN (1997) Calorimetric and spectroscopic studies of the thermotropic phase behavior of the *n*-saturated 1,2-diacylphosphatidylglycerols. Biophys J 72:779–793

Chapter 17
Pressure Effects on Artificial and Cellular Membranes

Roland Winter

Abstract We review the combined effect of temperature and pressure on the structure, dynamics and phase behavior of lipid bilayers, differing in chain length, headgroup structure and composition as revealed by thermodynamic, spectroscopic and scattering experiments. The effect of additives, such as ions, cholesterol, and anaesthetics is discussed as well. Our data include also reports on the effect of pressure on the lateral organization of heterogeneous lipid membranes and lipid extracts from cellular membranes, as well as the influence of peptide and protein incorporation on the pressure-dependent structure and phase behavior of lipid membranes. Moreover, the effects of pressure on membrane protein function are summarized. Finally, we introduce pressure as a kinetic variable for studying the kinetics of various lipid phase transformations.

Keywords Membranes • Lipid bilayers • Self-assembly • High pressure • Membrane proteins

17.1 Introduction

The amphiphilic nature of lipid molecules leads to their self-aggregation in water solution, so that the hydrocarbon chains are segregated away from contact with water, while the hydrophilic headgroups are hydrated. The main driving force behind this self-assembly is the hydrophobic effect, which on its own would lead to a macroscopic phase separation (one polar and one non-polar phase), but the requirement that the polar headgroups like to be in contact with water prevents such phase separation. Packing restrictions of the lipid molecules lead to different aggregate topologies. A useful concept for a qualitative understanding of the assembly structure and phase behavior of amphiphile systems is based on a consideration of the shape of the lipid molecules, which can be characterized by a

---

R. Winter (✉)
Physical Chemistry I – Biophysical Chemistry, TU Dortmund University, Otto-Hahn Str. 6, D-44227 Dortmund, Germany
e-mail: roland.winter@tu-dortmund.de

K. Akasaka, H. Matsuki (eds.), *High Pressure Bioscience*,
Subcellular Biochemistry 72, DOI 10.1007/978-94-017-9918-8_17

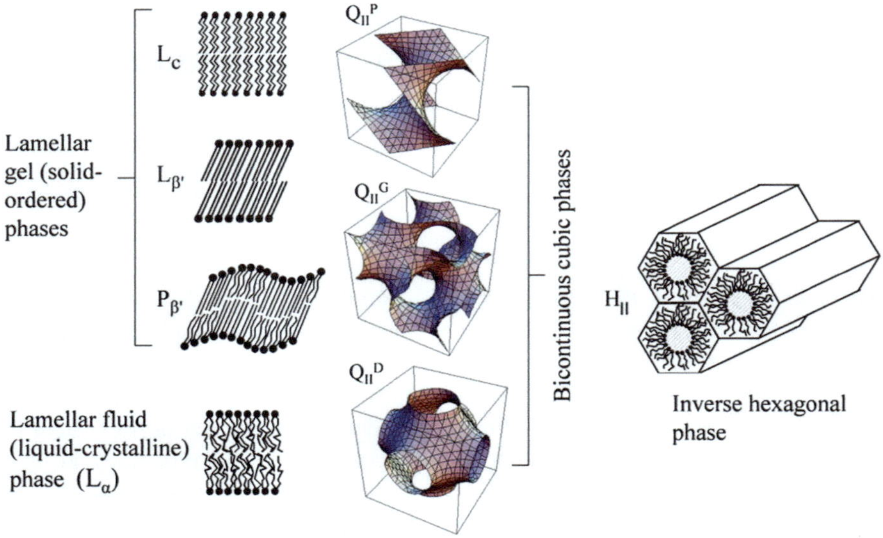

Fig. 17.1 Schematic drawing of various lamellar and non-lamellar lyotropic lipid mesophases adopted by membrane lipids (L_c, lamellar crystalline; $L_{\beta'}$, $P_{\beta'}$, lamellar gel; L_α, lamellar liquid-crystalline (fluid-like); $Q_{II}{}^G$ (space group Ia3d), $Q_{II}{}^P$ (space group Im3m), $Q_{II}{}^D$ (space group Pn3m) cubic phases – lipid bilayers are draped on the minimal surfaces shown; H_{II}, *inverse hexagonal*). Numerous factors determine the particular mesophase structure, e.g., the type of lipid, the lipid chain length and degree of unsaturation, the headgroup area and charge density, solvent properties, pH, the level of hydration, temperature and pressure (Modified from Winter and Jeworrek 2009; Meersman et al. 2013)

dimensionless packing parameter, P, defined by the ratio $\nu(a{\cdot}l)^{-1}$, where ν is the molecular volume, l is the molecular length, and a is the molecule's cross-sectional area (headgroup) at the hydrocarbon-water interface (Seddon and Templer 1993). When $P \approx 1$, i.e. for cylindrical-like molecules, there are optimal conditions for the formation of a lipid bilayer structure (Fig. 17.1). For $P > 1$, the molecules are wedge-shaped and the lipid monolayer prefers to curve towards the water region, and an inverse hexagonal (H_{II}) phase may form (see Fig. 17.1).

The conformation and shape of the lipid molecule depends not only sensitively on temperature but also on pressure, thereby rendering lipid mesophase structures also pressure sensitive. It has been found that pressure stress affects all levels of cellular physiology, including membrane physiology, transport, and signalling processes (Bartlett 2002; Daniel et al. 2006; Meersman et al. 2013; Winter 2013), and biological membranes seem to be one of the most pressure sensitive cellular components. Throughout most of the Earth's history, in particular microbial life has experienced a range of pressures, both dynamic pressure when the young Earth was heavily bombarded, and static pressure in subsurface environments that could have served as a refuge and where microbial life still flourishes nowadays (Picard and Daniel 2013). Here we discuss results of studies of the effects of pressure on the structure, dynamics and phase behavior of lyotropic lipid mesophases, model and natural membrane systems as well as pressure effects on the interaction of peptides with membranes. High-pressure biophysical techniques have largely been developed in recent years (Balny et al. 1992; Winter 2002, 2003a; Woenckhaus et al. 2000; Arnold et al. 2003; Jonas 1991; Akasaka 2006). In this review we focus on concepts and the presentation of selected characteristic results, only.

17.2 Pressure Effects on Lamellar Lipid Bilayer Phases

Biologically relevant lyotropic lipid mesophases are mostly formed by phospholipids. They exhibit a rich structural polymorphism, depending on their molecular structure, level of hydration, ionic strength, temperature and pressure. The basic structural element of biological membranes consists of lamellar lipid bilayers (Fig. 17.1). In addition to their interest as prototypical membranes, lipid bilayers have been fascinating objects of physical chemistry studies for many years. Most of their lipids possess two acyl-chains and a hyphrophilic headgroup. The composition (configuration, length, degree of unsaturation) of the lipid chains and the headgroup (e.g., charge) and hence the physical-chemical properties of the membrane can vary significantly in cellular membranes. Not only is the entire cell membrane very complex, containing a large variety (sometimes hundreds) of different lipid molecules and a large body (even up to about 50 %) of proteins performing versatile functions, but also the simplest lipid bilayer consisting of only one or two kinds of lipid molecules already exhibits a rather complex temperature and pressure dependent phase behavior.

Lipid bilayers display various phase transitions including a chain melting (gel-to-fluid) transition. Phospholipids often exhibit two thermotropic lamellar phase transitions, a gel-to-gel ($L_{\beta'}$-$P_{\beta'}$) pretransition and a gel-to-fluid (or liquid-crystalline) ($P_{\beta'}$-L_{α}) main (chain melting) transition at a higher temperature. Some phsopholipids, such as phosphatidylcholines, display a tilt angle of the lipid chains with respect to the lipid bilayer normal. In the fluid-like L_{α} phase, the acyl-chains of the lipid bilayers are conformationally disordered ("melted"), whereas in the gel phases, the chains are more extended, i.e. their order parameter is high (Fig. 17.1). The lipids in the ordered $L_{\beta'}$ phase are arranged on a two-dimensional triangular lattice in the membrane phase. This phase is also called solid-ordered (s_o) phase. Besides neutral or zwitterionic lipids, also negatively charged lipids are present in cellular membranes. The chain-melting temperature, T_m, of negatively charged lipid membranes generally increases when neutralizing the charges by proteins or divalent ions.

Upon application of hydrostatic pressure, a range of pressure-induced phase transformations has been observed as well (Landwehr and Winter 1994a, b; Hammouda et al. 1997; Czeslik et al. 1998; Winter 2001; Winter et al. 1989, 2000; Winter and Dzwolak 2004).

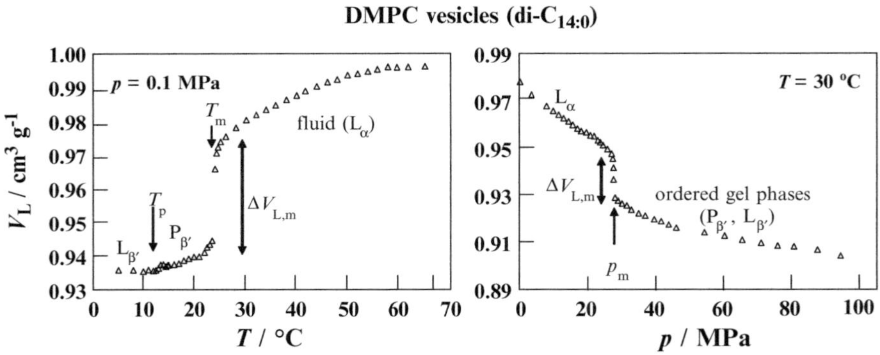

Fig. 17.2 Effect of temperature (at 1 bar = 0.1 MPa) (*left*) and pressure (at $T = 30$ °C) (*right*) on the partial lipid volume V_L of DMPC bilayers as obtained from densimetric measurements. The gel-to-gel ($L_{\beta'}$-to-$P_{\beta'}$) and gel-to-fluid ($P_{\beta'}$-to-L_α) lamellar phase transition appears at temperatures T_p and T_m, respectively. The pressure-induced fluid-to-gel transition at $T = 30$ °C occurs at pressure p_m (Modified from Böttner et al. 1994; Winter and Jeworrek 2009; Meersmann et al. 2013)

Owing to the fact that the average end-to-end distance of disordered hydrocarbon chains in the L_α-phase is smaller than that of ordered (all-trans) chains, the bilayer becomes thinner during melting at the $P_{\beta'}$-to-L_α-transition, even though the partial lipid volume increases. This is demonstrated in Fig. 17.2 which shows the temperature dependence of the partial specific lipid volume, V_L, of dimyristoyl-phosphatidylcholine (DMPC)[1] in water (Böttner et al. 1994). Measuring molecular volumes versus temperature or pressure reveals phase transitions, and comparison of the volume changes at phase transitions leads to estimates of the cohesive energy of the hydrocarbon chains (Hallinen et al. 2012; Seemann et al. 2001). The change of V_L near 14 °C corresponds to a small volume change in course of the $L_{\beta'}$-to-$P_{\beta'}$ transition. The main (chain-melting) transition at $T_m = 23.9$ °C is accompanied by a pronounced 3 % change in volume, which is essentially due to changes of the chain cross-sectional area, because the chain disorder increases drastically at the transition. Upon hydrostatic compression of the lipid bilayer, owing to its anisotropic nature, lateral shrinking is observed which is accompanied by an increase in bilayer thickness due to a straightening (decrease of gauche conformers and kinks) of the acyl-chains. Figure 17.2 exhibits the pressure dependence of

[1] Abbreviations: DMPC 1,2-dimyristoyl-sn-glycero-3-phosphatidylcholine (di-C14:0); DMPS 1,2-dimyristoyl-sn-glycero-3-phosphatidylserin (di-C14:0); DPPC 1,2-dipalmitoyl-sn-glycero-3-phosphatidylcholine (di-C16:0); DPPE 1,2-dipalmitoyl-sn-glycero-3-phosphatidylethanolamine (di-C16:0); DOPC 1,2-dioleoyl-sn-glycero-3-phosphatidylcholine (di-C18:1,cis); DOPE 1,2-dioleoyl-sn-glycero-3-phosphatidylethanolamine (di-C18:1,cis); POPC 1-palmitoyl-2-oleoyl-sn-glycero-3-phosphatidylcholine (C16:0,C18:1,cis); DLPC 1,2-dilauroyl-sn-glycero-3-phosphatidylcholine; MO monoolein; ME monoelaidin; PA palmitoylic acid; MA myristoylic acid; Chol cholesterol.

V_L at a temperature above T_m, 30 °C. Increasing pressure triggers the phase transformation from the L_α to the gel phase, as can be seen from the rather abrupt decrease of the lipid volume at 27 MPa.

Biological membranes can also melt. Typically, such melting transitions are found about 10 °C below the body or growth temperature. It has been speculated that biological membranes adapt their lipid composition such that the temperature distance to the melting transition is maintained. The same may hold true for adaptation to high-pressure conditions. In lipid bilayers, the fluctuations in enthalpy, volume and area are high close to the melting transition. High enthalpy fluctuations lead to high heat capacity, high volume fluctuations to a high volume compressibility, and high area fluctuations lead to a high area compressibility. In turn, area fluctuations lead to fluctuations in curvature and bending elasticity (Heimburg 2007). Such fluctuations may be required for various physiological functions, including membrane budding and fusion, and are expected to be attenuated upon pressurization.

A common slope of the order of 0.2 °C MPa^{-1} has been observed for the gel-fluid phase boundary of saturated phospholipids as shown in Fig. 17.3 (Winter 2001;

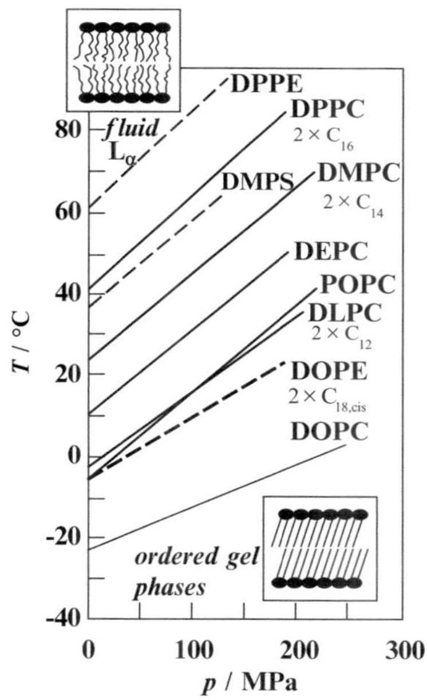

Fig. 17.3 *T*, *p*-phase diagram for the main (chain-melting) transition of different phospholipid bilayer systems. The fluid-like (liquid-crystalline) L_α-phase is observed in the low-pressure, high-temperature regions of the phase diagram; the gel (solid-ordered) phase regions appear at low temperatures and high pressures, respectively. The gel-to-fluid transition lines of the phosphatidylcholines are drawn as *solid lines*, those of the phospholipids with a different headgroup (PE, PS) as *dashed lines*. See footnote for abbreviations (Modified from Meersmann et al. 2013; Winter and Dzwolak 2004; Winter and Jeworrek 2009).

Winter et al. 2000; Winter and Dzwolak 2004). Assuming the validity of the Clapeyron relation describing first-order phase transitions for this single-lipid bilayer system, $\mathrm{d}T_\mathrm{m}/\mathrm{d}p = T_\mathrm{m}\Delta V_\mathrm{m}(T_\mathrm{m}, p)/\Delta H_\mathrm{m}(T_\mathrm{m}, p)$, the positive slope can be explained by an endothermic enthalpy change, ΔH_m, and a partial molar volume increase, ΔV_m, for the gel-to-fluid transition, in agreement with direct thermodynamic measurements (Janosch et al. 2004; Krivanek et al. 2008; Seemann et al. 2003). The transition enthalpy, ΔH_m, is 36 kJ mol^{-1} for DPPC at ambient pressure, and decreases slightly with increasing pressure ($(\mathrm{d}\Delta H_\mathrm{m}/\mathrm{d}p) = -0.034$ kJ·mol^{-1}·MPa^{-1}) (Potekhin et al. 2008). The drop of enthalpy change with pressure evidences a significant difference in the coefficients of thermal expansion of the two phases. Similarly, ΔV_m decreases linearly with increasing pressure (from 22.9 cm^3 mol^{-1} at 0.1 MPa to ~13 cm^3 mol^{-1} at 200 MPa, i.e., $\mathrm{d}\Delta V_\mathrm{m}/\mathrm{d}p = -0.0493$ cm^3·mol^{-1} MPa^{-1}) (Potekhin et al. 2008), which is due to the significant difference in the lipid's compressibility coefficients in the fluid and gel phase, respectively. The transition half-width ($\Delta T_{\mathrm{m},1/2}$), which can be estimated as the ratio of the calorimetric peak area $\Delta H_{\mathrm{m,cal}}$ to its amplitude $C_{p,\mathrm{max}}$, can be estimated from the van't Hoff enthalpy change by using $\Delta H_{\mathrm{m,vH}} = 4\mathrm{R}{T_\mathrm{m}}^2 C_{p,\mathrm{max}}/\Delta H_{\mathrm{m,cal}} = 4\mathrm{R}{T_\mathrm{m}}^2/\Delta T_{\mathrm{m},1/2}$. $\Delta T_{\mathrm{m},1/2}$ has been found to be essentially independent of pressure, and the average number of lipid molecules comprising the cooperative unit N of the transition (assuming $N = \Delta H_{\mathrm{m,cal}}/\Delta H_{\mathrm{m,vH}}$) grows slightly with increasing pressure and temperature.

Similar transition slopes have been determined for the mono-*cis*-unsaturated lipid POPC, the phosphatidylserine DMPS, and the phosphatidylethanolamine DPPE. Only the slopes of the di-*cis*-unsaturated lipids DOPC and DOPE have been found to be markedly smaller. The two *cis*-double bonds of DOPC and DOPE lead to very low transition temperatures and Clapeyron slopes, as they impose kinks in the linear conformations of the lipid acyl-chains, thereby creating significant free volume fluctuations in the bilayer and rendering the ordering effect of HHP less effective. To remain in a physiologically relevant, i.e. fluid-like state at high pressures, more of such *cis*-unsaturated lipids are incorporated into cellular membranes of deep-sea organisms (where temperatures as low as 2–4 °C and pressures up to the 100 MPa range are encountered), which constitutes an important example of homeoviscous adaptation (Yayanos 1998; Bartlett 2002, 2008; Allen et al. 1999), but other strategies might be followed as well (Abe 2013). For example, the ratio of unsaturated to saturated fatty acids of the piezophilic deep-sea bacterium CNPT3 is linearly dependent on the hydrostatic pressure at which the bacteria were cultivated (DeLong et al. 1985, 1987). The unsaturated to saturated ratio increases from 1.9 at ambient pressure to about 3 at 69 MPa at 2 °C. Interestingly, archaeal membranes under high-pressure stress seem to use similar strategies for homeoviscous adaptation as observed in bacteria and eucarya (Hanford et al. 2002; Oger et al. 2010, 2013). Whereas the latter essentially regulate the length and the satuation level of their lipid chains to adjust membrane fluidity, archaea, because of their specific lipids, have to adopt different strategies, such as the incorporation of cyclopentane rings into the isoprenoid chains or the modification of the proportions of diether vs. tetraether lipids (Oger et al. 2013).

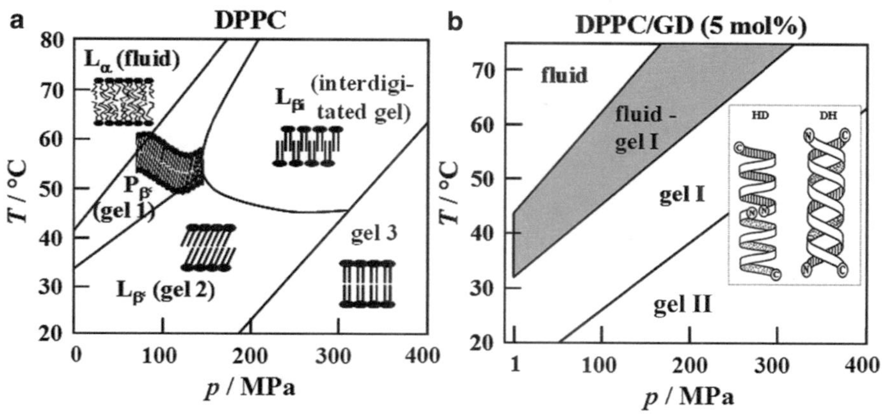

Fig. 17.4 (**a**) T, p-phase diagram of DPPC bilayers in excess water with ordered gel phases gel 1 ($P_{\beta'}$), gel 2 ($L_{\beta'}$) and the pressure-induced gel phases gel 3 and $L_{\beta i}$. (**b**) Phase diagram of DPPC-gramicidin D (GD) (5 mol%) in excess water as obtained from diffraction and spectroscopic data. The inset shows a schematic view of the helical dimer (HD) and double helix (DH) conformation of GD (Modified from Winter and Jeworrek 2009; Meersmann et al. 2013)

As seen in Fig. 17.3, pressure generally increases the chain-order of membranes, thus mimicking the effect of cooling. However, please note that applying high pressure can lead to the formation of additional ordered phases, which are not observed under ambient pressure conditions, such as a partially interdigitated high pressure gel phase, $L_{\beta i}$, which has been found for phospholipid bilayers with acyl-chain lengths $\geq C_{16}$ (Landwehr et al. 1994a, b; Hammouda et al. 1997). To illustrate this phase variety, the results of a detailed small-angle X-ray and neutron scattering and FTIR spectroscopy study of the p,T-phase diagram of DPPC in excess water are shown in Fig. 17.4a. At much higher pressures as shown here, even further ordered gel phases appear, which differ in the tilt angle of the acyl-chains and the level of hydration in the interlamellar headgroup region. Interestingly, even at pressures where bulk water freezes, the lamellar structure of the membrane is preserved (Czeslik et al. 1998). Further pressure-induced phase transitions of phospholipid bilayer systems are discussed by Matsuki in this volume (e.g., Matsuki et al. 2005, 2007, 2013).

17.3 Pressure Effects on Lipid Mixtures and Lipid-Peptide Interactions

The next level of complexity encompasses T, p-phase diagrams of binary mixtures of saturated phospholipids (Winter et al. 1999a, b, 2000; Winter and Dzwolak 2004; Winter 2001). They are characterized by lamellar gel phases at low temperatures, a lamellar fluid phase at high temperatures, and an intermediate fluid-gel coexistence

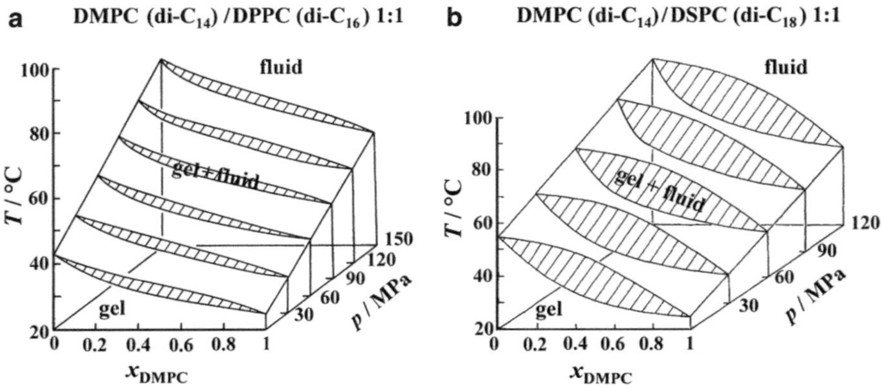

Fig. 17.5 *T, p, x* phase diagram of (**a**) equimolar DMPC/DPPC (di-C_{14}/di-C_{16}) and (**b**) equimolar DMPC/DSPC (di-C_{14}/di-C_{18}) multilamellar vesicles in excess water; x_L denotes the mole fraction of lipid L (Modified from Winter and Dzwolak 2004; Winter and Jeworrek 2009; Meersmann et al. 2013)

region (Fig. 17.5). The narrow fluid-gel coexistence region in the DMPC(di-C_{14}) – DPPC(di-C_{16}) system indicates a nearly ideal mixing behavior of the two components (a so-called isomorphous system). The coexistence region in the DMPC(di-C_{14}) – DSPC(di-C_{18}) system is broader and reveals pronounced deviations from ideality. As seen in Fig. 17.5, with increasing pressure the gel-fluid coexistence region of the binary lipid systems is shifted toward higher temperatures. A shift of about 0.22 °C MPa^{-1} is observed, which is similar to the slope of the gel-fluid transition line of the one-component lipid bilayer components (Landwehr et al. 1994a, b; Winter et al. 2000; Winter and Dzwolak 2004; Winter 2001).

Membranes also contain other components such as cholesterol, proteins and peptides. Cholesterol (Chol) constitutes up to ~30–50 % of the lipid content of animal cell membranes. Due to both its amphiphilic character and shape, it is inserted into phospholipid membranes, thickens fluid-like bilayers, increases the packing density of the lipid acyl-chains in a way that has been referred to as "condensing effect", increases the hydrophobicity and deceases the water permeability (Seemann et al. 2003; Krivanek et al. 2008). As revealed by fluorescence spectroscopic methods, an increase in pressure up of 100 MPa is much less effective in suppressing water permeability than cholesterol embedded in fluid DPPC bilayers at high concentration levels. For concentrations above ~30 mol% cholesterol, the main phase transition can hardly be detected any more. At a concentration of ~40 mol% cholesterol, the order parameter values are found to be almost independent of pressure. Hence, sterols like cholesterol can efficiently regulate the structure, motional freedom and hydrophobicity of lipid membranes, so that they can withstand even drastic changes in environmental conditions, such as in external pressure and temperature.

Membrane proteins can constitute about 30 % of the entire protein content of a cell and act as anchors, signaling modules, enzymes or transporters. Membrane lipids and proteins have been shown to influence each other directly and in response to environmental changes. Pressure studies on this interaction have been scant. One example is the channel peptide gramicidin D (GD) on the structure and phase behavior of phospholipid bilayers (Zein et al. 2000; Eisenblatter et al. 2005). Gramicidin is polymorphic, being able to adopt a range of structures with different topologies. Common forms are the dimeric single-stranded right-handed $\beta^{6.3}$-helix with a length of 24 Å, and the antiparallel double-stranded $\beta^{5.6}$-helix, being ~32 Å long. For comparison, the hydrophobic fluid bilayer thickness is about 30 Å for DPPC bilayers, and the hydrophobic thickness of the gel phases is larger by ~5 Å. Depending on the gramicidin concentration, significant changes of the lipid bilayer structure and phase behavior were observed. These include disappearance of certain gel phases formed by the pure DPPC system, and the formation of broad two-phase coexistence regions at higher gramicidin concentrations (Fig. 17.4b). Likewise, the peptide conformation is influenced by the lipid environment. Depending on the phase state and lipid acyl-chain length, GD adopts at least two different types of quaternary structures in the bilayer environment, a double helical pore (DH) and a helical dimer channel (HD) (inset Fig. 17.4b). When the bilayer thickness changes at the gel-to-fluid main phase transition of DPPC, the conformational equilibrium of the peptide also changes (Zein et al. 2000). Hence, not only the lipid bilayer structure and T, p-dependent phase behavior drastically depends on the polypeptide concentration, but also the peptide's conformational dynamics (and hence function) may be significantly influenced by the lipid environment. No pressure-induced unfolding of the polypeptide is observed up to 1.0 GPa. For large integral and peripheral proteins, however, pressure-induced changes in the physical state of the membrane may lead to a weakening of protein-lipid interactions as well as to protein dissociation (see below).

Studies were also carried out on the phase behavior of cholesterol-containing ternary lipid mixtures, containing an unsaturated phospholipid like DOPC or POPC and a saturated lipid like sphingomyelin (SM) or DPPC. Such lipid systems are supposed to mimic distinct liquid-ordered lipid regions (domains), coined "rafts", which seem to be also present in cell membranes and are thought to be important for cellular functions such as signal transduction and the sorting and transport of lipids and proteins (Janosch et al. 2004; Nicolini et al. 2005, 2006; Jeworrek et al. 2008; Weise et al. 2009; Kapoor et al. 2012a, b). Lipid domain formation can be influenced by temperature, pH, calcium ions, protein adsorption, as well as by pressure. We determined the liquid-disordered/liquid-ordered (l_d/l_o) phase coexistence region of canonical model raft mixtures such as POPC/SM/Chol (1:1:1), which extends over a rather wide temperature range. An overall fluid phase without domains is only reached at high temperatures, above ~50 °C (Nicolini et al. 2006). Upon pressurization at ambient temperatures (20–40 °C), an overall (liquid- and solid-) ordered state is reached at pressures of about 100–200 MPa. A similar behavior was observed for the model raft mixture DOPC/DPPC/Chol (1:2:1) (Fig. 17.6) (Kapoor et al. 2012a, b). Interestingly, in this pressure range of ~200 MPa, cessation of

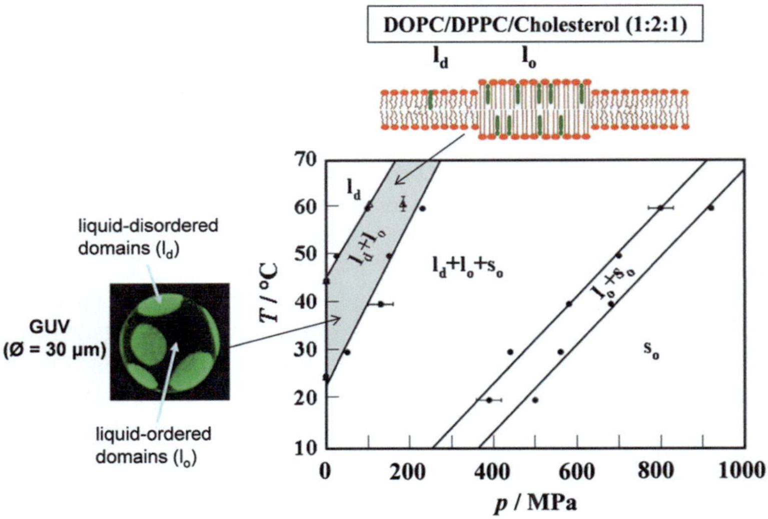

Fig. 17.6 *T, p*-phase diagram of the ternary lipid mixture DOPC/DPPC/Chol (1:2:1) in excess water as obtained from FTIR spectroscopy and small-angle X-ray scattering data. The $l_d + l_o$ two-phase coexistence region is marked in *grey*. The liquid-disordered (all-fluid) phase is represented by l_d, whereas l_o and s_o are the liquid- and solid-ordered phases, respectively. *Left*: Fluorescence microscopic visualization of l_o-l_d coexistence regions in a giant unilamellar vesicle (GUV) (Modified from Winter and Jeworrek 2009; Meersmann et al. 2013)

membrane protein function in natural membrane environments has been observed for a variety of systems (De Smedt et al. 1979; Chong et al. 1985; Kato et al. 2002; Ulmer et al. 2002; Powalska et al. 2007; Linke et al. 2008; Periasamy et al. 2009), which might be related to the membrane matrix reaching a physiologically unacceptable overall ordered state in this pressure range.

Finally, we have a short look at possible prebiotic lipid mixtures. Life very likely originated in the depth of the protoocean of the Hadean Earth, i.e., under high hydrostatic pressure conditions. In the prebiotic era, single-chain amphiphiles or fatty acids are at the focal point for research into the origin of life. They can be abiotically synthesized via Fisher-Tropsch type reactions under simulated geochemical conditions such as hydrothermal systems, and have also been detected in carbonaceous meteorites. Recently, we showed that hydrostatic pressure increases the amount of vesicular structures of such short-length fatty-acid systems, which could serve as prebiotic membrane envelopes (Kapoor et al. 2014). Lacking the complex, highly evolved protein machinery of present-day biomembranes, the dynamic nature of fluid fatty-acid-membranes seems particularly well suited for growth, division, and nutrient uptake. These studies also showed that pressures in the range encountered in the deep sea could serve as a multifunctional toggle: limiting their flexibility and permeability (and hence leakage rate) to solutes, while still keeping the membrane in a fluid-like state, which is required for function.

17.4 Dynamical Properties of Lipid Bilayers upon Compression

Rather little is known about pressure effects on the dynamical properties of lipid bilayers at elevated pressures (Winter 2003a, b; Jonas et al. 1991). Of particular interest is the effect of pressure on the lateral diffusion, which is related to biological functions such as transport and membrane-associated signaling processes. Pressure effects on the lateral self diffusion coefficient, D_{lat}, of DPPC and POPC vesicles have been studied by Jonas et al. (1991): D_{lat} of DPPC in the fluid-like phase decreases by about 30 % from 0.1 to 30 MPa at 50 °C. A further 70 % decrease in the D_{lat}-value occurs at the pressure-induced L_α to gel phase transition. Compared to the lateral diffusion, the rotational dynamics of lipids and small amphiphilic molecules in membranes seems to be less influenced by high pressure (Bernsdorff et al. 1996, 1997). Phospholipid flip-flop and lipid transfer between membranes is also slowed down by high pressure (Macdonald 1992; Mantulin et al. 1984).

17.5 Effects of Ionic Strength, Amphiphilic Drugs and Anaesthetics

The membrane's structure, dynamics and phase behavior is also influenced by the ionic strength of the surrounding solvent (in particular for lipids with charged headgroups) and the incorporation of amphiphilic molecules, such as anaesthetics and drugs (Winter 2003a, b; Witer and Czeslik 2000; Wlodarczyk et al. 2006; Böttner et al. 1993; Winter et al. 1991; Eisenblätter et al. 2000; Wong et al. 1988; Winter and Jonas 1999; Mountcastle et al. 1978).

The effect of inorganic ions on the melting transition temperature, T_m, of phospholipids depends on the nature of the ions and the surface charge of the lipid membrane. The effect is especially pronounced when Ca^{2+} ions are adsorbed on negatively charged membranes. To assess the effect of salts on the thermodynamic parameters of the gel-to-fluid transition of zwitterionic and anionic phospholipid bilayers, experiments were performed in the presence of mono- and divalent salts, such as NaCl and $CaCl_2$. The addition of 0.1 M NaCl to the zwitterionic DPPC lipid dispersion does not virtually change the transition temperature of DPPC. 20 mM $CaCl_2$ raises the transition temperature by about 1 °C, however. The dT_m/dp transition slope of the main transition is hardly affected by the addition of these salts, but the rates of the enthalpy and volume changes decrease slightly upon addition of the salts (Potekhin et al. 2008). Calorimetric measurements on DMPC/Ca^{2+} dispersions revealed that an increasing Ca^{2+} concentration leads to an increase in main transition temperatures with little change in transition enthalpy, and also to an increase of the $L_{\beta'}$-$P_{\beta'}$ gel-to-gel transition temperature, until both transitions merge at very high salt concentrations. The main transition is shifted towards smaller pressures with increasing temperature in comparison to that of pure

DMPC dispersions. Otherwise, the transition slope dT_m/dp is essentially parallel to that of pure DMPC, and the volume change ΔV_m at the main transition is of similar magnitude. A similar behavior has been observed for the negatively charged lipid DMPS upon addition of Ca^{2+} (Winter and Czeslik 2000).

The substances that cause general anaesthetic effects comprise a wide range of chemically and structurally dissimilar molecules (e.g., alcohols, ethers, inert gases). Interestingly, in 1950 Johnson and Flagler reported that tadpoles anaesthetized in 3–6 vol.% ethanol (EtOH) solution awake upon application of 14–35 MPa of hydrostatic pressure (Johnson and Flagler 1950), pointing to an antagonistic effect of HHP and anaesthesia. In fact, it has been demonstrated that general anaesthetics, such as halothane ($CF_3CHBrCl$) and enflurane (CHF_2OCF_2CHFCl), decrease the gel-to-liquid crystalline phase transition temperature of phospholipid bilayers and that this effect can be reversed by application of moderate pressures. The anaesthetic effects of chemically inert gases, such as xenon, nitrous oxide (N_2O), and nitrogen narcosis under hyperbaric conditions have long provided an astonishing phenomenon for neuroscience as well. Figure 17.7 shows the effect of various gases on the gel-to-fluid transition of DPPC bilayers as a function of pressure as revealed by high pressure calorimetric, light scattering and fluorescence anisotropy measurements (MacNaughtan and Macdonald 1980; Tamura et al. 2003; Danielowski 1997). For comparison, the effect of hydrostatic pressure is depicted as well. The least lipid-soluble of the inert gases, He, fails to narcotise animals, it merely acts as a pressure-transmitting fluid and exhibits a similar Clapeyron slope of the main transition as hydrostatic pressure ($dT_m/dp \approx 0.22$ °C MPa^{-1}) (MacNaughtan and Macdonald 1980; Tamura et al. 2003; Winter and Pilgrim 1989). With increasing solubility in the fluid lipid phase, a reduction on T_m

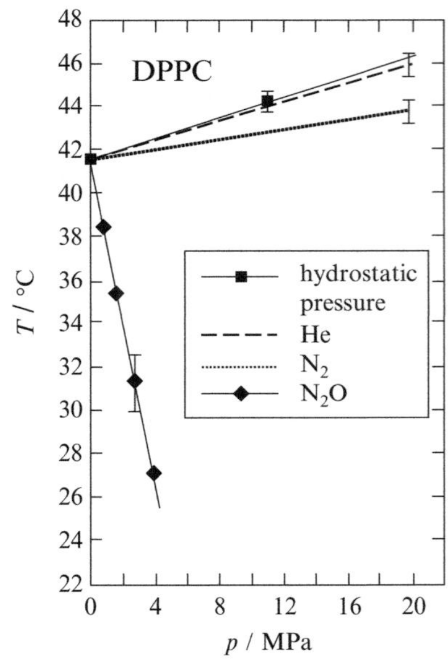

Fig. 17.7 Effect of the pressure of various gases on the chain-melting transition of DPPC vesicles. For comparison, the gel-to-fluid transition for pressurization under hydrostatic conditions is depicted (Modified from Winter and Jeworrek 2009)

should be expected (similar to a "freezing-point depression"). Nitrogen shows such a reduction in transition slope ($dT_m/dp \approx 0.06$ °C MPa^{-1}). The anaesthetic gas nitrous oxide lowers the T_m drastically with increasing gas pressure (at a rate of about −5 °C MPa^{-1}), indicating an increasing solubility in the apolar lipid phase. Also Ar has been shown to exhibit slightly negative slopes (MacNaughtan and Macdonald 1980). Inert gases, such as N_2O, may exert narcotic effects apparently by a mechanism similar to that of the more potent inhalational general anaesthetics.

The question of specific sites at which anaesthetics act is still a matter of debate. One of the earliest theories of anaesthesia is attributed to Meyer and Overton and based on their discovery that the potency of various anaesthetic species is generally proportional to their solubility in fatty acids (calibrated in olive oil at ambient pressure for gases at that time, in octane or lipid bilayers later on) (Meyer 1899; Overton 1909). Those finding led to the theory that anaesthetics dissolve in the lipid fraction of the cell membrane, thereby altering – by changes in membrane fluidity, volume expansion, and lateral structure (which could be reversed upon pressurization) – the physiological properties. It is now thought, however, that ion channels and neurotransmitter receptor sites formed within the neuronal cell membranes constitute essentially the primary sites of anaesthetic action (Wlodarczyk et al. 2006). But it is also likely that, depending on the concentration or gas pressure, both effects play a decisive role.

The influence of local anaesthetics, such as tetracaine (TTC), on the thermodynamic properties and the temperature and pressure dependent phase behavior of phospholipids has also been studied (Wlodarczyk et al. 2006; Böttner et al. 1993; Winter et al. 1991; Eisenblätter et al. 2000; Wong et al. 1988; Winter and Jonas 1999). From volumetric measurements it has been found that the main transition at ambient pressure shifts to a lower temperature and the isothermal compressibility, κ_T, increases by addition of TTC. Hence, the binding of the anaesthetic to membranes leads to a strong enhancement of the volume fluctuations in the neighborhood of T_m, and the addition of TTC shifts the pressure-induced fluid-to-gel phase transition towards somewhat higher pressures. Such findings might also be of relevance, as in lipid bilayer membranes, strong density or concentration fluctuations may be related to the transmembrane permeability of ions and small molecules. Also the biochemical action of local anaesthetics is still controversial as to whether or not the action is lipid mediated. One thing that is clear, however, is that they do strongly perturb the lipid bilayer system and change their thermomechanical properties.

17.6 Pressure Effects on Non-lamellar Lipid Phases

In aqueous solution, lipid molecules can self-assemble into a wide range of nanostructures that display one-, two- and three-dimensional morphologies. Corresponding examples are multilamellar phases, elongated tubes (hexagonal phases (H_{II})) and cubic phases (bicontinuous or micellar), and sometimes disordered lipid architectures (sponge or L_3 phases). Interestingly, besides the planar lipid bilayer matrix, also non-planar lyotropic lipid nanostructures and their hierarchically organized derivatives partly resemble biological structures like plasma membranes, vesicles, the Golgi apparatus, mitochondria and the endoplasmic reticulum (ER), and have probably also important implications in their functioning (Chandrashekhar et al. 2013). Inverse bicontinuous cubic phases (Q_{II}) are formed from a continuous fluid lipid bilayer three-dimensionally folded in a regular and well-defined morphology, separating two continuous interwoven networks of water channels. Cubic phase architectures are based on G (gyroid), D (double diamond) and P (primitive) types of infinite periodic minimal surfaces (IPMS). The corresponding phases have space groups Ia3d, Pn3m and Im3m, with 3, 4 and 6 water channels meeting at angles of 120°, 109.5°, and 90°, respectively. Lipids that can adopt such non-lamellar phases are present at substantial levels in biological membranes. Fundamental cell processes, such as endo- and exocytosis, fat digestion, membrane budding and fusion, involve a rearrangement of biological membranes where such non-lamellar highly curved lipid structures are involved, but probably also static cubic structures appear in biological cells. For example, the organized smooth ER has been shown to exhibit minimal surface topologies of gyroid and diamond types, along with the usual planar lamellae (Chandrashekhar et al. 2013).

The temperature- and pressure-dependent structure and phase behavior of a series of phospholipid systems, including phospholipid/fatty acid mixtures (e.g., DLPC/LA, DMPC/MA, DPPC/PA) and monoaclyglycerides (MO, ME), exhibiting non-lamellar phases have been studied (Erbes et al. 1996; Templer et al. 1998; Winter et al. 1999a, b, 2000; Winter and Dzwolak 2004). Contrary to DOPC, which shows a lamellar gel ($L_{\beta'}$)-to-L_α transition, the corresponding lipid DOPE with ethanolamine as (smaller) headgroup exhibits an additional phase transition from the lamellar L_α to the non-lamellar inverse hexagonal H_{II} phase at high temperatures

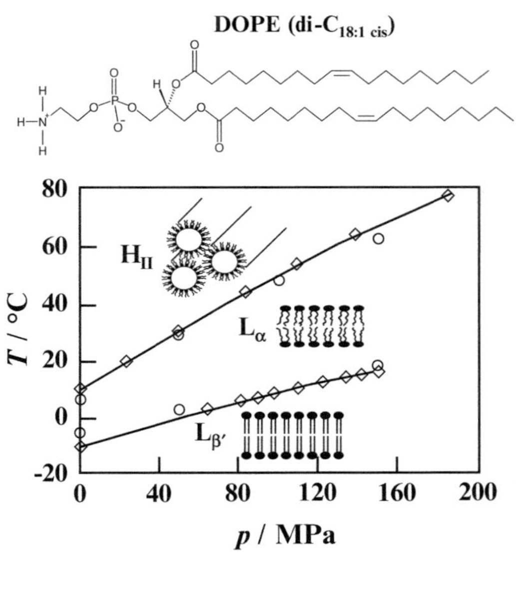

Fig. 17.8 *T*, *p*-phase diagram of DOPE in excess water. Lipid phases: $L_{\beta'}$, lamellar gel; L_α, lamellar liquid-crystalline (fluid-like) and H_{II}, inverse hexagonal (Modified from Winter and Jeworrek 2009; Meersmann et al. 2013)

(Fig. 17.8). As pressure forces a closer packing of the lipid chains, which results in a decreased number of *gauche* bonds and kinks in the chains, both transition temperatures, of the $L_{\beta'}$-L_α and the L_α-H_{II} transition, increase with increasing pressure. The L_α-H_{II} transition observed in DOPE/water and also in egg-PE/water (egg-PE is a natural mixture of different phosphatidylethanolamines) is the most pressure-sensitive lyotropic lipid phase transition found so far ($dT/dp \approx 0.40$ °C MPa^{-1}). The reason why this transition has such a strong pressure dependence is the strong pressure-dependence of the chain length and volume of its *cis*-unsaturated chains. Generally, at sufficiently high pressures, hexagonal and cubic lipid mesophases give way to lamellar structures as they exhibit smaller partial lipid volumes. Interestingly, in these systems inverse cubic phases $Q_{II}{}^D$ and $Q_{II}{}^P$ can be induced in the region of the L_α-H_{II} transition by subjecting the sample to extensive temperature or pressure cycles across the phase transition. For conditions that favor a spontaneous curvature of a lipid monolayer which is not too high, the topology of an inverse bicontinuous cubic phase can have a similar or even lower free energy than the lamellar or inverse hexagonal phase, as the cubic phases are characterized by a low curvature free energy and do not suffer the extreme chain packing stress predominant in the H_{II}-phase.

Incorporation of membrane proteins has been shown to affect also the barotropic phase behavior of curvature-inducing lipids. For example, incorporation of bacteriorhodopsin into monoolein (MO) has been found to stabilize more three-dimensional cubic than planar lamellar phases. Even the existence of a highly swollen gyroid cubic phase in the high water region could be found (Chandrashekhar et al. 2013). Such studies comprising non-lamellar lipid systems are promising for understanding complex biomembrane restructuring, including conditions how cells cope with extreme conditions of high pressures.

17.7 Kinetic Studies Using Pressure-Jump Relaxation Techniques

Phase transitions between lipid nanostructures are generally associated with deformations of the interfaces which, very often, imply also their fragmentation and fusion. Depending on the topology of the lipid structures involved, transition phenomena of different complexity may be observed. The pressure-jump technique in conjunction with synchrotron X-ray diffraction or spectroscopies has been used to study the time course of lipid phase transitions and to search for possible transient intermediate structures, with a view to unravelling the underlying transition mechanisms (Winter and Czeslik 2000; Winter and Köhling 2004; Kraineva et al. 2005; Squires et al. 2000, 2002; Köhling et al. 2002; Conn et al. 2006, 2008; Schiewek and Blume 2008). The pressure-jump technique offers several advantages over the temperature-jump approach: (1) Pressure propagates rapidly so that sample inhomogeneity is no problem. (2) Pressure-jumps can be performed bidirectionally, i.e., with increasing or decreasing pressure. (3) In the case of fully reversible structural changes of the sample, pressure-jumps can be repeated with identical amplitudes to allow for an averaging of the diffraction data over many jumps and hence an improvement of the counting statistics.

Here, we discuss one example, only, an interesting phase transition where membrane fusion occurs: a fluid lamellar to inverse bicontinuous cubic phase transition. Models for the process of membrane fusion rely – in cellular membranes with the help of various fusion proteins – on the formation of transient lipid contacts known as stalks, which subsequently break through to form the beginnings of tubular connections (so-called fusion pores) that are the fundamental connecting element in the inverse bicontinuous cubic phases, which consist of ordered arrays of such connections. As an example of a lamellar to cubic lipid phase transformation, Fig. 17.9 shows data on the MO-water system. A pressure-jump from 1,500 to 1 bar was used to induce the $L_\alpha \rightarrow$ Ia3d transition. The time-dependent SAXS patterns after the jump are shown in Fig. 17.9a. The first lamellar reflection (001) of the L_α phase and, at longer times, the reflections ($\sqrt{6}$, $\sqrt{8}$, $\sqrt{14}$, $\sqrt{16}$, ...) of the

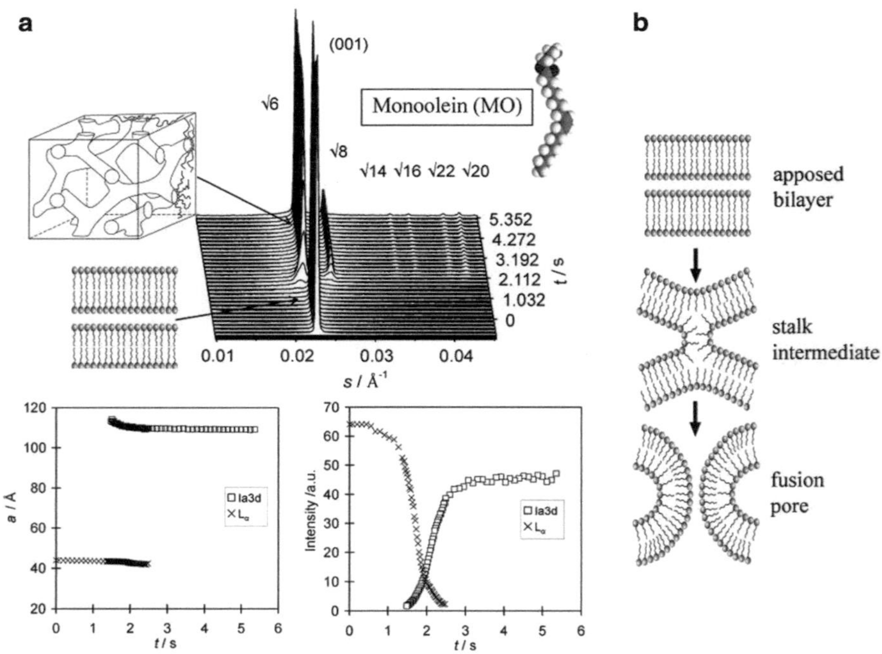

Fig. 17.9 (**a**) *Top*: Time-resolved SAXS data of monoolein (MO) in 20 wt% H_2O at 35 °C. At time zero, a fast (ms) pressure-jump from 1,500 to 1 bar induces a phase transition from the lamellar L_α to the Ia3d cubic phase (a schematic view of the two phases is added). Indexing of the Bragg reflections is indicated as well. *Bottom*: The corresponding intensities and lattice constants of the corresponding phases as a function of time. (**b**) Formation of a stalk intermediate between a lamellar bilayer and a fusion pore (Modified from Winter and Jeworrek 2009)

developing Ia3d phase are clearly visible. The time-evolution of the corresponding intensities and lattice constants is shown as well. The system starts in the lamellar phase with initial lattice parameter of $a = 44$ Å. The intensity of the L_α phase decays rapidly while the cubic phase Ia3d is formed concomitantly. The first peak of Ia3d appears after 1.5 s, and the intensity of the diffraction peaks of the L_α phase vanishes at ~2.5 s. The lamellar phase shrinks to a specific threshold interlamellar spacing, which is probably occurring at the same time that stalks are formed between bilayers (Fig. 17.9, right). In cellular membranes, membrane fusion-mediating proteins have to act by reducing the energy barrier to create these fusion intermediates.

Generally, the relaxation behavior and the kinetics of pressure-induced lipid phase transformations depend drastically on the topology of the lipid mesophase, and also on the temperature and the driving force, i.e., the applied pressure-jump amplitude, Δp (Erbes et al. 2000). Often multicomponent kinetic behavior has been observed, with short relaxation times (probably on the ns to μs time-scale), referring to the relaxation of lipid-chain conformations in response to the pressure change. Longer relaxation times are due to the kinetic trapping of the system. In most of such cases, the rate of the transition is limited by the transport and redistribution of water into and in the new lipid phase, and the obstruction factor of the different structures, especially in cases where non-lamellar (hexagonal and cubic) phases are involved. In addition, slow relaxation processes may originate from nucleation phenomena and domain growth. Maximum values of up to minutes have been observed at the melting transition of pure phospholipids (Heimburg 2007).

The pressure-jump relaxation technique may also be applied to more complex biochemical processes. For example, the technique has been applied to study the pressure dependence of the photocycle kinetics of bacteriorhodopsin (bR) from *Halobacterium salinarium* (Klink et al. 2002). Such time-resolved studies of lipid phase transitions and proteo-lipid systems are still at an early stage, but clearly have been shown already to be invaluable for helping explore transition pathways and mechanisms.

17.8 Reconstituted and Biological Membranes

Cellular membranes are complex heterogeneous aggregate structures, held together by the hydrophobic effect. The integrity and functionality of these membranes are vital for the cell, e.g., for energy production, transport, signaling processes, and maintenance of osmotic pressure and intracellular pH. Although some ion transporters are unaffected or even activated upon mild compression, certain other channels and pumps are inactivated at moderate to high pressures. It has often been observed that at pressures of several hundred MPa, membrane protein function ceases, and integral and peripheral proteins may even become detached from the membrane when its bilayer is sufficiently ordered by pressure, and depolymerization of attached cytoskeletal proteins may be involved as well. For example, the influence of hydrostatic pressure on the activity of Na^+, K^+-ATPase enriched in the plasma

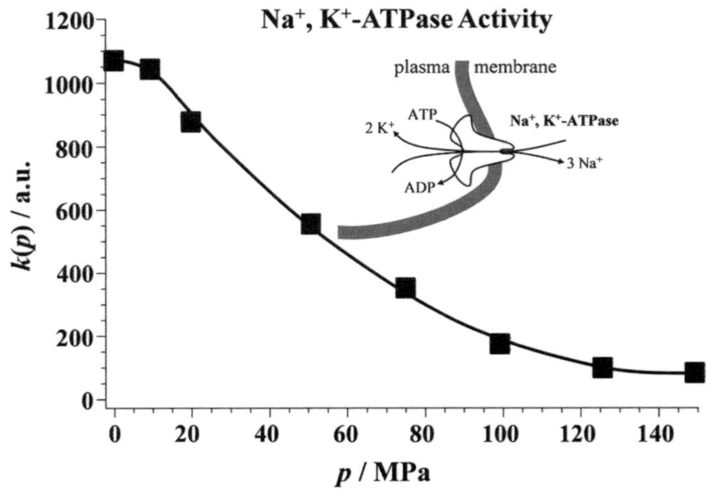

Fig. 17.10 Activity k (in arbitrary units) of Na^+, K^+-ATPase – as measured using an enzymatic assay – at selected pressures and $T = 37\ °C$. The free energy of hydrolysis of one ATP molecule is converted to up-hill transport by actively transporting 3 Na^+ ions out of and 2 K^+ into the cell (Modified from Winter and Jeworrek 2009; Meersmann et al. 2013)

membrane from rabbit kidney outer medulla was studied using a kinetic assay. The data shown in Fig. 17.10 reveal that the activity, k, of the Na^+, K^+-ATPase is retarded by pressures below 200 MPa. The plot of ln$k$ vs. p revealed an apparent activation volume of the pressure-induced inhibition reaction which amounts to $\Delta V^{\neq} = 47$ mL mol^{-1}. At higher pressures, exceeding 200 MPa, the enzyme is inactivated irreversibly (De Smedt et al. 1979; Chong et al. 1985). Kato et al. (2002) suggested that the activity of the enzyme shows at least three step changes induced by pressure: at pressures below and around 100 MPa, a decrease in the fluidity of the lipid bilayer and a reversible conformational change in the transmembrane protein is induced, leading to functional disorder of the membrane-associated ATPase activity. Pressures of 100–200 MPa cause a reversible phase transition and the dissociation or conformational changes in the protein subunits, and pressures exceeding 220 MPa irreversibly degrade the lipid-protein assembly due to protein unfolding and separation.

To be able to explore the effect of the lipid matrix itself on the enzyme activity, the Na^+, K^+-ATPase was also reconstituted into various lipid bilayer systems of different chain length, configuration, phase state and lateral organization including model raft mixtures. In the low-pressure region, around 10 MPa, a significant increase of the activity was observed for the enzyme reconstituted into DMPC and DOPC bilayers. Upon further compression, the enzyme activity decreases, reaching zero activity around 200 MPa for all reconstituted systems measured, similar to the natural system. A similar behavior has been found for the chloroplast ATP-synthase (Souza et al. 2004).

The effect of pressure was also determined for the HorA activity of the bacterium *Lactobacillus plantarum* (Ulmer et al. 2002), an ATP-dependent multidrug-resistance transporter of the ABC family. Changes were determined for the membrane composition of *L. plantarum* induced by different growth temperatures and their effect on the pressure inactivation, and a temperature-pressure phase diagram was constructed for the *L. plantarum* membranes that could be correlated with the respective kinetics of high-pressure inactivation. Upon pressure-induced phase transitions to rigid (i.e., gel-like) membrane structures, at pressures around 50–150 MPa for temperatures between 20 and 37 °C, fast inactivation of HorA was observed.

The effect of pressure and the influence of the lipid matrix on lipid-protein interactions was also studied for the multidrug resistance protein LmrA, which was expressed in the bacterium *Lactococcus lactis* and functionally reconstituted in different model membrane systems (Periasamy et al. 2009). The membrane systems chosen were composed of DMPC, DOPC, DMPC + 10 mol% Chol, and the heterogeneous model raft mixture DOPC:DPPC:Chol (1:2:1). Teichert (2008) showed that a sharp pressure-induced fluid-to-gel phase transition without the possibility for lipid sorting, such as in DMPC bilayers, has a drastic inhibitory effect on the LmrA activity. Inactivation of membrane protein function upon entering a rigid (gel-like or solid-ordered) membranous state seems to be a rather common phenomenon. Otherwise, an overall fluid-like membrane phase over the whole pressure-range covered, with suitable hydrophobic matching, such as for DOPC, prevents the membrane protein from total high pressure inactivation even at 200 MPa. Also the systems exhibiting thicker membranes with higher lipid order parameters, such as DMPC/10 mol% Chol and the model raft mixture, show remarkable pressure stabilities. The results also revealed that an efficient packing with optimal lipid adjustment to prevent (also pressure-induced) hydrophobic mismatch might be a particular prerequisite for the homodimer formation and hence function of LmrA.

Pressure has also been found to be of interest to modulate signaling events, such as ToxR/ToxS interaction and phospholipase activation by G-proteins (Linke et al. 2008; Scarlata 2005). High-pressure-induced dimer dissociation of membrane proteins *in vivo* was studied using the ToxR inner membrane-spanning transcription factor present in some piezosensitive and piezophilic bacteria. Analyses of ToxR derived from the mesophilic bacterium *Vibrio cholerae* were carried out by introducing protein variants in *E. coli* reporter strains carrying *a ToxR activatable reporter gene* fusion. Dimerization ceased at 20–50 MPa, depending on the nature of the transmembrane segment rather than as a result of changes in the pressure-induced lipid bilayer environment (Linke et al. 2008). Similar results were also obtained for ToxR derived from the piezophilic deep-sea bacterium *Photobacterium profundum* strain SS9 in both *E. coli* and SS9 backgrounds (Linke et al. 2009). Scarlata revealed perturbations of phospholipase PLβ-Gβγ association caused by the Gα(GDP) subunit in the membranous context that where not observable by atmospheric association measurements (Scarlata 2005). PLCβ membrane binding was stable throughout the 0.1-200 MPa range, Gβγ only at high concentrations, whereas Gα(GDP) dissociated from membranes above 100 MPa.

High pressure has also been used as a means to study the functioning of a number of ion channels in excitable cells (Conti et al. 1984; Wann et al. 1980; Heinemann et al. 1987; Meyer et al. 1997; Macdonald 1997, 2001, 2002a, b; Macdonald et al. 2005; Friedrich et al. 2006). Their structure, function, and regulation can be disturbed or damaged by HHP. The detailed mechanism of action is mostly unclear and might vary depending on the particular protein considered.

The BK (Big Potassium) channel, also called Maxi-K channel, in the plasmamembrane of bovine chromaffin cells is typical of a large group of high conductance channels, selective for K^+ ions, which are gated by voltage, Ca^{2+} concentration, and in some cases by fatty acids and membrane stretch (tension). High hydrostatic pressure was found to increase the probability of BK channels being open (90 MPa increasing the probability 30-fold). The pressure activation of the channel was reversed upon decompression. Channel conductance was unaffected, however. It was shown that pressure did not act by raising the order of the membrane bilayer, it was rather concluded that pressure most likely acted directly on the channel proteins and/or their modulating reactions (Macdonald 1997). An endogenous inhibitor was postulated to be dissociated by pressure (Macdonald 1997, 2001).

The effect of HHP was also studied on bacterial mechanosensitive channels of small conductance (MscS) in giant spheroplasts of *E. coli* using Heinemann's "flying-patch" patch-clamp technique. Mechanosensitive channels are found in a number of tissues and organisms and are thought to be sensors for a number of systems including the senses of touch, hearing and balance, as well as participating in cardiovascular regulation and osmotic homeostasis. Voltage has been used to activate the channel in a high-pressure chamber. It was found that pressure affects the channel kinetics but not the channel conductance. At negative pipette voltages (corresponding to membrane depolarization) the channel exhibited a reversible reduction in activity with increasing hydrostatic pressure between 0.1 and 90 MPa at room temperature. The reduced activity was characterized by a significant reduction in the channel-open probability resulting from shortening of the channel openings with increasing pressure, which indicated that HHP generally favored channel closing. The results suggested that lateral compression of the lipid bilayer and/or volume changes from conformational rearrangements of the channel protein under high-pressure conditions are responsible for the reduction of the channel activity (Petrov et al. 2007).

Recently, Petrov et al. (2011) have also investigated the effect of HHP on MscL G22E mutant channels reconstituted into azolectin liposome membranes and channels expressed in *E. coli* giant spheroplasts. Here, HHP in the range of 0.1 to 90 MPa was shown to increase the channel open probability by favoring the open state of the channel. Hydrostatic pressure affected the channel kinetics, as manifested by the propensity of the channel to gate at subconducting levels with an increase in pressure. It was proposed that the presence of water molecules around the hydrophobic gate of the G22E MscL channel induces hydration of the hydrophobic lock under HHP, causing frequent channel openings and preventing the channel closure in the absence of membrane tension. Generally, HHP perturbation was identified as a valuable experimental approach toward better understanding of the gating mechanism of complex channels such as MscL.

The study of high-pressure effects on membrane proteins, channels and membrane associated processes such as signaling events is still in its infancy. What is already clear, however, is that the membrane's physical-chemical effects (i.e., membrane composition, phase state, in-plane lateral organization, domain formation, hydrophobic matching, surface charge density, curvature elastic stress, and lateral pressure profile) markedly influence the lipid-protein interaction, activity and pressure stability of the embedded membrane protein. It seems also to be clear that the specific nature of the embedded membrane protein (e.g., its oligomeric assembly, a required dimerization reaction or clustering upon signaling, the formation of a signaling platforms, etc.) plays an important role in its pressure-dependent stability and activity as well.

Acknowledgements Financial support from the Deutsche Forschungsgemeinschaft (DFG, FOR 1979) is gratefully acknowledged.

References

Abe F (2013) Dynamic structural changes in microbial membranes in response to high hydrostatic pressure analyzed using time-resolved fluorescence anisotropy measurement. Biophys Chem 183:3–8

Akasaka K (2006) Probing conformational fluctuation of proteins by pressure perturbation. Chem Rev 106:1814–1835

Allen EE, Facciotti D, Bartlett DH (1999) Monounsaturated but not polyunsaturated fatty acids are required for growth of the deep-sea bacterium Photobacterium profundum SS9 at high pressure and low temperature. Appl Environ Microbiol 65:1710–1720

Arnold MR, Kalbitzer RH, Kremer W (2003) High-sensitivity sapphire cells for high pressure NMR spectroscopy on proteins. J Magn Reson 161:127–131

Balny C, Hayashi R, Heremans K, Masson P (eds) (1992) High pressure and biotechnology, vol 224. Colloque Inserm, John Libbey Eurotext Ltd, Montrouge, pp 195–208

Bartlett DH (2002) Pressure effects on in vivo microbial processes. Biochim Biophys Acta 1595:367–381

Bartlett DH, Ferguson G, Valle G (2008) Adaptations of the psychrotolerant piezophile Photobacterium profundum strain SS9. In: Michiels C, Bartlett D, Aertsen A (eds) High-pressure microbiology. ASM Press, Washington DC, pp 319–337

Bernsdorff C, Wolf A, Winter R (1996) The effect of temperature and pressure on structural and dynamic properties of phospholipid/sterol mixtures – a steady-state and time-resolved fluorescence anisotropy study. Z Phys Chem 193:151–173

Bernsdorff C, Wolf A, Winter R, Gratton E (1997) Effect of hydrostatic pressure on water penetration and rotational dynamics in phospholipid-cholesterol bilayers. Biophys J 72:1264–1277

Böttner M, Winter R (1993) Influence of the local anesthetic tetracaine on the phase behavior and the thermodynamic properties of phospholipid bilayers. Biophys J 65:2041–2046

Böttner M, Ceh D, Jacobs U, Winter R (1994) High pressure volumetric measurements on phospholipid bilayers. Z Phys Chem 184:205–218

Chandrashekhar VK, Ces O, Templer RH, Seddon JM (2013) Pressure effects on a protein-lipid model membrane. Soft Matter 9:6525–6531

Chong PL, Fortes PA, Jameson DM (1985) Mechanisms of inhibition of (Na, K)-ATPase by hydrostatic pressure studied with fluorescent probes. J Biol Chem 260:14484–14490

Conn CE, Ces O, Mulet X, Finet S, Winter R, Seddon JM, Templer RH (2006) Dynamics of structural transformations between lamellar and inverse bicontinuous cubic lyotropic phases. Phys Rev Lett 96:108102

Conn CE, Ces O, Squires AM, Mulet X, Winter R, Finet SM, Templer RH, Seddon JM (2008) A pressure-jump time-resolved X-ray diffraction study of cubic-cubic transition kinetics in monoolein. Langmuir 24:2331–2340

Conti F, Inoue I, Kukita F, Stühmer W (1984) Pressure dependence of sodium gating currents in the squid giant axon. Eur Biophys J 11:137–147

Czeslik C, Reis O, Winter R, Rapp G (1998) Effect of high pressure on the structure of dipalmitoylphosphatidylcholine bilayer membranes: a synchrotron-X-ray diffraction and FT-IR spectroscopy study using the diamond anvil technique. Chem Phys Lipids 91:135–144

Daniel I, Oger P, Winter R (2006) Origins of life and biochemistry under high pressure conditions. Chem Soc Rev 35:858–875

Danielowski K (1997) Einfluss von Lachgas, Argon und Stickstoff auf das Phasenverhalten von Phospholipiden. Diploma thesis, Dortmund University of Technology

De Smedt H, Borghgraef R, Ceuterick F, Heremans K (1979) Pressure effects on lipid-protein interactions in (Na^+, K^+)-ATPase. Biochim Biophys Acta 556:479–489

DeLong EF, Yayanos AA (1985) Adaptation of the membrane lipids of a deep-sea bacterium to changes in hydrostatic pressure. Science 228:1101–1103

DeLong EF, Yayanos AA (1987) Properties of the glucose transport system in some deep-sea bacteria. Appl Environ Microbiol 53:527–532

Eisenblatter J, Winter R (2005) Pressure effects on the structure and phase behavior of phospholipid polypeptide bilayers: a synchrotron small-angle X-ray scattering and ^{2}H-NMR spectroscopy study on DPPC gramicidin lipid bilayers. Z Phys Chem 219:1321–1345

Eisenblätter J, Zenerino A, Winter R (2000) High pressure ^{1}H-NMR on model biomembranes: a study of the local anaesthetic tetracaine incorporated into POPC lipid bilayers. Magn Reson Chem 38:662–667

Erbes J, Winter R, Rapp G (1996) Rate of phase transformations between mesophases of the 1:2 lecithin/fatty acid mixtures DMPC/MA and DPPC/PA - a time-resolved synchrotron X-ray diffraction study. Ber Bunsenges Phys Chem 100:1713–1722

Erbes J, Gabke A, Rapp G, Winter R (2000) Kinetics of phase transformations between lyotropic lipid mesophases of different topology: a time-resolved synchrotron X-ray diffraction study using the pressure-jump relaxation technique. Phys Chem Chem Phys 2:151–162

Friedrich O, Kress KR, Hartmann M, Frey B, Sommer K, Ludwig H, Fink RHA (2006) Prolonged high-pressure treatments in mammalian skeletal muscle result in loss of functional sodium channels and altered calcium channel kinetics. Cell Biochem Biophys 45:71–83

Hallinen KM, Tristam-Nagle S, Nagle JF (2012) Volumetric stability of lipid bilayers. Phys Chem Chem Phys 14:15452–15457

Hammouda B, Worcester D (1997) Interdigitated hydrocarbon chains in C20 and C22 phosphatidylcholines induced by hydrostatic pressure. Physica B Condens Matt 241–243:1175–1177

Hanford MJ, Peeples TL (2002) Archeal tetraether lipids: unique structures and applications. Appl Biochem Biotechnol 97:45–62

Heimburg T (2007) Thermal biophysics of membranes. Wiley-VCH, Weinheim

Heinemann SH, Conti F, Stühmer W, Neher E (1987) Effects of hydrostatic pressure on membrane processes. Sodium channels, calcium channels, and exocytosis. J Gen Physiol 90:765–778

Janosch S, Nicolini C, Ludolph B, Peters C, Volkert M, Hazlet TL, Gratton E, Waldmann H, Winter R (2004) Partitioning of dual-lipidated peptides into membrane microdomains: lipid sorting vs peptide aggregation. J Am Chem Soc 126:7496–7503

Jeworrek C, Pühse M, Winter R (2008) X-ray kinematography of phase transformations of three-component lipid mixtures: a time-resolved synchrotron X-ray scattering study using the pressure-jump relaxation technique. Langmuir 24:11851–11859

Johnson FH, Flagler EA (1950) Hydrostatic pressure reversal of narcosis in tadpoles. Science 112:91–92

Jonas J (ed) (1991) High pressure NMR. Springer-Verlag, Berlin

Kapoor S, Triola G, Vetter IR, Erlkamp M, Waldmann H, Winter R (2012a) Revealing conformational substates of lipidated N-Ras protein by pressure modulation. Proc Natl Acad Sci USA 109:460–465

Kapoor S, Weise K, Erlkamp M, Triola G, Waldmann H, Winter R (2012b) The role of G-domain orientation and nucleotide state on the Ras isoform-specific membrane interaction. Eur Biophys J 41:801–813

Kapoor S, Berghaus M, Suladze S, Prumbaum D, Grobelny S, Degen P, Raunser S, Winter R (2014) Prebiotic cell membranes that survive extreme environmental pressure conditions. Angew Chem Intern Ed 53:8397–8401

Kato M, Hayashi R, Tsuda T, Taniguchi K (2002) High pressure-induced changes of biological membrane. Study on the membrane-bound Na^{+}/K^{+}-ATPase as a model system. Eur J Biochem 269:110–118

Klink BU, Winter R, Engelhard M, Chizhov I (2002) Pressure dependence of the photocycle kinetics of bacteriorhodopsin. Biophys J 83:3490–3498

Köhling R, Woenckhaus J, Klyachko NL, Winter R (2002) Small-angle neutron scattering study of the effect of pressure on AOT-*n*-octane-water mesophases and the effect of α-chymotrypsin incorporation. Langmuir 18:8626–8632

Kraineva J, Narayanan RA, Kondrashkina E, Thiyagarajan P, Winter R (2005) Kinetics of lamellar-to-cubic and intercubic phase transitions of pure and cytochrome c containing monoolein dispersions monitored by time-resolved small-angle X-ray diffraction. Langmuir 21:3559–3571

Krivanek R, Okoro L, Winter R (2008) Effect of cholesterol and ergosterol on the compressibility and volume fluctuations of phospholipid-sterol bilayers in the critical point region: a molecular acoustic and calorimetric study. Biophys J 94:3538–3548

Landwehr A, Winter R (1994a) High-pressure differential thermal analysis of lamellar to lamellar and lamellar to non-lamellar lipid phase transitions. Ber Bunsenges Phys Chem 98:214–218

Landwehr A, Winter R (1994b) The T, x, p-phase diagram of binary phospholipid mixtures. Ber Bunsenges Phys Chem 98:1585–1589

Linke K, Periasamy N, Ehrmann M, Winter R, Vogel RF (2008) Influence of high pressure on the dimerization of ToxR, a protein involved in bacterial signal transduction. App Environ Microbiol 74:7821–7823

Linke K, Periasamy N, Eloe EA, Ehrmann M, Winter R, Barlett DH, Vogel RF (2009) Influence of membrane organization on the dimerization ability of ToxR from photobacterium profundum under high hydrostatic pressure. High Press Res 29:341–442

Macdonald AG (1992) Effects of high hydrostatic pressure on natural and artificial membranes. In: Balny C, Hayashi R, Heremans K, Mason P (eds) High pressure and biotechnology, vol 224, Colloque INSERM. John Libbey Eurotext Ltd., Montrouge, pp 67–75

Macdonald AG (1997) Effect of high hydrostatic pressure on the BK channel in bovine chromaffin cells. Biophys J 73:1866–1873

Macdonald AG (2001) High hydrostatic pressure activates BK channels. Biophys J 80:A888

Macdonald AG (2002a) Experiments on ion channels at high pressure. Biochim Biophys Acta 1595:387–389

Macdonald AG (2002b) Ion channels under high pressure. Comp Biochem Physiol A Mol Integr Physiol 131:587–593

Macdonald AG, Martinac B (2005) Effect of high hydrostatic pressure on the bacterial mechanosensitive channel MscS. Eur Biophys J 34:434–441

MacNaughtan W, MacDonald AG (1980) Effects of gaseous anaesthetics and inert gases on the phase transition in smectic mesophases of dipalmitoyl phosphatidylcholine. Biochim Biophys Acta 597:193–198

Mantulin WW, Gotto AM, Pownall HJ (1984) Effect of hydrostatic pressure on the transfer of a fluorescent phosphatidylcholine between apolipoprotein-phospholipid recombinants. J Am Chem Soc 106:3317–3319

Matsuki H, Goto M, Kusube M, Tamai N, Kaneshina S (2005) Barotropic phase transitions of 1-palmitoyl-2-stearoylphosphatidylcholine bilayer membrane. Chem Lett 34:270–271

Matsuki H, Miyazaki E, Sakano F, Tamai N, Kaneshina S (2007) Thermotropic and barotropic phase transitions in bilayer membranes of ether-linked phospholipids with varying alkyl chain lengths. Biochim Biophys Acta 1768:479–489

Matsuki H, Goto M, Tada K, Tamai N (2013) Thermotropic and barotropic phase behavior of phosphatidylcholine bilayers. Int J Mol Sci 14:2282–2302

Meersman F, Daniel I, Bartlett D, Winter R, Hazael R, McMillan PF (2013) High-pressure biochemistry and biophysics. Rev Mineral Geochem 75:607–648

Meyer HH (1899) Theorie der Alkoholnarkose. Arch Exp Path Pharmak 42:109–118

Meyer R, Heinemann SH (1997) Temperature and pressure dependence of Shaker K^+ channel N- and C-type inactivation. Eur Biophys J 26:433–445

Mountcastle DB, Biltonen RL, Halsey MJ (1978) Effect of anesthetics and pressure on the thermotropic behavior of multilamellar dipalmitoylphosphatidylcholine liposomes. Proc Natl Acad Sci USA 75:4906–4910

Nicolini C, Baranski J, Schlummer S, Palomo J, Lumbierres-Burgues M, Kahms M, Kuhlmann J, Sanchez S, Gratton E, Waldmann H, Winter R (2005) Visualizing association of N-Ras in lipid microdomains: influence of domain structure and interfacial adsorption. J Am Chem Soc 128:192–201

Nicolini C, Kraineva J, Khurana M, Periasamy N, Funari SS, Winter R (2006) Temperature and pressure effects on structural and conformational properties of POPC/SM/cholesterol model raft mixtures‘ a FT-IR, SAXS, DSC, PPC and Laurdan fluorescence spectroscopy study. Biochim Biophys Acta 1758:248–258

Oger PM, Cario A (2013) Adaptation of the membrane in archaea. Biophys Chem 183:3–8

Oger P, Jebbar M (2010) The many ways of coping with pressure. Res Microbiol 161:799–809

Overton E (1909) Studien über die Narkose. Gustav Fischer, Jena

Periasamy N, Teichert H, Weise K, Vogel RF, Winter R (2009) Effects of temperature and pressure on the lateral organization of model membranes with functionally reconstituted multidrug transporter LmrA. Biochim Biophys Acta 1788:390–401

Petrov E, Rohde PR, Macdonald AG, Martinac B (2007) Effect of high hydrostatic pressure and voltage on gating of the bacterial mechanosensitive channel of small conductance. Proc 4th Int Conf High Pressure Biosci Biotechol 1:20–27

Petrov E, Rohde PR, Martinac B (2011) Flying-patch patch-clamp study of G22E-MscL mutant under high hydrostatic pressure. Biophys J 100:1635–1641

Picard A, Daniel I (2013) Pressure as an environmental parameter for microbial life. Biophys Chem 183:30–41

Potekhin SA, Senin AA, Abdurakhmanov NN, Khusainova RS (2008) Thermodynamic invariants of gel to the liquid-crystal 1,2-diacylphosphatidylcholines transition. Biochim Biophys Acta 1778:2588–2593

Powalska E, Janosch S, Kinne-Saffran E, Kinne RKH, Fontes CFL, Mignaco JA, Winter R (2007) Fluorescence spectroscopic studies of pressure effects on Na+, K + -ATPase reconstituted into phospholipid bilayers and model raft mixtures. Biochemistry 46:1672–1683

Scarlata S (2005) Determination of the activation volume of PLCβ by Gβγ-subunits through the use of high hydrostatic pressure. Biophys J 88:2867–2874

Schiewek M, Blume A (2008) Pressure jump relaxation investigations of lipid membranes using FTIR spectroscopy. Eur Biophys J 38:219–228

Seddon JM, Templer RH (1993) Cubic phases of self-assembled amphiphilic aggregates. Phil Trans R Soc A 344:377–401

Seemann H, Winter R (2003) Volumetric properties, compressibilities and volume fluctuations in phospholipid-cholesterol bilayers. Z Phys Chem 217:831–846

Seemann H, Winter R, Royer CA (2001) Volume, expansivity and isothermal compressibility changes associated with temperature and pressure unfolding of staphylococcal nuclease. J Mol Biol 307:1091–1102

Souza MO, Creczynski-Pasa TB, Scofano HM, Graber P, Mignaco JA (2004) High hydrostatic pressure perturbs the interactions between CF0F1 subunits and induces a dual effect on activity. Int J Biochem Cell Biol 36:920–930

Squires A, Templer RH, Ces O, Gabke A, Woenckhaus J, Seddon JM, Winter R (2000) Kinetics of lyotropic phase transitions involving the inverse bicontinuous cubic phases. Langmuir 16:3578–3582

Squires AM, Templer RH, Seddon JM, Woenckhaus J, Winter R, Finet S, Theyencheri N (2002) Kinetics and mechanism of the lamellar to gyroid inverse bicontinuous cubic phase transition. Langmuir 18:7384–7392

Tamura K, Matsumoto T, Suzuki Y (2003) Effects of gas pressure on the phase transition temperature of liposomal bilayer membranes. In: Winter R (ed) Advances in high pressure bioscience and biotechnology II. Springer-Verlag, Heidelberg, pp 203–206

Teichert H (2008) Behaviour of membrane transport proteins under high hydrostatic pressure. M.Sc. Thesis, Technische Universität München

Templer RH, Seddon JM, Duesing PM, Winter R, Erbes J (1998) Modeling the phase behavior of the inverse hexagonal and inverse bicontinuous cubic phases in 2:1 fatty acid/phosphatidylcholine mixtures. J Phys Chem B 102:7262–7271

Ulmer HM, Herberhold H, Fahsel S, Gänzle MG, Winter R, Vogel RF (2002) Effects of pressure-induced membrane phase transitions on inactivation of HorA, an ATP-dependent multidrug resistance transporter, in lactobacillus plantarum. Appl Environ Microbiol 68:1088–1095

Wann KT, Macdonald AG (1980) The effects of pressure on excitable cells. Comp Biochem Phys A Phys 66:1–12

Weise K, Triola G, Brunsveld L, Waldmann H, Winter R (2009) Influence of the lipidation motif on the partitioning and association of N-Ras in model membrane subdomains. J Am Chem Soc 131:1557–1564

Winter R (2001) Effects of hydrostatic pressure on lipid and surfactant phases. Curr Opin Colloid Interface Sci 6:303–312

Winter R (2002) Synchrotron X-ray and neutron small-angle scattering of lyotropic lipid mesophases, model biomembranes and proteins in solution at high pressure. Biochim Biophys Acta 1595:160–184

Winter R (2003a) High pressure NMR studies on lyotropic lipid mesophases and model biomembranes. Ann Rep NMR Spectr 50:163–200

Winter R (ed) (2003b) High pressure bioscience and biotechnology II. Springer-Verlag, Heidelberg

Winter R (2013) Special issue: biomolecular systems under extreme environmental conditions. Biophys Chem 183:1–2

Winter R, Czeslik C (2000) Pressure effects on the structure of lyotropic lipid mesophases and model biomembrane systems. Z Kristallogr 215:454–474

Winter R, Dzwolak W (2004) Temperature-pressure configurational landscape of lipid bialyers and proteins. Cell Mol Biol 50:397–417

Winter R, Jeworrek C (2009) Effect of pressure on membranes. Soft Matter 5:3157–3173

Winter R, Jonas J (1999) High pressure molecular science. Kluwer Academic Publisher, Dordrecht, 358 of NATO Science Series E

Winter R, Köhling R (2004) Static and time-resolved synchrotron small-angle X-ray scattering studies of lyotropic lipid mesophases, model biomembranes and proteins in solution. J Phys Condens Matt 16:S327–S352

Winter R, Pilgrim WC (1989) A SANS study of high pressure phase transitions in model biomembranes. Ber Bunsenges Phys Chem 93:708–717

Winter R, Christmann MH, Böttner M, Thiyagarajan P, Heenan R (1991) The influence of the local anaesthetic tetracaine on the temperature and pressure dependent phase behaviour of model biomembranes. Ber Bunsenges Phys Chem 95:811–820

Winter R, Erbes J, Templer RH, Seddon JM, Syrykh A, Warrender NA, Rapp G (1999a) Inverse bicontinuous cubic phases in fatty acid/phosphatidylcholine mixtures: the effects of pressure and lipid composition. Phys Chem Chem Phys 1:887–893

Winter R, Gabke A, Czeslik C, Pfeifer P (1999b) Power-law fluctuations in phase-separated lipid membranes. Phys Rev E 60:7354–7359

Wlodarczyk A, McMillan PF, Greenfield SA (2006) High pressure effects in anaesthesia and narcosis. Chem Soc Rev 35:890–898

Woenckhaus J, Köhling R, Winter R, Thiyagarajan P, Finet S (2000) High pressure-jump apparatus for kinetic studies of protein folding reactions using the small-angle synchrotron X-ray scattering technique. Rev Sci Instrum 71:3895–3899

Wong PT, Siminovitch DJ, Mantsch HH (1988) Structure and properties of model membranes: new knowledge from high-pressure vibrational spectroscopy. Biochim Biophys Acta 47:139–171

Yayanos AA (1998) Empirical and theoretical aspects of life at high pressure in the deep sea. In: Horikoshi K, Grant WD (eds) Extremophiles: microbial life in extreme environments. Wiley-Liss, New York, pp 47–92

Zein M, Winter R (2000) Effect of temperature, pressure and lipid acyl chain length on the structure and phase behavior of phospholipid-gramicidin bilayers. Phys Chem Chem Phys 2:4545–4551

Chapter 18
Effects of High Hydrostatic Pressure on Microbial Cell Membranes: Structural and Functional Perspectives

Fumiyoshi Abe

Abstract Biological processes associated with dynamic structural features of membranes are highly sensitive to changes in hydrostatic pressure and temperature. Marine organisms potentially experience a broad range of pressure and temperature fluctuations. Hence, they have specialized cell membranes to perform membrane protein functions under various environmental conditions. Although the effects of high pressure on artificial lipid bilayers have been investigated in detail, little is known about how high pressure affects the structure of natural cell membranes and how organisms cope with pressure alterations. This review focused on the recent advances in research on the effects of high pressure on microbial membranes, particularly on the use of time-resolved fluorescence anisotropy measurement to determine membrane dynamics in deep-sea piezophiles.

Keywords Microbial cell membrane • *Saccharomyces cerevisiae* • Deep-sea piezophile • Eicosapentaenoic acid • High-pressure time-resolved fluorescence anisotropy measurement (HP-TRFAM)

18.1 Introduction

Oceans occupy 70 % of our planet and have an average depth of 3,800 m. In terms of volume, majority of the biosphere is made up of high-pressure environments. Therefore, marine organisms experience a broad range of hydrostatic pressures ranging from 0.1 to 100 MPa (0.1 MPa = 1 bar = 0.9869 atm = 1.0197 kg of force/cm^2; for clarity, "MPa" is used throughout), with varying temperatures. Although the effects of temperature have been widely investigated in various biological systems,

F. Abe (✉)
Department of Chemistry and Biological Science, College of Science and Engineering, Aoyama Gakuin University, 5-10-1 Fuchinobe, Chuo-ku, Sagamihara 252-5258, Japan

Institute of Biogeosciences, Japan Agency for Marine-Earth Science and Technology (JAMSTEC), Yokosuka, Japan
e-mail: abef@chem.aoyama.ac.jp

K. Akasaka, H. Matsuki (eds.), *High Pressure Bioscience*,
Subcellular Biochemistry 72, DOI 10.1007/978-94-017-9918-8_18

the effects of hydrostatic pressure have not been thoroughly analyzed. The primary effect of high pressure on lipid bilayers is tighter packing and restricting acyl chain motion. With increasing pressure, the gel-to-liquid crystalline coexistence region is shifted toward higher temperatures by approximately 22 C/100 MPa (Winter and Dzwolak 2005). For example, a pressure increase of 100 MPa increases the primary transition (L_β/L_α) temperature of stearoyl-oleyl phosphatidylcholine (SOPC) and dioleoylphosphatidylcholine (DOPC) membranes by 18.1 C and 23.3 C, respectively (Kaneshina et al. 1998). Hence at high pressure and room temperature, the membrane structure is analogous to that at low temperature and atmospheric pressure. Several excellent reviews on the effects of high pressure on artificial lipid bilayers should be consulted to understand the biophysical basis of membranes in detail (Winter 2002; Winter and Dzwolak 2005; Winter and Jeworrek 2009; Brooks et al. 2011; Matsuki et al. 2013). Piezophiles are microorganisms that prefer hydrostatic pressures higher than atmospheric pressure for their growth at least under laboratory conditions that mimic deep-sea, high-pressure environments. They are typically isolated from sediment samples of deep ocean floors. However, not all microbes inhabiting the deep sea are piezophiles; there are many mesophiles whose growth is impaired at high pressure relative to that at atmospheric pressure. These microorganisms retain the viability without exuberant growth but ensure the continuation of the species under high-pressure environments during evolution. Piezophiles have aroused the interest of researches not only because of their unique microbial adaptive strategy but also because of industrial applications. However, the basic understanding of piezo-adaptation is hampered because of the difficulty in genetic manipulation (e.g., gene disruption, overexpression, or mutagenesis) of these naturally occurring microorganisms. The use of model organisms such as enteric bacterium *Escherichia coli* or budding yeast *Saccharomyces cerevisiae* overcomes this difficulty in understanding the general traits of the whole organism. Thus, this review focuses on the effects of high pressure on membrane properties in both piezophiles and mesophiles.

18.2 Effects of High Pressure on Membranes in Yeast

High hydrostatic pressures higher than 100–150 MPa are generally considered lethal for most microorganisms; however, the lethal effects largely depend on the types of strains, culture media, or duration of pressure treatment. The effect of high pressure on the ultrastructure of *S. cerevisiae* cells was analyzed by immunoelectron microscopy using thin frozen sections. Pressure treatment of cells at 100 MPa for 10 min disrupts spindle pole bodies whereas treatment at 300 MPa disrupts subcellular structures, including the nuclear membrane (Kobori et al. 1995), and is accompanied by leakage of internal substrates such as amino acids or ions (Shimada et al. 1993). Compared with *S. cerevisiae*, the fission yeast *Schizosaccharomyces pombe* is more sensitive to high pressure (Sato et al. 1996). Physiological processes mediated by membrane proteins (e.g., respiration, nutrient uptake, ion flux, and signal transduction) are compromised by non-lethal levels of high pressure (<50–100 MPa) at least in part because of the perturbation of the membrane structure (de Smedt et al. 1979; Cossins and Macdonald 1989; Somero 1992; Bartlett 1999). Amino acid uptake by cells is mediated by permease/transporters in the plasma membrane. In *S. cerevisiae*, tryptophan uptake is one of the highest pressure-sensitive processes in yeast physiology. Experimental wild-type strains (e.g., YPH499 and W303-1A) usually carry several nutrient auxotrophic markers such as *trp1* (tryptophan), *leu2* (leucine), *lys2* (lysine), *his3* (histidine), *ade2* (adenine), or *ura3* (uracil) for the selection of plasmid-bearing transformants. High pressure of 15–25 MPa inhibits the activity of tryptophan permeases Tat1 and Tat2 and degrades these permease proteins. Because of reduced tryptophan availability under high pressure, growth of *trp1* cells is arrested in the G_1 phase of the cell cycle without any loss of viability (Abe and Horikoshi 2000). *trp1* cells also exhibit cold sensitivity at temperatures of 10 °C–15 °C. High pressure and low temperature are analogous in terms of increased packing and reduced fluidity of the membrane. In this regards, tryptophan permeases are sensitive to these structural alterations of the membrane caused by either high pressure or low temperature.

In the simplest kinetic mechanism in which the transition state presents the highest energy barrier and chemical transformation of substrate to product is the single rate-limiting step, the activation volume describes the direction and magnitude of hydrostatic pressure. The following equation represents a quantitative estimation of the effects of pressure:

$$(\partial \ln k/\partial p)_T = -\Delta V^{\neq}/RT \tag{18.1}$$

where k is the rate constant, p is the pressure (MPa), T is the absolute temperature (K), and R is the gas constant (mL MPa K^{-1} mol^{-1}). $\Delta V^{\neq}$ is the apparent volume change in activation (activation volume), representing the difference in volumes between the basal state and the transition state. Tryptophan uptake by Tat1 or Tat2 is accompanied by a large positive activation volume of 89.3 mL mol^{-1} or 50.8 mL mol^{-1}, respectively (Abe and Iida 2003). This indicates that the two permeases undergo a large volume increase during tryptophan translocation; hence, their activity is inhibited by increasing hydrostatic pressure. In contrast, leucine uptake mediated by leucine permease Bap2 is impaired by high pressure to a lesser extent (activation volume, 24 mL mol^{-1}) (Abe 2007). The difference in activation volumes between the homologous amino acid permeases Tat1, Tat2, and Bap2 could be attributable to volume changes associated with interactions of the substrate with the binding site or conformational change in permeases during substrate import.

Spontaneous mutants capable of growth under high pressure, namely, high-pressure growth (*HPG*) mutants, were isolated from *trp1* strain. This revealed regulatory components required for ubiquitin-dependent degradation of tryptophan permeases in response to high pressure. Tat1 and Tat2 are degraded when cells are exposed to a pressure of 25 MPa (Abe and Iida 2003). The *HPG* mutants were categorized into 4 linkage groups, *HPG1–HPG4*. *HPG1* mutation sites are located in the catalytic HECT domain of Rsp5 ubiquitin ligase (Abe and Iida

2003). In the absence of active Rsp5, Tat1 and Tat2 are markedly accumulated in the plasma membrane, thus allowing *trp1* cells to grow under high pressure. Particularly, Tat1 is highly stable at 0.1 MPa, with a very slow turnover rate (half-life ~ 6 h) but undergoes rapid Rsp5-dependent degradation at a pressure of 25 MPa (half-life < 3 h) (Suzuki et al. 2013). This implies that high pressure may induce conformational disorder in Tat1 proteins by directly or indirectly modulating the bilayer membrane structure; then, the disordered Tat1 proteins could be recognized by Rsp5 ubiquitin ligase for degradation.

Yeast plasma membrane is composed of glycerophospholipids, sphingolipids, and ergosterol. Ergosterol is a major constituent of the membrane and maintains plasma membrane rigidity by increasing packing efficiency of the interfacial regions of the membrane (Abe and Hiraki 2009). Genes encoding enzymes that catalyze the later steps of ergosterol biosynthesis, *ERG6*, *ERG2*, *ERG3*, *ERG4*, and *ERG5*, can be disrupted; however, the deletion mutants accumulate structurally abnormal sterols in the plasma membrane. Of these, *erg2*Δ and *erg6*Δ mutants show growth defect at 25 MPa even though they grow normally at 0.1 MPa (Abe and Minegishi 2008). It is still unclear whether ergosterol, but not the altered sterols, exerts stabilizing effects on membrane proteins or plays a role in the efficient trafficking of proteins to the plasma membrane when yeast cells are placed under high-pressure condition. The latter is likely because several genes involved in membrane trafficking (*VID24*, *VPS34*, *SEC22*, *PEP3*, *CHC1*, *PEP5*, *VPS45*, *ERG24*, *VPS54*, *AKR1*, and *SAC1*) are required for growth under high pressure (Abe and Minegishi 2008). The products of these genes might facilitate the delivery of newly synthesized proteins to appropriate locations (e.g., the bud neck, cell surface, or cell wall) under high-pressure condition.

Fatty acid composition affects cell survival under lethal pressure levels of 150–200 MPa. Supplementation with various fatty acids such as palmitoleic acid (C16:1), oleic acid (C18:1), linoleic acid (C18:2), or linolenic acid (C18:3) confers tolerance against high-pressure treatment at 150–200 MPa for 30 min to *ole1*Δ mutant that lacks membrane-bound Δ9 desaturase. The order of fatty acids for promoting survival is linolenic acid > linoleic acid > oleic acid > palmitoleic acid, indicating that higher proportion of unsaturated fatty acids helps maintain viability under high pressure (de Freitas et al. 2012). Whether the promoted unsaturation increases membrane fluidity and whether this allows the cells to survive under high pressure remain to be elucidated.

18.3 Effects of High Pressure on the Dynamic Structure of Bacterial Membranes

Numerous studies have elucidated the compositional alterations in membrane fatty acids of bacteria in response to environmental changes, including temperature, ions, pH, and chemicals, as an adaptive strategy employed by microorganisms (Denich et al. 2003; Trevors 2003). To perform physiological functions in hostile environments, bacteria potentially remodel the membrane by changing the ratio of (i) saturation to unsaturation, (ii) *cis* to *trans* unsaturation, (iii) branched to unbranched structure, and (iv) acyl chain length. Fatty acids containing single or more unsaturated bonds have more bulky conformation than their saturated counterparts do, thus allowing higher conformational freedom and lesser packing of the membrane. Therefore, it is commonly believed that organisms inhabiting cold or high-pressure environments have unsaturated bonds in cell membranes to compensate for the increased packing effect. This compensation mechanism is termed as "homeoviscous adaptation" (Sinensky 1974; Cossins and Macdonald 1989; Hazel 1995). However, very few examples of deep-sea microbes are available in which this adaptation has been proven experimentally. Furthermore, analysis of fatty acid species merely provides suggestive evidences about the dynamic structure of the membrane.

Fluorescence anisotropy measurement by using lipophilic fluorescence probes with polarized nature is a sensitive and useful tool to analyze the dynamic membrane structure. 1,6-Diphenyl-1,3,5-hexatriene (DPH) and its cationic derivative 1-[4-(trimethylamino)phenyl]-6-phenyl-1,3,5-hexatriene (TMA-DPH) are commonly used for such analyses. When added to model or natural membranes, DPH is primarily distributed perpendicular to the bilayer plane near the center of the membrane but is partially distributed parallel to the bilayer plane within acyl chain tails (van Langen et al. 1989). TMA-DPH is distributed near the lipid–water interface because of its charged moiety, thus reflecting only the interfacial region of the membrane. Time-resolved fluorescence anisotropy measurement based on time-correlated single-photon counting (TCSPC) provides quantitative information about membrane order and rotational motion of acyl chains in a single measurement. Details about the methodology of measuring microbial membranes are mentioned elsewhere (Abe 2013). Essentially, the dynamic membrane structure can be described by the order parameter S and the rotational diffusion coefficient D_w of the fluorescence probe embedded in the membrane. High hydrostatic pressure stiffens the membranes and restricts acyl chain motion, thereby increasing S and decreasing D_w. In our recent study, we created a system, namely, high-pressure time-resolved fluorescence anisotropy measurement (HP-TRFAM), to perform this analysis under high pressure. Analysis by using this system allows quantitative understanding of how high pressure affects membrane structure in deep-sea piezophiles as well as mesophiles (Usui et al. 2012; Abe and Usui 2013). The HP-TRFAM system comprises a high-pressure optical cell, a high-pressure pump, and a TCSPC device (Fig. 18.1).

The effect of high pressure on membrane properties of *E. coli* was analyzed using HP-TRFAM (Abe and Usui 2013). *E. coli* cells optimally grow at 37 °C and 0.1 MPa, but their growth is diminished at low temperature or high pressure. The order parameter S of the TMA-DPH-labeled membrane in *E. coli* grown at 37 °C was much higher at a lower temperature of 15 °C than at 37 °C. S linearly increased with increasing pressure at both the temperatures. Correspondingly, the rotational diffusion coefficient D_w was much higher at 37 °C than at 15 °C and linearly decreased with increasing pressure at both the temperatures (Abe and Usui 2013).

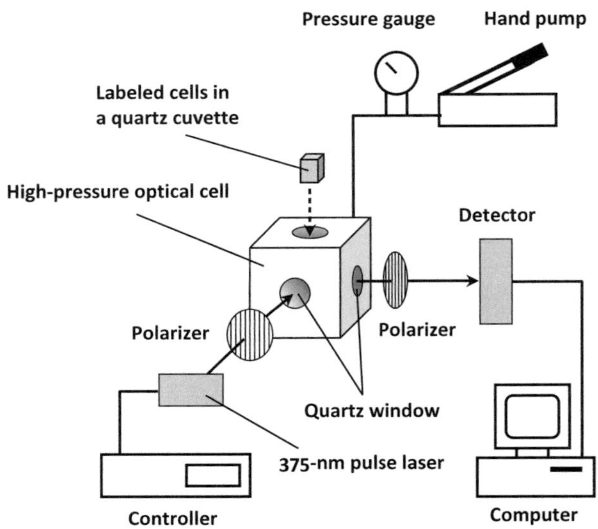

Fig. 18.1 Construction of a system employing HP-TRFAM. A high-pressure optical cell with quartz windows capable of applying hydrostatic pressures up to 200 MPa is mounted in a FluoroCube designed to perform TCSPC with a polarizing device and a 375-nm laser diode

Thus, *E. coli* cell membrane can be readily stiffened using high pressure. The degree of changes in S and D_w in the *E. coli* cell membrane with increasing pressure is analogous to those observed in 1-palmitoyl-2-oleoyl-sn-glycero-3-phosphocholine (POPC) membrane in the liquid-crystalline phase; however, the absolute values of S and D_w mutually differ between the natural and artificial membranes. In this regard, pressure dependence of the dynamic behavior is similar between the two completely different membranes as long as the POPC membrane is in the liquid-crystalline phase. However, it should be elucidated whether *E. coli* cell membrane responds to changes in pressure and temperature differentially when the cells are grown under high pressure or low temperature.

18.4 Role of Polyunsaturated Fatty Acids in Piezo-Adaptation

Polyunsaturated fatty acids (PUFAs) such as eicosapentaenoic acid (EPA; C20:5) and docosahexaenoic acid (DHA; C22:6) are commonly distributed in the membranes of deep-sea piezophiles (Fang et al. 2003; Valentine and Valentine 2004). Because of the bulky conformation, incorporation of these PUFAs into the membrane is assumed to compensate for membrane packing caused by high pressure or low temperature, thus aiding higher membrane fluidity. In some cases, the proportion of PUFAs in the cell membrane of piezophiles increases with pressure to allow their growth. However, incorporation of PUFAs into the membrane may not be always biologically relevant. The primary transition temperatures of distearoylphosphatidylcholine (DSPC), stearoyl-oleyl phosphatidylcholine (SOPC), and stearoyl-arachidonoyl phosphatidylcholine (SAPC) membranes are 55.6 °C, 6.7 °C, and −13.0 °C, respectively (Tada et al. 2010). According to the thermodynamic properties of phase transitions, volume changes (ΔV) associated with the gel-to-liquid crystalline phase are 31.6 mL mol^{-1}, 18.9 mL mol^{-1}, and 10.1 mL mol^{-1} for DSPC, SOPC, and SAPC membranes, respectively (Tada et al. 2010). The clear difference in the ΔV can be explained based on the large amount of free volumes of UFAs in the gel phase, without any significant difference in the volumes in the liquid crystalline phase. Because natural cell membranes are in liquid-crystalline-like phase, contribution of PUFAs to membranes fluidity might be small.

The role of monounsaturated fatty acids (MUFAs) and PUFAs was elucidated in *Photobacterium profundum* strain SS9, a piezophile that optimally grows at 28 MPa and 9 °C. Its membrane contains EPA at a proportion of 11 %. Whereas a mutant with defective EPA synthesis normally grows under high pressure, another mutant lacking *cis*-vaccenic acid (C18:1) exhibits marked growth defect under the same condition (Allen et al. 1999). Therefore, MUFAs but not PUFAs are required for the growth of *P. profundum* under high pressure. An EPA-deficient mutant was also created from psychrophilic but not piezophilic bacterium *Shewanella livingstonensis* Ac10 isolated from Antarctic seawater. The mutant becomes elongated at a low temperature of 4 °C and consequently shows cold-sensitive growth (Kawamoto et al. 2009). In this case, EPA plays a role in facilitating cell division at low temperature possibly by interacting with some proteins involved in cell division. Similarly, an EPA-deficient mutant was created from *Shewanella piezotolerans* WP3 isolated from the western Pacific Ocean, and the mutant showed growth defects at low temperature (4 °C and 0.1 MPa) and high pressure (20 °C and 20 MPa), indicating that EPA is required for growth under these conditions (Wang et al. 2009). *Shewanella violacea* strain DSS12 is a deep-sea bacterium isolated from the Ryukyu Trench at a depth of 5,110 m and exhibits piezophily, with optimal growth at 30 MPa and 8 °C (Nogi et al. 1998). *S. violacea* mutant lacking EPA shows growth defect at a pressure of 30 MPa and becomes elongated, indicating that EPA also plays a role in appropriate cell division under high pressure (Kawamoto et al. 2011).

The dynamic membrane structure was investigated in *S. violacea* using the HP-TRFAM system at pressures up to 150 MPa. It was shown that the membrane was extremely rigid, with high S and small D_w values even at 0.1 MPa and 10 °C (Usui et al. 2012). In a sharp contrast to the *E. coli* membrane, the S and D_w values remained almost unchanged in *S. violacea* membrane under varied pressure conditions (Usui et al. 2012; Abe and Usui 2013). This indicates that the *S. violacea* membrane is stable against changes in hydrostatic pressure while *E. coli* membrane is readily perturbed. This suggests modification of the general concept that cold- and high-pressure-adapted microorganisms have cell membranes that are more fluid. *S. violacea* mutant lacking EPA showed a compensatory increase in palmitoleic acid (C16:1) by two-fold. The cell membrane lacking EPA but containing a high proportion of palmitoleic acid was more disordered than a wild-type membrane,

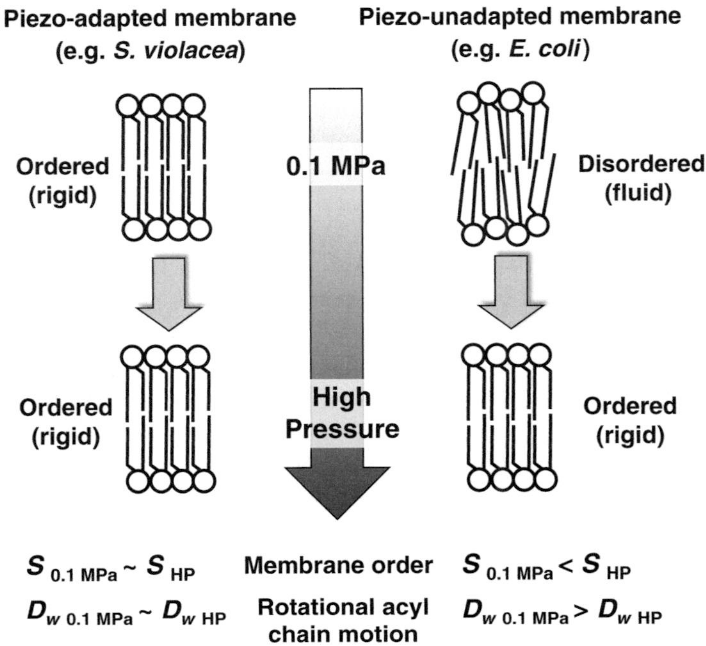

Fig. 18.2 The modified concept of piezo-adaptation in microbial cell membranes. The mechanism to compensate for the increased packing effect on the membranes is termed as "homeoviscous adaptation". However, maintaining physical properties of the membrane at a certain level, even highly rigid, under various pressure conditions could be more important for piezo-adaptation than simply increasing the fluidity of the membrane in a certain deep-sea bacteria

with reduced *S* and increase D_w values possibly occupying more spaces within the lipid bilayer (Usui et al. 2012). In addition, the EPA-lacking membrane is stiffened by increasing hydrostatic pressure, which is analogous to the *E. coli* membrane. In this situation, membrane-embedded proteins in the EPA-lacking membrane could be more perturbed by high pressure due to a structural change in the membrane. Accordingly, maintaining physical properties of the membrane at a certain level, even highly rigid, under a wide range of pressures could be more important for piezo-adaptation than simply increasing the fluidity of the membrane in a certain type of deep-sea bacteria (Fig. 18.2).

Membrane properties influence membrane-embedded proteins and vice versa. The effect of high pressure on the activity of multidrug transporter HorA in Lactobacillus plantarum was analyzed when the membrane composition, and hence the membrane physicochemical property, was controlled using varying temperatures (15 °C or 37 °C) for cell culture (Ulmer et al. 2002). In the analysis, the phase state of the membrane was determined using Fourier transform-infrared spectroscopy and 6-dodecanoyl-2-(dimethylamino)naphthalene (Laurdan) fluorescence spectroscopy. Laurdan and 6-propionyl-2-(dimethylamino)naphthalene (Prodan) are amphiphilic molecules with fluorescence excitation and emission spectra that are highly sensitive to the polarity and to the dipolar dynamic of the environment (Kusube et al. 2005). Hence, they can determine whether the membrane is in an ordered phase or in a fluid-like phase. It was observed that pressure denaturation of HorA was faster when pressure was applied to proteins embedded in a liquid-crystalline membrane than when pressure was applied to proteins in a gel-like membrane (Ulmer et al. 2002). This is in agreement with our results, which show that cell membranes of deep-sea piezophiles are rigid and imperturbable to high pressure (Usui et al. 2012). Model or natural membranes were reconstituted with multidrug transporter LmrA proteins of *Lactococcus lactis*, and the phase behavior was analyzed using the Laurdan fluorescence technique. While incorporation of LmrA increased the order parameter of dimyristoylphosphatidylcholine (DMPC, C14:0) membrane at pressures 0.1–200 MPa, it did not affect the DOPC (C18:1) membrane under the same pressure range (Periasamy et al. 2009). Particularly, incorporation of LmrA remarkably increased the generalized polarization function *GP* of natural lipids extracted from *L. plantarum*. Consequently, the reconstituted natural membrane with LmrA was much less sensitive to increasing hydrostatic pressure. The drastic change in membrane order was thought be caused by a hindering mismatch between the lipid hydrophobic region and the transmembrane segments of LmrA. Therefore, lipid–protein interaction is an important determinant of the phase behavior of the cell membrane as a function of hydrostatic pressure. Accordingly, how the membrane phase or fluidity affects the structure and function of membrane proteins and vice versa may differ with respect to the proteins, lipid composition, and pressure magnitude. Hence, there is no straightforward direction.

18.5 Conclusion

Natural cell membranes are a complex mixture of phospholipids, sterols, and numerous membrane proteins. Therefore, it is difficult to provide a straightforward account of the effect of high pressure on the phase behavior of the membranes, their structure, activity of membrane proteins, and cell growth and viability. Numerous studies on model membranes significantly contribute to the understandings of the effect of high pressure on natural cell membranes. Several excellent reviews may be consulted for further perspective (Winter and Dzwolak 2005; Winter and Jeworrek 2009; Brooks et al. 2011; Matsuki et al. 2013). In situ measurements of cell membranes under high pressure have not been properly investigated. Approaches by using HP-TRFAM or Laurdan and Prodan fluorescence technique may offer clues that link the structural aspects of membranes and membrane protein functions in a quantitative manner.

Acknowledgements This manuscript was prepared by supports of grants from the Japan Society for the Promotion of Science (No. 24580122) and the Program for the Strategic Research Foundation at Private Universities by the Ministry of Education, Culture, Sports, Science, and Technology (2013-2017).

References

Abe F (2007) Probing for dynamics of amino acid uptake in living yeast using hydrostatic pressure proceedings for the 4th international conference on high pressure bioscience and biotechnology (Eds Abe F and Suzuki A) 1: 134–138

Abe F (2013) Dynamic structural changes in microbial membranes in response to high hydrostatic pressure analyzed using time-resolved fluorescence anisotropy measurement. Biophys Chem 183:3–8

Abe F, Hiraki T (2009) Mechanistic role of ergosterol in membrane rigidity and cycloheximide resistance in *Saccharomyces cerevisiae*. Biochim Biophys Acta 1788:743–752

Abe F, Horikoshi K (2000) Tryptophan permease gene TAT2 confers high-pressure growth in *Saccharomyces cerevisiae*. Mol Cell Biol 20:8093–8102

Abe F, Iida H (2003) Pressure-induced differential regulation of the two tryptophan permeases Tat1 and Tat2 by ubiquitin ligase Rsp5 and its binding proteins, Bul1 and Bul2. Mol Cell Biol 23:7566–7584

Abe F, Minegishi H (2008) Global screening of genes essential for growth in high-pressure and cold environments: searching for basic adaptive strategies using a yeast deletion library. Genetics 178:851–872

Abe F, Usui K (2013) Effects of high hydrostatic pressure on the dynamic structure of living *Escherichia coli* membrane: a study using high-pressure time-resolved fluorescence anisotropy measurement. High Press Res 33:278–284

Allen EE, Facciotti D, Bartlett DH (1999) Monounsaturated but not polyunsaturated fatty acids are required for growth of the deep-sea bacterium *Photobacterium profundum* SS9 at high pressure and low temperature. Appl Environ Microbiol 65:1710–1720

Bartlett DH (1999) Microbial adaptations to the psychrosphere/piezosphere. J Mol Microbiol Biotechnol 1:93–100

Brooks NJ, Ces O, Templer RH, Seddon JM (2011) Pressure effects on lipid membrane structure and dynamics. Chem Phys Lipids 164:89–98

Cossins AR, Macdonald AG (1989) The adaptation of biological membranes to temperature and pressure: fish from the deep and cold. J Bioenerg Biomembr 21:115–135

de Freitas JM, Bravim F, Buss DS, Lemos EM, Fernandes AA, Fernandes PM (2012) Influence of cellular fatty acid composition on the response of *Saccharomyces cerevisiae* to hydrostatic pressure stress. FEMS Yeast Res 12:871–878

de Smedt H, Borghgraef R, Ceuterick F, Heremans K (1979) Pressure effects on lipid-protein interactions in ($Na^+ + K^+$)-ATPase. Biochim Biophys Acta 556:479–489

Denich TJ, Beaudette LA, Lee H, Trevors JT (2003) Effect of selected environmental and physico-chemical factors on bacterial cytoplasmic membranes. J Microbiol Methods 52:149–182

Fang J, Chan O, Kato C, Sato T, Peeples T, Niggemeyer K (2003) Phospholipid FA of piezophilic bacteria from the deep sea. Lipids 38:885–887

Hazel JR (1995) Thermal adaptation in biological membranes: is homeoviscous adaptation the explanation? Annu Rev Physiol 57:19–42

Kaneshina S, Ichimori H, Hata T, Matsuki H (1998) Barotropic phase transitions of dioleoylphosphatidylcholine and stearoyl-oleoylphosphatidylcholine bilayer membranes. Biochim Biophys Acta 1374:1–8

Kawamoto J, Kurihara T, Yamamoto K, Nagayasu M, Tani Y, Mihara H, Hosokawa M, Baba T, Sato SB, Esaki N (2009) Eicosapentaenoic acid plays a beneficial role in membrane organization and cell division of a cold-adapted bacterium, *Shewanella livingstonensis* Ac10. J Bacteriol 191:632–640

Kawamoto J, Sato T, Nakasone K, Kato C, Mihara H, Esaki N, Kurihara T (2011) Favourable effects of eicosapentaenoic acid on the late step of the cell division in a piezophilic bacterium, *Shewanella violacea* DSS12, at high-hydrostatic pressures. Environ Microbiol 13 :2293–2298

Kobori H, Sato M, Tameike A, Hamada K, Shimada S, Osumi M (1995) Ultrastructural effects of pressure stress to the nucleus in *Saccharomyces cerevisiae*: a study by immunoelectron microscopy using frozen thin sections. FEMS Microbiol Lett 132:253–258

Kusube M, Tamai N, Matsuki H, Kaneshina S (2005) Pressure-induced phase transitions of lipid bilayers observed by fluorescent probes prodan and laurdan. Biophys Chem 117:199–206

Matsuki H, Goto M, Tada K, Tamai N (2013) Thermotropic and barotropic phase behavior of phosphatidylcholine bilayers. Int J Mol Sci 14:2282–2302

Nogi Y, Kato C, Horikoshi K (1998) Taxonomic studies of deep-sea barophilic *Shewanella* strains and description of *Shewanella violacea* sp. nov. Arch Microbiol 170:331–338

Periasamy N, Teichert H, Weise K, Vogel RF, Winter R (2009) Effects of temperature and pressure on the lateral organization of model membranes with functionally reconstituted multidrug transporter LmrA. Biochim Biophys Acta 1788:390–401

Sato M, Kobori H, Ishijima SA, Feng ZH, Hamada K, Shimada S, Osumi M (1996) *Schizosaccharomyces pombe* is more sensitive to pressure stress than *Saccharomyces cerevisiae*. Cell Struct Funct 21:167–174

Shimada S, Andou M, Naito N, Yamada N, Osumi M, Hayashi R (1993) Effects of hydrostatic pressure on the ultrastructure and leakage of internal substances in the yeast *Saccharomyces cerevisiae*. Appl Microbiol Biotechnol 40:123–131

Sinensky M (1974) Homeoviscous adaptation–a homeostatic process that regulates the viscosity of membrane lipids in *Escherichia coli*. Proc Natl Acad Sci U S A 71:522–525

Somero GN (1992) Adaptations to high hydrostatic pressure. Annu Rev Physiol 54:557–577

Suzuki A, Mochizuki T, Uemura S, Hiraki T, Abe F (2013) Pressure-induced endocytic degradation of the *Saccharomyces cerevisiae* low-affinity tryptophan permease Tat1 is mediated by Rsp5 ubiquitin ligase and functionally redundant PPxY motif proteins. Eukaryot Cell 12:990–997

Tada K, Goto M, Tamai N, Matsuki H, Kaneshina S (2010) Pressure effect on the bilayer phase transition of asymmetric lipids with an unsaturated acyl chain. Ann N Y Acad Sci 1189:77–85

Trevors JT (2003) Fluorescent probes for bacterial cytoplasmic membrane research. J Biochem Biophys Methods 57:87–103

Ulmer HM, Herberhold H, Fahsel S, Ganzle MG, Winter R, Vogel RF (2002) Effects of pressure-induced membrane phase transitions on inactivation of HorA, an ATP-dependent multidrug resistance transporter, in *Lactobacillus plantarum*. Appl Environ Microbiol 68:1088–1095

Usui K, Hiraki T, Kawamoto J, Kurihara T, Nogi Y, Kato C, Abe F (2012) Eicosapentaenoic acid plays a role in stabilizing dynamic membrane structure in the deep-sea piezophile *Shewanella violacea*: a study employing high-pressure time-resolved fluorescence anisotropy measurement. Biochim Biophys Acta 1818:574–583

Valentine RC, Valentine DL (2004) Omega-3 fatty acids in cellular membranes: a unified concept. Prog Lipid Res 43:383–402

van Langen H, van Ginkel G, Shaw D, Levine YK (1989) The fidelity of response by 1-[4-(trimethylammonio)phenyl]-6-phenyl-1,3,5-hexatriene in time-resolved fluorescence anisotropy measurements on lipid vesicles. Effects of unsaturation, headgroup and cholesterol on orientational order and reorientational dynamics. Eur Biophys J 17:37–48

Wang F, Xiao X, Ou HY, Gai Y, Wang F (2009) Role and regulation of fatty acid biosynthesis in the response of *Shewanella piezotolerans* WP3 to different temperatures and pressures. J Bacteriol 191:2574–2584

Winter R (2002) Synchrotron X-ray and neutron small-angle scattering of lyotropic lipid mesophases, model biomembranes and proteins in solution at high pressure. Biochim Biophys Acta 1595:160–184

Winter R, Dzwolak W (2005) Exploring the temperature-pressure configurational landscape of biomolecules: from lipid membranes to proteins. Philos Transact A Math Phys Eng Sci 363:537–562, discussion 562–563

Winter R, Jeworrek C (2009) Effect of pressure on membranes. Soft Matter 5:3157–3173

Chapter 19
Homeoviscous Adaptation of Membranes in Archaea

Philippe M. Oger

Abstract Because membranes play a central role in regulating fluxes inward and outward from the cells, maintaining the appropriate structure of the membrane is crucial to maintain cellular integrity and functions. Microbes often face contrasted and fluctuating environmental conditions, to which they need to adapt or die. Membrane adaptation is achieved by a modification of the membrane lipid composition, a strategy termed homeoviscous adaptation. Homeoviscous adaptation in archaea involves strategies similar to that observed in bacteria and eucarya, such as the regulation of lipid chain length or saturation levels, as well as strategies specific to archaea, such as the regulation of the number of cycles along the isoprenoid chains or the regulation of the ratio between mono and bipolar lipids. Although not described yet described in hyperthermophilic bacteria, it is possible that these two strategies also apply to these latter organisms.

Keywords Archaeal lipids • Archaeol • Caldarchaeol • Homeoviscous adaptation

19.1 Introduction

As the first and ultimate barrier between the intracellular space and the wild-world, biological membranes play a fundamental role in the adaptation of microbes to their environments. Membranes are not just physical barriers to regulate inward and outward trafficking, they play a central role in energy storage and processing via the ion gradients, provide a matrix for environmental sensing, multicomponent metabolic and signaling pathways as well as motility. Maintaining optimal membrane function is therefore crucial for the cell. Thus, temperature-, hydrostatic pressure-, or pH-induced perturbations in membrane organization pose a serious challenge. Bacteria and eucarya regulate membrane fluidity by regulating the level of unsaturations

P.M. Oger (✉)
Laboratoire de Géologie de Lyon, UMR CNRS 5276, Ecole Normale Supérieure de Lyon, 46, Allée d'Italie, F-69342 Lyon, France
e-mail: poger@ens-lyon.fr

K. Akasaka, H. Matsuki (eds.), *High Pressure Bioscience*,
Subcellular Biochemistry 72, DOI 10.1007/978-94-017-9918-8_19

and the length of the lipid acyl chains or incorporate lipids with different polar head groups, an adaptation referred to as homeoviscous adaptation, which has been reviewed recently (Russell and Nichols 1999). This chapter presents evidence for membrane adaptation in archaea as a function of environmental constraints, and discuss about the possibility and limits of homeoviscous adaptation in archaea.

19.2 Structure and Properties of Archaeal Lipids

Archaeal membranes are essentially composed of phospho-, glyco- and phospho-glyco-lipids which differ substantially from their bacterial counterparts (Koga and Morii 2005; Koga 2011; Koga and Morii 2007; Matsumi et al. 2011). Archaeal polar lipids are composed of isoprenoid hydrocarbon chains bound by ether bonds to the *sn*-2 and *sn*-3 positions of glycerol (Fig. 19.1). Polar head groups consist of phosphatidyl or sugar moieties linked on the *sn*-1 position of the glycerol backbone (G-1 position). Thus, archaeal phospholipids have a G-1-P structure, which is a mirror image stereoisomer of the *sn*-glycerol-3-phosphate or G-3-P backbone of bacterial phospholipids. It is noteworthy that this enantiomeric difference, although central to the divergence between Archaea and the Bacteria/Eukarya in terms of evolution, e.g. the lipid divide (Koga and Morii 2007), is insignificant in terms of chemical behavior, since enatiomers have essentially equivalent physical behavior. Some archaeal lipids may have alternate backbone structures, in which the glycerol molecule is substituted to a branched nonitol molecule, such as calditol (Fig. 19.1d), onto which the two isoprenoid polar hydrocarbon chains are bound (De Rosa et al. 1980c; Nicolaus et al. 1990).

In addition to the G-1-P backbone structure, archaeal lipids have other unique characteristics. The core hydrocarbon chains of archaeal glycerol polar lipids are composed exclusively of isoprenoid hydrocarbons most often made up of a phytanyl chain (20 carbons, or C20, Fig. 19.1a, b) or a head-to-head condensation dimer of two phytanyl chains, e.g. a biphytanyl chain (40 carbons, C40, Fig. 19.1c-g). C20-based archaeal lipids are glycerol-dialkyl-glycerol-diether (GDGD or archaeol and archaeol derivatives), while C40-based lipids are glycerol-dialkyl-glycerol-tetraether (GDGT or caldarchaeol and derivatives) (Nishihara and Koga 1987). Archaeol and caldarchaeol and derivatives represent the vast majority of archaeal lipids (Table 19.1). Monopolar lipids organize in a bilayer similar in structure to that of bacteria or eucarya. The bipolar lipids form monolayers in which the molecule is fully stretched and spans the entire membrane thickness (De Rosa et al. 1983). Thermoacidophiles and especially the members of the *Sulfolobales*, present a mixture of GDGT and an asymmetric tetraether lipid which harbors a glycerol backbone at one end of the hydrophobic core and a calditol group at the other end, which is referred to as glycerol-dialkyl-nonitol-tetraether (GDNT, Fig. 19.1d) (De Rosa et al. 1980a; De Rosa et al. 1980c; Gulik et al. 1985). Archaeal lipids containing 1 or 2 C25 isoprenoid chains have been described in several species (De Rosa et al. 1982). Halobacteriales such as *Haloterrigena, Halococcus,*

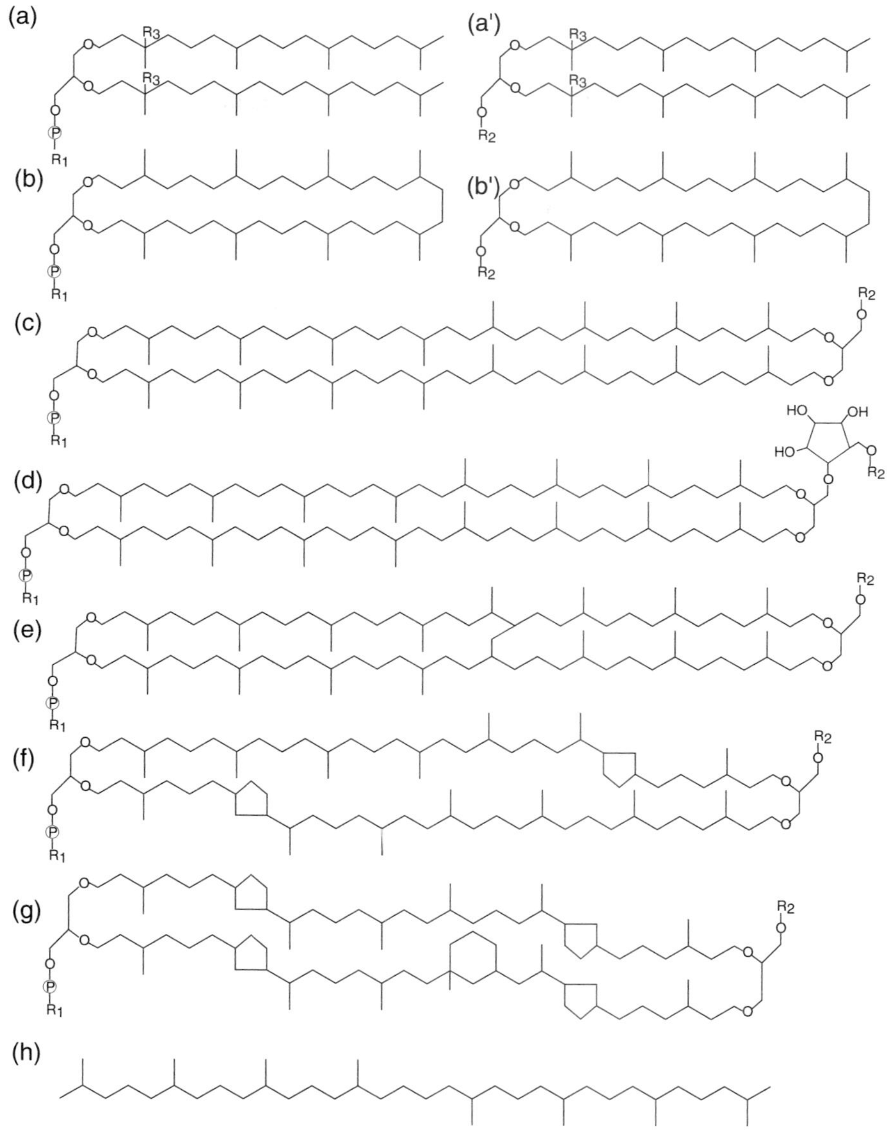

Fig. 19.1 Structures of archaeal lipids (**a**) diphytanylglycerol (archaeol: archaeal diether lipid); (**b**) cyclic archaeol; (**c**) caldarchaeol (archaeal tetraether lipid); (**d**) calditoglycerocaldarchaeol; (**e**) H-shaped caldarchaeol; (**f**) cyclopentane-containing caldarchaeol; (**g**) crenarchaeol; (**h**) lycopane-type isoprenoid hydrocarbons. R1, R2, R3: polar head groups. The headgroups of phospholipids can be a range of polar compounds – for example, glycerol, serine, inosine, ethanolamine, myo-inositol or aminopentanetetrols (see Table 19.1). Glycolipids also exhibit a range of sugar residues – for example, glucose, mannose, galactose, gulose, N-acetylglucosamine or combinations thereof (Adapted from Oger and Cario 2013)

Table 19.1 Temperature and pH optima, percentage or diether (C20), tetraether (C40) and H-shape tetra ether (H-C80) lipids, type of core lipids, polar head groups for the major groups of Crenarcheota and Euryarchaeota. Phosphatidyl headgroups : PI, phosphatidylinositol; PG, phosphatidylglycerol; PE, phosphatidylethanolamine; PGS, phosphatidylglycerylsulphate; PGP-Me, phosphatidylglycerylphosphate methyl ester (Oger and Cario 2013)

Class	pH	T	C20	C40	H-C80	Lipid type	Polar head
Acidolobales	4	75	traces	100		Cyclic and acyclic tetraethers	
Desulfurococcales	5.5–7.6	85–100	0–80	20–100		Tetraether with up to four cyclopentane rings	PI; galactose, mannose, glucose; mono and di glycosides
Sulfolobales	2–4	75–87	0	100		Caldarchaeol, calditoglycerocaldarchaeol; tetraethers with up to six cyclopentane rings	PG
Thermoproteales	4–7	75–100	0–5	95–100		Tetraether with up to four cyclopentane rings	PI; β-D-glucose
Methanobacteriales	6–7.5	65–85	9–17	83–91	31	31 % of an H-shaped caldarchaeol derivative	
Methanobacteriales	7–7.5	37–39	37.5–52	48–62.5			
Methanococcales	6	85	100	0		15 % Archaeol; 85 macrocyclic Archaeol	PE; glucopyranose
Methanococcales	7	25–35	100	0			
Methanomicrobiales	7,4	37	41	59			
Methanosarcinales	6.5–7.5	25–37	100	0		Archaeol	

(continued)

Table 19.1 (continued)

Class	pH	T	C20	C40	H-C80	Lipid type	Polar head
Halobacteriales	7–10	25–40	100	0		C20-C20, C20-C25	PG, PGS, PGP-Me; monoglycosyl, sulphate ester of adiglycosyl
Methanopyrales	6,5	97	100	0			mannose, glucose, galactose, N-acatylglucosamine
Thermococcales	6–8	83–101	0–100	0–100	0–48	Archaeol and caldarchaeol derivatives	PI
Thermoplasmatales	0.7–3	59–60	0	100		Tetraether with up to five cyclopen-tane rings	PG; glucose, gulose

Natronobacterium or Natronomonas produce C20–C25 diether lipids (Kamekura and Kates 1999), while the hyperthermophile *Aeropyrum pernix* (Morii et al. 1999; Sako et al. 1996) or methanogens such as Methanobacter *thermoautotrophicus* (Koga et al. 1993) produce C25-C25 diether lipids.

Additional modifications on the core isoprenoid chains include the presence of unsaturations at position C3 of the phytanyl chain of archaeol, the presence of hydroxy groups at the same position (hydroxyarchaeol) (Nichols et al. 2004), the presence of 1–4 cyclopentane rings along the biphytanyl chains of caldarchaeol (Fig. 19.1f) or the presence of cyclopentane and hexapentane rings as in crenarchaeol (Fig. 19.1g). The number of cyclopentane rings in the caldarchaeol derivatives will depend on the species. Tetraethers of *Thermoplasma* or *Sulfolobus* may contain up to four cyclopentane rings (De Rosa and Gambacorta 1988; Sugai et al. 1995) while those in *Archaeoglobus* contain only up to two cycles (Lai et al. 2008; Tarui et al. 2007). Unsaturated analogues of archaeol have been reported in several species, especially amongst the halophilic archaea (Franzmann et al. 1992; Moldoveanu and Kates 1988; Moldoveanu et al. 1990; Qiu et al. 2000; Sprott et al. 2003; Upasani et al. 1994). Although long thought to be the result of technical and analytical difficulties, the presence of these unsaturated lipids is now well established in at least three different types of archaea, a mesophilic methanogen, *Methanobacterium burtonii* (Franzmann et al. 1992), a hyperthermophilic methanogen *Methanopyrus kandleri* (Hafenbradl et al. 1993), and a member of the Thermococcales, *Thermococcus sp.* S557 (Gonthier et al. 2001). Macrocyclic archaeol (Fig. 19.1b), a derivative of archaeol in which the two phytanyl chains are bridged by a covalent bond was first described in *Methanocal-*

dococcus jannaschii, a thermophilic methanogen, in which it represents ca. half the membrane lipids (Comita and Gagosian 1983; Comita et al. 1984). Similarly, a H-shaped cross-linked derivative of caldarchaeol (Fig. 19.1e) was shown to represent 30 % of the core lipids in another methanogen, *Methanothermus fervidus* (Morii et al. 1998). Numerous derivatives of the above structures with cycles or bridges between the two isoprenoid chains have been identified in different Archaea. These include but are not limited to archaeol with cyclopentane rings or archaeol and macrocyclic archaeol with bridges at position C2 of the isoprenoid chains.

Two major types of polar headgroups are found in archaeal lipids. The first ones consist of phosphodiester-linked polar head groups, while the second only exhibit a sugar moiety. Bipolar archaeal lipids are asymmetrical, harboring a phosphatidyl moiety at one end of the lipid and a sugar moiety at the other end (Morii and Koga 1994). Phosphodiester-linked polar head groups are very similar to that observed in bacteria, with phosphatidylserine, phosphatidylcholine, phosphatidylethanolamine, phosphatidylinositol, phosphatidylglycerol, phosphatidylglycerophosphate, phosphatidylglycerosulfate, etc. (Koga and Morii 2005). There is a large diversity in sugar moieties, with molecules harboring one, two, or three units of different sugars such as gulose, glucose, galactose (Table 19.1) or other carbohydrates such as glycerol, inositol, glycerate, etc. The diversity of polar lipids in a single archaeon can be important, with species displaying more than 25 different lipid molecules, each of which may also present various unsaturation level or cyclopentane rings number (Nishihara and Koga 1987; Nishihara et al. 1987). Bipolar lipids, which harbor hydrophilic moieties on both ends and a long hydrophobic central domain can form monolayer membrane, which contrast with the typical lipid bi-layer observed in most organisms, including bacteria and eukarya. Bipolar lipids are oriented in the membranes with the phosphatidyl moiety inside the cell and the bulkier sugar moiety pointed towards the outside (Morii and Koga 1994).

Similar to what is observed in bacteria and eucarya, the lipid composition is typical of each archaeal family or species, although to date there is only limited information on this issue (Table 19.1). In halophiles, e.g. salt-adapted archaea, the polar lipids are mainly comprised of diether lipids with phytanyl (C20), or sesterterpenyl (C25) isoprenoid chains. In methanogens, acidophiles (low pH-adapted) and hyperthermophiles (high-temperature-adapted), the polar lipids typically consist of a mixture of diphytanylglycerol diether and dibiphytanyl tetraether lipids, with or without cyclopentane rings. It is clear that the diversity of archaeal lipids is actually largely unknown and might include a larger diversity of polar headgroups as well as modifications along the isoprenoid chains.

In addition to polar lipids, archaea were shown to produce several irregular isoprenoid hydrocarbons of the lycopene/squalene structure, with a chain length of 15–50 carbons (Fig. 19.1h) (Sprott et al. 1991; Tornabene et al. 1979; Manquin et al. 2004). Lycopene-type isoprenoid chains are derived from a tail-to-tail condensation of two poly-isoprenoid chains, which are not likely derived from the normal lipid synthesis pathway. The presence of straight chain fatty acid-based lipids in Archaea reported in several studies remains highly questionable because they can often be tracked down to the growth medium used for the cultures (Carballeira et al. 1997; Nishihara et al. 2000).

Most known Archaea have been isolated from the most extreme environments. It has been suggested that many of their cellular features have adapted to maintain the integrity and form of the cells to resist the extremes in pH, temperature, salinity or pressure experienced in these environments. Indeed, the ether linkages are more chemically stable than esther linkages to oxidation or pH (Mathai et al. 2001; Yamauchi et al. 1993). Archaeal diether lipid based bilayer membranes and tetraether monolayers exhibit a very high thermal stability (Elferink et al. 1994; Dante et al. 1995). The monolayer organization of the tetraether lipids provides extreme rigidity to these membranes. Archaeal lipid membranes have, in general, much lower phase transition temperature than fatty acyl ester lipids (Yamauchi et al. 1993). While membranes made of fatty acyl ester lipids are in the gel phase or in the liquid crystalline phase depending mostly on their fatty acid composition, archaeol- and caldarchaeol-based polar lipid membranes of archaea are assumed to be in the liquid crystalline phase over a wide temperature range, e.g. 0–100°C (Dannenmuller et al. 2000; Stewart et al. 1990). Small phase transitions have been observed in liposomes of *S. acidocaldarius* between 42°C and 69°C (Chong et al. 2005; Chong et al. 2003). These phase transitions are associated to very small volume changes, which suggests that the polar headgroup region of the lipid may still be rigid and tightly packed through hydrogen-bond network at elevated temperatures. Lateral mobility studies of bipolar tetraether lipids from *S. acidocaldarius* (Khan and Chong 2000; Morii et al. 1999) or *T. acidophilum* (Jarrell et al. 1998) demonstrate that lateral diffusion rate under physiological temperature is in the order of 2.10^{-8} cm s^{-1}, comparable to those of diacylglycerophospholipids in the liquid crystalline phase at 37°C for *E. coli*. Thus, despite the difference in lipids structure and composition and growth temperature, the two types of membranes display very similar properties.

19.3 Evidence of Homeoviscous Adaptation in Archaea

Because membranes act as physical barriers to regulate inward and outward trafficking, play a central role in energy storage and processing via the ion gradients, provide a matrix for environmental sensing, multicomponent metabolic and signaling pathways as well as motility, maintaining optimal membrane biological function is crucial for any organism. Based on the observation that the membrane lipids of *E. coli* cells grown under the contrasting temperatures of 43°C and 15°C were different (Marr and Ingraham 1962; Sinensky 1971), but the respective membranes displayed similar physical properties at their respective growth temperature (Table 19.2), Sinensky posed the basis of homeoviscous adaptation (Sinensky 1974). This theory states that membrane lipids composition variations in response to temperature (or other stresses) in a given organism favor the maintenance of the appropriate membrane fluidity for it to function optimally. Since its inception the notion of

Table 19.2 The viscosity and phase transition of *E. coli* lipid extracts and intact membranes from cells grown at different temperatures (Adapted from Sinensky 1974)

	Temperature of growth(T_G)	Temperature of analysis	rotational correlation time τ (ns)	coefficient of viscosity η (poise)	Phase transition Temperature (T_T)	$T_G - T_T$
Lipid extracts	15	15	2.8	1.8	-1 ± 1	16 ± 1
	30	30	2.7	1.9	16 ± 2	14 ± 1
	37	37	2.6	1.8		
	43	43	2.7	2.0	27 ± 1	16 ± 1
	43	15	13.8	15		
Membranes	23	23	3.5	2.5		
	23	37	1.6	1.0		
	37	37	3.3	2.5		

homeoviscous adaptation has been extended to include the notion of water and ion permeability and to accommodate for a dynamic rather than static model of the membrane (McElhaney 1984a, b; van de Vossenberg et al. 1999; van de Vossenberg et al. 1998). Homeoviscous adaptation of the membranes is made possible by the difference in physico-chemical behavior of membrane lipids which is dependent of several parameters such as the nature of the polar head groups, the length of the hydrocarbon chain length, the number of unsaturations along that chain. It is not the point of this chapter to address this issue in detail, and excellent reviews have been published to address the physical behavior of lipids as a function of composition, temperature or pressure conditions (Zein and Winter 2000; Winter 2002). For example, an increase in chain length by two carbons will increase the phase transition temperature of the lipid by 10–20°C, and decrease membrane permeability to proton and water (Winter 2002). The incorporation of a single unsaturation can shift the fluid/gel phase transition by 10 to 20°C (Russell and Nichols 1999; Winter 2002). Similarly, a change in the proportion of polar headgroups PE relative to PC lipids will increase the order in the membrane for lipids of similar acyl chain composition (Winter 2002; Winter and Jeworrek 2009). This is due in part to the reduced hydration and stearic bulk of the ethanolamine compared to choline, and the capacity of PE, and incapacity of PC, groups to form hydrogen bonds.

There is a body of evidence suggesting that archaeal membrane lipid compositions vary with environmental constraints, and that these variations may be interpreted within the framework of homeoviscous adaptation. The impact of temperature variations on the archaeal membrane composition was first studied in *Sulfolobus acidocaldarius* an hyperthermophilic acidophile. The lipids of *S. acidocaldarius* are limited to bipolar tetraethers of different head groups containing up to four cyclopentane rings (De Rosa et al. 1980a). The mean number of cyclopentane rings increases from 1.94 to 2.52 from 75°C to 89°C (Table 19.3).

Table 19.3 Effect of growth temperature, hydrostatic pressure, pH, and growth phase on the composition of archaeal membrane lipids. T_g: growth temperature in °C, P_g: hydrostatic pressure in MPa, pH_g: growth pH

Species	T_{OPT}(°C)	Stress	Range	Mechanism	Reference
Methanococcoides burtonii	23	T_g	4–23	Twofold increase of the mean number of unsaturationTwofold increase of the proportion of unsaturated lipids	(Nichols et al. 2004)
Archaeoglobus fulgidus	83	T_g	70–90	Increase in tetraether over diether lipids from 30 to 90 %Increase in the mean number of cyclopentane rings	(Lai et al. 2008)
Thermococcus kodakarensis	85	T_g	60–93	Increase in tetraether over diether lipids from 34 to 66 %	(Matsuno et al. 2009)
Methanocaldococcus jannaschii	85	T_g	47–75	Decrease of the tetraether/diether lipid ratioIncreased proportion of macrocyclic archaeol	(Sprott et al. 1991)
Sulfolobus acidocaldarius	87	T_g	75–89	1 unit increase of the mean number of cyclopentane rings	(De Rosa et al. 1980b)
Thermoplasma acidophilum	55	T_g	39–59	0.5 unit increase in the mean number of cyclopentane rings	(Ernst et al. 1998)
Pyrococcus horikoshii	98	T_g	82–103	0.2 unit increase of the mean number of cyclopentane rings	(Sugai et al. 2000; Chong 2010)
Picrophilus torridus	55	T_g	45–62	0.5 increase of the mean number of cyclopentane rings	(Uda et al. 2004)

(continued)

Table 19.3 (continued)

Species	T_{OPT}(°C)	Stress	Range	Mechanism	Reference
Thermoplasma vulcanium	60	T_g	45–62	1.2 units increase of the mean number of cyclopentane rings	(Uda et al. 2004)
Picrophilus oshimae	55	T_g	45–62	0.5 unit increase of the mean number of cyclopentane rings	(Uda et al. 2004)
Ferroplasma acidophilum	35	T_g	35–45	0.2 unit increase in the mean number of cyclopentane rings	(Uda et al. 2004)
Acidilobus sulfurireducens	81	T_g	65–81°C	0.5 unit increase of the mean number of cyclopentane rings	(Boyd et al. 2011)
Methanocaldococcus jannaschii	85	P_g	0.1–50 MPa	Decrease of the proportion of tetraether and diether from 46 to 36 %Increase in the proportion of macrocyclic diether from 36 to 64 %	(Kaneshiro and Clark 1995)
Thermoplasma acidophilum	59	pH_g	1.2–3.0	1.1 unit increase in the mean number of cyclopentane rings	(Shimada et al. 2008)
Acidilobus sulfurireducens	81	pH_g	3.0–5.0	0.5 unit increase of the mean number of cyclopentane rings	(Boyd et al. 2011)
Thermococcus kodakarensis	85	Growth phase	Log vs. stationary	Increase of the tetraether over diether ratio(Mean increase between 20 and 45 %)	(Matsuno et al. 2009)

Incorporating cyclopentane rings is expected to increase the gel/fluid phase transition temperature of the lipid (Gliozzi et al. 1983). It will increase the packing efficiency of the membrane lipids, which in turn increases membrane stability as a function of increasing temperature or salinity, and decreasing pressure or pH, and consequently lowers the permeability (Chong et al. 2012). In addition, each such cyclization reduces the available modes of movement of the lipids, further increasing membrane rigidity. Thus, it has been proposed that an increase in the number of cyclopentane rings is a homeoviscous strategy in archaea to help maintain membrane stability, help cope with the increase in molecule agitation and membrane expansion as the temperature increases (De Rosa et al. 1980b). Similar temperature-dependent increase in cyclopentane ring numbers have been reported for several archaeal species (Table 19.3). In *Thermoplasma acidophilum, Pycrophilus oshimae and Picrophilus torridus*, which grow optimally at 55°C, the mean number of cyclopentane rings increases by ca. 0.5 unit from 39°C to 60°C (Ernst et al. 1998; Uda et al. 2001; Uda et al. 2004). This increase reaches 1.2 units for *Thermoplasma volcanium*, which optimal growth temperature is 60°C, but is only of 0.2 units *Ferroplasma acidophilum* which optimum is the lowest at 35°C. In archaea harboring mixed diphytanyl diether and dibiphytanyl tetraether lipids, the tetraether-to-diether lipid ratio has been shown to vary with growth temperature (Lai et al. 2008; Matsuno et al. 2009; Sprott et al. 1991). In *Archaeoglobus fulgidus* the tetraether to diether core lipid ratio increases from 0.3 for cells grown at 70°C to 0.9 for cells grown at 89°C. It is expected that the increase in the tetraether lipid will stabilize the membranes by forming monolayer-type membranes or domains in the membrane. This putative homeoviscous adaptation strategy is not limited to *Archaeoglobus*, since in the *Thermococcales* species *Thermococcus kodakarensis*, the tetraether lipid composition of cells in the exponential phase is also shifted up from 36 % at 60°C to 66 % at 93°C with increasing temperature (Matsuno et al. 2009). It is thus very probable that the homeoviscous adaptation in archaea harboring mixed C20-C40 membrane lipids will involve the regulation of the ratio of these lipids to accommodate with the variations in temperature. Variations of the tetraether/diether ratio is also observed in the hyperthermophilic methanogen *Methanocaldococcus jannaschii*. In this later strain, tetraether proportion was shown to decrease with temperature rather than increase. However, the lipid composition of that strain is abruptly shifted towards another type of membrane lipid, the macrocyclic derivative of archaeol (Fig. 19.1b). This increase in macrocyclic archaeol proportions constitutes the third homeoviscous adaptive mechanism existing in Archaea. It has been shown that the covalent bound between the two free ends of the isoprenoid chains of the archaeol in macrocyclic archaeol reduces the motion of the molecule drastically, to values close to that observed for tetraether lipids (Dannenmuller et al. 2000). In addition, macrocyclic archaeol has been demonstrated to be an efficient barrier against proton and solute leakage under high temperature. Thus, increasing the proportion of macrocyclic archaeol is expected to have similar impacts on membrane stability and permeability as an increase in the tetraether/diether ratio.

Homeoviscous adaptation in archaea may proceed along a fourth route. *Methanobacterium burtonii* is the first archaeal eurypsychrophile described with a relative broad growth temperature range (–2.5 to 29°C) and an optimal growth temperature of 25°C. It has been isolated from cold (2°C), anaerobic, methane saturated waters from the antarctic Ace Lake (Franzmann et al. 1992). Membrane of *M. burtonii* contain 4 different diether lipids with a phosphatidylinositol (PI) or a phosphatidylglycerol (PG) polar head. The archaeol core lipid may contain an hydroxyl group (Ar-OH) at position C-3 of the isoprenoid chain on the *sn*-3 position of the glycerol backbone. In addition, these lipids are polyunsaturated with up to five unsaturations. Under a decrease in temperature from 23°C to 4°C, the mean number of unsaturation per lipid molecule is increasing almost two-fold, regardless of the lipid core (ArPI, ArPG, Ar-OHPI, Ar-OHPG). Thus the strategy adopted by *M. burtonii* is similar to that adopted by many psychrophilic bacteria (Russell and Nichols 1999). It is noteworthy that the presence of the hydroxyl group is also likely to modify the properties of the lipid in the membrane and contribute to increased fluidity. Indeed, the presence of the hydroxyl group in close proximity to the head group might increase the surface area and shorten the chain length (Sprott et al. 1990). Chain-length regulation is a common homeoviscous adaptive strategy in bacteria, in which a reduction in length is associated with increase membrane fluidity (Russell and Nichols 1999).

Membrane lipid composition in archaea not only varies with temperature conditions for growth. Indeed, shifting conditions of pressure, pH or salinity induce essentially the same cellular response. To date, there is only one report of homeoviscous adaptation in archaea in response to high hydrostatic pressure. It was shown that in *Methanocaldococcus jannaschii*, the proportion of tetra ether decreased from 46 to 36 % when the cells were grown under 50 MPa in comparison to cells grown under atmospheric pressure (Kaneshiro and Clark 1995). Macrocyclic archaeol is the most abundant lipid in this organisms under shifting pressure and temperature conditions. Concomitantly, there is an increase of macrocyclic diether from 36 to 64 %. Membrane lipid composition has also been shown to vary substantially in *Thermoplasma acidophilum* as a response to pH increase. In this strain, at 55°C growth temperature, each caldarchaeol molecule contains 5.1 cyclopentane rings at pH 3.0, 4.8 rings at pH 2.4, 4.1 rings at pH 1.8 and 4.0 at pH 1.2 (Shimada et al. 2008). This increase of cyclopentane rings in the tetraether lipids mirrors that observed for an increase in temperature (0.5 unit for 20°C increase). One should note that the homeoviscous adaptation strategies described above are not mutually exclusive. Indeed, the hyperthermophilic methanogenic archaeon, *Methanocaldococcus jannaschii*, controls both the proportions of its various di- and tetraether lipids, but also the degree of cyclization of its archaeol derivatives and the proportion of non polar lipids such as lycopene-type hydrocarbons in response to temperature and hydrostatic pressure variations (Sprott et al. 1991). In *Archaeoglobus fulgidus*, the drastic variation of the tetraether/diether ratio observed with a shift from 70°C to 90°C is accompanied by the increase of the mean number of cyclopentane rings in the tetraether lipids from one to two cycles (Lai et al. 2008).

19.4 The Limits of Homeoviscous Adaptation in the Archaea

Mechanistically, the archaeal homeoviscous adaptation resembles in part that of bacteria and eukarya, e.g. the regulation of the unsaturation level of membrane lipids in the psychrophile *Methanobacterium burtonii*. Other mechanisms are specific to Archaea because their specificity takes root in the nature of the archaeal lipids, e.g. the existence of bipolar ether-linked transmembrane lipids. All lipid composition fluctuations presented above can be explained within the framework of homeoviscous adaptation taken *sensu largo* to include the regulation of membrane permeability. Yet, when one considers the lipid compositions of the different archaeal species, several membrane lipid compositions may prove more difficult to interpret within this framework. We will use here four examples to illustrate this point.

As mentioned above, tetraether synthesizing archaea have been shown to regulate the number of the cyclopentane rings in their tetraether as a function of temperature or pH. Archaea that have cyclopentane rings containing tetraethers may grow over a large range of temperature, e.g. from 59°C for the thermoacidophile *Thermoplasma acidophilum* (Uda et al. 2001; Uda et al. 1999) up to 100°C for *Hyperthermus butylicus* (Zillig et al. 1990; Zillig et al. 1991) and a large range of pH, from 0.7 for *Picrophilus acidophilum* (Schleper et al. 1995) to 7 for *Hyperthermus butylicus* (Zillig et al. 1990; Zillig et al. 1991). Despite the observation of a temperature- or pH-induced increase in the number of cyclopentane rings, it is difficult to determine a clear correlation between the number of ring and the temperature or the pH. For example, *Hyperthermus butylicus*, which optimal growth temperature is 100°C, was shown to accumulate up to two cyclopentane rings, while others such as *Sulfolobus solfataricus* will accumulate lipids with up to six rings when growing optimally at 87°C. It is noteworthy that the difference between the lipid composition of these two species is actually more important than that observed in a given species to compensate a similar difference of temperature ($\Delta T = 20°C$, Table 19.3). Overall, the composition in tetraether lipids in each species seems to reflect its position in the taxonomic tree, rather than its optimal growth characteristics. This later view is supported by several reports on the presence of tetraether with or without cyclopentane rings in mesophilic and cold environments and cold-adapted archaea, and especially from marine environments (DeLong et al. 1998). The number of cyclopentane rings reported is in the same order, e.g. up to four cyclopentane rings, as those observed for many hyperthermophiles or thermoacidophiles. This composition would predict a very stable, but fairly rigid, and impermeable membrane. Thus it is yet not clear how membrane viscosity is regulated in these low temperature-adaptated organisms.

Similar conclusions can be drawn from the comparison of the growth optima of archaea that are able to synthesize a single type of ether lipid (Table 19.4). Considering the diversity of Archaea only able to synthesize monopolar diether lipids, one can see two populations exhibiting divergent temperature optima: the first set is mesophile, composed essentially of methanogens and halophiles with

Table 19.4 Temperature and pH optima for archaea harboring only diether lipids in their membranes (Oger and Cario 2013)

Genus	Species	pH	T	Ar	CalAr
Methanococcus	*vannielli*	7.2	25	100	0
Methanolobus	*tindarius*	6.5	25	100	0
Halorubrum	*lacuprofundi*	7.4	25	100	0
Methanococcus	*voltae*	7	35	100	0
Methanosaeta	*concilii*	7.5	35	100	0
Methanosarcina	*barkeri*	7	37	100	0
Methanohalophilus	*mahii*	7.5	37	100	0
Methanosaeta	*soehngenii*	7.5	37	100	0
Methanococcus	*jannaschii*	6	85	100	0
Thermococcus	*barophilus*	7	85	100	0
Thermococcus	*celer*	5.8	88	100	0
Aeropyrum	*pernix*	7	90	100	0
Methanopyrus	*kandleri*	6.5	97	100	0

optima ranging from 25 to 37°C, and the second is hyperthermophilic with optima ranging from 85 °C to 98 °C. Halophiles synthesize a mixture of C20-C20 and C20-C25 ether lipids with a particular polar headgroup (discussed below), *Methanocaldococcus jannaschii* synthesizes a macrocyclic derivative of archaeol (discussed above), while all other species synthesize archaeol. Archaeol, the C20-C25, or the C25-C25 diether lipids define membranes with higher permeability and fluidity at low temperature than tetraether lipids, which justifies their synthesis in mesophiles. It is more difficult to explain archaeol-based membranes in hyperthermophiles such as *Thermococcus barophilus*, *Thermococcus celer*, *Aeropyrum pernix* or *Methanopyrus kandleri*, which optimum growth temperature ranges from 85°C to 98°C, within the framework of the homeoviscous adaptation. It is also clear in this example that the lipid composition in these organisms reflects their taxonomic position as is observed with cyclic tetraether lipids, and that other yet undetermined factors are explaining the compatibility of the lipid composition with the optimal growth parameters.

If there is a correlation between the phylogenetic position and the membrane lipid composition, related organisms may yet exhibit a large diversity of lipid composition when grown in the same conditions, e.g. in the same medium, at the same pH and temperature (Sugai et al. 2004). Archaea from the *Thermococcales* were all isolated from hydrothermal systems, and exhibit very close physico-chemical growth requirements. Most *Thermococcus* species have growth temperature optima close to 85°C, while *Pyrococcus* usually grow at temperature close to 100°C. It has long been known that *Thermococcales* could synthesize both di- and tetraether lipids (Lattuati et al. 1998; Carballeira et al. 1997; Lanzotti et al. 1989; Sprott et al. 1997). As discussed above, homeoviscous adaptation in this group involves the regulation of the ratio of these two lipid types (Matsuno et al. 2009). However, species of this group exhibit a large range of lipid compositions (Table 19.5). Amongst the

Table 19.5 Membrane lipid composition of *Thermococcales* species grown at 85°C (Sugai et al. 2004)

		C20	C40	H-C80
Thermococcus	*aggregans*	42.1	57.9	
Thermococcus	*zilligii*	33.2	66.8	
Pyrococcus	*furiosus*	31.2	68.8	
Thermococcus	*fumicolans*	27.3	72.7	
Thermococcus	*aegeus*	27.2	72.8	
Thermococcus	*litoralis*	26.8	73.2	
Pyrococcus	*woesei*	26.5	73.5	
Thermococcus	*pacificus*	25.4	74.6	
Thermococcus	*gorgonarius*	16.8	83.2	
Thermococcus	*profundus*	16	84	
Thermococcus	*peptonophilus*	14.3	85.7	
Palaeococcus	*ferrophilus*	12.2	87.8	
Thermococcus	*stetteri*	9.6	90.4	
Thermococcus	*guaymasensis*	12.7	75.3	12
Thermococcus	*celer*	38.8	45.7	15.5
Pyrococcus	*horikoshii*	5.9	60.1	34
Thermococcus	*waiotapuensis*	15.2	35.9	48.9

17 strains tested, most strains synthesized archaeol and caldarchaeol, but 4 also synthesized an H-shaped caldarchaeol derivative (Sugai et al. 2004). In this study, the measured proportion of di-ether ranged from 42.1 % in *T. aggregans* to 5.9 % in *P. horikoshii*, while the proportion of H-shaped caldarchaeol reached 48.9 % in *T. waiotapuensis*. As noted above, the proportion of di-ether may even be higher, since it has been shown to be the only lipids found in *T. barophilus* (Marteinsson et al. 1999). Once again, it is striking that the variations observed in this group of phylogenetically related organisms grown in the same conditions are larger than those evidenced upon thermal, pH or high pressure treatment.

Last, another interesting archaeal lipid is the novel tetraether containing four cyclopentane rings and one cyclohexane ring, named crenarchaeol, which has been found exclusively in crenarchaea, and especially in marine planktonic cells (Schouten et al. 2000; Damste et al. 2002). These environments have very low temperature, ca 1–2 C in average. Thus it is quite difficult to understand why such low-temperature adapted organisms would form membrane with lipids containing five cycles. Molecular dynamics simulations suggested a possible explanation to this paradox, by showing that the cyclohexane ring decreases membrane density, which could be a cold adaptation strategy. The later discovery of the predominance of crenarchaeol in hot spring crenarchaea would argue against such hypothesis, and rather favor the synthesis of crenarchaeol as a crenarchaea specific lipid as seen for other Archaea (Pearson et al. 2004; Pitcher et al. 2010).

19.5 Conclusions

There is conclusive evidence that archaea use mechanisms resembling the homeoviscous adaptation observed in bacteria and eukarya. When these latter essentially regulate the length and the saturation level of their lipid chains to regulate membrane fluidity, archaea because of their specific lipids have to adopt different strategies, e.g. the incorporation of cyclopentane rings into the isoprenoid chains or the modification of the proportions of diether vs tetraether lipids. Despite these differences, the impact on membrane behavior is essentially the same, to regulate the packing of lipids as a function of fluctuations in temperature, or other environmental parameters. Homeoviscous adaptation offers a convenient and simple conceptualization to explain the variations of the composition of membrane lipids. Yet, several aspects remain difficult to explain within this framework. The most puzzling are linked to the composition in tetraether lipids of Archaea. These lipids form extremely rigid, packed and impermeable structures, which fit in the membranes of hyperthermophilic archaea from which they were first isolated. Yet these tetraether lipids are common in many mesophilic, and psychrophilic archaea, and are also abundant in samples of low-temperature environments, although it is difficult to understand why these low-temperature adapted microbes would incorporate these lipids in their membranes. Conversely, numerous hyperthermophilic archaea often capable of growth over 100°C do not have tetraether lipids in their membranes. Rather than suggesting that homeoviscous adaptation does not exist in Archaea, it suggests that we may have overlooked important parameters taking part in the adaptation of the Archaeal membrane in response to stress. Amongst these, four seem to be able to influence the behavior of the membrane: (1) The S-layer: archaeal cells are encased in a 2D protein matrix, anchored in the membrane which defines a pseudoperiplasm space around the cell, but which influence on membrane dynamics has not yet been evaluated; (2) The spatial organization of lipids in the membrane, especially in those composed of a mix of di- and tetraethers may influence local membrane dynamics; (3) The presence of non-polar membrane lipids, such as lycopene-type poly-isoprenoid hydrocarbons, in the structure of the membrane may modify substantially the behavior of the membrane, in a similar way that the presence of sterols in bacteria and eukarya modify the fluidity and biological function of membranes. (4) The nature of polar headgroups. There is a need to obtain precise descriptions of these polar headgroups and the relative proportions and structures of the lipids. Keeping in mind that lipid compositions in Archaea are strongly influenced by growth conditions (Matsuno et al. 2009), future research to address these questions will require the close cooperation between the biophysicists who can described the behavior of the biological systems at the molecular level, the biochemist who can provide detailed analyses of the composition of the membrane lipids and the microbiologists who can provide control on the cell cycle of the model organisms and prepare biological significant material.

References

Boyd ES, Pearson A, Pi Y, Li W-J, Zhang YG, He L, Zhang CL, Geesey GG (2011) Temperature and pH controls on glycerol dibiphytanyl glycerol tetraether lipid composition in the hyperthermophilic crenarchaeon *Acidilobus sulfurireducens*. Extremophiles 15:59–65

Carballeira NM, Reyes M, Sostre A, Huang HS, Verhagen M, Adams MWW (1997) Unusual fatty acid compositions of the hyperthermophilic archaeon *Pyrococcus furiosus* and the bacterium *Thermotoga maritima*. J Bacteriol 179:2766–2768

Chong PLG (2010) Archaebacterial bipolar tetraether lipids: physico-chemical and membrane properties. Chem Phys Lipids 163:253–265

Chong PLG, Zein M, Khan TK, Winter R (2003) Structure and conformation of bipolar tetraether lipid membranes derived from thermoacidophilic archaeon *Sulfolobus acidocaldarius* as revealed by small-angle X-ray scattering and high-pressure FT-IR spectroscopy. J Phys Chem B 107:8694–8700

Chong PLG, Ravindra R, Khurana M, English V, Winter R (2005) Pressure perturbation and differential scanning calorimetric studies of bipolar tetraether liposomes derived from the thermoacidophilic archaeon *Sulfolobus acidocaldarius*. Biophys J 89:1841–1849

Chong PLG, Ayesa U, Daswani V, Hur E (2012) On physical properties of tetraether lipid membranes: effects of cyclopentane rings. Archaea. doi:10.1155/2012/138439, 11 pages

Comita PB, Gagosian RB (1983) Membrane lipid from deep-sea hydrothermal vent methanogen – a new macrocyclic glycerol diether. Science 222:1329–1331. doi:10.1126/science.222.4630.1329

Comita PB, Gagosian RB, Pang H, Costello CE (1984) Structural elucidation of a unique macrocyclic membrane lipid from a new extremophilic, deep-sea hydrothermal vent archaebacterium, *Methanococcus jannaschii*. J Biol Chem 259:5234–5241

Damste JSS, Schouten S, Hopmans EC, van Duin ACT, Geenevasen JAJ (2002) Crenarchaeol: the characteristic core glycerol dibiphytanyl glycerol tetraether membrane lipid of cosmopolitan pelagic crenarchaeota. J Lipid Res 43:1641–1651

Dannenmuller O, Arakawa K, Eguchi T, Kakinuma K, Blanc S, Albrecht AM, Schmutz M, Nakatani Y, Ourisson G (2000) Membrane properties of archaeal macrocyclic diether phospholipids. Chem Eur J 6:645–654

Dante S, De Rosa M, Maccioni E, Morana A, Nicolini C, Rustichelli F, Troitsky VI, Yang B (1995) Thermal stability of bipolar lipid langmuir-blodgett films by X-ray diffraction. Mol Cryst Liq Cryst 262:191–207

De Rosa M, Gambacorta A (1988) The lipids of archaebacteria. Prog Lipid Res 27:153–175

De Rosa M, Esposito E, Gambacorta A, Nicolaus B, Bu'Lock J (1980a) Complex lipids of *Caldariella acidophila*, a thermoacidophile archaebacterium. Phytochemistry 19:821–826

De Rosa M, Esposito E, Gambacorta A, Nicolaus B, Bu'Lock JD (1980b) Effects of temperature on ether lipid composition of *Caldariella acidophila*. Phytochemistry 19:827–831

De Rosa M, De Rosa S, Gambacorta A, Bulock JD (1980c) Structure of calditol, a new branched-chain nonitol, and the derived tetraether lipids in thermoacidophile archaebacterial of the *Caldariella* group. Phytochemistry 19:249–254

De Rosa M, Gambacorta A, Nicolaus B, Ross HNM, Grant WD, Bulock JD (1982) An asymmetrical archaebacterial diether lipid from alkaliphilic halophiles. J Gen Microbiol 128:343–348

De Rosa M, Gambacorta A, Nicolaus B (1983) A new type of membrane in thermophilic archaebacteria based on bipolar ether lipids. J Membr Sci 16:287–294

DeLong E, King L, Massana R, Cittone H, Murray A, Schleper C, Wakeham S (1998) Dibiphytanyl ether lipids in nonthermophilic crenarchaeotes. Appl Environ Microbiol 64:1133–1138

Elferink MGL, Dewit JG, Driessen AJM, Konings WN (1994) Stability and proton permeability of liposomes composed of archaeal tetraether lipids. Biochim Biophys Acta-Biomembr 1193:247–254

Ernst M, Freisleben HJ, Antonopoulos E, Henkel L, Mlekusch W, Reibnegger G (1998) Calorimetry of archaeal tetraether lipid – indication of a novel metastable thermotropic phase in the main phospholipid from *Thermoplasma acidophilum* cultured at 59 degrees C. Chem Phys Lipids 94:1–12

Franzmann P, Springer N, Ludwig W, Conway de Macario E, Rohde M (1992) A methanogenic archaeon from Ace Lake, Antarctica: *Methanococcoides burtonii* sp. nov. Syst Appl Microbiol 15:573–581

Gliozzi A, Paoli G, De Rosa M, Gambacorta A (1983) Effect of isoprenoid cyclization on the transition temperature of lipids in thermophilic archaebacteria. Biochim Biophys Acta 735:234–242

Gonthier I, Rager MN, Metzger P, Guezennec J, Largeau C (2001) A di-O-dihydrogeranylgeranyl glycerol from *Thermococcus* S557, a novel ether lipid, and likely intermediate in the biosynthesis of diethers in Archaea. Tetrahedron Lett 42:2795–2797

Gulik A, Luzzati V, De Rosa M, Gambacorta A (1985) Structure and polymorphism of bipolar isoprenoyl ether lipids from archaebacteria. J Mol Biol 182:131–149

Hafenbradl D, Keller M, Thiericke R, Stetter KO (1993) A novel unsaturated archael ether core lipid from the hyperthermophile *Methanopyrus kandleri*. Syst Appl Microbiol 16:165–169

Jarrell HC, Zukotynski KA, Sprott GD (1998) Lateral diffusion of the total polar lipids from *Thermoplasma acidophilum* in multilamellar liposomes. BBA-Biomembr 1369:259–266

Kamekura M, Kates M (1999) Structural diversity of membrane lipids in members of *Halobacteriaceae*. Biosci Biotech Bioch 63:969–972

Kaneshiro SM, Clark DS (1995) Pressure effects on the composition and thermal behavior of lipids from the deep-sea thermophile *Methanococcus jannaschii*. J Bacteriol 177:3668–3672

Khan TK, Chong PLG (2000) Studies of archaebacterial bipolar tetraether liposomes by perylene fluorescence. Biophys J 78:1390–1399

Koga Y (2011) Early evolution of membrane lipids: how did the lipid divide occur? J Mol Evol 72:274–282

Koga Y, Morii H (2005) Recent advances in structural research on ether lipids from archaea including comparative and physiological aspects. Biosci Biotech Bioch 69:2019–2034

Koga Y, Morii H (2007) Biosynthesis of ether-type polar lipids in archaea and evolutionary considerations. Microbiol Mol Biol Rev 71:97–120

Koga Y, Nishihara M, Morii H, Akagawamatsushita M (1993) Ether polar lipids of methanogenic bacteria – structures, comparative aspects and biosyntheses. Microbiol Rev 57:164–182

Lai D, Springstead J, Monbouquette H (2008) Effect of growth temperature on ether lipid biochemistry in *Archaeoglobus fulgidus*. Extremophiles 12:271–278

Lanzotti V, Trincone A, Nicolaus B, Zillig W, De Rosa M, Gambacorta A (1989) Complex lipids of *Pyrococcus* and AN1, thermophilic members of Archaebacteria belonging to the *Thermococcales*. Biochim Biophys Acta 1004:44–48

Lattuati A, Guezennec J, Metzger P, Largeau C (1998) Lipids of *Thermococcus hydrothermalis*, an archaea isolated from a deep-sea hydrothermal vent. Lipids 33:319–326. doi:10.1007/s11745-998-0211-0

Manquin B, Morgan J, Ju J, Muller-Spath T, Clark D (2004) Production of C35 isoprenoids depends on H_2 availability during cultivation of the hyperthermophile *Methanococcus jannaschii*. Extremophiles 8:13–21

Marr AG, Ingraham JL (1962) Effect of temperature on composition of fatty acids in *Escherichia coli*. J Bacteriol 84:1260–1267

Marteinsson VT, Birrien JL, Reysenbach AL, Vernet M, Marie D, Gambacorta A, Messner P, Sleytr UB, Prieur D (1999) *Thermococcus barophilus* sp. nov., a new barophilic and hyperthermophilic archaeon isolated under high hydrostatic pressure from a deep-sea hydrothermal vent. Int J Syst Bacteriol 49:351–359

Mathai JC, Sprott GD, Zeidel ML (2001) Molecular mechanisms of water and solute transport across archaebacterial lipid membranes. J Biol Chem 276:27266–27271

Matsumi R, Atomi H, Driessen AJM, van der Oost J (2011) Isoprenoid biosynthesis in archaea – biochemical and evolutionary implications. Res Microbiol 162:39–52

Matsuno Y, Sugai A, Higashibata H, Fukuda W, Ueda K, Uda I, Sato I, Itoh T, Imanaka T, Fujiwara S (2009) Effect of growth temperature and growth phase on the lipid composition of the archaeal membrane from *Thermococcus kodakarensis*. Biosci Biotech Bioch 73:104–108

McElhaney RN (1984a) The relationship between membrane fluidity and phase state and the ability of bacteria and mycoplasma to grow and survive at various temperatures. In: Kates M, Manson L (eds) Membrane fluidity. Plenum Press, New York, pp 249–276

McElhaney RN (1984b) The structure and function of the *Acholeplasma laidlawii* plasma membrane. Biochim Biophys Acta 779:1–42

Moldoveanu N, Kates M (1988) Biosynthetic studies of the polar lipids of *Halobacterium cutirubrum* – formation of isoprenyl ether intermediates. Biochim Biophys Acta 960:164–182

Moldoveanu N, Kates M, Montero CG, Ventosa A (1990) Polar lipids of nonalkalophilic halococci. Biochim Biophys Acta 1046:127–135

Morii H, Koga Y (1994) Asymmetrical topology of diether- and tetraether-type polar lipids in membranes of *Methanobacterium thermoautotrophicum* cells. J Biol Chem 269:10492–10497

Morii H, Eguchi T, Nishihara M, Kakinuma K, Konig H, Koga Y (1998) A novel ether core lipid with H-shaped C-80-isoprenoid hydrocarbon chain from the hyperthermophilic methanogen *Methanothermus fervidus*. Biochim Biophys Acta-Lipid Lipid Met 1390:339–345

Morii H, Yagi H, Akutsu H, Nomura N, Sako Y, Koga Y (1999) A novel phosphoglycolipid archaetidyl (glucosyl)inositol with two sesterterpanyl chains from the aerobic hyperthermophilic archaeon *Aeropyrum pernix* K1. Biochim Biophys Acta-Mol Cell Biol L 1436:426–436

Nichols DS, Miller MR, Davies NW, Goodchild A, Raftery M, Cavicchioli R (2004) Cold adaptation in the antarctic archaeon *Methanococcoides burtonii* involves membrane lipid unsaturation. J Bacteriol 186:8508–8515

Nicolaus B, Trincone A, Esposito E, Vaccaro MR, Gambacorta A, De Rosa M (1990) Calditol tetraether lipids of the archaebacterium *Sulfolobus solfataricus* – biosynthetic studies. Biochem J 266:785–791

Nishihara M, Koga Y (1987) Extraction and composition of polar lipids from the archaebacterium, *Methanobacterium thermoautotrophicum* – effective extraction of tetraether lipids by an acidified solvent. J Biochem 101:997–1005

Nishihara M, Morii H, Koga Y (1987) Structure determination of a quartet of novel tetraether from *Methanobacterium thermoautotrophicum*. J Biochem 101:1007–1015

Nishihara M, Nagahama S, Ohga M, Koga Y (2000) Straight-chain fatty alcohols in the hyperthermophilic archaeon *Pyrococcus furiosus*. Extremophiles 4:275–277

Oger P, Cario A (2013) Adaptation of the archaeal membrane. Biophys Chem 15:42–56

Pearson A, Huang Z, Ingalls AE, Romanek CS, Wiegel J, Freeman KH, Smittenberg RH, Zhang CL (2004) Nonmarine crenarchaeol in Nevada hot springs. Appl Environ Microbiol 70:5229–5237

Pitcher A, Rychlik N, Hopmans EC, Spieck E, Rijpstra WIC, Ossebaar J, Schouten S, Wagner M, Damste JSS (2010) Crenarchaeol dominates the membrane lipids of candidatus *Nitrososphaera gargensis*, a thermophilic group I. 1b archaeon. ISME J 4:542–552

Qiu DF, Games MPL, Xiao XY, Games DE, Walton TJ (2000) Characterisation of membrane phospholipids and glycolipids from a halophilic archaebacterium by high-performance liquid chromatography/electrospray mass spectrometry. Rapid Commun Mass Sp 14:1586–1591

Russell NJ, Nichols DS (1999) Polyunsaturated fatty acids in marine bacteria – a dogma rewritten. Microbiology 145:767–779

Sako Y, Nomura N, Uchida A, Ishida Y, Morii H, Koga Y, Hoaki T, Maruyama T (1996) *Aeropyrum pernix* gen nov, sp nov, a novel aerobic hyperthermophilic archaeon growing at temperatures up to 100 degrees C. Int J Syst Bacteriol 46:1070–1077

Schleper C, Puehler G, Holz I, Gambacorta A, Janekovic D, Santarius U, Klenk HP, Zillig W (1995) *Picrophilus* gen. nov, fam. nov – a novel aerobic, heterotrophic, thermoacidophilic genus and family comprising archaea capable of growth around pH = 0. J Bacteriol 177:7050–7059

Schouten S, Hopmans EC, Pancost RD, Damste JSS (2000) Widespread occurrence of structurally diverse tetraether membrane lipids: evidence for the ubiquitous presence of low-temperature relatives of hyperthermophiles. Proc Natl Acad Sci U S A 97:14421–14426

Shimada H, Nemoto N, Shida Y, Oshima T, Yamagishi A (2008) Effects of pH and temperature on the composition of polar lipids in *Thermoplasma acidophilum* HO-62. J Bacteriol 190:5404–5411

Sinensky M (1971) Temperature control of phospholipid biosynthesis in *Escherichia coli*. J Bacteriol 106:449–455

Sinensky M (1974) Homeoviscous adaptation – homerostatic process that regulates viscosity of membrane lipids in *Escherichia coli*. Proc Natl Acad Sci U S A 71:522–525

Sprott GD, Ekiel I, Dicaire C (1990) Novel, acid-labile, hydroxyether lipid cores in methanogenic bacteria. J Biol Chem 265:13735–13740

Sprott GD, Meloche M, Richards JC (1991) Proportions of diether, macrocyclic diether, and tetraether lipids in *Methanococcus jannaschii* grown at different temperatures. J Bacteriol 173:3907–3910

Sprott GD, Agnew BJ, Patel GB (1997) Structural features of ether lipids in the archaeobacterial thermophiles *Pyrococcus furiosus*, *Methanopyrus kandleri*, *Methanothermus fervidus*, and *Sulfolobus acidocaldarius*. Can J Microbiol 43:467–476

Sprott GD, Larocque S, Cadotte N, Dicaire CJ, McGee M, Brisson JR (2003) Novel polar lipids of halophilic eubacterium *Planococcus* H8 and archaeon *Haloferax volcanii*. Biochim Biophys Acta-Mol Cell Biol L 1633:179–188

Stewart LC, Kates M, Ekiel IH, Smith ICP (1990) Molecular order and dynamics of diphytanylglycerol phospholipids – a ^{2}H NMR and ^{31}P NMR study. Chem Phys Lipids 54:115–129

Sugai A, Sakuma R, Fukuda I, Itoh YH, Kon K, Ando S, Itoh T (1995) The structure of the core polyol of the ether lipids from *Sulfolobus acidocaldarius*. Lipids 30:339–344

Sugai A, Masuchi Y, Uda I, Itoh T, Itoh Y (2000) Core lipids of hyperthermophilic archaeon. *Pyrococcus horikoshii* OT3. J Jpn Oil Chem Soc 49:695–700

Sugai A, Udai I, Itoh Y, Itoh H (2004) The core lipid composition of the 17 strains of hyperthermophilic archaea, *Thermococcales*. J Oleo Sci 53:41–44

Tarui M, Tanaka N, Tomura K, Ohga M, Morii H, Koga Y (2007) Lipid component parts analysis of the hyperthermophilic sulfate- reducing archaeon *Archaeoglobus fulgidus*. J UOEH 29:131–139

Tornabene TG, Langworthy TA, Holzer G, Oro J (1979) Squalenes, phytanes, and other isoprenoids as major neutral lipids of methanogenic and thermoacidophilic archaebacteria. J Mol Evol 13:73–83

Uda I, Sugai A, Kon K, Ando S, Itoh YH, Itoh T (1999) Isolation and characterization of novel neutral glycolipids from *Thermoplasma acidophilum*. Biochim Biophys Acta-Mol Cell Biol L 1439:363–370

Uda I, Sugai A, Itoh YH, Itoh T (2001) Variation in molecular species of polar lipids from *Thermoplasma acidophilum* depends on growth temperature. Lipids 36:103–105

Uda Y, Sugai A, Itoh Y, Itoh T (2004) Variation in molecular species of core lipids from the order *Thermoplasmales* strains depends on the growth temperature. J Oleo Sci 53:399–404

Upasani VN, Desai SG, Moldoveanu N, Kates M (1994) Lipids extremely halophilic archaebacteriaa from saline environments in India – a novel glycolipid in *Natronobacterium* strains. Microbiology 140:1959–1966

van de Vossenberg JLCM, Driessen AJM, Konings WN (1998) The essence of being extremophilic: the role of the unique archaeal membrane lipids. Extremophiles 2:163–170. doi:10.1007/s007920050056

van de Vossenberg J, Driessen AJM, da Costa MS, Konings WN (1999) Homeostasis of the membrane proton permeability in *Bacillus subtilis* grown at different temperatures. Biochim Biophys Acta-Biomembranes 1419:97–104

Winter R (2002) Effect of lipid chain length, temperature, pressure and composition on the lateral organisation and phase behavior of lipid bilayer/gramicidin mixtures. Biophys J 82:153A–153A

Winter R, Jeworrek C (2009) Effect of pressure on membranes. Soft Matter 5:3157–3173

Yamauchi K, Doi K, Yoshida Y, Kinoshita M (1993) Archaebacterial lipids: highly proton-impermeable membranes from 1,2-diphytanyl-sn-glycero-3- phosphocholine. Biochim Biophys Acta 1146:178–182

Zein M, Winter R (2000) Effect of temperature, pressure and lipid acyl chain length on the structure and phase behaviour of phospholipid-gramicidin bilayers. Phys Chem Chem Phys 2:4545–4551

Zillig W, Holz I, Janekovic D, Klenk HP, Imsel E, Trent J, Wunderl S, Forjaz VH, Coutinho R, Ferreira T (1990) *Hyperthermus butylicus*, a hyperthermophilic sulfur-reducing archaebacterium that ferments peptides. J Bacteriol 172:3959–3965

Zillig W, Holz I, Wunderl S (1991) *Hyperthermus butylicus* gen. nov, sp. nov, a hyperthermophilic, anaerobic, peptide-fermenting, facultatively H_2S generating archaebacterium. Int J Syst Bacteriol 41:169–170

Part VI
Pressure Adaptation and Tolerance of Proteins and Living Organisms

Editors' Foreword of Part VI

Hydrostatic pressure is known to affect cellular physiology differently, depending on the level of pressure applied. One of the puzzles in pressure effect on living cells yet to be solved is why the pressure of a few tens MPa decreases the growth rate, while that of a few hundred MPa decreases the cellular viability. To identify crucial proteins selectively impaired or damaged, Iwahashi in Chap. 20 employs the hierarchical cluster analysis of expression profiles in yeast cells that have been exposed to different levels of hydrostatic pressure. They found that proteins related to membrane and cell wall biosynthesis are likely the crucial targets of pressure-induced damage in the low-pressure range, whereas proteasome activity- and endoplasmic reticulum-localized proteins can be the crucial targets of pressure-induced damage in the high-pressure range. In Chap. 21, Ohmae et al. deal with the topic of pressure and environmental adaptation of protein structures and function for the case of a deep-sea bacterium, here dihydrofolate reductase (DHFR). To find mechanisms of adaptation or loss of viability of cells further on the molecular level, the next step would be to study details of structural and functional changes of some crucial proteins, such as dihydrofolate reductase (DHFR) as a function of pressure. By comparing DHFR from *Moritella profunda*, a deep-sea bactrerium, with DHFR from *Escherichia coli*, the authors show that the thermodynamic stabilities, volume differences, and enzyme activities under pressure are all different between the two proteins, while their backbone structures are almost identical in crystals. The result suggests a number of interesting possibilities as to the molecular design of this protein to adapt to the environment. Finally, in Chap. 22, Ono shows the extreme surprising observation that spores of a certain moss spore, exposed to pressures can easily endure hydrostatic pressures of up to 7.5 GPa, or a certain spore can germinate even after exposure to up to 20 GPa for a limited time, still germinates. The finding of this unusual tolerance of the moss spores to pressure may even suggest the possibility that a certain form of life might travel through the outer space with a meteorite.

Chapter 20	Pressure-Dependent Gene Activation in Yeast Cells	Hitoshi Iwahashi
Chapter 21	Environmental Adaptation of Dihydrofolate Reductase from Deep-Sea Bacteria	Eiji Ohmae, Kunihiko Gekko, and Chiaki Kato
Chapter 22	Moss Spores can Tolerate Ultra-high Pressure	Fumihisa Ono

Chapter 20
Pressure-Dependent Gene Activation in Yeast Cells

Hitoshi Iwahashi

Abstract Hydrostatic pressure is one of the physical factors affecting cellular physiology. Hydrostatic pressure of a few tens MPa decreases the growth rate and a few hundred MPa decreases the cellular viability. To find clues to understand how such pressure effects on living cells relating to damages on protein molecules, we employed yeast DNA microarrays and analyzed genome-wide gene-expression levels in yeast cells which have been exposed to different levels of hydrostatic pressure. These include the cells temporarily adapted to a high pressure (from 0.1 to 30 MPa) and to a low pressure (from 30 to 0.1 MPa). These conditions cause yeast cells decreases of growth rate. We also analyzed gene expression profiles from the cells recovering after the sublethal pressure treatment at 180 MPa at 4 °C for 0 min and at 40 MPa at 4 °C for 16 h. These conditions cause yeast cells decreases of cellular viability. The activated genes by the temporary adaptations to both of the high pressure and the low pressure suggest that proteins related to membrane biosynthesis and cell wall biosynthesis can be crucial targets of pressure-induced damages, whereas the activated genes under recovering conditions after exposure to the sublethal high pressure suggest that proteasome activity and proteins localized in endoplasmic reticulum can be the crucial targets or the essential factors to survive.

Keywords Genomics • Metabolomics • Proteasome • *Saccharomyces cerevisiae* • Yeast

20.1 Biology Under Hydrostatic Pressure

What happens on molecular levels in a biological cell when it is exposed to high hydrostatic pressure? This is a fundamental and difficult question in biology in view of the complex nature of molecular events governing the physiology even of a biological cell. It is well known that hydrostatic pressure causes growth

H. Iwahashi (✉)
Department of Applied Life Science, Faculty of Applied Biological Science, Gifu University, 1-1, Yanagido, Gifu, Gifu 501-1193, Japan
e-mail: h1884@gifu-u.ac.jp

K. Akasaka, H. Matsuki (eds.), *High Pressure Bioscience*, Subcellular Biochemistry 72, DOI 10.1007/978-94-017-9918-8_20

inhibition, growth arrest, cellular death and breakdown of cellular structures. Under a few tens of MPa, yeast grows temporarily adapted to altered hydrostatic pressure conditions. Growth arrest is a condition that yeast cells cannot grow but no death is observed. Hydrostatic pressure of a few hundred of MPa decreases the microbial surviving abilities (Yayanos and Pollard 1969; Iwahashi et al. 2003a, 2005). These observations suggest that different cellular processes may be involved in the pressure effects in different ranges of hydrostatic pressure.

In this chapter, we describe our efforts to try to identify, as much as possible, the main biological processes that are affected in each range of pressure using yeast *Saccharomyces cerevisiae*. This species is the first eukaryote whose whole genomic sequence is completed and the genome wide gene expression profiles can be evaluated using DNA microarray technology (Momose and Iwahashi 2001). Proteins, whether they are in membranes or in cellular fluids, are considered among the most crucial targets of pressure, as they are the "life machines", any damage of which would cause growth inhibition, growth arrest, and death of biological cell. To find out proteins that are affected or damaged most seriously by the pressure in each pressure range, we focus on proteins that cells try to recover during and after the "attack" of pressure. We do not exclude the possibility, however, that the breakdown of cellular systems cause cellular death. These cellular systems include programmed cellular death, proteasome functions, autophagy (Tsukada and Ohsumi 1993), and so on.

Our strategy is the following: once the cells are damaged, not to the fatal stage, but allowed to recover from the initial damage of pressure, they would try to recover proteins that have been damaged to the non-functional stage. By recovering proteins, yeast cells may temporarily adapt to high-pressure conditions (Fig. 20.1 from A to B) and may restore the damages after the sublethal high-pressure treatment (Fig. 20.1 from A to C and D). To find out major protein systems to be damaged as a result of applied pressure, we evaluated the gene-expression profiles using the cells temporarily adapted to high pressure (Fig. 20.1 from A to B) and the cells recovered from sublethal pressure conditions (Fig. 20.1 from A to C and D).

20.1.1 High Pressure Inhibits Cellular Growth

Escherichia coli and yeast *S. cerevisiae* grow under high pressure up to 40–50 MPa with other appropriate conditions (Tamura et al. 1992; Iwahashi et al. 2005). The growth rates are decreased with increasing pressure (Iwahashi et al. 2005). Growth limit up to approximately 50 MPa is shared by many microorganisms (Yayanos and Pollard 1969; Abe and Horikoshi 1995; Iwahashi et al. 2005). This pressure range is corresponding to the upper limit of macromolecular synthesis such as protein, mRNA, and DNA (Yayanos and Pollard 1969). These reports suggest that the microbes temporarily adapt to high pressure up to approximately 50 MPa and do increase their populations (Fig. 20.2).

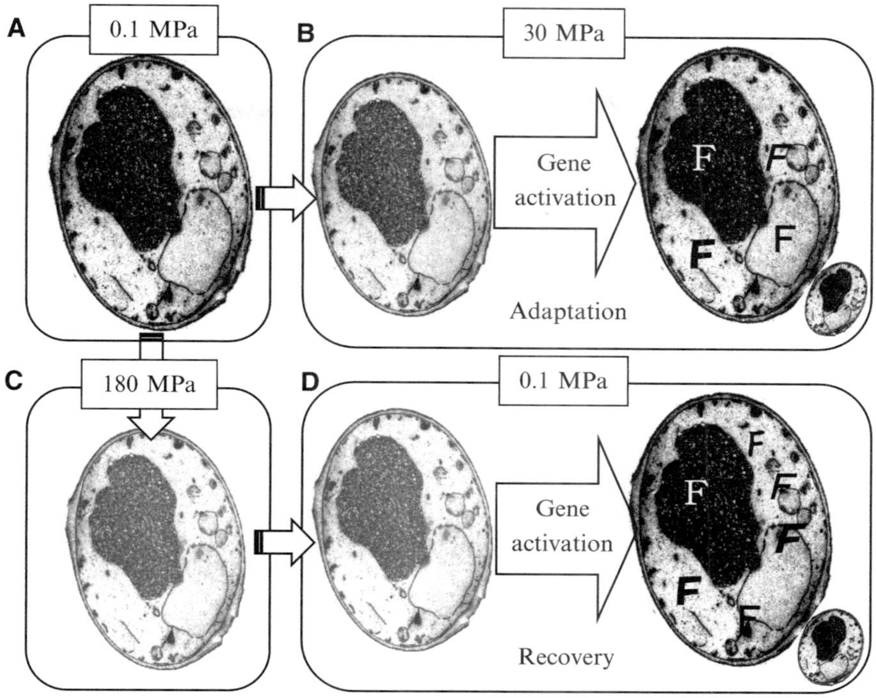

Fig. 20.1 Temporary adaptation under high pressure and recovery after high pressure conditions. (A) Normally growing yeast cells under 0.10 MPa. (B) Yeast cells under 30 MPa, temporarily adapt to high pressure. They have damages and they try to restore damages by activating specific genes. (C) Yeast cells under 180 MPa, have damages that may cause cellular death. (D) Yeast cells under 0.1 from 180 MPa. They try to restore damages by activating genes. The factors (F, F, ***F***, and **F**) can be produced from newly synthesized mRNAs and we may detect these mRNA by DNA microarray

Yeast cells grown under pressure apparently show that their cell density (absorbance at 660 nm/colony forming units (CFU)) is much higher than that grown under atmospheric pressure (Iwahashi et al. 2005). This suggests that cells become bigger immediately after decompression, which has been confirmed by flow cytometry analysis after decompression (Iwahashi et al. 2005). Thus, *S cerevisiae* temporarily adapts to high-pressure conditions by increasing density of cellular components (Iwahashi et al. 2005).

20.1.2 High Pressure Causes Growth Arrest

Growth arrest is observed from 50 to 100 MPa with yeast *S. cerevisiae* (Iwahashi et al. 2005), however physiology around these conditions is not well characterized.

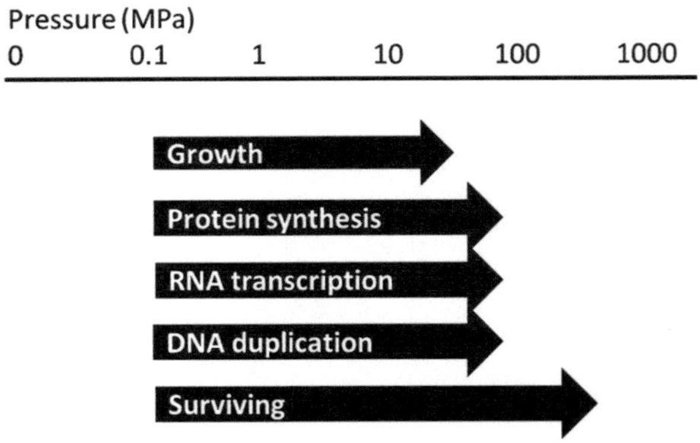

Fig. 20.2 High pressure decreases the ability of growth, protein synthesis, RNA transcription, DNA duplication, and surviving. The *arrows* show the upper limit of those abilities. For example, *Escherichia coli* and yeast *S. cerevisiae* grow under high pressure up to 40–50 MPa with appropriate other conditions (Tamura et al. 1992; Iwahashi et al. 2005). More than 100 MPa, DNA duplication, RNA translation, and protein synthesis can not be expected (Yayanos and Pollard 1969). A few hundred MPa decreases the microbial surviving abilities (Yayanos and Pollard 1969; Iwahashi et al. 2003a, 2005)

This is because of no biological alteration can be monitored and no macromolecular response can be expected because the stop of DNA duplication, RNA translation, and protein synthesis (Yayanos and Pollard 1969) (Fig. 20.2).

20.1.3 High Pressure Causes Cellular Death

In comparison to growth inhibition, cellular death by pressure treatment occurs in a relatively larger pressure range depending on species of microbes and also depends on temperature. For example, *Lactobacillus case*i is killed under 300 and 550 MPa is lethal to *E. coli* (Sonoike et al. 1993; O'Reilly et al. 2000). On the other hand, eucaryotic microorganisms are more sensitive to pressure. *S. cerevisiae* can be inactivated under around 150 MPa (Iwahashi et al. 1991) and *Penicillium roqueforti* of fungi is shown to be more sensitive than *Staphylococcus aureus* (Sonoike et al. 1993). Interestingly, the effect of pressure depends strongly on temperature. Sonoike et al. (1993) showed that *L. casei* is killed under 300 MPa with a death rate of 0.32 at 0 °C, 0.1 at 20 °C, and 0.32 at 60 °C. For yeast *S. cerevisiae*, 40 MPa is sublethal at 4 °C, but grows at 25 °C (Iwahashi et al. 2003a). It should be noted that *E. coli* inactivated by pressure comes back to life. Koseki and Yamamoto (2006) reported that completely inactivated *E. coli* confirmed by CFU becomes to show CFU under appropriate conditions. This is confirmed by Ohshima et al. (2013). This observation shows that microbes have a crucial damage by pressure treatment and this crucial damage can be restored under appropriate conditions.

20.1.4 High Pressure Causes Breakdown of Cellular Structures

Breakdown of cellular structures is observed at much high pressure than that causing cellular death. Osumi et al. (1996) showed the disruptions of subcellular structures at 300 MPa and inner structures at 400 MPa by electron microscope. The pressure conditions are much higher than the conditions that cause complete cellular death at 200 MPa. They also observed breakdown of spindle pole bodies at 100 MPa but this was repaired in 24 h. Thus, the spindle pole body cannot be a crucial factor for cellular death.

20.1.5 Gene Activation and Pressure Stress Response

DNA microarray technology enabled us to monitor almost all gene activations at transcription levels under stress conditions such as chemicals, temperature, and pressure (Momose and Iwahashi 2001; Murata et al. 2003). This technology makes us understand stress responses from genome wide functional information.

The relation between gene activation and biological response by the pressure treatment is summarized in Fig. 20.3. Microbes grow up to approximately 50 MPa by temporary adaptation to high pressure (Fig. 20.3 A). Once adapted to high pressure, decreased pressure conditions can be the stress and microbes have to adapt to low pressure (Fig. 20.3 B). Under the growth-arrested conditions (Fig. 20.3 C), microbes may be damaged and these damages can be completely repaired (Fig. 20.3 D). Under these conditions, the damages are not lethal and easily repaired. To understand crucial damages caused by pressure, we have to treat microbes under the pressure conditions that decreases surviving abilities (Fig. 20.3 E and H). Under these conditions we are able to monitor restoration from the pressure causing damages. However, we are not able to monitor the specific activation of genes as RNA transcription cannot be expected (Fig. 20.2). We need repairing response (Fig. 20.3 F) to understand the damages. Under the much higher-pressure conditions (Fig. 20.3 G) we may not expect the living microbes (Fig. 20.3 I) and no response can be expected. Thus, activation of genes relating to pressure stress can be monitored in the Steps A, B, and F in Fig. 20.3.

Based on these biological reasons, we evaluated the gene-expression levels using cells temporarily adapting to high pressure (Fig. 20.1 from A to B and Fig. 20.3 A) and readapting to low pressure (Fig. 20.1 from B to A and Fig. 20.3 B), and cells restoring the damages after sublethal pressure conditions (Fig. 20.1 from A to C and D and Fig. 20.3 F).

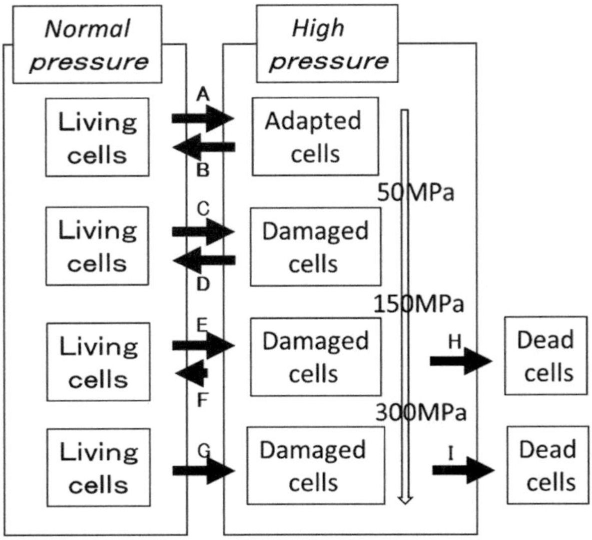

Fig. 20.3 Responses of Yeast *S. cerevisiae* under high-pressure conditions. Yeast grows up to approximately 50 MPa by temporary adaptation to high pressure (A). Once adapted to high pressure, decreased pressure conditions can be the stress and yeast has to adapt to lower pressure (B). Under the growth-arrested conditions (C), microbes may be damaged and these damages can be completely repaired (D). To understand lethal damages caused by pressure, we have to put yeast under the pressure that decreases surviving abilities (E and H). In these conditions we are able to monitor restoration from the pressure caused damages (F). Under the much higher-pressure conditions (G) we may not expect the living yeast (I) and no response can be expected

20.2 Temporary Adaptation of Yeast Cells to High Pressure and Low Pressure

20.2.1 Pressure-Dependent Gene Activation in Yeast Cells Temporarily Adapted to High Pressure

The purpose of this section is to understand the temporary adaptation mechanisms of yeast cells to high hydrostatic pressure conditions by pressure-dependent gene activation (Fig. 20.1 from A to B). We selected high-pressure condition of 30 MPa as this condition decreased the growth rate to 50 % compared to that of 0.1 MPa (Iwahashi et al. 2005).

We compared the expression profiles of yeast cells grown under a pressure of 30 MPa to those of yeast cells grown under 0.1 MPa using DNA microarrays (Iwahashi et al. 2005). This activation of genes corresponds to the temporary adaptation shown in Fig. 20.1 (from A to B). From approximately 6,000 genes, we found 366 genes were activated more than two times by the pressure dependent adaptation. While, 253 genes were inactivated to low than 0.5-fold (Iwahashi et al. 2005).

Table 20.1 List of activated genes in yeast cells temporarily adapted to high pressure (Fig. 20.1 from A to B) (Iwahashi et al. 2005)

	Name	Function of gene
1	<u>*PST1*</u>	Strong similarity to SPS2 protein
2	<u>*INO1*</u>	Myo-inositol-1-phosphate synthase
3	<u>*MET6*</u>	Homocysteine methyltransferase
4	RTA1	Involved in 7-aminocholesterol resistance
5	*DAN1*	Only expressed under anaerobic conditions
6	*GPG1*	Hypothetical protein
7	*PIR3*	Member of the Pir1p/Pir2p/Pir3p family
8	*HSP12*	Heat shock protein
9	<u>*OPI3*</u>	Methylene-fatty-acyl-phospholipid synthase
10	<u>*PTP2*</u>	Protein-tyrosine-phosphatase
11	*YPS3*	GPI-anchored aspartyl protease 3 (yapsin 3)
12	*STR3*	Similarity to cystathionine beta-lyase
13	*SRL3*	Similarity to YOR083w
14	*MOH1*	Similarity to hypothetical protein – human
15	*POX1*	Acyl-CoA oxidase
16	*PBI2*	Proteinase B inhibitor 2
17	*AFR1*	Involved in morphogenesis of the mating
18	*MET13*	Putative methylene tetrahydrofolate reductase
19	*NTH1*	Neutral trehalase (alpha,alpha-trehalase)
20	*SED1*	Abundant cell surface glycoprotein

Commonly activated genes among higher and lower pressure as shown by *underlines*

The top 20 activated genes that have common names are listed with functional information in Table 20.1. The most highly induced genes were *PST1*, followed by *INO1*. These genes were induced by more than 15-fold.

The 366 activated genes were characterized using the functional categories of MIPS (Munich Information Center for Protein Sequences) in The Institute of Bioinformatics and Systems Biology (http://www.helmholtz-muenchen.de/en/ibis). High numbers of genes were induced in the functional categories of "CELL CYCLE AND DNA PROCESSING", "CELL RESCUE, DEFENSE AND VIRULENCE", and "METABOLISM". In particular, a large proportion of genes in the category of "CELL RESCUE, DEFENSE AND VIRULENCE" (13.7 %) was observed. It was significant that genes in subcategories of "stress response" "amino acid metabolism", "nitrogen and sulfur metabolism", "C-compound and carbohydrate metabolism", "cell growth/morphogenesis", and "lipid, fatty-acid and isoprenoid metabolism" were activated (Iwahashi et al. 2005). Based on the functional information attached with the activated genes, it appears that 30 MPa pressure causes yeast cells to activate the stress response including heat shock proteins as well as pathways for the metabolism of carbon compounds, lipids, and amino acids (Iwahashi et al. 2005). These results and altered cell density suggest the possibility that high-pressure shock is related with membrane structure.

20.2.2 Pressure-Dependent Gene Activation in Yeast Cells Temporarily Adapted to Low Pressure

Once yeast cells adapted to high pressure, low pressure can be a stress. However, few researches were focused on the low-pressure shock response. In this section, pressure-dependent gene activation in yeast cells for adaptation to low pressure is described. We selected yeast cells growing under pressure of 30 MPa as the normally growing cells. These cells were transferred to 0.1 MPa by decompression. We could not harvest enough cells for DNA microarray analysis after the change of pressure from 40 to 0.1 MPa conditions. We compared the levels of mRNA production between the cells growing under pressure of 30 MPa and the cells incubated for 2 h after decreasing pressure. This activation of genes corresponds to the reactions shown in Fig. 20.1 (from B to A).

We select the genes that have two times higher than that of control as the activated genes (Iwahashi et al. 2003a). By this selection, we found 168 genes. On the other hand, we also selected the genes that have 0.5 time lower than that of control as the inactivated genes. We found 134 genes in this selection. The top 20 genes that have common names are listed with functional information in Table 20.2. The most highly induced gene was *PDC1*, followed by *MET6*. These genes were induced by more than 7-fold.

Those activated 168 genes were characterized according to the functional categories of MIPS (Iwahashi et al. 2003a). By this characterization, we found the genes that involved in functional categories of "CELLULAR COMMUNICATION" and "METABOLISM" were significantly activated. Within the group of genes of "METABOLISM", the genes that related to subcategories of "amino acid metabolism", "C-compound and carbohydrate metabolism", and "lipid, fatty-acid and isoprenoid metabolism" were activated. In addition, the genes whose products were localized in cell wall, secretion protein, mitochondrion, and nucleus were significantly activated. These results suggest that low-pressure shock caused the alteration in membrane structure.

20.2.3 Temporary Adaptation to Altered Pressure Conditions Needs the Activation of Membrane Biosynthesis and Cell Wall Biosynthesis

We compared activated genes between the conditions adapted to high pressure and low pressure. Surprisingly, 5 genes of *PST1, INO1, MET6, OPI3, and PTP2* were commonly activated genes (Table 20.1 and 20.2 shown by underlines). *PST1* encodes cell wall protein that contains a putative GPI-attachment site secreted by regenerating protoplasts (Brown et al. 2007). *INO1* encodes inositol-3-phosphate synthase involved in synthesis of inositol phosphates and inositol-containing phospholipids (Culbertson and Henry 1975). *MET6* encodes cobalamin-independent

Table 20.2 List of activated genes in yeast cells temporarily adapted to low pressure (Fig. 20.1 from B to A) (Iwahashi et al. 2005)

	Name	Function of gene
1	*PDC1*	Pyruvate decarboxylase
2	*MET6*	Isozyme of methionine synthase
3	*KTR2*	Putative mannosyltransferase
4	*OPI3*	Methylene-fatty-acyl-phospholipid synthase
5	*APN1*	AP endonuclease
6	*MET17*	O-acetylhomoserine sulfhydrylase
7	*ARG4*	Argininosuccinate lyase
8	*PTP2*	Protein tyrosine phosphatase
9	*MTD1*	Methylenetetrahydrofolate dehydrogenase
10	*LHS1*	Novel member of the Hsp70 family
11	*SAM2*	S-adenosylmethionine synthetase
12	*INO1*	L-myo-inositol-1-phosphate synthase
13	*FAT1*	Fatty acid transporter
14	*ERO1*	Required for protein disulfide bond formation in the ER
15	*PDI1*	Protein disulfide isomerase
16	*PTM1*	Putative membrane protein
17	*ILV2*	Acetolactate synthase
18	*ARO8*	Aromatic amino acid aminotransferase
19	*CHS1*	Chitin synthase 1
20	*PST1*	Strong similarity to SPS2 protein

Commonly activated genes among higher and lower pressure as shown by *underlines*

methionine synthase; involved in methionine biosynthesis and regeneration (Masselot and De Robichon-Szulmajster 1975). *OPI3* encodes Methylene- fatty-acyl-phospholipid synthase catalyzing the last two steps in phosphatidylcholine biosynthesis and also known as phospholipid methyltransferase (Kodaki and Yamashita 1987). *PTP2* encodes phosphotyrosine-specific protein phosphatase; involved in the inactivation of mitogen-activated protein kinase during cell wall organization (Mattison et al. 1999). Among these genes, there are four genes directly related to membrane or cell wall biosynthesis and methionine is required for lipid biosynthesis. Thus, the temporary adaptation to high pressure or to low pressure may need rearrangement of membrane or cell wall structures.

The activation of genes for membrane structure and cell wall structure caused by 30 MPa pressure is consistent with the observed cellular transformation confirmed by the flow cytometry analysis (Iwahashi et al. 2005). The effect of pressure on tryptophan permease in the membrane is under intense study by Abe et al. (Abe and Horikoshi 1995, 2000; Abe 2004). These researchers found that hydrostatic pressure in the range of 15–25 MPa caused arrest of the cell cycle in G1 phase in an exponentially growing culture of a *S. cerevisiae* tryptophan auxotroph. The study proved that G1 arrest was due to damage to tryptophan permease, which resulted in tryptophan starvation. They suggested that hydrostatic pressure might affect the activity of tryptophan permease through changes in the lipid bilayer structure (Abe and Horikoshi 2000; Abe 2004). The genome-wide mRNA expression profiles and the behavior of tryptophan permease under a pressure that caused growth inhibition suggest that pressure affects membrane structure. The membrane structure of eucaryotic microorganisms is more complex than that of procaryotic microorganisms and may be for the reason that eucaryotic microorganisms are more sensitive to hydrostatic pressure (Iwahashi et al. 1991; Sonoike et al. 1993).

20.3 Pressure-Dependent Gene Activation in Yeast Cells Recovering from the Damages After High-Pressure Treatment

20.3.1 Selection of Activated Genes During the Recovering Conditions After Pressure

In this section, we focus on the damages that lead to cellular death but may be restored, as they can be the primary and crucial damages by hydrostatic pressure treatment. To find out those responses and activation of genes, we selected the pressure conditions of 60 % viability compared with the control as we needed cells alive or cells trying to survive, namely 180 MPa at 4 °C for 0 min (instant) and 40 MPa at 4 °C for 16 h (Iwahashi et al. 2003b). The temperature was selected to arrest cells during high-pressure treatment. Yeast may adapt to the conditions during the pressure treatment at moderate temperatures. This is the reason why we arrest yeast cells at 4 °C. It should be noted that yeast cannot activate genes as the macromolecular biosynthesis is arrested under more than 100 MPa conditions (Fig. 20.2).

After the high-pressure treatment, we incubated yeast cells under 0.1 MPa at 25 °C for 1 h. This incubation allows yeast cells to start restoring the damages and we can monitor the responses from recovering yeast cells. This activation of genes corresponds to the reactions shown in Fig. 20.1 from A to C and D. We found 286 activated genes greater than two times with 180 MPa at 4 °C for 0 min and 218 those with 40 MPa at 4 °C for 16 h (Iwahashi et al. 2003b). We characterized these genes according to the functional categories of the MIPS. The proportion of activated genes (number of induced genes / number of genes in the category) was high in the categories of "ENERGY," "CELL RESCUE," and "PROTEIN DESTINATION". More than 100 induced genes fell into the category of "CELLULAR ORGANIZATION". In both pressure treatments, many genes that belong to the category of "CELLULAR ORGANIZATION" were activated. This category includes the genes whose localizations were characterized. Thus, we analyzed the localization of proteins encoded by activated genes by using the MIPS. After the both of pressure treatment, many genes whose products were localized in the ER (Endoplasmic Reticulum), mitochondria, nucleus, and cytoplasm were

Table 20.3 Commonly activated genes after high-pressure treatments under 180 and 40 MPa

	Code	Name	Function of gene
1	YGR142W	*BTN2*	v-SNARE binding protein
2	YBR072W	*HSP26*	Heat shock protein 26
3	YLL026W	*HSP104*	Heat shock protein 104
4	YLR216C	*CPR6*	Member of the cyclophilin family
5	YBL078C	*ATG8*	Essential for autophagy
6	YGR286C	*BIO2*	Biotin synthase//biotin synthetase
7	YGR180C	*RNR4*	Ribonucleotide reductase
8	YAL005C	*SSA1*	Heat shock protein of HSP70 family
9	YLL039C	*UBI4*	Ubiquitin

highly activated (Iwahashi et al. 2003b). This suggests that pressure damages, activates and repairs the ER, mitochondria, and nucleus.

Commonly activated genes by both treatments are listed in Table 20.3. Half the number of those activated genes related to protein metabolisms including *HSP104*. *HSP104* is a gene that was confirmed to contribute barotolerance and well known molecular chaperon (Iwahashi et al. 1997a, b, 1998, 2003b). This suggests that protein metabolism is essential to repair the damages after high-pressure treatment. Furthermore, molecular chaperons contribute to the function of ER (Iwahashi et al. 2003b). This agrees with the localization analysis, which suggests the importance of ER restoration.

20.3.2 Comparison of Pressure-Dependent Gene Activation and Other Stress Dependent Gene Activation

We have accumulated genome-wide expression profiles of genes activated by environmental stresses (GEO, http://www.ncbi.nlm.nih.gov/geo/). We compared the expression profiles of 3,875 genes after high-pressure treatment and other environmental stresses, using hierarchical cluster analysis (Murata et al. 2003). The hierarchical cluster analysis is based on correlation coefficient among expression profiles. The expression profiles those share high correlation coefficients make clusters of groups and we may understand similar responses make clusters together (Fig. 20.4). The selected 3,875 genes were those that had shown previous high intensity (greater than average) under at least one stress condition. This selection was carried out because genes with low expression are likely to yield erroneous results.

By the cluster analysis, it is well characterized that chemicals that cause soluble protein denaturation make one cluster (Kitagawa et al. 2001, 2003). Zineb, maneb, thiuram, TPN, and methyl mercury (II) chloride are chemicals that cause direct denaturation of soluble proteins. Any profile activated by pressure does not come to this cluster (Fig. 20.4). This strongly suggests that pressure do not primary cause

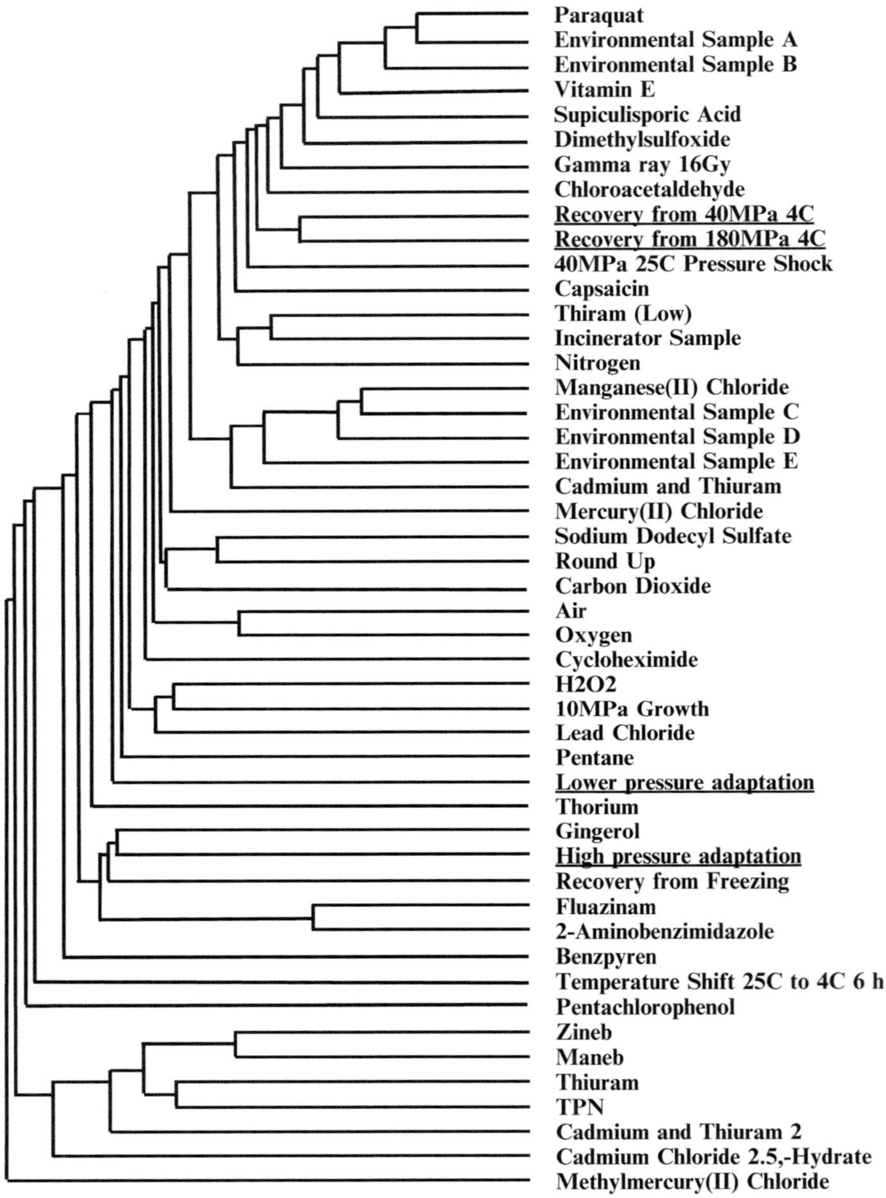

Fig. 20.4 Hierarchical cluster analysis of expression profiles among stress responses. The hierarchical cluster analysis is based on correlation coefficient among expression profiles. Here the expression profiles sharing high correlation coefficients make clusters of groups

soluble protein denaturation. However, molecular chaperons such as *HSP104* are activated and heat shock treatment was shown to make yeast barotolerant (Iwahashi et al. 1991). This means that protein structures are affected by pressure. This can be explained by the observation that membrane proteins have the damages by high-pressure treatment. A significant number of genes whose products are localized in the ER are activated after the high-pressure treatment (Iwahashi et al. 2003b) (Table. 20.3).

20.3.3 Proteasome and Its Function Could Be the Crucial Target of Damages Under High Pressure Conditions

Proteins can be grouped into essential proteins and nonessential proteins. Here essential proteins are those proteins of yeast that are thought to be critical for its growth under specific conditions and nonessential proteins is not required under those conditions. Thus, nonessential proteins can be excluded from the proteins as crucial target of pressure. These essential proteins are encoded by the corresponding essential genes. It is suggested that yeast cell growing in YPD (2 % polypeptone, 1 % yeast extract, and 2 % glucose) medium requires 1,156 essential genes from the total of approximately 6,000 genes (http://www.yeastgenome.org/). As described above, we found 286 activated genes greater than two times with 180 MPa at 4 °C for 0 min and 218 those with 40 MPa at 4 °C for 16 h (Iwahashi et al. 2003b). From the two gene groups (the essential genes and the activated genes), we selected 13 common genes as the pressure-activated essential genes (Tanaka et al. 2010). Those are *KAR2* (Nuclear fusion protein), *NCE103* (Involved in non-classical protein export pathway), *OLE1* (Stearoyl-CoA desaturase), *PRE1* (20S proteasome subunit C11 (beta4)), *PRE3* (20S proteasome subunit (beta1)), YFR050c *PRE4* (20S proteasome subunit (beta7)), *RPN11* (26S proteasome regulatory subunit), YPR108w *RPN7* (Subunit of the regulatory particle of the proteasome), *RPN8* (26S proteasome regulatory subunit), *RPT2* (26S proteasome regulatory subunit), *RPT5* (26S proteasome regulatory subunit), YGL048c *RPT6* (26S proteasome regulatory subunit), and *UBC1* (E2 ubiquitin-conjugating enzyme). Surprisingly, almost all proteins contribute on proteasome functions. Thus, the crucial target of pressure apparently falls into proteasome-related proteins.

We also carried out metabolomics analysis of yeast cells recovered from conditions after pressure treatment of 40 MPa at 4 °C for 16 h (Tanaka et al. 2010). Metabolomics is an emerging field in analytical biochemistry, and the development of this field for comprehensive and quantitative analysis of organic acids, carbohydrates, and nucleotides is a necessity in the field of functional genomics (Tanaka et al. 2008). We assessed the metabolites of the cells restoring the damages after high-pressure treatment using CE/MS methods. We focused on amino acids, since these metabolites often contribute to the homeostasis of cells and can be easily separated from cells (Tanaka et al. 2008). The metabolomics analysis suggested the significant accumulation of glycine, valine, isoleucine, leucine, asparagine, aspartic acid, and tyrosine in the cells (Tanaka et al. 2010). Among these amino acids, glycine, valine, isoleucine, leucine, and tyrosine are known to be dominant in the membrane-spanning domain of transmembrane proteins (Tanaka et al. 2010). The metabolomics analysis agrees with genomic analysis that proteasome-related proteins are activated in recovering yeast cells from pressure damages. The activated proteasome may breakdown denatured transmembrane proteins.

20.3.4 The Crucial Target of Pressure Under Sublethal Conditions

Pressure-dependent gene activation and related results in yeast cells imply three kinds of candidate for the most crucial target of pressure. One is proteasome-related proteins as these genes are essential to grow and activated during recovering conditions. However, activated proteasome may be induced to degrade other proteins damaged by the pressure. As a significant number of genes whose products were localized in ER are activated after the high-pressure treatment (Iwahashi et al. 1991), the crucial target can be found from the substrates of proteasomes that are localized in ER. This substrate can be the second candidate and further screenings are required. The last candidate is indirect events after pressure treatment. The huge amount of denatured proteins by the pressure may have to be degraded to start cellular growth. The over capacity of proteasome functions may lead to the stop of cell cycles as the proteasome function is essential to make yeast grow in YPD medium.

20.4 Summary

To answer the question of "What happens on molecular levels in a biological cell when it is exposed to high hydrostatic pressure?" we evaluated activated genes under the two types of conditions. One is temporary adaptation of yeast cells to high pressure (Fig. 20.1 from A to B) and low pressure (Fig. 20.1 from B to A). The other is recovering condition from the damages after sublethal high-pressure treatment (Fig. 20.1 from A to C and D). Under the temporary adaptation conditions, yeast cells activate genes related to membrane biosynthesis and cell wall biosynthesis (Fig. 20.5a). Under the recovery conditions, yeast cells activate essential genes related to proteasome functions and proteins localized in ER can be the crucial proteins (Fig. 20.5b). These findings may be one step and may lead to new experiments for answering the question.

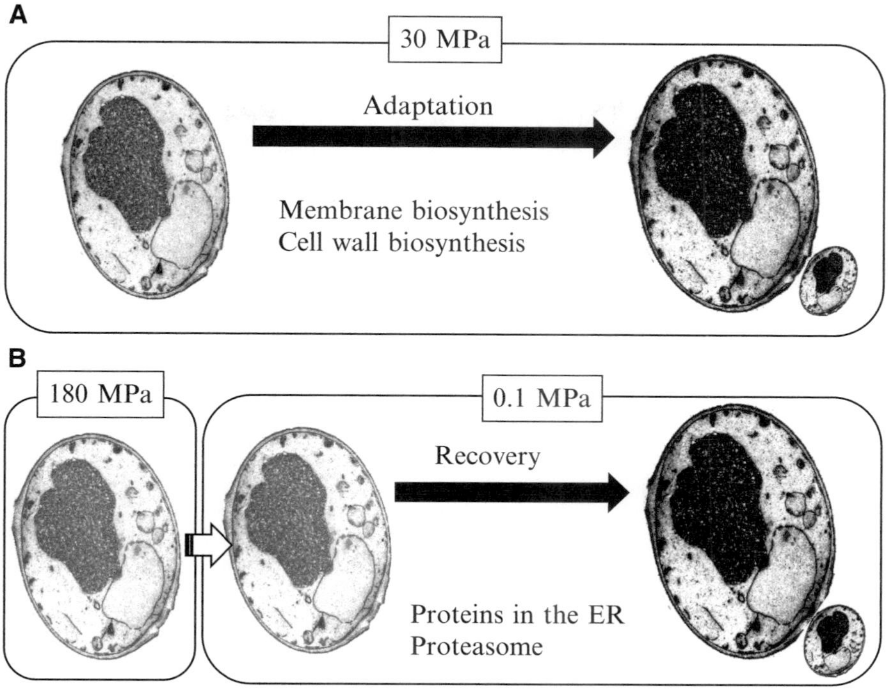

Fig. 20.5 What happens on molecular levels in a biological cell when it is exposed to high hydrostatic pressure? Temporary adaptation of yeast cells to high pressure (A). Recovery conditions after sublethal high pressure conditions (B)

References

Abe F (2004) Piezophysiology of yeast: occurrence and significance. Cell Mol Biol 50:437–445

Abe F, Horikoshi K (1995) Hydrostatic pressure promotes the acidification of vacuoles in *Saccharomyces cerevisiae*. FEMS Microbiol Lett 130:307–312

Abe F, Horikoshi K (2000) Tryptophan permease gene TAT2 confers high-pressure growth in *Saccharomyces cerevisiae*. Mol Cell Biol 20:8093–8102

Brown SL, Stockdale VJ, Pettolino F, Pocock KF, de Barros Lopes M, Williams PJ, Bacic A, Fincher GB, Høj PB, Waters EJ (2007) Reducing haziness in white wine by overexpression of *Saccharomyces cerevisiae* genes YOL155c and YDR055w. Appl Microbiol Biotechnol 73:1363–1376

Culbertson MR, Henry SA (1975) Inositol-requiring mutants of *Saccharomyces cerevisiae*. Genetics 80:23–40

Iwahashi H, Obuchi K, Kaul SC, Komatsu Y (1991) Induction of barotolerance by heat shock treatment in yeast. FEMS Microbiol Lett 80:325–328

Iwahashi H, Obuchi K, Fujii S, Komatsu Y (1997a) Barotolerance is dependent on both trehalose and hsp104 but essentially different from thermotolerance in *Saccharomyces cerevisiae*. Lett Appl Microbiol 25:43–47

Iwahashi H, Obuchi K, Fujii S, Komatsu Y (1997b) Effect of temperature on the role of Hsp104 and trehalose in barotolerance of *Saccharomyces cerevisiae*. FEBS Lett 416:1–5

Iwahashi H, Nwaka S, Obuchi K, Komatsu Y (1998) Evidence for the interplay between trehalose metabolism and heat shock proteins 104 in yeast. Appl Environ Microbiol 64:4614–4617

Iwahashi H, Shimizu H, Odani M, Komatsu Y (2003a) Piezophysiology of genome wide gene expression levels in the yeast *Saccharomyces cerevisiae*. Extremophile 7:291–298

Iwahashi H, Ishidou E, Odani M, Homma T, Oka S (2003b) Low pressure shock response of yeast. In: Winter R (ed) Advances in high pressure bioscience and biotechnology. Springer, Heidelberg, pp 275–278

Iwahashi H, Odani M, Ishidou E, Kitagawa E (2005) Adaptation of *Saccharomyces cerevisiae* to high hydrostatic pressure causing growth inhibition. FEBS Lett 579:2847–2852

Kitagawa E, Momose Y, Iwahashi H (2001) Correlation of the structures of agricultural fungicides to gene expression in *Saccharomyces cerevisiae* upon exposure to toxic doses. Environ Sci Tech 15:2788–2793

Kitagawa E, Momose Y, Iwahashi H (2003) Correlation of the structures of agricultural fungicides to gene expression in *Saccharomyces cerevisiae* upon exposure to toxic doses. Environ Sci Tech 15:2788–2793

Kodaki T, Yamashita S (1987) Yeast phosphatidylethanolamine methylation pathway. Cloning and characterization of two distinct methyltransferase genes. J Biol Chem 262:15428–15435

Koseki S, Yamamoto K (2006) Recovery of *Escherichia coli* ATCC 25922 in phosphate buffered saline after treatment with high hydrostatic pressure. Int J Food Microbiol 110:108–111

Masselot M, De Robichon-Szulmajster H (1975) Methionine biosynthesis in *Saccharomyces cerevisiae*. I. Genetical analysis of auxotrophic mutants. Mol Gen Genet 139:121–132

Mattison CP, Spencer SS, Kresge KA, Lee J, Ota IM (1999) Differential regulation of the cell wall integrity mitogen-activated protein kinase pathway in budding yeast by the protein tyrosine phosphatases Ptp2 and Ptp3. Mol Cell Biol 19:7651–7660

Momose Y, Iwahashi H (2001) Bioassay of cadmium using a DNA microarray: genome-wide expression patterns of *Saccharomyces cerevisiae* response to cadmium. Environ Toxicol Chem 20:2353–2360

Murata Y, Watanabe T, Sato M, Momose Y, Nakahara T, Oka S, Iwahashi H (2003) DMSO exposure facilitates phospholipid biosynthesis and cellular membrane proliferation in yeast cells. J Biol Chem 278:33185–33193

Ohshima S, Nomura K, Iwahashi H (2013) Clarification of the recovery mechanism of *Escherichia coli* after hydrostatic pressure treatment. High Pressure Res 33:308–314

O'Reilly CE, O'Connor PM, Kelly AL, Beresford TP, Murphy PM (2000) Use of hydrostatic pressure for inactivation of microbial contaminants in cheese. Appl Environ Microbiol 66:4890–4896

Osumi M, Sato M, Kobori H, Zha Hai feng, Ishijima AS, Hamada K, Shimada S (1996) Morphological effect of pressure stress on yeast. In: Hayashi R, Balony C (eds) High pressure bioscience and biotechnology. Elsevier, Amsterdam, pp 37–46

Sonoike K, Setoyama T, Kuma Y, Shinno T, Fukumoto K, Ishihara M (1993) Effects of pressure and temperature on the death rate of *Lactobacillus casei* and *Escherichia coli*. In: Hayashi R (ed) High pressure bioscience and food science. Sanei Press, Kyoto, pp 213–219

Tamura K, Shimizu T, Kourai H (1992) Effects of ethanol on the growth and elongation of *Escherichia coli* under high pressures up to 40 MPa. FEMS Microbiol Lett 78:321–324

Tanaka Y, Higashi T, Rakwal R, Wakida S, Iwahashi H (2008) Development of a capillary electrophoresis-mass spectrometry method using polymer capillaries for metabolomic analysis of yeast. Electrophoresis 29:2016–2023

Tanaka Y, Higashi T, Rakwal R, Shibato J, Wakida S, Iwahashi H (2010) The role of proteasome in yeast *Saccharomyces cerevisiae* response to sublethal high-pressure treatment. High Pres Res 30:519–523

Tsukada M, Ohsumi Y (1993) Isolation and characterization of autophagy-defective mutants of *Saccharomyces cerevisiae*. FEBS Lett 333:169–174

Yayanos AA, Pollard EC (1969) A study of the effects of hydrostatic pressure on macromolecular synthesis *in Escherichia coli*. Biophys J 9:1464–1482

Chapter 21
Environmental Adaptation of Dihydrofolate Reductase from Deep-Sea Bacteria

Eiji Ohmae, Kunihiko Gekko, and Chiaki Kato

Abstract In order to elucidate the molecular adaptation mechanisms of enzymes to the high hydrostatic pressure of the deep sea, we cloned, purified, and characterized more than ten dihydrofolate reductases (DHFRs) from bacteria living in deep-sea and ambient atmospheric pressure environments. The nucleotide and amino acid sequences of these DHFRs indicate the deep-sea bacteria are adapted to their environments after the differentiation of their genus from ancestors inhabiting atmospheric pressure environments. In particular, the backbone structure of the deep-sea DHFR from *Moritella profunda* (mpDHFR) almost overlapped with the normal homolog from *Escherichia coli* (ecDHFR). Thus, those of other DHFRs would also overlap on the basis of their sequence similarities. However, the structural stability of both DHFRs was quite different: compared to ecDHFR, mpDHFR was more thermally stable but less stable against urea and pressure unfolding. The smaller volume changes due to unfolding suggest that the native structure of mpDHFR has a smaller cavity and/or enhanced hydration compared to ecDHFR. High hydrostatic pressure reduced the enzymatic activity of many DHFRs, but three deep-sea DHFRs and the D27E mutant of ecDHFR exhibited pressure-dependent activation. The inverted activation volumes from positive to negative values indicate the modification of their structural dynamics, conversion of the rate-determining step of the enzymatic reaction, and different contributions of the cavity and hydration to the transition-state structure. Since the cavity and hydration depend on amino acid side chains, DHFRs would adapt to the deep-sea environment by regulating the cavity and hydration by substituting their amino

E. Ohmae (✉)
Department of Mathematical and Life Sciences, Graduate School of Science, Hiroshima University, Higashi-Hiroshima 739-8526, Japan
e-mail: ohmae@hiroshima-u.ac.jp

K. Gekko
Hiroshima Synchrotron Radiation Center, Hiroshima University, Higashi-Hiroshima 739-8526, Japan

C. Kato
Institute of Biogeosciences, Japan Agency for Marine-Earth Science and Technology (JAMSTEC), Yokosuka 237-0061, Japan

K. Akasaka, H. Matsuki (eds.), *High Pressure Bioscience*,
Subcellular Biochemistry 72, DOI 10.1007/978-94-017-9918-8_21

acid side chains without altering their backbone structure. The results of this study clearly indicate that the cavity and hydration play important roles in the adaptation of enzymes to the deep-sea environment.

Keywords Cavity and hydration • Deep-sea enzyme • Dihydrofolate reductase • Hydrostatic pressure • Molecular adaptation

Abbreviations

CSM	center of fluorescence spectral mass
DHF	dihydrofolate
DHFR	dihydrofolate reductase
THF	tetrahydrofolate

21.1 Introduction

The deep sea is an extreme environment characterized by darkness, low (or in some places, extremely high) temperature, and high hydrostatic pressure. Regardless of these unfavorable conditions for life, many organisms such as fishes, shellfishes, shrimps, crabs, and bacteria live in the deep sea. Since the intra- and extracellular temperature and pressure of deep-sea organisms should be the same, their enzymes—so-called "deep-sea enzymes"—must function under physiological conditions of the organisms. Furthermore, the mechanisms they use to adapt to such extreme conditions must involve the enzymes themselves. Understanding these environmental adaptation mechanisms at the molecular level could provide novel and useful knowledge that may elucidate the relationships among the structure, stability, and function of enzymes, leading to the industrial appreciations.

Dihydrofolate reductase (DHFR) [EC 1.5.1.3] is an indispensable enzyme for all organisms because of its role in the biosynthesis of purine nucleotides and some amino acids. Therefore, DHFR is regarded as a target enzyme for cancer, malaria, and infection drugs. Many studies have examined DHFRs from various organisms such as *Escherichia coli* (Schnell et al. 2004; Hammes et al. 2011; Lee and Goodey 2011), humans (Abali et al. 2008; Bertino 2009; Gonen and Assaraf 2012), *Candida albicans* (the causative microorganism of candidiasis) (Navarro-Martínez et al. 2006; Paulsen et al. 2009, 2011), the malaria parasite *Plasmodium falciparum* (Costanzo and Hartl 2011; Yuthavong et al. 2012), *Pneumocystis carinii* and *Pneumocystis jirovecii* (the causative microorganisms of Pneumocystis pneumonia) (Graffner-Nordberg et al. 2004; Cody et al. 2012, 2013), and methicillin-resistant *Staphylococcus aureus* (Oefner et al. 2009; Frey et al. 2010; Sunduru et al. 2011).

Furthermore, DHFR plays a role in the physiological conditions in which organisms live. Therefore, it is regarded as a model enzyme for studying the environmental adaptation of proteins from the hyperthermophilic bacterium *Thermotoga maritima* (Dams et al. 2000; Loveridge et al. 2009; Loveridge and Allemann 2010), the psychrophilic bacterium *Moritella profunda* (Xu et al. 2003a; Hay et al. 2009; Evans et al. 2010), and the extremely halophilic bacterium *Haloferax volcanii* (Pieper et al. 1998; Wright et al. 2002; Boroujerdi and Young 2009). The numerous studies on DHFR ensure that this enzyme is the first choice of protein to be examined in novel enzymatic studies. Accordingly, we selected DHFR as a model for studies on the functional characteristics and environmental adaptation mechanisms of deep-sea enzymes.

21.2 Isolation and Taxonomic Determination of Deep-Sea Bacteria

The development of full ocean depth submersibles has enabled the worldwide collection of deep-sea water or seafloor sediment samples including deep-sea bacteria (Fig. 21.1a). However, while many deep-sea samples are returned to the sea surface at their original temperature, their original pressure is rarely maintained, as this would require a specialized collection apparatus. Therefore, many obligatory piezophilic deep-sea bacteria may not be able to survive when collected and exposed to atmospheric pressure for a long time. However, some deep-sea bacteria can survive and grow under atmospheric pressure, although their growth rate is significantly reduced. Therefore, we usually cultivate and isolate deep-sea bacteria using a pressure vessel as follows (Kato et al. 1995; Kato 2011a). First, a small amount of deep-sea sample and marine broth 2216 (Difco Laboratories, Detroit, MI, USA) are sealed in a plastic bag and placed into a pressure vessel. Then, the vessel is placed in a refrigerator and incubated at 2–4 °C for 7–14 days under the same pressure at the depth at which the sample was collected. Next, a few grown bacteria are picked up by a disposable syringe and repeatedly cultivated using low-melting-point agar medium under the same pressure and temperature conditions to isolate single colonies. Electron microscopy images of some isolated deep-sea bacteria are shown in Fig. 21.1b–f.

The taxonomic determination of the isolated bacteria is performed on the basis of 5S and 16S ribosomal DNA sequences and other methods (Kato et al. 1998; Nogi et al. 2004). About a half of deep-sea microorganisms belong to archaea and others belong to various kinds of bacteria according to the 16S ribosomal DNA sequences (Kato et al. 2008a), but all cultivated deep-sea piezophilic bacteria included in the Gamma-proteobacteria subgroup (which includes well-known genera such as *Escherichia*, *Pseudomonas*, *Salmonella*, and *Vibrio*) belong to only five genera: *Shewanella*, *Moritella*, *Psychromonas*, *Photobacterium*, and *Colwellia* (Nogi et al. 2002; Nogi 2008; Kato 2011b). These five genera also contain species living in atmospheric pressure environments. Therefore, these deep-sea bacteria adapted to the high-pressure and low-temperature conditions of the deep sea after the differenti-

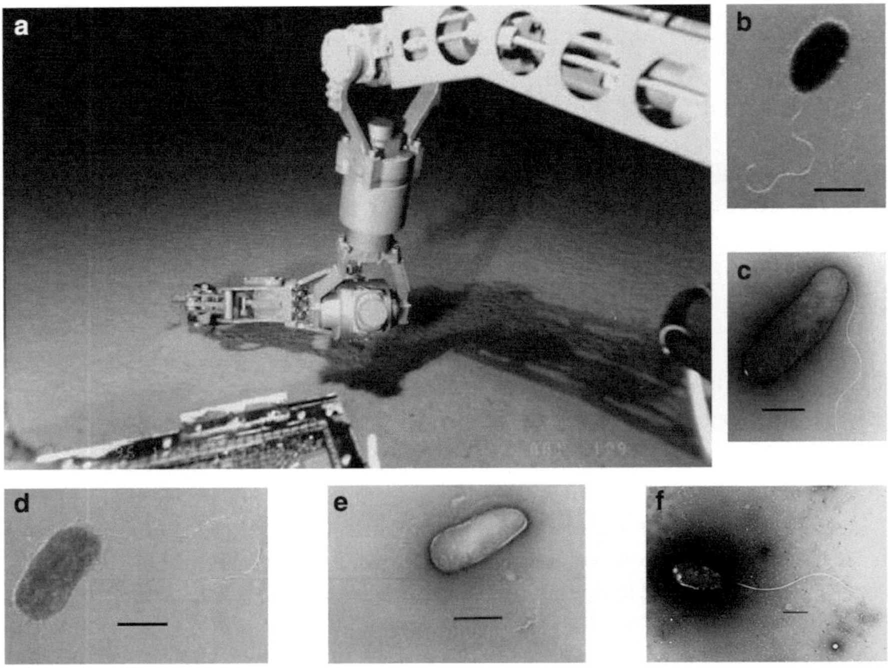

Fig. 21.1 Images of a manipulator of the KAIKO submersible collecting a deep-sea sediment sample in the Mariana Trench at a depth of 10,898 m by means of sterilized mud sample (**a**), and electron microscope images of the deep-sea bacteria *Shewanella violacea* strain DSS12 (**b**), *Moritella yayanosii* strain DB21MT-5 (**c**), *Moritella japonica* strain DSK1 (**d**), *Shewanella benthica* strain DB21MT-2 (**e**), and *Photobacterium profundum* strain DSJ4 (**f**). Scale bars in B–F: 1 μm (Pictures (a), (b), (e), and (f) are taken from Kato et al. (1997), Nogi et al. (1998b), Kato et al. (1998), and Nogi et al. (1998c), respectively, with permissions of publishers. Copyright holder of pictures (c) and (d) is JAMSTEC (1998–2014))

ation of their genera from atmospheric pressure environments. The bacterial species and their DHFR names used in this study are listed in Table 21.1.

21.3 Cloning and Overexpression of Deep-Sea DHFRs

Although the whole genome sequences of many bacteria are registered in the DDBJ/GenBank/EMBL sequence databases, only a few deep-sea bacteria have been sequenced. In addition, only the 16S ribosomal DNA sequences of some of the novel deep-sea bacteria used in this study are known. Taking advantage of the fact that the overexpression of DHFR protein confers a trimethoprim-resistant phenotype in *E. coli*, we initially cloned genomic DNA fragments obtained by restriction enzyme digestion into the pUC118 vector, introduced them into competent *E. coli*

Table 21.1 Descriptions of the bacterial species and DHFR names used in this article

Bacterial species	Isolation source or depth	Piezophilicity of bacteria	References	DHFR names
Shewanella benthica ATCC43992	4,575 m	Piezophilic	Deming et al. (1984), MacDonell and Colwell (1985)	sb43992DHFR
S. benthica DB21MT-2	10,898 m	Piezophilic	Kato et al. (1998)	sb21DHFR
S. benthica DB6705	6,356 m	Piezophilic	Kato et al. (1995), Nogi et al. (1998b)	sb6705DHFR
S. frigidimarina ACAM591	Antarctic sea ice	Piezosensitive	Bowman et al. (1997)	sfDHFR
S. gelidimarina ACAM456	Antarctic sea ice	Piezotolerant	Bowman et al. (1997)	sgDHFR
S. oneidensis MR-1	Oneida Lake	Piezosensitive	Venkateswaran et al. (1999)	soDHFR
S. putrefaciens IAM12079	Rancid butter	Piezosensitive	Owen et al. (1978)	spDHFR
S. violacea DSS12	5,110 m	Piezophilic	Kato et al. (1995), Nogi et al. (1998b)	svDHFR
Moritella abyssi 2693	2,815 m	Piezophilic	Xu et al. (2003b)	maDHFR
M. japonica DSK1	6,356 m	Piezophilic	Nogi et al. (1998a)	mjDHFR
M. marina	Sea water	Piezosensitive	Urakawa et al. (1998)	mmDHFR
M. profunda 2674	2,815 m	Piezophilic	Xu et al. (2003b)	mpDHFR
M. yayanosii DB21MT-5	10,898 m	Piezophilic	Kato et al. (1998)	myDHFR
Photobacterium phosphoreum	Sea water	Piezosensitive	Hastings and Nealson (1981)	ppDHFR
P. profundum DSJ4	5,110 m	Piezophilic	Nogi et al. (1998c)	ppr4DHFR
P. profundum SS9	2,551 m	Piezophilic	Bartlett et al. (1989)	ppr9DHFR
Psychromonas kaikoae JT7304	7,434 m	Piezophilic	Nogi et al. (2002)	pkDHFR
Escherichia coli		Piezosensitive		ecDHFR

cells, and selected colonies growing on a Luria broth plate containing 100 mg L^{-1} ampicillin and 5 mg L^{-1} trimethoprim (Murakami et al. 2010). The nucleotide sequences of the deep-sea DHFR genes located in the cloned DNA fragment, which were too long to read their overall nucleotide sequences, were determined using an inverted trimethoprim-sensitive mutant obtained by a transposon-based *in vitro* mutation method (Gene Jumper Kit; Invitrogen, Carlsbad, CA, USA) (Murakami et al. 2010). After the DNA sequences of several deep-sea DHFR genes were determined, degenerate PCR and a PCR-based chromosome walking technique

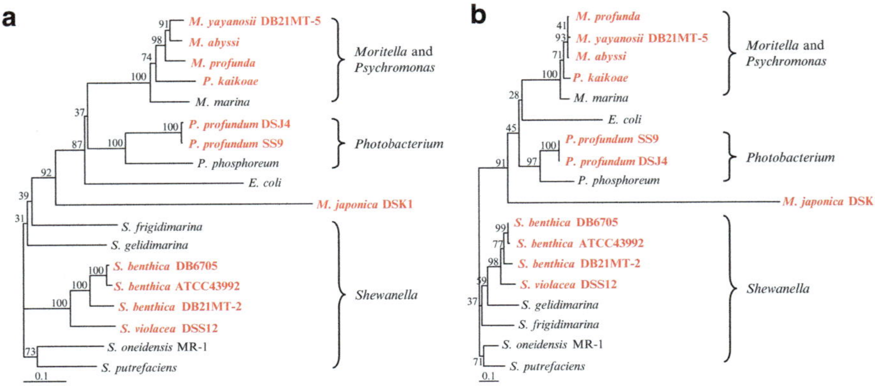

Fig. 21.2 Phylogenetic trees showing the relationships of deep-sea bacteria and their congeneric species constructed with the nucleotide (**a**) and amino acid (**b**) sequences of dihydrofolate reductases (DHFRs) using the neighbor-joining method. The scale represents the average number of substitutions per site. Bootstrap values (%) indicated at the branches were calculated from 1,000 trees. Deep-sea bacteria are indicated by *bold* letters

(DNA Walking Speed-up Kit; Seegene, Seoul, Korea) were used to clone other deep-sea DHFR genes (Murakami et al. 2011). A genomic DNA amplification method using a DNA polymerase from ϕ29 phage (GenomiPhi Kit; GE Healthcare UK Ltd., Buckinghamshire, UK) was also used to save the precious genomic DNA (Murakami et al. 2011). The development of these techniques enables the cloning and overexpression of deep-sea DHFRs from less than 1 ng genomic DNA from deep-sea bacteria. These techniques also enable the cloning and overexpression of DHFRs from uncultivable bacteria and metagenomic libraries.

Figure 21.2 shows the phylogenetic trees constructed using the determined nucleotide (a) and amino acid (b) sequences of DHFRs from bacteria living in deep-sea and ambient atmospheric pressure environments. As expected from the 16S ribosomal DNA sequences, each genus was essentially differentiated from the others except for one deep-sea bacterium: *M. japonica* strain DSK1, isolated from the Japan Trench at a depth of 6,356 m (Nogi et al. 1998a), was phylogenetically distant from the other *Moritella* species. However, as shown in Fig. 21.3, all DHFR proteins from these bacteria were successfully overexpressed in *E. coli* cells.

21.4 Primary and Tertiary Structures of Deep-Sea DHFRs

Figure 21.4 shows the primary structures of deep-sea DHFRs and their normal homologs from congeneric species living in ambient atmospheric pressure environments. The amino acid sequences of the N-terminal region were relatively conserved, but those of the C-terminal region were very variable. However, the length of the polypeptides was almost the same among species at approximately 160

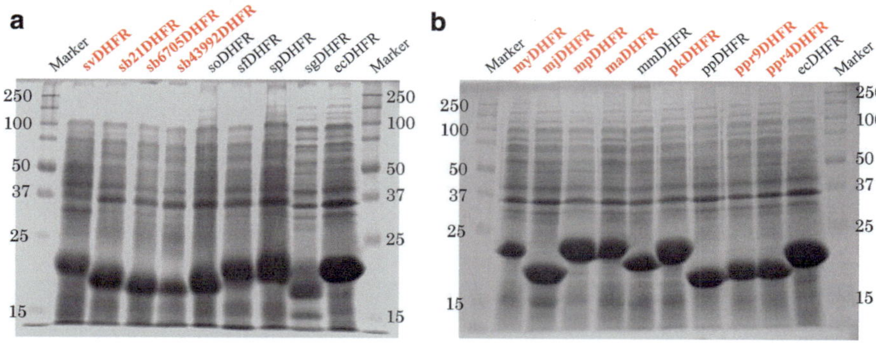

Fig. 21.3 SDS-PAGE of the overexpressed DHFRs from deep-sea and congeneric ambient atmospheric pressure bacteria in *Escherichia coli* cells. Deep-sea DHFRs are indicated by *bold* letters

residues. Furthermore, the active-site residue in the *Shewanella benthica* sequence, Asp28, was completely conserved in all DHFRs as an acidic amino acid residue, i.e., aspartic or glutamic acid. These results suggest that all DHFRs fold in a similar fashion to maintain enzyme function.

Figure 21.5a shows a superimposed drawing of the *E. coli* DHFR (ecDHFR, PDB ID: 1rx2) (Sawaya and Kraut 1997) and the deep-sea DHFR from *M. profunda* (mpDHFR, PDB ID: 2zza) (Ohmae et al. 2012) collected off the West African coast at a depth of 2,815 m (Xu et al. 2003b). As shown in Fig. 21.5a, the backbone structures of both DHFRs almost overlap even though their sequence similarity was only 55 % (Fig. 21.4). Since the other DHFRs except that of *M. japonica* (mjDHFR) have comparable sequence similarity (48–56 %), their backbone structures would also overlap with that of ecDHFR. Although the sequence similarity of mjDHFR to ecDHFR was only 32 %, similar values are reported for human and mouse DHFRs, whose backbone structures are very similar to that of ecDHFR (Oefner et al. 1988; Stammers et al. 1987). Therefore, the backbone structure of mjDHFR would also be close to that of ecDHFR.

Another deep-sea enzyme of known crystallographic structure is 3-isopropylmalate dehydrogenase from *S. benthica* strain DB21MT-2, which was isolated from the Mariana Trench at a depth of 10,897 m (Kato et al. 1998). As shown in Fig. 21.5b, the backbone structure of this enzyme (PDB ID: 3vmk) completely overlaps with that of the normal homolog from *S. oneidensis* strain MR-1 (PDB ID: 3vmj) (Nagae et al. 2012), which was collected from Oneida Lake (Venkateswaran et al. 1999). Similar structural conservation was also observed in other deep-sea enzymes such as aspartate carbamoyltransferase from *M. profunda* (De Vos et al. 2007), α-glucosidase from *Geobacillus* sp. strain HTA-462 (Shirai et al. 2008), Cu/Zn superoxide dismutase from the deep-sea yeast *Cryptococcus liquefaciens* strain N6 (Teh et al. 2008), and superoxide dismutase from the deep-sea worm *Alvinella pompejana* (Shin et al. 2009), as found for other hyperthermophilic enzymes (Vieille and Zeikus 2001). Deep-sea enzymes with unknown crystal structures

	10 20 ↓30 40 50 60 70 80 90	100 110 120 130 140 150 160	Length	Identity	Accession No.
sb43992DHFR	-MKIAMIVAMANNRVIGKDNKMPWHLPEDLKHFKAMTLGKPIVMGRKTYESIGRALPGRLNIVISRQEGLSIAGVTCVTSFESAVAAAGEC	EELVVIGGG-QLYAALLSQADKLYLTEINLDVDGDTHFPRWDDGSWQRISQETAINDKGLEYSFINLVKKC-----	160	49%	AB538271
sb6705DHFR	-MKIAMIAAMANNRVIGKDNKMPWHLPEDLKHFKAMTLGKPIVMGRKTYESIGRALPGRLNIVISRQEGLSIAGVTCVTSFESAVAAAGEC	EELVVIGGG-QLYAALLSQADKLYLTEINLDVDGDTHFPRWDDGSWQRISQETAINDKGLEYSFINLVKKC-----	160	50%	AB538272
sb21DHFR	-MKIAMIAAMANNRVIGKNNQMPWHLPEDLKHFKAMTLGKPIVMGRKTYESIGRALPGRLNIVISRQQGLTIPGVTCVTSFESAVAAAGEC	EELVVIGGG-QLYAALLSQADKLYLTEINLDVDGDTHFPQWDDGSWQRVSQETSINDKGVEYSFINLVKKC-----	160	51%	AB538270
svDHFR	-MKIAMIAAMANNRVIGKDNQMPWHLPEDLKHFKAMTLGKPIVMGRKTYDSIGRALPGRLNIVISRQEGLSIPGVTCVTSFEMAVEAAGEC	EELVVIGGG-QLYASLLSKADKLYLTEINLDVAGDTFFPQWDDGSWERISQEMLVNGAGLEYSFINLVKKC-----	160	52%	AB519243
sgDHFR	ILKIAMIAAMANNRVIGKDNKMPWHLPEDLRHFKAVTLGKPIVMGRKTYESIGRPLPGRHNIVISRQADLAIDGVTTVTCFEDAKLAAGDC	EELVVIGGG-QLYTSLLCLSDVLYLTEISLDVEGDTYFPHWDDGSWLEMSRDVAISGKQLEYSFIKMVKKC-----	161	48%	AB538273
spDHFR	-MRIAMIAAMANNRVIGKDNKMPWHLPEDLKHFKAMTLGKPVVMGRKTFESIGRPLPGRHNIVISRQADLVIEGVTSVTSFDAAKQAAGEC	EELVIMGGG-QLYRELLPQADILYLTQINLDVEGDTYFPEWDARDWQETESLSCINADGLEYRFINLTKKC-----	160	53%	CP000681
soDHFR	-MRIAMIAAMANNRVIGKDNKMPWHLPEDLRHFKAMTLGKPVVMGRKTFESIGRPLPGRHNIVISRQADLQIEGVTCVTSFEAAKRVAGDC	EELVVIGGG-QLYKQLLPQADRLYLTQINLDVDGDTFFPAWEDNEWQKTETQPSISADGLEYNFINLVKKC-----	160	48%	AE014299
sfDHFR	-MRIAMIAAMANNRVIGKDNKMPWHLPEDLRHFKAMTLSKPVVMGRKTYESIGRPLPGRHNIVISRNSQLSIEGVTCVKSFDEAITVAGDC	EEIVVIGGG-QLYQQLLPQADVLYLTLIDVDVDGDTVFPAWDDGSWDLQNSVTATNDKGLQYSFNTLVKKC-----	160	51%	CP000447
mmDHFR	-----MIAALANNRVIGLDNKMPWHLPAELQLFKRATLGKPIVMGRNTFESIGRPLPGRLNIVLSRQRDYVAEGVTVVATLEDAIVAAGDV	EELMIIGGA-TIYSQCLAAADRLYLTHIELTTAGDTWFPDYEQYNWQEIEQEYFVADDNNPHDYRFSLLERVK---	158	55%	AB862596
mjDHFR	-MKLSLMAAKSKNGVIGNGPDIPWQAKGEQLLFKAMTYNQWLLVGRKTFESMGTLPNRKYAVISRSDFVSEDENVMVFSSIESALRQLADI	TDHVIVSGGGEIYRSLISKVDTLHISTIDIETEGDIVFPDIPDSFSLAFEQKFESNINYCYQIWQNS---------	157	32%	AB505968
myDHFR	-----MIAALANNRVIGLDNKMPWHLPAELQLFKRATLGKPIVMGRNTFESIGRPLPGRLNIVLSRQTDYQPEGVTVVATLEDAVVAAGDV	EELMIIGGA-TIYNQYLAAADRLYLTHIELTIEGDTWFPDYEQYHWQEIEHESYAADDKNPHDYRFSLLERVK---	158	55%	AB505967
mpDHFR	-MIVSMIAALANNRVIGLDNKMPWHLPAELQLFKRATLGKPIVMGRNTFESIGRPLPGRLNIVLSRQTDYQPEGVTVVATLEDAVVAAGDV	EELMIIGGA-TIYNQCLAAADRLYLTHIELTTEGDTWFPDYEQYNWQEIEHESYAADDKNPHNYRFSLLERVK---	162	55%	AJ487535
maDHFR	-----MIAALANNRVIGLDNKMPWHLPAELQLFKRATLGKPIVMGRNTFESIGRPLPGRLNIVLSRQTDYQPEGVTVVATLEDAVVAAGDV	EELMIIGGA-TIYNQCLAAADRLYLTHIELTTEGDTWFPDYEQYNWQEIEHESYAADDKNPHDYHFSLLERVK---	158	55%	AB862595
pkDHFR	-----MIAALANNRVIGLDNKMPWHLPAELQLFKRATLGKPIVMGRNTFESIGRPLPGRLNIVLSRQTGYQPEGVTVVATLEDAVIAAGDV	EELMIIGGA-TIYQQCLAAADRLYLTHIELTTAGDTWFPDYEQYNWHEIVHENYSADDKNPYDYHFSLLERVK---	158	54%	AB862599
ppDHFR	-MKISMIAAMAHDRIIGKDNSMPWHLPADFAWFKRCTLGKPIVMGRKTFESIGRPLPGRLNIVISRNPNFTAEGITVVSDIAAAKLVAGDA	PELMIIGGG-SIYQACLPIADSLYLTFIDADLEGDTQFPDWGEH-WHRVESQRYNADDKNAYAMEFVILERS----	160	55%	AB862598
ppr4DHFR	-----MIAAMAKDRVIGKDNAMPWHLPADFAWFKKSTLGKPIVMGRKTFESIGRPLPGRLNIVISRNPEFSVEGVTVVADIEAAKHAAGDI	GELMVIGGG-SIYEACLPSADRLYLTFIDLDVDGDTCFPDWGEG-WIKSYTEHYSADEKNAHDMQFVILDRNVDLT	160	55%	AB862597
ppr9DHFR	-----MIAAMAKDRVIGKDNAMPWHLPADFAWFKKSTLGKPIVMGRKTFESIGRPLPGRLNIVISRNPEFSVEGVTVVADIEAAKHAAGDI	GELMVIGGG-SIYEACLPSADRLYLTFIDLDVDGDTCFPDWGEG-WIKSYTEHYSADEKNAHDMQFVILERNVDLT	160	56%	AB505966
ecDHFR	--MISLIAALAVDRVIGMENAMPWNLPADLAWFKRNTLNKPVIMGRHTWESIGRPLPGRKNIILSSQP-GTDDRVTWVKSVDEAIAACGDV	PEIMVIGGG-RVYEQFLPKAQKLYLTHIDAEVEGDTHFPDYEPDDWESVFSEFHDADAQNSHSYCFEILERR----	159	100%	AP009048

Fig. 21.4 Amino acid sequences of DHFRs from deep-sea bacteria and their congeneric species. Deep-sea DHFRs are indicated by *bold* letters. Residue numbering is based on the sequence of the DHFR from *S. benthica*. Conserved amino acid residues are indicated by *shadows* employing fully conserved residues as *bold* letters. The active site residue is indicated by an *arrow* on the numbering row. The sequence length, level of identity with ecDHFR, and accession numbers for the DDBJ/GenBank/EMBL sequence databases are also indicated at the end of each sequence

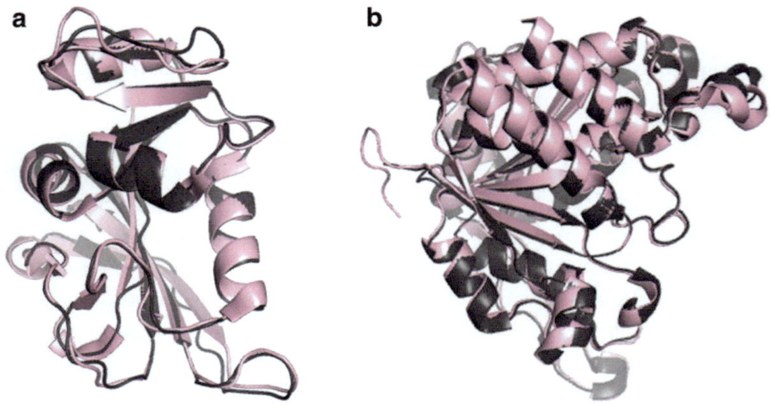

Fig. 21.5 Superimposed drawings of the crystal structures of deep-sea enzymes (*pale* color) and their normal homologs (*dark* color). (**a**) mpDHFR (PDB code: 2zza; *pale*) and ecDHFR (1rx2; *dark*). (**b**) 3-Isopropylmalate dehydrogenases (subunit of the homodimer) from *S. benthica* strain DB21MT-2 (3vmk; *pale*) and *S. oneidensis* strain MR-1 (3vmj; *dark*) (Figures drawn using PyMol software [http://www.pymol.org/])

generally have the same folded structures as their normal homologs in homology modeling (Purcarea et al. 1997; Saito et al. 2006; Kato et al. 2008b; Brindley et al. 2008; Xie et al. 2009). Taken together, these results indicate that the tertiary structures of deep-sea enzymes are essentially the same as those of their homologs from bacteria living in ambient atmospheric pressure environments.

21.5 Structural Stability of Deep-Sea DHFRs

It was previously believed that deep-sea enzymes are more thermally unstable than those from bacteria living in atmospheric pressure environments, because the deep sea is cold in addition to being a high-pressure environment. Indeed, many deep-sea enzymes exhibit maximum activity at lower temperatures than mesophilic enzymes (Zhang et al. 2007; Zhang and Zeng 2008; Murakami et al. 2010; Yang and Dang 2011). However, the temperature of maximum activity is not necessarily a measure of structural stability. In the present study, the thermal stability of mpDHFR was considerably higher than that of ecDHFR (Ohmae et al. 2012). Figure 21.6 shows the temperature dependence of the far-ultraviolet circular dichroism spectra of ecDHFR and mpDHFR. The ellipticities in the circular dichroisn spectrum of ecDHFR around 220 nm become less negative with increasing temperature (Fig. 21.6a), indicating the thermal unfolding of the secondary structure. In contrast, the ellipticities of mpDHFR become more negative with increasing temperature (Fig. 21.6b), suggesting an increase in secondary structures. Similar abnormal thermal stability was also observed in other deep-sea DHFRs from *S. violacea* (svDHFR) isolated from the Ryukyu Trench at a depth of 5,110 m (Kato et al. 1995) and *Photobacterium profundum* strain SS9 (ppr9DHFR) isolated from the Sulu Sea at a depth of 2,551 m (Bartlett et al. 1989), although these DHFRs aggregated at 65 °C and 70 °C, respectively (Fig. 21.6c). Therefore, the thermal stability of DHFRs is not necessarily proportional to the temperature at which organisms inhabit.

The structural stability of ecDHFR and mpDHFR against urea and pressure was monitored using fluorescence spectroscopy (Ohmae et al. 2012). The center of fluorescence spectral mass (CSM) was calculated using the following equation:

$$CSM = \Sigma\lambda_i \ F_i / \Sigma F_i \tag{21.1}$$

where λ_i and F_i are the wave number (in cm^{-1}) and fluorescence intensity at wavelength *i*, respectively (Vidugiris and Royer 1998). The resultant CSM data were plotted against urea concentration (Fig. 21.7a) and directly fitted to the two-state unfolding model—native (N) $\Longleftrightarrow$ unfolded (U)—by nonlinear least-squares regression analysis as follows:

$$CSM_{obs} = \{CSM_N + CSM_U \exp(-\Delta G_u/RT)\} / \{1 + \exp(-\Delta G_u/RT)\} \tag{21.2}$$

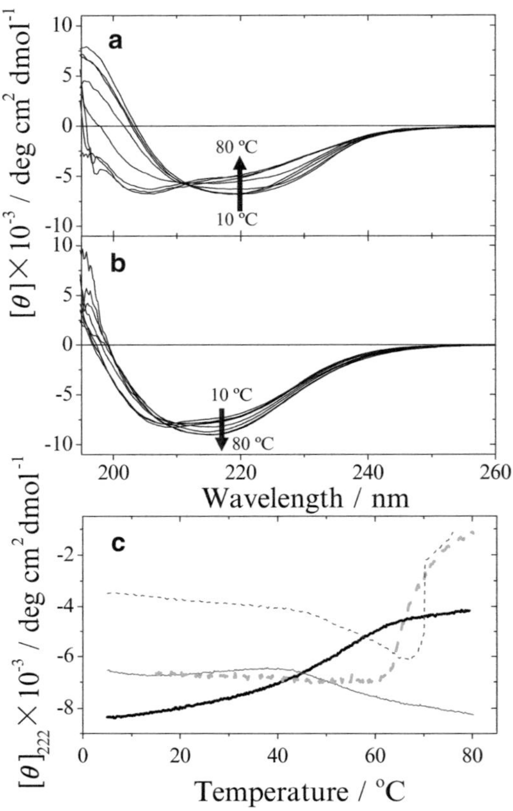

Fig. 21.6 Temperature dependence of the far-ultraviolet circular dichroism spectra of ecDHFR (**a**) and mpDHFR (**b**) at pH 8.0. The buffer used was 20 mM Tris-hydrochloride containing 0.1 mM EDTA and 0.1 mM dithiothreitol. (**c**) Temperature dependence of the molar ellipticities at 222 nm of ecDHFR (*thick continued line*) and three deep-sea DHFRs: mpDHFR (*thin continued line*), svDHFR (*thick dashed line*), and ppr9DHFR (*thin dashed line*). The buffer used was the same as in panels a and b, except that the pH was 7.0 for svDHFR and ppr9DHFR (Figure reproduced from Ohmae et al. (2012) and additional data)

where R is the gas constant, T is temperature, ΔG_u is the change in Gibbs free energy due to unfolding, and CSM_N and CSM_U are the CSM values of the native and unfolded forms, respectively. CSM_N and CSM_U at a given urea concentration were estimated by assuming the same linear dependence in the transition region as the pure native (i.e., pre-transition region) and unfolded states (i.e., post-transition region). The urea concentration dependence of ΔG_u was calculated as follows (Pace 1985):

$$\Delta G_u = \Delta G_U^o - m\,[\text{urea}] \qquad (21.3)$$

where $\Delta G°_U$ is the Gibbs free energy change due to urea unfolding in the absence of denaturant, and the slope m is the parameter reflecting the cooperativity of the transitions. The urea concentration at the midpoint of the transition (i.e., $\Delta G_u = 0$) was defined as C_m. The calculated values of $\Delta G°_U$, m, and C_m are listed in Table 21.2. The $\Delta G°_U$ value of mpDHFR was approximately one-third of ecDHFR, indicating that mpDHFR is significantly less stable than ecDHFR. The DHFRs from six *Shewanella* species living in both deep-sea and ambient atmospheric pressure

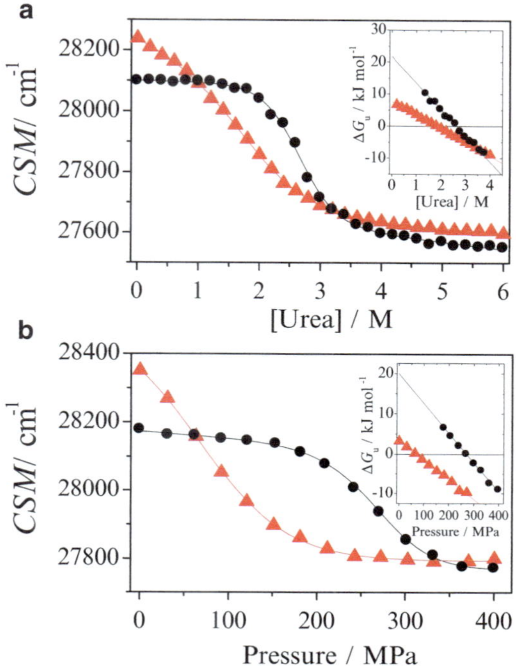

Fig. 21.7 Urea concentration (**a**) and pressure (**b**) dependences of the CSM of mpDHFR (*triangle*) and ecDHFR (*circle*) at pH 8.0. The experimental temperatures were 25.0 °C (a) and 20.4 °C (b). *Lines* indicate the non-linear least-squares fits for Eqs. 21.2, 21.3, and 21.4 (see Sect. 21.5). *Insets* show urea concentration (a) and pressure (b) dependence of the changes in Gibbs free energy due to protein unfolding (Figure reproduced from Ohmae et al. (2012))

Table 21.2 Thermodynamic parameters for the urea denaturation of ecDHFR and seven DHFRs at pH 8.0 determined by fluorescence spectra[a]

DHFR	Temperature (°C)	ΔG^{o}_{U} (kJ mol^{-1})	m (kJ mol^{-1} M^{-1})	C_m (M)
ecDHFR[b]	25	21.8 ± 1.8	8.2 ± 0.7	2.7 ± 0.3
ecDHFR[c]	15	26.9 ± 3.7	9.4 ± 1.2	2.9 ± 0.5
mpDHFR[b]	25	7.9 ± 0.6	4.3 ± 0.2	1.8 ± 0.2
svDHFR[c]	15	8.0 ± 0.5	3.5 ± 0.2	2.3 ± 0.2
sb21DHFR[c]	15	8.7 ± 1.2	4.6 ± 0.3	1.9 ± 0.3
sb6705DHFR[c]	15	7.9 ± 0.9	4.7 ± 0.3	1.7 ± 0.2
sfDHFR[c]	15	8.3 ± 0.9	4.0 ± 0.2	2.1 ± 0.2
soDHFR[c]	15	6.7 ± 1.0	5.9 ± 0.3	1.1 ± 0.2
spDHFR[c]	15	8.3 ± 1.4	6.9 ± 0.6	1.2 ± 0.2

[a]The buffer used was 20 mM Tris-hydrochloride containing 0.1 mM EDTA and 0.1 mM dithiothreitol
[b]Ohmae et al. (2012)
[c]Murakami et al. (2011)

environment were also less stable than ecDHFR (Table 21.2) (Murakami et al. 2011). As these *Shewanella* DHFRs share more than 80 % sequence similarity with each other and all of them exhibited abnormal thermal stability (unpublished data), it appears that the ancient *Shewanella* species already possessed a heat-resistant but urea-vulnerable DHFR before adapting to the deep-sea environment.

The pressure unfolding data could also be fitted to Eq. 21.2 (Fig. 21.7b). The pressure dependence of ΔG_u was calculated as follows:

$$\Delta G_u = \Delta G^o_p + P\Delta V_P \qquad (21.4)$$

where ΔG°_P is the Gibbs free energy change due to pressure unfolding at 0 MPa, P is the pressure, and ΔV_P is the partial molar volume change due to pressure unfolding. ΔG°_P can be regarded as the Gibbs free energy change at atmospheric pressure, because the pressure difference of 0.1 MPa is negligible in this experiment. The calculated ΔG°_P values at 20.4 °C for ecDHFR and mpDHFR were 20.0 ± 2.6 and 3.3 ± 0.3 kJ mol^{-1}, respectively, further indicating the lower structural stability of mpDHFR (Ohmae et al. 2012).

Interestingly, the ΔV_P of mpDHFR (−49 ± 3 mL mol^{-1}) was smaller (i.e., less negative) than that of ecDHFR (−74 ± 11 mL mol^{-1}). The volume change due to the urea unfolding of mpDHFR (−53 ± 7 mL mol^{-1}), which was obtained from the pressure dependence of the ΔG°_U values, was also smaller (i.e., less negative) than that of ecDHFR (−85 ± 7 mL mol^{-1}) (Ohmae et al. 2012). These volume changes provide useful information about the mechanism of pressure adaptation of DHFR structure (see Sect. 21.7).

21.6 Enzymatic Function of Deep-Sea DHFRs

The most interesting characteristic of deep-sea enzymes is the pressure-induced activation of their enzymatic activity. The enzymatic activity of three deep-sea DHFRs—svDHFR, mpDHFR, and DHFR from *S. benthica* strain DB21MT-2 (sb21DHFR)—increased as the pressure was increased up to 50 MPa and then gradually decreased with additional pressurization (Fig. 21.8a) (Murakami et al. 2010, 2011; Ohmae et al. 2012). Conversely, the enzyme activities of all other DHFRs from bacteria living in both deep-sea and ambient atmospheric pressure environments did not exhibit such pressure activation. Therefore, it appears that some pressure adaptation mechanisms are involved in the activities of these three deep-sea DHFRs. However, such pressure adaptation mechanisms are not specific to these deep-sea enzymes. We recently found that a single methylene insertion at the active site, D27E, can substantially invert the pressure dependence of the enzyme activity of ecDHFR (Fig. 21.8a) (Ohmae et al. 2013a). Since the structural stability of ecDHFR against pressure is higher than that of mpDHFR (Fig. 21.7b), this mutant could adapt to and functions sufficiently in the deep-sea environment.

From the results of the enzymatic activity, we calculated the activation free energy (ΔG^*) and activation volume (ΔV^*) of the enzymatic reaction (Fig. 21.8b) using the following equation, which is applicable at saturating substrate concentrations:

$$\Delta V^* = \partial \Delta G^* / \partial P = \partial \left(-RT \ln k_{cat}\right) / \partial P = \partial \left(-RT \ln v\right) / \partial P \qquad (21.5)$$

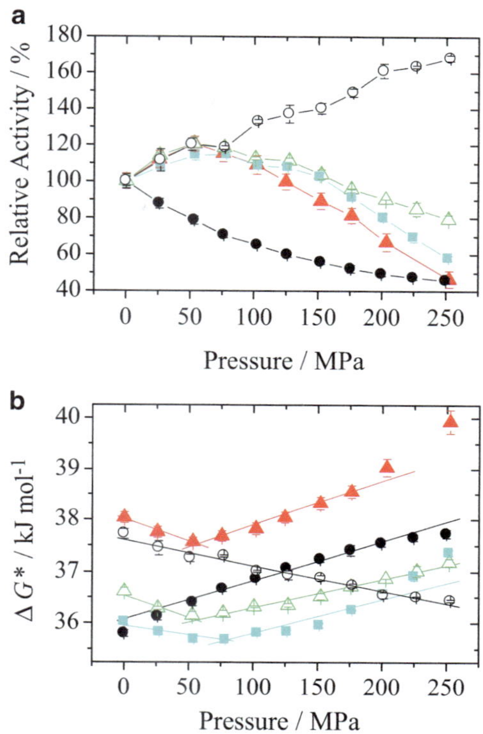

Fig. 21.8 Pressure dependence of the relative activity (**a**) and activation free energy (**b**) of three deep-sea DHFRs—mpDHFR (*closed triangle*), svDHFR (*open triangle*), and sb21DHFR (*closed square*)—compared with wild-type ecDHFR (*closed circle*) and its D27E mutant (*open circle*). The experimental temperature was 25.0 °C. *Lines* in panel b indicate linear fits (Figure reproduced from published data (Murakami et al. 2011; Ohmae et al. 2012; 2013a))

where v is the initial velocity (Ohmae et al. 2008). Table 21.3 lists the calculated activation volumes of the enzymatic reaction for the deep-sea and homologous DHFRs. The activation volume ranged from -8.6 mL mol^{-1} for svDHFR and mpDHFR to 30.5 mL mol^{-1} for sfDHFR obtained from *S. frigidimarina* strain ACAM591 collected from Antarctic sea ice (Bowman et al. 1997). Since the activation volume reflects the volume changes of the cavity and hydration between the ground and transition states in the rate-determining step of the enzymatic reaction as discussed in Sect. 21.7, the variations of this value indicate variations in the rate-determining step.

The enzymatic reaction of DHFR includes at least five steps: two binding steps of dihydrofolate (DHF) and NADPH, a chemical oxidation–reduction step, and two releasing steps of tetrahydrofolate (THF) and $NADP^+$; the rate-determining step of ecDHFR at neutral pH and atmospheric pressure is the THF-releasing step (Fierke et al. 1987). Thus, the positive activation volumes for sb21DHFR (6.5 ± 0.1 mL mol^{-1}) and mpDHFR (8.6 ± 0.9 mL mol^{-1}) at the higher pressure regions, which are comparable to that of ecDHFR (7.5 ± 0.2 mL mol^{-1}), indicate that the THF-releasing step is also the rate-determining step in the enzymatic reaction of both DHFRs. In addition, the much larger positive activation volumes for sfDHFR at the lower (14.0 ± 0.1 mL mol^{-1}) and higher

Table 21.3 Activation volumes at 25 °C and pH 7.0 for the enzymatic reaction of DHFRs obtained from bacteria living in deep-sea and ambient atmospheric pressure conditions[a]

DHFR	ΔV^* (mL mol^{-1})	
ecDFHR (wild-type)[b]	7.5 ± 0.2 (0.1–250 MPa)	
ecDHFR (D27E mutant)[c]	−4.8 ± 0.1 (0.1–250 MPa)	
svDHFR[d]	−8.6 ± 1.9 (0.1–50 MPa)	5.6 ± 0.1 (50–250 MPa)
sb21DHFR[d]	−3.5 ± 0.6 (0.1–75 MPa)	6.5 ± 0.1 (75–250 MPa)
sb6705DHFR[d]	2.0 ± 0.1 (0.1–25 MPa)	29.0 ± 0.3 (25–250 MPa)
sfDHFR[d]	14.0 ± 0.1 (0.1–125 MPa)	30.5 ± 0.2 (125–250 MPa)
soDHFR[d]	4.1 ± 1.4 (0.1–50 MPa)	13.1 ± 0.2 (50–250 MPa)
spDHFR[d]	11.5 ± 0.2 (0.1–125 MPa)	23.3 ± 1.0 (125–250 MPa)
mjDHFR[b]	38.7 ± 0.3 (0.1–250 MPa)	
myDHFR[b]	1.7 ± 0.6 (0.1–75 MPa)	16.5 ± 0.6 (75–250 MPa)
mpDHFR[e]	−8.6 ± 2.5 (0.1–50 MPa)	8.6 ± 0.9 (50–250 MPa)
ppr9DHFR[b]	13.8 ± 0.4 (0.1–250 MPa)	

Values in *parentheses* indicate the pressure range used for the calculation
[a]The buffer used was 20 mM Tris-hydrochloride containing 0.1 mM EDTA, 0.1 mM dithiothreitol, 250 μM NADPH, and 250 μM DHF
[b]Murakami et al. (2010)
[c]Ohmae et al. (2013a)
[d]Murakami et al. (2011)
[e]Ohmae et al. (2012)

(30.5 ± 0.2 mL mol^{-1}) pressure regions suggest that the $NADP^+$-releasing step and conformational changes such as partial unfolding, respectively, are the rate-determining steps, because the two binding steps cannot be the rate-determining steps under saturated concentrations of DHF and NADPH (both 250 μM). Conversely, when the activation volume is negative, the transition state has a smaller cavity and/or enhanced (i.e., higher) hydration than the ground state; this is the case when the chemical oxidation–reduction step is the rate-determining step (Northrop and Cho 2000; Quirk and Northrop 2001), because partial charges would be stabilized by the hydration or condensation of the hydrated water in the transition-state structure.

21.7 Adaptation Mechanisms of DHFR to the Deep-Sea Environment

As the structural differences between deep-sea enzymes and their normal homologs are small (Fig. 21.5), the pressure adaptation mechanisms of the deep-sea enzymes must originate from other sources. Since hydrostatic pressure affects the volume of a system, volume changes due to protein unfolding or during the enzymatic reaction are important for the molecular adaptation mechanisms of enzymes to the deep-sea environment. The partial molar volume ($V°$) of a protein in water is determined by the following three factors: the constitutive atomic volume (V_c), cavity resulting from imperfect atomic packing (V_{cav}), and volume change as a result of hydration (ΔV_{sol}) (Kauzman 1959; Ohmae et al. 2013b):

$$V^{o} = V_{c} + V_{cav} + \Delta V_{sol} \tag{21.6}$$

Since the constitutive atomic volume does not change during protein unfolding, i.e., the constitutive atomic volume is independent of the conformation of the protein, the volume change due to unfolding (ΔV_P) is a product of the modified internal cavity and hydration as follows:

$$\Delta V_P = V^{o}\,(\text{unfolded state}) - V^{o}\,(\text{native state}) = \Delta V_{cav} + \Delta\Delta V_{sol} \tag{21.7}$$

Cavities contribute positively and hydration contributes negatively to $V°$; hence, the experimentally determined negative value of ΔV_P can be ascribed to decreased cavity volume and/or increased hydration due to unfolding. However, the effect of hydration, $\Delta\Delta V_{sol}$, originates from the volume changes of external voids surrounding the protein molecule, because the constitutive atomic volume of water molecules also does not change. Furthermore, the internal cavities and external voids are readily exchanged, because the conformation of the protein fluctuates rapidly in solution. Therefore, the individual contributions of the cavity and hydration to ΔV_P cannot be discriminated by isothermal compression experiments, although they can be discriminated quantitatively by adiabatic compression experiments in which the protein compressed faster than its conformational fluctuation.

As discussed in Sect. 21.5, mpDHFR has smaller ΔV_P values hence it is expected to take on a loosely packed and highly hydrated native structure in solution, compared with ecDHFR. There are some indirect evidences for this hypothesis. Significant pressure and urea-concentration dependences of the fluorescence spectra of mpDHFR in the native state (Fig. 21.7) suggest easy accession of solvent molecules to the tryptophan side chains located on the inside of the protein molecule. Abnormal thermal stability of deep-sea DHFRs (Fig. 21.6) also supports the enhanced hydration in the native structure, which reduces the heat capacity change due to thermal unfolding. Since the cavity and hydration depend on the amino acid side chains, enzymes could adapt to the deep-sea environment by substituting the amino acid side chains without changing their backbone structure.

Equation 21.7 is also applicable for ΔV^* as follows:

$$\Delta V^{*} = V^{o}\,(\text{transition state}) - V^{o}\,(\text{ground state}) = \Delta V_{cav} + \Delta\Delta V_{sol} \tag{21.8}$$

Therefore, the experimentally observed negative and positive values of ΔV^* can be ascribed to decreased cavity volume and/or increased hydration between the transition- and ground-state structures in the rate-determining step and vice versa, respectively. Since the change of the protein structure during the enzymatic reaction is small compared to that during unfolding, ΔV^* can be attributed to the local change of the cavity and hydration around the active site of the enzyme. In the case of DHFR, as discussed in Sect. 21.6, the negative ΔV^* values could be explained by the stabilization of partial charges due to the hydration or condensation of the hydrated water in the transition-state structure. Another possible origin of the negative ΔV^* value is loop or domain movement such as the "open" conformation, which enhances hydration in the transition-state structure. Conversely, the positive ΔV^* value of DHFRs could be attributed to the release of a product or cofactor, or dehydration due to the conformational "closing" of the transition-state structure. In any case, enzymes functionally adapt to the high-pressure environment of the deep sea by modifying their structural dynamics as well as the cavity and hydration in their transition-state structure without substantially altering their ground-state structure.

21.8 Conclusions

The structural differences between deep-sea DHFRs and their normal homologs are small. However, they have rather variable structural stability; deep-sea DHFRs are thermally stable but are unstable with respect to urea and pressure unfolding. The smaller volume changes due to unfolding suggest that the native structures of deep-sea DHFRs have a smaller cavity and/or enhanced hydration compared to ecDHFR. Since the cavity and hydration depend on the amino acid side chains, DHFRs could adapt to the deep-sea environment by substituting amino acid side chains without altering their backbone structure. The inverted and pressure-dependent activation volumes of deep-sea DHFRs indicate the conversion of the rate-determining step of the enzymatic reaction. Thus, enzymes can adapt to the high-pressure environment of the deep sea by modifying the structural dynamics as well as the cavity and hydration of the transition-state structure without substantially altering their ground-state structure. The findings of the present work clearly indicate cavities and hydration play critical roles in the adaptation of enzymes to the deep-sea environment. Regardless, additional studies on other deep-sea enzymes are required to confirm the molecular-adaptation mechanisms of deep-sea enzymes.

Acknowledgements We thank Dr. Yuichi Nogi of JAMSTEC for donating the bacterial genomic DNAs. We also thank Profs. Kazuyuki Akasaka, Ryo Kitahara, Kaoru Nakasone, Shin-ichi Tate, Kunihiro Kuwajima, and Fumiyoshi Abe as well as Drs. Kazumi Hata, Chiho Murakami, and Takako Sato for their helpful discussions and experimental contributions. This work was financially supported by JSPS KAKENHI (Grant Number: 24570186 to E.O.).

References

Abali EE, Skacel NE, Celikkaya H, Hsieh YC (2008) Regulation of human dihydrofolate reductase activity and expression. Vitam Horm 79:267–292

Bartlett D, Wright M, Yayanos AA, Silverman M (1989) Isolation of a gene regulated by hydrostatic pressure in a deep-sea bacterium. Nature 342:572–574

Bertino JR (2009) Cancer research: from folate antagonism to molecular targets. Best Pract Res Clin Haematol 22:577–582

Boroujerdi AF, Young JK (2009) NMR-derived folate-bound structure of dihydrofolate reductase 1 from the halophile *Haloferax volcanii*. Biopolymers 91:140–144

Bowman JP, McCammon SA, Nichols DS, Skerratt JH, Rea SM, Nichols PD, McMeekin TA (1997) *Shewanella gelidimarina* sp. nov. and *Shewanella frigidimarina* sp. nov., novel Antarctic species with the ability to produce eicosapentaenoic acid (20:5 omega 3) and grow anaerobically by dissimilatory Fe(III) reduction. Int J Syst Bacteriol 47:1040–1047

Brindley AA, Pickersgill RW, Partridge JC, Dunstan DJ, Hunt DM, Warren MJ (2008) Enzyme sequence and its relationship to hyperbaric stability of artificial and natural fish lactate dehydrogenases. PLoS One 3:e2042

Cody V, Pace J, Stewart E (2012) Structural analysis of *Pneumocystis carinii* dihydrofolate reductase complexed with NADPH and 2,4-diamino-6-[2-(5-carboxypent-1-yn-1-yl)-5-methoxybenzyl]-5-methylpyrido[2,3-d]pyrimidine. Acta Crystallogr Sect F Struct Biol Cryst Commun 68:418–423

Cody V, Pace J, Queener SF, Adair OO, Gangjee A (2013) Kinetic and structural analysis for potent antifolate inhibition of *Pneumocystis jirovecii*, *Pneumocystis carinii*, and human dihydrofolate reductases and their active-site variants. Antimicrob Agents Chemother 57:2669–2677

Costanzo MS, Hartl DL (2011) The evolutionary landscape of antifolate resistance in *Plasmodium falciparum*. J Genet 90:187–190

Dams T, Auerbach G, Bader G, Jacob U, Ploom T, Huber R, Jaenicke R (2000) The crystal structure of dihydrofolate reductase from *Thermotoga maritima*: molecular features of thermostability. J Mol Biol 297:659–672

Deming JW, Hada H, Colwell RR, Luehrsen KR, Fox GE (1984) The ribonucleotide sequence of 5s rRNA from two strains of deep-sea barophilic bacteria. J Gen Microbiol 130:1911–1920

De Vos D, Xu Y, Hulpiau P, Vergauwen B, Van Beeumen JJ (2007) Structural investigation of cold activity and regulation of aspartate carbamoyltransferase from the extreme psychrophilic bacterium *Moritella profunda*. J Mol Biol 365:379–395

Evans RM, Behiry EM, Tey LH, Guo J, Loveridge EJ, Allemann RK (2010) Catalysis by dihydrofolate reductase from the psychropiezophile *Moritella profunda*. Chembiochem 11:2010–2017

Fierke CA, Johnson KA, Benkovic SJ (1987) Construction and evaluation of the kinetic scheme associated with dihydrofolate reductase from *Escherichia coli*. Biochemistry 26:4085–4092

Frey KM, Lombardo MN, Wright DL, Anderson AC (2010) Towards the understanding of resistance mechanisms in clinically isolated trimethoprim-resistant, methicillin-resistant *Staphylococcus aureus* dihydrofolate reductase. J Struct Biol 170:93–97

Gonen N, Assaraf YG (2012) Antifolates in cancer therapy: structure, activity and mechanisms of drug resistance. Drug Resist Updat 15:183–210

Graffner-Nordberg M, Kolmodin K, Aqvist J, Queener SF, Hallberg A (2004) Design, synthesis, and computational affinity prediction of ester soft drugs as inhibitors of dihydrofolate reductase from *Pneumocystis carinii*. Eur J Pharm Sci 22:43–54

Hammes GG, Benkovic SJ, Hammes-Schiffer S (2011) Flexibility, diversity, and cooperativity: pillars of enzyme catalysis. Biochemistry 50:10422–10430

Hastings JW, Nealson KH (1981) The symbiotic luminous bacteria. In: Starr M, Stolp H, Trüper H, Balows A, Schlegel H (eds) The prokaryotes: a handbook on habitats, isolation, and identification of bacteria. Springer-Verlag, Berlin, pp 1322–1345

Hay S, Evans RM, Levy C, Loveridge EJ, Wang X, Leys D, Allemann RK (2009) Are the catalytic properties of enzymes from piezophilic organisms pressure adapted? Chembiochem 10:2348–2353

Kato C (2011a) Cultivation methods for piezophiles. In: Horikoshi K, Antranikian G, Bull A, Robb F, Stetter K (eds) Extremophiles handbook. Springer-Verlag, Tokyo, pp 719–726

Kato C (2011b) Distribution of piezophiles. In: Horikoshi K, Antranikian G, Bull A, Robb F, Stetter K (eds) Extremophiles handbook. Springer-Verlag, Tokyo, pp 643–655

Kato C, Sato T, Horikoshi K (1995) Isolation and properties of barophilic and barotolerant bacteria from deep-sea mud samples. Biodivers Conserv 4:1–9

Kato C, Li L, Tamaoka J, Horikoshi K (1997) Molecular analyses of the sediment of the 11000-m deep Mariana Trench. Extremophiles 1:117–123

Kato C, Li L, Nogi Y, Nakamura Y, Tamaoka J, Horikoshi K (1998) Extremely barophilic bacteria isolated from the Mariana Trench, Challenger Deep, at a depth of 11,000 meters. Appl Environ Microbiol 64:1510–1513

Kato C, Arakawa S, Sato T, Xiao X (2008a) Culture-independent characterization of microbial diversity in selected deep-sea sediments. In: Michiels C, Bartlett DH, Aertsen A (eds) High-pressure microbiology. ASM Press, Washington, DC, pp 219–236

Kato C, Sato T, Abe F, Ohmae E, Tamegai H, Nakasone K, Siddiqui KS, Thomas T (2008b) Protein adaptation to high-pressure environments. In: Siddiqui KS, Thomas T (eds) Protein adaptation in extremophiles. Nova Science Publisher, New York, pp 167–191

Kauzman W (1959) Some factors in the interpretation of protein denaturation. Adv Protein Chem 14:1–63

Lee J, Goodey NM (2011) Catalytic contributions from remote regions of enzyme structure. Chem Rev 111:7595–7624

Loveridge EJ, Allemann RK (2010) The temperature dependence of the kinetic isotope effects of dihydrofolate reductase from *Thermotoga maritima* is influenced by intersubunit interactions. Biochemistry 49:5390–5396

Loveridge EJ, Rodriguez RJ, Swanwick RS, Allemann RK (2009) Effect of dimerization on the stability and catalytic activity of dihydrofolate reductase from the hyperthermophile *Thermotoga maritima*. Biochemistry 48:5922–5933

MacDonell MT, Colwell RR (1985) Phylogeny of the *Vibrionaceae*, and recommendation for two new genera, *Listonella* and *Shewanella*. Syst Appl Microbiol 6:171–182

Murakami C, Ohmae E, Tate S, Gekko K, Nakasone K, Kato C (2010) Cloning and characterization of dihydrofolate reductases from deep-sea bacteria. J Biochem 147:591–599

Murakami C, Ohmae E, Tate S, Gekko K, Nakasone K, Kato C (2011) Comparative study on dihydrofolate reductases from *Shewanella* species living in deep-sea and ambient atmospheric-pressure environments. Extremophiles 15:165–175

Nagae T, Kato C, Watanabe N (2012) Structural analysis of 3-isopropylmalate dehydrogenase from the obligate piezophile *Shewanella benthica* DB21MT-2 and the nonpiezophile *Shewanella oneidensis* MR-1. Acta Crystallogr Sect F Struct Biol Cryst Commun 68:265–268

Navarro-Martínez MD, García-Cánovas F, Rodríguez-López JN (2006) Tea polyphenol epigallocatechin-3-gallate inhibits ergosterol synthesis by disturbing folic acid metabolism in *Candida albicans*. J Antimicrob Chemother 57:1083–1092

Nogi Y (2008) Bacteria in the deep sea: psychropiezophiles. In: Margesin R, Schinner F, Marx JC, Gerday C (eds) Psychrophiles: from Biodiversity to Biotechnology. Springer, Berlin, pp 73–82

Nogi Y, Kato C, Horikoshi K (1998a) *Moritella japonica sp. nov.,* a novel barophilic bacterium isolated from a Japan Trench sediment. J Gen Appl Microbiol 44:289–295

Nogi Y, Kato C, Horikoshi K (1998b) Taxonomic studies of deep-sea barophilic *Shewanella* strains and description of *Shewanella violacea* sp. nov., a new balophilic bacterial species. Arch Microbiol 170:331–338

Nogi Y, Masui N, Kato C (1998c) *Photobacterium profindum* sp. nov., a new, moderately barophilic bacterial species isolated from a deep-sea sediment. Extremophiles 2:1–7

Nogi Y, Kato C, Horikoshi K (2002) *Psychromonas kaikoae* sp. nov., a novel from the deepest piezophilic bacterium cold-seep sediments in the Japan Trench. Int J Syst Evol Microbiol 52:1527–1532

Nogi Y, Hosoya S, Kato C, Horikoshi K (2004) *Colwellia piezophila* sp. nov., a novel piezophilic species from deep-sea sediments of the Japan Trench. Int J Syst Evol Microbiol 54:1627–1631

Northrop DB, Cho YK (2000) Effects of pressure on deuterium isotope effects of yeast alcohol dehydrogenase: evidence for mechanical models of catalysis. Biochemistry 39:2406–2412

Oefner C, D'Arcy A, Winkler FK (1988) Crystal structure of human dihydrofolate reductase complexed with folate. Eur J Biochem 174:377–385

Oefner C, Parisi S, Schulz H, Lociuro S, Dale GE (2009) Inhibitory properties and X-ray crystallographic study of the binding of AR-101, AR-102 and iclaprim in ternary complexes with NADPH and dihydrofolate reductase from *Staphylococcus aureus*. Acta Crystallogr Sect F Struct Biol Cryst Commun 65:751–757

Ohmae E, Tatsuta M, Abe F, Kato C, Tanaka N, Kunugi S, Gekko K (2008) Effects of pressure on enzyme function of *Escherichia coli* dihydrofolate reductase. Biochim Biophys Acta 1784:1115–1121

Ohmae E, Murakami C, Tate S, Gekko K, Hata K, Akasaka K, Kato C (2012) Pressure dependence of activity and stability of dihydrofolate reductases of the deep-sea bacterium *Moritella profunda* and *Escherichia coli*. Biochim Biophys Acta 1824:511–519

Ohmae E, Miyashita Y, Tate S, Gekko K, Kitazawa S, Kitahara R, Kuwajima K (2013a) Solvent environments significantly affect the enzymatic function of *Escherichia coli* dihydrofolate reductase: comparison of wild-type protein and active-site mutant D27E. Biochim Biophys Acta 1834:2782–2794

Ohmae E, Miyashita Y, Kato C (2013b) Thermodynamic and functional characteristics of deep-sea enzymes revealed by pressure effects. Extremophiles 17:701–709

Owen RJ, Legors RM, Lapage SP (1978) Base composition, size and sequence similarities of genoma deoxyribonucleic acids from clinical isolates of *Pseudomonas putrefaciens*. J Gen Microbiol 104:127–138

Pace CN (1985) Determination and analysis of urea and guanidine hydrochloride denaturation curves. In: Hirs CHW, Timasheff SN (eds) Methods in enzymology, vol 131. Academic Press, New York, pp 267–280

Paulsen JL, Liu J, Bolstad DB, Smith AE, Priestley ND, Wright DL, Anderson AC (2009) In vitro biological activity and structural analysis of 2,4-diamino-5-(2′-arylpropargyl)pyrimidine inhibitors of *Candida albicans*. Bioorg Med Chem 17:4866–4872

Paulsen JL, Bendel SD, Andereson AC (2011) Crystal structures of *Candida albicans* dihydrofolate reductase bound to propargyl-linked antifolates reveal the flexibility of active site loop residues critical for ligand potency and selectivity. Chem Biol Drug Des 78:505–512

Pieper U, Kapadia G, Mevarech M, Herzberg O (1998) Structural features of halophilicity derived from the crystal structure of dihydrofolate reductase from the Dead Sea halophilic archaeon, *Haloferax volcanii*. Structure 6:75–88

Purcarea C, Hervé G, Ladjimi MM, Cunin R (1997) Aspartate transcarbamylase from the deep-sea hyperthermophilic archaeon *Pyrococcus abyssi*: genetic organization, structure, and expression in *Escherichia coli*. J Bacteriol 179:4143–4157

Quirk DJ, Northrop DB (2001) Effect of pressure on deuterium isotope effects of formate dehydrogenase. Biochemistry 40:847–851

Saito R, Kato C, Nakayama A (2006) Amino acid substitutions in malate dehydrogenases of piezophilic bacteria isolated from intestinal contents of deep-sea fishes retrieved from the abyssal zone. J Gen Appl Microbiol 52:9–19

Sawaya MR, Kraut J (1997) Loop and subdomain movements in the mechanism of *Escherichia coli* dihydrofolate reductase: crystallographic evidence. Biochemistry 36:586–603

Schnell JR, Dyson HJ, Wright PE (2004) Structure, dynamics, and catalytic function of dihydrofolate reductase. Annu Rev Biophys Biomol Struct 33:119–140

Shin DS, DiDonato M, Barondeau DP, Hura GL, Hitomi C, Berglund JA, Getzoff ED, Cary SC, Tainer JA (2009) Superoxide dismutase from the eukaryotic thermophile *Alvinella pompejana*: structures, stability, mechanism, and insights into amyotrophic lateral sclerosis. J Mol Biol 385:1534–1555

Shirai T, Hung VS, Morinaka K, Kobayashi T, Ito S (2008) Crystal structure of GH13 α-glucosidase GSJ from one of the deepest sea bacteria. Proteins 73:126–133

Stammers DK, Champness JN, Beddell CR, Dann JG, Eliopoulos E, Geddes AJ, Ogg D, North AC (1987) The structure of mouse L1210 dihydrofolate reductase-drug complexes and the construction of a model of human enzyme. FEBS Lett 218:178–184

Sunduru N, Gupta L, Chauhan K, Mishra NN, Shukla PK, Chauhan PM (2011) Synthesis and antibacterial evaluation of novel 8-fluoro Norfloxacin derivatives as potential probes for methicillin and vancomycin-resistant *Staphylococcus aureus*. Eur J Med Chem 46:1232–1244

Teh AH, Kanamasa S, Kajiwara S, Kumasaka T (2008) Structure of Cu/Zn superoxide dismutase from the heavy-metal-tolerant yeast *Cryptococcus liquefaciens* strain N6. Biochem Biophys Res Commun 374:475–478

Urakawa H, Kita TK, Steven SE, Ohwada K, Colwell RR (1998) A proposal to transfer *Vibrio marinus* (Russell 1891) to a new genus *Moritella* gen. nov. as *Moritella marina* comb. nov. FEMS Microbiol Lett 165:373–378

Venkateswaran K, Moser DP, Dollhopf ME, Lies DP, Saffarini DA, MacGregor BJ, Ringelberg DB, White DC, Nishijima M, Sano H, Burghardt J, Stackebrandt E, Nealson KH (1999) Polyphasic taxonomy of the genus *Shewanella* and description of *Shewanella oneidensis* sp nov. Int J Syst Bacteriol 49:705–724

Vidugiris GJA, Royer CA (1998) Determination of the volume changes for pressure-induced transitions of apomyoglobin between the native, molten globule, and unfolded states. Biophys J 75:463–470

Vieille C, Zeikus GJ (2001) Hyperthermophilic enzymes: sources, uses, and molecular mechanisms for thermostability. Microbiol Mol Biol Rev 65:1–43

Wright DB, Banks DD, Lohman JR, Hilsenbeck JL, Gloss LM (2002) The effect of salts on the activity and stability of *Escherichia coli* and *Haloferax volcanii* dihydrofolate reductases. J Mol Biol 323:327–344

Xie BB, Bian F, Chen XL, He HL, Guo J, Gao X, Zeng YX, Chen B, Zhou BC, Zhang YZ (2009) Cold adaptation of zinc metalloproteases in the thermolysin family from deep sea and arctic sea ice bacteria revealed by catalytic and structural properties and molecular dynamics. J Biol Chem 284:9257–9269

Xu Y, Feller G, Gerday C, Gransdorff N (2003a) *Moritella* cold-active dihydrofolate reductase: are there natural limits to optimization of catalytic efficiency at low temperature? J Bacteriol 185:5519–5526

Xu Y, Nogi Y, Kato C, Liang Z, Rüger HJ, De Kegel D, Glansdorff N (2003b) *Moritella profunda* sp. nov. and *Moritella abyssi* sp. nov., two psychropiezophilic organisms isolated from deep Atlantic sediments. Int J Syst Evol Microbiol 53:533–538

Yang J, Dang H (2011) Cloning and characterization of a novel cold-active endoglucanase establishing a new subfamily of glycosyl hydrolase family 5 from a psychrophilic deep-sea bacterium. FEMS Microbiol Lett 325:71–77

Yuthavong Y, Tarnchompoo B, Vilaivan T, Chitnumsub P, Kamchonwongpaisan S, Charman SA, McLennan DN, White KL, Vivas L, Bongard E, Thongphanchang C, Taweechai S, Vanichtanankul J, Rattanajak R, Arwon U, Fantauzzi P, Yuvaniyama J, Charman WN, Matthews D (2012) Malarial dihydrofolate reductase as a paradigm for drug development against a resistance-compromised target. Proc Natl Acad Sci U S A 109:16823–16828

Zhang JW, Zeng RY (2008) Molecular cloning and expression of a cold-adapted lipase gene from an Antarctic deep sea psychrotrophic bacterium *Pseudomonas* sp. 7323. Mar Biotechnol (NY) 10:612–621

Zhang J, Lin S, Zeng R (2007) Cloning, expression, and characterization of a cold-adapted lipase gene from an Antarctic deep-sea psychrotrophic bacterium, *Psychrobacter* sp 7195. J Microbiol Biotechnol 17:604–610

Chapter 22
Moss Spores Can Tolerate Ultra-high Pressure

Fumihisa Ono

Abstract It was shown by the present authors' group that tardigrade can survive under very high hydrostatic pressure range of 7.5 GPa. In the case of small land plants, the tolerance to high pressure seems stronger than such a small animal, tardigrade. We have extended our experiments to two popular mosses, *Ptychomitrium* and *Venturiella* searching for lives under very high pressure range. Spore placentas of moss *Ptychomitrium* and *Venturiella* were sealed in a small Teflon capsule together with a liquid pressure medium. The capsule was put in the center of a pyrophillite cube, and the maximum pressure of 7.5 GPa was applied using a 250-t cubic anvil press. The pressure was kept constant at the maximum pressure for various duration of time between 6 and 144 h. After the pressure was released, the spores were seeded on agar medium, and incubated for one week and longer at 25 °C with white light of 2,000–2,500 Lx. It was proven that 80–90 % of the spores of moss *Ptychomitrium* were alive and germinated after being exposed to the very high pressure of 7.5 GPa for up to 48 h. Furthermore, a relatively high germination rate of about 35 % was retained even after exposure to 7.5 GPa for 144 h. It was also proven that 70–90 % of the spores of moss *Venturiella* were alive and germinated after exposure to 7.5 GPa for up to 72 h. However, after exposed to 7.5 GPa for 144 h, only four individuals in a hundred were germinated. The pressure tolerances of the mosses *Ptychomitrium* and *Venturiella* were found to be much stronger than tardigrade. The experiments were extended to the ultra-high pressure range up to 20 GPa, and a few spores of *Venturiella* were germinated up to the length of 30 μm. The mechanisms for such strong tolerance of moss spores will be discussed on the bases of the quality of the applied hydrostatic pressure, degree of the physical deformation and extremophiles.

Keywords Moss spores under high pressure • High pressure tolerance • *Ptychomitrium* • *Venturiella*

F. Ono (✉)
Department of Applied Science, Okayama University of Science, 1-1 Ridaicho, Kitaku, Okayama 700-0005, Japan

Department of Chemistry, University College London, 20 Gordon Street, London WC1H 0AJ, UK
e-mail: fumihisa@das.ous.ac.jp

K. Akasaka, H. Matsuki (eds.), *High Pressure Bioscience*,
Subcellular Biochemistry 72, DOI 10.1007/978-94-017-9918-8_22

22.1 Overview

Nobody would have believed that small creatures can withstand against such a very high pressure of several GPa order for such a long time of more than a few hours until a several experiments were conducted concerning the search for life under very high hydrostatic pressure (Ono et al. 2008; Horikawa et al. 2009; Nishihira et al. 2010). In our experiments all kinds of living samples, both animals and plants tested were proven to keep their lives after exposure to such a very high hydrostatic pressure of 7.5 GPa. Among the candidates for these experiments, the samples that showed the strongest tolerance against the very high pressure were moss spores of *Ptychomitrium* and *Venturiella*. It was shown (Ono et al. 2010a; Nishihira et al. 2010) that the spores of these mosses can survive under very high pressure, as high as 7.5 GPa, for up to 144 h. Then, a big question arose that what mechanism made the living creatures so strong against very high pressure. To make a contribution to open up this question, we have made a closer investigation to the moss spores from a biophysical and biochemical point of view.

22.2 Strong Viability of Mosses

Mosses (Bryophyta) are non-vascular plants, small and simple land plants with leaves and a stem without roots. They do not have flowers or seeds, but in certain seasons they produce ten- to hundred-thousands of spores in a spore case. They have been living on the Earth since a few billion years ago, and their spores are strong against extreme environmental conditions such as long-term dried weather, in vacuum and at high temperatures (Kenrick and Crane 1997). There are approximately 12,000 species, and mostly found in areas of humid atmosphere with cold to moderately warm climate.

Ptychomitrium sinense belongs to Grimmiaceae, and are mostly found on rocks at sunshiny places in Japan, Korea, China, Vietnam and other countries including north western America. In a dried atmosphere, their leaves shrink and their color turns brown. On the other hand, when they have certain amount of moisture, they become straight and green, again. A spore case of *Ptychomitrium* contains several tens of thousands of spores. Their spores are spherical in shape with a diameter of 20–30 μm. Once spores are dried, they become strong against extreme environmental conditions and remain alive for a long period. When they are seeded on a humid culture medium, they begin to germinate within 1 week. They do not have proper roots, but rhizoids which are thread-like and anchor them to the surface of a big stone. This moss, therefore, would be a good candidate for the search of plant life that would survive under very high pressure. Figure 22.1a shows a wild gregarious of the *Ptychomitrium* and (b) an individual with a spore case, stem and leaves which was picked up from a field in the city of Okayama.

Fig. 22.1 Moss *Ptychomitrium sinense*; The photo (**a**) was taken at a field in the city of Okayama, and (**b**) shows a spore case with a stem and leaves

Venturiella sinensis belongs to *Erpodiaceae*, also being one of the popular mosses seen in moderate climate areas such as in Japan. The spores of *Venturiella* are spherical in shape with a diameter of about 20–30 μm. A spore case, being in the size of 0.5 mm in diameter and 1 mm in length contains several ten thousands of spores. Once spores are put in a dried atmosphere, they become strong against extreme environmental conditions and remain alive for long years. When they are seeded on a humid medium, they begin to germinate within 1 week. This moss is also a good candidate that can tolerate very high hydrostatic pressure of GPa order. Figure 22.2 shows spore cases of the moss *Venturiella sinensis* picked up from a field in the city of Okayama.

22.3 Generation of High Hydrostatic Pressure up to Several GPa

To generate a hydrostatic pressure of several GPa, a cubic anvil press operated by a 250-t press was used. Between the two pistons of the press, six anvils made of tungsten-carbide alloy were arranged to compress a cubic high-pressure cell made of pyrophillite. To keep a good hydrostaticity, a teflon capsule with liquid pressure medium was adopted to contain biological specimens.

Specimens of spore placentas of the moss, *Ptychomitrium* collected from the field of the city of Okayama were dehydrated by keeping them in a low humidity container of about 20 % for more than 1 week. Then, two or three of them were sealed in a small teflon capsule with an inner diameter and length of 1.6 and 1.8 mm, respectively. Fluorinert (perfluoricarbon, $C_8H_{16}F$; PC77, Sumitomo 3 M) was used as the liquid pressure medium, and put into the teflon capsule together with the moss spore placentas. It was proven that soaking in Fluorinert has no practical effect on

Fig. 22.2 (**a**) Moss *Venturiella* collected from a field in the city of Okayama, (**b**) in an expanded scale and (**c**) a spore case with full of spores picked up from the moss shown in (a). Each ripe spore case has a light-brown colored cap

the life of tardigrades (Ono et al. 2008) and *Artemia* eggs (Ono et al. 2010b; Minami et al. 2010) at the cryptobiotic state. The capsule was put in the center of a cube which was made of pyrophillite with an edge length of 6.0 mm. An illustration of the high pressure cell is shown in Fig. 22.3.

The pyrophillite cube was compressed by six tungsten-carbide anvils with a front edge length of 4.0 mm. These anvils were compressed by a 250-t press. The arrangement of the set of anvils of the present cubic anvil press is shown in Fig. 22.4.

The pressure was determined by using a calibrated curve of the relation between the press load and the actual pressure established before the experiment (Matsushita et al. 2003, 2005). Estimated error of the intensity of pressure was smaller than 0.3 GPa. In this apparatus, the press load was controlled automatically to keep the intensity of the pressure constant during the operation at the maximum pressure.

The pressure was increased from ambient to the maximum pressure of 7.5 GPa at a rate of 0.3 GPa min^{-1}. The pressure was kept constant at the maximum pressure for various duration of time between 6 and 144 h, and then, decreased down to the ambient pressure with the same rate as on increasing. The high-pressure experiments were made at room temperature.

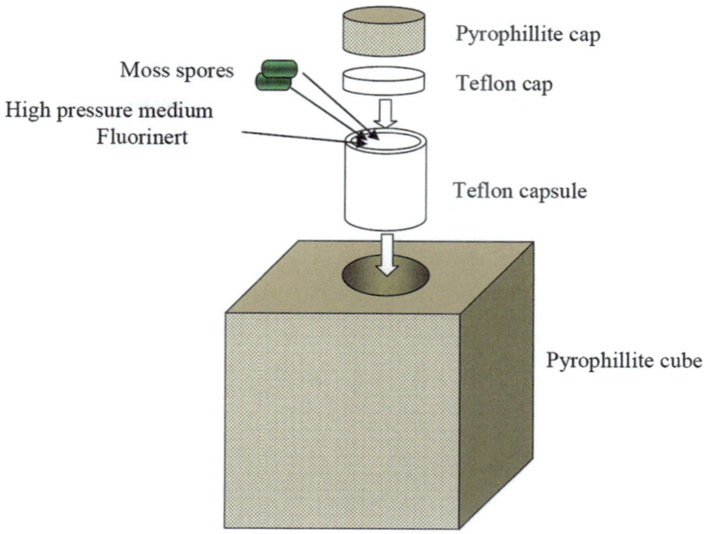

Fig. 22.3 An assembly of the high pressure cell; the teflon capsule which contains a few moss spore cases with liquid pressure medium fluorinert is inserted in the pyrophillite cube. The edge length of the cube is 6.0 mm

Fig. 22.4 An illustration of the six tungsten-carbide anvils compressing a pyrophillite cube put in the center of the anvils. The set of the anvils is operated by a 250-t press

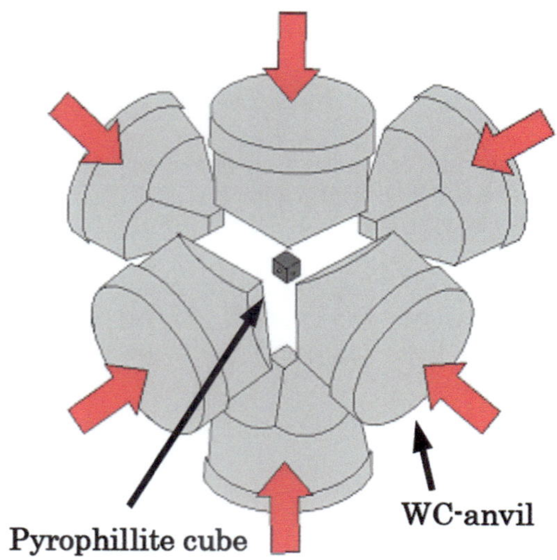

Figure 22.5a and b shows the tungsten-carbide anvils with the pyrophillite cubic cell before and after the pressure of 7.5 GPa was generated and the pressure was kept for 144 h. Immediately after the pressure was released, the spore placentas were brought out from the teflon capsule and the spores were seeded on a humid medium. The seeded medium was kept in an incubator for 1–4 weeks or more under white light with an intensity of 2,500 Lx. These spores were investigated under an

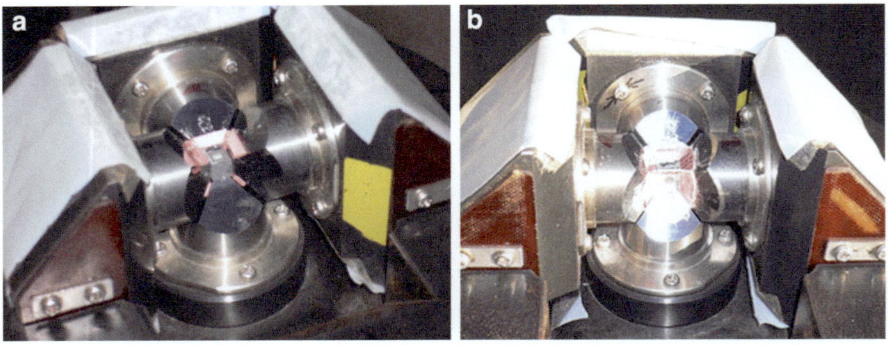

Fig. 22.5 (**a**) Four of the six WC-anvils and the pyrophillite cube before, and (**b**) after the exposure to the high pressure of 7.5 GPa

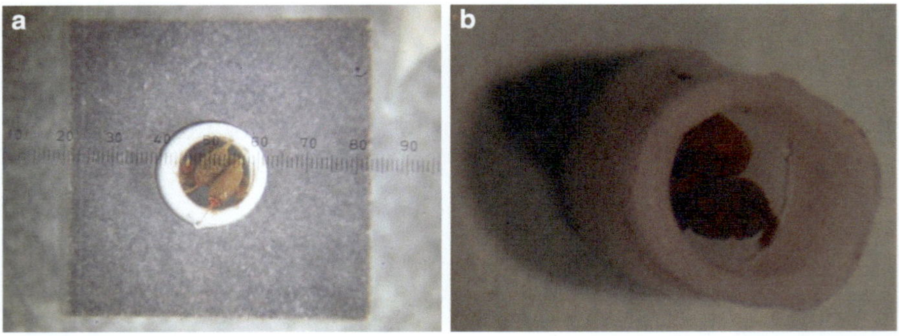

Fig. 22.6 (**a**) The Teflon capsule containing three spore placentas before exposure to high pressure, and (**b**) brought out from the pyrophillite cube after exposure to 7.5 GPa (Nishihira et al. 2010; Ono et al. 2010a)

optical microscope. Figure 22.6a shows the teflon capsule containing three spore placentas before exposure to high pressure, and (b) shows after the pressure was released.

22.4 Results for *Ptychomitrium*

After exposures to the very high pressure of 7.5 GPa for various duration of time between 6 and 144 h, the spores of moss *Ptychomitrium* were seeded on an agar and cultivated in an incubator. The temperature in the incubator was set constant at 25 °C with white light of the intensity 2,500 Lx (Nishihira et al. 2010).

Up to 89 % of *Ptychomitrium* spores exposed to 7.5 GPa for 6 h germinated. This germination rate is almost as high as that of the control group which was not exposed to high pressure. This high germination rate was preserved up to the exposure to the

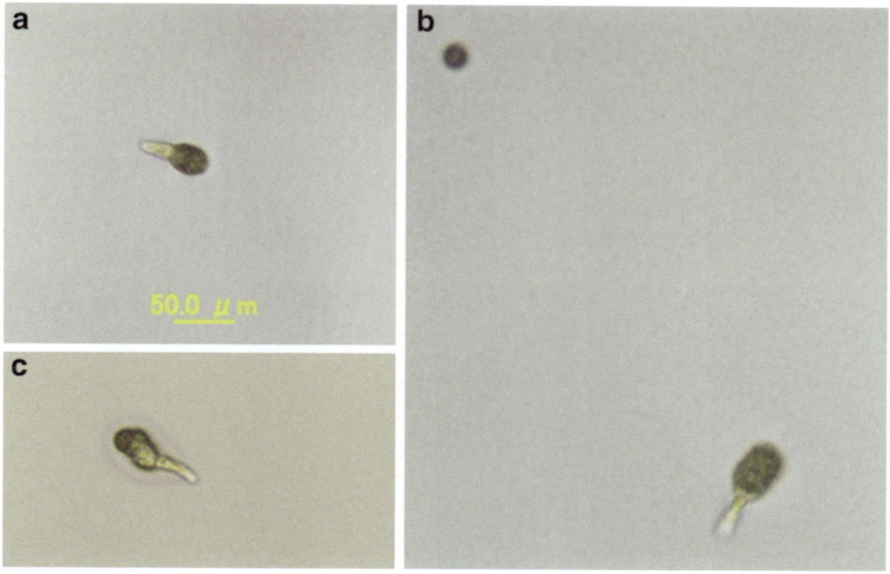

Fig. 22.7 (**a**) Germinated spore of *Ptychomitrium* after exposure to 7.5 GPa for 6 h, (**b**) 12 h and (**c**) control. These photos were taken 5 days after the spores were seeded on an ager

high pressure for 48 h. Germinated spores exposed to 7.5GPa for 6 and 12 h are shown in Fig. 22.7a and b, respectively with the control in (c). The spore seen at the upper left hand side in (b) is not germinated, yet. These photos were taken 5 days after seeded. As seen in these photos, the length of the germinated protonemata was about 30 μm, and no difference was seen between the high pressure exposed group and the control.

On the other hand, the germination rate for the spores exposed to high pressure for longer than 48 h decreased, down to 29–38 % after the exposure for 144 h. The spores exposed to high pressure for 144 h were shown in Fig. 22.8. In this figure germination of some of the spores is clearly seen, and this fact means that they can tolerate the high hydrostatic pressure of 7.5 GPa for such a long time duration of 144 h (Nishihira et al. 2010).

It was found (Nishihira et al. 2010) that all the germinated spore protonemata exposed to 7.5 GPa for less than 24 h frizzled, which is a characteristic of *Ptychomitrium*. On the other hand, for the spores exposed to high pressure for more than 24 h, all the protonemata grew straight. The mechanism of frizzling of the moss *Ptychomitrium* was damaged during the exposure to 7.5 GPa for more than 24 h. A typical example of frizzled protonemata is shown in Fig. 22.9. The spores shown in (a) germinated from those exposed to 7.5 GPa for 6 h, and the one shown in (b) germinated from the spore exposed to high pressure for 144 h.

The number of germinated spores of *Ptychomitrium* after exposure to 7.5 GPa for various duration of time are shown in Table 22.1. The moss *Ptychomitrium* has

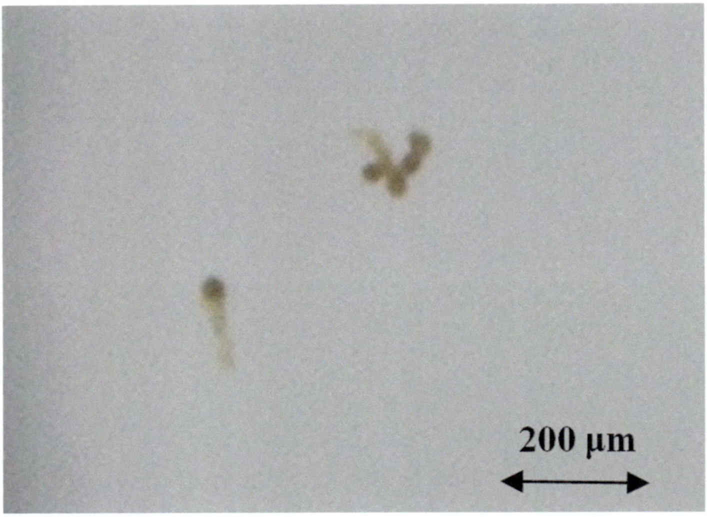

Fig. 22.8 Spores of *Ptychomitrium* exposed to 7.5 GPa for 144 h and seeded on an agar. The photo was taken 3 weeks after seeded. In this photograph, two spores germinated, while the other three did not germinate

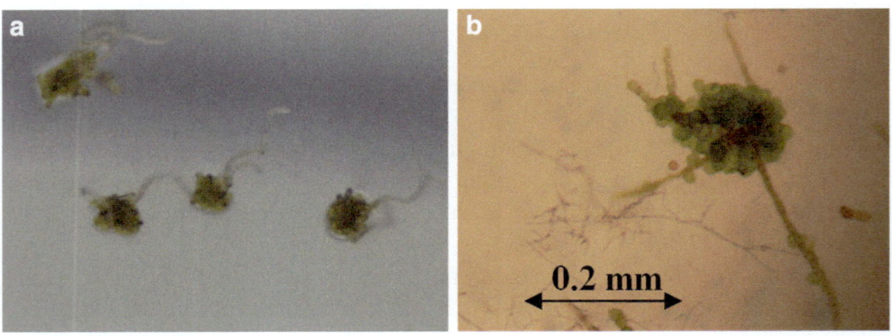

Fig. 22.9 (**a**) Germinated spores of *Ptychomitrium* exposed to 7.5 GPa for 6 h, and (**b**) for 144 h. These photographs were taken 3 weeks after seeded. It is clearly seen in (a) that germinated protonemata frizzled, while no such frizz was observed in (b) (Nishihira et al. 2010)

greater high pressure tolerance than the tardigrade (Ono et al. 2008) and Artemia (Ono et al. 2010b; Minami et al. 2010).

From this experiment, we confirmed that moss spores are still alive after exposure to the very high hydrostatic pressure of 7.5 GPa, which corresponds to the pressure at the depth of about 180 km below the surface of the Earth. This depth is equal to 1/35 of the Earth's radius, and is in the middle part of the upper mantle (Stacy and Davis 2008).

Table 22.1 Number of living and germinated spores of *Ptychomitrium* after exposure to 7.5 GPa tabulated with the exposure time to the maximum hydrostatic pressure of 7.5 GPa (Nishihira et al. 2010)

Exposed time (h)	Control	6	48	72	144	144
Germinated	66	239	334	115	335	267
Not germinated	12	56	51	116	552	640
Total	78	295	385	231	887	907
Germination rate (%)	84.6	81.0	86.8	49.8	37.8	29.4

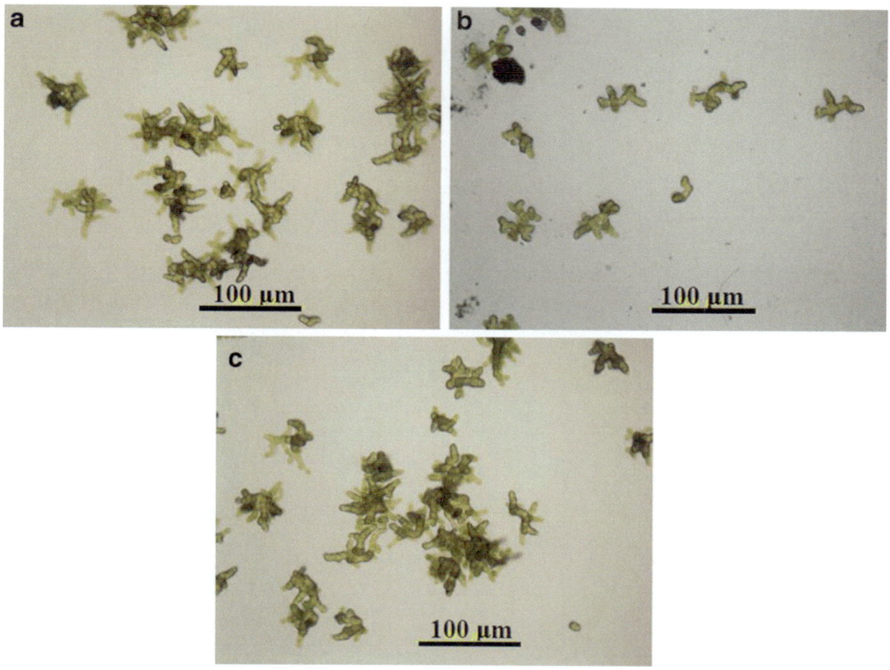

Fig. 22.10 Germinated spores of the moss *Venturiella*; (**a**) exposed to 7.5 GPa for 24 h, (**b**) exposed to 7.5 GPa for 72 h and (**c**) control. These photographs were taken 3 weeks after they were seeded

22.5 Results for *Venturiella*

After exposures to 7.5 GPa for various duration of time between 24 and 144 h, the spores of moss *Venturiella* were seeded on an agar and cultivated in an incubator with the same condition as of *Ptychomitrium*.

Germinated spores of moss *Venturiella* after exposure to 7.5 GPa for 24 and 72 h are shown in Fig. 22.10a and b, respectively. As seen in this figure, the spores began to germinate in both high pressure exposed groups. For comparison, germinated spores of the control group are shown in Fig. 22.10c. The germination rates of these

Fig. 22.11 The moss *Venturiella* exposed to the high hydrostatic pressure of 7.5 GPa for 72 h. The photo was taken 5 months after the spores were seeded on an ager medium (Ono et al. 2010a)

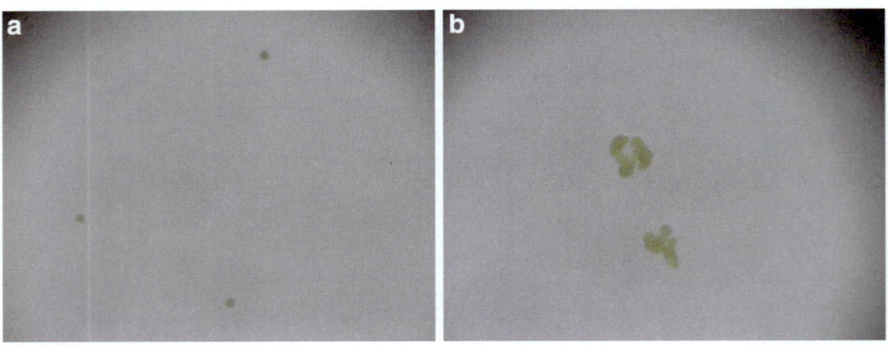

Fig. 22.12 (**a**) Spores of *Venturiella* exposed to the high hydrostatic pressure of 7.5 GPa for 144 h, and (**b**) control. The spores in (a) have not germinated yet. These photos were taken 2 weeks after seeded (Ono et al. 2010a)

high-pressure exposed groups were 70–90 %, which were as high as the control. After being incubated for 5 months, these spores grew up to have green leaves as seen in Fig. 22.11 (Ono et al. 2010a).

On the other hand, the spores exposed to 7.5GPa for 144 h, only four individuals in a hundred germinated. Furthermore, the germination delayed about 1 week than those exposed to high pressure for 72 h. The spores of *Venturiella* exposed to high pressure for 144 h and 2 weeks after seeded are shown in Fig. 22.12. As seen in this figure, the spores have not germinated yet. This exposure time of 144 h seems almost the limit of the tolerance for the moss *Venturiella* to the high pressure of

Table 22.2 Number of living and germinated spores of *Venturiella* after exposure to 7.5 GPa tabulated with the exposed time to the maximum hydrostatic pressure of 7.5 GPa

Exposed time (h)	Control	24	48	72	72	144
Germinated	107	844	334	202	95	4
Not germinated	2	20	51	53	43	100
Total	109	864	385	255	138	104
Germination rate (%)	98.2	97.7	86.8	79.2	68.8	4

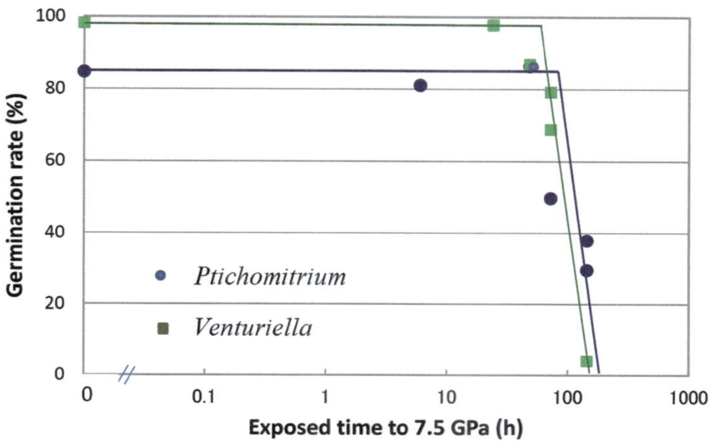

Fig. 22.13 Germination rate of *Ptychomitrium* and *Venturiella* spores plotted as a function of exposure time to 7.5 GPa (Ono et al. 2010a; Nishihira et al. 2010)

7.5 GPa. This pressure tolerance of moss is found to be one order stronger than the small animal, tardigrade (Ono et al. 2008). The percentages of the number of germinated spores of *Venturiella* after exposure to 7.5 GPa for various duration of time are shown in Table 22.2.

Observed germination rates of the mosses Ptichomitrium and Venturiella are plotted in Fig. 22.13 as a function of the exposure time to high pressure. As seen in this figure, the tolerances of the two mosses are almost the same strength, and they can preserve their lives under 7.5 GPa for up to 144 h.

From these experiments shown above the effect of high hydrostatic pressure of 7.5 GPa on moss spores and their survival limit were made clear. The survival limit of the exposure time to the high pressure, 7.5 GPa is determined to be around 144 h. We convinced that moss spores are still alive after being exposed to such a very high hydrostatic pressure of 7.5 GPa. Considering the fact that synthesis of diamond from graphite was made under the pressure of 5.5 GPa at 1,800 °C (Bovenkerk et al. 1959), the present results that moss spores can tolerate the high hydrostatic pressure of 7.5 GPa are un-understandable. The protein molecules and genes of the moss spores would have damaged at least to some extent. But the present results did not show any damages in the spores.

The results of the present investigation may give the possibility (Ono et al. 2010a; Nishihira et al. 2010) that the spores of mosses, *Ptychomitrium* and *Venturiella* can travel through the space in a large meteorite, and reach the earth alive from other planet or galaxy.

22.6 Further Exposure to Ultra High Pressure Range

As it was shown from the study mentioned in Sect. 22.3 that the pressure generated and the duration of time, 7.5 GPa for up to 144 h, are almost the mechanical limit of the cubic anvil press, it would be necessary to adopt a multi-stage multi-anvil type high pressure equipment to extend the investigation to a higher pressure range. An octahedral anvil press was used to extend the high pressure study to an ultra-high pressure range up to 20 GPa (Ono et al. 2013).

Several spore cases of moss *Venturiella* were collected again from the field in the city of Okayama, Japan. Two of them were sealed in the same small teflon capsule as used in the cubic anvil press described in Sect. 22.3 with a liquid pressure medium fluorinert (PC77, Sumitomo 3 M). The inner diameter and the height of the capsule were 1.6 mm and 1.8 mm, respectively. This capsule was put in the center of an octahedron made of MgO with an edge length of 8.0 mm. To generate a pressure up to 20 GPa a two-stage octahedral anvil press installed at the Institute for Study of the Earth's Interior (ISEI) of Okayama University was adopted. The pressure was increased to 20 GPa at the rate of 40 GPa h^{-1}, kept for 30 min at the maximum pressure, and then decreased down to ambient pressure at the rate of 20 GPa h^{-1}. The pressure was determined by using a calibrated curve of the relation between the press load and the actual pressure established before the experiment. Figure 22.14

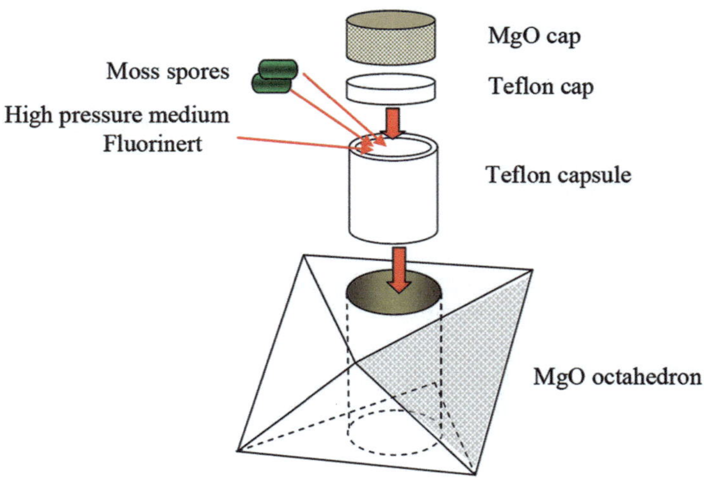

Fig. 22.14 An illustration of the octahedral high pressure cell made of MgO and teflon capsule with moss spore samples

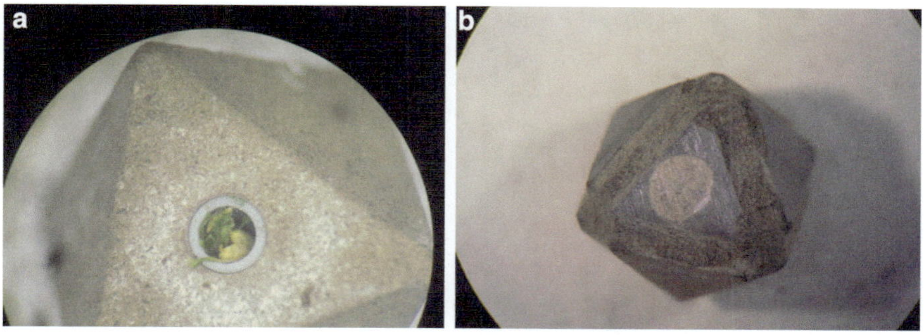

Fig. 22.15 (**a**) An octahedral high pressure cell made of MgO with a tefron capsule containing two moss spores before exposure to high pressure, (**b**) the MgO high pressure cell brought out from the octahedral anvil press after the maximum pressure of 20 GPa was generated for 30 min (Ono et al. 2013)

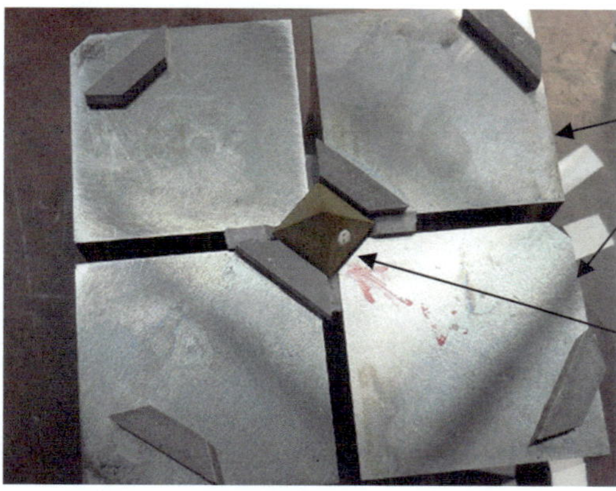

Fig. 22.16 Four of the 8 WC- anvils and the MgO-octahedron high pressure cell before exposure to high pressure

shows an illustration of the high-pressure cell and the teflon capsule with two moss spore cases set into the center of the octahedron.

Figure 22.15a shows an octahedron high pressure cell with a tefron capsule containing two moss spores before exposure to high pressure, and (b) the MgO octahedron after exposed to 20 GPa. Figure 22.16 shows 4 of the 8 anvils with the octahedral high-pressure cell.

After the pressure was released, the teflon capsule was brought out from the MgO high pressure cell, and the moss spores were investigated under an optical microscope. Figure 22.17a shows the capsule with samples before exposure to high pressure, and (b) the octahedron with the teflon capsule after the pressure was released. In this figure it is seen that the capsule was distorted reflecting the shape of the octahedral high pressure cell.

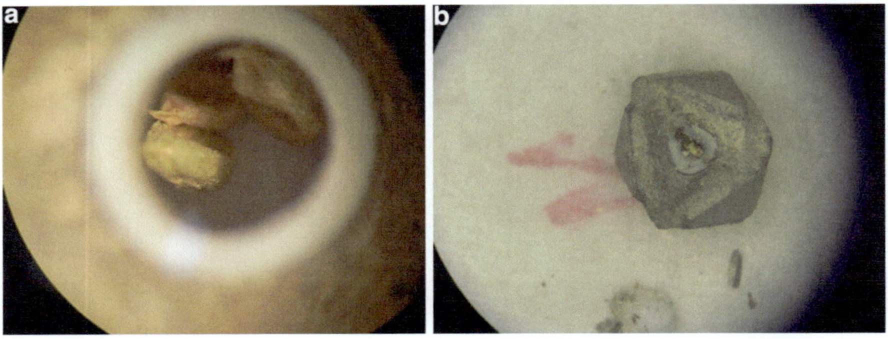

Fig. 22.17 (**a**) Teflon capsule with samples before exposure to high pressure and (**b**) after exposure to 20 GPa for 30 min (Ono et al. 2013)

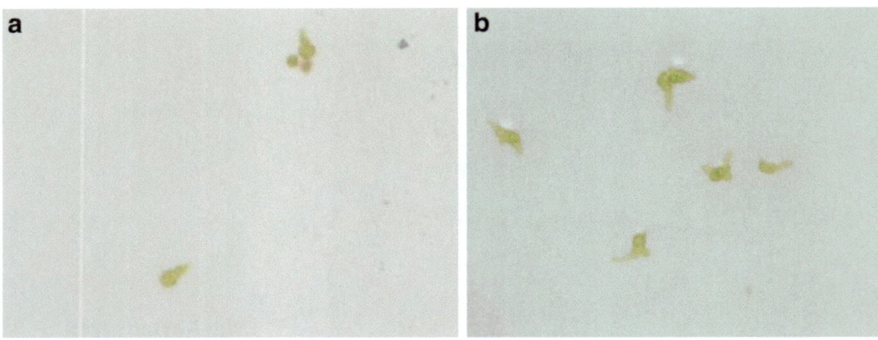

Fig. 22.18 (**a**) A germinated spore protonema of *Venturiella* after exposure to 20 GPa for 30 min and seeded on an agar, (**b**) germinated spores of the control group. These photos were taken 9 days after seeded

The spores of the moss *Venturiella* exposed to 20 GPa for 30 min were seeded on an agar. Just after seeded, the spores seemed normal and their color was green. They were kept at 298 K in an incubator for several weeks or more under white light with an intensity of 2,500 Lx. They became a little swelled at 4 days after seeded.

Figure 22.18a shows a germinated spore protonema of *Venturiella* after exposure to 20 GPa for 30 min (Ono et al. 2013). In this photograph one more germinated spore is seen at the upper right side. This fact means that the two spores were alive even after they were exposed to such an extremely high pressure of 20 GPa. The length of the protonema of a few spores reached up to 30 μm. However, these protonemata did not grow any longer with further incubation, faded gradually and disappeared completely after further incubation for 3 months. For comparison, control moss spores which were not exposed to high pressure are shown in Fig. 22.18b. In this figure germination of protonema is clearly seen in almost all the spores. The germination rate of the spores of the high pressure-exposed group was

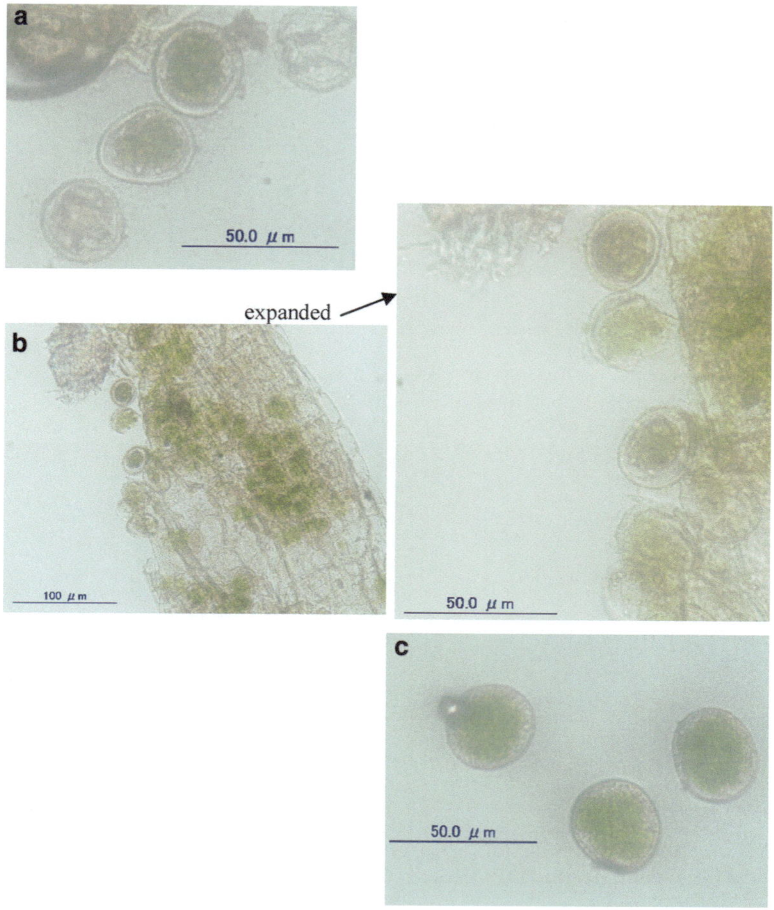

Fig. 22.19 Microscope images of spores of moss *Venturiella*, (**a**) and (**b**) exposed to 20 GPa for 30 min, and (**c**) the control group (Ono et al. 2013)

25 % of the total counted number of the spores. On the other hand, the germination rate for the control was 98 % (Ono et al. 2013).

Changes in physical properties in moss spores exposed to ultra-high pressures were investigated under a microscope by comparing the images for the high pressure-exposed group with those of the control. It is seen in Fig. 22.19a and b that some of the spores exposed to 20 GPa for 30 min were broken and the inner organisms were pushed out of the spore membrane completely. In some other spores, damages were not so serious, but there remains a large plastic deformation. These facts indicate that there exists a certain amount of pressure gradience in the high pressure medium. Under the ultra-high pressure of 20 GPa, the pressure medium fluorinert is no more liquid, and pressure gradience would be generated even by adopting an octahedral anvil press. The diameters of the inner organisms

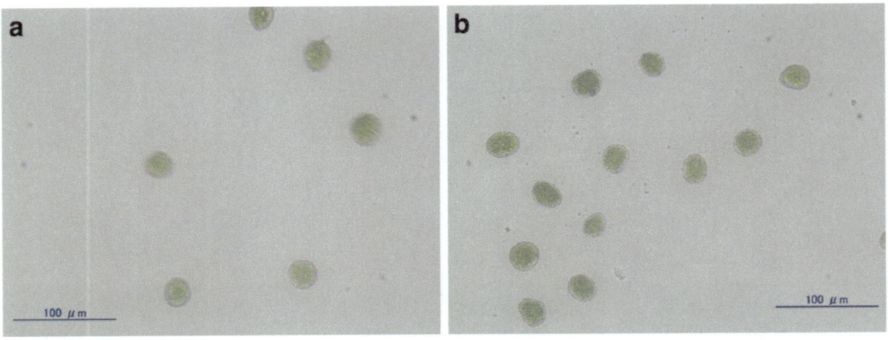

Fig. 22.20 Microscope photographs of spores of moss *Venturiella*, (**a**) exposed to 7.5 GPa for 30 min, and (**b**) the control group (Ono et al. 2013)

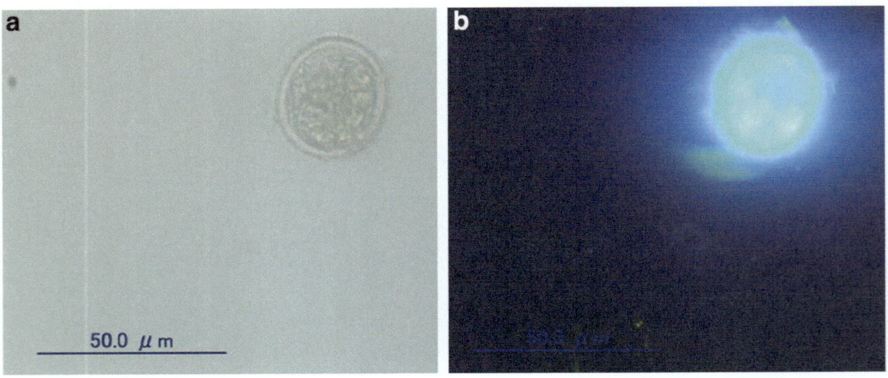

Fig. 22.21 (**a**) Microscope images of a spore of moss *Venturiella* exposed to 20 GPa for 30 min, and (**b**) the same spore after the DAPI-treatment (Ono et al. 2013)

became smaller than those of the control shown in Fig. 22.19c. As a result, a large gap appeared between the spore membranes and the inner organisms. The average outer diameter of the spores decreased about 16 %, which corresponds to a 40 % decrease in volume (Ono et al. 2013).

Plastic deformations of spores exposed to 7.5 and 15 GPa were also investigated using an optical microscope. Observed microscope images for moss spores after exposure to 7.5 GPa for 30 min and control are shown in Fig. 22.20a and b, respectively. The average volume contraction in the spores exposed to 7.5 GPa for 30 min was as small as about 10 %. With this amount of plastic deformation, the germination rate did not decrease and almost all of the spores in high-pressure exposed group were alive.

To make sure whether the high pressure exposed spores of *Venturiella* were alive or not, a DAPI (4′,6-Diamidino-2-phenylindole) treatment was adopted. If a spore is alive, its color appears in red after the DAPI-treatment, while if it is dead its color appears in light-blue. Figure 22.21 shows a spore exposed to 20 GPa for 30 min (a)

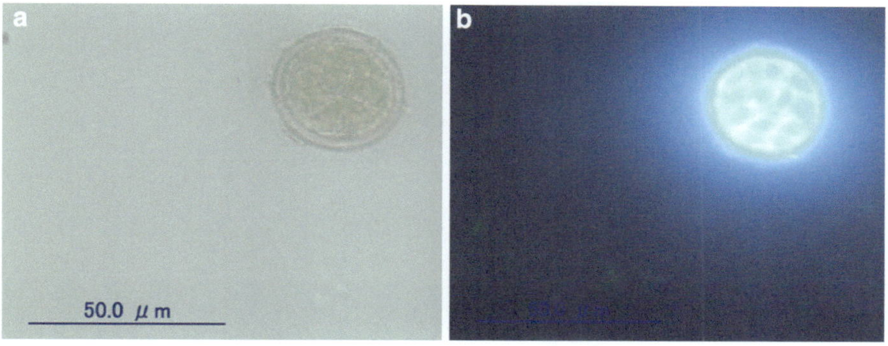

Fig. 22.22 (**a**) Microscope images of a spore of moss *Venturiella* exposed to 15 GPa for 30 min, and (**b**) the same spore after the DAPI-treatment

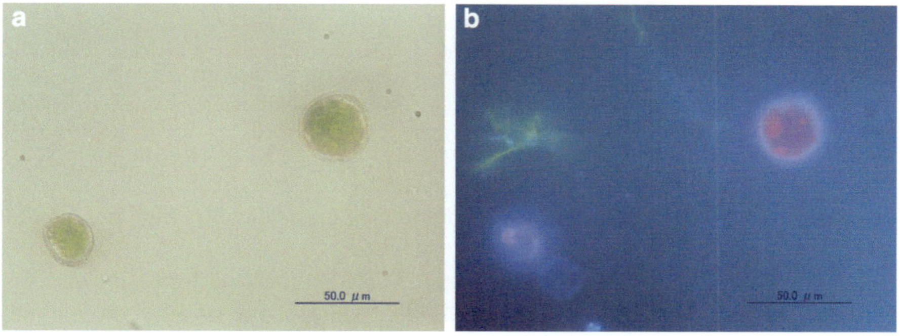

Fig. 22.23 (**a**) Microscope images of two spores of moss *Venturiella* exposed to 7.5 GPa for 30 min, and (**b**) the same spores after the DAPI-treatment (The figures are reproduced from the reference Ono et al. (2013), http://www.tandfonline.com) with the permission of Taylor & Francis Ltd)

before and (b) after the DAPI-treatment. Figure 22.22 also shows a spore exposed to 15 GPa for 30 min (a) before and (b) after the DAPI-treatment. From these figures it is clearly seen that these spores were not alive.

On the other hand, as shown in Fig. 22.23, the color of the two spores exposed to 7.5 GPa for 30 min both turned red after the DAPI-treatment. This means that these two spores were alive after the exposure to 7.5 GPa.

Observed volume contraction of spores of *Venturiella* was plotted in Fig. 22.24 as a function of exposed pressure for 30 min. In this figure the volume of spores under pressure, estimated by extrapolating the average values of compressibility data for various proteins (Gekko and Hasegawa 1986), was also plotted. As we have no compressibility data available for the present samples, we adopted the average values of the data for proteins up to 0.5 GPa, for water (ice) up to 1 GPa (Fine and Millero 1973) and for a solid compound of FeS up to 20 GPa (Kusaba et al. 1998). Though this is merely a rough estimation, it seems worth estimating up to what

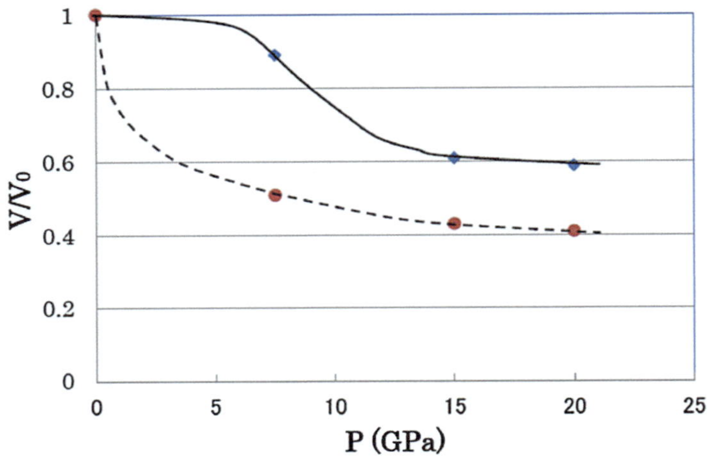

Fig. 22.24 Change of the volume of spores of moss *Venturiella* plotted as a function of exposed pressure. Estimated volume under pressure was also shown by the *broken curve* (The figure is reproduced from the reference Ono et al. (2013), http://www.tandfonline.com) with the permission of Taylor & Francis Ltd)

extent the biological samples would be compressed under the ultra high pressure of 20 GPa.

It was shown from the present investigation that the plastic deformation of moss spores of *Venturiella* after exposure to 20 GPa was as large as 40 % in volume. Nevertheless, a few spores germinated into protonemata to the length of about 30 μm. On the other hand, almost all the spores were found alive after exposure to 7.5 GPa for 30 min, and their plastic deformation was only 10 % in volume. The large difference in the effect of pressure on moss spores between exposure to 7.5 and 20 GPa may be due to freezing of the pressure medium, Fluorinert by pressure. It was reported that fluorinert began to freeze at around 1 GPa, while a good hydrostaticity was preserved up to 10 GPa (Uwatoko et al. 2008). This report may support the results of our experiments showing survival of almost all the animals and plants tested under 7.5 GPa and only 25 % of the moss spores surviving after exposure to 20 GPa.

The effect of hydrostatic pressure on proteins has been studied by a few authors (Gekko and Hasegawa 1986) up to a pressure range around 0.5 GPa. They analyzed partial volume and partial compressibility of some proteins and found that these two parameters were sensitive to the structures of proteins due to hydration and atomic packing within cavities. They pointed out that the pressure dependence of volume gave important information on the flexibility of the native structure and the conformation of denatured states of protein molecules. An analysis of the structure of some proteins under high pressure was made (Akasaka 2003) up to 0.3 GPa by using a pressure variable NMR system. The structural deformation and conformational equilibrium were determined in some proteins under high pressure.

However, all of these detailed analyses of proteins under pressure were made up to 0.5 GPa, and no information was available in the ultra high pressure range of 7.5–20 GPa until the experiment shown above was carried out.

The results of the investigation mentioned in the above section and of the series of the experiments of searching life of small animals and plants under extremely high pressure may give a hint to the hypothetical imagination that life can travel through space (Guidetti et al. 2012) in a large meteorite, and reach the earth alive from another planet or galaxy. It may also give support to the report on findings of some DNA building blocks in a meteorite found in the Arctic (Callahan et al. 2011).

The finding that a plant can stand against very high pressure of 20 GPa may open up other possibilities in the search for very strong species that can withstand very hard environmental conditions and to improve breed of vegetables (Mori et al. 2012; Mustey et al. 2014) and fruits for a better taste and greater yield.

22.7 Discussion

Effects of hydrostatic pressure on biological samples have mostly been investigated to the order of 0.5 GPa. Proteins began to unfold around 0.3 GPa, where most bacteria and many other organisms die (Suzuki and Taniguchi 1972; Weber and Drickamer 1983; Heremans 1987; Silva et al. 1992). High-pressure sterilizing techniques have been used industrially to keep foods fresh. However, in the series of experiments made by the present authors' group (Ono et al. 2008, 2010a, b, 2012, 2013; Minami et al. 2010; Nishihira et al. 2010, 2012; Shibata et al. 2014), all the living samples tested were proven to keep their lives after exposure to such a very high pressure of 7.5 GPa for the duration of at least more than a few hours. Among the candidates for these experiments, the samples that showed the strongest tolerance against the very high pressure were moss spores of *Ptychomitrium* and *Venturiella*. The spores of these mosses can survive after exposure to 7.5 GPa, for up to 144 h. It seems worthwhile to consider that what mechanism made the living creatures so strong against very high pressure.

First, we consider the quality of the hydrostatic pressure generated in these experiments. As we have been using a cubic anvil press and an octahedral anvil press, the uniformity of the generated pressure is much higher than that using a piston-cylinder type high-pressure equipment at above the pressure of freezing point of the liquid high-pressure medium. It was reported (Uwatoko et al. 2008) that fluorinert began to freeze at the pressure above 1 GPa. Therefore, a multi-anvil type press is especially useful at the pressure range of several GPa order where the pressure medium freezes. However, even using a multi-anvil press a certain amount of pressure-gradience may arise above the freezing point of the liquid high-pressure medium.

To estimate the effect of pressure-gradience, we simply assume a cylindrical shape with the length l and the cross-sectional area S for the specimen. Considering an existence of a pressure-gradience along the cylindrical axis (x-direction), $\mathrm{d}p/\mathrm{d}x$, forces exerted to the both end of the specimen along x-direction are written as,

$$F = l\ S\mathrm{d}p/\mathrm{d}x. \tag{22.1}$$

By taking into account the physical fact that this force propagates everywhere in the biological sample along x-direction, and that the effect of this force to the cells and molecules of the living sample can be scaled as the force exerted per unit area, the intensity I in terms of the seriousness of the damage caused in the sample can be considered, to the first approximation, to be proportional to this force per unit area. Thus,

$$I = A\ F/S = A\ l\ \mathrm{d}p/\mathrm{d}x, \tag{22.2}$$

where A is a constant. We define, here, the high-pressure tolerance T as the duration of time for which the living sample could tolerate under high pressure. As a result, the high-pressure tolerance can be expressed to the first approximation as,

$$T = B(\mathrm{d}p/\mathrm{d}x)^{-1}\ l^{-1} \tag{22.3}$$

where B is a constant which is proportional to A^{-1}. When p is fixed at 7.5 GPa, as in the present case, $(\mathrm{d}p/\mathrm{d}x)^{-1}$ is also a constant. From this equation, the strongest tolerance of moss spores can be explained, as they are the smallest among the specimens investigated. Recently, we extended our study to a further smaller sized sample, blue-green alga *Microcystis flos-aquae* of which the diameter of a membrane is about 2–4 μm. It was shown that the pressure tolerance was enormously strong and survived and multiplicated after exposure to 20 GPa for 30 min (Ono et al. 2014). However, there seems to be an exception. In the case of yeast (Shibata et al. 2014), the diameter of a membrane is smaller than a moss spore. Never the less, the high-pressure tolerance is much weaker than that of mosses. In the case of yeast, we must consider the size of the sample not as a cell, but as a dry-yeast granule.

Next, we consider the effect of plastic deformation caused in a biological sample by the heavy compression. It was shown (Ono et al. 2013) that in the moss spores the plastic deformation caused by the application of 7.5 GPa for 30 min was only 10 % in volume, while after exposure to 20 GPa it was as much as 40 %. Therefore, it was confirmed that a plastic deformation of 10 % in volume can keep the moss spores alive, while that of 40 % would kill most of the spores. The high plasticity of proteins and genes also seems to play an important role in the strong tolerance against heavy deformation. Owing to this unique character that proteins and genes possess, most part of the strain that proteins and genes receive under heavy compression could be elastic deformation.

Thirdly, we consider the tolerance as time-dependent high-pressure resistivity of biological samples. Considering the experimental fact that all the biological samples that have been tested were alive after exposure to 7.5 GPa for at least 1 h, and that the tolerance defined as the tolerated duration of time is roughly in inverse proportion to the size of the sample, there are some chemical reactions or time-dependent phase transformations taking place during the exposure to ultra-high pressure. However, it is difficult to point out, at this moment, what kind of chemical reaction or phase transformation is involved in the high-pressure tolerance of biological samples.

Finally, we extend our discussion to extremophiles that active living samples gain just before they are dehydrated. It is well known that once an animal or plant is brought into metabolically inactive state, it acquires very strong resistivity against severe environmental conditions such as high temperatures, high radioactive beam irradiations and high pressures. Primary metabolites such as amino-acids, sugars, vitamins etc. play an important role to the strong resistivity of plants and animals against severe environmental conditions. Among these metabolites, sugars such as trehalose (Higashiyama 2002) seem to be one of the most important substances. It is known that animals and plants gain a certain amount of trehalose just before they become metabolically inactive.

Amount of residual water also seems important. However, a plant seed usually contains a certain amount of water, as much as about 10 %, which would freeze under high pressure to cause serious damages in the organs and cells of the seed. On the other hand, our experiment showed that seeds of white clover tolerate the high pressure of 7.5 GPa for more than 1 h (Nishihira et al. 2012). This result proves that residual water is not an important factor responsible for the strong resistivity of the living samples against very high pressure.

In the end of discussion, we emphasize the fact that, just before brought into cryptobiotic state, both animals and plants acquire much stronger extremophiles than we had expected before we began these high-pressure experiments. The high plasticity of proteins also seems to play an important role in the strong tolerance of living samples against ultra-high pressure.

22.8 Conclusion

By using a cubic anvil press it was shown that 80–90 % of spores of moss *Ptychomitrium* were alive and germinated after exposure to an extremely high hydrostatic pressure of 7.5 GPa for up to 48 h. It was also shown that 70–90 % of the spores of the moss *Venturiella* were alive and germinated after exposure to the high pressure of 7.5 GPa for up to 72 h. However, after exposure to 7.5 GPa for 144 h, the germination rate of *Ptychomitrium* spores decreased to 32 %. Interestingly, the result reveal that at least about one third of the spores of moss *Ptychomitrium* were still alive and germinated after being exposed to 7.5 GPa for such a long period of 144 h. In the case of *Venturiella*, however, only four individuals in a hundred were alive and germinated after exposure to 7.5 GPa for 144 h. Compared with tardigrades, moss spores are about five times stronger against high pressure.

The investigation was extended to the ultra-high pressure range up to 20 GPa by using an octahedral anvil press. It was shown that 25 % of the spores of moss Venturiella were alive and germinated after exposure to 20 GPa for 30 min.

However, germinated protonemata grew only up to the length of 30 μm. They did not grow any further and faded away after incubation for 3 months. However, it was really a great surprise that moss spores germinated protonema to the length of 30 μm after exposure to such an ultra-high pressure of 20 GPa.

As the mechanisms responsible for the very strong tolerance of living creatures in a cryptobiotic state against ultra-high pressure, primary metabolites such as amino-acids, sugars, vitamins etc. play an important role. Among these metabolites, sugars such as trehalose seem to be one of the most important substances. Animals and plants gain a certain amount of trehalose just before they become metabolically inactive. It was proven that residual water is not an important factor responsible for the strong resistivity of the living samples against ultra-high pressure.

Considering the experimental fact that in all the biological samples that have been tested the tolerance depends on the duration of time exposed to high pressure, there are some chemical reactions or phase transformations taking place during the exposure to ultra-high pressure. However, it is difficult to point out what kind of chemical reaction or phase transformation is involved.

The effect of plastic deformation caused in a biological sample by the heavy compression seems an important factor to the strong tolerance. The plastic deformation caused by the application of 7.5 GPa was only 10 % in volume, while after exposure to 20 GPa it was as much as 40 %. Therefore, a plastic deformation of 10 % in volume can keep the moss spores alive, while that of 40 % would kill most of the spores. The high plasticity of proteins also seems to play an important role in the strong tolerance against ultra-high pressure.

The results of the present investigation suggest the possibility of moss spores traveling through outer space in a large meteorite, and reaching the earth alive from other planet or galaxy.

Acknowledgements The author of this chapter wishes to express his sincere thanks to Mrs. N. Nishihira of Okayama Ichinomiya Senior High School for incubation of mosses and kind help throughout the experiment. He also thanks to Professors Y. Mori, K. Takarabe and Y. Hada of Okayama University of Science, Dr. M. Saigusa and Dr. Y. Matsushima of Okayama University and Prof. N. Saini of University of Rome for valuable discussion. Thanks are extended to Prof. E. Ito and Dr. D. Yamazaki of Institute for Study of the Earth's Interior (ISEI) of Okayama University for their help operating the octahedral anvil press.
This chapter presents a result of a joint research program carried out at the Institute for Study of the Earth's Interior, Okayama University.

References

Akasaka K (2003) Highly fluctuating protein structures revealed by variable-pressure nuclear magnetic resonance. Biochemistry 42:10875–10885

Bovenkerk HP, Bundy FP, Hall HT, Strong HM, Wentorf RH (1959) Preparation of diamond. Nature 184:1094–1098

Callahan MP, Smith KE, Cleaves HJ, Ruzicka J, Stern JC, Glavin DP, House CH, Dworkin JP (2011) Carbonaceous meteorites contain a wide range of extraterrestrial nucleobases. Proc Natl Acad Sci 108:13995–13998

Fine RA, Millero FJ (1973) Compressibility of water as a function of temperature and pressure. J Chem Phys 59:5529–5536

Gekko K, Hasegawa Y (1986) Compressibility-structure relationship of globular proteins. Biochemistry 25:6563–6571

Guidetti R, Rizzo AM, Altiero T, Rebecchi L (2012) What can we learn from the toughest animals of the Earth? Water bears (tardigrades) as multicellular model organisms in order to perform scientific preparations for lunar exploration. Planet Space Sci 74:97–102

Heremans K (1987) In: Jonnasch WM, Marquis RE, Zimmerman AM (eds) Current perspectives in high pressure biology. Academic, London, pp 224–244

Higashiyama T (2002) Novel functions and applications of trehalose. Pure Appl Chem 74:1263–1269

Horikawa DD, Iwata K, Kawai K, Koseki S, Okuda T, Yamamoo K (2009) High hydrostatic pressure tolerance of four different anhydrobiotic animal species. Zoolog Sci 26:238–242

Kenrick P, Crane PR (1997) The origin and early evolution of plants on land. Nature 389:33–39

Kusaba K, Syono Y, Kikegawa T, Shimomura O (1998) Structure and phase equilibria of FeS under high pressure and temperature. In: Manghnini MH, Yagi T (eds) Properties of earth and planetary materials at high pressure and temperature. American Geophysical Union, Washington, DC, pp 297–306

Matsushita M, Endo S, Miura K, Ono F (2003) Pressure induced magnetic phase transition in Fe-Ni Invar alloy. J Magn Magn Mater 265:352–356

Matsushita M, Miyoshi M, Endo S, Ono F (2005) Pressure-induced magnetic phase transitions in Fe-based Invar alloys. Phys Rev B 72:214404-1-5

Minami K, Ono F, Mori Y, Takarabe K, Saigusa M, Matsushima Y, Saini NL, Yamashita M (2010) Strong environmental tolerance of *Artemia* under very high pressure. J Phys Conf Ser 215:012164-4p

Mori Y, Yokota S, Ono F (2012) Germination of vegetable seeds exposed to very high pressure. J Phys Conf Ser 377:012055-4p

Mustey A, Leighs JA, Appleby-Thomas GJ, Wood D, Hazael R, McMillan PF, Hazell PJ (2014) The effect of hydrostatic vs. shock pressure treatment of plant seeds. J Phys Conf Ser 500:182025-6p

Nishihira N, Shindou A, Saigusa M, Ono F, Matsushima Y, Mori Y, Takarabe K, Saini NL, Yamashita M (2010) Preserving life of moss *Ptychomitrium* under very high pressure. J Phys Chem Solid 71:1123–1126

Nishihira N, Iwasaki T, Shinpou R, Hara A, Ono F, Hada Y, Mori Y, Takarabe K, Saigusa M, Matsushima Y, Saini NL, Yamashita M (2012) Maintaining viability of white clover under very high pressure. J Appl Phys 111:112619-4p

Ono F, Saigusa M, Uozumi T, Matsushima Y, Ikeda H, Saini NL, Yamashita M (2008) Effect of high hydrostatic pressure on to life of the tiny animal tardigrade. J Phys Chem Solid 69:2297–2300

Ono F, Mori Y, Takarabe K, Nishihira N, Shindo A, Saigusa M, Matsushima Y, Saini NL, Yamashita M (2010a) Strong environmental tolerance of moss *Venturiella* under very high pressure. J Phys Conf Ser 215:012165-4p

Ono F, Minami K, Saigusa M, Matsushima Y, Mori Y, Takarabe K, Saini NL, Yamashita N (2010b) Life of Artemia under very high pressure. J Phys Chem Solid 71:1127–1130

Ono F, Mori Y, Sougawa M, Takarabe K, Hada Y, Nishihira N, Motose H, Saigusa M, Matsushima Y, Yamazaki D, Ito E, Saini NL (2012) Effect of very high pressure on life of plants and animals. J Phys Conf Ser 377:012053-6p

Ono F, Nishihira N, Sougawa M, Hada Y, Mori Y, Takarabe K, Saigusa M, Matsushima Y, Yamazaki D, Ito E, Saini NL (2013) Distortion o spores of moss Venturiella under ultra high pressure. High Pres Res 33:362–368

Ono F, Nishihira N, Hada Y, Mori Y, Takarabe K, Saigusa M, Matsushima Y, Yamazaki D, Ito E (2014) Strong tolerance of blue-green alga *Microcystis flos-aquae* to very high pressure. J Phys Chem Solids. doi:10.1016/j.jpcs.2014.08.008

Shibata M, Torigoe M, Matsumoto Y, Yamamoto M, Takizawa N, Hada Y, Mori Y, Takarabe K, Ono F (2014) Tolerance of budding yeast Saccharomyces cerevisiae to ultra high pressure. J Phys Conf Ser 500:102004-5p

Silva JL, Luan P, Glaser M, Voss EW, Weber G (1992) Effects of hydrostatic pressure on a membrane-enveloped virus; high immunogenicity of the pressure-inactivated virus. J Virol 66:2111–2117

Stacy FD, Davis PM (2008) Physics of the Earth, 4th edn. Cambridge University Press, Cambridge, p 280

Suzuki K, Taniguchi Y (1972) Effect of pressure on biopolymers and model systems. Symp Soc Exp Biol 26:103–124

Uwatoko Y, Matsubayashi K, Matsumoto T, Aso M, Nishi N, Fujiwara T, Hedo M, Tabata S, Takagi K, Tado M, Kagi H (2008) Development of palm cubic anvil apparatus for low temperature physics. Rev High Pres Sci Tech 18:230–236 (in Japanese)

Weber G, Drickamer HG (1983) The effect of high pressure upon proteins and other biomolecules. Q Rev Biophys 16:89–112

Part VII
High Pressure Food Processing and Sterilization

Editors' Foreword of Part VII

In this part, recent developments in two major areas of pressure application to food materials are described. One area of application is the efficient use of pressure in sterilizing food materials, particularly spores which are known to be particularly pressure-tolerant. Lenz and Vogel in Chap. 23 and Yamazaki and collaborators in Chap. 26 report examples of new approach to this problem by combining pressure and temperature treatments, with which efficient sterilization is attainable. Although the mechanism is yet to be clarified, the method represents a new trend in the use of pressure for sterilization. In Chap. 25, Izumi reports on their new attempts to sterilize raw agricultural products with pressure, and reports that basically pressure at 400 MPa gives a sufficient level of sterilization on fresh-cut vegetables.

The other area is the use of pressure to improve the quality of food product. Shigematsu in Chap. 24 as well as Yamazaki in Chap. 26 reports the utility of relatively low pressures in increasing food qualities in terms of nutrition and taste. The relatively low range of pressure needed to improve qualities of food or other biomaterials will promise its utility in a wide range of applications in near future.

Chapter 23
Pressure-Based Strategy for the Inactivation of Spores

Christian A. Lenz and Rudi F. Vogel

Abstract Since the first application of high hydrostatic pressure (HHP) for food preservation more than 100 years ago, a wealth of knowledge has been gained on molecular mechanisms underlying the HHP-mediated destruction of microorganisms. However, one observation made back then is still valid, i.e. that HHP alone is not sufficient for the complete inactivation of bacterial endospores. To achieve "commercial sterility" of low-acid foods, i.e. inactivation of spores capable of growing in a specific product under typical storage conditions, a combination of HHP with other hurdles is required (most effectively with heat (HPT)). Although HPT processes are not yet industrially applied, continuous technical progress and increasing consumer demand for minimally processed, additive-free food with long shelf life, makes HPT sterilization a promising alternative to thermal processing.

In recent years, considerable progress has been made in understanding the response of spores of the model organism *B. subtilis* to HPT treatments and detailed insights into some basic mechanisms in *Clostridium* species shed new light on differences in the HPT-mediated inactivation of *Bacillus* and *Clostridium* spores. In this chapter, current knowledge on sporulation and germination processes, which presents the basis for understanding development and loss of the extreme resistance properties of spores, is summarized highlighting commonalities and differences between *Bacillus* and *Clostridium* species. In this context, the effect of HPT treatments on spores, inactivation mechanism and kinetics, the role of population heterogeneity, and influence factors on the results of inactivation studies are discussed.

Keywords Bacterial endospores • Food safety • Germination • High hydrostatic pressure (HHP) • High pressure thermal (HPT) processing • Inactivation mechanism • Pressure-assisted thermal processing (PATP) • Resistance • Sporulation • Sterilization

---

C.A. Lenz • R.F. Vogel (✉)
Lehrstuhl für Technische Mikrobiologie, Technische Universität München,
85354 Freising, Germany
e-mail: rudi.vogel@wzw.tum.de

K. Akasaka, H. Matsuki (eds.), *High Pressure Bioscience*,
Subcellular Biochemistry 72, DOI 10.1007/978-94-017-9918-8_23

23.1 Introduction

23.1.1 History of Spores and HHP

In addition to salting, drying, smoking or fermenting food, the application of heat treatments has a long history in the preservation of food and dates back to not later than the last decade of the seventeenth century (in: (Gould 2006)). More than 100 years later, a method to thermally sterilize food using hermetically sealed containers was described by Appert (1810). This was long before the microbiological origin of food spoilage was identified (Pasteur 1866) and bacterial spores were studied. Although the presence of refractile bodies in bacterial cells was already reported in the first half of the nineteenth century (Ehrenberg 1838), it took more than 30 years until the term "spores" appeared within the context of "permanent" cells of bacteria (*Bacillus subtilis* (Cohn 1872)), which was followed by first studies on bacterial spores conducted by Cohn (Cohn 1876) and Koch (*Bacillus anthracis* (Koch 1876)).

The application of high hydrostatic pressure (HHP) treatments to preserve food was first described only 23 years later. In 1899, Hite tried to find alternative preservation methods for heat sensitive food and was able to demonstrate that HHP treatments at approx. 700 MPa can significantly increase the shelf-life of milk with less detrimental effects on sensorial properties compared to heat treatments (Hite 1899). However, relatively soon it became clear that bacterial spores are highly resistant to HHP treatments (Chlopin and Tammann 1903) and even pressures levels of about 1,200 MPa are not sufficient to achieve an effective inactivation (*B. subtilis* (Larson et al. 1918)). This clearly indicated that such treatments are not suitable for an application in food sterilization processes. In the late 1960s some attempts were made to utilize HHP-triggered germination at low pressure levels to establish a "tyndallization"-like two-step HHP process, in which germinated, i.e. heat sensitive spores were subsequently killed by mild heat treatments (Clouston and Wills 1969; Gould and Sale 1970). However, the great heterogeneity in germination requirements and pressure sensitivity of spores from different species and single spores within a population resulting in non-log-linear germination kinetics made such approaches too unreliable for an application in commercial food sterilization processes (Gould 2006).

In the 1980s, technological progress and the increasing consumer demand for fresh-like food with increased shelf-life resulted in first commercial applications of HHP pasteurization processes for the production of acidic ($pH < 4.5$) food, where spores from pathogenic species are not able to initiate growth. Thereafter, the number of pressure treated, chilled, low-acid food products ($pH \geq 4.5$) steadily increased. This group of foods nowadays presents the majority of pressure treated products on the market. Commonly, pressure levels of up to 600 MPa and initial temperatures below (or at) room temperature are used in the production of meat products and vegetables each accounting for approx. 27 % of HHP treated food, followed by (fruit) juices and beverages (approx. 14 %), and seafood and fish products (approx. 13 %) (Tonello-Samson 2014).

However, such commonly applied processes are not sufficient to inactivate spores effectively. Spore inactivation by HHP can be enhanced by applying a two-step process or by adding synergistically acting substances to food. Both approaches have some disadvantages and have not yet been shown to be capable of complete spore inactivation in a broad range of food products (see Sect. 23.3.5). An effective way to inactivate spores using HHP-based processes requires a combination of HHP and heat, where the detrimental effect of heat on food quality is still lower compared to conventional retorting, and process control can be improved.

In such processes, the adiabatic heat of compression can be exploited to fast and homogenously reach the sterilization temperature of 121.1 °C in a preheated product, which reduces the heating time and therefore the thermal load and avoids over-/underprocessing of the edges/interior of a food product (pressure-assisted thermal sterilization, PATS). Alternatively, high pressure at temperatures below the sterilization temperature can be used to achieve commercial sterility utilizing the synergistic effect between pressure and temperature on spore inactivation while significantly reducing the thermal load. Acronyms for such processes are numerous and include PATP (pressure-assisted thermal processing, e.g. Park et al. 2014), HP/HT (e.g. (Shpigelman et al. 2014)) or HPHT (high pressure high temperature, e.g. (Kebede et al. 2013)), HPS (high-pressure sterilisation treatments combined with elevated starting temperatures, e.g. (Krebbers et al. 2003)), HPTP (high pressure–thermal processing, e.g. (Devi et al. 2015)), HPTS (high pressure thermal sterilization, e.g. (Reineke et al. 2013b)), and THP (combined thermal-high pressure processing, e.g. (Fraeye et al. 2010)). Throughout this chapter, the acronym HPT (high pressure thermal, e.g. HPT (Knoerzer and Chapman 2011; Delgado et al. 2013; García-Parra et al. 2014)) is used whenever, high pressure is combined with temperatures above ambient temperatures. HPT is used since it is relatively short, can be used to describe treatments/processing/inactivation/resistance, does not predetermine the desired inactivation result (such as sterilization), and is not judgmentally with respect to the contribution of pressure and temperature to the inactivation result (such as PATP).

HPT processes are not yet applied in an industrial scale, which is presumably due to two major reasons: (i) Technical constraints including the deficiency in robustness and increased fatigue of industrial HHP equipment at elevated temperatures impede commercial application. (ii) Despite of the discovery of bacterial spores and the first experiments on high pressure treatment of food more than 100 years ago and the wealth of knowledge that has been gained since then, some basic mechanisms including the HHP/HPT resistance of spores still evade complete understanding.

However, HPT processing still has the advantage to enable the production of stable and safe food while nutritionally valuable molecules (e.g. vitamins), the appearance (color, texture), and the taste are affected to a lower extent as compared to thermal processing (Heinz and Buckow 2010; Knorr et al. 2011). Due to continuous technical progress, increasing consumer demands for minimally processed, healthy food with long shelf life, and considerable efforts made by legal authorities to decrease the levels of traditional food preservatives (e.g. salt), HHP-based sterilization processes increasingly appear to present a promising alternative to thermal processing.

23.1.2 Important Spore-Forming Species

Many members of the orders *Bacillales* and *Clostridiales* can produce spores. The ubiquitous presence of bacterial spores in nature, their dormancy and extreme resistance to chemical and physical stress conditions, and the ability to grow out under favorable conditions prevalent in many (mainly low-acid) food products makes them a key problem in food industry.

Various genera such as *Alicyclobacillus* (*A.*), *Desulfotomaculum* (*D.*), *Thermoanaerobacterium* (*T.*) and *Geobacillus* (*G.*) include certain species that are associated with food spoilage. Examples include *A. acidoterrestris* (Steyn et al. 2011), *D. nigrificans*, *T.* (formally *Clostridium* (*C.*)) *thermosaccharolyticum* and *G.* (formally *Bacillus* (*B.*)) *stearothermophilus*.

The genera, which include species that are most frequently associated with problems in food preservation (spoilage and safety), are *Bacillus* and *Clostridium*. Typical spoilage organisms can be found in the species *B. coagulans* (flat-sour), *B. thermoacidurans* (acid spoilage), and occasionally *B. licheniformis*, *B. pumilus*, and *B. subtilis* (Oomes et al. 2007). Additionally, various *Clostridium* species can be involved in spoilage including *C. bifermentans* (sulfide spoilage, rotten egg odor, black color), *C. sporogenes* (putrefactive fermentation), *C. butyricum*, *C. tertium* (butyric odor) and various psychrophilic and psychrotrophic *Clostridium* species primarily associated with spoilage of chilled red meat (Adam et al. 2010).

Major spore-forming, food-associated, pathogenic species include *B. cereus* (emetic toxin cereulide and diarrheal toxins), *C. perfringens* (intoxication, diarrhea/vomiting, enterotoxin; infection, gas gangrene, alpha toxin) and *C. botulinum* (intoxication, flaccid paralysis, botulinum neurotoxin; infection, infant botulism) (Scallan et al. 2011). *C. difficile* (infection, diarrhea, enterotoxin A, cytotoxin B) can also present a problem for food safety. In addition to species of significance in food, the genera *Bacillus* and *Clostridium* also comprise clinically important species (e.g. *B. anthracis* (anthrax) and *C. tetani* (tetanus)) and strains of importance in industrial biotechnology (e.g. *B. amyloliquefaciens* (alpha amylase, BamH1 restriction enzyme), *B. thuringiensis* (insecticidal toxin), *C. phytofermentans*, *C. thermocellum* (cellulose degradation), *C. acetobutylicum* and *C. beijerinckii* (solvent production)).

However, in the past decades, *B. subtilis* served as model organism for studying basic mechanisms such as the sporulation cascade, resistance and germination mechanisms. Although this organism can be occasionally involved in food spoilage, it certainly doesn't play any important role in food spoilage and does not present any threat to food safety. Reasons for this choice may be found in its non-pathogenic nature, its natural transferability and the ready availability of genetic and molecular biological data, e.g. its complete genome sequence. Although basic mechanisms are still not understood completely, a wealth of knowledge has been obtained on *B. subtilis* spores (Setlow and Johnson 2013) including insights into HHP- and HPT-mediated inactivation mechanisms (Reineke et al. 2013a).

For *Clostridium* species, *C. acetobutylicum*, has been suggested to present a suitable model organism studying basic sporulation and germination mechanisms due to its interesting solventogenic properties, its non-pathogenicity, and a high degree in genetic homology compared to many other *Clostridium* species (Paredes et al. 2005). Additionally, a considerable amount of research on basic mechanisms has been conducted using *C. perfringens* (Paredes-Sabja et al. 2011). However, there are still huge differences in the state of knowledge on *B. subtilis* compared to other spore-forming bacteria.

Although considerable variations in some key proteins and signal transduction pathways can impede direct transferability of results, basic elements involved in sporulation (Paredes et al. 2005; de Hoon et al. 2010; Li and McClane 2010; Wörner et al. 2006) and germination (Paredes-Sabja et al. 2011) are highly conserved among various spore-forming species. Additionally, the basic structure and composition of spores of various species appears to be very similar (Setlow and Johnson 2013). Thus, available data for model organisms present a valuable basis for understanding basic mechanisms in other spore formers.

Knowledge on the developmental program leading to spore formation, i.e. sporulation, presents an important basis for understanding the development of the extremely high resistance of spores to physical treatments such as HHP or HPT processing. Since the loss of these resistance properties plays a crucial role in HHP-mediated spore inactivation, an in-depth understanding of this process requires a detailed picture of molecular events during spore germination. Therefore, the following paragraphs provide an overview of the spore cycle and describe events during sporulation and germination considering major differences between the genera *Bacillus* and *Clostridium*. Especially genus-specific differences in germination pathways directly explain some major difference in HHP and HPT inactivation mechanisms of *Bacillus* and *Clostridium* spores.

23.2 The Spore Cycle

In response to unfavorable conditions, which can be typically found during the stationary phase, cells from spore-forming species can initiate sporulation (Fig. 23.1). However, the entry into stationary phase does not obligatory lead to immediate sporulation. Typical stationary phase events, which are not necessarily required for but often regulated by factors involved in sporulation include that cells upregulate genes associated with nutrient transport and metabolism (carbohydrates, amino acids and inorganic ions), try to make additional nutrient sources accessible, and to evade or somehow to adapt to the new situation. The induction of alternative metabolic substrate pathways can occur due to nutrient depletion (e.g. arginine depletion, arginine biosynthesis), but does not necessarily require complete depletion (e.g. >40 % of initial glucose remaining, *C. acetobutylicum* (Jones et al. 2008)). Strategies to improve nutrient accessibility include the secretion of degradative enzymes such as amylases and proteases (e.g. *B. subtilis*, Phillips and Strauch 2002) and harming/killing of other organism including sister cells (e.g. *B. subtilis* (Gonzalez-Pastor et al. 2003)), the competitive microbiota (antibiotics)

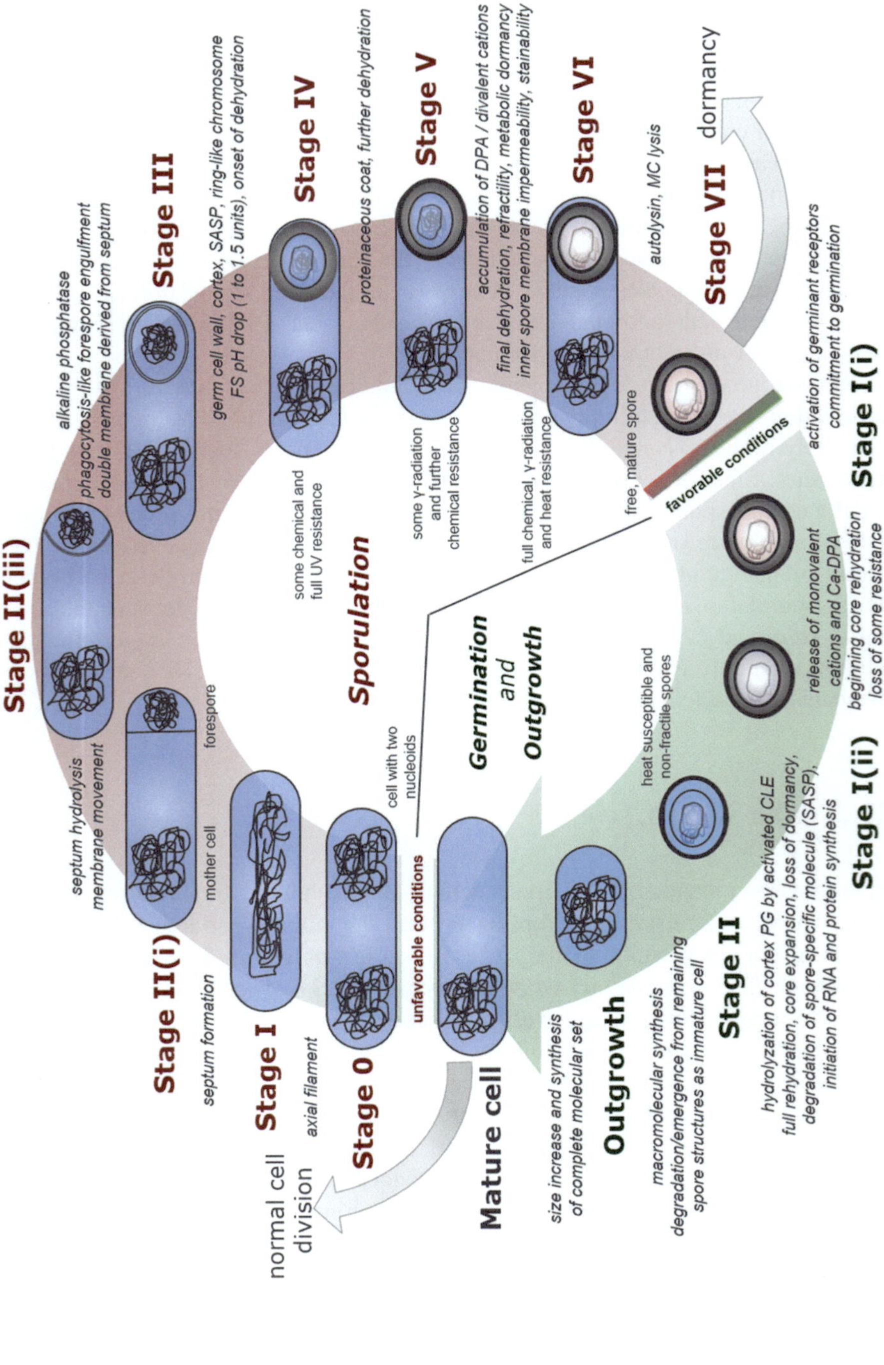
Stage II(iii)
septum hydrolysis
membrane movement
alkaline phosphatase
phagocytosis-like forespore engulfment
double membrane derived from septum
Stage III
germ cell wall, cortex, SASP, ring-like chromosome
FS pH drop (1 to 1.5 units), onset of dehydration
Stage IV
proteinaceous coat, further dehydration
Stage V
accumulation of DPA / divalent cations
final dehydration, refractility, metabolic dormancy
inner spore membrane impermeability, stainability
Stage VI
autolysin, MC lysis
Stage VII
dormancy
Stage II(i)
septum formation
mother cell
forespore
Stage I
axial filament
Stage 0
cell with two nucleoids
unfavorable conditions
Sporulation
some chemical and full UV resistance
some γ-radiation and further chemical resistance
full chemical, γ-radiation and heat resistance
free, mature spore
favorable conditions
Germination and Outgrowth
heat susceptible and non-fractile spores
Mature cell
normal cell division
activation of germinant receptors
commitment to germination
Stage I(i)
release of monovalent cations and Ca-DPA
beginning core rehydration
loss of some resistance
Stage I(ii)
hydrolyzation of cortex PG by activated CLE
full rehydration, core expansion, loss of dormancy,
degradation of spore-specific molecule (SASP),
initiation of RNA and protein synthesis
Stage II
macromolecular synthesis
degradation/emergence from remaining
spore structures as immature cell
Outgrowth
size increase and synthesis
of complete molecular set

(e.g. *B. subtilis* (Irigul-Sonmez et al. 2014)), or higher organisms such as insects (e.g. *B. thuringensis*) or animals (protein toxins) (e.g. *C. botulinum*, *C. tetani* (Connan et al. 2013)). Motility and chemotaxis can present a stationary phase response to evade the nutrient-poor environment (e.g. *C. difficile* (Saujet et al. 2011)) with the underlying machinery being well conserved in *Bacillus* and *Clostridium* species and closely linked to sporulation (e.g. *B. subtilis* (Aizawa et al. 2002); *C. acetobutylicum* (Tomas et al. 2004; Alsaker and Papoutsakis 2005; Zhao et al. 2005)). In contrast, competence development to adapt to the new situation presents a relatively uncommon stationary phase response (e.g. *B. subtilis* (Stiegelmeyer and Giddings 2013)).

Finally, cells can initiate sporulation. The decision for a single cell within a population to form a spore depends on the success of the strategies mentioned above, starvation conditions (Chubukov and Sauer 2014) and extracellular signals (e.g. cell density-dependent (Lazazzera et al. 1999; Bischofs et al. 2009; Schultz et al. 2009; Lopez and Kolter 2010)). Quorum-sensing (QS) mechanisms mediated by signal peptides are thought to play an important role in both *Bacillus* and *Clostridium* species. Prominent examples are the Phr-Rap system in *Bacillus* species (peptides as phosphatase regulators-receptor aspartyl phosphatases; *B. subtilis* (Bischofs et al. 2009)) and the Agr-like system in *Clostridium* species (Accessory gene regulator (Myers et al. 2006); *C. perfringens* (Li et al. 2011); *C. sporogenes*, *C. botulinum* type A (Cooksley et al. 2010)). Although various environmental factors, which push a cell towards or delay the decision to initiate sporulation have been identified and the mechanism of sporulation induction is quite well characterized (at least in *B. subtilis*; see Sect. 23.2.1.1.1), single cell fate in the context of the entire population is largely unexplored. However, it seems likely that differences in the fate of single cells at this stage influence the heterogeneity of a spore population (see Sect. 23.3.3), which has a significant impact on the success of HHP-based inactivation strategies (see Sect. 23.3.5).

Mature spores are metabolically dormant and differ significantly from vegetative cells in their composition, structure and resistance properties enabling spore-forming organisms to survive long periods without the availability of nutrients under harsh environmental conditions. Under certain environmental conditions (e.g. in the presence of nutrient germinants) spores can germinate, grow out and convert back to a growing cell.

Fig. 23.1 General scheme of the spore cycle. Stages 0 through VII indicate major morphological stages during sporulation. Stages I(i) through II indicate germination stages. Major events between the stages (italics) and important spore properties at a specific stage (plain text inside of the circle) are described. Black coils in the oval cells indicate DNA. Spore layers: thin black line = inner membrane; thick gray layer = cortex; outer black layer = coat. Notably, spore-formers are not always rod-shaped, the sporangium can swell during sporulation, spore formation does not necessarily occur at a terminal position, and spores can contain a large exosporium as outermost layer, which is not depicted. *MC* mother cell, *SASP* small acid soluble protein, *FS* forespore, *DPA* dipicolinic acid, *PG* peptidoglycan, *CLE* cortex lytic enzymes (Data for figure adapted from Setlow and Johnson (2013) and Setlow (2003))

The spore cycle of *B. subtilis* has been most extensively studied (Setlow and Johnson 2013). A simplified scheme of the spore cycle of such a rod shaped bacterium without terminal swelling of the forespore compartment or elongation of the mother cell is depicted in Fig. 23.1. The spore cycle in *Clostridium* species is presumed to include essentially the same steps with the difference that the sporangium swells during the course of sporulation. Based on genomic information, *Bacillus* and *Clostridium* sporulation programs show many parallels and many components of the spore germination machinery are highly conserved among spore-forming members of the *Bacillales* and *Clostridiales* orders. However, although orchestration of differentiation programs is less understood in *Clostridium* species and far from being completely elucidated at the molecular level, studies on the transcriptional level have contributed to better understanding of molecular events that underlie sporulation and suggest that there are some major differences in sporulation (see Sect. 23.2.1.1) (Paredes et al. 2005; Jones et al. 2008). Additionally, both proteins and signal transduction pathways involved in germination are substantially different between *Bacillus* and *Clostridium* species (see Sect. 23.2.2) (Paredes-Sabja et al. 2011).

The following paragraph summarizes important morphological events during sporulation, which putatively occur in a similar manner in *Bacillus* and *Clostridium* species and lead to a stepwise development of the extreme resistance of spores.

23.2.1 Sporulation

Although sporulation is a continuous process and intermediate forms do not present discrete entities in which a sporulating cell remains, this developmental process is commonly divided into eight stages (stage 0 through VII) based on observable morphological changes and helpful in describing mutant strains (*spo* mutants), in which sporulation is blocked in a particular stage (Setlow and Johnson 2013) (Fig. 23.1).

In a vegetative, growing cell, two nucleotides are formed (stage 0) that align in an axial filament (stage I). Transition to stage II is characterized by typical stationary phase responses as mentioned above (e.g. secretion of degradative enzymes). In stage II, the first distinct morphological change becomes visible. A septum is formed leading to an asymmetric division that separates the sporulating cell into two unequal cellular compartments, i.e. a large mother cell (MC) and a smaller forespore (FS) compartment (stage II(i)). This is followed by septum hydrolysis, membrane movement and curvature (stage II(iii)). At this point, first scaffold proteins crucial to later coat assembly are attached to the outer surface of the membrane that encases the forespore compartment. A phagocytosis-like process leads to the complete engulfment of the forespore, which now presents an intracellular protoplast surrounded by a double membrane (two membranes with opposite polarities) (stage III). The transition to the next stage is characterized by significant changes in the forespore core region including a decrease in volume and pH, the onset of dehydration, and synthesis of small acid soluble proteins (SASP), which are involved in arranging the chromosome in ring-like structure. Additionally, the germ cell wall (essentially identical peptidoglycan (PG) structure compared to growing cells) and the cortex (similar components but unique PG structure) are formed between the inner and the outer forespore membranes (stage IV). The assembly of unique coat proteins, which is a continuative process beginning shortly after asymmetric cell division, is then intensified. This leads to the encasement of the spore by several proteinaceous coat layers (stage V). A large amount of pyridine-2,6-dicarboxylic acid (dipicolinic acid, DPA) is synthesized in the mother cell and accumulated in the spore compartment. This is paralleled by the uptake of divalent cations, i.e. predominately Ca^{2+}, but also high amounts of Mg^{2+} and Mn^{2+}. Final dehydration of the forespore interior is crucial to the development of metabolic dormancy and causes increased refractility resulting in the typical phase-bright appearance of spores. The permeability of the inner spore membrane decreases, which impedes the access of stain to the spore interior. Spore encasement by the outermost layers (coat and exosporium) is completed, which marks the end of the maturation process (stage VI). Finally, the mother cell lyses and releases the mature spore into the environment (stage VII) (Hoch 1993; Sonenshein 2000; Eichenberger et al. 2004; Piggot and Hilbert 2004; Paredes et al. 2005; Setlow and Johnson 2013).

23.2.1.1 The Developmental Program Underlying Sporulation

The initiation of sporulation requires synthesis and activation of the master regulator Spo0A (see Sect. 23.2.1.1.1). The exceedance of a threshold level of activated Spo0A triggers a cascade in which the four sporulation sigma factors, σ^F, σ^G, σ^E, and σ^K play key roles (see Sect. 23.2.1.1.2). This ensures the timely and spatially ordered stepwise development of morphological spore characteristics, which determine spore properties including their extreme resistance. Key findings on the developmental program underlying sporulation and how this process is regulated are summarized in the following paragraphs considering differences between *Bacillus* and *Clostridium* species (Paredes et al. 2005; Jones et al. 2008; Setlow and Johnson 2013).

Initiation of Sporulation Via the Activation of the Master Regulator Spo0A

Prerequisites for the initiation of sporulation are a complete chromosome replication (Sonenshein 2000), DNA damage repair (Rowland et al. 2004) and induction of tricarboxylic acid cycle enzymes (Phillips and Strauch 2002). Sporulation triggers during stationary phase can include nutrient limitation (starvation from carbon and/or nitrogen) (*B. subtilis* (Setlow and Johnson 2013)), a decrease in (intracellular) pH (*C. acetobutylicum* (Jones and Woods 1986)), and increased levels of intracellular reduction energy (NAD(P)H) (*C. acetobutylicum* (Meyer and Papoutsakis 1989)). Furthermore, the addition of various substances can promote sporulation, e.g. decoyinine (inhibitor of guanine nucleotide synthesis, *B. subtilis* (Setlow

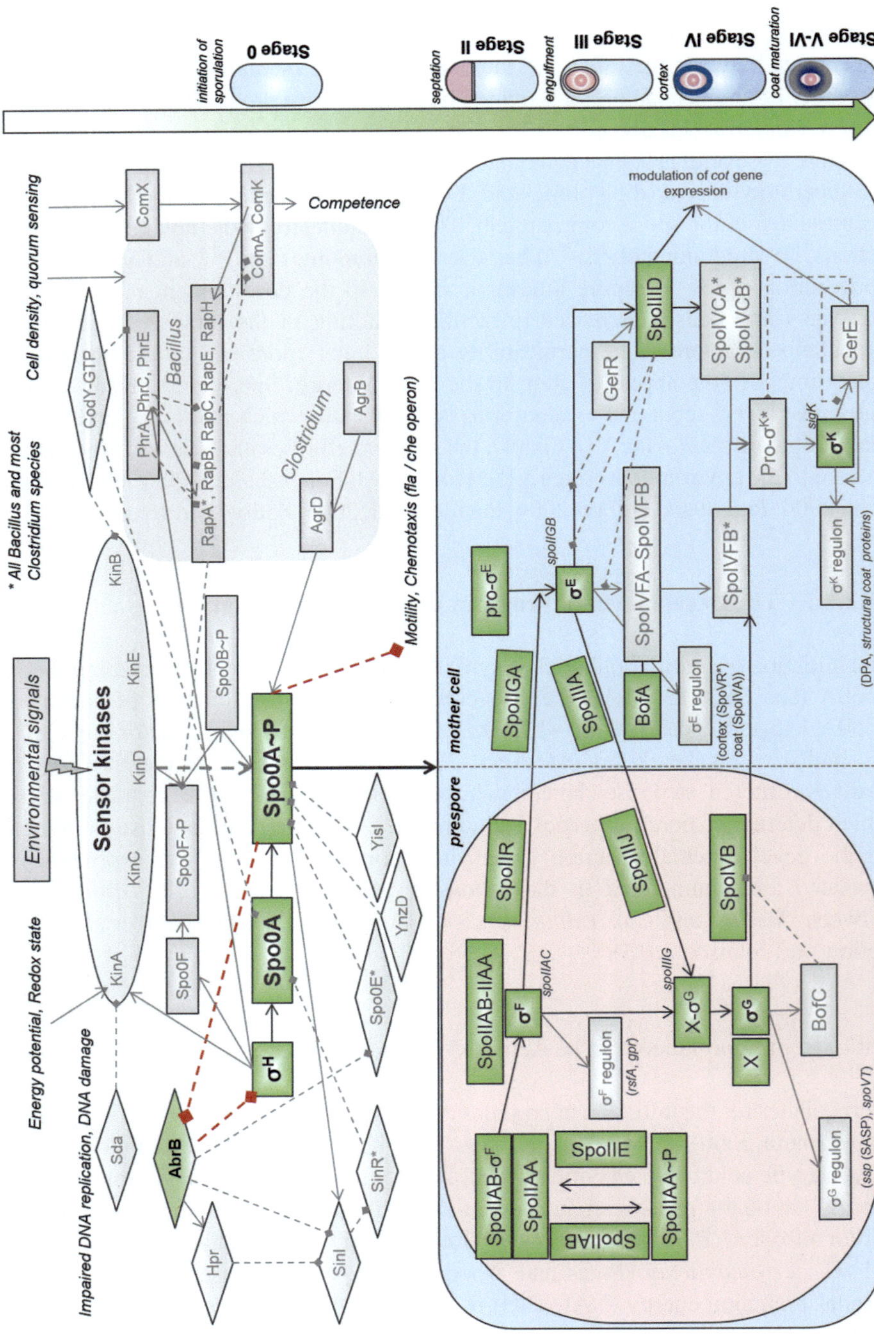
initiation of sporulation
Stage 0
septation
Stage II
engulfment
Stage III
cortex
Stage IV
coat maturation
Stage V-VI
* All Bacillus and most Clostridium species
Environmental signals
Energy potential, Redox state
Impaired DNA replication, DNA damage
Cell density, quorum sensing
Sensor kinases
KinA
KinB
KinC
KinD
KinE
Sda
CodY-GTP
PhrA, PhrC, PhrE
Bacillus
RapA*, RapB, RapC, RapE, RapH
ComX
ComA, ComK
Competence
Clostridium
AgrD
AgrB
Spo0F
Spo0F~P
Spo0B~P
Spo0A
Spo0A~P
σH
AbrB
Hpr
SinI
SinR*
Spo0E*
YnzD
YisI
Motility, Chemotaxis (fla / che operon)
prespore
mother cell
SpoIIGA
SpoIIIA
BofA
σE regulon
(cortex (SpoVR*) coat (SpoIVA))
pro-σE
spoIIGB
σE
SpoIVFA-SpoIVFB
SpoIVFB*
GerR
SpoIIID
modulation of cot gene expression
SpoIVCA* SpoIVCB*
Pro-σK*
sigK
σK
GerE
σK regulon
(DPA, structural coat proteins)
SpoIIR
SpoIIIJ
SpoIVB
SpoIIAB-IIAA
σF
spoIIAC
σF regulon
(rsfA, gpr)
spoIIIG
X-σG
σG
X
BofC
σG regulon
(ssp (SASP), spoVT)
SpoIIAB-σF
SpoIIAA
SpoIIE
SpoIIAB
SpoIIAA~P

and Johnson 2013)), carboxylic acids (membrane uncouplers, *C. acetobutylicum* (Husemann and Papoutsakis 1988)), high carbon source and ATP levels (mimicking phosphate or nitrogen limitation, *C. acetobutylicum* (Meyer and Papoutsakis 1989)), butyrate, and carbon monoxide (inhibitors of H_2 formation, Bahl et al. 1995; Woods 1995; Jones and Woods 1986; Girbal and Soucaille 1998).

The master regulator Spo0A presents a highly sensitive, self-reinforcing switch determining whether a single cell initiates sporulation or not. Once Spo0A levels in its activated, i.e. phosphorylated form exceed a certain threshold level, sporulation is initiated and the sporulation cascade proceeds (Fig. 23.2).

Spo0A Synthesis

The transcription of *spo0A* begins no later than during the transition between exponential and stationary phase (*B. subtilis* (Piggot and Hilbert 2004); *C. acetobutylicum* (Jones et al. 2008)) and is aided by the earliest sporulation sigma factor, σ^H. Despite of some differences in the time-dependent expression pattern (Weir et al. 1991; Durre and Hollergschwandner 2004), σ^H is highly conserved among different spore formers (e.g. *B. subtilis* (Stragier and Losick 1996); *C. acetobutylicum* (Jones et al. 2008)). A burst of Spo0A synthesis at the onset of sporulation can be observed in various spore formers (e.g. *C. acetobutylicum* (Harris et al. 2002); *C. perfringens* (Huang et al. 2004); *C. beijerinckii* (Wilkinson et al. 1995); *C. cellulolyticum* (Ravagnani et al. 2000)). This occurs due to a just-in-time control by multiple regulatory mechanisms (*B. subtilis* (Chastanet and Losick 2011)). Whereas the Spo0A amino acid sequence is highly conserved in *Bacillus* and *Clostridium* species (Stephenson and Lewis 2005), signaling used is different between spore-forming species (Stephenson and Hoch 2002; Durre and Hollergschwandner 2004). For example, Spo0A of a proteolytic *C. botulinum* type A strain can repress growth genes in *B. subtilis* but fails to induce sporulation (Wörner et al. 2006), and *C. acetobutylicum* or *C. tetani* but not *C. botulinum* or *C. difficile* Spo0A can complement a *C. perfringens spo0A* mutant (low transcription instead of general incompatibility is possibly involved in the latter effect) (Sarker and Huang 2006).

Fig. 23.2 Sporulation cascade. *Top*: initiation of sporulation; *Bottom left*: forespore-specific regulation mechanisms; *Bottom right*: mother cell-specific regulation mechanisms. On the *right* an approximate time line of corresponding events during sporulation is depicted. Diamond fields indicate repressors of sporulation during stage 0. *Dashed lines* connect repressors and their target. The target is marked by a small diamond at the end of this line. *Straight arrows* indicate activation pointing on the activated target. *Dark boxes* indicate parts of the cascade conserved in *B. subtilis* (*Bacillus* species) and *Clostridium* spores. Asterisk indicates parts that are conserved in *B. subtilis* (*Bacillus* species) and only a few *Clostridium* species. *Gray dashed lines, arrow* and *boxes* indicate mechanisms identified in *B. subtilis* (except for the Agr system from *Clostridium* species labeled explicitly) (Figure adapted from Paredes et al. (2005) and modified according to Jones et al. (2008), and Setlow and Johnson (2013). Note that there are some more sporulation genes conserved among *Bacillus* and *Clostridium* species (Galperin et al. 2012), which might be not directly involved in the cascade or play uncertain roles)

Spo0A Activation
Initiation of sporulation requires Spo0A activation, i.e. phosphorylation, which presents a highly conserved mechanism among various spore formers. In both *Bacillus* and *Clostridium* species, the first step towards Spo0A phosphorylation is the activation of sensor kinases. Although the Spo0A activation mechanism is barely understood in many organisms, spore-forming species are thought to generally possess multiple orphan kinases allowing initiation of sporulation in response to a variety of environmental conditions (Brunsing et al. 2005), e.g. two major (KinA, B) and three minor kinases (KinC, D, E) in *B. subtilis* (Jiang et al. 2000; Stephenson and Hoch 2002). In addition to various candidates, which have not yet been proven to lead to Spo0A activation, several sensor histidine kinases can directly phosphorylate Spo0A in *Clostridium* species, e.g. CBO1120 in *C. botulinum* (Wörner et al. 2006), HK CD2492 in *C. difficile* (Underwood et al. 2009), and CAC0323, CAC2730 and CAC0903 in *C. acetobutylicum* (Alsaker and Papoutsakis 2005; Jones et al. 2008; Steiner et al. 2011). However, just like the strong species dependency of extracellular signals that drive sporulation, the functionality of such kinases can vary considerably between different species, which is thought to reflect the adaptation to different habitats.

A major difference between *Bacillus* and *Clostridium* species can be found in intermediate steps leading to Spo0A activation. In *Bacillus* species, activated kinases phosphorylate the sporulation initiation phosphotransferase, Spo0F, followed by the single-domain response regulator/phosphotransferase, Spo0B, which activates Spo0A (*B. subtilis* KinA can also activate Spo0A directly with low efficiency (Trach and Hoch 1993)). In contrast, Spo0A is directly activated by sensor kinases in *Clostridium* species, which generally lack Spo0F and Spo0B phosphotransferases (Sebaihia et al. 2007).

Regulation of Spo0A Activation
Spo0A ~ P modulates the expression of over 100 genes directly or indirectly by binding to sites upstream of several key genes (Fujita et al. 2005). Since a threshold level of Spo0A ~ P needs to be exceeded for the initiation of sporulation (Fujita and Losick 2005), (i) a decrease in *spo0A* expression levels, (ii) inhibition of sensor kinases, or (iii) dephosphorylation of Spo0A (or one of the phosphotransferases Spo0F or Spo0B in *Bacillus* species) represses the initiation of sporulation. Indeed, a complex regulatory network employing such mechanisms (of which some are well characterized in *B. subtilis*) regulates and fine-tunes the initiation of sporulation in response to various intracellular and extracellular conditions (Jiang et al. 2000; Piggot and Hilbert 2004; Fujita and Losick 2005; Fujita et al. 2005).

(i) An example for a direct repressor of *spo0A* is *sinR*, which directly binds to the promoter sequences of the *spo0A* operon (also to *sigF* and *sigE*, *B. subtilis* (Mandicmulec et al. 1995)) and putatively plays a similar role in some *Clostridium* species (*C. acetobutylicum*, *C. tetani* (Paredes et al. 2005; Scotcher et al. 2005; Jones et al. 2008)). In contrast, the SinR repressor, SinI (stimulated by Spo0A ~ P and probably repressed by AbrB and Hpr), is absent in *Clostridium* species (Mandicmulec et al. 1995; Paredes et al. 2005). A conserved (Mandicmulec et al. 1995; Nolling et al. 2001; Scotcher and Bennett 2005; Jones et al. 2008) indirect mechanism regulating *spo0A* expression levels is the transition-state regulator AbrB, which negatively controls σ^H leading to a decrease in *spo0A* transcription and activates other repressors. Near the onset of sporulation, high Spo0A ~ P levels decrease intracellular AbrB levels caused by the high affinity of the gene *abrB* for Spo0A ~ P (*B. subtilis* (Mandicmulec et al. 1995)).

(ii) Kinase inhibitors are well characterized in *B. subtilis*. For example, Sda inhibits KinA in response to impaired DNA replication or DNA damage (Rowland et al. 2004) and CodY (with its corepressor GTP) is involved in the regulation of Spo0A ~ P levels putatively via inhibiting KinB (Molle et al. 2003; Ratnayake-Lecamwasam et al. 2001). Although kinase inhibitors are likely to be also present in other spore-formers, knowledge on such inhibitors is scarce, which is putatively related to their kinase specificity and the low degree of conservation of certain kinases among spore-forming species.

(iii) Sporulation repressors acting antagonistic to the histidine kinases by dephosphorylating Spo0A ~ P identified in *B. subtilis* include the aspartyl phosphate phosphatases Spo0E (Diaz et al. 2008) and YnzD and YisI (highly homologous to Spo0E) (Perego 2001). In contrast to YnzD and YisI, Spo0E-like aspartyl phosphate phosphatases are conserved in spore-forming *Bacillus* species (e.g. *B. subtilis* (Diaz et al. 2008), *B. anthracis* (Grenha et al. 2006)) and some *Clostridium* species (Paredes et al. 2005). Additionally, Spo0F ~ P dephosphorylation (e.g. by RapA, RapB) plays a role in modulating the phosphorylation state of Spo0A in *B. subtilis*. However, both Spo0F and Rap phosphatases are absent in *Clostridium* species (Sebaihia et al. 2007).

The Sporulation Cascade

Sporulation is regulated by a complex network, in which gene expression is temporally and spatially (mother cell/forespore) different, and highly ordered (Fig. 23.2). In this process, increased levels of σ^H and Spo0A ~ P activate the expression of several key sporulation genes, most importantly four sporulation-specific sigma factors. Associated with RNA polymerase, these sigma factors drive different patterns of gene expression in a compartment-specific manner, i.e. the early and the late sigma factors σ^F and σ^G in the prespore and the early and the late sigma factors σ^E and σ^K in the mother cell. An extensive 'cross-talk' between the mother cell and the forespore compartment and developmental checkpoints ensure that gene expression remains temporally coordinated in the two compartments. Many target genes (i.e. sigma factor regulons) described in detail in *B. subtilis* are different or absent in *Clostridium* species (Paredes et al. 2005). However, the four sigma factors appear to be highly conserved in *Bacillus* (Stragier and Losick 1996) and several *Clostridium* species (Jones et al. 2008). This and the high degree of conservation of other key sporulation genes suggests that, in addition to the basic sporulation stages, the genetic background of the sporulation cascade is similar in various spore-forming organisms.

Stage II Prespore Sigma F

The forespore chromosome is transported across the sporulation septum, i.e. from the mother cell to the forespore compartment, via a SpoIIIE-mediated channel formation (*B. subtilis* (Becker and Pogliano 2007; Fleming et al. 2010); SpoIIIE conserved (Galperin et al. 2012)), which is followed by septal membrane fission, i.e. the final step in unequal cell division.

In *B. subtilis*, activated Spo0A leads to an upregulation of the *spoIIA* operon coding for SpoAA, SpoAB and the first forespore-specific sigma factor σ^F (Schmidt et al. 1990). Before septum formation is completed, σ^F is present in an inactive form σ^F-SpoIIAB (promoted by high ATP/ADP ratio). After septum formation, the anti-anti-σ factor SpoIIAA is dephosphorylated by the phosphatase SpoIIE present in the sporulation septum (Clarkson et al. 2004; Yudkin and Clarkson 2005), SpoIIAA interacts with the anti-σ factor SpoIIAB, and σ^F is released in its active form (Hilbert and Piggot 2004). Genes involved (Paredes et al. 2005) and the mechanism of σ^F activation (*C. acetobutylicum* (Jones et al. 2008, 2011; Bi et al. 2011)) are thought to be conserved in *Clostridium* species.

In *B. subtilis*, $E\sigma^F$ transcribes numerous genes in the forespore including genes coding for the DNA binding protein RsfA (modifies $E\sigma^F$ specifity), SpoIIR (compartmental cross-talk necessary for σ^E activation), Gpr (protease acting on SASP in the early germination process), and the late forespore-specific sigma factor σ^G (*spoIIIG*) (Wang et al. 2006). Although *spoIIR* along with other sporulation related genes (such as *gpr*, *spoIIP*, *sigG*, *lonB*) are conserved in *Clostridium* species (Paredes et al. 2005; Galperin et al. 2012) they are not controlled by σ^F in such organisms (*C. acetobutylicum* (Jones et al. 2011)).

Stage II Mother Cell Sigma E

In the mother cell compartment, the *spoIIG* operon coding for both the protease SpoIIGA and the first mother cell-specific sigma factor, σ^E, is transcribed early by $E\sigma^A$ (*B. subtilis* (Piggot and Hilbert 2004)). Activation of σ^E from its inactive precursor, pro-σ^E, requires septation and the physical interaction between SpoIIGA and the σ^F-regulated SpoIIR (Labell et al. 1987).

Active σ^E has been reported to control the expression of over 250 genes in the mother cell (Eichenberger et al. 2004) including SpoIIIA (involved in compartmental cross talk and activation of the late forespore-specific σ^G), genes required for cortex formation (e.g. *pdaA*) (Fukushima et al. 2002) and some genes involved in coat biosynthesis and assembly (e.g. *cot* genes) (Eichenberger et al. 2004). Additionally, σ^E-containing RNA polymerase is responsible for the transcription of *spoIIID* (transcriptional regulator modulating $E\sigma^E$ action), *gerR* (requiring SpoIIID; directly controlling late sporulation genes *cotB*, *cotU* and *spoVIF* (Cangiano et al. 2010)), the *spoIVF* operon (required for activation of the late mother cell-associated sigma factor σ^K), and the *spoIVC* operon (precursor protein for σ^K).

Genes coding for σ^E and SpoIIGA (Paredes et al. 2005) and the mechanism of σ^E activation are conserved in other spore-forming organisms (*C. acetobutylicum* (Jones et al. 2011)). Additionally, σ^E-dependent operons *spoIIIAA-AH* (required for the activation of σ^G), *spoVR* (involved in cortex synthesis), and *spoIVA* (involved in cortex formation and spore coat assembly) are similar in *Bacillus* and *Clostridium* species (Jones et al. 2008). GerR, the SpoIVF and the SpoIVC operon are not conserved in *Clostridium* species (Paredes et al. 2005).

Stage III-V Prespore Sigma G
Activation of the late, sporulation stage III-associated (Cutting et al. 1990) forespore-specific sigma factor σ^G (*spoIIIG*) requires the products of the $E\sigma^E$ transcribed *spoIIIA* operon in the mother cell and the action of SpoIIIJ in the prespore compartment (*B. subtilis* (Eichenberger et al. 2004)). However, regulation of σ^G activity appears to be multilayered and macromolecular synthesis and σ^G-directed gene activation in the forespore depend on mother cell-forespore channels (possibly gap junction-like feeding tubes) through which the mother cell nurtures the developing spore by providing small molecules needed for biosynthetic activity (Camp and Losick 2009). SpoIIIAH is targeted specifically to the membrane surrounding the forespore, through an interaction of its C-terminal extracellular domain with the C-terminal extracellular domain of the forespore membrane protein SpoIIQ and forms part (maybe by forming a ring structure) of such a channel that is required for the activation of σ^G (Meisner et al. 2008).

Genes coding for SpoIIIA, SpoIIIJ, σ^G are conserved in *Clostridium* species (Paredes et al. 2005). In contrast to *B. subtilis*, adjacent *sigE* and *sigG* do not have a σ^F promoter between them (Paredes et al. 2004) and are expressed as a single transcript *spoIIGA-sigE-sigG* (*spoIIGA-sigE* and *sigG* transcripts also detected (Harris et al. 2002)).

σ^G controls the expression of over 110 genes in the forespore (*B. subtilis* (Steil et al. 2005)) including *spoIVB* (required for the expression of the late mother cell-specific sigma factor, σ^K. (Piggot and Hilbert 2004)), the gene coding for BofC (SpoIVB inhibitor), *ssp* genes (coding for SASP (Setlow 1994, 1995)), and *spoVT* (coding for a DNA binding protein modulating transcription of genes by $E\sigma^G$ (Wang et al. 2006) including nutrient germinant receptor genes (Ramirez-Peralta et al. 2012)).

In *Clostridium* species, genes under the control of σ^G also include those coding for SpoIVB, SpoVT and SASP (*C. acetobutylicum* (Jones et al. 2008)) and σ^G is also required for the expression of the late mother cell-specific σ factor σ^K (*C. difficile* (Piggot and Hilbert 2004)). BofC is not conserved in *Clostridium* species (Paredes et al. 2004).

Stage III-V Mother Cell Sigma K
In *B. subtilis*, σ^K is formed by splicing together two genes (*spoIVCB* and *spoIIIC*), both under the control of σ^E and SpoIIID, separated by a skin element (Stragier et al. 1989). Synthesized in an inactive form (pro-σ^K), σ^K activation requires the expression of *spoIVB*, which is under the control of the forespore-specific σ^G and the σ^E-controlled *spoIVF* operon (protease processing pro-σ^K to σ^K, SpoIVFB and its inhibitor, SpoIVFA) in the mother cell.

In conjunction with the σ^E controlled SpoIIID, $E\sigma^K$ transcribes genes responsible for DPA synthesis, *cot* genes, and *gerE* (transcriptional activator modulating $E\sigma^K$

action, leading to altered *cot* gene expression patterns (Cangiano et al. 2010)). Among the last genes transcribed by Eσ^K are genes involved in mother cell lysis (e.g. *cwlC*) (Piggot and Hilbert 2004).

Interruption of the *sigK* gene and the requirement of the excision of a skin (*sigK* intervening sequence) element are very uncommon among spore-formers and were only found in *B. subtilis* and *C. difficile* (Haraldsen and Sonenshein 2003). In other spore-formers there is a single gene encoding σ^K (e.g. *C. acetobutylicum* (Sauer et al. 1994; Nolling et al. 2001)) likely under the control of *spoIIID* and σ^E. The genes coding for σ^K and SpoIIID but not GerE, SpoIVF and SpoIVC are conserved in *Clostridium* species (Paredes et al. 2005). However, σ^K but not its important role has been proposed to be conserved among *Bacillus* and *Clostridium* species (Jones et al. 2008, 2011).

Although there are some exceptions, e.g. σ^K controlled phosphatase *yisI* acting on Spo0A$\sim$P or late sporulation genes (*yusW* and *yhbA*) under the control of the early sporulation sigma factor σ^F (*B. subtilis* (Oomes et al. 2009)), σ^F, σ^E and σ^G, and σ^K controlled genes are primarily (but not exclusively) involved in early, middle, and late sporulation, respectively. Generally, the *Bacillus* model describing the temporal orchestration of known sporulation-related transcription factors with the cascade progressing in the order σ^H, Spo0A, σ^F, σ^E, and σ^G appears to hold true in *Clostridium* species. In addition to the key sigma factors, major activating/processing proteins involved in sigma factor activation in *B. subtilis* are likely to play a similar role in *C. acetobutylicum* (Jones et al. 2008, 2011). Perturbations in the ordered course of the sporulation phosphorelay can cause problems in proper spore assembly and alterations in structural features of spores (Veening et al. 2006). This putatively plays a major role in altered sporulation conditions affecting resistance properties of spores from both genera (see Sect. 23.3.4.1).

Apart from these general commonalities there are major differences in the time span required for sporulation (e.g. 2–3 h in *B. subtilis* and 8–20 h between sporulation initiation and initial forespore formation in solventogenic *Clostridium* species (Long et al. 1984; Santangelo et al. 1998)) and in the regulons of the specific sporulation sigma factors in *Bacillus* and *Clostridium* species. These major differences during sporulation, i.e. where spore properties determining their resistance, dormancy, and germinability are developed, certainly contribute to difficulties in transferring HPT inactivation strategies between these genera and add food for thought to the discussion on possible "surrogate" organisms for food safety studies.

23.2.2 Germination and Outgrowth

Various environmental conditions can cause the conversion of a dormant spore back to an actively growing and doubling cell. Since the required apparatus is already present in mature dormant spores, germination is an essentially biophysical process, which does not require any new macromolecule synthesis (Moir 2006). The complete process can be divided into three main events, i.e. germination, outgrowth, and doubling. The times a spore requires to pass through each of the individual stages (i) vary considerably (e.g. 11, 71 and 16 % of the total lag time for non-proteolytic *C. botulinum* type B spores (Stringer et al. 2011)), (ii) represent independent events (e.g. impossible to predict total lag from germination times), (iii) are not correlated for single spores within a population, (iv) are differentially affected by environmental sporulation and recovery conditions (e.g. presence of NaCl or temperature), and (v) can be dramatically changed by physical treatments such as heat and HHP (Webb et al. 2007; Stringer et al. 2011). The high heterogeneity in the time span, a single spore within a population requires from the presence of suitable conditions to the first cell doubling, i.e. total lag time, and its high dependence on environmental factors and the history of a spore has major implications for food safety risk assessment (Augustin 2011). This heterogeneity in germination times in particular presents a major reason impeding a successful implementation of (two-step) spore inactivation processes relying on spore germination and aiming at complete inactivation of spore populations (see Sects. 23.3.3 and 23.3.5).

Generally, germination can be triggered by the activation of germinant receptors, directly by exogenous enzymes, and under certain physical conditions including HHP/HPT. However, pathways involved vary considerably and the efficiency of a certain substance or physical condition to trigger germination is highly organism-dependent.

Since germination plays a key role in the inactivation of bacterial endospores by high pressure treatments and this process can follow pathways involving the activation of germinant receptors and germinant receptor-independent pathways, a detailed picture of molecular events during such germination processes is important to understand how HHP/HPT treatments provoke spore inactivation. The following paragraphs provide an overview of germination pathways dependent (Sect. 23.2.2.1) and independent (Sect. 23.2.2.2) of germinant receptors considering differences between *Bacillus* and *Clostridium*.

23.2.2.1 Major Events in nGR-Mediated Germination

Germinant receptors responding to nutrients (nGRs) are highly conserved among spore-forming species (Paredes-Sabja et al. 2011). Although this class of receptors can also respond to non-nutrient low-molecular-weight germinants and physical conditions, the term nGRs is used here. The nGR-mediated germination process comprises two stages (Setlow et al. 2001; Setlow 2003; Setlow and Johnson 2013). In the beginning of stage I (stage I(i) in Fig. 23.1), specific chemical compounds, i.e. nutrient or non-nutrient germinants (see Sect. 23.2.2.1.1), have to penetrate outer spore layers (exosporium, coat and cortex), bind to and activate receptor proteins localized in the inner spore membrane, i.e. different substrate-specific nGRs (see Sect. 23.2.2.1.2). This finally leads to the commitment of spores to germination. This first step, i.e. the germination initiation time, represents a major variable in kinetics of nGR-triggered spore germination and, at least in *B. subtilis*, largely accounts for the superdormancy of spores (Zhang et al. 2012). After commitment removal of the germinant or reversal of germinant binding can no longer block germination. Signal transduction (see Sect. 23.2.2.1.3) initiates a generally irreversible cascade of degradative processes. Later in stage I (stage I(ii) in Fig. 23.1) monovalent cations are released from the spore core (approx. 80 % of Na^+ and K^+ in *B. megaterium* (Rode and Foster 1966)) via an energy-independent process (Setlow 2003). The release of H^+ elevates the core pH from ~6.5 to 7.7, which is essential for the initiation of spore metabolism later in germination. Furthermore, activated nGRs trigger the release of over 90 % of the spore core's large depot of DPA present in the core as a 1:1 chelate with divalent cations, predominantly Ca^{2+} (>90 % Ca release, *B. megaterium*, Rode and Foster 1966; Paidhungat and Setlow 2000, 2001; Wang et al. 2011). The rapid loss of Ca-DPA marks the beginning of partial core rehydration (from approx. 35 to 45 % water in *B. subtilis* (Setlow et al. 2001)), putatively requiring an increase of the inner spore membrane permeability. Partial rehydration at this point is not sufficient for protein mobility or enzyme activity (Cowan et al. 2003) but already causes some decrease in spore wet-heat resistance. This and the initiation of further degradative steps are the major reasons for Ca-DPA release being a crucial and rate-liming step in the inactivation of *Bacillus* spores by HHP and HPT treatments (see Sect. 23.3.2.1)

In germination stage II (Fig. 23.1) cortex lytic enzymes (CLEs) activated by the DPA release from the core hydrolyze cortex peptidoglycan (Setlow et al. 2001) (see Sect. 23.2.2.1.4), which presents a prerequisite for complete core rehydration. Since spores contain preformed enzyme-substrate pairs with the substrates protected from degradation by regulatory mechanisms in the developing forespore and by dehydration in the resting spore, rehydration enables the rapid resumption of enzyme activity in the spore protoplast. For example, the PGM/3PGA (phosphoglycerate mutase/3-phosphoglyceric acid) and the GPR/SASP (germination protease/α-/β-type SASP) substrate/enzyme pairs, which are well characterized in *B. subtilis*, are stable for months or years but interact in the first 15–30 min after rehydration (Setlow 1994). The core expands and rehydration allows for to the initiation of RNA and protein synthesis. Germinated spores completely lost spore-specific properties including dormancy, refractility and their extreme resistance to physical stress.

During outgrowth the fully germinated spore emerges from remaining spore structures (i.e. the coat layers) and turns into an immature cell. This is followed by a maturation process, in which the complete macromolecular set of a normal vegetative cell is restored finally leading to active cell division.

Interestingly, the time until the first cell division occurs after outgrowth differs from that, a normal cell needs for doubling and contributes to the total lag time between nGR activation and growth (Stringer et al. 2011).

Germinants Activating nGRs

The nGR-mediated germination pathway (Fig. 23.3) can be triggered by numerous nutrients and non-nutrient agents. Nutrient germinants include a wide variety of amino acids (most prominently L-alanine, effective in *Bacillus* and *Clostridium* spores (Peck 2009)), sugars (e.g. glucose and fructose), and purine ribonucleosides (inosine and adenosine). Non-nutrient germinants that are capable of inducing physiologic germination via the activation of nGRs include exogenous Ca-DPA or inorganic ions (e.g. K^+).

Generally, a single compound can be sufficient to trigger germination, e.g. L-alanine in *B. subtilis* (Moir and Smith 1990) and *B. anthracis* (Luu et al. 2011), L-cysteine and L-threonine in *B. cereus* (Hornstra et al. 2006b), L-Asparagin, KCl or Ca-DPA in *C. perfringens* (Paredes-Sabja et al. 2008c), glucose in proteolytic *C. botulinum* (Foegeding and Busta 1983; Chaibi et al. 1996), inosine in *B. cereus* (Abel-Santos and Dodatko 2007), and salts such as KBr in *B. megaterium* (Setlow and Setlow 1977). Additionally, combinations of germinants can be required for the induction of sporulation presenting either an alternative to germination induced by single components, e.g. inosine together with amino acids in *B. anthracis* (Fisher and Hanna 2005) and the AGFK mixture (L-asparagine, glucose, fructose and potassium ions) in *B. subtilis* (Setlow 2003), or the only way to induce germination, e.g. amino acids together with L-lactate (or other substance) in non-proteolytic *C. botulinum* (Plowman and Peck 2002).

The presence of co-germinants, i.e. substance that are not essential for the induction of germination but can stimulate the activity of another co-germinant or germinant, can influence (enhance) germination (e.g. sodium bicarbonate or sodium thioglycollate, non-proteolytic *C. botulinum* (Plowman and Peck 2002); inorganic phosphate and sodium ions in *C. perfringens* (Paredes-Sabja et al. 2009c); inorganic ions (Ca^{2+}) assist in low level inosine germination, *B. cereus* T (Shibata et al. 1993)). On the other hand, different nutrient combinations can result in a negative effect on the induction of germination, e.g. L-methionine can impede L-serine- and L-valine-mediated germination in *B. anthracis* (Luu et al. 2011). The finding that further addition of L-histidine can counteract the negative effect on L-serine-, but not L-valine-mediated germination suggests that there can exist a complex interplay between the different germination pathways (Luu et al. 2011).

In addition to co-germinants, a low environmental pH, contact with reducing agents, and, most prominently, a sublethal heat shock can activate spores, i.e. facilitate subsequent nutrient germination. Although heat shock requirements (effective temperatures) are species-specific, sublethal heat appears to be a common activator of both *Bacillus* and *Clostridium* spores (Plowman and Peck 2002; Garcia et al. 2010). This effect might be related to changes in nGR proteins altering receptor responsiveness or a facilitated access of germinants to their receptors (Alimova et al. 2006). Although it was demonstrated that a sublethal heat shock leads to

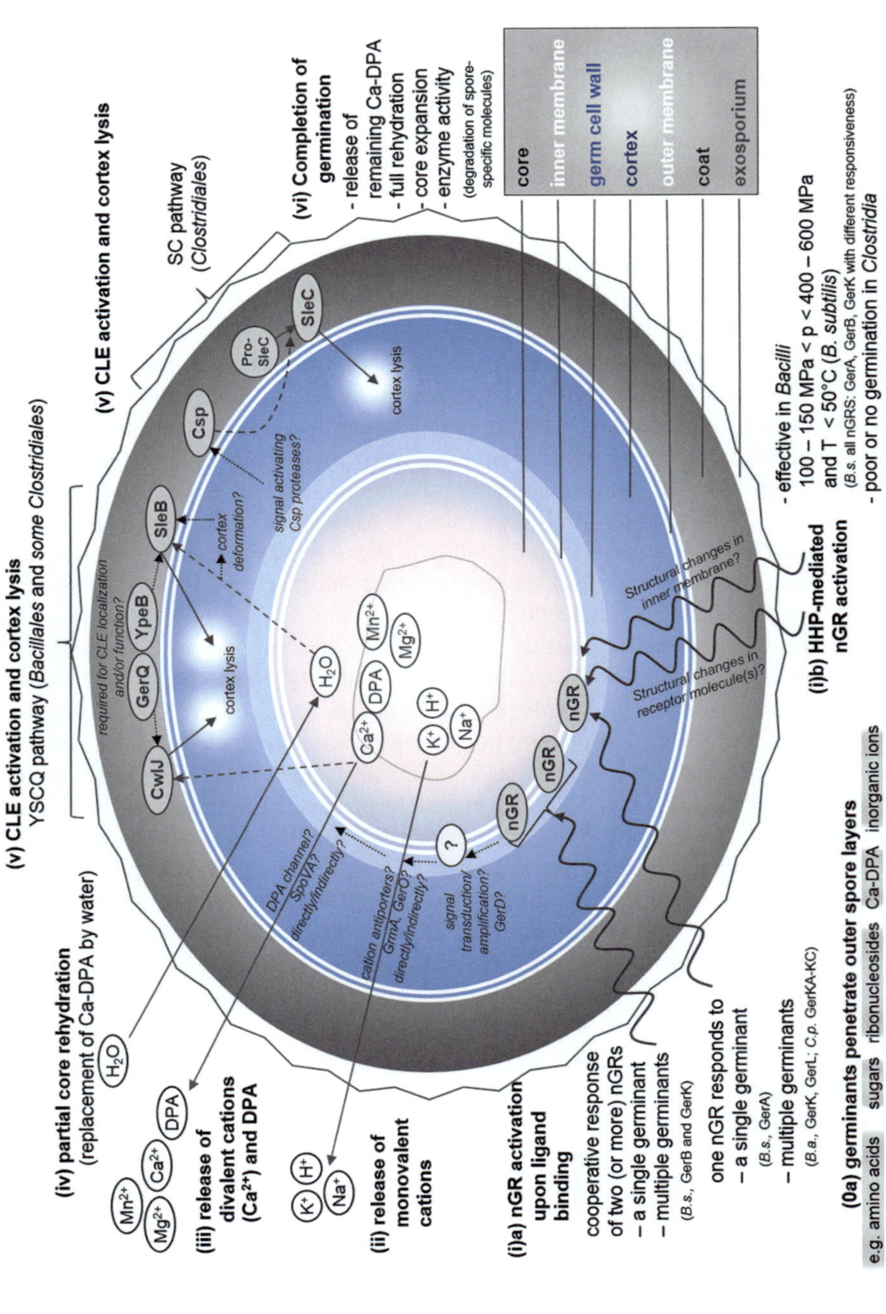
(v) CLE activation and cortex lysis
YSCQ pathway (Bacillales and some Clostridiales)
required for CLE localization and/or function?
CwlJ
GerQ
YpeB
SleB
Csp
Pro-SleC
SleC
cortex lysis
cortex lysis
cortex deformation?
signal activating Csp proteases?
cortex lysis
(v) CLE activation and cortex lysis
SC pathway (Clostridiales)
(vi) Completion of germination
- release of remaining Ca-DPA
- full rehydration
- core expansion
- enzyme activity
(degradation of spore-specific molecules)
core
inner membrane
germ cell wall
cortex
outer membrane
coat
exosporium
(iv) partial core rehydration
(replacement of Ca-DPA by water)
H_2O
Mn^{2+} Ca^{2+} DPA Mg^{2+}
(iii) release of divalent cations (Ca^{2+}) and DPA
K^+ H^+ Na^+
(ii) release of monovalent cations
DPA channel? SpoVA? directly/indirectly?
cation antiporters? GrmA, GerO? directly/indirectly?
signal transduction/ amplification? GerD?
?
nGR
nGR
nGR
(i)a nGR activation upon ligand binding
cooperative response of two (or more) nGRs
– a single germinant
– multiple germinants
(B.s., GerB and GerK)
one nGR responds to
– a single germinant
(B.s., GerA)
– multiple germinants
(B.a., GerK, GerL; C.p. GerKA-KC)
Structural changes in inner membrane?
Structural changes in receptor molecule(s)?
(i)b HHP-mediated nGR activation
- effective in Bacilli
100 – 150 MPa < p < 400 – 600 MPa
and T < 50°C (B. subtilis)
(B.s. all nGRS: GerA, GerB, GerK with different responsiveness)
- poor or no germination in Clostridia
(0a) germinants penetrate outer spore layers
e.g. amino acids
sugars
ribonucleosides
Ca-DPA
inorganic ions

the denaturation of some spore proteins, which appears to be a largely reversible process, the underlying mechanism is still not completely clear (Zhang et al. 2009).

The striking differences in the ability of bacterial spores to sense different nutrients appears to reflect their adaptation to different environmental niches (Paredes-Sabja et al. 2009c). In addition to the species dependence, germination requirements can significantly vary among different strains of a species (e.g. *B. cereus* (van der Voort et al. 2010)) and between individual spores within a population (Ghosh et al. 2009). Whereas the latter might be the result of a stochastic process (Ghosh and Setlow 2009), organism-dependent differences in the specificity for germinants initiating germination via the activation of germinant receptors is due to the differences in such highly specific receptors.

nGRs

nGRs are thought to be synthesized in the forespore compartment of the developing spore (in *B. subtilis* transcribed under the control of the late sporulation sigma factor σ^G and regulated by transcription factors SpoVT and YlyA). nGrs are commonly incorporated into the inner spore membrane (e.g. *C. perfringens* (Banawas et al. 2013); *C. botulinum* (Alberto et al. 2005)) in low numbers, where they potentially co-localize (Setlow and Johnson 2013) and associate with other small proteins modulating their function (Ramirez-Peralta et al. 2013).

Orthologues of the well-characterized *B. subtilis* nGR, GerA are highly conserved within spore-forming species (except *C. difficile* (Ramirez et al. 2010; Paredes-Sabja et al. 2011)). However, the organization of genes encoding nGR subunits is heterogeneous in different spore-formers, where mono- through quattrocistronic *gerA* family operons can be found (approx. 50 % tricistronic operons as in *B. subtilis*) (Paredes-Sabja et al. 2011). Although individual nGR specificities for nutrients can broadly overlap (Luu et al. 2011), nGRs are generally highly substrate-specific, e.g. inosine analogs fail to induce germination and even small modifications of nucleobase or sugar moieties of inosine impede germination of *B. cereus* spores (Abel-Santos and Dodatko 2007). Substrate-specificity frequently even includes stereospecificity (e.g. L-alanine might trigger but D-alanine can

---

Fig. 23.3 Nutrient germinant receptor (nGR)-dependent germination pathways in *Bacillus* and *Clostridium* spores. The different spore compartments are labeled on the right. Note that the actual size of spore layers in relation to each other can vary considerably. The coat comprises multiple (commonly up to four) distinct layers. Spores of some species possess no, some a very large exosporium. The position of the proteins involved in germination might not be depicted precisely and some of them are additionally present in other layers. Germination steps, which are not yet completely understood are described in italics. *Curled arrows* = exogenous activation events initiating germination; *straight solid arrows* = efflux or influx of molecules and enzymatic action on the cortex; *dashed arrows* = activation within the degradative cascade; *dotted arrows* = interaction mechanism not completely clear. *B.s. B. subtilis, B.a. B. anthracis, C.p. C. perfringens* (Data for figure adapted from Paredes-Sabja et al. (2011))

inhibit germination), which is determined by only a few amino acids within a germination protein (*B. subtilis* (Paidhungat and Setlow 1999)). This substrate specificity of nGRs considerably varies between different species and strains, which is a major reason for differences in germination requirements.

Despite of the specificity of nGRs, the effectiveness of a certain germinant or mixture of germinants, i.e. germination rate (time required for germination) and germination efficiency (spore fraction within a population, which will germinate), depends on several other factors. Such factors include the germinant concentration, the number of functional nGRs in a spore, and their accessibility, i.e. permeability of outer spore layers for germinants. The passage of germinants through outer layers might be facilitated by specific coat proteins (e.g. GerP, *B. subtilis* and *B. cereus* (Behravan et al. 2000); *B. anthracis* (Carr et al. 2010)). Additionally, the presence of specific enzymes in the outer spore layers might be involved in modulating germination, i.e. converting substrates to effective or ineffective germinant molecules. For example, purine-specific nucleoside hydrolase present in the outer layers of various *Bacillus* spores (*B. cereus* (Hornstra et al. 2006a); *B. anthracis* (Redmond et al. 2004)) was reported to suppress inosine- or adenosine-induced germination of *B. thuringiensis* spores (Liang et al. 2008), and spore-specific alanine racemase catalyzing the interconversion of D- and L-alanine can modulate L-alanine-induced germination in *B anthracis* (Chesnokova et al. 2009).

Generally, one nGR can respond to a single germinant (e.g. *B. subtilis* GerA, L-alanine (Moir and Smith 1990)) or multiple germinants (e.g. *C. perfringens* GerKA–GerKC, L-asparagine, KCl, Ca-DPA (Paredes-Sabja et al. 2008c)). Additionally, multiple nGRs can respond cooperatively to a single or multiple germinants (e.g. *B. subtilis* GerB and GerK, AGFK (L-asparagine, glucose, fructose and K^+ ions (Setlow 2003))). In addition to nutrient–triggered germination, nGRs, appear to be also required for outgrowth in some organisms (e.g. *C. perfringens* GerKA–KC and GerKB (Paredes-Sabja et al. 2008c, 2011)). However, the underlying mechanism is not clear.

Signal Transduction

The exact process of nGR activation and the nature of signal transduction after their activation, i.e. the spore's commitment to germination, both remain to be elucidated even in well-characterized model organisms. The lipoprotein GerD, which is localized in the spore inner membrane of *Bacillales* (Pelczar et al. 2007; Pelczar and Setlow 2008) in discrete clusters with nGRs (*B. subtilis* (Griffiths et al. 2011)), but absent in *Clostridiales* species, has been discussed as being involved in signal transduction from activated nGRs to proteins acting downstream in the germination process (Paredes-Sabja et al. 2011; Chen et al. 2014). For the release of monovalent cations, cation antiporters could play a role (e.g. GrmA in *B. megaterium* (Tani et al. 1996); GerO (and GerQ) in *C. perfringens* (Paredes-Sabja et al. 2009a)). However, this role is questionable as it might be only indirect (during sporulation), appears to be species- and even strain-dependent (*B. megaterium* (Christie and Lowe 2007)), and not involved in germination triggered by various germinants (inosine but not

L-alanine germination in *B. cereus* (Senior and Moir 2008)). Additionally, proteins encoded by the *spoVA* operon that are involved in the uptake of Ca-DPA into the developing spore (Li et al. 2012) can putatively physically interact with nGRs and might be involved in the release of Ca-DPA (e.g. by DPA channel formation) at least in some *Bacillus* species (Paredes-Sabja et al. 2011). However, the exact mechanism and if SpoVA proteins are directly involved in germination or contribute indirectly via their role during sporulation is not clear.

Cortex Hydrolysis

Cortex lytic enzymes (CLEs) are synthesized during sporulation in the forespore, the mother cell, or both compartments, are localized in proximity to the inner and outer cortex surfaces, and can have different binding specificities, but typically recognize and cleave cortex PG exclusively, which is thought to be due to the presence of cortex PG-specific muramic acid δ-lactam (Makino and Moriyama 2002) (Fig. 23.3).

Two distinct pathways for cortex hydrolysis are highly conserved in *Bacillales* and *Clostridiales* species, i.e. YpeB–SleB–CwlJ–GerQ (YSCQ) and the Csp–SleC (CS) pathway. *Bacillales* species commonly rely on the YSCQ pathway, while the CS system appears to be generally absent. In contrast, *Clostridiales* can utilize either one or both pathways (Paredes-Sabja et al. 2011). Additionally, other CLEs can be present in spore-forming species, which might be not essential for germination (e.g. SleM present in *B. weihenstephanensis* and some *Clostridium* species (Paredes-Sabja et al. 2011)). In a few species lacking obvious orthologues of the proteins involved in cortex hydrolysis via the YSCQ or CS pathway, various different potential CLEs are present, which might be involved in different mechanisms of cortex hydrolysis (Paredes-Sabja et al. 2011). This, however, remains to be elucidated.

In the YSCQ pathway, either one of the two CLEs, SleB and CwlJ is sufficient to complete germination. CwlJ is thought to be primarily located in the spore coat, possibly associated with other coat proteins (Bagyan and Setlow 2002), and requires GerQ, which is cross-linked into high-molecular-mass complexes in the spore coat by a transglutaminase (Tgl) (Monroe and Setlow 2006). Cortex hydrolysis by CwlJ can be triggered by both Ca–DPA released from the spore core during germination and by exogenous Ca-DPA. SleB is thought to be localized in the coat, the cortex and putatively also in the inner membrane (*B. subtilis* (Chirakkal et al. 2002)). Proper localization has been demonstrated to require the interaction between the SleB peptidoglycan-binding motif and the δ-lactam structure of the cortex (Masayama et al. 2006). Additionally, YpeB, which can be found in the same locations as SleB (Chirakkal et al. 2002), is required for proper SleB localization, stabilization, activation, and/or function. SleB activation is putatively triggered by partial core rehydration and the accompanying deformation of the cortex.

In the Csp–SleC (CS) pathway, the CLE SleC is essential for cortex hydrolysis during spore germination (Paredes-Sabja et al. 2009b). SleC is an exo-acting lytic transglycosylase possessing a peptidoglycan-binding domain (Gutelius et al. 2014),

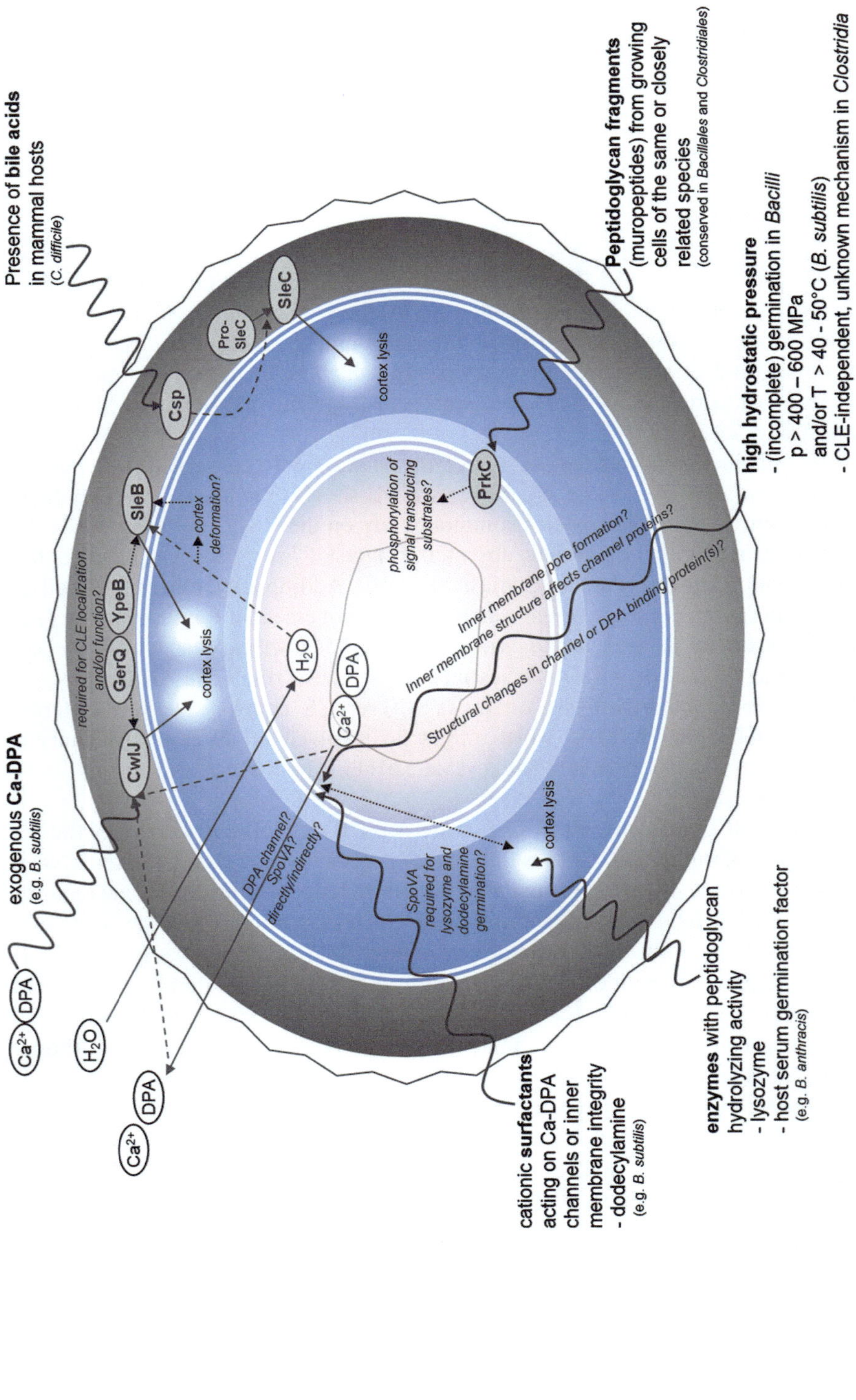
Presence of bile acids
in mammal hosts
(C. difficile)
exogenous Ca-DPA
(e.g. B. subtilis)
Ca^{2+}
DPA
H_2O
CwlJ
GerQ
YpeB
SleB
Csp
Pro-SleC
SleC
PrkC
required for CLE localization
and/or function?
cortex deformation?
cortex lysis
cortex lysis
cortex lysis
cortex lysis
phosphorylation of
signal transducing
substrates?
Inner membrane pore formation?
Inner membrane structure affects channel proteins?
Structural changes in channel or DPA binding protein(s)?
DPA channel?
SpoVA?
directly/indirectly?
SpoVA
required for
lysozyme and
dodecylamine
germination?
Peptidoglycan fragments
(muropeptides) from growing
cells of the same or closely
related species
(conserved in Bacillales and Clostridiales)
high hydrostatic pressure
- (incomplete) germination in Bacilli
p > 400 – 600 MPa
and/or T > 40 - 50°C (B. subtilis)
- CLE-independent, unknown mechanism in Clostridia
enzymes with peptidoglycan
hydrolyzing activity
- lysozyme
- host serum germination factor
(e.g. B. anthracis)
cationic surfactants
acting on Ca-DPA
channels or inner
membrane integrity
- dodecylamine
(e.g. B. subtilis)

which might be localized in the spore similar to SleB. In contrast to SleB and CwlJ, SleC is synthesized as a precursor and exists as an inactive zymogen (pro-SleC) in dormant spores. Thus, SleC requires activation by germination-specific serine proteases (Csp proteases, e.g. encoded by the *cspABC* operon in *C. difficile* (Adams et al. 2013)) to initiate cortex hydrolysis (Paredes-Sabja et al. 2011). However, the CS germination pathway is not activated upon Ca-DPA release (as it is the case for CwlJ) or partial core rehydration and cortex deformation (as it is the case for SleB), and it is not clear how the activity of Csp proteases is regulated in nGR-mediated germination.

23.2.2.2 nGR-Independent Germination

In addition to the initiation of germination via the activation of nGRs in response to favorable nutrient conditions, specific nGR-independent germination pathways allow spores to respond to the presence of host-specific factors or sense germination and vegetative growth occurring in their environment (Fig. 23.4).

For example, in addition to germination via the activation of nGRs (e.g. in *C. perfringens* (Paredes-Sabja et al. 2011)), exogenous Ca-DPA can directly activate the CLE CwlJ in some species resulting in cortex hydrolysis and complete germination (e.g. *B. subtilis* (Setlow 2003)). Although concentrations of Ca-DPA released by naturally occurring spore populations might be low and possibly too low to efficiently trigger germination of spores, exogenous Ca-DPA could play a role in sensing spore germination in their environment.

The germination-specific protease CspC can act as germinant receptor, which is activated in the presence of bile acids in *C. difficile* spores (Francis et al. 2013), in which orthologues to common nGRs have not yet been identified (Paredes-Sabja et al. 2011). This leads to the initiation of germination via proteolytic activation of the cortex hydrolase, SleC, specifically in mammalian intestines (Adams et al. 2013).

Furthermore, enzymes with peptidoglycan hydrolyzing activity e.g. host serum germination factor (lysozyme) can trigger sporulation by directly degrading the PG of spores (*B. anthracis* (Giebel et al. 2009)). Since many spores are well protected against cortex lysis by lysozyme due to the impermeability of their proteinaceous coat layers, lysozyme might only be active in initiating germination, when these layers are severely damaged (Peck and Fernandez 1995). Since Ca-DPA release

Fig. 23.4 Nutrient germinant receptor (nGR)-independent germination pathways in *Bacillus* and *Clostridium* spores. The different spore compartments are identical to those described in Fig. 23.3. The position of the proteins involved in germination might not be depicted precisely. Events, which are not yet completely understood, are described in italics. *Curled arrows* = exogenous activation events initiating germination; *straight solid arrows* = efflux or influx of molecules and enzymatic action on the cortex; *dashed arrows* = activation within the degradative cascade; *dotted arrows* = interaction mechanism not completely clear (Data for figure adapted from Paredes-Sabja et al. (2011), Adams et al. (2013), Francis et al. (2013), and Chen et al. (2014))

triggered by lysozyme treatment of decoated *B. subtilis* spores requires SpoVA proteins, it has been speculated, that Ca-DPA channel opening presents a step in lysozyme-induced germination (Vepachedu and Setlow 2007).

A highly conserved mechanism (including non-spore-formers) (Paredes-Sabja et al. 2011), which is (putatively) completely independent from nGRs and proteins involved in the nGR-mediated initiation of germination, involves PrkC as germinant receptor. PrkC is a eukaryotic-like protein kinase, which is localized in the inner spore membrane, phosphorylates serine/threonine residues, and enables spores to sense growth of cells of the same or closely related species via the detection of PG fragments from growing but not from lysed cells (Shah et al. 2008). Such muropeptides functioning as germinants are thought to activate PrkC by physically binding to the extracellular region of PrkC through interactions mediated by their mesodiaminopimelic acid (DAP) moiety (Squeglia et al. 2011). Divalent cations (Zn^{2+}) were shown to influence the activity of PrkC and its cognate phosphatase PrpC in *B. anthracis* (Arora et al. 2013) and possible substrates phosphorylated by PrkC have been identified including the ribosomal GTPases EF-G and CpgA, and the elongation factor EF-Tu (Shah and Dworkin 2010). However, the exact mechanism by which PrkC triggers spore germination remains to be elucidated.

Finally, germination can also be induced by cationic surfactants such as dodecylamine (Rode and Foster 1960). The role of surfactants as non-nutrient germinants in spore germination in their natural environment is not clear. Dodecylamine bypasses nGRs to trigger spore germination, most likely by causing DPA release in some fashion (Setlow et al. 2003). In this process it has been speculated that either dodecylamine directly opens Ca-DPA channels, where SpoVA might participate in channel formation, or alters inner membrane properties resulting in the release of Ca-DPA (*B. subtilis* (Vepachedu and Setlow 2007)). Ca-DPA release then triggers subsequent germination events similar to the nGR-mediated germination pathway.

In addition to responding to various nutrient and non-nutrient germinants described above, spore germination can also be triggered by physical treatments. It has been reported that violent agitation in the presence of glass beads, i.e. mechanical abrasion, can trigger germination of *B. subtilis* spores, which might occur due to physical damage of the cortex structure inducing CLE activation, i.e. either CwlJ or SleB (Jones et al. 2005). More relevant to food processing, HHP can trigger germination of spores from various species, which plays an important role in HHP-mediated spore inactivation (at least in *Bacillus* species).

23.3 High Pressure Inactivation of Bacterial Endospores

23.3.1 Possible Molecular Targets for HHP

HHP can induce morphological changes in spore structures. Pressurization of *G.* (*B.*) *stearothermophilus* (Hayakawa et al. 1994) and *B. cereus* spores (Gola et al. 1996) was reported to result in lengthening and flattening when treated at 300 MPa (20 °C,

5 min) and "breaking" of spores treated at 500 MPa (20 °C, 5 min) (Gola et al. 1996). Accordingly, pressure treatments of *B. anthracis* spores were reported to result in physicochemical protein modifications, increased membrane permeability, and flattening and lengthening of *B. anthracis* spores (Clery-Barraud et al. 2004). However, in addition to physicochemical alterations of spore components, the initiation of germination might play a role in some of the observable morphological changes after HHP treatments.

Generally, high pressure induces physical and chemical changes, according to Le Chatelier's principle, shifting the equilibrium to a lower total volume to counteract the imposed change, i.e. favoring biochemical and enzymatic reactions where the total volume is reduced, whereas processes leading to an increase in the total volume are retarded. Thus, the formation or disruption of non-covalent bonds, the degree of hydration or conformational changes of proteins that lead to total volume changes are likely to be affected by high pressure. Primarily disrupting the structure of secondary and tertiary bonded molecules while covalent bonds remain largely unaffected presents one of the major advantages for the application of HHP treatments for the production of safe, high quality food products (Knorr et al. 2011).

HHP stabilizes DNA hydrogen bonds, i.e. the melting temperature for duplex–single strand transition increases upon pressurization. Although HHP-mediated destabilization of DNA molecules can occur depending on their melting temperature, DNA helices are thought to be stabilized by HHP in most cases (Winter and Dzwolak 2005). Accordingly, DNA was shown to be stable at up to 1 GPa in vegetative cells (Cheftel 1995). Since spore DNA is present in the highly dehydrated environment within the spore core and additionally protected against heat stress by spore-specific molecules (e.g. SASP), it is unlikely that DNA presents a primary target in spore inactivation by HHP or HPT processes.

In contrast, lipid systems are way more sensitive to HHP and easily undergo phase transformations upon alterations in environmental conditions, which is dependent on their molecular structure, water content, the pH, ionic strength, temperature and pressure (Winter and Dzwolak 2005; Winter and Jeworrek 2009). Accordingly, membranes present a major target in the HHP-mediated inactivation of vegetative cells (Gänzle and Vogel 2001). In spores, the outer spore membrane plays a role primarily during sporulation, whereas the inner spore membrane is additionally important for spore dormancy and resistance, which is due to its unique structure (density) and composition (membrane proteins). The inner membrane of the developing forespore is a functional membrane with a phospholipid content similar to that of growing cells, but the volume surrounded by this membrane in mature spores is smaller than it would be expected from the phospholipid content and largely expends during germination without membrane synthesis, which indicates that it is densely packed (Cowan et al. 2004). This makes the inner membrane an extremely strong permeability barrier for almost all molecules including water, confers resistance to DNA damaging chemicals (Westphal et al. 2003; Cortezzo and Setlow 2005) and protects the core from rehydration, which is crucial to spore dormancy (Sunde et al. 2009). Since the ability to maintain the dehydrated state and retain DPA in the spore core presents an important factor for HHP resistance, the barrier function of the inner membrane is likely to be important for the resistance against such treatments. The phase transformation behavior of membranes in response to HHP is highly organism-specific (extreme case: piezophilic deep-sea organisms (Abe 2013)) and commonly adjusted by organisms in response to altered environmental conditions via modulating chain length, chain unsaturation or head group structure (homeoviscous adaptation) (Winter and Jeworrek 2009). Thus, differences in the HHP resistance observed between different species or in response to different sporulation conditions (e.g. incubation temperature), might be, at least partially related to differences in inner membrane composition. Although much work has been conducted to elucidate the phase behavior of model and vegetative cell membranes, knowledge on the exact effect of HHP on the inner spore membrane is scarce and further research is necessary to elucidate the role of this spore layer in HHP inactivation.

Additionally, the inner membrane harbors various proteins important for germination (e.g. nGRs (Setlow 2003), SpoVA (Li et al. 2012)). Although large integral and peripheral proteins can dissociate and their interactions with the lipid phase can be weakened by HHP, membrane proteins can be very pressure-stable (no unfolding up to 1 GPa) (Winter and Jeworrek 2009). However, the membrane structure and T,p-dependent phase behavior is influenced by membrane proteins and, in turn, protein conformation (and function) can be influenced by the lipid environment (Ulmer et al. 2002; Winter and Jeworrek 2009). Thus, HHP, can generally act on proteins or protein complexes directly to alter their properties and function or act via changing the structure of the surrounding lipid phase. Considering that large parts of the germination machinery are localized in the inner membrane and that germination is the first step in the HHP-mediated inactivation of spores from various species, this layer is likely to present a molecular target for HHP-mediated inactivation.

HHP favors hydrogen bond formation (protein secondary structure), and is unfavorable for hydrophobic interactions (protein tertiary structure). Generally, the effect of HHP on proteins depends on protein properties and the presence of salts and osmolytes (e.g. sucrose, sorbitol, glycerol). HHP favors a decrease in the total volume of protein–water systems. This can be achieved by filling the protein's void volumes when they become accessible to solvent upon dissociation, solvent-exposure of charged groups that have been involved in stabilization of protein assemblies (e.g. salt bridges; electrostriction effect), or potentially hydration of hydrophobic residues. The conformation of proteins denatured by HHP often resembles compact 'molten globule' type structures (Winter and Dzwolak 2005). Damage to proteins in the core (e.g. preformed enzymes, translation machinery) would putatively have consequences during late germination, outgrowth, and protein de novo synthesis rather than direct implications. Although proteins in the core are thought to be highly stabilized, especially by the extreme dehydration in this spore area, the germination protease Gpr, responsible for degrading α-/β-type SASP during late germination, appears to be not functional after HPT treatments at high pressure levels and/or temperatures (Wuytack et al. 1998), which might be due to its pressure-/temperature-mediated inactivation (Reineke et al. 2013a). Thus, inactivation of other important core proteins might play a role in spore inactivation by HPT processing. Although the spore coat has been mentioned in the context of HPP resistance (Reddy et al. 2006), the role of this thick proteinaceous layer is non-existent in heat inactivation (Setlow 2006) and far from clear in HHP-mediated inactivation of spores. If the coat would be involved in spore resistance to HHP, proteins in this layer might also present targets for HHP-mediated spore inactivation.

However, similar to heat inactivation it is difficult to spot a single target for the inactivation of microorganisms by HHP/HPT treatments and it is likely to be a composition of different engaged processes. Once spores are germinated by high pressure, molecular targets become similar to that for the inactivation of vegetative cells.

23.3.2 HHP Inactivation Mechanism

In contrast to direct inactivation by heat, high pressure-mediated spore inactivation of *Bacillus* spores has been proposed to be caused by an indirect mechanism involving germination (Wuytack et al. 1998; Margosch et al. 2006; Mathys et al. 2007a; Reineke et al. 2012). However, there seem to exist major differences in the pathways involved in *Bacillus* and *Clostridium* spores (Sarker et al. 2013).

23.3.2.1 *Bacillus* Spores

The majority of data on HHP-mediated germination and inactivation is available for the model organism *B. subtilis*. For spores of this organism, the release of DPA presents an early and rate limiting step of HHP-mediated inactivation (Reineke et al. 2013b). Consequently, the ability of spores to retain DPA largely determines their resistance to high pressure treatments. This putatively accounts for the different resistance properties of spores to HPT treatments and heat alone (Margosch et al. 2004b) and the fact that high wet heat resistance does not necessarily correlate with high pressure resistance (Margosch et al. 2004a; Olivier et al. 2011). Generally, the effectiveness and the underlying mechanism of HHP-mediated inactivation of bacterial endospores is greatly dependent on the temperature and pressure applied. According to the results from a number of studies, which investigated the HHP-mediated inactivation of *Bacillus*, predominantly *B. subtilis* spores, the underlying mechanism of HHP-mediated inactivation varies between three more or less concrete ranges of pressure-temperature combinations. (i) Moderate high pressure (approx. 100 up to 600 MPa) combined with moderate temperature (up to 50 °C), (ii) moderate high pressure (up to 500–600 MPa) combined with elevated temperature (above 50 °C) (iii) very high pressure (above 600 MPa) (Reineke et al. 2012, 2013b) (Fig. 23.5).

(i) Pressure levels that are sufficient to trigger nutrient-like germination already begin around 100 MPa (Paidhungat et al. 2002; Margosch et al. 2006; Torres and Velazquez 2005). Typically, pressures ranging between approx. 150 and

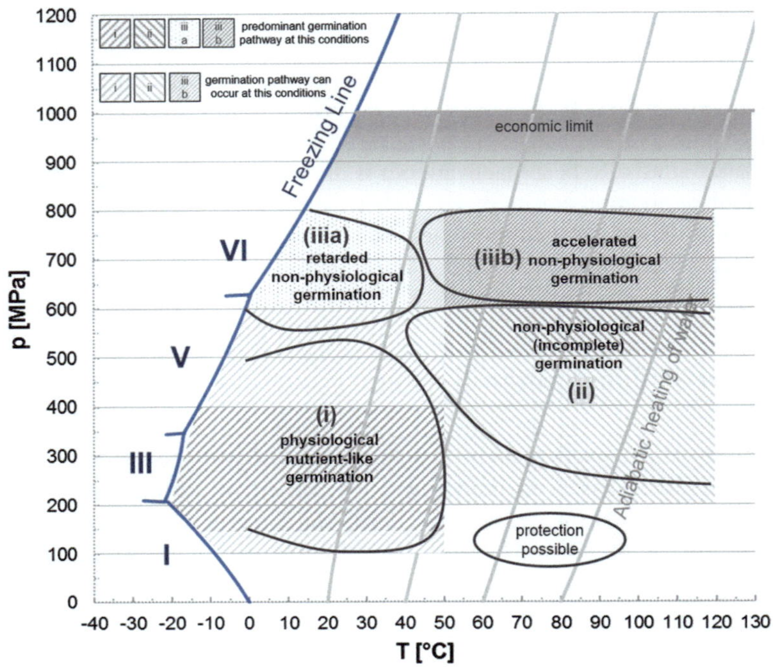

Fig. 23.5 Pressure/temperature-dependence of HHP-induced germination pathways in *B. subtilis*. *Black solid lines* roughly indicate areas, where different pathways are likely to be induced. *Dark shaded* areas mark core regions where a certain germination pathway is thought to dominate. *Light shaded* areas mark p/T combinations, where a certain pathway can potentially occur (Data for figure adapted from Reineke et al. (2013a), Reineke et al. (2012), Reineke et al. (2010), Black et al. (2005), and Black et al. (2007b))

400 MPa can effectively trigger germination of different *Bacillus* spores in the absence of nutrients and in a temperature and pH-dependent manner (Gould and Sale 1970; Reineke et al. 2012). Notably, a short (30 s) HHP pulse at 150 MPa can be sufficient to commit *B. subtilis* spores to germination, although this commitment by a pressure pulse is, in contrast to nutrient germination, more variable among individual spores and frequently reversible at ambient pressure (Kong et al. 2014). This is thought to occur due to the activation of nGRs (putatively any of the nGRs present in *B. subtilis* (Black et al. 2005) and *B. cereus* (Wei et al. 2009), but with different responsiveness (*B. subtilis* (Doona et al. 2014))). Requirements for HHP-mediated nGR activation at low pressure levels (e.g. 150 MPa) have been reported to be similar, though not identical to requirements for nutrient activation of these receptors (*B. subtilis* (Black et al. 2005)). For example, all *B. subtilis* nGRs (GerA, GerB, GerK) require diacylglycerylation to induce nutrient germination, whereas this appears to be the case only for GerA to induce germination by low pressures (Wuytack et al. 1998; Black et al. 2005). However, the exact mechanism of nGR activation

is known neither for nutrient- nor for HHP-triggered germination. For the latter, structural changes of the receptor molecules and/or structural changes in the inner membrane, where nGRs are situated, seem to be possible (Black et al. 2007a). Although there is some evidence that non-nutrient germination can occur at pressures as low as 200 MPa (30–60 °C) presumably via direct Ca-DPA-release (*B. subtilis* Δ*gerABK*, Δ*sleB* spores (Reineke et al. 2012)), nGR-mediated germination appears to be clearly dominating at moderate pressures/temperatures and essentially follows the nutrient-triggered germination pathway (Ca-DPA release, partial core rehydration, CLE activation, cortex hydrolysis, full germination (Setlow 2003; Black et al. 2005)).

Similar to nutrient-induced germination, HHP-mediated activation of nGRs triggers a relatively slow release of Ca-DPA. Results from measured DPA-release and flow cytometry experiments indicate that *B. subtilis* spore germination via nGR activation can be triggered by up to 600 MPa (threshold pressure) and up to 50 °C (Reineke et al. 2013b). Accordingly, pressure-induced germination of *B. subtilis* at up to 600 MPa has been suggested earlier to follow the physiological germination pathway, since inhibitors of nutrient-induced germination ($HgCl_2$ and N_α-*p*-Tosyl-L-arginine methyl ester (TAME)) also impeded HHP-induced germination at 600 MPa (Wuytack et al. 1998). In accordance with the proposed germination mechanism, nGR levels are the major factor determining *B. subtilis* germination rates at low pressure levels (150 MPa), whereas other germination related proteins play a minor role (Doona et al. 2014).

Since nGRs are highly species-specific proteins, which respond to a variety of different nutrients with different efficiency, it is reasonable that they also have different properties including their responsiveness to HHP and heat stability. Thus, it is not surprising that HHP-induced germination by pressure in the discussed range (100–150 MPa up to 400–600 MPa) does not occur with exactly the same efficiency in other spore-formers as it does in *B. subtilis*, especially when considering that events downstream of nGR activation also influence the germination efficiency. For example, the optimal temperature for the completion of spore germination at 150 MPa is around 40 and 65 °C for *B. subtilis* (Black et al. 2007a) and *B. cereus* spores (Wei et al. 2009), respectively. However, nGR-mediated germination triggered by such pressure levels is thought to be a common pathway in spores of various *Bacillus* species and might occur in a very similar fashion compared to *B. subtilis* especially in species with high similarity in *ger* operon organization and nGR structure.

(ii) Higher inactivation levels of *B. subtilis* spores at pressures up to 600 MPa can be achieved by additionally increasing the treatment temperature above 40–50 °C. In contrast to treatments at lower temperatures in the same pressure range, such parameter combinations lead to a rapid DPA release (up to 90 % in less than 5 min), which does not continue after pressurization (Reineke et al. 2013b). This is indicative of a non-physiological release of DPA. Although pressure levels at which nGR-independent germination primarily occurs have been reported in some studies to begin around 550 MPa or slightly lower (Margosch et al. 2006;

Torres and Velazquez 2005), there is some evidence that such a mechanism starts to occur at pressure levels as low as 200 MPa and temperatures at or even below 50 °C in *B. subtilis* spores (Reineke et al. 2012). When the pressure is decreased to a very low level, e.g. 61 MPa combined with 93 °C, this combination can even result in retarded inactivation of *B. subtilis* as compared to ambient pressure (Johnson and Zobell 1949). At such treatment conditions, the activity of nGRs might be reduced (nGR activity decreases as temperature increases (Paidhungat et al. 2002)) and pressures applied might be too low to promote significant damage of molecular targets (e.g. inner membrane or proteins).

The exact mechanism of a non-physiologic DPA release triggered by HHP is not clear. Effects on the solubility of DPA seem unlikely to be involved since the DPA concentration of the core exceeds the solubility limit by far. Thus, the formation of pores due to inner membrane damage might be a possible mechanism. Additionally, DPA-channel opening (Paidhungat et al. 2002; Wilson et al. 2008) due to changes in inner membrane organization or alterations in the structure of channel proteins (Black et al. 2005) or associated DPA binding proteins (SpoVAD (Li et al. 2012)) might be involved.

This direct and rapid DPA-release can occur independently from the presence of functional nGRs, is not limited by the activity of CLEs, is followed by partial core rehydration, but leads to an incomplete germination process in *B. subtilis* (Wuytack et al. 1998; Paidhungat et al. 2002; Setlow 2003; Black et al. 2007b; Reineke et al. 2011, 2012, 2013b). In comparison to pressure treatments at a low pressure (100 MPa), elevated pressure levels (500 MPa, 40 °C) (Wuytack et al. 1998) also trigger Ca-DPA release and cortex degradation, but spores remain more resistant against a second HHP treatment, UV light, and hydrogen peroxide, which was suggested to be due to 500 MPa treated spores failing in degrading their SASPs (Wuytack et al. 1998). Additionally, such spores fail to induce ATP generation indicating that germination is arrested at some stage (Wuytack et al. 1998).

Notably, the presence of Ca-DPA in the core is not required for resistance to HHP and its release does not results in inactivation at 550 MPa and 37 °C, which indicates that genetic material and essential enzymes inside a partially hydrated spore core lacking DPA are not damaged by such a treatment (Reineke et al. 2011). Accordingly, inactivation via this incomplete germination has been previously described as occurring due to a two-stage mechanism. In the first stage, HPP and heat treatment act to generate sub-lethally injured, DPA free and phase bright spores. In the second stage, these spores are inactivated by moderate heat putatively largely independent of the pressure level (Margosch et al. 2004a, b). Later, this model has been extended as for HPP treated *G. stearothermophilus* spores (600 MPa at 77 °C), four distinct spore populations were detected (Mathys et al. 2007a). For these populations, an at least three step inactivation mechanism was suggested involving a germination step following hydrolysis of the spore cortex, an unknown step, and finally an inactivation step with physical compromise of the spore's inner membrane (Mathys et al. 2007a).

(iii) Studies comparing the heat and the combined high pressure and heat resistances of bacterial spores showed that pressure and heat act synergistically to deliver spore inactivation (Ahn et al. 2007; Bull et al. 2009). However, above pressure levels of 600 MPa the synergism between pressure and temperature appears to diminish and DPA release and subsequent inactivation is thought to be governed by the treatment temperature as major influence factor (Reineke et al. 2013b). Depending on the treatment temperature this has two major consequences.

(iiia) At 550 MPa, pressure-induced germination primarily occurs via a non-nutrient pathway and SpoVA protein levels rather than nGR levels in a spore influence germination rates (Doona et al. 2014). However, increased pressure levels (over 400–500 MPa) combined with ambient temperatures (up to 40 °C) have been reported to stabilize spores due to a retarded germination resulting in low or no inactivation, e.g. of *B. subtilis* (Wuytack et al. 1998; Reineke et al. 2012), *G. (B.) stearothermophilus* (Ardia 2004; Mathys et al. 2009) and *B. amyloliquefaciens* spores (Rajan et al. 2006).

(iiib) When pressure levels above 400–600 MPa are combined with elevated temperatures above 50–60 °C, an acceleration of both germination and inactivation rates can be observed, which indicates the important role of high treatment temperatures and has been suggested to be useful if germination should be induced in a relatively short process time (Reineke et al. 2011; Reineke et al. 2012). Since the temperature is the governing factor for spore inactivation it is likely that at some point, germination is bypassed completely or at least plays a diminished role and a direct spore inactivation occurs similar to heat inactivation (de Heij et al. 2003). However, although increasing pressure levels commonly accelerate spore inactivation strongly and steadily, a protective effect of extremely high pressure levels combined with high temperatures was also reported. For example, treatments at 800–1,200 MPa and 120 °C result in a lower inactivation of *B. amyloliquefaciens* and *C. botulinum* spores as compared to heat alone, and a pronounced pressure-dependent tailing effect can lead to a small spore fraction surviving conditions up to 120 °C and 1.4 GPa during isothermal treatments (Margosch et al. 2006). However, such pressure levels exceed the limit of industrially available HHP units.

The mechanism of direct Ca-DPA release appears to be a common response of *Bacillus* spores to elevated pressure and/or temperature, CLEs are commonly involved in the subsequent germination process, and the cortex lytic machinery is highly conserved among Bacillales (Paredes-Sabja et al. 2011). Thus, it is likely that mechanisms underlying spore germination and inactivation of various *Bacillus* spores under such conditions are similar to that described for *B. subtilis*. Obvious differences in the HPT resistance of different *Bacillus* spores might occur due to differences in composition and structure of spore components.

23.3.2.2 *Clostridium* Spores

The mechanism by which HHP inactivates *Clostridium* spores is largely unknown, but appears to exhibit some striking differences compared to *Bacillus* spores, which impedes the transferability of results obtained for *B. subtilis*. Furthermore, HHP/HPT resistance of spores from different *Clostridium* species appears to vary considerably. With respect to differences in the behavior of different types of *C. botulinum* described in the following text it should be considered that the species *C. botulinum* consists of four phylogenetically distinct lineages comprising several different species including representatives of *C. sporogenes* sharing the genes for neurotoxin production (Collins and East 1998).

The majority of *Bacillus* and *Clostridium* spores have considerable differences in nutrient germination requirements, which is putatively related to the different genetic architecture (Paredes-Sabja et al. 2011) and different conformation of their nGRs. Hence, it is not surprising that *Clostridium* nGRs also harbor a different responsiveness to HHP and that HHP levels triggering nGR-mediated germination of *B. subtilis* spores do not provoke identical effects in *Clostridium* spores. Indeed, although there exists a huge gap of knowledge and it cannot be excluded that low pressure levels might be effective in inducing nGR-mediated germination in spores of some *Clostridium* species, available data indicate that low pressure levels at moderate temperatures have no or a very small effect on *Clostridium* spores. For example, 100–200 MPa, 7 min treatments do not induce germination in *C. perfringens* spores within 60 min after pressure treatment (Akhtar et al. 2009) and pressure cycling (60 MPa followed by 400 MPa at 60 °C) reduces *C. sporogenes* viable spore counts by less than three log cycles (Mills et al. 1998).

Additionally, CLEs, which are thought to play a major role in pressure-induced germination via the non-nutrient pathway in *Bacillus* spores, are different in *Clostridium* spores. Many *Clostridium* species rely on the CS machinery for spore cortex lysis during nutrient-triggered germination, which is completely absent in *Bacillus* spores, and of which their activation mechanism during nGR-mediated germination is not clear (Paredes-Sabja et al. 2011). Since the majority of *Clostridium* species lack the CLE CwlJ (Paredes-Sabja et al. 2011), which is thought be activated by endogenous or exogenous DPA, it is unlikely that the direct release of Ca-DPA caused by treatments at elevated HHP levels triggers germination as it is proposed to occur in *Bacillus* spores. Although spores of many *Clostridium* species possess SleB orthologues, YpeB, which is required for proper localization and/or function of SleB and conserved in *Bacillus* spores, is absent in the majority of *Clostridium* species (Paredes-Sabja et al. 2011). Results obtained for *C. difficile* suggest that spores of this organism rely solely on their CS system and that SleB is not functional possibly due to the absence of YpeB (Burns et al. 2010; Cartman and Minton 2010). Thus, the requirement of YpeB for functional SleB might be conserved among *Bacillus* and *Clostridium* spores, which suggests that it is unlikely that the HHP-mediated release of Ca-DPA leading to partial core rehydration and cortex deformation causes germination via the action of SleB in *Clostridium* spores.

Additionally, *C. perfringens* spores, which possess the CS system and SleB, but not YpeB, are stable and germinate normally even when they lack DPA (Paredes-Sabja et al. 2008b). This makes it unlikely, that the CS cortex lysis machinery is activated by HHP through a DPA-dependent pathway. Furthermore, the *C. perfringens* SleC, CspB, and the Ca-DPA release process were reported to present possible targets for damage by wet heat (Wang et al. 2012), which suggests that SleC activation somehow triggered by the release of Ca-DPA is unlikely to be involved in germination in response to high pressure at elevated temperatures.

Accordingly, it has been speculated that only *Clostridium* species that possess the complete YpeB-SleB cortex lytic machinery such as *C. sporogenes* and at least one *C. botulinum* type A strain (Paredes-Sabja et al. 2011) can germinate effectively in response to high pressure treatments at elevated temperatures via Ca-DPA release, SleB activation, and cortex lysis (Sarker et al. 2013). Indeed, pressure-induced germination is very poor in spores of various *Clostridium* species, e.g. max. 52 % germinated *C. perfringens* spores after 483 MPa, 50 °C, 5 min treatments (heat susceptible fraction 24 h after pressure treatment) (Kalchayanand et al. 2004). *C. sporogenes* germinates slightly better (82 %) under the same conditions (Kalchayanand et al. 2004) and a further prolongation of the dwell time (15–120 min) is necessary for an effective pressure-induced germination, which then occurs at an optimum pressure of around 200 MPa combined with mild heat (40–70 °C) (Ishimori et al. 2012). Notably, the DPA release of a pressure-resistant, proteolytic *C. botulinum* type B strain lacking YpeB can be observed quantitatively only at conditions where five log inactivation is reached. This indicates that DPA is released by a physicochemical rather than a physiological process and that the ability to retain DPA plays a differently important role for the HPT resistance of such spores compared to *Bacillus* spores (Margosch et al. 2004a; Margosch 2004).

Although the exact mechanism of inactivation of *Clostridium* spores is not known, available data suggest that germination induced by low pressure and moderate temperatures via the activation of nGRs, occurs only poorly or very slowly in comparison to *Bacillus* spores. Since nGRs are highly species- and sometimes strain-specific molecules, HHP-triggered germination via the activation of nGRs could potentially work effectively in some *Clostridium* species or strains, which have not been examined so far. Furthermore, the release of Ca-DPA at elevated pressures and temperatures appears to occur slower than in *Bacillus* spores and a different mechanism might be involved. Additionally, pressure-mediated inactivation at elevated temperatures is likely to occur independently from known CLEs (maybe except of a few YpeB-SleB positive strains) via physicochemical processes. The HHP resistance of spores varies greatly between genera and species.

Differences in the proposed inactivation pathways discussed above are likely to account for large differences in the HHP resistance of *Bacillus* and *Clostridium* spores. As an example, three log inactivation isorate curves for the *Bacillus* model organism, *B. subtilis*, and for the food intoxicator, *C. botulinum* type E, are shown in Fig. 23.6. The markedly different shapes of the curves reflect differences in the resistance of these two organisms against HPT treatments, i.e. *C. botulinum* type E,

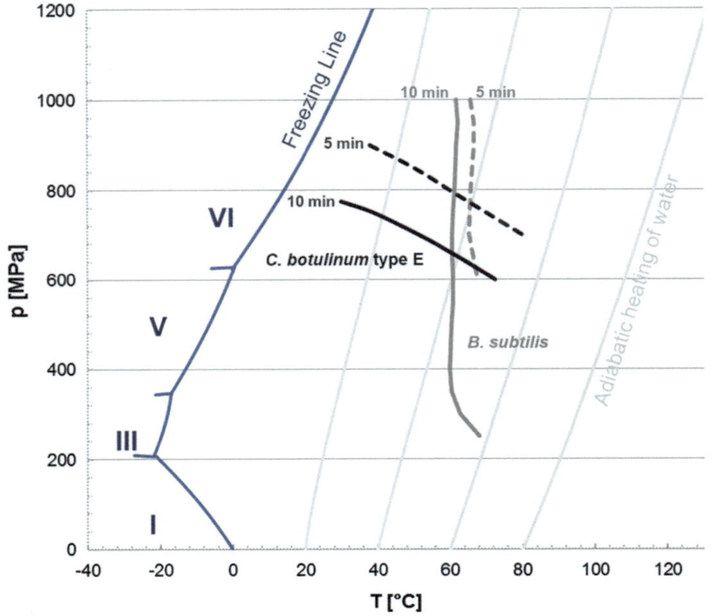

Fig. 23.6 Isorate lines for a three log inactivation of *B. subtilis* strain PS832 (*gray* (Reineke et al. 2012)) and non-proteolytic *C. botulinum* type E strain TMW 2.990 (*black*, unpublished data). p/T combinations required to achieve a three log inactivation after 5 min HPT treatments are depicted as *dashed lines*. A three log inactivation after a holding time of 10 min requires p/T combinations marked by the *solid lines*

which tends to be less resistant compared to other *C. botulinum* types, requires much higher pressure levels at high temperatures to be inactivated by three log cycles. At lower temperatures (below approx. 60 °C) in combination with high pressure levels, *B. subtilis* but not *C. botulinum* type E inactivation is retarded. Isorate curves for other *Bacillus* or *Clostridium* species can vary significantly from those depicted, which likely reflects species-specific differences in resistance factors (e.g. inner membrane composition) and the germination machinery. Although there can also exist huge differences in the HPT resistance between different strains of a single species, such strains usually do not follow different inactivation pathways, i.e. the curve might be shifted but would have a similar shape.

23.3.3 Inactivation Kinetics and Population Heterogeneity

Inactivation kinetics of bacterial endospores often do not follow first order reaction kinetics. An initial lag phase in inactivation kinetics, i.e. shoulder, and flattening of survivor curves at longer treatment times, i.e. tailing, can be frequently observed.

Especially HHP-treatments at ambient to moderate temperatures can result in a pronounced shoulder formation, i.e. an initial increase in the cell count from treated compared to untreated spore samples (mainly *Bacillus* spores) (Reineke et al. 2011). For example, an approx. 0.3 log (initially 8 log) increase for *B. subtilis* spores treated below 600 MPa at 30 °C and 60 °C has been observed (Reineke et al. 2012). There are two main reasons discussed to account for a shoulder formation including the pressure-induced germination of superdormant spores (e.g. *B. cereus*, *B. subtilis* (Wei et al. 2010)) and the disassembly of spore agglomerates (e.g. *G. (B.) stearothermophilus* (Mathys et al. 2007b)). Additionally, tailing of inactivation curves was frequently described and is related to the heterogeneity of spore populations with respect to various resistance properties and dormancy. Notably, a pronounced tailing effect was observed for proteolytic *C. botulinum* spores where a highly resistant fraction within the spore population survived even at extreme pressure-temperature combination (1.4 GPa, 120 °C) (Margosch et al. 2006). However, the shape of the inactivation curves can also follow log linear kinetics, which is certainly organism- and frequently treatment intensity-dependent. Examples for such (almost) log linear inactivation kinetics obtained for *C. botulinum* type E spores at 60 °C for 1 s up to 10 min are depicted in Fig. 23.7 (inactivation kinetics A 750 MPa, B 900 MPa). Although there is a slight trend towards a tailed shape at 900 MPa and above, this organism generally does not show pronounced tailing effects at any treatment conditions that are capable to efficiently induce inactivation (>2 log cycles) (unpublished data).

Four distinct physiological states of pressure treated spores have been previously described, i.e. spores in a dormant, germinated, an unknown physiology state after germination, and an inactivated state (inner membrane disrupted) (*B. licheniformis* (Mathys et al. 2007a)). In addition to the differences between these physiological states, especially the still dormant spore fraction is likely to be highly heterogeneous within itself as it may contain spores, which are damaged to a specific degree. Differently severe damage of the germination machinery after HPT treatments can be observed monitoring growth from single spores. For example, 200 MPa, 70 °C treatments (16 min, *B. subtilis*) slightly and 800 MPa, 70 °C treatments (4 min, *B. licheniformis*) strongly prolong detection times of single *Bacillus* spores and widen associated distributions (Margosch et al. 2004b). Similarly, HPT treatments at 600 MPa, 80 °C and 800 MPa, 60 °C markedly extended detection times and increased their heterogeneity in *C. botulinum* type E spores (Lenz et al. 2014). This heterogeneity underlines the presence of spores that are differently equipped, e.g. with functional nGRs, before and after pressure treatments. Notably, variations in the expression levels of nGRs were shown to be not the primary factor that controls spore germination heterogeneity in untreated *B. subtilis* spores (Zhang et al. 2013) indicating that other spore components modulating the germination or involved in the ability to retain Ca-DPA play an important role. Furthermore, low pressure levels at ambient temperatures (100 MPa, 20 °C) left the detection times of *B. licheniformis* largely unaltered (Margosch et al. 2004b), whereas 200 MPa, 20 °C treatments tended to decrease *C. botulinum* type E detection times and their

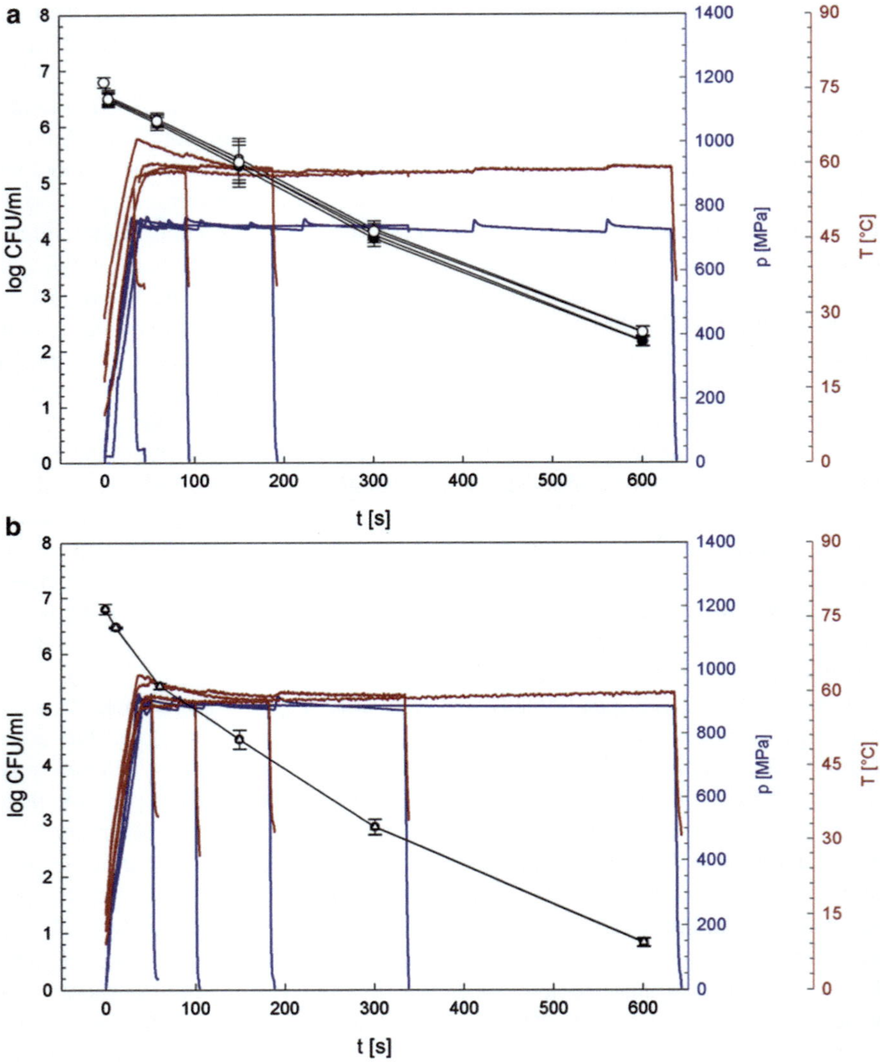

Fig. 23.7 Inactivation kinetics (1 s–10 min) of *C. botulinum* type E strain TMW 2.990 at 750 MPa, 60 °C (**a**) and 900 MPa, 60 °C (**b**). *Black lines* with data points and standard deviation indicate survivor curves (log CFU/mL) as determined by plate counts after a 7 d incubation under anaerobic conditions. The lower curves in graphs **a** and **b** show the pressure levels recorded during the different treatments. The upper graphs refer to the sample temperature monitored during the treatments (Note that pressure and temperature curves are shifted on the time axis by the time needed for the pressure build-up)

variability (Lenz et al. 2014). This is indicative of an activation of spores, e.g. involving increased responsiveness or accessibility of nGRs or changes in spore layers modulating germination.

This has two major practical implications: extended times, which spores need to induce growth can lead to an overestimation of spore lethality in inactivation studies when incubation times after HPT treatments are too short. Secondly, reduced times and narrowed distributions at low pressures and ambient temperatures indicate an enhanced chance of spores growing out and, e.g. in the case of *C. botulinum*, produce toxin within a distinct period of time (e.g. product shelf life).

23.3.4 Influence Factors on HHP-Mediated Spore Inactivation

In addition to species- and strain-specific differences in HHP-mediated germination requirements and HPT resistance, there exists a variety of factors influencing the outcome of spore inactivation studies. Such influence factors can either affect the resistance properties of spores (e.g. sporulation conditions), act synergistically or antagonistically with HHP (e.g. the matrix surrounding spores), or influence the recovery and detection of treated spores (germination conditions). Additionally, the design of HPT experiments (equipment, parameter control) can significantly alter the outcome of spore inactivation studies. Different factors are summarized in the following paragraphs and Fig. 23.8 highlighting putative major influence factors.

23.3.4.1 Sporulation Conditions

Two major factors affecting the resistance of spores against HHP-mediated inactivation include the sporulation medium composition in general and the mineral content of spores in particular as well as the temperature at which spores are produced. Generally, the modification of sporulation conditions can lead to perturbations in the ordered course of the sporulation phosphorelay (Veening et al. 2006) (Fig. 23.2), which can cause problems in proper spore assembly and alterations in structural features of spores, such as protoplast composition, membrane fluidity, and coat structure, which can affect spore germination and resistance properties.

For example, sporulation media resembling the natural habitat of *C. botulinum* can yield highly HPT-resistant spores (*C. botulinum* type A (Margosch et al. 2004a, 2006), *C. botulinum* type E (Lenz and Vogel 2014)). This effect is likely to be conferred, at least partially, by the amounts of divalent cations (Ca^{2+}, Mg^{2+} and Mn^{2+}) present in the sporulation medium and shown to be required for HHP and HPT resistance in a mineral type and concentration-dependent manner (*C. botulinum* type E (Lenz and Vogel 2014)).

For *Bacillus* spores it was shown that, in contrast to heat resistance, HHP resistance increases when spores are demineralized by acid titration and decreases again upon remineralization with single cation types. Such experiments suggested a

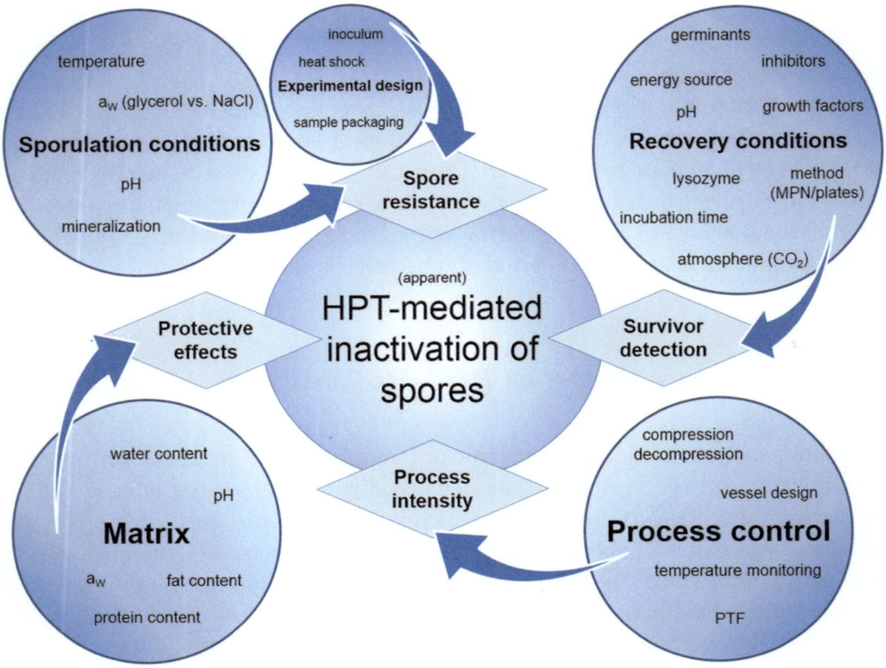

Fig. 23.8 Overview of factors that can influence the outcome of HPT spore inactivation studies via their effect on spore resistance properties, the detection of survivors from HPT treatments, the intensity of the HPT process, and the protection of spores from their inactivation by HPT treatments

general order of resistance of $H^+ > K^+ > Ca^{2+} = Mg^{2+} >$ native spores (49 MPa, *B. megaterium* (Bender and Marquis 1982), 100–300 MPa, *B. subtilis* (Igura et al. 2003)). This effect has been reported to occur in a similar manner in *B. subtilis* spores grown in medium supplemented with different minerals and subjected to 800 MPa, 70 °C treatments (Margosch et al. 2004b).

The putative involvement of the cortex lytic machinery (cation-dependent activities of CLEs) in the effect observed for *B. subtilis* together with the differences in HHP-mediated germination pathways between *Bacillus* and *Clostridium* species suggests that the presence of cations does not necessarily affect the HHP-mediate inactivation in an identical way in spores of different spore-forming organisms. Potential factors leading to an altered HPT resistance in response to the presence of different cations include the accumulation of cations in the core and protective effects on macromolecules, effects on the core water content, and changes in the spore structure mediated by alterations in enzyme activity or gene expression patterns during sporulation (Oomes et al. 2009). However, underlying mechanisms are not clear in either species.

The sporulation temperature is putatively one of the most important and certainly one of the best characterized influence factors during sporulation and was shown to affect a large variety of properties of both *Bacillus* and *Clostridium* spores.

In spores of different *Bacillus* species, sporulation temperature has been reported to negatively correlate with (i.e. a decrease in sporulation temperature increases . . .) surface roughness (Lindsay et al. 1990), spore size (Baweja et al. 2008; Garcia et al. 2010), wet density (Lindsay et al. 1990), the percentage of L-alanine substitution in the cortex peptidoglycan (at least slightly (Melly et al. 2002; Atrih and Foster 2001)), the level of at least one coat protein (Melly et al. 2002), and the overall size and integrity of the exosporium (Faille et al. 2007). For both *Bacillus* and *Clostridium* species it was shown that spore properties negatively correlating with the sporulation temperature include the core water content (Beaman and Gerhardt 1986; Popham et al. 1995; Melly et al. 2002) (not significantly (Atrih and Foster 2001; Paredes-Sabja et al. 2008a)), and the degree of unsaturation of the inner membrane (Gaughran 1947; Peck et al. 1995; Aguilar et al. 1998; Evans et al. 1998) (not significant (Planchon et al. 2011)).

Furthermore, sporulation temperature has been reported to positively correlate with (i.e. a decrease in sporulation temperature decreases . . .) the core/(core + cortex) volume ratio (Lindsay et al. 1990), the DPA content (no linear relationship (Lindsay et al. 1990; Planchon et al. 2011)) (not significant (Melly et al. 2002)) (only at elevated temperatures (Baweja et al. 2008)), the mineral contents (not linear (Lindsay et al. 1990; Igura et al. 2003)) (not significant (Atrih and Foster 2001)), and cross-linked muramic acid, tripeptide side chains and tetrapeptides in the cortex (slight and sometimes not significant (Atrih and Foster 2001; Melly et al. 2002)). Additionally, a decrease in the sporulation temperature was shown to decrease the average acyl chain length in the inner membrane of *Clostridium* spores (Peck et al. 1995; Evans et al. 1998).

In contrast, α/β-type SASP levels have been reported to remain unaffected by the sporulation temperature (Movahedi and Waites 2000; Melly et al. 2002) and the degree of δ-lactam residues in the cortex tends to peak at the optimum growth temperature (Atrih and Foster 2001; Melly et al. 2002).

Additionally, the HPT resistance of *Bacillus* and *Clostridium* spores was shown to be influenced by the sporulation temperature (summarized in Table 23.1). Results from various studies indicate that this effect is frequently treatment intensity-, i.e. pressure level and temperature-dependent, but largely independent from the treatment duration. Although there can be exceptions (Black et al. 2007b) and the sporulation temperature does not significantly influence HHP-mediated inactivation at many process conditions, spore resistance generally tends to decrease with increasing sporulation temperatures (Raso et al. 1998a, b; Igura et al. 2003; Margosch 2004; Black et al. 2005; Garcia et al. 2010; Nguyen Thi Minh et al. 2011; Olivier et al. 2012; Lenz and Vogel 2015).

Since the homoviscous adaptation of membrane properties is a conserved mechanism among various bacteria, it seems likely that alterations in the inner membrane are involved in sporulation temperature-mediated effects. Furthermore, cold stress induces σ^B (Mendez et al. 2004), which activates germinant receptor operons (*B. weihenstephanensis* (Garcia et al. 2010)) and contributes to proper germination (*B. cereus* (van Schaik and Abee 2005)). Additionally, effects of the sporulation temperature and divalent cations appear to be interconnected, at least in some *Bacillus* (Igura et al. 2003) and *Clostridium* (Lenz and Vogel 2014) spores.

Table 23.1 Summary of available data on sporulation temperature effects on the HHP resistance. Unless mentioned explicitly sporulation temperature is indicated on top of each column. (ST) = standard temperature used in the respective study; ↑ = HHP resistance increased; ↓ = HHP resistance decreased; (≈) = HHP resistance unaltered; *B. w.* = *B. weihenstephanensis*, *B. a.* = *B. amyloliquefaciens*, *B. s.* = *B. sporothermodurans*, *B. c.* = *B. coagulans*; *initial temperature prior to pressurization; **Pressure-induced germination measured. Especially at low pressures at ambient temperatures, sensitivity to pressure-induced germination does not necessarily correlate with HHP inactivation as germination is generally higher, but not all germinated spores are inactivated

Organism	p [MPa]	T [°C]	Dwell [min]	Matrix	Initial [spores/mL]	Sporulation temperature [°C]: 12 (or 13)	20 (or 19 or 18)	23	30 (or 28)	37 (or 33)	44 (or 38)	48 - 50	Reference
B. subtilis	150	37*	1 – 7	Tris-HCl pH 7.5	OD_{600} = 1.0			↑ >20% ger. @ 7 min (23–30: consistently lower germ. for 1.5 - 7 min)	↑ >40% ger. @ 7 min	(ST) >80% germinated after 7 min**	↓ 90% ger. @ 7 min consis. 1.5 - 7 min		(Black et al., 2005)**
	100	55	30	Dest. water	10^6				(ST) ≈ 1 log reduction	↓ >2 log reduction = lowest resistance	↓ ≈ 1.3 log reduction		
	200	55	30	Dest. water	10^6				(ST) ≈ 1.7 log reduct.	↓ ≈ 3.5 log reduct. (37–44: similar resist. of 37 and 44°C spores)	↓ ≈ 3.5 log reduct.		(Igura et al., 2003)
	300	55	30	Dest. water	10^6				(ST) ≈ 2 log reduction	↓ < 4.5 log reduct. (37–44: slight difference of 37 vs. 44°C spores)	↓ > 4.5 log reduct.		
	350	40	60	Dest. water	10^9 (spray-dried)		↓ (19°C) 3.37 log red.		(ST) ≈ 2.3 log red. (not mentioned)			(≈) (49°C) 2.31 log red.	(Nguyen Thi Minh et al., 2011)
	500	50*	0.5 - 5	Tris–HCl pH 7.5	OD_{600} = 1.0			↓ ≈ 95% germinated after 2.5 min (23–44: consistently between 0.5 and 2.5 min, ≈ identical germination after 5 min)	↓ >90% germinated after 2.5 min	(ST) 75% germinated after 2.5 min**	↑ 50% germinated after 2.5 min		(Black et al., 2007b)**
	800	70	1s -16 min	mashed carrot pH 5.15	10^7				(ST) 2.5 log reduction after 1s		↓ 5 log red. after 1s	↓ (48°C) 6 log red. after 1s (= detection limit)	(Margosch, 2004)

B. cereus	**55 - 400**	**25, 40**	15	Mac-Ilvaine buffer pH7	10^8		(≈) log inactivation 0 – 0.5 at 25°C 0 – 2.5 at 40°C		(ST)	(≈) log inactivation 0 – 0.5 at 25°C 0 – 2.5 at 40°C			(Raso et al., 1998a)
							inact. level HHP but not spo. temp.-dependent, consistent for 250 MPa, 25°C, 0.5 – 30 min						
	550, 690	**25**	15	Mac-Ilvaine buffer pH7	10^8		(≈) slightly less than 1 and 1.5 log reduct.		(ST) Slightly over 1 and 1.3 log reduction	↓ Approx. 1.4 and 2.3 log inactivation			
							resistance of 20°C spores marginally higher vs. 30°C spores						
	550, 690	**40**	15	Mac-Ilvaine buffer pH7	10^8		↑ ≈ 3.1 - 3.4 log red.		(ST) ≈ 4.4 - 5.7 log red.	↓ ≈ 5.3 - 6.5 log red.			
	55 - 150	**60**	15	Mac-Ilvaine buffer pH7	10^8		(≈) < 0.5 – 2.8 log red.		(ST) < 0.5 – 2.8 log reduction	↓ 1.8 – 3.8 log red.			
	200, 250, 300	**60**	15	Mac-Ilvaine buffer pH7	10^8		↑ ≈ 3 - 3.2 log red.		(ST) 3.5, 4 and 5.2 log reduction	↓ 3.9, 4.6, 5.4 log red.			
							general trends confirmed for 690 MPa, 40°C, 0.5 – 30 min						
	400 - 690	**60**	15	Mac-Ilvaine buffer pH7	10^8		↑ 3.7 – 6 log red.		(ST) 6 – over 7 log reduction (detect limit)	↓ 6 – over 7 log red. (below detect. limit)			
									400 MPa: resistance 30 ≈ 37°C spores				
							general trends confirmed for 690 MPa, 60°C, 0.5 – 30 min						
B. w.	**150**	**27**	0.5	Phosp. buffer pH 7.4	OD_{600} = 0.6-1.0	↑ 50% germination	↑ 15% germination		(ST) >99% germination**				(Garcia et al., 2010)**
	500	**34**	2	Phosp. buffer pH 7.4	OD_{600} = 0.6-1.0	(≈)	(≈)		(ST) 35% germinated**				
B. a.	**600**	**110**	2 - 3	Nutrient agar	10^5 to 10^6 /g				↑ D = 4.0 (3 min)	(ST) D value 0.58 (2 min dwell)			(Olivier et al., 2012)
B. s.	**600**	**110**	0.75	Nutrient agar	10^5 to 10^6 /g				↑ D value 0.24	(ST) D value 0.14			
B. c.	**600**	**110**	4 - 6	Nutrient agar	10^5 to 10^6 /g					(ST) D value 0.5 6 min dwell		↓ (50°C) D = 0.24 (4 min)	
	200	**40**	10	IPB buffer pH 7	5 × 10^7	(≈) (13°C) (0.5 log inactivation)	(≈) (18°C) (0.5 log inactivation)	(≈) (23°C) (0.6 log inactivation)	(ST) 28°C (0.6 log inactivation)	(≈) (33°C) (0.9 log inactivation)	(≈) (38°C) (1 log inactivation)		(Lenz and Vogel, 2015)
	200	**80**	10	IPB buffer pH 7	5 × 10^7	(≈) (13°C) (2.4 log inactivation)	(≈) (18°C) (2.4 log inactivation)	(≈) (23°C) (2.5 log inactivation)	(ST) 28°C (2.6 log inactivation)	(≈) (33°C) (2.8 log inactivation)	(≈) (38°C) (2.7 log inactivation)		
	800	**40**	10	IPB buffer pH 7	5 × 10^7	(≈) (13°C) (2.7 log inactivation)	(≈) (18°C) (2.6 log inactivation)	(≈) (23°C) (2.7 log inactivation)	(ST) 28°C (3 log inactivation)	↓ (33°C) (3.6 log inactivation)	↓ (38°C) (4.6 log inactivation)		
	800	**80**	1 - 10	IPB buffer pH 7	5 × 10^7	↑ (13°C) (4.9 log inactivation)	↑ (18°C) (4.6 log inactivation)	↑ (23°C) (4.7 log inactivation)	(ST) 28°C (6.5 log inactivation)	↓ (33°C) (7.5 log = complete)	↓ (38°C) (7.5 log = complete)		

In addition to the sporulation temperature-dependent uptake of minerals during sporulation, both influence factors may act on an overlapping subset of spore resistance factors. However, underlying mechanisms are far from being completely understood and many other sporulation temperature-mediated alterations in spore properties mentioned above could play a role.

Generally, any other factors during sporulation, which affect the germination machinery, spore layers it is situated in, or, generally, the ability of spores to retain Ca-DPA in the core could influence the HPT resistance of spores. Some other factors, which exert an (at least slight) effect on the spore resistance to HHP-mediated germination of *B. subtilis* include aeration, the pH, and the water activity (different effect of glycerol and NaCl) of the sporulation medium (Nguyen Thi Minh et al. 2011). The latter effect might be related to osmotic stress conditions provoking alterations in the inner membrane (levels of individual fatty acids and phospholipids (Lopez et al. 2002)). Both the effect of pH (*B. cereus* (Oh and Moon 2003)) and NaCl concentration (*B. subtilis* (Black et al. 2007b)) appear to be dependent on the HPT process intensity. However, not much is known about the precise mechanism and such effects in other spore-formers.

Notably, sporulation conditions such as variable nutrient availability and sub-optimal temperatures, which are commonly prevalent in soil or sediment, the natural habitat of many spore-formers, can affect (increase) the resistance properties of spores from various species. This can have direct implications on the evaluation of food safety as it may lead to the underestimation of risk originating from pathogenic spore-formers.

23.3.4.2 Germination Conditions

Conditions at which spores are incubated after HPT treatments can significantly influence their recovery (germination, outgrowth, growth) and thus the determined inactivation efficiency of the treatment.

Influence factors include the composition of the recovery medium, where suitable concentrations of potent germinants, co-germinants or mixtures thereof can compensate for low numbers of functional nGRs (Clements and Moir 1998; Hornstra et al. 2006b). Additionally, the medium pH has to be suitable for spore recovery, and compounds required during outgrowth and growth need to be present in suitable amounts (e.g. metal ions (Rode and Foster 1966), amino acids (Hawirko et al. 1979), energy source). However, a suitable pH and required substances are usually present in rich growth media commonly used for the cultivation of spore-forming species and an increased amount of germinants or co-germinants is not necessarily required to detect increased numbers of survivors after HPT treatments (proteolytic *C. botulinum* type A, non-proteolytic type B (Reddy et al. 2010)). Additionally, some other substance were suggested to aid in the recovery of injured spores (e.g. starch (Foegeding and Busta 1981)), or inhibit germination of (injured) spores (e.g. NaCl, heat treated *B. subtilis*, *B. licheniformis*, *B. coagulans* (Cazemier et al. 2001); unsaturated fatty acids, untreated proteolytic *C. botulinum* types A and B (Foster and Wynne 1948)).

The germination temperature can generally influence spore germinability (Evans et al. 1997). Lower temperatures were shown to prolong germination times and increase the associated heterogeneity within a population without changing the final extent of germination (non-proteolytic *C. botulinum* type B (Stringer et al. 2009)). Although this might be different for HPT treated spores, recovery temperatures below the optimum growth temperature have been frequently reported to enhance the recovery of physically (heat) damaged spores (Williams and Reed 1942; Sugiyama 1951; Condón et al. 1996).

Additionally, the environmental gas atmosphere plays a role in germination and outgrowth of *Clostridium* spores and low levels of CO_2 promote germination and outgrowth, e.g. of *C. botulinum* (proteolytic (Wynne and Foster 1948) and non-proteolytic types (Gibson et al. 2000; Fernandez et al. 2001)). However, whether this plays a role in facilitating recovery of damaged spores has not yet been determined.

Furthermore, the addition of lysozyme can enhance the recovery of thermally injured spores penetrating through heat damaged coat layers and degrading the PG (*C. perfringens* (Ando and Tsuzuki 1983); non-proteolytic *C. botulinum* type B, E, F (Peck et al. 1992); *Clostridium* sp. RKD (Dixit et al. 2005)). However, this effect might be strain- and treatment intensity-dependent or influenced by the recovery time, since it was not detected for other heat or HPT-damaged *C. botulinum* strains (proteolytic type A, non-proteolytic type B (Reddy et al. 2010)).

Finally, the incubation time is an important factor, which can largely influence the outcome of spore inactivation studies. This is primarily related to the heterogeneity in spore populations, which is commonly increased after lethal HPT treatments (see Sect. 23.3.3). For example, at least 5 days of incubation were shown to be required to obtain reliable plate count results for heat or HPT treated *C. botulinum* spores (proteolytic type A, non-proteolytic type B (Reddy et al. 2010)). On the other hand, using most probable number methodology for estimating the number of surviving spores an incubation period of four month in liquid recovery medium was required and resulted in slightly higher recovery compared to plate counts (Reddy et al. 2010).

Neglecting the possibility that certain recovery conditions can increase the number of surviving spores detected can lead to inaccurate results of inactivation studies and an over-estimation of spore lethality. From a practical point of view, this is particularly important where conditions enhancing spore recovery are prevalent during food storage but (frequently) not considered in laboratory inactivation studies presenting the basis for food safety risk assessment. Possible examples include long incubation times during extended shelf-life, low recovery temperatures during refrigerated storage, CO_2 in modified atmosphere packaged products, and the presence of lysozyme e.g. in fish.

23.3.4.3 Matrix

Environmental characteristics can strongly affect the survival of spores during HPT treatments. One of the most important influence factors is the pH of the matrix surrounding the spores, which plays an important role in spore inactivation not only in food. Due to the shift of the dissociation equilibrium caused by alterations in both pressure and temperature, it is important to consider this factor even when conducting spore inactivation studies in simple aqueous model systems. The use of buffer solutions that are not pressure and temperature stable, i.e. not suitable to largely maintain their pH during HPT treatments, can result in highly accelerated spore inactivation rates (Mathys et al. 2008). Generally, a low pH is thought to increase HPT spore inactivation (e.g. *B. coagulans* (Roberts and Hoover 1996); *B. cereus* (Gao and Ju 2010)). However, this effect may be species-dependent and putatively occurs in a process intensity-dependent manner, since variations in the pH (3–8) did not result in significant alterations in *B. subtilis* spore inactivation by HPT treatments at 100 and 600 MPa, 40 °C (Wuytack and Michiels 2001). Additionally, it was shown that the pH value and the inactivation of proteolytic *C. botulinum* type B spores do only correlate within certain limits, i.e. no significant difference between inactivation at pH 5.15 and 6, but markedly increased inactivation rate at pH 4 (Margosch et al. 2004a). Another important role of the pH can be found in its synergistic action with some (antimicrobial) substances such as citric and acetic acids (*B. amyloliquefaciens* (Ratphitagsanti et al. 2010)) or nisin (*B. coagulans* (Gao and Ju 2011)). A temporally decrease of the pH of food products under pressure likely contributes to spore inactivation. However, pressure-induced changes in the pH of complex food matrices cannot be easily determined and food matrices generally exert protective rather than synergistic effects on the HPT inactivation of bacterial endospores.

Additional food properties, which can influence the HPT resistance of spores include the total water content, water activity, protein and fat content. For example, soybean protein and sucrose significantly protect *B. cereus* spores against HPT inactivation (540 MPa, 71 °C, 16.8 min (Gao and Ju 2010)). In contrast, soybean oil showed only a slight and less significant protective effect (Gao and Ju 2010).

The protective effect of the fat content is still a matter of discussion. Although spore protection can be often observed in fat-rich environments, it has been frequently attributed to the low water activity of the product due to the concentration of solutes in the aqueous phase rather than to the fat itself (e.g. *B. stearothermophilus* in cocoa mass (Ananta et al. 2001)).

Similar to the HPT resistance of spores, which does generally not correlate with their heat resistance when spore inactivation is assessed in model buffer suspensions, protection conferred by the food matrix against HPT and heat do not necessarily correlate with each other. For example, this was shown for proteolytic *C. botulinum* types A and B and *C. sporogenes* spores in Bolognese sauce, cream sauce and rice water agar (Bull et al. 2009).

For practical considerations on food safety, the variety of different effects exerted by various food components and their interconnection in complex food matrices, which may additionally be species- and HPT treatment intensity-dependent indicate that it is extremely difficult to precisely predict the efficacy of HPT processes to inactivate spores in food. Thus, modeling approaches can help to characterize matrix effects and roughly estimate possible inactivation levels of a certain spore type in a narrow range of products. However, a final, reliable evaluation of the reachable degree of food safety has to be made conducting challenge tests in a specific food, under specific process conditions, and with spores of relevant spore-forming species.

23.3.4.4 Experimental Design

There is a multitude of studies that dealt with the inactivation of bacterial endospores by HPT treatments and numerous spore inactivation kinetics have been recorded using model spore suspensions and in various food products. However, there are some factors related to the experimental design and the control of process parameters, which can influence inactivation results and impede the comparability of results.

Effective temperatures activating spores can be strongly species-dependent, e.g. as low as 40 °C for psychrophilic *C. frigoris* (Spring et al. 2003), around 60 °C for *C. botulinum* type E (Juneja et al. 1995), and around 65–85 °C for many *Bacillus* spores such as *B. subtilis* and *B. cereus* (Evans and Curran 1943; Sogin et al. 1972; Leuschner and Lillford 1999). A sublethal heat shock in the latter temperature range, has been frequently used in spore inactivation studies either with the aim to inactivate remaining vegetative cells in a spore suspension or to activate spores (e.g. to reduce shoulder formation in inactivation kinetics). However, heat shocked spores can germinate better and can be much more susceptible to subsequent HPT treatments (e.g. *C. perfringens* (Akhtar et al. 2009)). Since it is unlikely, that spore populations present in their natural environment such as soil or sea sediment are situated in a uniformly activated state, a sublethal heat shock before HPT treatments can lead to a drastically overestimated spore lethality, which should be avoided.

Furthermore, it is commonly necessary to conduct inactivation experiments with high inoculation levels to demonstrate an effective inactivation by a certain number of log cycles, e.g. six log inactivation depending on the target organism, the food composition, the intended storage temperature, and the presence of additional hurdles. However, a higher initial inoculum can result in lower spore inactivation levels detected after HPT treatments (10^4–10^9 *B. subtilis* spores, 100 MPa, 45–75 °C, 10–120 min (Furukawa et al. 2002)), which was suggested to occur due to increased spore clump formation in highly concentrated spore suspension. Additionally, an opposite effect could also occur when very high inoculum levels are used, since clustered spores might germinate more rapidly due to nGR-independent germination pathways allowing spores to sense germination in their environment (see Sect. 23.2.2.2 and Fig. 23.4). Although this was shown to be not the case for non-proteolytic *C. botulinum* type B (Webb et al. 2012), it cannot be excluded that such an effect occurs in highly concentrated spore suspensions when spores of other species are investigated. This could be the case, e.g. for *Bacillus* spores, which readily release Ca-DPA under pressure and possess the Ca-DPA-responsive CLE, CwlJ. Notably, the latter mechanism is highly speculative, but would result in the same conclusion compared to clump formation at high spore densities, i.e. that inoculation levels for inactivation kinetics should be chosen as low as possible (but high enough to demonstrate the stipulated log reduction). Additionally, this makes the conduction of challenge tests using low spore inocula indispensable.

Another influence factor, which was shown to influence spore inactivation by HPT processes could be the type of sample packaging (Patazca et al. 2013). This has been attributed to differences in temperatures, samples are subjected to at different places in flexible plastic pouches (Patazca et al. 2013), which could be related to differences in the heat transfer between the sample and the pressure transmitting fluid (PTF). In the simplest case, packaging systems can affect the apparent spore resistance due to adhesion effects of spores to the packaging material.

23.3.4.5 Process Control

While the critical process parameters pressure level and treatment time can be usually held constant easily, the control of constant treatment temperatures has been frequently neglected in the past. Missing information on temperature profiles and/or high fluctuations in the sample temperature during HHP/HPT treatments can impede the comparability of results from inactivation studies. This is due to the temperature increase during pressurization caused by the compression work against intermolecular forces, i.e. compression heating or adiabatic heating (in the narrow sense only applying to compression heating in an insulated environment). Fluctuations in the treatment temperature play an important role especially when high pressure levels are combined with high temperatures, e.g. to achieve an effective inactivation of bacterial endospores.

Influence factors, which should be considered include: (i) Compression and decompression rates (Ratphitagsanti et al. 2009; Syed et al. 2012). Among other possible effects, these rates obviously influence the total process time and thereby the total "pressure and thermal load" a sample is subjected to. Additionally, the maximum temperature reach at the end of the pressure build-up can be influenced depending on how precisely HPT processes are controlled (vessel volume, thermostating jackets, etc.). (ii) The type (thermodynamic properties) of pressure transmitting fluid used (Balasubramaniam and Balasubramanian 2003; Scurrah et al. 2006; Patazca et al. 2013). (iii) The vessel design (material properties and shape), which can influence the temperature distribution in HPT systems (applies mainly to larger or industrial HHP units) and heat transfer between vessel wall and PTF and (Juliano et al. 2009; Koutchma et al. 2010; Knoerzer and Chapman 2011). (iv) The position of monitoring the process temperature in a HHP vessel or directly in the sample.

Since a number of reports lack information on one or more of these possible influence factors, it is often difficult to compare data obtained. For detailed studies on endospore inactivation it can be advantageous to use relatively short compression and decompression rates and taking into account only the inactivation effect occurring during the holding time (i.e. inactivation after total treatment time minus inactivation during compression and decompression (0 or 1 s treatments)). Additionally, a PTF with fast temperature response (conductive heat transfer) but low compression heating, relatively small and precisely tempered pressure vessels, monitoring the process temperature inside of the sample, and setting the sample starting temperature to a suitable level are certainly advantageous. The application of pressure treatments with (close to) isobaric, isothermal holding times can drastically enhance data transferability between different HHP systems (Reineke et al. 2012). However, in any case process parameters should be at least specified as detailed as possible to facilitate the comparability of results.

Accordingly, for food processing, practical challenges in the design of HHP or HPT processes and the control of process parameters are mainly related to an inhomogeneous temperature distribution during the treatment. There are a number of factors influencing the temperature profile of a food product during the treatment, which are primarily related to the adiabatic heat of compression of all materials inside of the pressure chamber, i.e. food product, packaging material, product carrier and PTF. This adiabatic heating effect is generally strongly dependent on the material properties, the starting temperature, the compression rate, and the target pressure of the pressurization process. Especially the adiabatic heating of the food product strongly varies with its composition and can reach heating rates of approx. 6 up to almost 9 °C per 100 MPa in pure oil or fat (Rasanayagam et al. 2003). Since industrial HHP systems are usually not temperature controlled, the actual temperature profile of a product also depends on heat transfer occurring between the warm product, the colder PTF and the even colder wall of the pressure chamber. This is generally a function of heat transfer properties of the different material involved and the surface available for heat transfer. Adiabatic heating properties of the product carrier and insulating properties of the packaging material can slow down this process (Knoerzer and Chapman 2011). Furthermore, heat convection resulting in a vertical temperature gradient can occur, which have to be considered especially when large (high) vessels and extended pressure dwell times are used.

Additionally, there exists the possibility of a matrix-related effect, which may occur during HHP processing of large solid food resulting in pressure non-uniformity. For example, the pressure in the center of 1.4–2.2 kg pieces of ham treated at 400–600 MPa, 7–24 °C, was reported to be up to 9 MPa lower compared to the pressure in the pressure transmitting fluid (Minerich and Labuza 2003). This effect, however, could be frequently neglected due to a smaller size and higher compressibility of many other food products. Last but not least, this effect is drastically smaller compared to other effects affecting spore inactivation such as temperature inhomogeneity.

23.3.5 *Strategies to Enhance HHP-Mediated Spore Inactivation*

Strategies to inactivate bacterial endospores by HHP without the need for its combination with high temperatures include the introduction of two-step processes or the addition of synergistically acting compounds.

Two-step processes can consist of a mild pressure treatment to induce spore germination followed by a treatment at higher pressure and moderate temperature or by a mild heat treatment to inactivate germinated spores. Alternatively, a heat activation step at a moderate temperature followed by a high pressure treatment (Mills et al. 1998) and combinations with irradiation treatments before or after pressurization (Crawford et al. 1996) have been explored in some studies. Although optimization of the process parameters of the two steps can result in the successful inactivation of *Bacillus* spores by several log cycles (Raso et al. 1998b), reliable sterilization processes completely inactivating such spores were not established. Furthermore, an application of such processes to effectively inactivate *Clostridium* spores appears to be even more difficult. Reasons can be found in the great heterogeneity in pressure sensitivity and the effectiveness of HHP-induced germination among spores of different species and within a population (Heinz and Knorr 1996; Wuytack et al. 1998). Accordingly, complete germination can commonly not be achieved (Aouadhi et al. 2012). Furthermore, germination does not follow log-linear kinetics and dwell times (*B. subtilis* optimum for germination around 300 MPa, 30 °C, 60 min) that may be required for triggering the germination of a large portion of the spore population can be very long (Reineke et al. 2012). Finally, germination induced by low pressure levels does not (or only poorly) occur in the majority of *Clostridium* spores, which altogether makes such two-step approaches unsuitable for an industrial application.

A possibility to enhance the induction of germination and inactivation of *Clostridium* (and *Bacillus*) spores is to add suitable nutrient germinants, e.g. to a food, combined with a heat activation step and a subsequent HPT treatment, e.g. L-asparagine, KCl, 80 °C for 10 min and 586 MPa, 73 °C for 10 min to inactivate *C. perfringens* (Akhtar et al. 2009). Although such an approach enhanced inactivation levels at relatively mild treatment conditions, it also failed to provoke complete (>6 log) reduction in viable cell counts. Additionally, it might not be desired to alter food product formulations, requirements for nutrient-triggered germination can vary considerably between species and within populations, and it is questionable whether the presence of nutrients for an optimal germination, e.g. of pathogenic spore-formers, presents a feature a food product should have. All this makes it difficult to use this approach for a broad spectrum of foods and against spores from various species. Whereas significant spore reduction can be achieved, additional strategies are required to ensure food safety upon storage.

An effective way to decrease the temperature required for effective spore inactivation during HPT processes is the addition of antimicrobial compounds such as essential oils (Gayan et al. 2012) or sucrose laurate (only effective at pH $\leq$ 6.0; (Stewart et al. 2000)). Especially nisin has been demonstrated to be very effective in combination with HHP treatments, e.g. in inactivating *C. perfingens* (Gao et al. 2011), *A. acidoterrestris* in apple juice (Sokolowska et al. 2012), or *B. subtilis* and *B. cereus* in milk, although failing to provoke complete inactivation (Black et al. 2008). Such a combination of antimicrobial substances and HHP processing might be reasonable in some cases. However, economical, legal and food product-specific limiting factors have to be considered, e.g. that preservatives are often not desired or permitted in food products, especially not in high amounts, and that the use of certain substances that affect sensorial food properties might be limited to only a few food products. Additionally, an important marketing argument for HHP-treated food, i.e. the possibility of labeling food as 'additive-free', cannot be used anymore.

23.4 Conclusions

Considerable progress made in recent years in understanding the basic mechanism during sporulation and germination of both *Bacillus* and *Clostridium* spores presents a valuable basis for explaining how the extreme resistance properties of spores develop and how they are lost in consequence of physical stress, e.g. during HPT treatments.

Although there are some gaps of knowledge that need to be closed, differences in currently proposed germination mechanisms of *Bacillus* and *Clostridium* spores appear to be large, and can at least to some extent explain the significant differences observable in their response to HHP and HPT treatments. However, in contrast to *Bacillus* spores, which might respond to HPT treatments in a similar way as *B. subtilis* does, the exact inactivation mechanism of *Clostridium* spores is still largely unknown.

Major other points that still need to be elucidated include the identification of molecular targets, which are affected by HPT and relevant to spore inactivation, and how they are affected, e.g. by sporulation conditions, or protected, e.g. by the food matrix. Additionally, detailed reasons for the frequently large population heterogeneity in germination and resistance properties are not known in detail, especially not in *Clostridium* spores. In-depth knowledge on such relations and reasons would facilitate future risk assessment and could lead to the identification of process combinations or hurdle strategies to increase the effectivity of spore inactivation.

Generally, additional comparable data such as inactivation kinetics obtained under accurately controlled process conditions can enable the refinement of risk assessment and might give some additional hints on underlying inactivation mechanisms in different organisms. With a view to the described differences between *Bacillus* and *Clostridium* and the heterogeneity of spore properties even within species, in particular within the *C. botulinum* species-complex, the use of any surrogate organisms should be avoided in high pressure challenge testing or data generation for safety considerations.

Finally, various strategies for the HHP-mediated inactivation of bacterial endospores have been proposed in the past. However, their suitability for an industrial application largely depends on the type of food product and its intended storage conditions. For the production of food products, in which pathogenic/food intoxicating spore-formers are not able to grow under normal conditions of distribution and storage, i.e. acidic ($pH < 4.5$), low a_W (<0.93), and/or cold-stored (<3 °C when cold storage alone) food, HHP processes are currently applied to inactivate vegetative cells and increase the shelf-life of such products. An increased HHP process intensity (e.g. ambient to moderate temperatures), the application of two-step processes (any combinations of mild heat, HHP or HPT), or the combination of HHP or mild HPT processes with antimicrobial or other synergistically acting substances can be used to additionally target bacterial endospores (see Sect. 23.3.5). Inactivation of spores, e.g. of the acid-tolerant organisms *C. pasteurianum* and *B. coagulans* by such treatments, can lead to a further increase in shelf-life of acidic/low a_W/cold-stored products. However, since the intensity of HPT processes at ambient or moderate temperatures is generally too low, two-step inactivation processes mainly rely on the statistical event of spore germination, and HHP/antimicrobial combinations are highly food matrix-dependent, complete spore activation is generally not achieved by such processes. This makes such processes suitable for serving as an additional hurdle in food processing, but not as preservation technique for the production of low-acid ($pH > 4.5$), shelf-stable food with a water activity above 0.93.

For a successful implementation of HHP-based approaches for the production of such foods, it is necessary to demonstrate that a specific process can reliably achieve spore inactivation irrespectively of the heterogeneity of spore populations and statistical germination events. HPT processes at temperatures above a critical inactivation threshold, where the process temperature governs inactivation, appear likely to be suitable for this purpose. Although this technology is currently not applied industrially, such processes may enable the production of shelf-stable and safe food with less detrimental effects on sensorial and nutritional characteristics as compared to thermal processing. Together with the fact that consumers increasingly demand for minimally processed, healthy food with long shelf-life, and legal authorities continuously try to decrease the levels of traditional food preservatives (e.g. salt), HPT processing appears to be a promising alternative to conventional thermal processing.

References

Abe F (2013) Dynamic structural changes in microbial membranes in response to high hydrostatic pressure analyzed using time-resolved fluorescence anisotropy measurement. Biophys Chem 183:3–8. doi:10.1016/j.bpc.2013.05.005

Abel-Santos E, Dodatko T (2007) Differential nucleoside recognition during *Bacillus cereus* 569 (ATCC 10876) spore germination. New J Chem 31:748–755. doi:10.1039/B616695d

Adam KH, Flint SH, Brightwell G (2010) Psychrophilic and psychrotrophic *clostridia*: sporulation and germination processes and their role in the spoilage of chilled, vacuum-packaged beef, lamb and venison. Int J Food Sci Technol 45:1539–1544

Adams CM, Eckenroth BE, Putnam EE, Doublié S, Shen A (2013) Structural and functional analysis of the CspB protease required for *Clostridium* spore germination. Plos Pathog 9:e1003165. doi:10.1371/journal.ppat.1003165

Aguilar PS, Cronan JE Jr, de Mendoza D (1998) A *Bacillus subtilis* gene induced by cold shock encodes a membrane phospholipid desaturase. J Bacteriol 180:2194–2200

Ahn J, Balasubramaniam VM, Yousef AE (2007) Inactivation kinetics of selected aerobic and anaerobic bacterial spores by pressure-assisted thermal processing. Int J Food Microbiol 113:321–329. doi:10.1016/j.ijfoodmicro.2006.08.012

Aizawa S-I, Zhulin IB, Márquez-Magaña L, Ordal GW (2002) Chemotaxis and Motility. In: Sonenshein AL, Hoch JA, Losick R (eds) *Bacillus subtilis* and its closest relatives. FSA – Food Standards Agency, ASM Press, Washington, DC, pp 437–452

Akhtar S, Paredes-Sabja D, Torres JA, Sarker MR (2009) Strategy to inactivate *Clostridium perfringens* spores in meat products. Food Microbiol 26:272–277. doi:10.1016/j.fm.2008.12.011

Alberto FO, Botella L, Carlin F, Nguyen-The C, Broussolle W (2005) The *Clostridium botulinum* GerAB germination protein is located in the inner membrane of spores. FEMS Microbiol Lett 253:231–235. doi:10.1016/j.femsle.2005.09.037

Alimova A, Katz A, Gottlieb P, Alfano RR (2006) Proteins and dipicolinic acid released during heat shock activation of *Bacillus subtilis* spores probed by optical spectroscopy. Appl Opt 45:445–450

Alsaker KV, Papoutsakis ET (2005) Transcriptional program of early sporulation and stationary-phase events in *Clostridium acetobutylicum*. J Bacteriol 187:7103–7118. doi:10.1128/Jb.187.20.7103-7118.2005

Ananta E, Heinz V, Schlüter O, Knorr D (2001) Kinetic studies on high-pressure inactivation of *Bacillus stearothermophilus* spores suspended in food matrices. Innovative Food Sci Emerg Technol 2:261–272

Ando Y, Tsuzuki T (1983) Mechanism of chemical manipulation of the heat-resistance of *Clostridium perfringens* spores. J Appl Bacteriol 54:197–202

Aouadhi C, Simonin H, Prevost H, de Lamballerie M, Maaroufi A, Mejri S (2012) Optimization of pressure-induced germination of *Bacillus sporothermodurans* spores in water and milk. Food Microbiol 30:1–7. doi:10.1016/J.Fm.2011.12.006

Appert N (1810) L'art de conserver pendant plusiers annes toutes les substances animales et vegetables. Patris & Co., Quay Napoleon, Paris

Ardia A (2004) Process considerations on the application of high pressure treatment at elevated temperature levels for food preservation. Dissertation, Technical University of Berlin

Arora G, Sajid A, Arulanandh MD, Misra R, Singhal A, Kumar S, Singh LK, Mattoo AR, Raj R, Maiti S, Basu-Modak S, Singh Y (2013) Zinc regulates the activity of kinase-phosphatase pair (BasPrkC/BasPrpC) in *Bacillus anthracis*. Biometals 26:715–730. doi:10.1007/s10534-013-9646-y

Atrih A, Foster SJ (2001) Analysis of the role of bacterial endospore cortex structure in resistance properties and demonstration of its conservation amongst species. J Appl Microbiol 91:364–372, doi:jam1394 [pii]

Augustin JC (2011) Challenges in risk assessment and predictive microbiology of foodborne spore-forming bacteria. Food Microbiol 28(2):209–213. doi:10.1016/j.fm.2010.05.003

Bagyan I, Setlow P (2002) Localization of the cortex lytic enzyme CwlJ in spores of *Bacillus subtilis*. J Bacteriol 184:1219–1224

Bahl H, Müller H, Behrens S, Joseph H, Narberhaus F (1995) Expression of heat-shock genes in *Clostridium acetobutylicum*. FEMS Microbiol Rev 17:341–348

Balasubramaniam VM, Balasubramanian S (2003) Compression heating influence of pressure transmitting fluids on bacteria inactivation during high pressure processing. Food Res Int 36:661–668. doi:10.1016/S0963-9969(03)00014-0

Banawas S, Paredes-Sabja D, Korza G, Li Y, Hao B, Setlow P, Sarker MR (2013) The *Clostridium perfringens* germinant receptor protein GerKC is located in the spore inner membrane and is crucial for spore germination. J Bacteriol 195:5084–5091. doi:10.1128/JB.00901-13

Baweja RB, Zaman MS, Mattoo AR, Sharma K, Tripathi V, Aggarwal A, Dubey GP, Kurupati RK, Ganguli M, Chaudhury NK, Sen S, Das TK, Gade WN, Singh Y (2008) Properties of *Bacillus anthracis* spores prepared under various environmental conditions. Arch Microbiol 189:71–79. doi:10.1007/s00203-007-0295-9

Beaman TC, Gerhardt P (1986) Heat resistance of bacterial spores correlated with protoplast dehydration, mineralization, and thermal adaptation. Appl Environ Microbiol 52:1242–1246

Becker EC, Pogliano K (2007) Cell-specific SpoIIIE assembly and DNA translocation polarity are dictated by chromosome orientation. Mol Microbiol 66:1066–1079. doi:10.1111/j.1365-2958.2007.05992.x

Behravan J, Chirakkal H, Masson A, Moir A (2000) Mutations in the gerP locus of *Bacillus subtilis* and *Bacillus cereus* affect access of germinants to their targets in spores. J Bacteriol 182:1987–1994

Bender GR, Marquis RE (1982) Sensitivity of various salt forms of *Bacillus megaterium* spores to the germinating action of hydrostatic-pressure. Can J Microbiol 28:643–649

Bi CH, Jones SW, Hess DR, Tracy BP, Papoutsakis ET (2011) SpoIIE is necessary for asymmetric division, sporulation, and expression of sigma(F), sigma(E), and sigma(G) but does not control solvent production in *Clostridium acetobutylicum* ATCC 824. J Bacteriol 193:5130–5137. doi:10.1128/Jb.05474-11

Bischofs IB, Hug JA, Liu AW, Wolf DM, Arkin AP (2009) Complexity in bacterial cell-cell communication: quorum signal integration and subpopulation signaling in the *Bacillus subtilis* phosphorelay. Proc Natl Acad Sci USA 106:6459–6464. doi:10.1073/pnas.0810878106

Black EP, Koziol-Dube K, Guan D, Wei J, Setlow B, Cortezzo DE, Hoover DG, Setlow P (2005) Factors influencing germination of *Bacillus subtilis* spores via activation of nutrient receptors by high pressure. Appl Environ Microb 71:5879–5887

Black EP, Setlow P, Hocking AD, Stewart CM, Kelly AL, Hoover DG (2007a) Response of spores to high-pressure processing. Compr Rev Food Sci Food Saf 6:103–119. doi:10.1111/j.1541-4337.2007.00021.x

Black EP, Wei J, Atluri S, Cortezzo DE, Koziol-Dube K, Hoover DG, Setlow P (2007b) Analysis of factors influencing the rate of germination of spores of *Bacillus subtilis* by very high pressure. J Appl Microb 102:65–76

Black EP, Linton M, McCall RD, Curran W, Fitzgerald GF, Kelly AL, Patterson MF (2008) The combined effects of high pressure and nisin on germination and inactivation of *Bacillus* spores in milk. J Appl Microbiol 105:78–87. doi:10.1111/j.1365-2672.2007.03722.x

Brunsing RL, La Clair C, Tang S, Chiang C, Hancock LE, Perego M, Hoch JA (2005) Characterization of sporulation histidine kinases of *Bacillus anthracis*. J Bacteriol 187:6972–6981. doi:10.1128/Jb.187.20.6972-6981.2005

Bull MK, Olivier SA, van Diepenbeek RJ, Kormelink F, Chapman B (2009) Synergistic inactivation of spores of proteolytic *Clostridium botulinum* strains by high pressure and heat is strain and product dependent. Appl Environ Microbiol 75:434–445. doi:10.1128/AEM.01426-08

Burns DA, Heap JT, Minton NP (2010) SleC is essential for germination of *Clostridium difficile* spores in nutrient-rich medium supplemented with the bile salt taurocholate. J Bacteriol 192:657–664. doi:10.1128/JB.01209-09

Camp AH, Losick R (2009) A feeding tube model for activation of a cell-specific transcription factor during sporulation in *Bacillus subtilis*. Genes Dev 23:1014–1024. doi:10.1101/Gad.1781709

Cangiano G, Mazzone A, Baccigalupi L, Isticato R, Eichenberger P, De Felice M, Ricca E (2010) Direct and indirect control of late sporulation genes by GerR of *Bacillus subtilis*. J Bacteriol 192:3406–3413. doi:10.1128/Jb.00329-10

Carr KA, Janes BK, Hanna PC (2010) Role of the gerP operon in germination and outgrowth of *Bacillus anthracis* spores. PloS One 5:e9128. doi:10.1371/journal.pone.0009128

Cartman ST, Minton NP (2010) A mariner-based transposon system for in vivo random mutagenesis of *Clostridium difficile*. Appl Environ Microbiol 76:1103–1109. doi:10.1128/AEM.02525-09

Cazemier AE, Wagenaars SFM, ter Steeg PF (2001) Effect of sporulation and recovery medium on the heat resistance and amount of injury of spores from spoilage *bacilli*. J Appl Microbiol 90:761–770

Chaibi A, Ababouch LH, Busta FF (1996) Inhibition by monoglycerides of L-alanine-triggered Bacillus cereus and *Clostridium botulinum* spore germination and outgrowth. J Food Prot 59:832–837

Chastanet A, Losick R (2011) Just-in-time control of Spo0A synthesis in *Bacillus subtilis* by multiple regulatory mechanisms. J Bacteriol 193:6366–6374. doi:10.1128/Jb.06057-11

Cheftel JC (1995) Review: high-pressure, microbial inactivation and food preservation. Food Sci Technol Int 1:75–90. doi:10.1177/108201329500100203

Chen Y, Ray WK, Helm RF, Melville SB, Popham DL (2014) Levels of germination proteins in *Bacillus subtilis* dormant, superdormant, and germinating spores. PloS One 9:e95781. doi:10.1371/journal.pone.0095781

Chesnokova ON, McPherson SA, Steichen CT, Turnbough CL Jr (2009) The spore-specific alanine racemase of *Bacillus anthracis* and its role in suppressing germination during spore development. J Bacteriol 191:1303–1310. doi:10.1128/JB.01098-08

Chirakkal H, O'Rourke M, Atrih A, Foster SJ, Moir A (2002) Analysis of spore cortex lytic enzymes and related proteins in *Bacillus subtilis* endospore germination. Microbiology 148:2383–2392

Chlopin GW, Tammann G (1903) Über den Einfluß hoher Drücke auf Mikroorganismen. Z Hyg Infektionskr 45:171–204

Christie G, Lowe CR (2007) Role of chromosomal and plasmid-borne receptor homologues in the response of *Bacillus megaterium* QM B1551 spores to germinants. J Bacteriol 189:4375–4383. doi:10.1128/JB.00110-07

Chubukov V, Sauer U (2014) Environmental dependence of stationary-phase metabolism in *Bacillus subtilis* and *Escherichia coli*. Appl Environ Microbiol 80:2901–2909. doi:10.1128/AEM.00061-14

Clarkson J, Campbell ID, Yudkin MD (2004) Efficient regulation of sigmaF, the first sporulation-specific sigma factor in *B. subtilis*. J Mol Biol 342:1187–1195. doi:10.1016/j.jmb.2004.07.090

Clements MO, Moir A (1998) Role of the gerI operon of *Bacillus cereus* 569 in the response of spores to germinants. J Bacteriol 180:6729–6735

Clery-Barraud C, Gaubert A, Masson P, Vidal D (2004) Combined effects of high hydrostatic pressure and temperature for inactivation of *Bacillus anthracis* spores. Appl Environ Microbiol 70:635–637

Clouston JG, Wills PA (1969) Initiation of germination and inactivation of *Bacillus pumilus* spores by hydrostatic pressure. J Bacteriol 97:684–690

Cohn FJ (1872) Untersuchungen über Bakterien. In: Cohn F J (Hrsg) Beiträge zur Biologie der Pflanzen Band I, Heft 2. J U Kers's Verlag, Breslau, p 145

Cohn FJ (1876) Untersuchungen über Bakterien. IV. In: Cohn F J (Hrsg) Beiträge zur Biologie der Pflanzen Band II, Heft 2. J U Kers's Verlag, Breslau, pp 249–276

Collins MD, East AK (1998) Phylogeny and taxonomy of the food borne pathogen *Clostridium botulinum* and its neurotoxins. J Appl Microbiol 84:5–17

Condón S, Palop A, Raso J, Sala FJ (1996) Influence of the incubation temperature after heat treatment upon the estimated heat resistance values of spores of *Bacillus subtilis*. Lett Appl Microbiol 22:149–152

Connan C, Deneve C, Mazuet C, Popoff MR (2013) Regulation of toxin synthesis in *Clostridium botulinum* and *Clostridium tetani*. Toxicon 75:90–100. doi:10.1016/j.toxicon.2013.06.001

Cooksley CM, Davis IJ, Winzer K, Chan WC, Peck MW, Minton NP (2010) Regulation of neurotoxin production and sporulation by a putative agrBD signaling system in proteolytic *Clostridium botulinum*. Appl Environ Microbiol 76:4448–4460. doi:10.1128/Aem.03038-09

Cortezzo DE, Setlow P (2005) Analysis of factors that influence the sensitivity of spores of *Bacillus subtilis* to DNA damaging chemicals. J Appl Microbiol 98:606–617. doi:10.1111/j.1365-2672.2004.02495.x

Cowan AE, Koppel DE, Setlow B, Setlow P (2003) A soluble protein is immobile in dormant spores of *Bacillus subtilis* but is mobile in germinated spores: implications for spore dormancy. Proc Natl Acad Sci U S A 100:4209–4214. doi:10.1073/pnas.0636762100

Cowan AE, Olivastro EM, Koppel DE, Loshon CA, Setlow B, Setlow P (2004) Lipids in the inner membrane of dormant spores of Bacillus species are largely immobile. Proc Natl Acad Sci U S A 101:7733–7738. doi:10.1073/pnas.0306859101

Crawford YJ, Murano EA, Olson DG, Shenoy K (1996) Use of high hydrostatic pressure and irradiation to eliminate *Clostridium sporogenes* spores in chicken breast. J Food Prot 59:711–715

Cutting S, Oke V, Driks A, Losick R, Lu SJ, Kroos L (1990) A forespore checkpoint for mother cell gene-expression during development in *Bacillus subtilis*. Cell 62:239–250

de Heij WBC, van Schepdael LJMM, Moezelaar R, Hoogland H, Matser AM, van den Berg RW (2003) High-pressure sterilization: maximizing the benefits of adiabatic heating. Food Technol-Chicago 57:37–41

de Hoon MJ, Eichenberger P, Vitkup D (2010) Hierarchical evolution of the bacterial sporulation network. Curr Biol 20:735–745. doi:10.1016/j.cub.2010.06.031

Delgado FJ, Contador R, Álvarez-Barrientos A, Cava R, Delgado-Adámez J, Ramírez R (2013) Effect of high pressure thermal processing on some essential nutrients and immunological components present in breast milk. Innovative Food Sci Emerg Techol 19:50–56. doi:10.1016/j.ifset.2013.05.006

Devi AF, Buckow R, Singh T, Hemar Y, Kasapis S (2015) Colour change and proteolysis of skim milk during high pressure thermal–processing. J Food Eng 147:102–110. doi:10.1016/j.jfoodeng.2014.09.017

Diaz AR, Stephenson S, Green JM, Levdikov VM, Wilkinson AJ, Perego M (2008) Functional role for a conserved aspartate in the Spo0E signature motif involved in the dephosphorylation of the Bacillus subtilis sporulation regulator Spo0A. J Biol Chem 283:2962–2972. doi:10.1074/jbc.M709032200

Dixit A, Alam SI, Dhaked RK, Singh L (2005) Sporulation and heat resistance of spores from a *Clostridium* sp RKD. J Food Sci Technol 70:367–373

Doona CJ, Ghosh S, Feeherry FF, Ramirez-Peralta A, Huang Y, Chen H, Setlow P (2014) High pressure germination of *Bacillus subtilis* spores with alterations in levels and types of germination proteins. J Appl Microbiol 117:711–720. doi:10.1111/jam.12557

Durre P, Hollergschwandner C (2004) Initiation of endospore formation in *Clostridium acetobutylicum*. Anaerobe 10:69–74. doi:10.1016/j.anaerobe.2003.11.001

Ehrenberg CG (1838) Die Infusionstierchen als vollkommene Organismen. Verlag Leopold Voss, Leipzig

Eichenberger P, Fujita M, Jensen ST, Conlon EM, Rudner DZ, Wang ST, Ferguson C, Haga K, Sato T, Liu JS, Losick R (2004) The program of gene transcription for a single differentiating cell type during sporulation in *Bacillus subtilis*. PloS Biol 2:1664–1683. doi:10.1371/journal.pbio.0020328

Evans FR, Curran HR (1943) The accelerating effect of sublethal heat on spore germination in mesophilic aerobic bacteria. J Bacteriol 46:513–523

Evans RI, Russell NJ, Gould GW, McClure PJ (1997) The germinability of spores of a psychrotolerant, nonproteolytic strain of *Clostridium botulinum* is influenced by their formation and storage temperature. J Appl Microbiol 83:273–280

Evans RI, McClure PJ, Gould GW, Russell NJ (1998) The effect of growth temperature on the phospholipid and fatty acyl compositions of non-proteolytic *Clostridium botulinum*. Int J Food Microbiol 40:159–167

Faille C, Tauveron G, Gentil-Lelievre CL, Slomianny C (2007) Occurrence of *Bacillus cereus* spores with a damaged exosporium: consequences on the spore adhesion on surfaces of food processing lines. J Food Prot 70:2346–2353

Fernandez PS, Baranyi J, Peck MW (2001) A predictive model of growth from spores of non-proteolytic *Clostridium botulinum* in the presence of different CO_2 concentrations as influenced by chill temperature, pH and NaCl. Food Microbiol 18:453–461. doi:10.1006/fmic.2001.0425

Fisher N, Hanna P (2005) Characterization of Bacillus anthracis germinant receptors in vitro. J Bacteriol 187:8055–8062. doi:10.1128/JB.187.23.8055-8062.2005

Fleming TC, Shin JY, Lee SH, Becker E, Huang KC, Bustamante C, Pogliano K (2010) Dynamic SpoIIIE assembly mediates septal membrane fission during *Bacillus subtilis* sporulation. Genes Dev 24:1160–1172. doi:10.1101/Gad.1925210

Foegeding PM, Busta FF (1981) Bacterial spore injury – an update. J Food Prot 44:776–786

Foegeding PM, Busta FF (1983) Proposed role of lactate in germination of hypochlorite-treated *Clostridium botulinum* spores. Appl Environ Microbiol 45:1369–1373

Foster JW, Wynne ES (1948) Physiological studies on spore germination, with special reference to *Clostridium botulinum*: IV. Inhibition of germination by unsaturated C(18) fatty acids. J Bacteriol 55:495–501

Fraeye I, Knockaert G, Van Buggenhout S, Duvetter T, Hendrickx M, Van Loey A (2010) Enzyme infusion prior to thermal/high pressure processing of strawberries: mechanistic insight into firmness evolution. Innovative Food Sci Emerg Technol 11:23–31. doi:10.1016/j.ifset.2009.06.007

Francis MB, Allen CA, Shrestha R, Sorg JA (2013) Bile acid recognition by the *Clostridium difficile* germinant receptor, CspC, is important for establishing infection. PloS Pathog 9:e1003356. doi:10.1371/journal.ppat.1003356

Fujita M, Losick R (2005) Evidence that entry into sporulation in *Bacillus subtilis* is governed by a gradual increase in the level and activity of the master regulator Spo0A. Genes Dev 19:2236–2244. doi:10.1101/Gad.1335705

Fujita M, Gonzalez-Pastor JE, Losick R (2005) High- and low-threshold genes in the Spo0A regulon of *Bacillus subtilis*. J Bacteriol 187:1357–1368. doi:10.1128/Jb.187.4.1357-1368.2005

Fukushima T, Yamamoto H, Atrih A, Foster SJ, Sekiguchi J (2002) A polysaccharide deacetylase gene (pdaA) is required for germination and for production of muramic delta-lactam residues in the spore cortex of *Bacillus subtilis*. J Bacteriol 184:6007–6015

Furukawa S, Noma S, Shimoda M, Hayakawa I (2002) Effect of initial concentration of bacterial suspensions on their inactivation by high hydrostatic pressure. Int J Food Sci Technol 37:573–577. doi:10.1046/j.1365-2621.2002.00586.x

Galperin MY, Mekhedov SL, Puigbo P, Smirnov S, Wolf YI, Rigden DJ (2012) Genomic determinants of sporulation in Bacilli and Clostridia: towards the minimal set of sporulation-specific genes. Environ Microbiol. doi:10.1111/j.1462-2920.2012.02841.x

Gänzle MG, Vogel RF (2001) On-line fluorescence determination of pressure mediated outer membrane damage in *Escherichia coli*. Syst Appl Microbiol 24:477–485. doi:10.1078/0723-2020-00069

Gao YL, Ju XR (2010) Modelling the effects of food ingredients and pH on high-pressure processing inactivation of Bacillus cereus spores: a laboratorial study. Int J Food Sci Technol 45:1862–1869. doi:10.1111/j.1365-2621.2010.02350.x

Gao YL, Ju XR (2011) Inactivation of *Bacillus coagulans* spores subjected to combinations of high-pressure processing and nisin. Trans ASABE 54:385–392

Gao YL, Qiu WF, Wu D, Fu Q (2011) Assessment of *Clostridium perfringens* spore response to high hydrostatic pressure and heat with nisin. Appl Biochem Biotechnol 164:1083–1095. doi:10.1007/s12010-011-9196-0

Garcia D, van der Voort M, Abee T (2010) Comparative analysis of *Bacillus weihenstephanensis* KBAB4 spores obtained at different temperatures. Int J Food Microb 140:146–153

García-Parra J, González-Cebrino F, Cava R, Ramírez R (2014) Effect of a different high pressure thermal processing compared to a traditional thermal treatment on a red flesh and peel plum purée. Innovative Food Sci Emerg Technol. doi:10.1016/j.ifset.2014.08.002

Gaughran ERL (1947) The saturation of bacterial lipids as a function of temperature. J Bacteriol 53:506

Gayan E, Torres JA, Paredes-Sabja D (2012) Hurdle approach to increase the microbial inactivation by high pressure processing: effect of essential oils. Food Eng Rev 4:141–148

Ghosh S, Setlow P (2009) Isolation and characterization of superdormant spores of *Bacillus* species. J Bacteriol 191:1787–1797. doi:10.1128/Jb.01668-08

Ghosh S, Zhang PF, Li YQ, Setlow P (2009) Superdormant spores of *Bacillus* species have elevated wet-heat resistance and temperature requirements for heat activation. J Bacteriol 191:5584–5591. doi:10.1128/Jb.00736-09

Gibson AM, Ellis-Brownlee RC, Cahill ME, Szabo EA, Fletcher GC, Bremer PJ (2000) The effect of 100% CO_2 on the growth of nonproteolytic *Clostridium botulinum* at chill temperatures. Int J Food Microbiol 54:39–48

Giebel JD, Carr KA, Anderson EC, Hanna PC (2009) The germination-specific lytic enzymes SleB, CwlJ1, and CwlJ2 each contribute to Bacillus anthracis spore germination and virulence. J Bacteriol 191(18):5569–5576. doi:10.1128/JB.00408-09

Girbal L, Soucaille P (1998) Regulation of solvent production in *Clostridium acetobutylicum*. Trends Biotechnol 16:11–16

Gola S, Foman C, Carpi G, Maggi A, Cassara A, Rovere P (1996) Inactivation of bacterial spores in phosphate buffer and in vegetable cream treated with high pressures. Prog Biotechnol 13:253–259

Gonzalez-Pastor JE, Hobbs EC, Losick R (2003) Cannibalism by sporulating bacteria. Science 301:510–513. doi:10.1126/science.1086462

Gould GW (2006) History of science-spores. J Appl Microbiol 101:507–513. doi:10.1111/j.1365-2672.2006.02888.x

Gould GW, Sale AJ (1970) Initiation of germination of bacterial spores by hydrostatic pressure. J Gen Microbiol 60:335–346

Grenha R, Rzechorzek NJ, Brannigan JA, de Jong RN, Ab E, Diercks T, Truffault V, Ladds JC, Fogg MJ, Bongiorni C, Perego M, Kaptein R, Wilson KS, Folkers GE, Wilkinson AJ (2006) Structural characterization of Spo0E-like protein-aspartic acid phosphatases that regulate sporulation in bacilli. J Biol Chem 281:37993–38003. doi:10.1074/jbc.M607617200

Griffiths KK, Zhang J, Cowan AE, Yu J, Setlow P (2011) Germination proteins in the inner membrane of dormant *Bacillus subtilis* spores colocalize in a discrete cluster. Mol Microbiol 81:1061–1077. doi:10.1111/j.1365-2958.2011.07753.x

Gutelius D, Hokeness K, Logan SM, Reid CW (2014) Functional analysis of SleC from *Clostridium difficile*: an essential lytic transglycosylase involved in spore germination. Microbiology 160:209–216. doi:10.1099/mic.0.072454-0

Haraldsen JD, Sonenshein AL (2003) Efficient sporulation in *Clostridium difficile* requires disruption of the sigma(K) gene. Mol Microbiol 48(3):811–821

Harris LM, Welker NE, Papoutsakis ET (2002) Northern, morphological, and fermentation analysis of spo0A inactivation and overexpression in *Clostridium acetobutylicum* ATCC 824. J Bacteriol 184:3586–3597. doi:10.1128/Jb.184.13.3586-3597.2002

Hawirko RZ, Naccarato CA, Lee RP, Maeba PY (1979) Outgrowth and sporulation studies on *Clostridium botulinum* type E: influence of isoleucine. Can J Microbiol 25:522–527

Hayakawa I, Kanno T, Tomita M, Fujio Y (1994) Application of high-pressure for spore inactivation and protein denaturation. J Food Sci 59:159–163. doi:10.1111/j.1365-2621.1994.tb06923.x

Heinz V, Buckow R (2010) Food preservation by high pressure. J Verbr Lebensm 5:73–81. doi:10.1007/s00003-009-0311-x

Heinz V, Knorr D (1996) High pressure inactivation kinetics of *Bacillus subtilis* cells by a three-state-model considering distributed resistance mechanisms. Food Biotechnol 10:149–161

Hilbert DW, Piggot PJ (2004) Compartmentalization of gene expression during *Bacillus subtilis* spore formation. Microbiol Mol Biol Rev 68:234–262. doi:10.1128/Mmbr.68.2.234-262.2004

Hite BH (1899) The effect of pressure in the preservation of milk. A preliminary report, vol 58, Bulletin West Virginia University Agriculture Experiment Station. West Virginia Agricultural Experiment Station, Morgantown, pp 15–35

Hoch JA (1993) Regulation of the phosphorelay and the initiation of sporulation in *Bacillus subtilis*. Annu Rev Microbiol 47:441–465

Hornstra LM, de Vries YP, de Vos WM, Abee T (2006a) Influence of sporulation medium composition on transcription of ger operons and the germination response of spores of *Bacillus cereus* ATCC 14579. Appl Environ Microbiol 72:3746–3749. doi:10.1128/Aem.72.5.3746-3749.2006

Hornstra LM, de Vries YP, Wells-Bennik MHJ, de Vos WM, Abee T (2006b) Characterization of germination receptors of *Bacillus cereus* ATCC 14579. Appl Environ Microbiol 72:44–53. doi:10.1128/Aem.72.1.44-53.2006

Huang IH, Waters M, Grau RR, Sarker MR (2004) Disruption of the gene (spoOA) encoding sporulation transcription factor blocks endospore formation and enterotoxin production in enterotoxigenic *Clostridium perfringens* type A. FEMS Microbiol Lett 233:233–240. doi:10.1016/j.femsle.2004.02.014

Husemann MHW, Papoutsakis ET (1988) Solventogenesis in *Clostridium acetobutylicum* fermentations related to carboxylic-acid and proton concentrations. Biotechnol Bioeng 32:843–852

Igura N, Kamimura Y, Islam MS, Shimoda M, Hayakawa I (2003) Effects of minerals on resistance of *Bacillus subtilis* spores to heat and hydrostatic pressure. Appl Environ Microb 69:6307–6310

Irigul-Sonmez O, Koroglu TE, Ozturk B, Kovacs AT, Kuipers OP, Yazgan-Karatas A (2014) In *Bacillus subtilis* LutR is part of the global complex regulatory network governing the adaptation to the transition from exponential growth to stationary phase. Microbiology 160:243–260. doi:10.1099/mic.0.064675-0

Ishimori T, Takahashi K, Goto M, Nakagawa S, Kasai Y, Konagaya Y, Batori H, Kobayashi A, Urakami H (2012) Synergistic effects of high hydrostatic pressure, mild heating, and amino acids on germination and inactivation of *Clostridium sporogenes* spores. Appl Environ Microbiol 78:8202–8207. doi:10.1128/Aem.02007-12

Jiang M, Shao W, Perego M, Hoch JA (2000) Multiple histidine kinases regulate entry into stationary phase and sporulation in *Bacillus subtilis*. Mol Microbiol 38:535–542

Johnson FH, Zobell CE (1949) The retardation of thermal disinfection of *Bacillus subtilis* spores by hydrostatic pressure. J Bacteriol 57:353–358

Jones DT, Woods DR (1986) Acetone-butanol fermentation revisited. Microbiol Rev 50:484–524

Jones CA, Padula NL, Setlow P (2005) Effect of mechanical abrasion on the viability, disruption and germination of spores of *Bacillus subtilis*. J Appl Microbiol 99:1484–1494. doi:10.1111/j.1365-2672.2005.02744.x

Jones SW, Paredes CJ, Tracy B, Cheng N, Sillers R, Senger RS, Papoutsakis ET (2008) The transcriptional program underlying the physiology of clostridial sporulation. Genome Biol 9:R114. doi:10.1186/gb-2008-9-7-r114

Jones SW, Tracy BP, Gaida SM, Papoutsakis ET (2011) Inactivation of sigmaF in *Clostridium acetobutylicum* ATCC 824 blocks sporulation prior to asymmetric division and abolishes sigmaE and sigmaG protein expression but does not block solvent formation. J Bacteriol 193:2429–2440. doi:10.1128/JB.00088-11

Juliano P, Knoerzer K, Fryer PJ, Versteeg C (2009) *C. botulinum* inactivation kinetics implemented in a computational model of a high-pressure sterilization process. Biotechnol Prog 25:163–175. doi:10.1002/btpr.116

Juneja VK, Eblen BS, Marmer BS, Williams AC, Palumbo SA, Miller AJ (1995) Thermal-resistance of nonproteolytic type B and type E *Clostridium botulinum* spores in phosphate buffer and turkey slurry. J Food Prot 58:758–763

Kalchayanand N, Dunne CP, Sikes A, Ray B (2004) Germination induction and inactivation of *Clostridium* spores at medium-range hydrostatic pressure treatment. Innovative Food Sci Emerg Technol 5:277–283. doi:10.1016/j.ifset.2004.02.004

Kebede BT, Grauwet T, Tabilo-Munizaga G, Palmers S, Vervoort L, Hendrickx M, Van Loey A (2013) Headspace components that discriminate between thermal and high pressure high temperature treated green vegetables: identification and linkage to possible process-induced chemical changes. Food Chem 141:1603–1613. doi:10.1016/j.foodchem.2013.05.097

Knoerzer K, Chapman B (2011) Effect of material properties and processing conditions on the prediction accuracy of a CFD model for simulating high pressure thermal (HPT) processing. J Food Eng 104:404–413. doi:10.1016/j.jfoodeng.2011.01.003

Knorr D, Reineke K, Mathys A, Heinz V, Buckow R (2011) High-pressure-induced effects on bacterial spores, vegetative microorganisms, and enzymes. Food engineering interfaces. In: 10th International congress on engineering and food, Vina del Mar, Apr 2008, pp 325–340. doi: 10.1007/978-1-4419-7475-4_14

Koch R (1876) Untersuchungen über Bakterien. V. Die Aetiologie der Milzbrand-Krankheit, begründet auf die Entwicklungsgeschichte des *Bacillus Anthracis*. In: Cohn F (ed) Beiträge zur Biologie der Pflanzen Band 2, Heft 2. J.U. Kern's Verlag, Breslau, pp 277–310

Kong LB, Doona CJ, Setlow P, Li YQ (2014) Monitoring rates and heterogeneity of high-pressure germination of *Bacillus* spores by phase-contrast microscopy of individual spores. Appl Environ Microbiol 80:345–353. doi:10.1128/Aem.03043-13

Koutchma T, Song Y, Setikaite I, Juliano P, Barbosa-Canovas GV, Dunne CP, Patazca E (2010) Packaging evaluation for high-pressure high-temperature sterilization of shelf-stable foods. J Food Process Eng 33:1097–1114. doi:10.1111/j.1745-4530.2008.00328.x

Krebbers B, Matser AM, Hoogerwerf SW, Moezelaar R, Tomassen MMM, van den Berg RW (2003) Combined high-pressure and thermal treatments for processing of tomato puree: evaluation of microbial inactivation and quality parameters. Innovative Food Sci Emerg Technol 4:377–385. doi:10.1016/S1466-8564(03)00045-6

Labell TL, Trempy JE, Haldenwang WG (1987) Sporulation-specific sigma-factor sigma-29 of *Bacillus subtilis* is synthesized from a precursor protein, P-31. Proc Natl Acad Sci USA 84:1784–1788

Larson WP, Hartzell TB, Diehl HS (1918) The effect of high pressures on bacteria. J Infect Dis 22:271–279

Lazazzera BA, Kurtser IG, McQuade RS, Grossman AD (1999) An autoregulatory circuit affecting peptide signaling in *Bacillus subtilis*. J Bacteriol 181:5193–5200

Lenz CA, Vogel RF (2014) Effect of sporulation medium and its divalent cation content on the heat and high pressure resistance of *Clostridium botulinum* type E spores. Food Microbiol 44:156–167. doi:10.1016/j.fm.2014.05.010

Lenz CA, Vogel RF (2015) Differential effects of sporulation temperature on the high pressure resistance of *Clostridium botulinum* type E spores and the interconnection with sporulation medium cation contents. Food Microb 46:434–442

Lenz CA, Schnabel J, Vogel RF (2014) The variability of times to detect growth from individual *Clostridium botulinum* type E endospores is differentially affected by high pressure treatments. High Pressure Res 34:1–7. doi:10.1080/08957959.2014.975699

Leuschner RG, Lillford PJ (1999) Effects of temperature and heat activation on germination of individual spores of *Bacillus subtilis*. Lett Appl Microbiol 29:228–232

Li JH, McClane BA (2010) Evaluating the involvement of alternative sigma factors SigF and SigG in *Clostridium perfringens* sporulation and enterotoxin synthesis. Infect Immun 78:4286–4293. doi:10.1128/IAI.00528-10

Li JH, Chen J, Vidal JE, McClane BA (2011) The Agr-like quorum-sensing system regulates sporulation and production of enterotoxin and Beta2 toxin by *Clostridium perfringens* type A non-food-borne human gastrointestinal disease strain F5603. Infect Immun 79:2451–2459. doi:10.1128/Iai.00169-11

Li YF, Davis A, Korza G, Zhang PF, Li YQ, Setlow B, Setlow P, Hao B (2012) Role of a SpoVA protein in dipicolinic acid uptake into developing spores of Bacillus subtilis. J Bacteriol 194:1875–1884. doi:10.1128/Jb.00062-12

Liang L, He X, Liu G, Tan H (2008) The role of a purine-specific nucleoside hydrolase in spore germination of Bacillus thuringiensis. Microbiology 154:1333–1340. doi:10.1099/mic.0.2007/014399-0

Lindsay JA, Barton LE, Leinart AS, Pankratz HS (1990) The effect of sporulation temperature on sporal characteristics of *Bacillus subtilis* A. Curr Microbiol 21:75–79

Long S, Jones DT, Woods DR (1984) Initiation of solvent production, clostridial stage and endospore formation in *Clostridium acetobutylicum* P262. Appl Microbiol Biotechnol 20:256–261

Lopez D, Kolter R (2010) Extracellular signals that define distinct and coexisting cell fates in *Bacillus subtilis*. FEMS Microbiol Rev 34:134–149. doi:10.1111/j.1574-6976.2009.00199.x

Lopez CS, Garda HA, Rivas EA (2002) The effect of osmotic stress on the biophysical behavior of the *Bacillus subtilis* membrane studied by dynamic and steady-state fluorescence anisotropy. Arch Biochem Biophys 408:220–228

Luu H, Akoachere M, Patra M, Abel-Santos E (2011) Cooperativity and interference of germination pathways in Bacillus anthracis spores. J Bacteriol 193:4192–4198. doi:10.1128/JB.05126-11

Makino S, Moriyama R (2002) Hydrolysis of cortex peptidoglycan during bacterial spore germination. Med Sci Monit 8:119–127

Mandicmulec I, Doukhan L, Smith I (1995) The *Bacillus subtilis* Sinr protein is a repressor of the key sporulation gene-Spo0a. J Bacteriol 177:4619–4627

Margosch D (2004) Behaviour of bacterial endospores and toxins as safety determinants in low acid pressurized food. Dissertation, Technische Universitaet Muenchen

Margosch D, Ehrmann MA, Gänzle MG, Vogel RF (2004a) Comparison of pressure and heat resistance of *Clostridium botulinum* and other endospores in mashed carrots. J Food Prot 67:2530–2537

Margosch D, Ganzle MG, Ehrmann MA, Vogel RF (2004b) Pressure inactivation of Bacillus endospores. Appl Environ Microbiol 70:7321–7328. doi:10.1128/Aem.70.12.7321-7328.2004

Margosch D, Ehrmann MA, Buckow R, Heinz V, Vogel RF, Ganzle MG (2006) High-pressure-mediated survival of *Clostridium botulinum* and Bacillus amyloliquefaciens endospores at high temperature. Appl Environ Microbiol 72:3476–3481. doi:10.1128/AEM.72.5.3476-3481.2006

Masayama A, Fukuoka H, Kato S, Yoshimura T, Moriyama M, Moriyama R (2006) Subcellular localization of a germination-specific cortex-lytic enzyme, SleB, of *Bacilli* during sporulation. Genes Genet Syst 81:163–169

Mathys A, Chapman B, Bull M, Heinz V, Knorr D (2007a) Flow cytometric assessment of *Bacillus* spore response to high pressure and heat. Innovative Food Sci Emerg Technol 8:519–527. doi:10.1016/j.ifset.2007.06.010

Mathys A, Heinz V, Schwartz FH, Knorr D (2007b) Impact of agglomeration on the quantitative assessment of *Bacillus stearothermophilus* heat inactivation. J Food Eng 81:380–387. doi:10.1016/j.jfoodeng.2006.11.012

Mathys A, Kallmeyer R, Heinz V, Knorr D (2008) Impact of dissociation equilibrium shift on bacterial spore inactivation by heat and pressure. Food Control 19:1165–1173

Mathys A, Reineke K, Heinz V, Knorr D (2009) High pressure thermal sterilization – development and application of temperature controlled spore inactivation studies. High Pressure Res 29:3–7. doi:10.1080/08957950802567240

Meisner J, Wang X, Serrano M, Henriques AO, Moran CP (2008) A channel connecting the mother cell and forespore during bacterial endospore formation. Proc Natl Acad Sci USA 105:15100–15105. doi:10.1073/pnas.0806301105

Melly E, Genest PC, Gilmore ME, Little S, Popham DL, Driks A, Setlow P (2002) Analysis of the properties of spores of *Bacillus subtilis* prepared at different temperatures. J Appl Microbiol 92:1105–1115

Mendez MB, Orsaria LM, Philippe V, Pedrido ME, Grau RR (2004) Novel roles of the master transcription factors Spo0A and sigma(B) for survival and sporulation of *Bacillus subtilis* at low growth temperature. J Bacteriol 186:989–1000. doi:10.1128/Jb.186.4.989-1000.2004

Meyer CL, Papoutsakis ET (1989) Increased levels of ATP and NADH are associated with increased solvent production in continuous cultures of *Clostridium acetobutylicum*. Appl Microbiol Biotechnol 30:450–459

Mills G, Earnshaw R, Patterson MF (1998) Effects of high hydrostatic pressure on *Clostridium sporogenes* spores. Lett Appl Microbiol 26:227–230

Minerich PL, Labuza TP (2003) Development of a pressure indicator for high hydrostatic pressure processing of foods. Innovative Food Sci Emerg Technol 4:235–243

Moir A (2006) How do spores germinate? J Appl Microbiol 101:526–530. doi:10.1111/j.1365-2672.2006.02885.x

Moir A, Smith DA (1990) The genetics of bacterial spore germination. Annu Rev Microbiol 44:531–553. doi:10.1146/annurev.mi.44.100190.002531

Molle V, Nakaura Y, Shivers RP, Yamaguchi H, Losick R, Fujita Y, Sonenshein AL (2003) Additional targets of the *Bacillus subtilis* global regulator CodY identified by chromatin immunoprecipitation and genome-wide transcript analysis. J Bacteriol 185:1911–1922

Monroe A, Setlow P (2006) Localization of the transglutaminase cross-linking sites in the *Bacillus subtilis* spore coat protein GerQ. J Bacteriol 188:7609–7616. doi:10.1128/JB.01116-06

Movahedi S, Waites W (2000) A two-dimensional protein gel electrophoresis study of the heat stress response of *Bacillus subtilis* cells during sporulation. J Bacteriol 182:4758–4763

Myers GSA, Rasko DA, Cheung JK, Ravel J, Seshadri R, DeBoy RT, Ren QH, Varga J, Awad MM, Brinkac LM, Daugherty SC, Haft DH, Dodson RJ, Madupu R, Nelson WC, Rosovitz MJ, Sullivan SA, Khouri H, Dimitrov GI, Watkins KL, Mulligan S, Benton J, Radune D, Fisher DJ, Atkins HS, Hiscox T, Jost BH, Billington SJ, Songer JG, McClane BA, Titball RW, Rood JI, Melville SB, Paulsen IT (2006) Skewed genomic variability in strains of the toxigenic bacterial pathogen, *Clostridium perfringens*. Genome Res 16:1031–1040. doi:10.1101/Gr.5238106

Nguyen Thi Minh H, Durand A, Loison P, Perrier-Cornet JM, Gervais P (2011) Effect of sporulation conditions on the resistance of *Bacillus subtilis* spores to heat and high pressure. Appl Microb Biotechnol 90:1409–1417

Nolling J, Breton G, Omelchenko MV, Makarova KS, Zeng QD, Gibson R, Lee HM, Dubois J, Qiu DY, Hitti J, Wolf YI, Tatusov RL, Sabathe F, Doucette-Stamm L, Soucaille P, Daly MJ, Bennett GN, Koonin EV, Smith DR, Finishing GSCP (2001) Genome sequence and comparative analysis of the solvent-producing bacterium *Clostridium acetobutylicum*. J Bacteriol 183:4823–4838

Oh S, Moon MJ (2003) Inactivation of *Bacillus cereus* spores by high hydrostatic pressure at different temperatures. J Food Prot 66:599–603

Olivier SA, Bull MK, Stone G, van Diepenbeek RJ, Kormelink F, Jacops L, Chapman B (2011) Strong and consistently synergistic inactivation of spores of spoilage-associated Bacillus and Geobacillus spp. by high pressure and heat compared with inactivation by heat alone. Appl Environ Microbiol 77:2317–2324. doi:10.1128/Aem.01957-10

Olivier SA, Bull MK, Chapman B (2012) *Bacillus* spp. spores produced at lower temperatures are more resistant to high pressure thermal processes but mineralization does not predict relative resistance. Innovative Food Sci Emerg Technol 16:96–101

Oomes SJ, van Zuijlen AC, Hehenkamp JO, Witsenboer H, van der Vossen JM, Brul S (2007) The characterisation of *Bacillus* spores occurring in the manufacturing of (low acid) canned products. Int J Food Microbiol 120:85–94. doi:10.1016/j.ijfoodmicro.2007.06.013

Oomes SJ, Jonker MJ, Wittink FR, Hehenkamp JO, Breit TM, Brul S (2009) The effect of calcium on the transcriptome of sporulating *B. subtilis* cells. Int J Food Microbiol 133:234–242. doi:10.1016/j.ijfoodmicro.2009.05.019

Paidhungat M, Setlow P (1999) Isolation and characterization of mutations in *Bacillus subtilis* that allow spore germination in the novel germinant D-alanine. J Bacteriol 181:3341–3350

Paidhungat M, Setlow P (2000) Role of Ger proteins in nutrient and nonnutrient triggering of spore germination in *Bacillus subtilis*. J Bacteriol 182:2513–2519

Paidhungat M, Setlow P (2001) Localization of a germinant receptor protein (GerBA) to the inner membrane of *Bacillus subtilis* spores. J Bacteriol 183:3982–3990

Paidhungat M, Setlow B, Daniels WB, Hoover D, Papafragkou E, Setlow P (2002) Mechanisms of induction of germination of *Bacillus subtilis* spores by high pressure. Appl Environ Microbiol 68:3172–3175. doi:10.1128/Aem.68.6.3172-3175.2002

Paredes CJ, Rigoutsos I, Papoutsakis ET (2004) Transcriptional organization of the *Clostridium acetobutylicum* genome. Nucleic Acids Res 32:1973–1981. doi:10.1093/Nar/Gkh509

Paredes CJ, Alsaker KV, Papoutsakis ET (2005) A comparative genomic view of clostridial sporulation and physiology. Nat Rev Microbiol 3:969–978. doi:10.1038/nrmicro1288

Paredes-Sabja D, Sarker N, Setlow B, Setlow P, Sarker MR (2008a) Roles of DacB and spm proteins in clostridium perfringens spore resistance to moist heat, chemicals, and UV radiation. Appl Environ Microbiol 74:3730–3738. doi:10.1128/AEM.00169-08, AEM.00169-08 [pii]

Paredes-Sabja D, Setlow B, Setlow P, Sarker MR (2008b) Characterization of *Clostridium perfringens* spores that lack spoVA proteins and dipicolinic acid. J Bacteriol 190:4648–4659. doi:10.1128/Jb.00325-08

Paredes-Sabja D, Torres JA, Setlow P, Sarker MR (2008c) *Clostridium perfringens* spore germination: characterization of germinants and their receptors. J Bacteriol 190:1190–1201. doi:10.1128/JB.01748-07

Paredes-Sabja D, Setlow P, Sarker MR (2009a) GerO, a putative $Na^{+}/H^{+}-K^{+}$ antiporter, is essential for normal germination of spores of the pathogenic bacterium *Clostridium perfringens*. J Bacteriol 191:3822–3831. doi:10.1128/Jb.00158-09
Paredes-Sabja D, Setlow P, Sarker MR (2009b) SleC is essential for cortex peptidoglycan hydrolysis during germination of spores of the pathogenic bacterium *Clostridium perfringens*. J Bacteriol 191:2711–2720. doi:10.1128/Jb.01832-08
Paredes-Sabja D, Udompijitkul P, Sarker MR (2009c) Inorganic phosphate and sodium ions are cogerminants for spores of *Clostridium perfringens* type A food poisoning-related isolates. Appl Environ Microbiol 75:6299–6305. doi:10.1128/AEM.00822-09
Paredes-Sabja D, Setlow P, Sarker MR (2011) Germination of spores of *Bacillales* and *Clostridiales* species: mechanisms and proteins involved. Trends Microbiol 19:85–94. doi:10.1016/j.tim.2010.10.004
Park SH, Balasubramaniam VM, Sastry SK (2014) Quality of shelf-stable low-acid vegetables processed using pressure-ohmic-thermal sterilization. Lwt-Food Sci Technol 57:243–252. doi:10.1016/j.lwt.2013.12.036
Pasteur L (1866) Etudes sur le vin. Des Seances del L'Academie Des Sciences 63:509–510
Patazca E, Morrissey TR, Loeza V, Reddy NR, Skinner GE, Larkin JW (2013) Effect of packaging systems and pressure fluids on inactivation of *Clostridium botulinum* spores by combined high pressure and thermal processing. J Food Prot 76:448–455. doi:10.4315/0362-028X.JFP-12-181
Peck MW (2009) Biology and genomic analysis of Clostridium botulinum. Adv Microb Physiol 55(183–265):320. doi:10.1016/S0065-2911(09)05503-9
Peck MW, Fernandez PS (1995) Effect of lysozyme concentration, heating at 90 degrees C, and then incubation at chilled temperatures on growth from spores of non-proteolytic Clostridium botulinum. Lett Appl Microbiol 21:50–54
Peck MW, Fairbairn DA, Lund BM (1992) The effect of recovery medium on the estimated heat-inactivation of spores of nonproteolytic *Clostridium botulinum*. Lett Appl Microbiol 15:146–151
Peck MW, Evans RI, Fairbairn DA, Hartley MG, Russell NJ (1995) Effect of sporulation temperature on some properties of spores of non-proteolytic *Clostridium botulinum*. Int J Food Microbiol 28:289–297
Pelczar PL, Setlow P (2008) Localization of the germination protein GerD to the inner membrane in *Bacillus subtilis* spores. J Bacteriol 190:5635–5641. doi:10.1128/JB.00670-08
Pelczar PL, Igarashi T, Setlow B, Setlow P (2007) Role of GerD in germination of *Bacillus subtilis* spores. J Bacteriol 189:1090–1098. doi:10.1128/JB.01606-06
Perego M (2001) A new family of aspartyl phosphate phosphatases targeting the sporulation transcription factor Spo0A of *Bacillus subtilis*. Mol Microbiol 42:133–143
Phillips ZE, Strauch MA (2002) *Bacillus subtilis* sporulation and stationary phase gene expression. Cell Mol Life Sci 59:392–402
Piggot PJ, Hilbert DW (2004) Sporulation of *Bacillus subtilis*. Curr Opin Microbiol 7:579–586. doi:10.1016/j.mib.2004.10.001
Planchon S, Dargaignaratz C, Levy C, Ginies C, Broussolle V, Carlin F (2011) Spores of *Bacillus cereus* strain KBAB4 produced at 10 degrees C and 30 degrees C display variations in their properties. Food Microbiol 28:291–297. doi:10.1016/J.Fm.2010.07.015
Plowman J, Peck MW (2002) Use of a novel method to characterize the response of spores of non-proteolytic *Clostridium botulinum* types B, E and F to a wide range of germinants and conditions. J Appl Microbiol 92:681–694
Popham DL, Sengupta S, Setlow P (1995) Heat, hydrogen peroxide, and UV resistance of *Bacillus subtilis* spores with increased core water content and with or without major DNA-binding proteins. Appl Environ Microbiol 61:3633–3638
Rajan S, Ahn J, Balasubramaniam VM, Yousef AE (2006) Combined pressure-thermal inactivation kinetics of *Bacillus amyloliquefaciens* spores in egg patty mince. J Food Prot 69:853–860
Ramirez N, Liggins M, Abel-Santos E (2010) Kinetic evidence for the presence of putative germination receptors in *Clostridium difficile* spores. J Bacteriol 192:4215–4222. doi:10.1128/JB.00488-10

Ramirez-Peralta A, Stewart KAV, Thomas SK, Setlow B, Chen Z, Li YQ, Setlow P (2012) Effects of the SpoVT regulatory protein on the germination and germination protein levels of spores of Bacillus subtilis. J Bacteriol 194:3417–3425

Ramirez-Peralta A, Gupta S, Butzin XY, Setlow B, Korza G, Leyva-Vazquez MA, Christie G, Setlow P (2013) Identification of new proteins that modulate the germination of spores of *bacillus* species. J Bacteriol 195:3009–3021. doi:10.1128/JB.00257-13

Rasanayagam V, Balasubramaniam VM, Ting E, Sizer CE, Bush C, Anderson C (2003) Compression heating of selected fatty food materials during high-pressure processing. J Food Sci 68:254–259. doi:10.1111/j.1365-2621.2003.tb14148.x

Raso J, Barbosa-Canovas G, Swanson BG (1998a) Sporulation temperature affects initiation of germination and inactivation by high hydrostatic pressure of *Bacillus cereus*. J Appl Microb 85:17–24

Raso J, Gongora-Nieto MM, Barbosa-Canovas GV, Swanson BG (1998b) Influence of several environmental factors on the initiation of germination and inactivation of *Bacillus cereus* by high hydrostatic pressure. Int J Food Microbiol 44:125–132

Ratnayake-Lecamwasam M, Serror P, Wong KW, Sonenshein AL (2001) *Bacillus subtilis* CodY represses early-stationary-phase genes by sensing GTP levels. Genes Dev 15:1093–1103. doi:10.1101/gad.874201

Ratphitagsanti W, Ahn J, Balasubramaniam VM, Yousef AE (2009) Influence of pressurization rate and pressure pulsing on the inactivation of *Bacillus amyloliquefaciens* spores during pressure-assisted thermal processing. J Food Prot 72:775–782

Ratphitagsanti W, De Lamo-Castellvi S, Balasubramaniam VM, Yousef AE (2010) Efficacy of pressure-assisted thermal processing, in combination with organic acids, against *Bacillus amyloliquefaciens* spores suspended in deionized water and carrot puree. J Food Sci 75:M46–M52

Ravagnani A, Jennert KCB, Steiner E, Grunberg R, Jefferies JR, Wilkinson SR, Young DI, Tidswell EC, Brown DP, Youngman P, Morris JG, Young M (2000) Spo0A directly controls the switch from acid to solvent production in solvent-forming *clostridia*. Mol Microbiol 37:1172–1185

Reddy NR, Tetzloff RC, Solomon HM, Larkin JW (2006) Inactivation of *Clostridium botulinum* nonproteolytic type B spores by high pressure processing at moderate to elevated high temperatures. Innovative Food Sci Emerg Technol 7:169–175. doi:10.1016/j.ifset.2006.03.002

Reddy NR, Tetzloff RC, Skinner GE (2010) Effect of media, additives, and incubation conditions on the recovery of high pressure and heat-injured *Clostridium botulinum* spores. Food Microbiol 27:613–617. doi:10.1016/J.Fm.2010.02.004

Redmond C, Baillie LW, Hibbs S, Moir AJ, Moir A (2004) Identification of proteins in the exosporium of *Bacillus anthracis*. Microbiology 150:355–363

Reineke K, Doehner I, Mathys A, Knorr D (2010) Pressure induced germination of *Bacillus subtilis* spores – a mechanistic study. In: 6th international conference on high pressure bioscience and biotechnology (HPBB2010), Freising-Weihenstephan, p 43 (Lecture L410)

Reineke K, Mathys A, Knorr D (2011) The impact of high pressure and temperature on bacterial spores: inactivation mechanisms of *Bacillus subtilis* above 500 MPa. J Food Sci 76:189–197. doi:10.1111/j.1750-3841.2011.02066.x

Reineke K, Doehner I, Schlumbach K, Baier D, Mathys A, Knorr D (2012) The different pathways of spore germination and inactivation in dependence of pressure and temperature. Innovative Food Sci Emerg Technol 13:31–41. doi:10.1016/j.ifset.2011.09.006

Reineke K, Mathys A, Heinz V, Knorr D (2013a) Mechanisms of endospore inactivation under high pressure. Trends Microbiol 21:296–304. doi:10.1016/j.tim.2013.03.001

Reineke K, Schlumbach K, Baier D, Mathys A, Knorr D (2013b) The release of dipicolinic acid — the rate-limiting step of *Bacillus* endospore inactivation during the high pressure thermal sterilization process. Int J Food Microbiol 162:55–63

Roberts CM, Hoover DG (1996) Sensitivity of *Bacillus coagulans* spores to combinations of high hydrostatic pressure, heat, acidity and nisin. J Appl Bacteriol 81:363–368

Rode LJ, Foster JW (1960) The action of surfactants on bacterial spores. Arch Mikrobiol 36:67–94

Rode LJ, Foster JW (1966) Quantitative aspects of exchangeable calcium in spores of *Bacillus megaterium*. J Bacteriol 91:1589–1593

Rowland SL, Burkholder WF, Cunningham KA, Maciejewski MW, Grossman AD, King GF (2004) Structure and mechanism of action of Sda, an inhibitor of the histidine kinases that regulate initiation of sporulation in *Bacillus subtilis*. Mol Cell 13:689–701

Santangelo JD, Kuhn A, Treuner-Lange A, Durre P (1998) Sporulation and time course expression of sigma-factor homologous genes in *Clostridium acetobutylicum*. FEMS Microbiol Lett 161:157–164

Sarker MR, Huang IH (2006) Complementation of a *Clostridium perfringens* spo0A mutant with wild-type spo0A from other *Clostridium* species. Appl Environ Microbiol 72:6388–6393. doi:10.1128/Aem.02218-05

Sarker MR, Akhtar S, Torres JA, Paredes-Sabja D (2013) High hydrostatic pressure-induced inactivation of bacterial spores. Crit Rev Microbiol. doi:10.3109/1040841X.2013.788475

Sauer U, Treuner A, Buchholz M, Santangelo JD, Durre P (1994) Sporulation and primary sigma-factor homologous genes in *Clostridium acetobutylicum*. J Bacteriol 176:6572–6582

Saujet L, Monot M, Dupuy B, Soutourina O, Martin-Verstraete I (2011) The key sigma factor of transition phase, SigH, controls sporulation, metabolism, and virulence factor expression in *Clostridium difficile*. J Bacteriol 193:3186–3196. doi:10.1128/JB.00272-11

Scallan E, Hoekstra RM, Angulo FJ, Tauxe RV, Widdowson MA, Roy SL, Jones JL, Griffin PM (2011) Foodborne illness acquired in the United States-major pathogens. Emerg Infect Dis 17:7–15. doi:10.3201/eid1701.091101p1

Schmidt R, Margolis P, Duncan L, Coppolecchia R, Moran CP, Losick R (1990) Control of developmental transcription factor-rho-F by sporulation regulatory proteins *spoIIAA* and *spoIIAB* in *Bacillus subtilis*. Proc Natl Acad Sci USA 87:9221–9225

Schultz D, Wolynes PG, Ben Jacob E, Onuchic JN (2009) Deciding fate in adverse times: sporulation and competence in *Bacillus subtilis*. Proc Natl Acad Sci USA 106:21027–21034. doi:10.1073/pnas.0912185106

Scotcher MC, Bennett GN (2005) SpoIIE regulates sporulation but does not directly affect solventogenesis in *Clostridium acetobutylicum* ATCC 824. J Bacteriol 187:1930–1936. doi:10.1128/Jb.187.6.1930-1936.2005

Scotcher MC, Rudolph FB, Bennett GN (2005) Expression of abrB310 and sinR, and effects of decreased abrB310 expression on the transition from acidogenesis to solventogenesis, in *Clostridium acetobutylicum* ATCC 824. Appl Environ Microbiol 71:1987–1995. doi:10.1128/Aem.71.4.1987-1995.2005

Scurrah KJ, Robertson RE, Craven HM, Pearce LE, Szabo EA (2006) Inactivation of *Bacillus* spores in reconstituted skim milk by combined high pressure and heat treatment. J Appl Microbiol 101:172–180. doi:10.1111/j.1365-2672.2006.02897.x

Sebaihia M, Peck MW, Minton NP, Thomson NR, Holden MTG, Mitchell WJ, Carter AT, Bentley SD, Mason DR, Crossman L, Paul CJ, Ivens A, Wells-Bennik MHJ, Davis IJ, Cerdeno-Tarraga AM, Churcher C, Quail MA, Chillingworth T, Feltwell T, Fraser A, Goodhead I, Hance Z, Jagels K, Larke N, Maddison M, Moule S, Mungall K, Norbertczak H, Rabbinowitsch E, Sanders M, Simmonds M, White B, Whithead S, Parkhill J (2007) Genome sequence of a proteolytic (Group I) *Clostridium botulinum* strain Hall A and comparative analysis of the clostridial genomes. Genome Res 17:1082–1092. doi:10.1101/gr.6282807

Senior A, Moir A (2008) The *Bacillus cereus* GerN and GerT protein homologs have distinct roles in spore germination and outgrowth, respectively. J Bacteriol 190:6148–6152. doi:10.1128/Jb.00789-08

Setlow P (1994) Mechanisms which contribute to the long-term survival of spores of *Bacillus* species. Soc Appl Bacteriol Symp Ser 23:49–60

Setlow P (1995) Mechanisms for the prevention of damage to DNA in spores of *Bacillus* species. Annu Rev Microbiol 49:29–54. doi:10.1146/annurev.mi.49.100195.000333

Setlow P (2003) Spore germination. Curr Opin Microbiol 6:550–556. doi:10.1016/j.mib.2003.10.001

Setlow B, Cowan AE, Setlow P (2003) Germination of spores of Bacillus subtilis with dodecylamine. J Appl Microb 95(3):637–648. doi:10.1046/j.1365-2672.2003.02015

Setlow P (2006) Spores of *Bacillus subtilis*: their resistance to and killing by radiation, heat and chemicals. J Appl Microbiol 101:514–525. doi:10.1111/j.1365-2672.2005.02736.x

Setlow P, Johnson EA (2013) Spores and their significance. In: Doyle MP, Buchanan RL (eds) Food microbiology: fundamentals and frontiers, 4th edn. American Society of Microbiology Press, Washington, DC, pp 45–79

Setlow B, Setlow P (1977) Levels of oxidized and reduced pyridine nucleotides in dormant spores and during growth, sporulation, and spore germination of Bacillus megaterium. J Bacteriol 129:857–865

Setlow B, Melly E, Setlow P (2001) Properties of spores of *Bacillus subtilis* blocked at an intermediate stage in spore germination. J Bacteriol 183(16):4894–4899. doi:10.1128/JB.183.16.4894-4899.2001

Shah IM, Dworkin J (2010) Induction and regulation of a secreted peptidoglycan hydrolase by a membrane Ser/Thr kinase that detects muropeptides. Mol Microbiol 75:1232–1243. doi:10.1111/j.1365-2958.2010.07046.x

Shah IM, Laaberki MH, Popham DL, Dworkin J (2008) A eukaryotic-like Ser/Thr kinase signals bacteria to exit dormancy in response to peptidoglycan fragments. Cell 135:486–496. doi:10.1016/j.cell.2008.08.039

Shibata H, Adachi S, Hirose Y, Ike M, Tani I, Hashimoto T (1993) Role of calcium in biphasic germination of *Bacillus cereus* T-spores sensitized to lysozyme. Microbiol Immunol 37:187–194

Shpigelman A, Kyomugasho C, Christiaens S, Van Loey AM, Hendrickx ME (2014) Thermal and high pressure high temperature processes result in distinctly different pectin non-enzymatic conversions. Food Hydrocoll 39:251–263

Sogin ML, McCall WA, Ordal JZ (1972) Effect of heat activation conditions on the germinal response of *Bacillus cereus* T spores. In: Halvorson HO, Hanson R, Campbell LL (eds) Spores, vol 5. American Society for Microbiology, Washington, DC, pp 363–367

Sokolowska B, Skapska S, Fonberg-Broczek M, Niezgoda J, Chotkiewicz M, Dekowska A, Rzoska S (2012) The combined effect of high pressure and nisin or lysozyme on the inactivation of *Alicyclobacillus acidoterrestris* spores in apple juice. High Pressure Res 32:119–127

Sonenshein AL (2000) Control of sporulation initiation in *Bacillus subtilis*. Curr Opin Microbiol 3:561–566

Spring S, Merkhoffer B, Weiss N, Kroppenstedt RM, Hippe H, Stackebrandt E (2003) Characterization of novel psychrophilic clostridia from an Antarctic microbial mat: description of Clostridium frigoris sp. nov., Clostridium lacusfryxellense sp. nov., Clostridium bowmanii sp. nov. and Clostridium psychrophilum sp. nov. and reclassification of Clostridium laramiense as Clostridium estertheticum subsp. laramiense subsp. nov. Int J Syst Evol Microbiol 53:1019–1029

Squeglia F, Marchetti R, Ruggiero A, Lanzetta R, Marasco D, Dworkin J, Petoukhov M, Molinaro A, Berisio R, Silipo A (2011) Chemical basis of peptidoglycan discrimination by PrkC, a key kinase involved in bacterial resuscitation from dormancy. J Am Chem Soc 133:20676–20679. doi:10.1021/ja208080r

Steil L, Serrano M, Henriques AO, Volker U (2005) Genome-wide analysis of temporally regulated and compartment-specific gene expression in sporulating cells of *Bacillus subtilis*. Microbiol-Sgm 151:399–420

Steiner E, Dago AE, Young DI, Heap JT, Minton NP, Hoch JA, Young M (2011) Multiple orphan histidine kinases interact directly with Spo0A to control the initiation of endospore formation in *Clostridium acetobutylicum*. Mol Microbiol 80:641–654. doi:10.1111/j.1365-2958.2011.07608.x

Stephenson K, Hoch JA (2002) Evolution of signalling in the sporulation phosphorelay. Mol Microbiol 46:297–304

Stephenson K, Lewis RJ (2005) Molecular insights into the initiation of sporulation in Gram-positive bacteria: new technologies for an old phenomenon. FEMS Microbiol Rev 29:281–301. doi:10.1016/j.femsre.2004.10.003

Stewart CM, Dunne CP, Sikes A, Hoover DG (2000) Sensitivity of spores of *Bacillus subtilis* and *Clostridium sporogenes* PA 3679 to combinations of high hydrostatic pressure and other processing parameters. Innovative Food Sci Emerg Technol 1:49–56

Steyn CE, Cameron M, Witthuhn RC (2011) Occurrence of *Alicyclobacillus* in the fruit processing environment – a review. Int J Food Microbiol 147:1–11

Stiegelmeyer SM, Giddings MC (2013) Agent-based modeling of competence phenotype switching in *Bacillus subtilis*. Theor Biol Med Model 10:23. doi:10.1186/1742-4682-10-23

Stragier P, Losick R (1996) Molecular genetics of sporulation in *Bacillus subtilis*. Annu Rev Genet 30:297–341

Stragier P, Kunkel B, Kroos L, Losick R (1989) Chromosomal rearrangement generating a composite gene for a developmental transcription factor. Science 243:507–512

Stringer SC, Webb MD, Peck MW (2009) Contrasting effects of heat treatment and incubation temperature on germination and outgrowth of individual spores of nonproteolytic Clostridium botulinum bacteria. Appl Environ Microb 75(9):2712–2719. doi:10.1128/Aem.02572-08

Stringer SC, Webb MD, Peck MW (2011) Lag time variability in individual spores of *Clostridium botulinum*. Food Microbiol 28:228–235. doi:10.1016/j.fm.2010.03.003

Sugiyama H (1951) Studies on factors affecting the heat resistance of spores of Clostridium botulinum. J Bacteriol 62:81–96

Sunde EP, Setlow P, Hederstedt L, Halle B (2009) The physical state of water in bacterial spores. Proc Natl Acad Sci U S A 106:19334–19339. doi:10.1073/pnas.0908712106

Syed QA, Reineke K, Saldo J, Buffa M, Guamis B, Knorr D (2012) Effect of compression and decompression rates during high hydrostatic pressure processing on inactivation kinetics of bacterial spores at different temperatures. Food Control 25:361–367. doi:10.1016/j.foodcont.2011.10.061

Tani K, Watanabe T, Matsuda H, Nasu M, Kondo M (1996) Cloning and sequencing of the spore germination gene of *Bacillus megaterium* ATCC 12872: similarities to the NaH-antiporter gene of *Enterococcus hirae*. Microbiol Immunol 40:99–105

Tomas CA, Beamish J, Papoutsakis ET (2004) Transcriptional analysis of butanol stress and tolerance in *Clostridium acetobutylicum*. J Bacteriol 186:2006–2018. doi:10.1128/Jb.186.7.2006-2018.2004

Tonello-Samson C (2014) Industrial application of high pressure processing in food industry. In: 8th international conference on high pressure bioscience and biotechnology, Nantes, 15–18 July 2014. Book of Abstracts, p 48 and personal communication

Torres JA, Velazquez G (2005) Commercial opportunities and research challenges in the high pressure processing of foods. J Food Eng 67:95–112

Trach KA, Hoch JA (1993) Multisensory activation of the phosphorelay initiating sporulation in *Bacillus subtilis* – Identification and sequence of the protein-kinase of the alternate pathway. Mol Microbiol 8:69–79

Ulmer HM, Herberhold H, Fahsel S, Gänzle MG, Winter R, Vogel RF (2002) Effects of pressure-induced membrane phase transitions on inactivation of HorA, an ATP-dependent multidrug resistance transporter, in *Lactobacillus plantarum*. Appl Environ Microbiol 68:1088–1095

Underwood S, Guan S, Vijayasubhash V, Baines SD, Graham L, Lewis RJ, Wilcox MH, Stephenson K (2009) Characterisation of the sporulation initiation pathway of *Clostridium difficile* and the role in toxin production. J Bacteriol 191:7296–7305. doi:10.1128/Jb.00882-09

van der Voort M, Garcia D, Moezelaar R, Abee T (2010) Germinant receptor diversity and germination responses of four strains of the *Bacillus cereus* group. Int J Food Microbiol 139:108–115. doi:10.1016/j.ijfoodmicro.2010.01.028

van Schaik W, Abee T (2005) The role of sigma(B) in the stress response of Gram-positive bacteria – targets for food preservation and safety. Curr Opin Biotechnol 16:218–224. doi:10.1016/j.copbio.2005.01.008

Veening JW, Kuipers OP, Brul S, Hellingwerf KJ, Kort R (2006) Effects of phosphorelay perturbations on architecture, sporulation, and spore resistance in biofilms of *Bacillus subtilis*. J Bacteriol 188:3099–3109. doi:10.1128/JB.188.8.3099-3109.2006

Vepachedu VR, Setlow P (2007) Role of SpoVA proteins in release of dipicolinic acid during germination of *Bacillus subtilis* spores triggered by dodecylamine or lysozyme. J Bacteriol 189:1565–1572. doi:10.1128/JB.01613-06

Wang ST, Setlow B, Conlon EM, Lyon JL, Imamura D, Sato T, Setlow P, Losick R, Eichenberger P (2006) The forespore line of gene expression in *Bacillus subtilis*. J Mol Biol 358:16–37. doi:10.1016/j.jmb.2006.01.059

Wang G, Zhang P, Paredes-Sabja D, Green C, Setlow P, Sarker MR, Li YQ (2011) Analysis of the germination of individual *Clostridium perfringens* spores and its heterogeneity. J Appl Microbiol 111:1212–1223. doi:10.1111/j.1365-2672.2011.05135.x

Wang G, Paredes-Sabja D, Sarker MR, Green C, Setlow P, Li YQ (2012) Effects of wet heat treatment on the germination of individual spores of *Clostridium perfringens*. J Appl Microbiol 113:824–836. doi:10.1111/j.1365-2672.2012.05387.x

Webb MD, Pin C, Peck MW, Stringer SC (2007) Historical and contemporary NaCl concentrations affect the duration and distribution of lag times from individual spores of nonproteolytic *Clostridium botulinum*. Appl Environ Microbiol 73:2118–2127. doi:10.1128/AEM.01744-06

Webb MD, Stringer SC, Le Marc Y, Baranyi J, Peck MW (2012) Does proximity to neighbours affect germination of spores of non-proteolytic *Clostridium botulinum*? Food Microbiol 32:104–109. doi:10.1016/j.fm.2012.04.015

Wei J, Setlow P, Hoover DG (2009) Effects of moderately high pressure plus heat on the germination and inactivation of *Bacillus cereus* spores lacking proteins involved in germination. Lett Appl Microbiol 49:646–651. doi:10.1111/j.1472-765X.2009.02721.x

Wei J, Shah IM, Ghosh S, Dworkin J, Hoover DG, Setlow P (2010) Superdormant spores of *Bacillus* species germinate normally with high pressure, peptidoglycan fragments, and bryostatin. J Bacteriol 192:1455–1458. doi:10.1128/Jb.01497-09

Weir J, Predich M, Dubnau E, Nair G, Smith I (1991) Regulation of Spo0h, a gene coding for the *Bacillus subtilis* sigma-H factor. J Bacteriol 173:521–529

Westphal AJ, Price PB, Leighton TJ, Wheeler KE (2003) Kinetics of size changes of individual *Bacillus thuringiensis* spores in response to changes in relative humidity. Proc Natl Acad Sci USA 100:3461–3466. doi:10.1073/pnas.232710999

Wilkinson SR, Young DI, Morris JG, Young M (1995) Molecular-genetics and the initiation of solventogenesis in *Clostridium berijerinckii* (formerly *Clostridium acetobutylicum*) Ncimb-8052. FEMS Microbiol Rev 17:275–285

Williams OB, Reed JM (1942) The significance of the incubation temperature of recovery cultures in determining spore resistance to heat. J Infect Dis 71:225–227

Wilson DR, Dabrowski L, Stringer S, Moezelaar R, Brocklehurst TF (2008) High pressure in combination with elevated temperature as a method for the sterilisation of food. Trends Food Sci Technol 19:289–299. doi:10.1016/j.tifs.2008.01.005

Winter R, Dzwolak W (2005) Exploring the temperature-pressure configurational landscape of biomolecules: from lipid membranes to proteins. Philos Transact A Math Phys Eng Sci 363:537–562. doi:10.1098/rsta.2004.1507, discussion 562–533

Winter R, Jeworrek C (2009) Effect of pressure on membranes. Soft Matter 5:3157–3173. doi:10.1039/B901690B

Woods DR (1995) The genetic-engineering of microbial solvent production. Trends Biotechnol 13:259–264

Wörner K, Szurmant H, Chiang C, Hoch JA (2006) Phosphorylation and functional analysis of the sporulation initiation factor Spo0A from *Clostridium botulinum*. Mol Microbiol 59:1000–1012. doi:10.1111/j.1365-2958.2005.04988.x

Wuytack EY, Michiels CW (2001) A study on the effects of high pressure and heat on Bacillus subtilis spores at low pH. Int J Food Microbiol 64:333–341, doi:S0168160500004785 [pii]

Wuytack EY, Boven S, Michiels CW (1998) Comparative study of pressure-induced germination of *Bacillus subtilis* spores at low and high pressures. Appl Environ Microbiol 64:3220–3224

Wynne ES, Foster JW (1948) Physiological studies on spore germination, with special reference to *Clostridium botulinum*: III. Carbon dioxide and germination, with a note on carbon dioxide and aerobic spores. J Bacteriol 55:331–339

Yudkin MD, Clarkson J (2005) Differential gene expression in genetically identical sister cells: the initiation of sporulation in *Bacillus subtilis*. Mol Microbiol 56:578–589. doi:10.1111/j.1365-2958.2005.04594.x

Zhang P, Setlow P, Li Y (2009) Characterization of single heat-activated *Bacillus* spores using laser tweezers Raman spectroscopy. Opt Express 17:16480–16491. doi:10.1364/OE.17.016480

Zhang P, Kong L, Wang G, Scotland M, Ghosh S, Setlow B, Setlow P, Li YQ (2012) Analysis of the slow germination of multiple individual superdormant *Bacillus subtilis* spores using multifocus Raman microspectroscopy and differential interference contrast microscopy. J Appl Microbiol 112:526–536. doi:10.1111/j.1365-2672.2011.05230.x

Zhang JQ, Griffiths KK, Cowan A, Setlow P, Yu J (2013) Expression level of *Bacillus subtilis* germinant receptors determines the average rate but not the heterogeneity of spore germination. J Bacteriol 195:1735–1740. doi:10.1128/JB.02212-12

Zhao YS, Tomas CA, Rudolph FB, Papoutsakis ET, Bennett GN (2005) Intracellular butyryl phosphate and acetyl phosphate concentrations in *Clostridium acetobutylicum* and their implications for solvent formation. Appl Environ Microbiol 71:530–537. doi:10.1128/Aem.71.1.530-537.2005

Chapter 24
Use of Pressure Activation in Food Quality Improvement

Toru Shigematsu

Abstract Beside intensive studies on inactivation microorganisms by high hydrostatic pressure (HP) for food storage, pressure effects on property of food materials have also been studied based on knowledge in pressure effect on biomolecules. Pressure effects on biological membranes and mass transfer in cellular biological materials and on enzyme activity would give an idea that HP treatment can introduce two types of activations into food materials: improved mass transfer and enzyme activity. Studies focusing on these pressure activations on food materials were then reviewed. Rice flour with an exclusively fine mean particle size and small starch damage was obtained due to improved water absorption properties and/or enzyme activity by HP. HP treatment increased of free amino acids and γ-aminobutyric acid (GABA) in rice and soybeans due to improved proteolysis and amino acid metabolism. Improvement of antioxidant activity and alteration of polyphenolic-compounds composition in food materials were also demonstrated by HP treatment. The HP-induced activations on food materials could contribute towards processing technologies for food quality improvement.

Keywords Biological membranes • Mass transfer • Enzyme activity • Food quality improvement

24.1 Overview

Since the earliest experiments for inactivation of *Escherichia coli* and *Staphylococcus aureus* under 2,900 atm at 25 °C by Roger H. in 1895, a number of studies have been accumulated on inactivation of microorganisms by high hydrostatic pressure (HP) (ZoBell 1970). This phenomenon was also expected to apply pressure as non-thermal techniques for preservation and extension of shelf life of foods and agro products. In 1899, Hite demonstrated the preservation period of milk could

T. Shigematsu (✉)
Faculty of Applied Life Sciences, Niigata University of Pharmacy and Applied Life Sciences (NUPALS), 265-1 Higashijima, Akiha-ku, Niigata 956-8603, Japan
e-mail: shige@nupals.ac.jp

K. Akasaka, H. Matsuki (eds.), *High Pressure Bioscience*,
Subcellular Biochemistry 72, DOI 10.1007/978-94-017-9918-8_24

be extended by HP treatment at 650 MPa for 10 min (Hite 1899). Hite et al. then demonstrated that HP treatment could also extend preservation period of vegetables in 1914 (Hite et al. 1914). Their results indicated that among spoilage microorganisms yeast showed to be more sensitive against pressure than bacterial strains. Bacterial strains, which showed relatively pressure resistance, cannot grow under acidic conditions. Thus, anti-spoilage effect of HP treatment was concluded. However, some pressure-resistant bacteria, such as spores of *Bacillus* sp., have been demonstrated to remain after HP treatment. So, the foodstuffs contaminated by the bacterial spores are sterilized with the combined pressure and thermal treatment (Wuytack and Michiels 2001; Aoyama et al. 2005).

Beside those studies on HP inactivation of microorganisms for food storage, pressure effects on physical property of food have also been studied. In 1914, Bridgman applied 500–600 MPa on egg and demonstrated that egg yolk and egg white coagulated by HP treatment but no effect on egg shell (Bridgman 1914). He indicated that pressure caused denaturation of the egg proteins by a different manner from heat treatment. Payens and Heremans (1969) reported the influence of pressure on β-casein molecule in milk. Macfarlane (1973) demonstrated meat could be softened by pressure under certain conditions. In Japan, knowledge concerning effects of HP on organisms and biomaterials has been accumulated since around 1950s (Murakami 1970). After Hayashi proposed advanced applications of HP on food processing (Hayashi 1987), research and development for establishment of HP food processing technologies was accelerated. His idea was that use of pressure (100–1,000 MPa) instead of heat treatment can allow manufacturing high quality foods without the change in the flavor and nutrient component in comparison with the heated one.

In general, HP treatment above approximately 100 MPa at ambient temperature on a cellular biological material has led to damage of biological membranes caused by phase transition of the lipid bilayers (Meersman and Heremans 2008). When biological membranes are damaged, mass transfer inside and between cells can be promoted. On the other hands, proteins are relatively resistant against pressure compared with lipids. Pressures at approximately 200–300 MPa are known to dissociate protein oligomers to their monomers. Monomeric proteins are denatured between 400 and 800 MPa. Certain enzymes were reportedly still active even under 600 MPa (Knorr et al. 2006). The pressure tolerance of enzymes indicates that HP treatment at approximately 100–400 MPa could increase apparent enzyme activities by improved association between enzymes and their substrates due to improved mass transfer by damage of biological membranes. Moreover, conformational equilibrium shift of proteins under HP might also increase enzyme activities.

These findings could lead us to an idea that HP treatment can introduce two types of activations into food materials: improved mass transfer and enzyme activity. In this chapter, studies focusing on pressure activations on food materials towards processing technologies for food quality improvement were reviewed.

24.2 Pressure Effects on Biomembranes and Mass Transfer in Cellular Biological Materials

Based on the phase transition of lipid bilayer by HP above 100 MPa at ambient temperature, HP was assumed to damage the cell structure in the plant tissue, especially affecting the plasma membrane (Kato and Hayashi 1999). Transmission electron microscopy revealed that HP at more than 100 MPa affected cellular membrane systems, especially in nuclear membrane, in yeast *Saccharomyces cerevisiae* (Osumi 1990; Shimada et al. 1993). The dielectric property analysis was also successfully used to evaluate damage of cellular membrane systems. The Cole-Cole arc decreased with increase in hydrostatic pressure applied for onion (Ueno et al. 2009a), turnip root (Ueno et al. 2009b), Japanese radish (Ueno et al. 2009c) and soybean (Ueno et al. 2010).

Damages in internal cell structures and membranes of food materials could lead to changes in food texture and physicochemical properties (Tangwongchai et al. 2000; Islam et al. 2003), as well as improved mass transfer of materials within cells and/or tissues. Then, drying, absorption, extraction and so on are expected and reported to accelerate (Eshtiaghi et al. 1994). The rate of mass transfer depends on the damage of internal cell structure. For example, the drying rate of Japanese radish treated by HP at 400 MPa (ambient temperature) for 5 min was faster than that of the untreated product, slower than those treated by both freeze-thaw and heat, and close to that treated by chloroform vapor (Ueno et al. 2009c). The water absorption of rice grains soaked in distilled water also improved under HP of 100–300 MPa at 40 °C (Fig. 24.1) (Kido et al. 2013). The improved mass transfer caused by HP could lead to improvement of association between certain enzymes and their substrates in cells and/or tissues of food materials. Thus HP treatment would have a potential to initiate and enhance certain enzymatic reactions in food materials, if the enzymes are still active during and after HP treatment.

24.3 Pressure Effects on Enzyme Activity

When HP applied for a protein, its conformational equilibrium between native and denatured states (N $\leftrightarrows$ D) would shift towards a state with the smallest volume according to Le Chatelier's principle. The conformational change by pressure is considered to mainly due to decrease in volume of cavities in the protein and to the shift of the hydration state (Fourme et al. 2012). This reversible change of the population of proteins with each conformational state could cause to the change of apparent activity of an enzyme by HP. Kido et al. (2013) analyzed pectinase (Sumizyme PTE; Shin Nihon Chemical) activity using pectin as a substrate under pressure at 100–300 MPa (Fig. 24.2). The initial reaction rate increased from 0.1 to 200 MPa with increase in the pressure level, but decreased from 200 to 300 MPa. The maximum initial reaction rate was obtained at 200 MPa. These

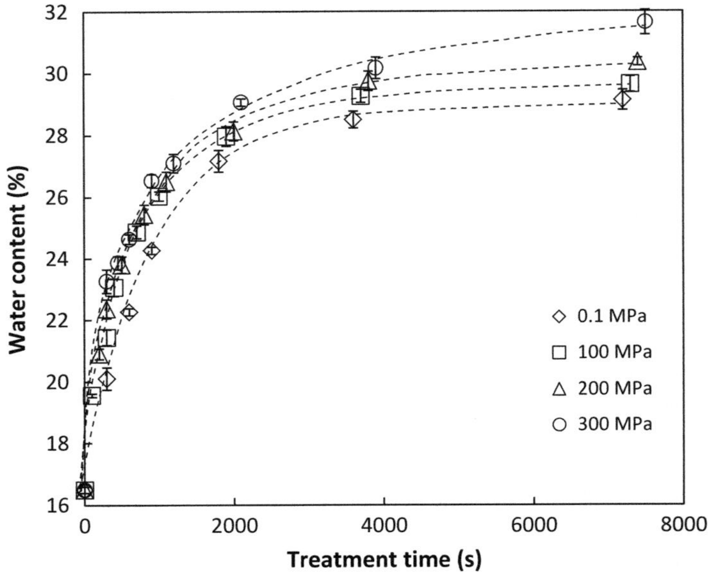

Fig. 24.1 Water absorption curves of rice grains soaked in distilled water under 0.1 MPa (atmospheric pressure; *diamonds*), 100 MPa (*squares*), 200 MPa (*triangles*) or 300 MPa (*circles*) at 40 °C. Error bars represent the standard deviations from at least three experiments. *Dotted lines* represent approximated curves based on the water absorption equation (Kido et al. 2013)

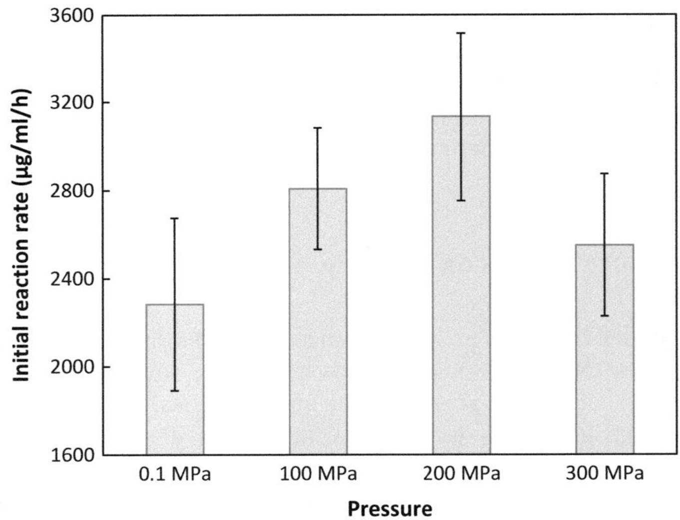

Fig. 24.2 Initial reaction rate of pectinase (Sumizyme PTE) using pectin as the substrate under HP treatment. Error bars represent the standard deviations from at least three experiments

results suggest that pressure effect on the conformational states of the pectinase gives the apparent optimum activity under pressure condition at 200 MPa, and that the conformational equilibrium shifts to the denatured state under a high-pressure levels such as 300 MPa. Although the pressure resistance against conformational equilibrium shift to their denatured states varies depending on the proteins, certain enzymes, such as α- and β-amylase and β-glucanase, were reportedly still active even over 600 MPa at ambient temperature (Knorr et al. 2006). HP roughly 100–600 MPa could not only suppress but also improve the activity of enzymes due to the conformational equilibrium shift.

Interestingly, Asaka and Hayashi (1991) applied HP treatment of 100–600 MPa (25 °C) for 10 min on the crude extract containing polyphenol oxidase in pear fruits, and subsequently analyzed its activity under atmospheric pressure at 30 °C. As the results, the activity of polyphenol oxidase increased with increase in the pressure level applied with the maximum activity at 400 and 500 MPa. HP treatment at 600 MPa relatively decreased the activity than that at the 400 and 500 MPa. The results indicate that the improvement of enzymatic activity of the polyphenol oxidase by the HP treatment for 10 min still remain even after depressurization. The conformation shift caused by pressure is considered as a reversible process. Depressurization makes the proteins with the denatured state to irreversible hydrophobic aggregation, which is pressure-gelatinization of proteins. The results by Asaka and Hayashi provided a possibility that certain enzymes are activated by irreversible conformation shift during high-pressure treatment. These results allow us to have an insight that HP can be used for improved enzymatic reactions in food materials.

24.4 Super-Fine Rice Flour Production by Enzymatic Treatment with High Hydrostatic Pressure

Rice flour is used for the production of a number of foods, particularly in Asian countries. In response to the recent expansion of rice-flour use, new and improved rice-flour production methods should be developed in order to meet this rise in demand. In general, rice flours with fine particle size and low starch damage are suitable for processing rice-flour-based bread and Western-style sweets, i.e. as a substitute for wheat flour. Particle size can be controlled by pulverization conditions. However, particle size and starch damage tend to have a trade-off relationship; fine rice flour obtained through harder pulverization tends to show higher starch damage. Pectinase used in the enzyme-treated milling method (Morohashi et al. 1998) only mildly damages the tissue structure of rice grains before subsequent mild wet pulverization, which leads to the production of fine rice flour with low starch damage. Based on the HP effect that improved enzyme activity and mass transfer, we applied HP for the enzymatic treatment to improve the enzyme-treated milling method (Kido et al. 2013; Homma et al. 2013).

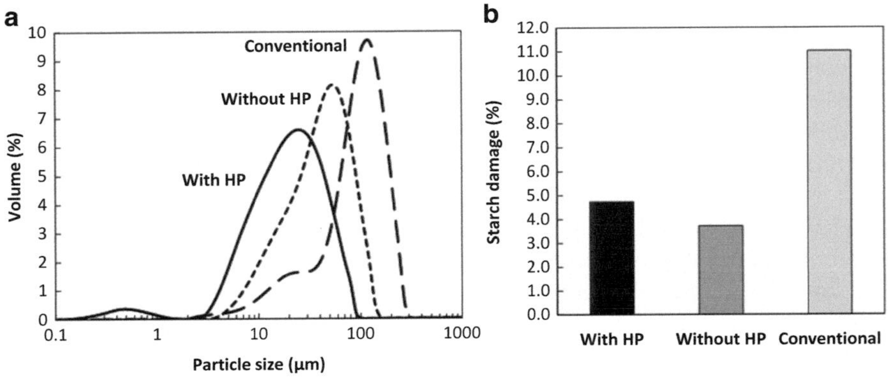

Fig. 24.3 Volume-based particle size distribution (**a**) and starch damage (**b**) of rice flour manufactured by conventional, enzyme-treated milling method without HP and enzyme-treated milling method with HP at 200 MPa (40 °C)

The apparent pectinase activity showed the highest under HP at 200 MPa. The water absorption ability also showed to be improved under HP. We therefore construct a process, which consists of HP treatment at 200 MPa (40 °C) for 1 h and subsequent wet pulverization. Using this process, rice flour with an exclusively fine mean particle size less than 20 μm and starch damage less than 5 % was obtained (Fig. 24.3). The resulting super-fine rice flour has a superior processing property suitable for bread, pasta, noodles and Western-style sweets. A pressure level of 200 MPa would give pectinase the optimal conformational equilibrium for activity and/or optimal association with pectin. The improved water absorption would contribute to tissue damage in the rice grain, thus reducing particle size in the resulting rice flour. The production process strongly correlates to improved water absorption properties and/or enzyme activity as a result of HP treatment.

24.5 Increase of Free Amino Acids and γ-aminobutyric Acid in Rice and Soybeans by Hydrostatic Pressure

Especially in Asian countries, rice is one of the most important crops. A number of researches and developments have been conducted for improvement of nutrition values and functions of rice. We investigated the possibility that improvement of proteolysis and amino acid-metabolism in brown rice by HP treatment (Shigematsu et al. 2010). Brown rice grains were soaked in water and subjected to HP treatment at 200 MPa (20 °C) for 10 min. After the treatment, brown rice grains were swept with paper towel to remove water and stored at 25 °C for 4 days. The free amino acid distribution of HP- treated samples, just after HP treatment, showed no apparent difference from that of untreated control without HP treatment. However,

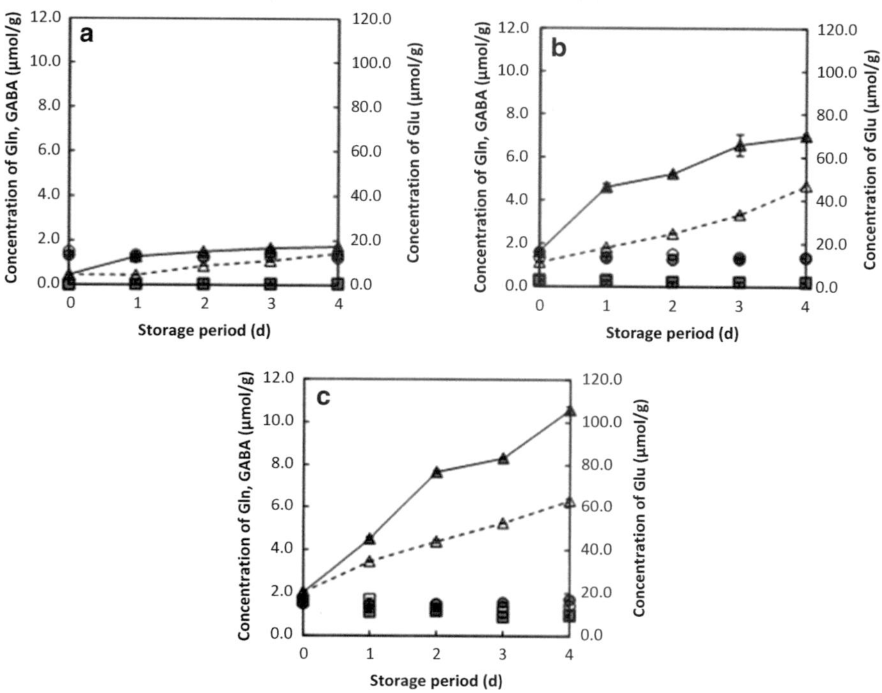

Fig. 24.4 Time course of GABA (*triangles*), Glu (*boxes*), and Gln (*circles*) concentrations in HP-treated (*closed symbols*) and HP-untreated (*open symbols*) samples, which had been soaked in Glu concentration of 0 (**a**), 0.01 (**b**), 0.05 g ml^{-1} (**c**) during storage. Error bars representing the standard deviations from at least three experiments

during 4 days storage at 25 °C after HP treatment, certain amino acids including γ-aminobutyric acid (GABA) in the HP-treated samples increased with time and showed higher concentrations than those in untreated samples (Fig. 24.4a). To investigate the feasibility for use of HP-treated brown rice grains as a bioreactor producing GABA, glutamic acid (Glu) was supplied into brown rice grains during water soaking and subjected to HP treatment. The GABA concentrations during storage increased with the increase in the Glu concentrations in the soaking solutions (Fig. 24.4b, c). The initial GABA production rate was accelerated by HP treatment.

In soybeans, which contain a large amount of proteins, the effect of HP on the free amino acids was analysed (Ueno et al. 2010). Water-soaked soybeans were subjected to HP treatment at 200 MPa (20 °C) for 10 min. After the treatment, soybeans were swept with paper towel to remove water and stored at 25 °C for 4 days. As in the brown rice, during 4 days storage at 25 °C after HP treatment, certain amino acids including GABA in the HP-treated soybean samples increased with time and showed higher concentrations than those in untreated samples. These results provide feasibility for a novel use of HP technology to alter the metabolic pathways of food materials, such as proteolysis and amino acid metabolism in brown rice grains and soybeans, and to accumulate useful metabolites.

24.6 Improvement of Antioxidant Activity of Food Materials by High Hydrostatic Pressure

Ueno et al. (2009a) applied HP treatments on onion with 200 and 400 MPa (20 °C) for 5 min. During storage at 25 °C after the HP treatment, the antioxidant activity evaluated as DPPH radical scavenging activity increased, whereas no alteration was observed for onion without HP treatment. Flavonoids were extracted from the onion samples and analyzed by high performance liquid chromatography (HPLC). The concentration of quercetin, which is the major flavonoid and antioxidant in onion, in the onion with HP treated increased during the storage. As the results, HP treatment could improve the antioxidant activity of food materials via alteration of antioxidants composition possibly caused by internal enzymatic reactions triggered by HP.

Then we analyzed effect of HP and subsequent preservation on the antioxidant activities of food materials (Shigematsu et al. 2012). Total 51 food materials were subjected to HP treatment at 200 MPa (20 °C) for 10 min and subsequently storage at 25 °C for 2 days. The HP-untreated control samples for each food material were also prepared. The antioxidant activity, evaluated as oxygen radical scavenging capacity (ORAC), of 26 products increased by HP-treatment. In the 26 products, 11 products showed HP-dependent increase of ORAC value during storage after HP-treatment (results of 23 products were shown in Fig. 24.5). Edible part of *Petroselinum crispum* (parsley) showed the highest value of ORAC in the 11 products.

P. crispum was then subjected to HP treatment at 100–600 MPa (20 °C) for 10 min and subsequently storage for 1 h. At each pressure level, the ORAC value was higher than that in the HP untreated control, with the highest value in HP treatment at 300 MPa. The analysis on polyphenolic compounds in *P. crispum*, which was HP-treated at 300–600 MPa (20 °C) for 10 min, showed that the concentration of malonylapiin [apigenin 7-*O*-(6-*O*-malonylglucoside)] decreased, but apiin [apigenin 7-*O*-glucoside] increased during storage for 30 min, resulting in the increase in the malony apiin/apiin ratio (Fig. 24.6). This result suggested that conversion of malonylapiin to apiin was triggered by the HP treatment. The alteration of the apigenin composition would be related to the HP-dependent increase in antioxidant activity in *P. crispum*.

An *in vitro* reconstruction experiment was carried out to evaluate the enzymatic conversion of the apigenins. The methanol extract of pre-boiled parsley and soluble extract of parsley were prepared as the crude extracts containing substrate and enzyme, respectively. After mixing the substrate and enzyme extracts, increase of apiin and decrease of malonylapiin were observed during incubation. In intact cells of parsley, malonylapiin and esterases are separately localized in vacuoles and cytoplasm, respectively (Lechtenberg et al. 2007; Luthria 2008). Our results

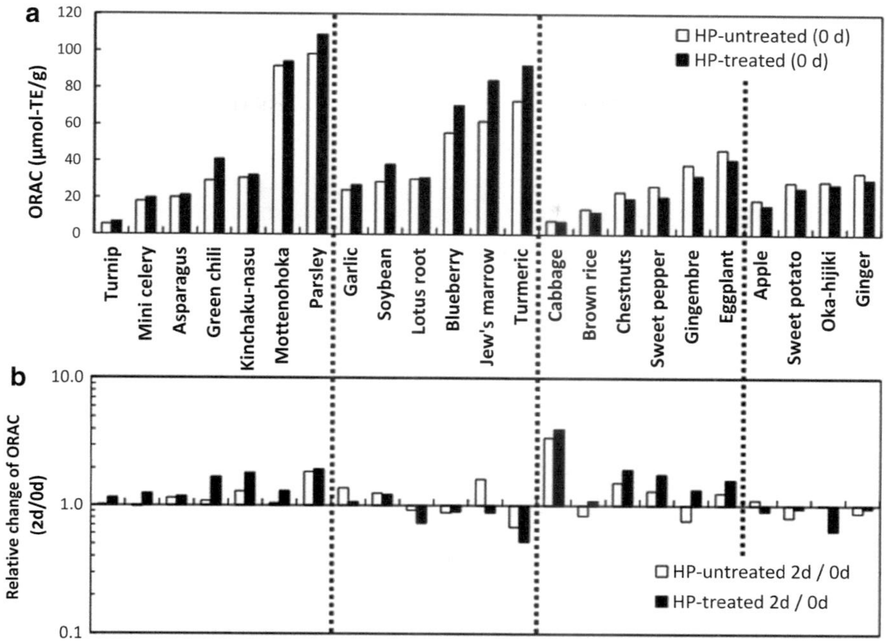

Fig. 24.5 The ORAC values, evaluated as Trolox equivalent, of HP-treated (*closed bars*) and untreated (*open bars*) food materials just after treatment (**a**) and relative change of ORAC value during 2 days storage (**b**)

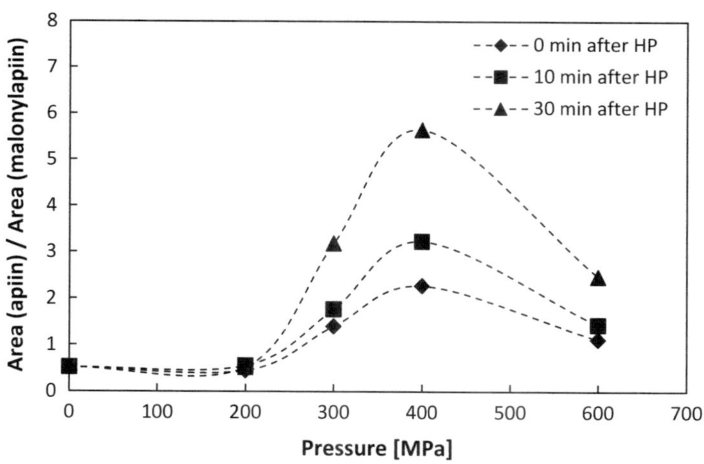

Fig. 24.6 Effect of pressure levels and time after HP treatment on malonylapiin/apiin ratio, evaluated as peak area values in HPLC analyses

suggest that damage of vacuole by HP treatment allowed improved the association between malonylapiin and esterases, which leads the HP dependent conversion of malonylapiin to apiin.

24.7 Summary

Pressure effects on biological membranes and mass transfer in cellular biological materials and on enzyme activity resulted in the two types of activations into food materials by HP treatment approximately at 100–400 MPa: improved mass transfer and enzyme activity. Rice flour with an exclusively fine mean particle size and small starch damage was obtained due to improved water absorption properties and/or enzyme activity by HP at 200 MPa. HP treatments could also cause alteration of biochemical properties or composition of compounds in food materials via the two types of activations. HP treatment at 200 MPa for 10 min increased of free amino acids and γ-aminobutyric acid (GABA) in rice and soybeans during subsequent storage under atmospheric pressure, which would due to proteolysis and amino acid metabolism by improved mass transfer and apparent activities of internal enzymes. In this case, HP caused alteration of cells in food materials, where biological membrane systems were damaged but certain enzymes remain still active. This HP-induced transformation of food materials or "Hi-Pit (Ueno et al. 2009a)" could thus change the component composition via progress of internal enzymatic reactions during subsequent storage after HP treatment. Improvement of antioxidant activity and alteration of polyphenolic-compounds composition in food materials by HP treatment would be also explained by "Hi-Pit" theory.

The HP-induced activations on food materials could be applied for alteration of physiochemical properties, as well as biochemical properties of the food materials. For example, accumulation of useful compounds via improved enzymatic reactions, activated by HP, should contribute towards processing technologies for food quality improvement.

References

Aoyama Y, Shigeta Y, Okazaki T, Hagura Y, Suzuki K (2005) Germination and inactivation of *Bacillus subtilis* spores under combined conditions of hydrostatic pressure and medium temperature. Food Sci Technol Res 11:101–105

Asaka M, Hayashi R (1991) Activation of polyphenoloxidase in pear fruits by high pressure treatment. Agric Biol Chem 55:2439–2440

Bridgman PW (1914) The coagulation of albumen by pressure. J Biol Chem 19:511–512

Eshtiaghi MN, Stute R, Knorr D (1994) High-pressure and freezing pretreatment effects on drying, rehydration, texture and color of green beans, carrots and potatoes. J Food Sci 59:1168–1170

Fourme R, Eric Girard E, Akasaka K (2012) High-pressure macromolecular crystallography and NMR: status, achievements and prospects. Curr Opin Struct Biol 22:636–642

Hayashi R (1987) Food processing and ingredients (in Japanese). Shokuhin to kaihatsu 22:55–62

Hite BH (1899) The effect of pressure in the preservation of milk. West Virginia Univ Agric Exp Stat Bull 54:15–35

Hite BH, Giddings NJ, Weakly CE (1914) The effects of pressure on certain microorganisms encountered in the preservation of fruits and vegetables. West Virginia Univ Agric Exp Stat Bull 146:2–67

Homma N, Nishiwaki T, Kobayashi K, Kido M, Yamamoto K, Shigematsu T, Suzuki A (2013) Japan patent 5,326,147

Islam MS, Igura N, Shimoda M, Hayakawa I (2003) Effects of low hydrostatic pressure and moderate heat on texture, pectic substances and color of carrot. Eur Food Res Technol 217:34–38

Kato M, Hayashi R (1999) Effects of high pressure on lipids and biomembranes for understanding high-pressure-induced biological phenomena. Biosci Biotechnol Biochem 63:1321–1328

Kido M, Kobayashi K, Chino S, Nishiwaki T, Homma N, Hayashi M, Yamamoto K, Shigematsu T (2013) Super-fine rice flour production by enzymatic treatment with high hydrostatic pressure processing. High Pres Res 33:237–244

Knorr D, Heinz V, Buckow R (2006) High pressure application for food biopolymers. Biochim Biophys Acta 1764:619–631

Lechtenberg M, Zumdick S, Gerhards C, Schmidt TJ, Hensel A (2007) Evaluation of analytical markers characterising different drying methods of parsley leaves (*Petroselinum crispum* L.). Pharmazie 62:949–954

Luthria DL (2008) Influence of experimental conditions on the extraction of phenolic compounds from parsley (*Petroselinum crispum*) flakes using a pressurized liquid extractor. Food Chem 107:745–752

Macfarlane JJ (1973) Pre-rigor pressurization of muscle: effects on pH, shear value and taste panel assessment. J Food Sci 38:294–298

Meersman F, Heremans K (2008) High hydrostatic pressure effects in the biosphere: from molecules to microbiology. In: Michiels C, Barlett DH, Aertsen A (eds) High-pressure microbiology. ASM Press, Washington, DC, pp 1–17

Morohashi K, Nabeya T, Yoshii Y, Egawa K (1998) Japan patent 3,076,552

Murakami TH (1970) Japanese studies on hydrostatic pressure. In: Zimmerman AM (ed) High pressure effects on cellular processes. Academic, New York, pp 131–138

Osumi M (1990) Effects of hydrostatic pressure to ultrastructure of yeast cells (in Japanese). In: Hayashi R (ed) Pressure-processed food – research and development. Sanei Shuppan, Kyoto, pp 157–164

Payens TAJ, Heremans K (1969) Effect of pressure on the temperature-dependent association of β-casein. Biopolymers 8:335–345

Shigematsu T, Murakami M, Nakajima K, Uno Y, Sakano A, Narahara Y, Hayashi M, Ueno S, Fujii T (2010) Bioconversion of glutamic acid to γ-aminobutyric acid (GABA) in brown rice grains induced by high pressure treatment. Jpn J Food Eng 11:189–199

Shigematsu T, Nakajima K, Inagaki K, Kawamura T, Nakamura M, Hayashi M, Kumakura S, Iguchi A, Hirayama M, Ueno S, Fujii T (2012) Effect of high hydrostatic pressure and subsequent preservation on the antioxidant activities of agricultural products. In: 7th international conference on High Pressure Bioscience and Biotechnology (HPBB 2012) book of abstracts, p 41

Shimada S, Andou M, Naito N, Yamada N, Osumi M, Hayashi R (1993) Effects of hydrostatic pressure on the ultrastructure and leakage of internal substances in the yeast *Saccharomyces cerevisiae*. Appl Microbiol Biotechnol 40:123–131

Tangwongchai R, Ledward DA, Ames JM (2000) Effect of high-pressure treatment on the texture of cherry tomato. J Agric Food Chem 48:1434–1441

Ueno S, Shigematsu T, Kuga K, Saito M, Hayashi M, Fujii T (2009a) High-pressure induced transformation of onion (in Japanese). Jpn J Food Eng 10:37–43

Ueno S, Hayashi M, Shigematsu T, Fujii T (2009b) Formation of green-blue compounds in *Brassica rapa* root by high pressure processing and subsequent storage. Biosci Biotechnol Biochem 73:943–945

Ueno S, Izumi T, Fujii T (2009c) Estimation of damage to cells of Japanese radish induced by high pressure with drying rate as index. Biosci Biotechnol Biochem 73:1699–1703

Ueno S, Shigematsu T, Watanabe T, Nakajima K, Murakami M, Hayashi M, Fujii T (2010) Generation of free amino acids and γ-aminobutyric acid in water-soaked soybean by high-hydrostatic pressure processing. J Agric Food Chem 58:1208–1213

Wuytack EY, Michiels CW (2001) A study on the effects of high pressure and heat on *Bacillus subtilis* spores at low pH. Int J Food Microbiol 64:333–341

ZoBell CE (1970) Pressure effects on morphology and life processes of bacteria. In: Zimmerman AM (ed) High pressure effects on cellular processes. Academic, New York, pp 85–130

Chapter 25
Use of Pressure for Improving Storage Quality of Fresh-Cut Produce

Hidemi Izumi

Abstract The microflora of fresh-cut produce is comprised primarily of phytopathogenic and soilborne organisms, but the product could be contaminated with foodborne pathogens. Populations of bacteria, molds, and yeasts associated with fresh-cut produce decreased to non-detectable levels following a high pressure (HP) treatment of 400 MPa for 10 min at room temperature, except for spore-forming bacteria such as *Bacillus* spp. which were inactivated when subjected to 600 MPa at 60 °C for 10 min. The HP treatment of 400 MPa for 5–10 min at room temperature for fresh-cut lotus root and pineapple may be commercially feasible as an alternative to chemical sterilization and thermal blanching, respectively. The HP treatment reduced the epiphytic microorganisms of the products to non-detectable levels, and the microbial counts remained at the initial levels during storage at 1 °C with minimal changes in physicochemical and visual quality of the products. However, the HP treatment induced cellular disruption in plant tissue that contributed to the changes in appearance of several fresh-cut vegetables. To improve storage quality, combining lower pressures with complementary technologies should be useful for successful application of HP for other fresh-cut produce.

Keywords Fresh-cut produce • High pressure • Microbiological quality • Physicochemical quality • Storage stability

25.1 Introduction

The United States Food and Drug Administration (FDA) defines fresh-cut produce as fresh fruits and vegetables for human consumption that have been minimally processed and altered in form by peeling, slicing, chopping, coring, or trimming with or without washing, prior to being packaged for use by the consumer or a retail establishment (Anonymous 2008). The market for fresh-cut produce continues to

H. Izumi (✉)
Faculty of Biology-Oriented Science and Technology, Kinki University,
Kinokawa 649-6493, Japan
e-mail: izumi@waka.kindai.ac.jp

K. Akasaka, H. Matsuki (eds.), *High Pressure Bioscience*,
Subcellular Biochemistry 72, DOI 10.1007/978-94-017-9918-8_25

grow in both retail and food service sectors in the U.S., the European Union, and Asian countries due to convenience, freshness, and nutrition of fresh-cut produce for consumers. However, the number of foodborne outbreaks and illnesses due to fresh and fresh-cut produce has been increasing with the increased production of fresh-cut produce. Indeed, fresh produce and produce dishes were responsible for 696 outbreaks and 25,222 illnesses in the U.S. between 2001 and 2010, a larger number than for seafood, poultry, beef, pork, and dairy categories (DeWaal and Glassman 2013). Fresh produce becomes contaminated with spoilage microorganisms and foodborne pathogens during production, harvesting, and processing, and the damaged cells and tissues that results from fresh-cut operations render the products more conducive to microbial growth. In the fresh-cut industry, product washing with sanitizers has been an important step for reducing the microbial population during processing, but the chemical treatments did not assure complete elimination of foodborne pathogens. Thus, physical treatment without any damage for the product should be considered as an alternative to chemical treatments for controlling microbial quality and safety of fresh-cut produce.

Since Hite et al. (1914) has been successful inactivating some microorganisms in culture media and sanitizing some fruits and fruit juices by high pressure (HP) technology, a number of studies on HP effects on microorganisms and fresh and processed fruits and vegetables have been conducted and reviewed by Hoover et al. (1989), Patterson (2005), and Gonzalez and Barrett (2010). They indicated that HP ranging from 100 to 900 MPa induced microbial inactivation by affecting the cell morphology, cell membrane, cell wall, biochemical reactions, and genetic mechanisms of microorganisms. The HP between 400 and 700 MPa is available in commercial systems for fruit jams and sauces, guacamole, sliced cooked hams, oysters, meal kits that contain meat, and salsa to reduce microorganism and inactivate harmful enzymes (San Martín et al. 2002; Patterson 2005). With liquid food products of fruits and vegetables, the HP of 300–600 MPa for 3–20 min at 20–25 °C has been reported to be beneficial in reducing the epiphytic microorganisms to non-detectable levels in apple juice (Muñoz et al. 2007; McKay 2009; Donsì et al. 2010), orange juice (Doğan and Erkmen 2003; Muñoz et al. 2007; Donsì et al. 2010), peach juice (Doğan and Erkmen 2003), cashew apple juice (Lavinas et al. 2008), tomato juice (Porretta et al. 1995), low-alcohol grape wine (Mok et al. 2006), vegetable soup (Muñoz et al. 2007), and ranch dressing including vegetables (Waite et al. 2009) with only minimal effects on the product itself. The HP treatment also eliminated *Escherichia coli* (Doğan and Erkmen 2003; Muñoz et al. 2007; Lavinas et al. 2008), *E. coli* O157 (Jordan et al. 2001; Whitney et al. 2008), and *Listeria monocytogenes* (Jordan et al. 2001) inoculated into the fruit juices. However, there are few studies on successful application of HP to fresh vegetables (Arroyo et al. 1997; Gonzalez et al. 2010), and little information exists on the effect of HP on the microbiological safety and shelf life of fresh-cut produce.

This chapter focuses on the influence of HP on inactivation of microorganisms associated with fresh-cut produce and microbiological and physicochemical quality of fresh-cut produce during storage. This will contribute to a better understanding of HP technology for fresh-cut produce. This technology appears to be commercially viable for the fresh-cut industry.

25.2 *In Vitro* Study on Microorganisms Associated with Fresh-Cut Produce

The microflora characteristic of fresh produce during its growth will persist after harvest and are therefore commonly found on fresh-cut produce. The microbial count is generally higher on fresh vegetables than fresh fruits, and approximately 80 % of the total isolates are bacteria in vegetables and molds in fruits (Izumi 2010). These epiphytic bacteria and fungi are mostly phytopathogenic and soilborne organisms that do not usually represent a public health concern.

Bacteria, molds, and yeasts associated with fresh-cut produce were processed at HP of 200, 400, and 600 MPa for 10 min at room temperature (≈23 °C) (Table 25.1). The treatment time and temperature were fixed at 10 min and room temperature, respectively, based on a preliminary study for practical use. The initial population of all microorganisms ranging from 4.8 to 7.0 CFU ml^{-1} were reduced to levels below the lower limit of detection or non-detectable levels at 400 MPa, except for spore-forming bacteria such as *Bacillus* spp. The population of *Bacillus cereus* and *B. thuringiensis* decreased by only 2.4 and 1.7 logs at 600 MPa, respectively. When

Table 25.1 Population of microorganisms associated with fresh-cut produce after HP treatment of 200, 400, and 600 MPa for 10 min at room temperature

	Mean ± SE population (Log CFU ml^{-1})			
Microorganism	Nontreatment	200 MPa	400 MPa	600 MPa
Gram negative bacteria				
Enterobacter hormaechei	5.1 ± 0.1^{a}	2.0 ± 0.1^{b}	ND	ND
Enterobacter kobei	5.7 ± 0.0^{a}	2.8 ± 0.2^{b}	ND	ND
Pseudomonas fulva	4.8 ± 0.0^{a}	2.8 ± 0.2^{b}	ND	ND
Pseudomonas nitroreducens	5.7 ± 0.0^{a}	5.1 ± 0.1^{b}	ND	ND
Escherichia coli	6.1 ± 0.0^{a}	4.9 ± 0.0^{b}	ND	ND
Gram positive bacteria				
Lactococcus lactis subsp. *lactis*	4.8 ± 0.0^{a}	4.8 ± 0.0^{a}	ND	ND
Lactobacillus curvatus	5.3 ± 0.1^{a}	5.2 ± 0.0^{a}	4.4 ± 0.1^{b}	ND
Bacillus cereus	5.5 ± 0.0^{a}	4.3 ± 0.0^{b}	4.1 ± 0.0^{c}	3.1 ± 0.0^{d}
Bacillus thuringiensis	5.2 ± 0.0^{a}	4.4 ± 0.1^{b}	4.2 ± 0.1^{c}	3.5 ± 0.0^{d}
Staphylococcus aureus	5.7 ± 0.1^{a}	5.3 ± 0.0^{b}	<1.4	ND
Molds				
Fusarium tricinctum	5.3 ± 0.0	ND	ND	ND
Penicillium solitum	6.3 ± 0.0^{a}	5.5 ± 0.1^{b}	ND	ND
Yeasts				
Candida rugopelliculosa	6.4 ± 0.1^{a}	5.5 ± 0.0^{b}	ND	ND
Pichia membranifaciens	7.0 ± 0.0^{a}	6.4 ± 0.0^{b}	ND	ND

[abcd]: Means with different letters within each microorganisms in detectable levels are significantly different ($p < 0.05$)
ND: No colonies detected on a plate of the original culture
<1.4: Below the detection level (fewer than 30 colonies on a plate of the original culture)

these bacteria were subjected to 600 MPa at 60 °C for 10 min, the population (5.1 log CFU ml^{-1}) decreased to non-detectable levels (data not shown). This result was in confirmation of other reports on effect of HP in combination with heat (95–105 °C) on inactivation of spores of *B. stearothermophilus* (Furukawa and Hayakawa 2000) and *B. amyloliquefaciens* (Ratphitagsanti et al. 2009), although mild pressure treatments (60 MPa at 40 °C) induced spore germination of *Bacillus* spp. through activation of the spore's nutrient receptors (Shigeta et al. 2007). In addition to these spoilage microorganisms, *E. coli* O157 (Gill and Ramaswamy 2008), *Salmonella typhimurium* (Arroyo et al. 1997), *L. monocytogenes* (Préstamo et al. 1999), and *Aeromonas hydrophila* (Durães-Carvalho et al. 2012) were completely inactivated by HP of 350–600 MPa for 3–10 min at room temperature, although the occurrence of pressure-resistant strains has been reported in *E. coli* O157:H7 (Jordan et al. 2001) and *L. monocytogenes* (Van Boeijen et al. 2008).

The resistance of microorganisms varies depending on the pressure applied, holding time, temperature, and type of microorganisms. Generally, gram-positive bacteria are more resistant to pressure than gram-negative bacteria, yeasts and molds; the most resistant are bacterial spores, as shown in Table 25.1. The cell membrane of microorganisms is known to be a primary site of pressure damage relative to cell walls and nucleic acids, although the mechanisms of microbial inactivation are still not fully understood (Hoover et al. 1989; Patterson 2005). The difference in pressure resistance between gram-positive and gram-negative bacteria may be attributed to the difference in structure of cell wall, and the more complex structure of spores than vegetative cells may contribute to the higher resistance of spores to pressure. Volume changes in proteins by HP also may be involved in the microbial inactivation, because high-pressure treatment favours biochemical reactions that lead to a volume decrease while it can inhibit or retard reactions that lead to a volume increase (Patterson 2005). It is likely that the resistance of microorganisms to HP will vary greatly due to the results of a combination of these factors.

25.3 *In Vivo* Study on Fresh-Cut Vegetables

25.3.1 Microbiological and Physicochemical Quality

When several fresh-cut vegetables (lettuce, carrots, cucumber, edible burdock root, and lotus root) were pressurized at 400 MPa for 10 min at room temperature, the treatment resulted in a large reduction of mesophilic aerobic bacteria (mesophiles), coliform groups (coliforms), and lactic acid bacteria to levels below the limit of detection or non-detectable levels, except for the mesophile counts in cucumber that contains *Bacillus* spp. (Table 25.2). When fresh-cut lettuce and carrots inoculated with *E. coli* and *Staphylococcus aureus* were subjected to the HP treatment, the microbial counts decreased from 5.2 to 5.4 log CFU g^{-1} for *E. coli* and 5.4 to

Table 25.2 Counts of mesophilic aerobic bacteria (mesophiles), coliform groups (coliforms), and lactic acid bacteria of fresh-cut vegetables either untreated or treated with HP at 400 MPa for 10 min at room temperature

	Mean ± SE population (Log CFU ml^{-1})					
	Mesophiles		Coliforms		Lactic acid bacteria	
Fresh-cut vegetables	Nontreatment	HP	Nontreatment	HP	Nontreatment	HP
Chopped lettuce	5.3 ± 0.1	ND	2.7 ± 0.1	ND	ND	ND
Carrot shreds	3.6 ± 0.0	<2.4	2.8 ± 0.1	ND	<2.4	ND
Carrot sticks	3.6 ± 0.1	ND	2.9 ± 0.1	ND	<2.4	ND
Cucumber slices	5.7 ± 0.1	3.6 ± 0.2 *	3.6 ± 0.1	ND	2.9 ± 0.0	<2.4
Edible burdock shreds	4.8 ± 0.2	<2.4	3.2 ± 0.2	ND	ND	ND
Lotus root slices	6.7 ± 0.0	<2.4	4.5 ± 0.0	ND	2.9 ± 0.1	ND

ND: No colonies detected on a plate of the original culture
<2.4: Below the detection level (fewer than 30 colonies on a plate of the original culture)
*Significant at $p < 0.05$ between paired nontreatment and HP treatment in detectable levels

5.8 log CFU g^{-1} for *S. aureus* to below the level of detection (data not shown). The initial population of fungi in fresh-cut lettuce (5.3 log CFU g^{-1}), carrots (3.5 log CFU g^{-1}), and cucumber (4.0 log CFU g^{-1}) also became undetected after the HP treatment (Wendakoon et al. 2010). These results indicate that the microbicidal effect of HP on fresh-cut vegetables is similar to that on fruit juices (Muñoz et al. 2007; Lavinas et al. 2008; McKay 2009; Donsì et al. 2010) and vegetable juices (Porretta et al. 1995; Muñoz et al. 2007).

Chopped lettuce, carrot sticks, and edible burdock shreds became brown as indicated by the decreasing L* value and carrot shreds softened as indicated by the decreasing texture expressed as a maximum shear force (N) after HP treatment of 400 MPa for 10 min at room temperature (Table 25.3). The HP treatment increased the rate of electrolyte leakage from tissue slices to 80–97 % with all fresh-cut vegetables. Since electrolyte leakage rate is an index of cell membrane permeability (King and Ludford 1983), the HP treatment may have induced cell membrane rupture and/or cell death of fresh-cuts. A scanning electron microscope (SEM) study of lotus root tissues revealed severe damage to the cell wall after the HP treatment (Fig. 25.1). Besides these undesirable effects, the HP treatment of lotus root reduced the activity of phenyl alanine ammonia lyase (PAL) that is a key enzyme in phenolic synthesis and responsible for oxidizing browning (López-Gálvez et al. 1996) as well as polyphenol oxidase (PPO) (Couture et al. 1993) (data not shown). Miao et al. (2011) also reported that the PAL activity of fresh-cut water bamboo shoot was inhibited by the HP treatment of 600 MPa for 10 min at room temperature. Although the effects of HP on enzymes are dependent on type of enzyme, source, substrate, pressure, time, and temperature (San Martín et al. 2002), model studies in a phosphate buffer showed the following rank of the enzymes according to their pressure induced inactivation: lipoxygenase, lactoperoxidase, pectinesterase, lipase, phosphatase, catalase, polyphenol oxidase, peroxidase (Seyderhelm et al. 1996).

Table 25.3 Lightness (L* value), texture, and electrolyte leakage rate of fresh-cut vegetables either untreated or treated with HP at 400 MPa for 10 min at room temperature

	Mean ± SE value								
	L* value			Texture (N)			Electrolyte leakage (%)		
Fresh-cut vegetables	Nontreatment	HP		Nontreatment	HP		Nontreatment	HP	
Chopped lettuce	66.3 ± 0.3	59.6 ± 2.5	*	2.0 ± 0.1	1.8 ± 0.2		18.5 ± 0.3	79.6 ± 9.2	*
Carrot shreds	47.4 ± 0.9	47.4 ± 1.7		0.6 ± 0.0	0.5 ± 0.0		44.2 ± 1.5	96.5 ± 0.1	*
Carrot sticks	64.3 ± 1.0	46.6 ± 0.6	*	5.1 ± 0.4	2.3 ± 0.3	*	8.2 ± 0.2	91.7 ± 0.5	*
Edible burdock shreds	40.1 ± 3.2	28.7 ± 1.1	*	2.5 ± 0.1	2.3 ± 0.3		31.8 ± 3.4	89.5 ± 1.8	*
Lotus root slices	63.7 ± 1.0	60.1 ± 3.9		8.6 ± 0.8	8.5 ± 1.5		25.8 ± 0.7	93.0 ± 0.6	*

*Significant at $p < 0.05$ between paired nontreatment and HP treatment

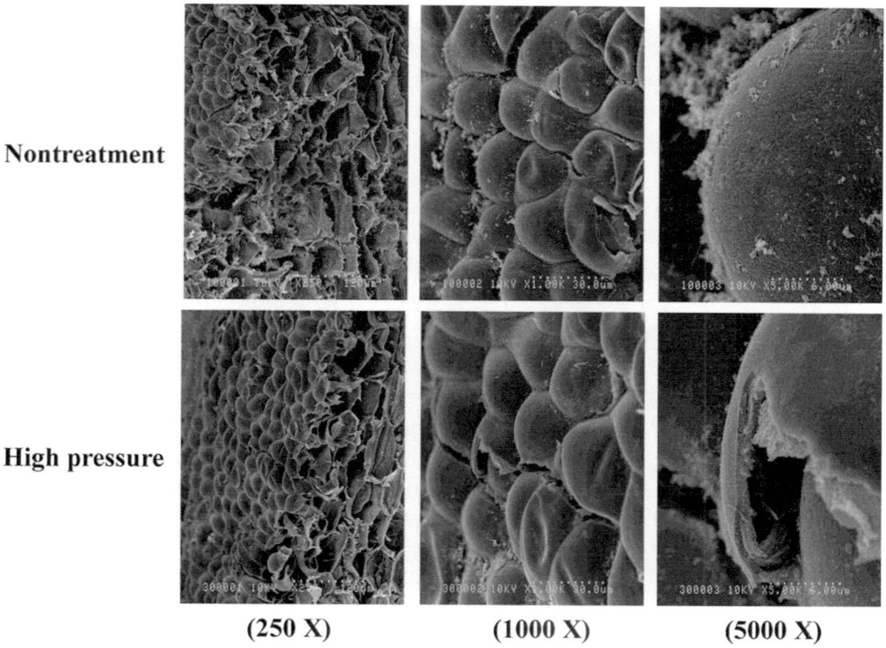

Fig. 25.1 Scanning electron microscopy photographs (250×, 1,000×, and 5,000×) of fresh-cut lotus root either untreated or treated with HP at 400 MPa for 10 min in film packages filled with water

It is known that HP affects not only microorganisms but also plant cells and tissues (Hoover et al. 1989; Gonzalez and Barrett 2010). In the HP treatments, the loss of membrane integrity commenced at 200–220 MPa due to protein unfolding and interface separation and total loss of membrane integrity occurred at 300 MPa and above in a model system of protein and lipid membrane (Kato et al. 2002) and onion cells (Gonzalez et al. 2010). With fresh-cut vegetables, the application of HP at 400 MPa caused cell wall disruption and intercellular component dispersion throughout the plant tissue, which would expose phenolic compounds to PPO and be associated with subsequent tissue browning in chopped lettuce, carrot shreds, and edible burdock shreds. The cellular disruption in relation to pectin degradation may have resulted in textural changes in carrot sticks as noted with carrot disks (Kato et al. 1997) and onion slices (Gonzalez et al. 2010). Therefore, HP treatment of some fresh-cut vegetables seems to be problematic due to the physical disruption of the cell, as shown with whole tomatoes and baby lettuce (Arroyo et al. 1997).

25.3.2 Storage Stability

Generally, fresh-cut vegetables are placed in a package and hermetically sealed for their distribution, which is called modified atmosphere packaging (MAP). The shelf life of some products is limited by the occurrence of browning discoloration during storage and transportation. In Japan, fresh-cut lotus root is usually distributed in film packages filled with water, in which the development of oxidizing browning could be reduced as compared to MAP. Hence storage quality of fresh-cut lotus root was evaluated after HP treatment with 400 MPa for 10 min at room temperature, held in film packaged with water and stored at 10 °C, 5 °C, and 1 °C. Although the transportation temperature of fresh-cut produce is commercially 10 °C in Japan, temperatures lower than 10 °C were included to extend the shelf life. The HP treatment reduced the counts of mesophiles (5.1 log CFU g^{-1}), coliforms (3.8 log CFU g^{-1}), and lactic acid bacteria (<2.4 log CFU g^{-1}) to non-detectable levels on fresh-cut lotus root and the bacterial counts remained below the limit of detection until day 2 at 10 °C, day 4 at 5 °C, and day 10 at 1 °C (Table 25.4). This trend was similar to other reports that recovery of bacteria during storage at 4 °C was not observed with HP treated fresh-cut mung bean sprouts (Muñoz et al. 2006) and pineapple (Alemán et al. 1998).

When surface color index (a* value) representing the green-red spectrum was used to interpret color differences of reddish browning among the pressurized slices stored at 10 °C, 5 °C, and 1 °C, the brown discoloration was less at 5 °C than at 10 °C and the fresh-cut lotus root did not develop browning during storage at 1 °C (Table 25.5). The rate of electrolyte leakage from tissue slices and the texture of fresh-cut lotus root remained at the initial level during storage at all temperatures (data not shown). These results indicate that HP treatment of 400 MPa for 10 min would be useful as a physical sterilization of fresh-cut lotus root when stored in film packages filled with water at 1 °C after treatment. Careful consideration for storage technology should be taken into account for using HP treatment, because different products react differently to HP in their physicochemical quality with correspondingly different storage effects.

25.4 *In Vivo* Study on Fresh-Cut Fruits

Generally, fresh-cut vegetables and fruits are washed after and before cutting, respectively, but not submerged in hot water during the processing operations. Only pineapple cubes are commercially submitted to hot water blanching for 1–3 min to inactivate deleterious enzymes and microorganisms. However, heating sometimes causes loss of color, texture, nutrition, and antioxidant capacity of some produce and juices (Rattanathanalerk et al. 2005; Castro et al. 2008; Zheng and Lu 2011; Jaiswal et al. 2012). In contrast, HP treatment is expected to be less detrimental than thermal treatment to many of the covalent compounds responsible for the sensory and nutritional quality of fruits and vegetables (Sancho et al. 1999; Butz et al. 2002, 2003; Roldán-Marín et al. 2009), because most covalent bonds participating in the protein structure are not affected by the pressure used in food processing (Balny 2004). The HP of fresh-cut pineapple may become a potential alternative to the conventional thermal blanching.

Table 25.4 Counts of mesophilic aerobic bacteria (mesophiles), coliform groups (coliforms), and lactic acid bacteria of fresh-cut lotus root either untreated or treated with HP at 400 MPa for 10 min and stored in film packages filled with water at 10 °C, 5 °C, and 1 °C

Storage temperature	Days in storage	Mean ± SE population (Log CFU ml^{-1})							
		Mesophiles			Coliforms			Lactic acid bacteria	
		Nontreatment	HP		Nontreatment	HP		Nontreatment	HP
10 °C	0	5.1 ± 0.1	ND		3.8 ± 0.2	ND		<2.4	ND
	2	7.4 ± 0.1	<2.4		6.3 ± 0.1	ND		<2.4	ND
	4	8.4 ± 0.1	3.5 ± 0.4	*	8.2 ± 0.3	4.4 ± 0.2	*	3.0 ± 0.1	ND
5 °C	4	6.0 ± 0.1	<2.4		5.7 ± 0.1	<2.4		3.0 ± 0.2	<2.4
	6	7.2 ± 0.2	2.7 ± 0.3	*	6.6 ± 0.3	<2.4		4.8 ± 0.3	<2.4
	8	7.3 ± 0.1	3.7 ± 1.0	*	7.2 ± 0.1	2.7 ± 0.2	*	5.2 ± 0.3	<2.4
1 °C	4	5.0 ± 0.0	<2.4		4.8 ± 0.1	ND		<2.4	ND
	6	5.0 ± 0.5	ND		4.5 ± 0.5	<2.4		<2.4	ND
	8	4.9 ± 0.0	ND		5.7 ± 0.0	ND		<2.4	<2.4
	10	5.1 ± 0.7	ND		5.6 ± 0.6	ND		3.1 ± 0.2	<2.4

ND: No colonies detected on a plate of the original culture
<2.4: Below the detection level (fewer than 30 colonies on a plate of the original culture)
*Significant at $p < 0.05$ between paired nontreatment and HP treatment in detectable levels

Table 25.5 Color index (a* value) of fresh-cut vegetables either untreated or treated with HP at 400 MPa for 10 min and stored in film packages filled with water at 10 °C, 5 °C, and 1 °C

		Mean ± SE value		
Storage temperature	Days in storage	Nontreatment	HP	
10 °C	0	−1.8 ± 0.8	−3.5 ± 0.6	
	2	2.7 ± 1.3	11.4 ± 0.5	*
	4	7.0 ± 2.7	17.5 ± 2.1	*
5 °C	4	1.7 ± 1.6	4.3 ± 0.5	
	6	2.8 ± 0.6	4.7 ± 4.1	
	8	3.1 ± 0.7	6.4 ± 0.5	*
1 °C	4	2.9 ± 0.5	3.7 ± 0.9	
	6	2.3 ± 0.8	4.5 ± 0.6	
	8	4.4 ± 0.8	6.0 ± 0.7	
	10	5.2 ± 1.0	6.2 ± 0.4	

*Significant at $p < 0.05$ between paired nontreatment and HP treatment

Table 25.6 Microbiological and physicochemical quality attributes of fresh-cut pineapple cubes either untreated or treated with HP at 400 MPa for 10 min or thermal blanching in 97 °C water for 3 min

		Mean ± SE		
Quality attributes		Nontreatment	High pressure	Blanching
Microbial counts (Log CFU g^{-1})				
Mesophiles		4.5 ± 0.0	ND	ND
Coliforms		ND	ND	ND
Fungi		4.9 ± 0.0	ND	ND
Physicochemical quality				
Color index	L^*value	66.5 ± 1.5	57.2 ± 4.7	56.1 ± 1.8
	b^*value	40.8 ± 1.3^a	39.0 ± 2.6^{ab}	31.6 ± 2.4^b
Texture (N)		11.6 ± 1.2	13.5 ± 0.8	14.0 ± 1.3
Electrolyte leakage (%)		85.0 ± 1.1^b	89.3 ± 3.1^{ab}	96.1 ± 1.6^a

ND: No colonies detected on a plate of the original culture
^{ab}: Means with different letters within each quality attribute are significantly different ($P < 0.05$)

Thus, the quality of fresh-cut pineapple cubes was assessed following treatments of either HP of 400 MPa for 5 min at room temperature or thermal blanching in 97 °C water for 3 min and subsequent storage. Both treatments reduced the microbial counts from 4.5 log CFU g^{-1} for mesophiles and 4.9 log CFU g^{-1} for fungi to non-detectable levels on pineapple cubes (Table 25.6). Coliforms were not detected on all samples. Alemán et al. (1998) also reported that the HP treatment of 270 and 340 MPa for fresh-cut pineapple reduced the bacterial and yeast counts to levels below the limit of detection. In terms of physicochemical quality, thermally blanched cubes presented a lower color index (b* value) for the blue-yellow spectrum and a higher rate of electrolyte leakage from tissue slices that is an index of cell membrane permeability as compared to the untreated cubes, whereas pressure treated cubes showed a similar color index (L* and b* values),

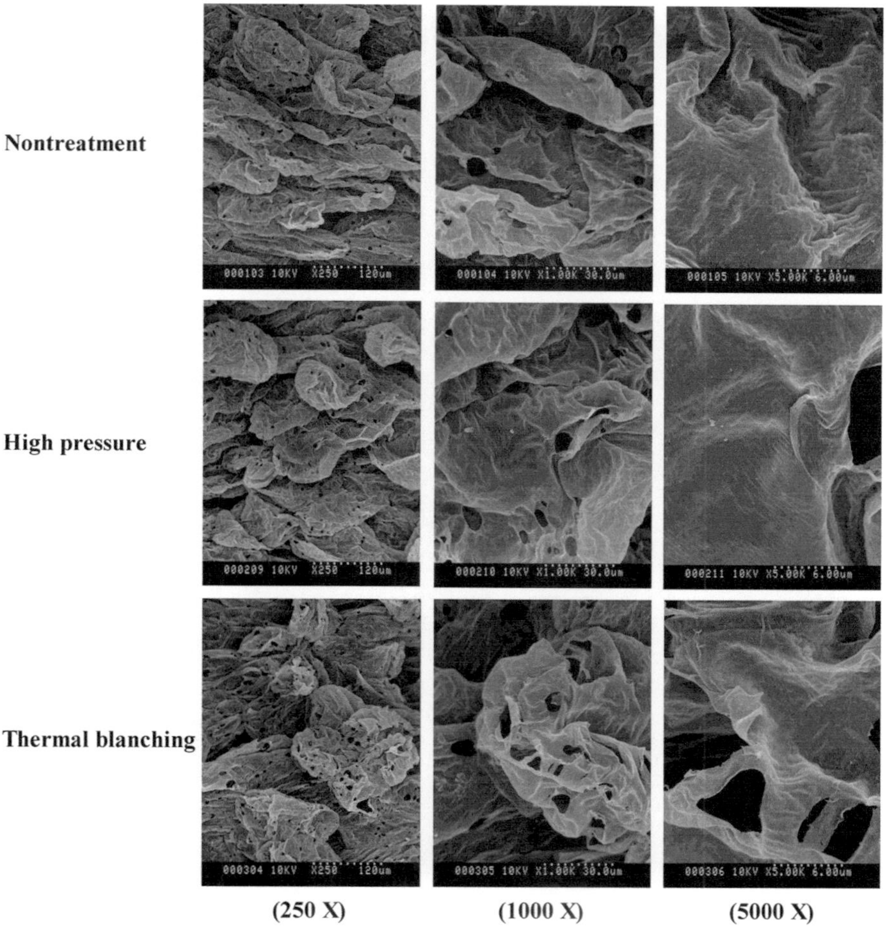

Fig. 25.2 Scanning electron microscopy photographs (250×, 1,000×, and 5,000×) of fresh-cut pineapple either untreated or treated with HP at 400 MPa for 10 min or thermal blanching in 97 °C water for 3 min

electrolyte leakage rate, and texture as the untreated cubes. The result of electrolyte leakage rate suggests that the thermal blanching may have induced severe damage to the cell membrane of pineapple as compared to the HP. A microstructural study revealed that both treatments did not cause either cell disruption or transformation in the tissues of fresh-cut pineapple (Fig. 25.2). Interestingly, holes of approximately 3 μm in diameter were found on the surface of cells, and the numbers of holes per cell were similar among untreated (17.2 ± 1.5), pressured (16.1 ± 1.5), and blanched (14.1 ± 0.5) samples. The cause for this phenomenon is unknown.

During storage of pineapple cubes in MAP at 1 °C for 6 days, counts of mesophiles and fungi increased to approximately 5 log CFU g^{-1} on untreated

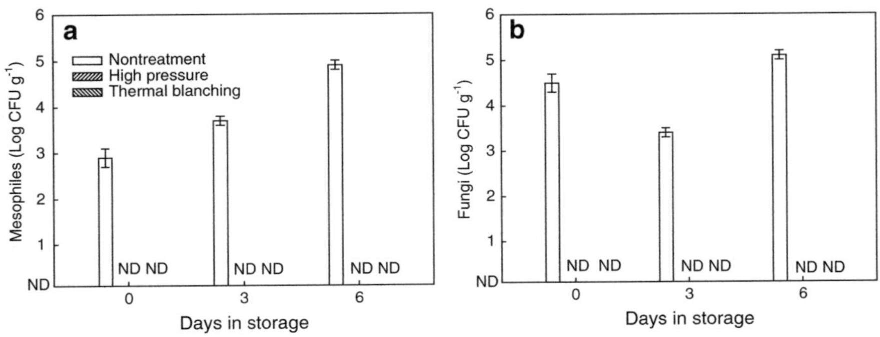

Fig. 25.3 Counts of mesophilic aerobic bacteria (mesophiles) (**a**) and fungi (**b**) of fresh-cut pineapple cubes either untreated or treated with HP at 400 MPa for 10 min or thermal blanching in 97 °C water for 3 min and stored in MAP at 1 °C

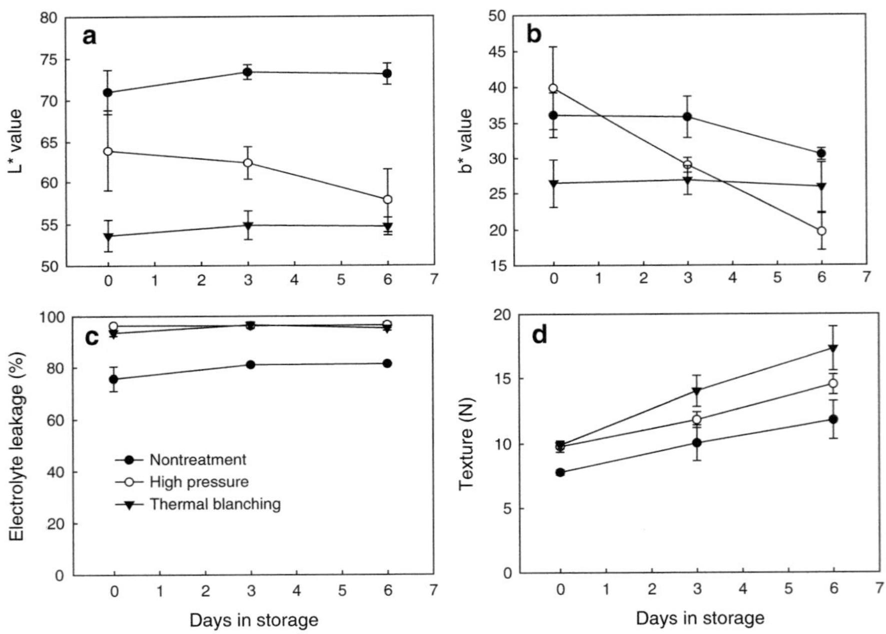

Fig. 25.4 Lightness (L* value) (**a**), color index (b* value) (**b**), electrolyte leakage rate (**c**), and texture (**d**) of fresh-cut pineapple cubes either untreated or treated with HP at 400 MPa for 10 min or thermal blanching in 97 °C water for 3 min and stored in MAP at 1 °C

samples but remained at non-detectable levels on either pressurised or blanched samples (Fig. 25.3). These results agree with other reports that no surviving microorganisms on fresh-cut pineapple treated by HP (Alemán et al. 1998) and thermal blanching combined with chemical treatments (López-Malo and Palou 2008) were detected during storage at low temperature. The HP treatment induced color deterioration with the decrease in L* and b* values, showed a higher rate of electrolyte leakage, and increased the texture of pineapple cubes relative to untreated samples during storage, but these quality changes were less when compared with blanched samples, except for the electrolyte leakage rate (Fig. 25.4). A reason for the color differences in L* and b* values may have been a non-enzymatic browning (Rattanathanalerk et al. 2005) or translucent appearance due to the diffusion of water and solutes across cell membranes (Montero-Calderón et al. 2008) of pineapple tissues. These results suggest that HP treatment of 400 MPa for 5 min would be a possible alternative to the conventional thermal blanching of fresh-cut pineapple.

25.5 Conclusions

The application of HP is uniform and instantaneous for all liquid and solid foods. The HP technology described in this chapter is directed towards inactivating microorganisms of fresh-cut produce to ensure the safety of the product to consumers. The HP treatment of fresh-cut lotus root and pineapple is considered a particularly promising alternative to chemical sterilization and thermal blanching, respectively. However, the HP induced changes in the appearance and delicate structure of some other vegetables. It renders them less suitable for consumption, because most of fresh-cut vegetables are used to be eaten raw as salads. Combining lower pressures with complementary technologies for quality control of fresh-cut produce is anticipated to reduce the microbial populations while avoiding the detrimental effects on organoleptic quality. For example, the use of various chemical antimicrobials (Whitney et al. 2008; Waite and Yousef 2008) and supercritical CO_2 (Spilimbergo et al. 2002) is being done for successful application of HP technology. Cost-effective HP treatment could be realized for fresh-cut produce in combination with other technologies.

References

Alemán GD, Ting EY, Farkas DF, Mordre SC, Hawes ACO, Torres JA (1998) Comparison of static and step-pulsed ultra-high pressure on the microbial stability of fresh cut pineapple. J Sci Food Agric 76:383–388

Anonymous (2008) Guidance for industry: guide to minimize microbial food safety hazards of fresh-cut fruits and vegetables. Center for Food Safety and Applied Nutrition, Washington, DC

Arroyo G, Sanz PD, Préstamo G (1997) Effect of high pressure on the reduction of microbial populations in vegetables. J Appl Microbiol 82:735–742

Balny C (2004) Pressure effects on weak interactions in biological systems. J Phys Condens Matter 16:S1245–S1253

Butz P, Edenharder R, Fernández GA, Fister H, Merkel C, Tauscher B (2002) Changes in functional properties of vegetables induced by high pressure treatment. Food Res Int 35:295–300

Butz P, Fernández GA, Lindauer R, Dieterich S, Bognár A, Tauscher B (2003) Influence of ultra high pressure processing on fruit and vegetable products. J Food Eng 56:233–236

Castro SM, Saraiva JA, Lopes-da-Silva JA, Delgadillo I, Van Loey A, Smout C, Hendrickx M (2008) Effect of thermal blanching and of high pressure treatments on sweet green and red bell pepper fruits (*Capsicum annuum* L.). Food Chem 107:1436–1449

Couture R, Cantwell MI, Ke D, Saltveit ME (1993) Physiological attributes related to quality attributes and storage life of minimally processed lettuce. HortScience 28:723–725

DeWaal CS, Glassman M (2013) Outbreak Alert! 2001–2010. Center for Science in the Public Interest. http://cspinet.org/new/pdf/outbreak_alert_2013_final.pdf. Accessed 10 Sept 2013

Doğan C, Erkmen O (2003) Ultra high hydrostatic pressure inactivation of *Escherichia coli* in milk, and orange and peach juices. Food Sci Tech Int 9:403–407

Donsì G, Ferrari G, Maresca P (2010) Pasteurization of fruit juices by means of a pulsed high pressure process. J Food Sci 75:E169–E177

Durães-Carvalho R, Souza AR, Martins LM, Sprogis ACS, Bispo JAC, Bonafe CFS, Yano T (2012) Effect of high hydrostatic pressure on *Aeromonas hydrophila* AH 191 growth in milk. J Food Sci 77:M417–M424

Furukawa S, Hayakawa I (2000) Investigation of desirable hydrostatic pressure required to sterilize *Bacillus stearothermophilus* IFP 12550 spores and its sterilization properties in glucose, sodium chloride and ethanol solutions. Food Res Int 33:901–905

Gill AO, Ramaswamy HS (2008) Application of high pressure processing to kill *Escherichia coli* O157 in ready-to-eat meats. J Food Prot 71:2182–2189

Gonzalez ME, Barrett DM (2010) Thermal, high pressure, and electric field processing effects on plant cell membrane integrity and relevance to fruit and vegetable quality. J Food Sci 75:R121–R130

Gonzalez ME, Jernstedt JA, Slaughter DC, Barrett DM (2010) Influence of cell integrity on textural properties of raw, high pressure, and thermally processed onions. J Food Sci 75:E409–E416

Hite BH, Giddings NJ, Weakley CE (1914) The effect of pressure on certain microorganisms encountered in the preservation of fruits and vegetables. Bull West Virginia Univ Agric Exp Sta 146:1–67

Hoover DG, Metrick C, Papineau AM, Farkas DF, Knorr D (1989) Biological effects of high hydrostatic pressure on food microorganisms. Food Technol 43:99–107

Izumi H (2010) Microbiological safety of fresh produce from the farm-to-table food chain. Mem Faculty Biol Oriented Sci Technol Kinki Univ 26:3–11

Jaiswal AK, Gupta S, Abu-Ghannam N (2012) Kinetic evaluation of colour, texture, polyphenols and antioxidant capacity of Irish York cabbage after blanching treatment. Food Chem 131:63–72

Jordan SL, Pascual C, Bracey E, Mackey BM (2001) Inactivation and injury of pressure-resistant strains of *Escherichia coli* O157 and *Listeria monocytogenes* in fruit juices. J Appl Microbiol 91:463–469

Kato N, Teramoto A, Fuchigami M (1997) Pectic substance degradation and texture of carrots as affected by pressurization. J Food Sci 62:359–362, 398

Kato M, Hayashi R, Tsuda T, Taniguchi K (2002) High pressure-induced changes of biological membrane – study on the membrane-bound Na^{+}/K^{+}-ATPase as a model system. Eur J Biochem 269:110–118

King MM, Ludford PM (1983) Chilling injury and electrolyte leakage in fruit of different tomato cultivars. J Am Soc Hort Sci 108:74–77

Lavinas FC, Miguel MAL, Lopes MLM, Mesquita VLV (2008) Effect of high hydrostatic pressure on cashew apple (*Anacardium occidentale* L.) juice preservation. J Food Sci 73:M273–M277

López-Gálvez G, Saltveit M, Cantwell M (1996) Wound-induced phenylalanine ammonia lyase activity: factors affecting its induction and correlation with the quality of minimally processed lettuces. Postharvest Biol Technol 9:223–233

López-Malo A, Palou E (2008) Storage stability of pineapple slices preserved by combined methods. Int J Food Sci Technol 43:289–295

McKay AM (2009) Inactivation of fungal spores in apple juice by high pressure homogenization. J Food Prot 72:2561–2564

Miao M, Wang Q, Zhang T, Jiang B (2011) Effect of high hydrostatic pressure (HHP) treatment on texture changes of water bamboo shoots cultivated in China. Postharvest Biol Technol 59:327–329

Mok C, Song K-T, Park Y-S, Lim S, Ruan R, Chen P (2006) High hydrostatic pressure pasteurization of red wine. J Food Sci 71:M265–M269

Montero-Calderón M, Rojas-Graü MA, Martín-Belloso O (2008) Effect of packaging conditions on quality and shelf-life of fresh-cut pineapple (*Ananas comosus*). Postharvest Biol Technol 50:182–189

Muñoz M, De Ancos B, Sánchez-Moreno C, Cano MP (2006) Evaluation of chemical and physical (high-pressure and temperature) treatments to improve the safety of minimally processed mung bean sprouts during refrigerated storage. J Food Prot 69:2395–2402

Muñoz M, De Ancos B, Sánchez-Moreno C, Cano MP (2007) Effects of high pressure and mild heat on endogenous microflora and on the inactivation and sublethal injury of *Escherichia coli* inoculated into fruit juices and vegetable soup. J Food Prot 70:1587–1593

Patterson MF (2005) Microbiology of pressure-treated foods. J Appl Microbiol 98:1400–1409

Porretta S, Birzi A, Ghizzoni C, Vicini E (1995) Effects of ultra-high hydrostatic pressure treatments on the quality of tomato juice. Food Chem 52:35–41

Préstamo G, Sanz PD, Fonberg-Broczek M, Arroyo G (1999) High pressure response of fruit jams contaminated with *Listeria monocytogenes*. Lett Appl Microbiol 28:313–316

Ratphitagsanti W, Ahn J, Balasubramaniam VM, Yousef AE (2009) Influence of pressurization rate and pressure pulsing on the inactivation of *Bacillus amyloliquefaciens* spores during pressure-assisted thermal processing. J Food Prot 72:775–782

Rattanathanalerk M, Chiewchan N, Srichumpoung W (2005) Effect of thermal processing on the quality loss of pineapple juice. J Food Eng 66:259–265

Roldán-Marín E, Sánchez-Moreno C, Lloría R, De Ancos B, Cano MP (2009) Onion high-pressure processing: flavonol content and antioxidant activity. LWT- Food Sci Technol 42:835–841

San Martín MF, Barbosa-Cánovas GV, Swanson BG (2002) Food processing by high hydrostatic pressure. Crit Rev Food Sci Nutr 42:627–645

Sancho F, Lambert Y, Demazeau G, Largeteau A, Bouvier J-M, Narbonne J-F (1999) Effect of ultra-high hydrostatic pressure on hydrosoluble vitamins. J Food Eng 39:247–253

Seyderhelm I, Boguslawski S, Michaelis G, Knorr D (1996) Pressure induced inactivation of selected food enzymes. J Food Sci 61:308–310

Shigeta Y, Aoyama Y, Okazaki T, Hagura Y, Suzuki K (2007) Hydrostatic pressure-induced germination and inactivation of *Bacillus* spores in the presence or absence of nutrients. Food Sci Technol Res 13:193–199

Spilimbergo S, Elvassore N, Bertucco A (2002) Microbial inactivation by high-pressure. J Supercritical Fluids 22:55–63

Van Boeijen IKH, Moezelaar R, Abee T, Zwietering MH (2008) Inactivation kinetics of three *Listeria monocytogenes* strains under high hydrostatic pressure. J Food Prot 71:2007–2013

Waite JG, Yousef AE (2008) Enhanced inactivation of foodborne pathogenic and spoilage bacteria by FD&C Red No. 3 and other xanthene derivatives during ultrahigh pressure processing. J Food Prot 71:1861–1867

Waite JG, Jones JM, Turek EJ, Dunne CP, Wright AO, Yang TCS, Beckwitt R, Yousef AE (2009) Production of shelf-stable ranch dressing using high-pressure processing. J Food Sci 74:M83–M93

Wendakoon SK, Matsuo H, Yamamoto H, Izumi H (2010) Effects of high pressure treatment on microbiological and organoleptic quality of fresh-cut vegetables. Acta Hort 875:297–301

Whitney BM, Williams RC, Eifert J, Marcy J (2008) High pressures in combination with antimicrobials to reduce *Escherichia coli* O157:H7 and *Salmonella* Agona in apple juice and orange juice. J Food Prot 71:820–824

Zheng H, Lu H (2011) Effect of microwave pretreatment on the kinetics of ascorbic acid degradation and peroxidase inactivation in different parts of green asparagus (*Asparagus officinalis* L.) during water blanching. Food Chem 128:1087–1093

Chapter 26
Application of High-Pressure Treatment to Enhancement of Functional Components in Agricultural Products and Development of Sterilized Foods

Eri Ohara, Mariko Kawamura, Miyuki Ogino, Eri Hoshino, Atsushi Kobayashi, Jun Hoshino, Akira Yamazaki, and Tadayuki Nishiumi

Abstract In 1988, the late Prof. Rikimaru Hayashi had first proposed "Use of High Pressure in Food", introducing his views, i.e., "heat and pressure are independently capable of transforming the state of a substance, and such state transforming factors are only heat and pressure in nature." Sc. D. Masaru Nakahara stated in his note that he had been impressed by the unique starting point of Prof. Hayashi's idea. Prof. Hayashi had explored some good method for food processing without using heat, so he alternatively thought of high-pressure treatment (Hayashi R (1989) Use of high pressure to food processing and preservation. In: Hayashi R (ed) Use of high pressure in food. San-Ei Publishing Co, Kyoto, pp 1–30; Nakahara M (1990) Water and ions at high pressure: their fundamental properties relevant to the pressure treatment to food. In: Hayashi R (ed) Pressure-processed food – research and development. San-Ei Publishing Co, Kyoto, pp 3–21). Since the start-up of Japanese research group of high pressure in biological field (the present "Japanese Research Group of High Pressure Bioscience and Biotechnology (JHPBB)") and "International Association of High Pressure Bioscience and Biotechnology (IAHPBB)", we have continued to research into the industrial use of high-pressure treatment over a period of 25 years to realize our dream, that is, the same as Prof. Hayashi's dream. Although heat and pressure were found to be independent factors capable of transforming the state of a substance, use of heat has been overwhelmingly more frequent in food processing up to now. However, the pressure treatment has the advantages of instantaneous transmission, uniform distribution in vessels, and ability of inducing uniform change in quality. The high-pressure treatment does not cause cleavage of the covalent bond in the substance, thereby lessening the

E. Ohara • M. Kawamura • E. Hoshino • A. Kobayashi • J. Hoshino • A. Yamazaki (✉)
Research Institute, Echigo Seika Co., Ltd, 1003-1, Takanashi, Ojiya, Niigata 947-0193, Japan
e-mail: a-yamazaki@echigoseika.co.jp

M. Ogino • T. Nishiumi
Department of Applied Biological Chemistry, Faculty of Agriculture, Niigata University, 8050, Ikarashi 2-nocho, Nishi-ku, Niigata 950-2181, Japan

K. Akasaka, H. Matsuki (eds.), *High Pressure Bioscience*,
Subcellular Biochemistry 72, DOI 10.1007/978-94-017-9918-8_26

decomposition of nutrients, the generation of offensive smell, and the production of abnormal materials when compared with the heat application. In addition, energy consumption in the high-pressure treatment is less than that in the heat treatment. For the reasons mentioned above, the high-pressure treatment has thus been regarded as suitable for future food processing, and much attention has been paid to the researches of high-pressure treatment again. Then, we reviewed the previous researches in which little interest had been taken because of imperfectness of non-heat sterilization. Surprisingly, we discovered some novel findings about the effect of high-pressure treatment, that is, pressure history on the subsequent event. Then, we decided to present two theses on the themes, "Application of High-pressure Treatment to Enhancement of Functional Components in Agricultural Products" and "Application of High-pressure Treatment to Development of Sterilized Foods".

Keywords High-pressure treatment • Sterilized foods • Functional components • Pressure history • Hi-Pit

26.1 Application of High-pressure Treatment to Enhancement of Functional Components in Agricultural Products

26.1.1 Introduction

The problem of food shortage has been raised worldwide. It is therefore an urgent and unavoidable issue to promote the efficient use of agricultural products. In this chapter, we present a surprising discovery that various functional components contained in agricultural products can be increased after the harvest to enhance the values of the agricultural products.

Agricultural products such as cereals, vegetables and fruits photosynthetically produce carbohydrates from water and carbon dioxide in the air, and generate functional components including amino acids and vitamins by the actions of endogenous enzymes. Such biosynthetic reactions are dominantly determined by the factors of the growing environment, for example, light, water and the concentration of oxygen, and the biosynthesis is mostly completed during the growing process of the agricultural products. Accordingly, the functional components have been considered to be unlikely to increase in the harvested products.

We found that the phenomena induced by a high-pressure treatment are combined with other physical, chemical or biological processing to cause another new phenomenon. We named this process "high-pressure induced transformation (Hi-Pit) (Yamazaki 2006)". In 1999, we discovered that gamma-aminobutyric acid (GABA) in brown rice can be increased by taking advantage of a change induced by the Hi-Pit effect. This discovery became a starting point for finding any other agricultural products enriched with the functional components.

Conventionally, when the high-pressure treatment is used for food processing, the optimal conditions for achieving the purposes such as sterilization, modification and the like, have been explored by controlling the three parameters, that is, the magnitude of pressure and the temperature in the course of the treatment, and the holding time of pressure application. When some biochemical reactions via enzymes are involved in the food processing, however, the steps before and after the high-pressure treatment should be regarded as part of the food processing and the optimal conditions should be explored from various angles to obtain desired results.

In the first part of this chapter, we introduce specific examples, i.e., the contents of GABA in brown rice, buckwheat and soybean, the contents of vitamin C in potato and avocado, and the behaviors of the lycopene content and the antioxidant capacity in tomato after the high-pressure treatment.

26.1.2 Destruction of Tissues and Acceleration of Enzyme Reactions by High-Pressure Treatment

When the agricultural products were soaked in water and subjected to high-pressure treatment, destruction of the cell walls occurred depending upon the magnitude of applied pressure. In Fig. 26.1, SEM photographs show the surfaces of polished rice grains. The rice grains in water were subjected to the high-pressure treatment at 25 °C (Sasagawa et al. 1994). By the application of a pressure of 300 MPa, the cells of the rice grain looked isotropically compressed and the starch granules were found to come out of the cells. When the pressure applied to the rice grains was increased to 500 MPa or more, the cells were compressed too strongly to recognize the original shape of the cell walls and the beginning of gelatinization was found in the starch granules coming out of the cells.

As stated above, due to destruction of the cell walls of the agricultural products by the high-pressure treatment, the water penetration into the cells was found to improve (Yamazaki et al. 1996) and the enzyme reaction between the endogenous enzyme in the cells of the agricultural products and the corresponding substrate was considered to accelerate, thereby increasing the functional components such as GABA, reducing sugars that are decomposition products of starches, vitamin, lycopene, allicin and the like.

26.1.3 Enhancement of GABA in Grains by High-Pressure Treatment

GABA, which is one of the functional components contained in the vegetables and cereals has recently attracted attention because GABA is considered to have

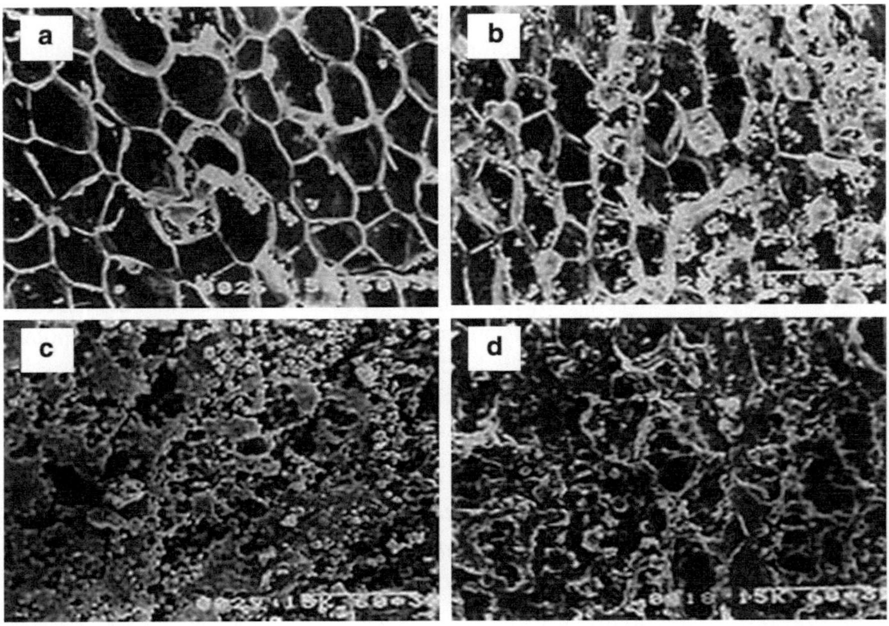

Fig. 26.1 SEM photographs showing the surface of polished rice grains subjected to high-pressure treatment at 20 °C for 10 min. (**a**) Control (not subjected to high-pressure treatment), (**b**) 300 MPa, (**c**) 500 MPa, (**d**) 700 MPa

not only a blood-pressure lowering effect, but also brain metabolism promoting effect, tranquilizing effect and renal function enhancement effect. It is known that GABA is produced from glutamic acid by the action of glutamic acid decarboxylase (GAD). Intensive studies have been conducted to increase the contents of GABA, for example, in a method of germinating brown rice (Saikusa et al. 1994) and a method of anaerobically treating tea leaves (Omori et al. 1987).

We have already reported that when cereals such as brown rice, buckwheat and soybean in water are subjected to high-pressure treatment and then allowed to stand under normal pressures, the resultant cereals can be enriched with GABA accumulation very efficiently (Kinefuchi et al. 1999a, b; Sasagawa et al. 2006). Figure 26.2 is a graph showing the contents of GABA in brown rice after high-pressure treatment. Each brown rice sample was drained after high-pressure treatment, and then allowed to stand for 18 h at 25 °C and a saturated humidity. The GABA content became larger as the magnitude of pressure applied in the treatment was increased up to 200 MPa. The higher pressure was considered to damage the cell walls more heavily because of an increase in the contact volume of the substrate with the enzyme. However, the pressure was set to 400 MPa, the enzyme itself responsible for generation of the GABA was supposed to be influenced by the pressure. This was considered to be a reason why the GABA content obtained after the high-pressure treatment of 400 MPa was lower than that after the treatment of 200 MPa.

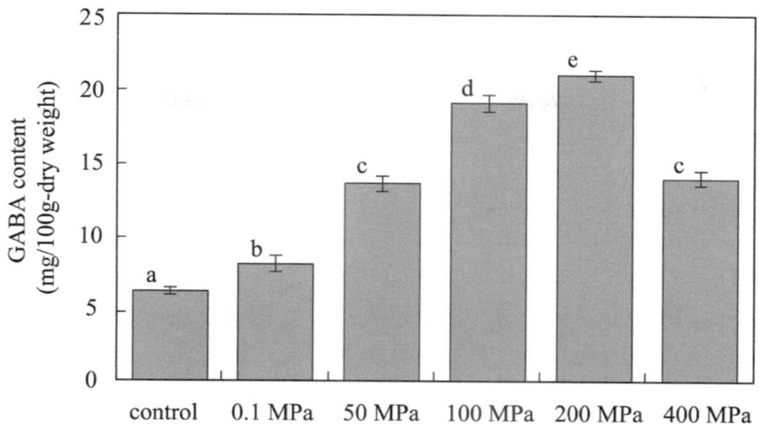

Fig. 26.2 Change in GABA content contained in brown rice after high-pressure treatment. Control is raw brown rice milled to 99 %. Other samples were soaked in water for 1 h, subjected to high-pressure treatment for 10 min at 25 °C, drained and then kept for 18 h at 25 °C and saturated humidity. GABA contents were determined after 18 h storage. 0.1 MPa (non-pressure treatment) was treated same as mentioned above, except for application of high-pressure treatment. There is a significant difference between the columns with the different letters ($p < 0.01$)

Similar to the brown rice, soybeans, black soybeans and buckwheat were also enriched with GABA accumulation. In any case, the GABA contents successfully increased about 20 times as much as those not subjected to the high-pressure treatment (Table 26.1).

To enrich the cereals with GABA accumulation, soaking of the cereals in water at 30 °C for 24 h or at 4 °C for 30 days is proposed, for example. In any case, however, special attention must be paid to the risk of GABA dissolving in water while the cereals are soaked in water, and the risk of propagation of microorganisms. In contrast to this, use of the high-pressure treatment has the advantage that glutamic acid and GABA can be prevented from dissolving in water only if water is strained out immediately after the high-pressure treatment. In addition, the time when the grains should be left to stand can be curtailed, so that the method using the high-pressure treatment can more efficiently increase the GABA content in the grains.

26.1.4 Enhancement of Vitamin C in Potato and Avocado by High-Pressure Treatment

Vitamin C (i.e., ascorbic acid, hereinafter referred to as AsA) contained in plants is biosynthesized via nine enzyme reactions using glucose as a substrate (Wheeler et al. 1998). Our experiments showed that potato and avocado, which are plants having relatively low AsA contents in nature can be enriched with AsA when subjected to a high-pressure treatment to promote the enzyme reactions (Sasagawa 2006).

Table 26.1 Increased functional component in food by taking advantage of Hi-Pit effect

			Content of functional component		High-pressure treatment		Left at normal pressures	
Food	Substrate	Functional component	Before treatment → after treatment	Multiplication factor	Optimal pressure (MPa)	Holding time (min)	Temperature (°C)	Duration
Brown rice	Glutamic acid	GABA	6.0 → 21 mg/100 g	3.5	200	2	25	15 h
Buckweat	Glutamic acid	GABA	2.7 → 50 mg/100 g	18.5	200	5	25	12 h
Soybean	Protein	Peptide	101 → 661 mg/100 ml	6.6	400	10	40	8 h
	Glutamic acid	GABA	8.2 → 161 mg/100 g	19.6	200	10	25	6 h
Black soybean	Glutamic acid	GABA	3.8 → 88 mg/100 g	23.2	200	10	25	6 h
Potato	Glucose	Vitamin C	35 → 70 mg/100 g	2.0	100	5	10	9 days
Avocado	Glucose	Vitamin C	15 → 48 mg/100 g	3.2	700	5	10	2 days
Tomato	Isopentenyl diphosphate	Lycopene	0.36 → 1.1 mg/100 g	3.1	400	10	20	24 h
Garlic	Alliin	Allicin	0 → 77.4 mg/100 g	–	400	5	10	9 days
Shiitake mushroom	Ribonucleic acid	Nucleic acid	CMP[a]: 21 → 36 mg/g	1.7	50	10	60	24 h
			IMP[b]: 4.5 → 20 mg/g	4.5				
Konbu (kelp)	Protein	Amino acid	Asp[c]: 9.5 → 12.2 mg/g	1.3	50	10	4	2 h
Onion[d]	Dihydroquercetin	Quercetin	92 → 206 μg/g	2.2	200	5	25	4 days

[a]Cytidine monophosphate
[b]Inosinic acid
[c]Aspartic acid
[d]Ueno et al. (2009)

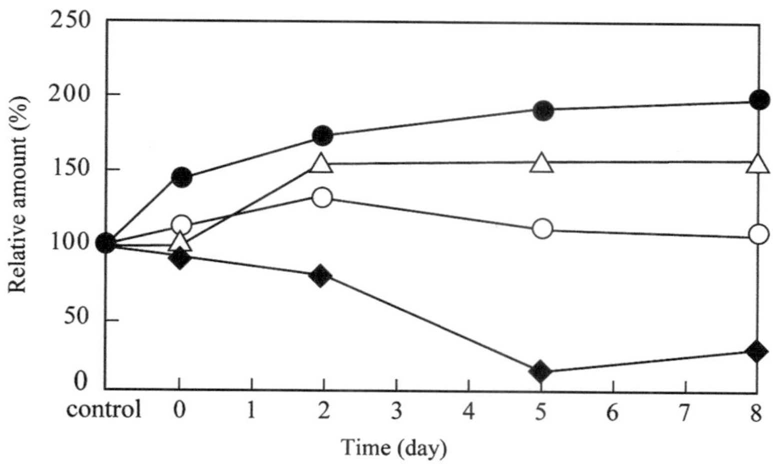

Fig. 26.3 Changes in vitamin C (AsA) contents in potatoes with time after high-pressure treatment. △; control (untreated), ○; 50 MPa, ●; 100 MPa, ♦; 700 MPa. Potato samples were subjected to high-pressure treatment for 5 min at 10 °C and then kept at 10 °C

Figure 26.3 is a graph showing changes in the AsA contents of potatoes with time after the potatoes were separately subjected to the respective high-pressure treatments and then refrigerated at 10 °C. The AsA content increased in the control group (not subjected to high-pressure treatment). The reason for this was considered to be as follows: starch was acceleratedly decomposed during the storage of potato at low temperatures, to generate glucose capable of serving as the substrate for the enzyme reactions. The increase of AsA content in the potato subjected to a high-pressure treatment of 100 MPa and the decrease of AsA content in the potato subjected to a high-pressure treatment of 700 MPa were both considered to be the results of accelerated enzyme reactions due to destruction of the cell walls by the high-pressure treatment. The AsA is known to be subject to oxidative decomposition in the presence of ascorbic acid oxidase (hereinafter referred to as AAO). In light of the above, both the reactions of generation and decomposition of AsA were considered to occur at the same time. Namely, each value of the AsA content shown in Fig. 26.3 was said to be a consequence of both the reactions of generation and decomposition of the AsA. In the high-pressure treatment of 700 MPa, the decrease in AsA content implied deactivation of the enzyme responsible for the enzyme reactions for biosynthesis and accelerated decomposition of the AsA due to destruction of the cell walls. In the high-pressure treatment of 50 MPa where destruction of the cell walls scarcely occurred, the AsA content became lower than that of the untreated potato. The reason for this was, we considered, that the stress of the pressure application produced active hydrogen in the cells, and the AsA is consumed to eliminate the active hydrogen.

Figure 26.4 is a graph showing changes of the AsA contents in avocados by the high-pressure treatment (50 MPa) depending on the degrees of ripeness. The

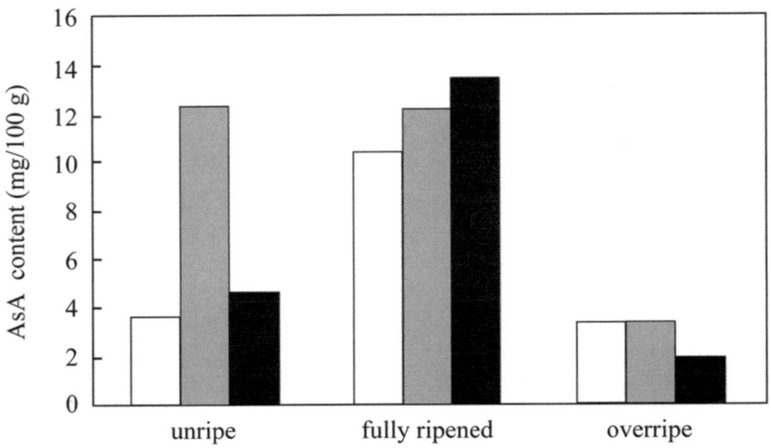

Fig. 26.4 Changes of vitamin C (AsA) contents after high-pressure treatment in avocados depending on degree of ripeness. □; control (untreated), ▨; immediately after high-pressure treatment, ■; 2 h later after high-pressure treatment. Avocado samples were subjected to high-pressure treatment of 50 MPa for 5 min at 10 °C and then kept at 10 °C

change in AsA content after the high-pressure treatment varied depending upon the degree of ripeness. In a fully ripened avocado, the AsA content continuously showed an increasing tendency for at least 2 h after completion of the high-pressure treatment. Then, in order to determine the AsA contents of fully ripened avocados, we further tried to subject the fully ripened avocados to the high-pressure treatments and refrigerate them. The results are shown in Fig. 26.5. When the pressure of 700 MPa was applied, the AsA content reached the maximum value on the second day after the high-pressure treatment, and then decreased. We inferred from such a tendency that destruction of the cell walls promoted the enzyme reactions, but the generated glucose available for the reactions were mostly consumed during the storage of 2 days, and after the second day, the decomposition reaction of AsA became predominant. In the case of potato, as mentioned above, the high-pressure treatment of 700 MPa was supposed to deactivate the enzyme responsible for the biosynthesis of AsA. In avocado, on the other hand, the AsA-producing enzyme was not deactivated because abundant lipid contained in the cells was considered to serve to protect the enzyme from being denatured due to pressure application. Generally, denaturation of protein due to the pressure application takes place in such a manner that the hydrophobic amino acid side chain of the protein is allowed to expose to water to cause hydration for compensating the decreased volume due to the pressure application. In the case of avocado, however, it is supposed that the lipids such as linoleic acid and oleic acid, and the hydrophobic group of enzyme exposed to water cause hydrophobic interaction, so that the hydration of enzyme is prevented and therefore the denaturation of the enzyme can be reduced. The increase of AsA content by the application of a pressure of 50 MPa was considered to be

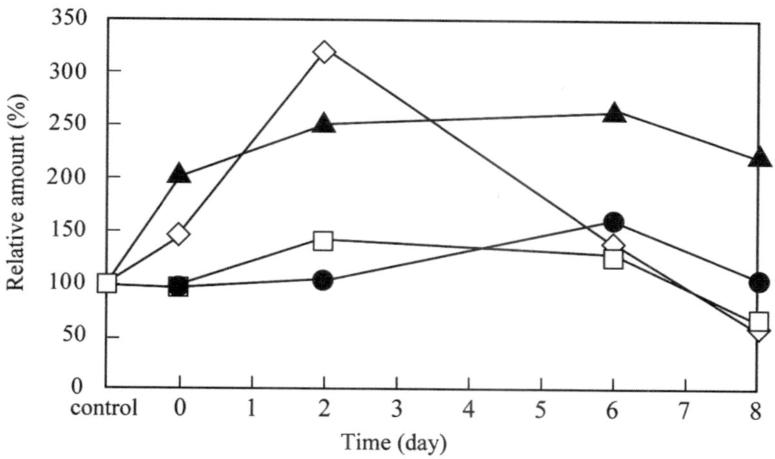

Fig. 26.5 Changes in vitamin C (AsA) contents in fully ripened avocados with time after high-pressure treatment. ●; control (untreated), ▲; 50 MPa, □; 100 MPa, ◇; 700 MPa. Fully ripened avocado samples were subjected to high-pressure treatment for 5 min at 10 °C and then kept at 10 °C

related to biodefense against the stress. When vegetables are exposed to the stress of injure such as cutting or the like, active hydrogen is generated. It has been reported that the AsA is consumed for trapping the active hydrogen, so that the synthesis of AsA is promoted (Oba 2002; Yamamoto and Oba 1999). We inferred from the above that the stress of pressure application activated the generation of AsA.

26.1.5 Enhancement of ORAC Value and Lycopene Content in Tomato by High-Pressure Treatment

The antioxidants contained in foods exhibit antioxidant effect by trapping active hydrogen and free radicals and inhibiting them from generating. Typical examples of the antioxidant include polyphenols such as anthocyanin and catechin, vitamin C, vitamin E, carotenoid and the like. It is common that foods contain two or more kinds of antioxidants, so that the oxygen radical absorbance capacity (ORAC) has been widely used as the indicator for expressing the total antioxidant capacity of the food. It is possible to determine the ORAC value including both L-ORAC value by the action of lipophilic antioxidants and H-ORAC value by the action of hydrophilic antioxidants.

Figure 26.6 shows changes in the L-ORAC value of tomatoes with time while tomatoes were kept at 20 °C after high-pressure treatment. Regardless of the magnitude of pressure at the high-pressure treatment, the L-ORAC values were found to increase after the high-pressure treatment. There was no particular change

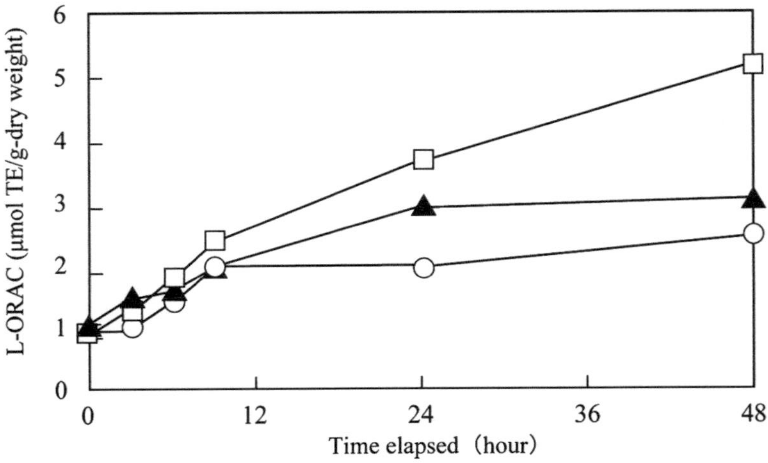

Fig. 26.6 Changes in L-ORAC value of tomatoes with time after high-pressure treatment. ○; control (untreated), ▲; 200 MPa, □; 400 MPa. *TE* Trolox equivalent. Tomato samples were subjected to high-pressure treatment for 5 min at 20 °C and then kept at 20 °C

in the H-ORAC value, the data of which is not shown in any figures. In view of the fact that the L-ORAC value of tomato increased after the high-pressure treatment, the content of lycopene in tomato was determined. The lycopene is the major lipophilic antioxidant contained in tomato. The results are shown in Fig. 26.7. The lycopene content also showed an increasing tendency after the high-pressure treatment. In particular, the lycopene content in the tomato sample group subjected to the high-pressure treatment of 400 MPa increased 3.1 times 24 h later. The reason for this was considered to be the accelerated enzyme reaction which was induced by the destruction of the cell walls of tomato when the pressure of 400 MPa was applied. However, the lycopene content decreased 48 h later, as shown in Fig. 26.7. The reactions for producing and decomposing lycopene were supposed to take place in tomato at the same time, and the decrease of lycopene content implied that the high-pressure treatment of 400 MPa affected some enzymes involved in the biosynthetic route to interrupt the biosynthesis of the substrates or intermediates. In the tomato sample group subjected to the high-pressure treatment of 400 MPa, the L-ORAC value significantly increased, though. Namely, the behavior of the L-ORAC value was not corresponding to that of the lycopene content in this tomato sample group. This suggested that other components contained in tomato, including polyphenols such as rutin and fat-soluble vitamins such as vitamin E and vitamin B should also contribute to the increase of the L-ORAC value. The increase of the ORAC value which is used as the total indicator of the antioxidant effect implied that a plurality of functional components concerned with the antioxidant capacity should be increased by the Hi-Pit effect.

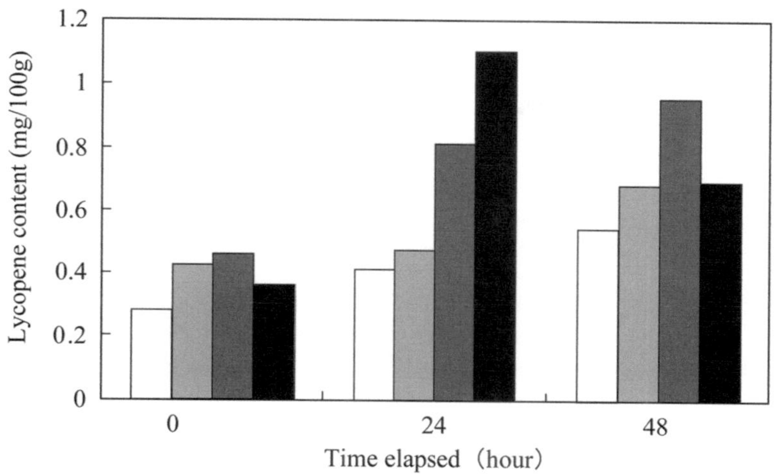

Fig. 26.7 Changes of lycopene contents in tomatoes with time after high-pressure treatment. □; control (untreated), ■; 100 MPa, ■; 200 MPa, ■; 400 MPa. Tomato samples were subjected to high-pressure treatment for 10 min at 20 °C and then kept at 20 °C

26.1.6 Enhancement of Allicin in Garlic by High-Pressure Treatment

U.S. National Cancer Institute concluded from the large-scale research that garlic may be the most potent food having cancer-preventive properties among foods such as vegetables and fruits (Caragay 1992). The scheme of how the functional components are produced in garlic are complicated, but can be summarized as follows (Fenwick and Hanley 1985): injury to the cells of garlic activates the enzyme allinase, which metabolizes alliin to allicin. Since allicin is so unstable that biosynthesis or thermolysis further converts allicin into another physiologically active substances such as ajoene, vinyldithiines, sulfides and the like. In view of the above-mentioned scheme, we assumed that the functional components derived from the precursor allicin can be increased if garlic is enriched with allicin that is produced from alliin via the enzyme reaction. Then, by subjecting garlic to a high-pressure treatment of 400 MPa to cause destruction of the cell walls, the amount of allicin generated after the high-pressure treatment was examined. Although no allicin was detected immediately after the high-pressure treatment, the amount of allicin attained to 77.4 mg 100 g^{-1} on the ninth day after the treatment, and then decreased (Table 26.1). It is commonly believed that allicin is produced when the garlic bulb is injured, for example, by grating. Advantageously, our experiments indicated that allicin can be produced with the whole bulb retaining its original shape, in virtue of the high-pressure treatment.

In addition to the foods specifically described in this chapter, various kinds of functional components in the harvested agricultural products, such as peptide in soybean, amino acid in shiitake mushroom, quercetin in onion and the like were successfully increased by the high-pressure treatment, as shown in Table 26.1.

26.1.7 Conclusions

In this chapter, we presented that the functional components in agricultural products can be increased after harvest by taking advantage of the Hi-Pit effect. When high pressure was applied to rice soaked in water, destruction of the cell walls of rice occurred depending upon the magnitude of pressure, and the increase of GABA content was observed during the storage after the high-pressure treatment. The increase in the GABA content was considered to be the result of accelerated enzyme reaction of the endogenous glutamic acid with GAD. Likewise, in each of other ten kinds of agricultural products, the increase in the content of functional component was confirmed. By the application of high pressure, non-covalent bond in the molecule of proteins and starches is subject to change, and the higher order structure of the molecule is decomposed, with the result that the area of part of the molecule in contact with water is increased and the hydration is therefore promoted (Gekko 1990). Most enzymes are composed of protein. Therefore, we considered that the enzyme begins to dissolve in water penetrating into the molecule of enzyme when the high-pressure treatment destroys the cell walls, and then the enzyme can come in contact with the substrate to accelerate the reaction. The functional components were considered to be thus increased. In light of the above, such biochemical reactions are deeply related to the presence of water. It is therefore necessary to pay attention to soaking of the target material in water prior to the high-pressure treatment in order to avoid the gas in the target material from being compressed by the high-pressure treatment.

The type of targeted functional component and the desired content of the functional component have been changing in accordance with the social demands and the users' requests. So, we should constantly try to formulate a new concept and explore the manufacturing conditions from various angles, as a matter of course, from economical viewpoint (Sasagawa and Yamazaki 2002). As shown in this chapter, the methods for increasing the functional components make use of the components originally contained in the materials themselves, so that the methods are advantageous in the safety of food and the manufacturing cost. Recently, there is increasing demands for health and long life. The way of increasing the functional components in the harvested agricultural products by taking advantage of the Hi-Pit effect as mentioned above is expected to propagate more and more as a promising and innovative method for enhancing the functions of food products without any addition.

26.2 Application of High-Pressure Treatment to Development of Sterilized Foods

26.2.1 Introduction

Since the late Prof. Rikimaru Hayashi first proposed "Use of High Pressure in Food" in 1988, application of high-pressure treatment to non-heat sterilization has become a big object to be investigated in Japan. The Japanese R & D Association for High Pressure Technology in Food Industry was established under the direction of Ministry of Agriculture, Forestry & Fisheries in 1989, and the research results were published in 1993. Along with the progress of the study, however, it has been found difficult to attain the commercial-level sterilization by the high-pressure treatment. Also, it has been demonstrated that the conditions for sterilization by high-pressure treatment are closely related to various ingredients and the properties in target foods (sugar, glucose, salt concentration, pH, water activity, fat globules and the like) (Taki et al. 1990, 1991; Hayakawa et al. 1995; Koseki and Yamamoto 2006). Especially against the heat-resistant spores the sterilization effect cannot readily be obtained even when the pressure of 400 MPa or more was applied for 1 h or more (Sonoike 1997). Consequently, it was impossible to find the conditions of high-pressure treatment which are industrially practicable and take advantage of non-heat sterilization.

In pursuit of the industrial application of the high-pressure treatment, we have come to indicate a production model of sterilized packaged rice by using a high-pressure treatment of 400 MPa for 10 min, which conditions are within the industrially practicable range (Sasagawa and Yamazaki 2002). We discussed comprehensive combinations of the pressure of the high-pressure treatment and the temperature of the heat sterilization to explore the possibility of sterilization, with the pressure being set to as low as possible within the range of 400 MPa or less. Yamazaki found that the phenomena induced by the high-pressure treatment are combined with other physical, chemical or biological processing to cause another new phenomenon. This process was named "High-pressure induced transformation (Hi-Pit)" by Yamazaki and Prof. Tomoyuki Fujii (Yamazaki 2006). As a result of extensive research, we discovered a surprisingly remarkable fact that the high-pressure treatment conducted on various kinds of microorganisms prior to heat sterilization brought the Hi-Pit effect, to reduce the heat resistance of the microorganisms. This discovery is introduced in the last part of this chapter.

26.2.2 Review of Sterilization Effect of High-Pressure Treatment on Microorganisms

The sterilization effect of high-pressure treatment against microorganisms largely varies depending on whether the microorganism is a heat-resistant spore or not.

Table 26.2 Sterilization effect of high-pressure treatment on various microorganisms

Microorganisms	Survival spore counts (cfu g^{-1})		
	Number of initial viable spores	High-pressure treatment (400 MPa, 10 min)	
		20 °C	45 °C
Penicillium verrucosum	1.5×10^4	2.0×10^4	0
Rhizopus oryzae	1.5×10^4	1	0
Cladosporium sphaerospermum	6.3×10^3	0	0
Fusarium oxysporum	5.4×10^5	2	0
Pseudomonas fluorescens	1.5×10^4	5	0
Streptococcus Cp22B	9.3×10^4	1.4×10^4	0

Table 26.2 shows the results of experiments to examine the effect of the high-pressure treatment on non-heat resistant microorganisms. When application of a high pressure of 400 MPa was held for 10 min at 20 °C, the microorganisms other than *Penicillium* and *Streptococcus* showed a drastic decrease in survival spore counts. In addition, when the temperature was increased to 45 °C during the high-pressure treatment, the survival spore counts of all the microorganisms used as the indicators decreased to undetectable levels. These results indicated that the effect of the high-pressure treatment on the non-heat resistant microorganisms is remarkable, and that the effect can be further enhanced by increasing the temperature in the course of the high-pressure treatment.

Sonoike organized the previous reports on the sterilization effect of high-pressure treatment on microorganisms published during the past 100 years (Sonoike 1997). According to those reports, most nutritive cells can be killed by the application of 400 MPa for 20 min or more. Among Gram-negative bacteria, fungi and yeasts, there are many kinds of spores that can be killed under the conditions milder than those mentioned above. Many reports prove the Gram-negative bacteria to be more sensitive to pressure and therefore get inactivated under the application of relatively lower pressure than Gram-positive bacteria. However, even by the application of high pressure of 400 MPa or more, it is considered very difficult to kill the heat resistant spores. Namely, the non-heat resistant microorganisms can be killed by the high-pressure treatment of 400 MPa; while it is found difficult to kill the heat resistant microorganisms only by the high-pressure treatment.

26.2.3 Sterilization Against Heat-Resistant Spores and Production of Sterilized Foods by Taking Advantage of the Hi-Pit Effect

We tried not only to explore the optimal conditions of the pressure, temperature and holding time of the high-pressure treatment to aim at perfect sterilization only

Table 26.3 Combination of pressure and temperature (holding time = 30 min) effective for reducing survival spore counts of *B. coagulans* spores

High-pressure treatment conditions		Survival spore counts (cfu g^{-1})				
Pressure (MPa)	Temperature (°C)	Fo = 1 120 °C, 1 min	Fo = 2 120 °C, 2 min	Fo = 4 120 °C, 4 min	Fo = 8 120 °C, 8 min	Fo = 12 120 °C, 12 min
Control (subjected to only heat sterilization)		9.2×10^5	8.0×10^5	8.0×10^5	2.9×10^3	2.4×10^2
100	30	9.0×10^5	8.3×10^5	6.7×10^4	3.3×10^3	5.3×10^2
	50	8.7×10^5	7.7×10^5	5.4×10^4	7.7×10^3	1.0×10^1
	70	7.6×10^5	6.7×10^5	3.3×10^4	6.7×10^2	6.2×10^1
	85	7.2×10^5	5.9×10^5	1.7×10^4	3.5×10^2	4.1×10^0
200	30	9.1×10^5	8.1×10^5	6.4×10^4	6.4×10^2	9.1×10^0
	50	8.4×10^5	1.2×10^5	5.4×10^4	9.3×10^1	ND
	70	5.9×10^4	1.1×10^4	9.9×10^3	1.1×10^0	ND
	85	4.3×10^4	2.0×10^4	8.1×10^2	ND	ND
400	30	9.5×10^4	6.7×10^4	7.1×10^3	6.7×10^1	7.2×10^0
	50	8.6×10^4	3.5×10^4	3.3×10^3	4.4×10^1	ND
	70	5.5×10^3	1.1×10^3	2.2×10^1	ND	ND
	85	3.8×10^3	1.0×10^3	ND	ND	ND

ND not detected. The spores were initially inoculated into sterile precooked food at a concentration of 10^6 cfu g^{-1}

with the high-pressure treatment, but also to take advantage of the Hi-Pit effect of the high-pressure treatment in combination with the subsequent heating treatment for killing the spores. Table 26.3 shows the results where the survival spore counts of *B. coagulans* obtained after a heat sterilization alone are compared with those obtained by conducting a high-pressure treatment (400 MPa or less) prior to the heat sterilization. The *B. coagulans* spores are extremely heat-resistant and often chosen as the indicator for determining the sterilization conditions for retort food products. As can be seen in Table 26.3, with the increases of the pressure and the temperature in the course of the high-pressure treatment conducted prior to the heat sterilization, the survival spore counts were largely decreasing. When only the heat sterilization was conducted, the number of viable cells was found to be 2.4×10^2 cfu g^{-1} even under the conditions of 120 °C for 12 min (Fo = 12). In contrast to this, when the high-pressure treatment of 200 MPa or more and 50 °C or more was conducted prior to the heat sterilization, the spores initially inoculated at a concentration of 10^6 cfu g^{-1} could be reduced to undetectable levels in any case. These results demonstrated that even though there remain some viable spores under heat sterilization conditions, the same conditions can be used for the heat sterilization if the high-pressure treatment of 200 MPa is conducted prior to the heat sterilization.

For the purpose of industrialization, we used two typical heat-resistant spores, *B. subtilis* and *B. cereus* to carry out experiments of sterilization by conducting a high-pressure treatment of 200 MPa prior to a heating treatment at 100 °C or less. Figure 26.8 shows a survival curve of spores *B. subtilis* subjected to a heating

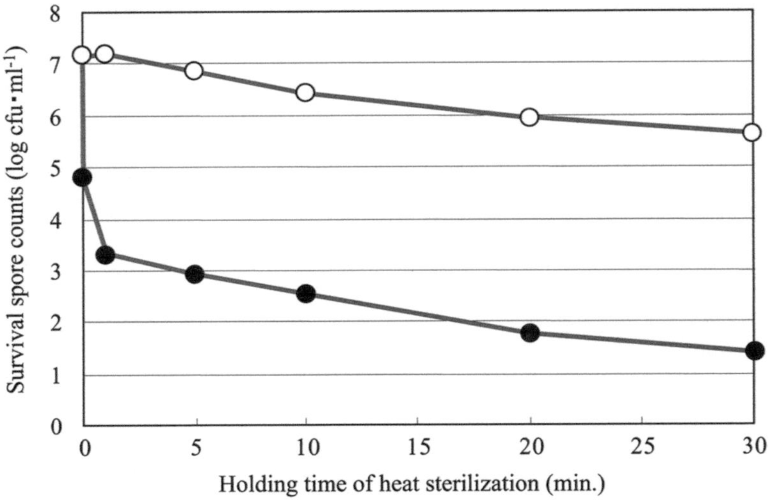

Fig. 26.8 Relationship between holding time of heat sterilization (90 °C) and survival spore counts of *B. subtilis* suspended in a 0.067 M suspension of phosphate buffer. Data: means ± SE (n = 3). -○-; control samples only subjected to heat sterilization at 90 °C. -●-; samples subjected to high-pressure treatment (200 MPa, 50 °C, 10 min) before heat sterilization at 90 °C

treatment of 90 °C after a high-pressure treatment of 200 MPa for 10 min at 50 °C. When the heating treatment was conducted alone, the survival spore counts moderately decreased as the holding time of heating treatment was extended. In the case of *B. subtilis*, the decimal reduction time (D-value: the time required at a certain temperature to kill 90 % of the microorganisms) was 18 min at 90 °C. In contrast to this, when the high-pressure treatment was conducted prior to the heating treatment, only 1 min heating treatment allowed the survival spore counts to decrease to about 1/100 of that obtained immediately after completion of the high-pressure treatment. Thereafter, the survival spore counts continued to decrease with almost the same gradient as that of the survival spore counts decreasing up to the D-value after only the heating treatment of 90 °C was conducted. After a lapse of 30 min, a decrease from the initial spore counts by 5.8 log was confirmed. In consideration of the D-value of 18 min at 90 °C, it will be necessary to conduct the heating treatment for about 100 min to decrease the survival spore counts by 5.8 log if the heating treatment is conducted alone. It can be said, however, that heating treatment can be curtailed to 30 min to obtain the same results by taking advantage of the Hi-Pit effect of the high-pressure treatment (200 MPa) conducted prior to the heating treatment.

Next, using the *B. cereus* spores which are known to cause deterioration of rice processed foods, we discussed the sterilization method based on the Hi-Pit effect. The conditions for high-pressure treatment were set to 200 MPa, 55 °C and 2 min in consideration of industrial practicability; and the temperature of the heating treatment was set to 100 °C that is the temperature for cooking rice. Figure 26.9 shows survival curves of *B. cereus* spores. When the heating treatment

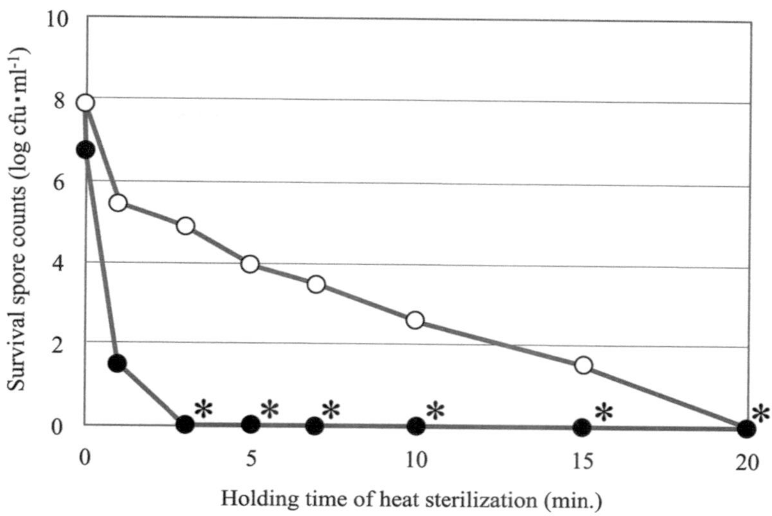

Fig. 26.9 Relationship between holding time of heat sterilization (100 °C) and survival spore counts of *B. cereus* suspended in a 0.067 M suspension of phosphate buffer. -○-; control samples only subjected to heat sterilization at 100 °C. -●-; samples subjected to high-pressure treatment (200 MPa, 55 °C, 2 min) before heat sterilization at 100 °C. *; not detected

was conducted alone, the survival spore counts linearly decreased with the passage of time, and finally attained to the undetectable level after 20 min. In contrast to this, when the high-pressure treatment was done prior to the heating treatment, the survival spore counts drastically decreased by 1 min heating treatment and attained to the undetectable level only in 3 min.

Generally, sterilized packaged rice products are prepared by cooking polished rice and hermetically sealing the package of cooked rice under an aseptic condition, and the resultant packaged rice products can be kept at room temperature for a half year to about 1 year. The sterilized packaged rice products have the advantages that they can be used as preserved food and the original taste of cooked rice can be maintained because the cooking temperature is set to about 100 °C, so that the sterilized packaged rice products are popular and familiar in Japan. Unlike retort food products, the above-mentioned packaged rice products are not exposed to high temperature and pressure for sterilization after the production. Therefore, the sterilized packaged rice is not subject to browning because the sugar, amino acid, protein, lipid and the like contained in rice do not cause any chemical change. That is why the sterilized packaged rice products have been exported recently to the Asian countries where rice products are evaluated based on the smell of cooked rice.

Figure 26.10 is a graph showing the survival spore counts at every step in the production process of the sterilized packaged rice products by making use of the high-pressure treatment. This production process is characterized in that microorganisms originating from the raw rice are removed and reduced step by step, and the rice subjected to the high-pressure treatment is cooked and the package of the cooked rice is finally sealed in the aseptic state.

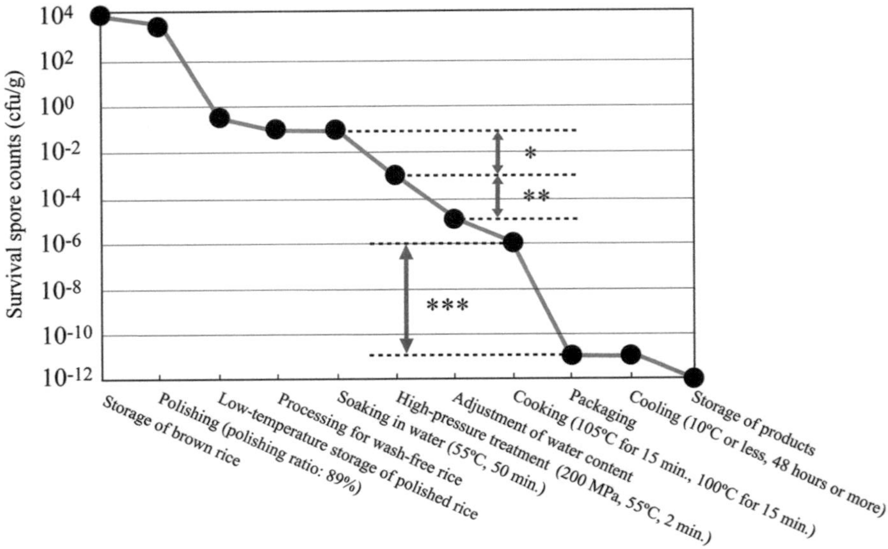

Fig. 26.10 Survival spore counts at every step in production process of cooked high-pressure processed rice. *; sterilization effect by soaking in water of 55 °C for 50 min. **; sterilization effect by high-pressure treatment (200 MPa, 55 °C, 2 min). ***; sterilization effect by heating treatment taking advantage of Hi-Pit effect

Our production method for sterilized packaged rice products based on the Hi-Pit effect has established high reliability, which is demonstrated by the actual output of 200,000,000 or more products for the past 20 years with no complaint about the commodity. The above-mentioned production method can thoroughly ensure the microbiological safety, i.e., a sterile state of the products although no chemical is added, and as a result, the products can be kept at room temperature for 5 years and therefore preserved for emergency. Further, when produced through the step of high-pressure treatment, it was found that the resultant sterilized packaged rice has the advantages of high degree of gelatinization, good digestibility, and increased contents of functional components such as GABA and the like (Yamazaki et al. 1996; Yamazaki and Sasagawa 1998; Kinefuchi et al. 1999c). Such a method designed by taking advantage of the Hi-Pit effect is considered to be promising as a new food processing method for production of various sterilized foods in the future.

26.2.4 Increase in Water Wettability of Surface of Spores and Decrease in Heat Resistance of Spores by High-Pressure Treatment

From the production process of the sterilized packaged rice products, we discovered the fact that the sterilizing efficiency can be improved by providing the step of

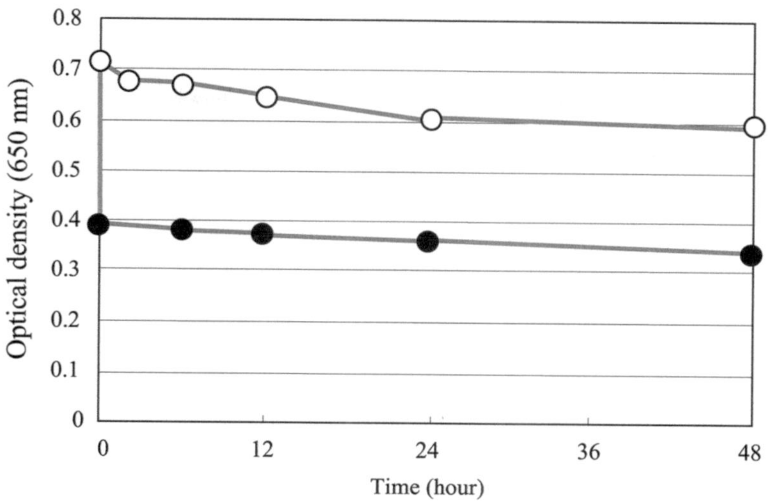

Fig. 26.11 Change in optical density at 650 nm of *B. cereus* spores suspended in a 0.067 M suspension of phosphate buffer left at 25 °C. -○-; control samples not subjected to high-pressure treatment. -●-; samples subjected to high-pressure treatment (200 MPa, 25 °C, 10 min)

high-pressure treatment prior to the heating treatment. Then, we demonstrated that the survival spore counts of *B. coagulans*, *B. subtilis* and *B. cereus* belonging to the genus *Bacillus* remarkably decrease in the case where the spores are subjected to a high-pressure treatment of 200 MPa and subsequently a heating treatment. The phenomenon of the decrease in heat resistance of the spores after the high-pressure treatment was assumed to result from the increase in heat conductivity toward the inside of the spores. Then, we observed the spore cells using a microscope. The spores before subjected to the high-pressure treatment were not stained with methylene blue, while the spores subjected to the high-pressure treatment became stained with methylene blue. Further, to discuss the affinity of the spore surface for water, the change in turbidity of a *B. cereus* suspension in phosphate buffer is shown in Fig. 26.11. Immediately after the high-pressure treatment the turbidity of the spore suspension sharply decreased. After that, however, the turbidity did not change while the suspension was allowed to stand at 25 °C. Generally, the decrease of turbidity is considered to indicate germination of the spores, and this type of decrease in turbidity appears a short time after the substantial germination (Levinson and Hyatt 1966). However, in our experiments, the decrease in turbidity appeared immediately after completion of the high-pressure treatment. Further, a drastic decrease or increase in turbidity did not appear in the control spore suspension (not subjected to the high-pressure treatment), which proved that the spores were not able to germinate or grow in the phosphate buffer solution. In light of the above, the decrease in turbidity of the suspension observed after the high-pressure treatment was considered to result from a certain physical action exerted

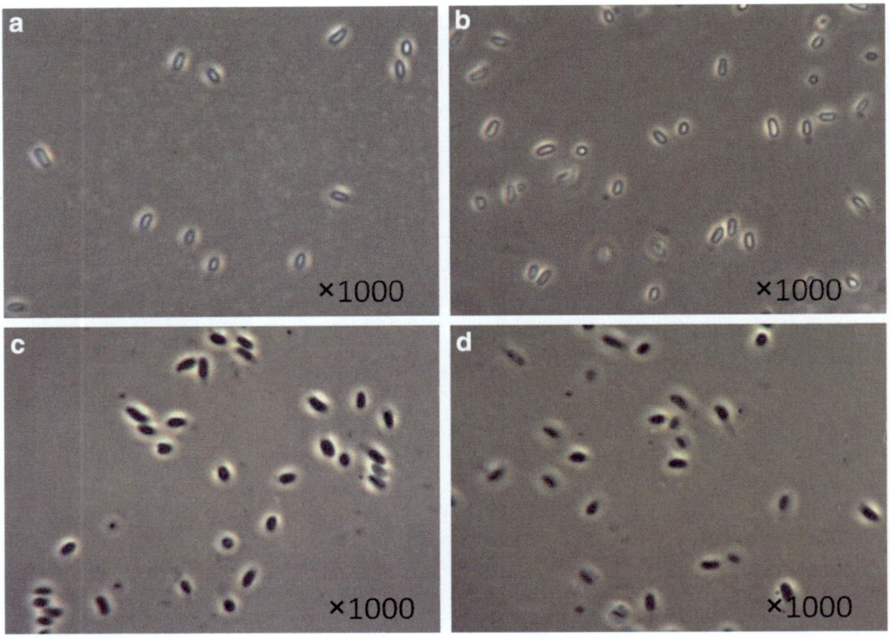

Fig. 26.12 Views of *B. cereus* spores suspended in a 0.067 M suspension of phosphate buffer taken by a phase contrast microscope. (**a**) Control, (**b**) control (left at 25 °C for 48 h), **c**) high-pressure treated spores (immediately after high-pressure treatment at 200 MPa, 25 °C, 10 min), **d**) high-pressure treated spores (left at 25 °C for 48 h after high-pressure treatment)

on the spores. The decrease in the heat resistance of spores after the high-pressure treatment was considered to take place by a mechanism utterly different from that of the physiological germination.

Figure 26.12 indicates views of *B. cereus* spores taken by a phase contrast microscope. The control spores not subjected to high-pressure treatment were observed as light images because of their high refractive index. In contrast to this, the spores taken immediately after the high-pressure treatment were observed as dark images. After the high-pressure treated spores were allowed to stand at 25 °C for 48 h, no particular change in the shape was observed as compared with the spores immediately after the high-pressure treatment. Namely, the spores assumed like a darkening state after the high-pressure treatment, but they did not change into nutritive cells after that. The microscopic observation proved the phenomenon induced by the high-pressure treatment to be different from physiological germination of the spores.

Figure 26.13 shows the results of quantitative assay of surface hydrophobicity of the spores. The spore surface showed a decreasing tendency of hydrophobicity and an increasing tendency of water wettability as the pressure applied to the spores increased. It was considered that the high-pressure treatment should improve the wettability of the spore surface by water, thereby allowing the spores to be more heat

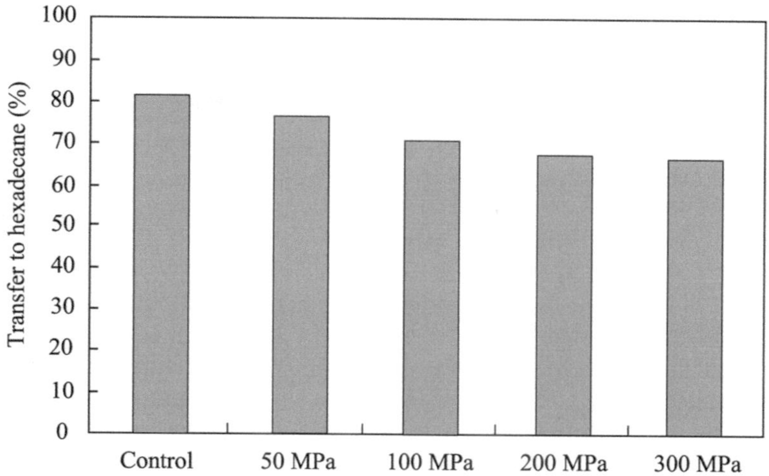

Fig. 26.13 Hydrophobicity of surface of *Bacillus cereus* spores suspended in a sterile distilled water after high-pressure treatment (at 20 °C for 10 min)

conductive and less heat resistant in the heating treatment and therefore realizing satisfactory sterilization. The effect of high-pressure treatment that substantially remains for a while and has a great influence on the subsequent step such as a heating treatment is referred to as "static pressure history". It will be one of the most important problems to clarify the duration of time when the "static pressure history" keeps effective for the subsequent step.

26.2.5 Conclusions

In the nineteenth century, Louis Pasteur, a French microbiologist developed the technique of sterilization at low temperatures of 62–65 °C, i.e., pasteurization. It has been 100 years since the pasteurization was developed, but pasteurization has been still now used for milk and Japanese traditional sake because the chemical makeup of foods can be left unaltered and the taste and nutrition values are not impaired. On the other hand, high-pressure treatment has attracted attention as a means of non-heat sterilization. When the pressure treatment was used, however, the pressure level was sometimes unreasonably increased, or the high-pressure treatment was used in combination with a heating treatment of 60 °C or more. In other words, the conditions of high-pressure treatment have been explored in such a manner that the original advantages of the high-pressure treatment cannot be exhibited. When the heating treatment of 60 °C or more is used together, the resultant sterilization method is no longer categorized into the non-heat sterilization. In the high-pressure treatment of 100 MPa, a change in volume of water is as small as 4 % and the energy generated by the pressure application is only 0.02 kcal mol^{-1}. In light of the above, it is very important to recognize the most significant benefit of the high-pressure treatment as causing no drastic change.

From the viewpoint of industrial practicability, we have continued the studies on high-pressure treatment for 25 years to put it to practical use, provided that the pressure of 400 MPa or less is applied within 10 min. As a result, we have come to the conclusion that perfect sterilization cannot be attained only by the high-pressure treatment, and the static pressure history can bring about an effect on the subsequent step (Hi-Pit effect). Currently, in our production process for sterilized packaged rice products based on the Hi-Pit effect, the pressure of 200 MPa is applied for only 2 min at the high-pressure treatment. The combination of the high-pressure treatment and the subsequent heating step of 100 °C can complete a sterile state by killing all the microorganisms including the heat-resistant spores originating from raw rice. The safety of our production method has been proved by the actual results of many years' production. We assume the decrease in heat-resistance of spores induced by the high-pressure treatment to be the result of a physical action, to construct a theory of the Hi-Pit effect. By clarifying the principle of the Hi-Pit effect, we believe that we can make the fullest use of the Hi-Pit effect.

We established an organization, "Research Association of High Pressure Technologies for the Creation of Future Industries" in 1999, having Prof. Rikimaru Hayashi as an advisor, for the purpose of making innovations in the fields of food, machinery, agriculture and medicine. Now, more than 100 companies have joined the association to make practical use of the high-pressure treatment, develop numerous kinds of high-pressure processed food products and accumulate the technologies related to the high-pressure treatment. In particular, we expect the application of the Hi-Pit effect to the production of various sterilized food products. The way of sterilization against the heat-resistant spores by taking advantage of the Hi-Pit effect as introduced in this paper is supposed to wipe and change the conventional views on high-pressure treatment. We confidently believe that the high-pressure treatment will develop into a technology useful for the human as a new sterilization means without impairing the quality of food, just like Pasteur's discovery.

References

Caragay AB (1992) Cancer-preventive foods and ingredients. Food Technol 46:65–68

Fenwick GR, Hanley AB (1985) The genus *Allium*. Part 2. Crit Rev Food Sci Nutr 22:273–377

Gekko K (1990) Hydration of biopolymers and pressure effects. In: Hayashi R (ed) Pressure-processed food – research and development. San-Ei Publishing Co, Kyoto, pp 23–35

Hayakawa K, Ueno Y, Kawamura S, Miyano Y, Kikushima S, Shou S, Hayashi R (1995) Preparation of salt-free *Miso* and its high-pressure treatment for preservation. Agric Biol Chem 69:1021–1026

Kinefuchi M, Sekiya M, Yamazaki A, Yamamoto K (1999a) Accumulation of GABA in brown rice by high-pressure treatment. J Jpn Soc Food Sci 46:323–328

Kinefuchi M, Sekiya M, Yamazaki A, Yamamoto K (1999b) Change in viable bacteria count in brown rice containing accumulated GABA by high-pressure treatment, and properties of processed brown rice. J Jpn Soc Food Sci 46:329–333

Kinefuchi M, Watanabe K, Komiya S, Yamazaki A, Yamamoto K (1999c) Characteristic of retrogradation of pressure-treated cooked rice. J Appl Glycosci 46:31–38

Koseki S, Yamamoto K (2006) pH and solute concentration of suspension media affect the outcome of high hydrostatic pressure treatment of *Listeria monocytogenes*. Int J Food Microbiol 111:175–179

Levinson HS, Hyatt MT (1966) Sequence of events during *Bacillus megaterim* spore germination. J Bacteriol 91:1811–1818

Oba K (2002) Enzymatic studies on the chemical changes of food materials in cookery science. J Home Econ Jpn 53:869–876

Omori M, Yano T, Okamoto J, Tsushida T, Murai T, Higuchi M (1987) Effect of anaerobically treated tea (Gabaron tea) on blood pressure of spontaneously hypertensive rats. Agric Biol Chem 61:1449–1451

Saikusa T, Horino T, Mori Y (1994) Distribution of free amino acids in the rice kernel and kernel fractions and the effect of water soaking on the distribution. J Agric Food Chem 42:1122–1125

Sasagawa A (2006) Development of food products using high-pressure induced transformation (Hi-Pit). Food Pack 47:504–509

Sasagawa A, Yamazaki A (2002) Development and industrialization of pressure-processed foods. In: Hayashi R (ed) Trends in High Pressure Bioscience and Biotechnology. Elsevier Science BV, Amsterdam, pp 375–384

Sasagawa A, Kinefuchi M, Yamazaki A, Yamada A (1994) The method of processing rice cakes with high pressure. In: Hayashi R, Kunugi S, Shimada S, Suzuki A (eds) High pressure bioscience. San-Ei Publishing Co, Kyoto, pp 336–343

Sasagawa A, Naiki Y, Nagashima S, Yamakura M, Yamazaki A, Yamada A (2006) Process for producing brown rice with increased accumulation of GABA using high-pressure treatment and properties of GABA-increased brown rice. J Appl Glycosci 53:27–33

Sonoike K (1997) High pressure sterilization technology-subject for the application to food. J Jpn Soc Food Sci 44:522–530

Taki Y, Awao T, Mitsuura N, Takagaki Y (1990) Sterilization of *Bacillus* sp. spores by hydrostatic pressure. In: Hayashi R (ed) Pressure processed food – research and development. San-Ei Publishing Co, Kyoto, pp 143–155

Taki Y, Awao T, Toba S, Mitsuura N (1991) Sterilization of *Bacillus* sp. spores by hydrostatic pressure, Part 2. In: Hayashi R (ed) High pressure science for food. San-Ei Publishing Co, Kyoto, pp 217–224

Ueno S, Shigematsu T, Kuga K, Saito M, Hayashi M, Fujii T (2009) High-pressure induced transformation of onion. Jpn J Food Eng 10:37–43

Wheeler GL, Jones MA, Smirnoff N (1998) The biosynthetic pathway of vitamin C in higher plants. Nature 393:365–369

Yamamoto A, Oba K (1999) Changes in the vitamin C content and enzyme activities involved in the synthesis and oxidation of ascorbic acid in cut vegetables. J Home Econ Jpn 50:1015–1020

Yamazaki A (2006) Development of food products using high-pressure induced transformation (Hi-Pit). Rev High Press Sci Technol 16:4–10

Yamazaki A, Sasagawa A (1998) Development of rice food products by high-pressure treatment. J Jpn Soc Food Sci 45:526–532

Yamazaki A, Kinefuchi M, Yamamoto K, Yamada A (1996) Physical properties and fine structure of grains of high-pressure-treated rice after cooking. Rev High Press Sci Technol 5:168–178

Part VIII
Pressure Effects on Motility, Physiology and Health

Editors' Foreword of Part VIII

This part deals with a rapidly developing new field of science-a frontier in pressure bioscience, namely the area of pressure-dependent physiology in living systems including humans. Chapter 27 by Nishiyama provides a prologue to this field, a clear visual example showing that pressure alters a physiology, a motility of a bacterial cell. In Chap. 28, Sokabe shows how mechanical forces can indeed activate physiology in bacterial and eukaryotic cells through the conformational change of the receptor protein. Here, pressure may take the form of uni-axial stress mediated via the fibrillar network of the cell rather than the isotropic hydrostatic pressure. Finally, in Chap. 29, Atomi discusses the physiological significance of a gravitational force to human physiology. She shows that, indeed, genes are activated to express a protein crystalline through some yet unclarified mechanisms upon applying the gravitational force to a human body, which appears to be crucial in human health. Contents in these chapters show indeed that pressure is not a variable foreign to life, but rather an intimate part of it.

Chapter 27
High-Pressure Microscopy for Studying Molecular Motors

Masayoshi Nishiyama

Abstract Movement is a fundamental characteristic of all living things. This biogenic function is carried out by various nanometer-sized molecular machines. Molecular motor is a typical molecular machinery in which the characteristic features of proteins are integrated; these include enzymatic activity, energy conversion, molecular recognition and self-assembly. These biologically important reactions occur with the association of water molecules that surround the motors. Applied pressures can alter the intermolecular interactions between the motors and water. In this chapter we describe the development of a high-pressure microscope and a new motility assay that enables the visualization of the motility of molecular motors under conditions of high-pressure. Our results demonstrate that applied pressure dynamically changes the motility of molecular motors such as kinesin, F_1-ATPase and bacterial flagellar motors.

Keywords High-pressure microscopy • Motility • Molecular motor • Kinesin • F_1-ATPase • Bacterial flagellar motor

27.1 Introduction

Water is the most abundant substance in cells and its weight accounts for 70 % of the total cell component. Most intracellular reactions occur in an aqueous environment and the unique properties of aqueous solutions can be attributed to the network of hydrogen bonds among water molecules. The hydrogen bond forms between a molecule of water and the electric-charged residue of the protein. Water molecules energetically optimize protein structure and are involved in the molecular recognition process (Chaplin 2006). In the absence of water molecules, proteins lack activity. Application of pressure is a powerful method for modulating intermolecular interactions between protein and water molecules. High-pressure technique has been

---

M. Nishiyama (✉)
The Hakubi Center for Advanced Research/Institute for Integrated Cell-Material Sciences, Kyoto University, Kyoto 606-8501, Japan
e-mail: mnishiyama@icems.kyoto-u.ac.jp

K. Akasaka, H. Matsuki (eds.), *High Pressure Bioscience*,
Subcellular Biochemistry 72, DOI 10.1007/978-94-017-9918-8_27

used as a thermodynamic tool to explore the biophysical properties of proteins, lipids, nucleic acids and other macromolecules. In general, the application of pressure of several hundred MPa does not seriously affect primary and secondary structures of protein molecules, but it does increase their structural fluctuation (Akasaka 2006). It also weakens protein-protein and protein-ligand interactions in solutions (Boonyaratanakornkit et al. 2002). These pressure-induced effects are thought to be caused by enhancement of the clustering of water molecules around hydrophobic and hydrostatic residues on the protein surface. This means that applied pressure is able to modulate the structure and function of protein molecules, without requiring the use of any chemical materials other than water molecules.

High-pressure techniques have been used in various types of spectroscopies, e.g. conventional spectroscopy (Kimura et al. 2006; Watanabe et al. 2013), Fourier transform infrared spectroscopy (Dzwolak et al. 2002), nuclear magnetic resonance (Kamatari et al. 2004), small-angle neutron scattering (Winter 2002; Filabozzi et al. 2010), small-angle X-ray scattering (Fujisawa et al. 2000; Brooks et al. 2010), X-ray crystallography (Colloc'h et al. 2006; Nagae et al. 2012), time-resolved spectroscopy (Hoshihara et al. 2008), and time-resolved fluorescence anisotropy (Usui et al. 2012). These methods are suitable for measuring the pressure-induced changes in the structure of protein molecules. In contrast to these spectroscopic techniques, optical microscopy is suitable for the observation of relatively large objects in the order of sub-micrometers or higher. Many high-pressure microscopy chambers have been developed to monitor the morphology and activity of cells.

In this chapter we describe a motility assay for investigating the mechanism of interaction between molecular motors and water molecules. We start by describing the development of a high-pressure microscope (Sect. 27.2) and its subsequent performance check (Sect. 27.3). We then show that applied pressure works directly to weaken the intermolecular interaction between tubulin molecules (Sect. 27.4). We describe the experimental results of a motility assay of several molecular motors including kinesin (Sect. 27.5), F_1-ATPase (Sect. 27.6) and flagellar motor (Sects. 27.7 and 27.8). Finally, we summarize the contents in this chapter (Sect. 27.8).

27.2 High-Pressure Microscope

'Seeing is believing'. As this proverb indicates, it is important to observe targets *in situ* using direct methods. This is why optical microscopy is a powerful method and is used in a wide range of research areas. It has also been used for many years in the area of high-pressure research. Applied pressures can induce significant changes in the morphology and activity of living cells of microorganisms (Kitching 1957; Otter and Salmon 1985; Abe 2007), sea urchin eggs (Inoue et al. 1975; Salmon 1975a), tissue cells (Frey et al. 2006), and muscle fibers (Ranatunga et al. 1990).

Many microscopes have been developed and constructed to observe microscopic objects under high-pressure conditions. A list of these high-pressure microscopes has been compiled by Vass and co-workers (Vass et al. 2010). The most important aspect in the development of the apparatus is the design of the high-pressure chamber that encloses the target object. In previous studies, a variety of chambers have been designed for microscopy, but most of them have focused on the aspect of pressure resistance and as a consequence the performance quality of the microscope is substandard. The reason for this is that the optical window of these chambers is frequently made of quartz ($n_e = 1.458$), sapphire ($n_e = 1.771$) or diamond ($n_e = 2.424$). Most commercially available objective lenses are not designed for the acquisition of images using these materials and the incompatibility leads to a decrease in the resolving power of images.

We have constructed a high-pressure chamber for acquiring images with high resolution even under high-pressure conditions (Nishiyama and Sowa 2012) (Fig. 27.1). The high-pressure chamber is equipped with two optical windows (BK7, $n_e = 1.519$). The refractive index is well-matched to the value of commercially available objective lenses. The aperture diameter and critical angle were 1.5 mm and 76°, respectively. Microscopic observations in the chamber were carried out through the observation window (thickness = 1.5 mm). The numerical apertures

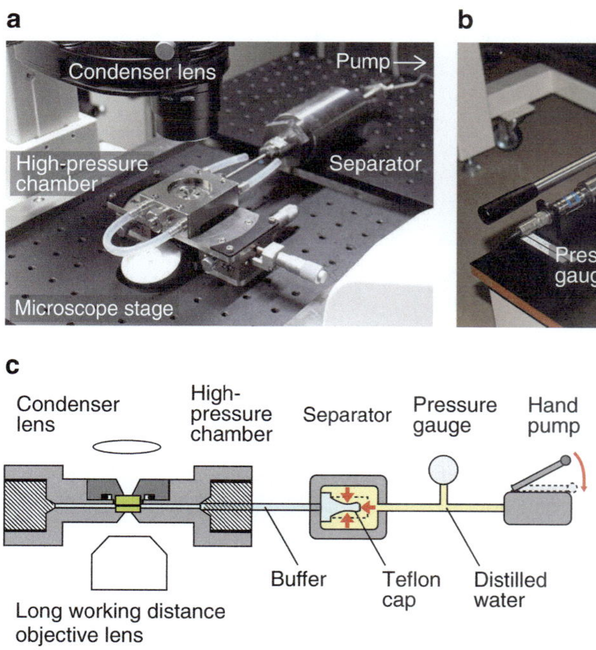

Fig. 27.1 High-pressure microscope. (**a**) High-pressure chamber and separator. Full details of the microscope stage are available to interested parties upon request. (**b**) High-pressure pump. (**c**) Schematic drawing of the experimental setup (not to scale)

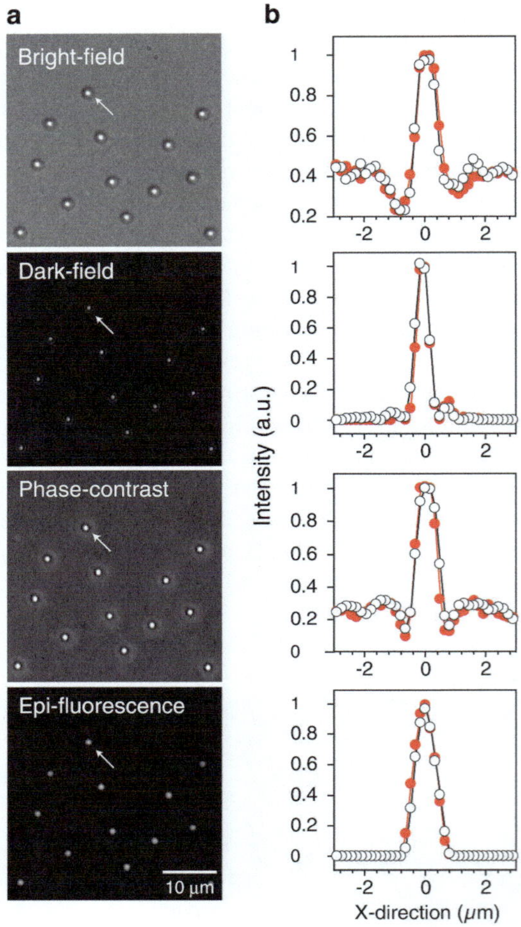

Fig. 27.2 (**a**) Various microscopic images of the same 1-μm beads at 100 MPa. These images were recorded at 30 frames s^{-1} and displayed without processing contrast enhancement and brightness offset. (**b**) Intensity profile of the bead at 0.1 (*open circles*) and 100 MPa (*solid circles*). The intensity profile of the same bead was plotted at each pixel in the x direction. The *arrows* in (a) indicate the corresponding original images (Modified from reference Nishiyama and Sowa 2012)

at the objective-lens and condenser-lens sides were 0.6 and 0.55, respectively. This design allowed us to acquire bright-field, dark-field, phase-contrast and epi-fluorescence microscopic images (Fig. 27.2).

The high-pressure chamber was connected to a separator, pressure gauge and high-pressure pump (Fig. 27.1). The separator conferred the advantage of reducing the total dead volume of buffer solution in the pressure line (Nishiyama et al. 2009). Hydrostatic pressure was applied to the pressure line using a hand pump. The inside of the Teflon cap was filled with buffer solution and connected to the chamber. The water pressure was transduced to the buffer solution by deformation of a thin Teflon cap in the separator and the pressure was then transmitted to the chamber. The hydrostatic pressure in the pressure line was measured using a pressure gauge.

These pressure devices were combined with a commercially available inverted microscope on a vibration-free table. Microscopic observation was carried out using a long-working-distance objective lens (NA 0.6, working distance ~3 mm; CFI ELWD ADM40 × C, Nikon). The images were acquired by different cameras, taking into account the sensitivity and frame rate required.

27.3 Performance Check

The withstanding pressure of our pressure apparatus was examined (Nishiyama and Sowa 2012). Using the hand pump, hydrostatic pressure was could be increased by several dozen MPa within a few seconds without any overshoot. Our apparatus was able to apply pressure up to 150 MPa, a limit imposed by the capabilities of the hand pump. This maximum pressure is ~1.5-fold higher than that of the deepest part of the Mariana Trench (10,924 m in depth), which is the highest pressure found outside the crust of the earth. This ability to withstand pressure at such a high level is sufficient for studying almost all biological activity on earth. High-pressure environment in the chamber was maintained appropriately without the pressure leakage. Applied pressure could be released nearly instantaneously by opening a valve.

The optical performance of the apparatus was then examined in detail (Nishiyama and Sowa 2012). Beads were fixed onto the surface of the observation window in the chamber. When pressure was applied to the chamber, the microscopic image of the beads moved out of focus because applied pressure deforms the pressure apparatuses. After readjusting the position of the objective lens, the microscopic images became normally focused again. The x-and y-positions of the bead did not shift largely, and it meant that we could track the same target under various pressure conditions. Figure 27.2a displays four different microscopic images of the same beads at 100 MPa. The intensity profile of a bead was independent of the applied pressure, suggesting that applied pressure did not affect image formation.

Finally, we confirmed that applied pressure significantly changed the physical characteristics of the buffer solutions (Nishiyama et al. 2009; Nishiyama and Sowa 2012). Our results showed that the application of 100 MPa of pressure increased the viscous drag by +5 %, temperature by +0.1 °C, and pH by +0.1. These results confirm that this system is useful for studying the effects of high pressure in our experiments.

In summary, we have succeeded in developing a high-pressure microscope that enables us to acquire various microscopic images with high-resolution and sensitivity, regardless of pressures. Recently, we were successful in expanding the numerical aperture of the opening on the objective–lens side (Nishiyama 2013; Vass et al. 2013). This achievement allows us to improve the sensitivity and resolution of microscopic observations so that highly-developed microscope techniques (Ishii et al. 2004b) could be introduced in the field of high-pressure research.

27.4 Microtubule Depolymerization from Both Ends

The microtubule is an important component of the cellular cytoskeleton (Nogales and Wang 2006). When microtubules polymerize and depolymerize, they push and pull against the plasma membrane from the inside of the cell and this results in changes in cell shape. The change in length of microtubules has to be precisely controlled as its regulation affects specific cellular processes such as the extension of growth cones in neurons.

The filamentous structure of microtubules is thermodynamically unstable and the length changes through the polymerization and depolymerization of tubulin molecules (Horio and Hotani 1986; Walker et al. 1988). These dynamic properties are strongly affected by physical conditions such as temperature and hydrostatic pressures. Salmon successfully developed a high-pressure microscope (Salmon and Ellis 1975) and under several dozen MPa of pressure, was able to visualize the induction of microtubule spindles to depolymerize immediately and reversibly in vitro (Salmon 1975c) and vivo (Salmon 1975a, b). The results indicated that pressure affects the polymerization-depolymerization dynamics of microtubules. We monitored individual microtubules in association with kinesin molecules at high-pressure and then examined the effects of pressure on the structure and function of microtubules (Nishiyama et al. 2009, 2010).

Polarity-marked microtubules were prepared by polymerizing dimly labeled tubulin molecules onto the ends of brightly labeled microtubule seeds (Fig. 27.3a). The individual microtubules were sparsely tethered to kinesin-motors on the observation window of a high-pressure chamber in the presence of 10 μM of the mitotic inhibitor taxol (paclitaxel) and 100 μM of the ATP analogue adenosine 5′-(ß, γ-imido) triphosphate (AMPPNP). The pressure-induced changes in the filamentous structure of the microtubules were examined by time-lapse microscopy. When pressure was applied to the sample solution, all the microtubules started to shorten from both ends. The changes in length were constant over time, irrespective of the microtubule polarity. At 150 MPa, the shortening rates were 1.0 and 1.1 $\mu m\ min^{-1}$ at the plus and minus ends, respectively. There was no evidence of either a "rescue"-like transition from shrinkage to growth of a microtubule or "catastrophe"-like rapid shrinkage. The shortening rates increased exponentially with pressure and were independent of the microtubule polarity (Fig. 27.3b).

The proto-filament at the ends of microtubules spontaneously adopt an outer-curved conformation (Nogales and Wang 2006). The tubulin molecules at the ends bind to the filament with a single tubulin–tubulin interaction in the axial direction. Therefore, microtubule shortening is caused by dissociation of one or several tubulin molecules in order. In contrast, microtubules are hardly severed following application of pressure (Nishiyama et al. 2009). Tubulin molecules embedded in the filamentous structure are well stabilized by tubulin–tubulin interactions not only in the axial direction, but also in the lateral direction. These multiple intermolecular interactions protect microtubules against being severed in the current experimental system.

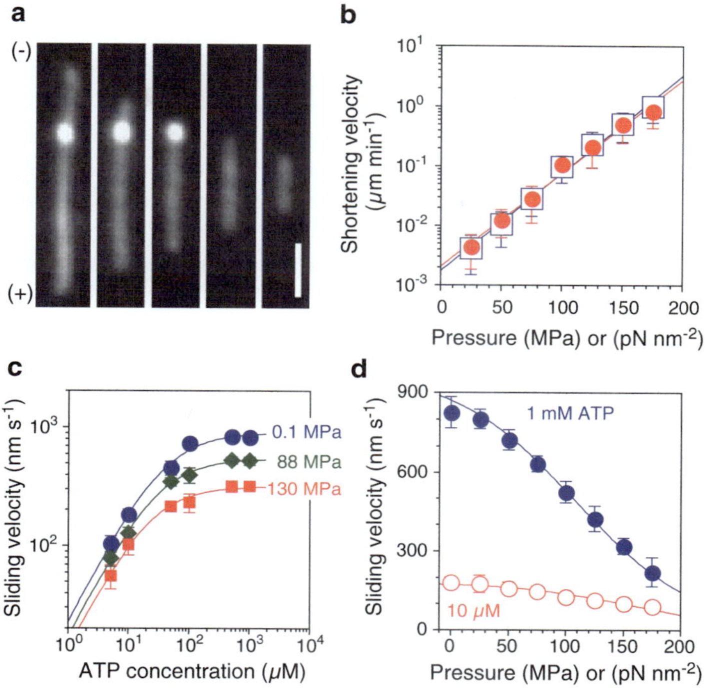

Fig. 27.3 Kinesin-microtubule complex. (**a**) Sequential fluorescence images of the same polarity-marked microtubule recorded at 2-min intervals. A pressure of 150 MPa was applied to a polarity-marked microtubule in the presence of 10 μM paclitaxel and 100 μM AMP-PNP. Scale bar, 5 μm. (**b**) Pressure dependence of shortening rates at plus (*open squares*) and minus (*solid circles*) ends (mean ± SD). (**c**) Michaelis-Menten kinetics, $v = v_{max} \times [ATP]/([ATP] + K_m)$, where v_{max} is the sliding velocity at saturation concentrations of ATP, and K_m is the Michaelis constant. (**d**) Pressure dependence of sliding velocity

27.5 Sliding Motion of a Microtubule Driven by Multiple Kinesin Molecules

Kinesin-1 is an ATP-driven molecular motor that moves stepwise in regular 8 nm steps along a microtubule (Svoboda et al. 1993; Kojima et al. 1997; Nishiyama et al. 2001). This movement is explained by a hand-over-hand model in which the two heads of kinesin work in a coordinated manner (Rice et al. 1999). One head remains bound to the microtubule while the other steps from the αβ-tubulin dimer behind the attached head to the dimer in front. The overall movement is 8 nm per ATPase cycle (Kojima et al. 1997; Schnitzer and Block 1997; Nishiyama et al. 2002). The stepping motion is strongly dependent on physical and chemical conditions, such as applied force (Kojima et al. 1997; Nishiyama et al. 2002), temperature (Kawaguchi and Ishiwata 2000; Taniguchi et al. 2005) and anesthetic agents (Miyamoto et al. 2000). Application of pressure is also expected to change the stepping motion but has not been studied. We studied pressure-induced effects on microtubule-based kinesin motility by performing microtubule-sliding assays under high-pressure conditions (Nishiyama et al. 2009).

Single microtubules were associated with multiple kinesin-1 molecules and attached to the surface of the observation window of the high-pressure chamber. In this experimental system, the motility of kinesin motors could be measured by the sliding motion of microtubules. When pressure was applied to the sample, the sliding velocity of microtubules immediately decreased from 790 nm s^{-1} at 0.1 MPa to 340 nm s^{-1} at 130 MPa. Most of the microtubules moved smoothly and continuously, even under pressurized conditions. The sliding velocity was constant over time for each pressure. After the release of pressure, the sliding velocity immediately returned to the original value. Thus the application of pressure acts as an inhibitor that directly and reversibly alters microtubule-based kinesin motility.

Systematic analysis of the sliding velocity of kinesin motors was then performed. Figure 27.3c shows the mean sliding velocities for each pressure and ATP concentration. The sliding velocity followed normal Michaelis-Menten kinetics. When the pressure was increased from 0.1 to 130 MPa, v_{max} decreased from 900 to 320 nm s^{-1}, whereas the K_m slightly decreased from 39 to 23 μM. Figure 27.3d shows the pressure-velocity relationships at ATP concentrations of 1 mM and 10 μM. The sliding velocity decreased monotonically with increases in pressure. Further analysis showed that the pressure mainly affects the stepping motion and not the ATP-binding reaction.

Interestingly, the pressure-velocity relationship (Fig. 27.3d) was very close to the force-velocity relationship of single kinesin molecules (Nishiyama et al. 2002). The application of either ~100 MPa (=1 pN $Å^{-2}$) of pressure or ~3 pN of force decreased the sliding velocity at 1 mM ATP by half. Our results suggest a similar inhibitory mechanism on kinesin motility. The application of pressure is thought to enhance the structural fluctuation and/or association of water molecules with the exposed regions of the kinesin head and microtubule. These pressure-induced effects could prevent kinesin motors from completing the stepping motion.

27.6 Single Molecule Analysis of the Rotation of F_1-ATPase

F_1-ATPase is the water-soluble part of ATP synthase and is an ATP-driven rotary molecular motor that rotates the rotary shaft against the surrounding stator ring, hydrolyzing ATP (Noji et al. 1997). An elementary step of the mechanochemical reaction of F_1 is 120° rotation that is coupled with a single turnover of ATP hydrolysis (Yasuda et al. 1998). The 120° step rotation can be divided into 80° substep triggered after ATP binding, and 40° substep triggered after ATP hydrolysis (Yasuda et al. 2001). The angular positions before the 80° and 40° substeps are therefore termed the 'binding angle' and the 'catalytic angle', respectively. Although the mechanochemical coupling mechanism of F_1-ATPase has been well studied, the molecular details of individual reaction steps remain unclear. We conducted a single-molecule rotation assay of F_1 from thermophilic bacteria under various pressures (Okuno et al. 2013).

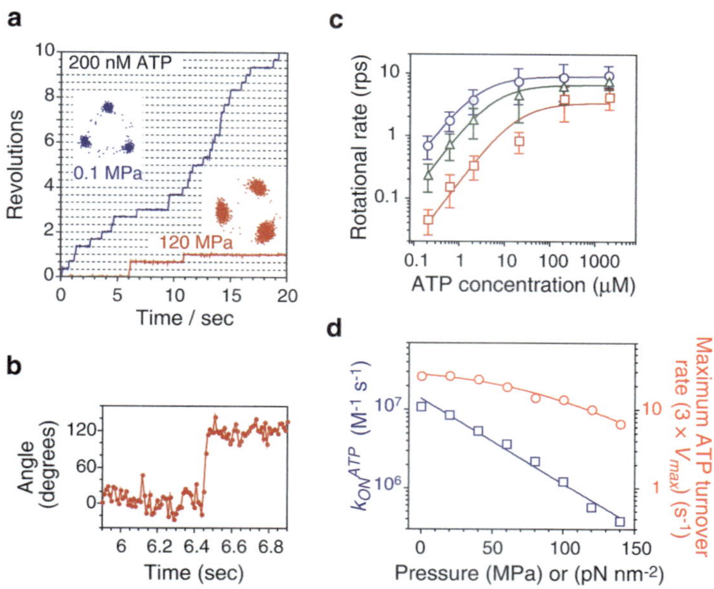

Fig. 27.4 Stepping rotation of F_1-ATPase. (**a**) Time courses and their centroid traces of the rotation of F_1-ATPase at 200 nM ATP. (**b**) Rising phase of single 120° step on an expanded scale at 120 MPa and 200 nM ATP. (**c**) Rotational rates as a function of ATP concentration at 0.1 (*circles*), 60 (*triangles*) and 120 MPa (*squares*). (**d**) Pressure dependence of the apparent ATP binding rate, k_{on}^{ATP} (*squares*) and maximum ATP turnover rate, $3 \times V_{max}$ (*circles*) (Modified from reference Okuno et al. 2013)

Figure 27.4a displays the time courses and xy plots of rotation of single wild-type F_1 molecules at an ATP-limiting concentration (200 nM). When 120 MPa pressure was applied to the system, the rotary motion drastically changed. The stepping rate significantly decreased, indicating applied pressure decreased the catalytic reactions of ATP molecules. However, applied pressure did not change the step size and directionality of the rotation. Most steps took place rapidly (Fig. 27.4b) and were independent of pressure. This means that the application of pressure did not seriously affect the angle velocity at the rising phase of the 120° step. It suggests that even at high-pressure conditions, F_1 retains high conformational stability and its native properties. An attempt was made to find the rotation of 80° and 40° substeps under high-pressure conditions. However, the data showed no evidence of substeps, suggesting that one or more pressure-sensitive reactions occurred at ATP binding angles, but not at catalytic angles. This is strongly supported by results of a rotation assay using a mutant F_1 (βE190D) (Okuno et al. 2013).

Systematic analysis of the rotational rate of single wild-type F_1 motors was performed under various concentrations of ATP and hydrostatic pressures (Fig. 27.4c). The rotational rate gradually decreased with increased pressure at each concentration of ATP. The rotational rate followed normal Michaelis-Menten kinetics. When the pressure was increased from 0.1 to 120 MPa, v_{max} decreased from 8.5 to 3.4 s^{-1}, whereas the K_m increased from 2.4 to 18 μM. The apparent ATP binding rate, $k_{ON}{}^{ATP}$ ($=3 \times V_{max} / K_m$), and maximum ATP turnover rate ($=3 \times V_{max}$) decreased with increased with pressure (Fig. 27.4d). The activation volumes determined from the pressure dependence of the rate constants were +100 $Å^3$ for ATP binding and +88 $Å^3$ for the other pressure-sensitive reaction. More detailed mechanisms can be elucidated by comparison with molecular dynamic simulations that include water molecules.

27.7 Swimming Motility of *Escherichia coli* Cells

Escherichia coli is a bacterium commonly used in research to study the effects of hydrostatic pressure on the activity of biological systems. In particular, *E. coli* has been used to investigate the effects of pressure on cell growth and morphological deformation (Zobell and Cobet 1964). At 30–50 MPa, cell division is inhibited leading to filamentous elongation of the cell body (Welch et al. 1993; Ishii et al. 2004a; Black et al. 2013). Cell growth is abolished at >60 MPa (Welch et al. 1993; Ishii et al. 2004a), and cell death results at >150 MPa (Kawarai et al. 2004; Manas and Mackey 2004; Moussa et al. 2007). These studies demonstrate that application of pressure inhibits the activity of living *E. coli* cells. Flagellar motility has been considered one of the most pressure-sensitive cellular processes (Meganathan and Marquis 1973; Eloe et al. 2008), but the detailed mechanisms that contribute to its inhibition by hydrostatic pressure remain unsolved.

Most wild-type strains of *E. coli* swim in solution using long (~10 μm) and thin (~20 nm) helical filaments (Namba and Vonderviszt 1997). The flagella vary in number (typically about 5–10 per cell) and originate at random points around the cell surface, a pattern called peritrichous flagellation (Fig. 27.5a). The cell turns its filament like a screw in either the counterclockwise (CCW, viewed from filament to motor) or clockwise (CW) direction. CCW rotation allows the left-handed helical filaments to form a bundle that propels the cell smoothly in a run. In contrast, CW rotation of a filament forces it out of the bundle and leads to a change in swimming direction called a tumble. The switching between CCW and CW rotation enables bacteria to migrate to more favorable environments (Porter et al. 2011). For the following experiments, we used the smooth-swimming strain RP4979 (*ΔcheY*) to study the pressure dependence of the motor function in swimming *E. coli* cells (Nishiyama and Kojima 2012). RP4979 cells lack the switch-inducing CheY protein which means that their flagellar motors rotate exclusively in the CCW direction, and cells can swim smoothly without tumbling (Scharf et al. 1998). Cells were cultured from frozen stocks to late logarithmic phase at 30 °C in tryptone broth (1 % Bacto tryptone, 0.5 % NaCl), a medium conventionally used for studying the motility of *E. coli* cells. The grown cells were suspended in a solution of 10 mM Tris, pH 7, 0.1 mM EDTA and were enclosed in the chamber. The medium does not contain

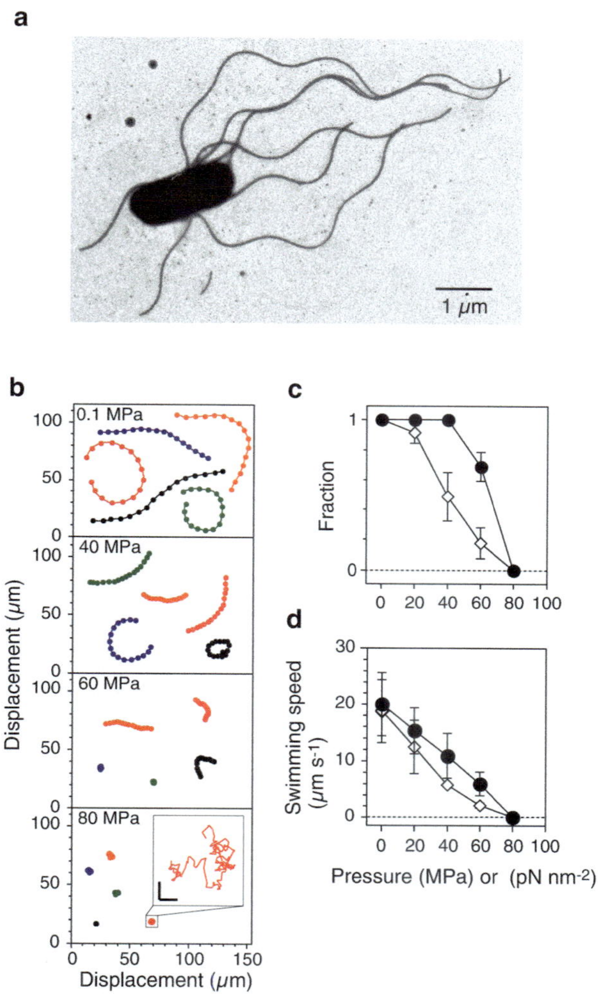

Fig. 27.5 Reversible inhibition of the swimming motility of *E. coli* cells. (**a**) Electron micrograph of a cell of *E. coli* showing peritrichous arrangement of flagella. (**b**) Trajectories of RP4979 cells. The positions of the cell were plotted every 10th frame for 5 s. (*Inset*) Trajectory of a cell at 80 MPa for 5 s on an expanded scale. Scale bar, 2 μm. (**c** and **d**) Swimming fraction and speed during the pressurization (*solid circles*) and depressurization processes (*open diamonds*). Swimming fractions were based on the number of cells that swam with a speed of >2 μm s^{-1} at each pressure. The speed was the average value of the swimming cells in (c). Error bars represent the SD (Modified from reference Nishiyama and Sowa 2012)

any sources of carbon and the cell condition was constantly maintained over our experimental period (typically and approximately 2 h).

The pressure was increased in a stepwise manner from 0.1 to 80 MPa and the motility of RP4979 cells measured (Fig. 27.5b). Under ambient conditions (0.1 MPa and 23 °C), RP4979 cells swam smoothly with a speed of ~20 μm s^{-1}. At 40 MPa, most cells still swam smoothly but the average speed decreased to ~50 % of its initial value at 0.1 MPa. At 60 MPa, many cells still swam, though swimming speed had drastically decreased and the some cells vibrated without showing any translational motion. At 80 MPa, most cells stopped directional swimming and diffused freely in the translational directions. This was confirmed by the fact that 2D mean-square displacement at 80 MPa increased linearly with time and demonstrates the cells only underwent random Brownian motion. Thus, the motility of swimming cells was partially inhibited at lower pressures and completely inhibited at 80 MPa. Figure 27.5c and d summarizes the swimming fraction and speed of the cells during the pressurization and depressurization processes.

Why do *E. coli* cells stop swimming at 80 MPa? The application of pressure may seriously damage the motility machinery, particularly flagellar motors and filaments. The damaged parts may be replaced by the protein only after the release of pressure. We therefore examined the motility of RP4979 cells just after release of 80 MPa of pressure and found that the swimming fraction and speed of cells instantaneously increased with time and returned almost to the initial value (Nishiyama and Sowa 2012). This rapid recovery of the motility of the swimming cells indicated that the cellular components responsible for motility, such as the flagellar filaments and motors, were not seriously damaged, even at 80 MPa. As shown in the next section, flagellar motors generate torque that is sufficient to join rotating filaments in a bundle. Thus, the applied pressure seems to inhibit the formation of rotating filament bundles that can propel the cell body in an aqueous environment.

27.8 Reverse Rotation of Flagellar Motors in *E. coli* Cells

The *E. coli* flagellar motor is one of the largest molecular machines in bacteria, with a molecular mass of ~11 MDa, approximately 13 different component proteins, and a further 25 or so proteins required for its expression and assembly (Macnab 1996). As shown in Fig. 27.6, it appears as a series of rings mounted on a rod surrounded by an array of studs embedded in the membrane (Minamino et al. 2008; Stock et al. 2012). The motor consists of a rotor and multiple stator units. Each stator is a torque-generating unit that constitutes two transmembrane ion channels. The motor is powered by protons that flow through these channels from outside of the cell to the inside (Berg 2003; Kojima and Blair 2004; Sowa and Berry 2008). The rotor spins relative to the cell body and its rotation transduces to the flagellum, whereas the stator units are anchored to the cell wall. Torque is generated by intermolecular interactions between the rotor and stator units. The chemomechanical energy conversion process is affected by physical and chemical conditions, such as viscous load (Ryu et al. 2000), temperature (Chen and Berg 2000; Inoue et al. 2013), pH (Manson et al. 1980; Shioi et al. 1980) and solvation (Yuan and Berg 2010). The application of pressure is also expected to modulate the

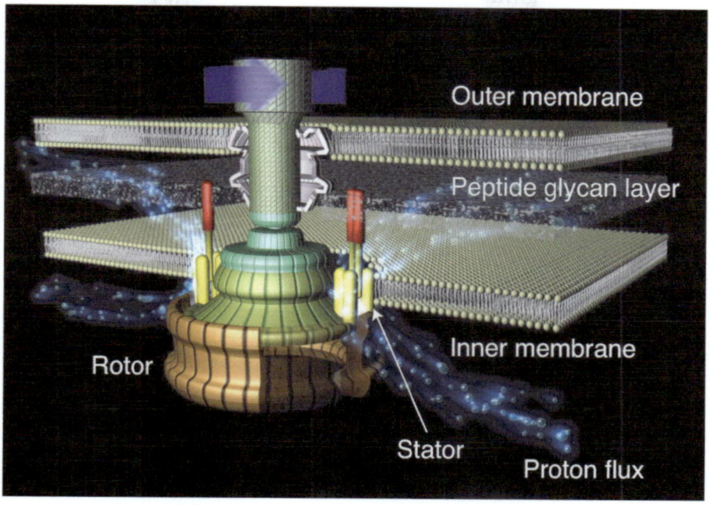

Fig. 27.6 Flagellar motor

torque generation process to the extent that *E. coli* cells stop swimming (Fig. 27.6). To study the effects of high pressure in detail, it is critical to monitor the motility of *E. coli* cells under these conditions.

We performed a tethered-cell assay using strain YS1326 (Nishiyama and Sowa 2012), which lacks switch-inducing protein CheY and expresses FliC-sticky filaments (Kuwajima 1988). Figure 27.7a displays the sequential phase-contrast images of the same single cell at 0.1, 80 and 120 MPa. In this experimental system (Fig. 27.7b), a single flagellar filament protruding from each cell was attached to the observation window in the high-pressure chamber, thus enabling rotation of the cell body, driven by the motor, to be monitored (Silverman and Simon 1974).

The time courses of the rotation of the cell at various pressures are displayed in Fig. 27.7c. Below 40 MPa, the motor rotated smoothly in the CCW direction, and upon application of pressure, the rotational speed did not change significantly. At 80 MPa, the rotational speed decreased sharply, but the motor still rotated in the CCW direction. At 120 MPa, the motor rotated still more slowly, but in the CW direction. Likewise, we found that many other cells still rotated in the CCW direction, changed direction frequently ('fluctuation'), or stopped rotation completely at 120 MPa.

The cell behavior was classified into CCW, fluctuation, CW, and stop states. The fractions were then calculated for each pressure and temperature condition (Fig. 27.7d). At 20 °C, the fraction of cells in the CCW state was almost equal to 1 when the pressure was less than 60 MPa. At more than 80 MPa, the fraction of cells in the CCW state decreased and reached ~0.1 at 140 MPa. On the other hand, the fraction of cells in the fluctuation state increased with pressure. After reaching a peak at 100 MPa, the fraction decreased with pressure. The number of cells in the stop and CW states monotonically increased with pressure.

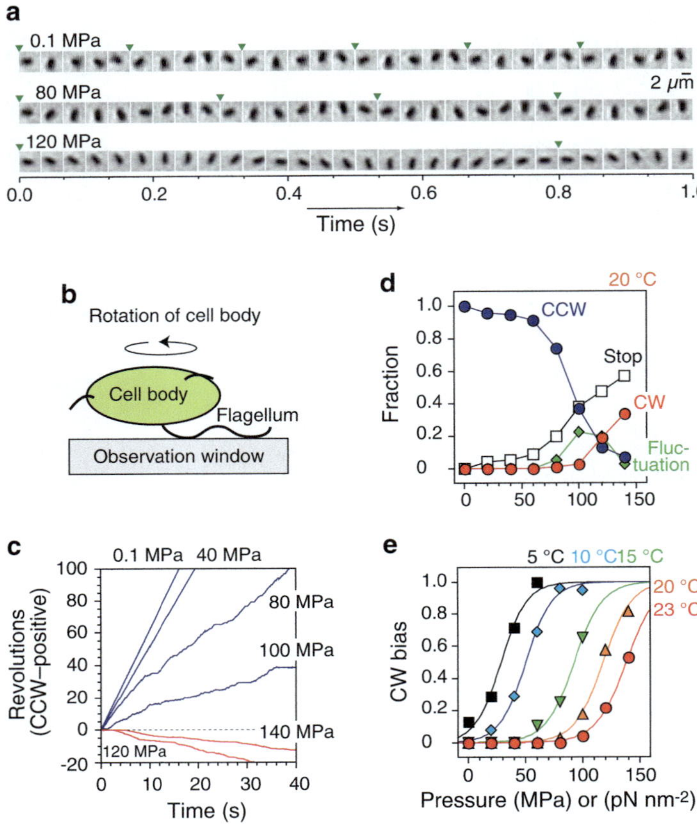

Fig. 27.7 Reverse rotation of *E. coli* flagellar motors. (**a**) Sequential phase-contrast images of the same tethered cell at 33-ms intervals. The pressures were 0.1 (*top*), 80 (*middle*) and 120 MPa (*bottom*). The *arrowheads* indicate completion of a turn. (**b**) Schematic drawing of the experimental system (not to scale). (**c**) Time course of rotations of the same rotating tethered cell in a. The experimental temperature was 20 °C, and the pressure was increased from 0.1 to 140 MPa. (**d**) Fraction of cells in the CCW (*blue*), fluctuation (*green*), CW (*red*), and stop (*white*) states at 20 °C. (**e**) CW bias. The plot at each temperature could be fitted by a sigmoidal curve (Modified from reference Nishiyama et al. 2013)

The pressure-induced change in the rotational direction was characterized by calculating the probability that a motor rotates in the CW direction (CW bias) and the results are summarized in Fig. 27.7e. The relationship between CW bias and applied pressure had similar sigmoidal curves, with the shift dependent on temperature, indicating motor direction is sensitive to pressure and temperature. The CW bias value at 20 °C steeply increased with pressure and reached 0.5 at ~120 MPa. At lower temperatures, pressure-induced changes in direction were observed at pressures of <120 MPa. The reversal in rotational direction occurred at a higher temperature than previously observed in a study in which flagellar motors rotated exclusively CW at −2 °C and 0.1 MPa (Turner et al. 1996). This suggests that increased pressure and decreased temperature act in a complementary fashion to reverse motor rotation from CCW to CW.

In wild type *E. coli* cells, CW rotation is normally caused by binding of CheY-P molecules to the switch complex of the motor. The CW bias increases with the intracellular concentration of CheY-P in a sigmoidal fashion (Cluzel et al. 2000). Interestingly, the applied pressure also changed CW bias in a sigmoidal fashion (Fig. 27.7e). Application of pressure generally promotes the formation of clusters of ordered water molecules on the surfaces of proteins. It is possible that hydration of the switch complex at high pressures induces structural changes similar to those caused by the binding of CheY-P.

27.9 Conclusion

We have demonstrated novel motility assays that change the structure and function of molecular motors. The pressure-induced changes could be caused by modifications of intermolecular interactions between protein and water molecules. The results described here suggest that pressure-induced effects on cell morphology and activity are directly caused by enhancement of hydration by water molecules. Most cellular processes are carried out by biomolecules located in an aqueous environment and the present techniques could be extended to investigate how the dynamic properties of these processes are produced in conjunction with water molecules.

Acknowledgments We would like to thank Tomoko Miyata for providing an Electron micrograph of *E. coli* (Fig. 27.5a), Akihiko Ishijima for illustrating a computer graphics of flagellar motor (Fig. 27.6). Yoshifumi Kimura for developing high-pressure chamber for microscopy, Daichi Okuno and Hiroyuki Noji for the measurement of F_1-ATPase, Yoshiyuki Sowa for the measurement of flagellar motors, Michio Homma, Akihiko Ishijima, Masahide Terazima and Yoshie Harada for technical support and discussion. This work was supported by PRESTO from JST, Grants-in-Aid for Scientific Research from MEXT and Shimadzu Science Foundation.

References

Abe F (2007) Exploration of the effects of high hydrostatic pressure on microbial growth, physiology and survival: perspectives from piezophysiology. Biosci Biotechnol Biochem 71:2347–2357

Akasaka K (2006) Probing conformational fluctuation of proteins by pressure perturbation. Chem Rev 106:1814–1835. doi:10.1021/cr040440z

Berg HC (2003) The rotary motor of bacterial flagella. Annu Rev Biochem 72:19–54. doi:10.1146/annurev.biochem.72.121801.161737

Black SL, Dawson A, Ward FB, Allen RJ (2013) Genes required for growth at high hydrostatic pressure in *Escherichia coli* K-12 identified by genome-wide screening. PLoS One 8:e73995. doi:10.1371/journal.pone.0073995

Boonyaratanakornkit BB, Park CB, Clark DS (2002) Pressure effects on intra- and intermolecular interactions within proteins. Biochim Biophys Acta 1595:235–249

Brooks NJ, Gauthe BL, Terrill NJ, Rogers SE, Templer RH, Ces O, Seddon JM (2010) Automated high pressure cell for pressure jump x-ray diffraction. Rev Sci Instrum 81:064103. doi:10.1063/1.3449332

Chaplin M (2006) Do we underestimate the importance of water in cell biology? Nat Rev Mol Cell Biol 7:861–866. doi:10.1038/nrm2021

Chen X, Berg HC (2000) Torque-speed relationship of the flagellar rotary motor of *Escherichia coli*. Biophys J 78:1036–1041. doi:10.1016/S0006-3495(00)76662-8

Colloc'h N, Girard E, Dhaussy AC, Kahn R, Ascone I, Mezouar M, Fourme R (2006) High pressure macromolecular crystallography: the 140-MPa crystal structure at 2.3 A resolution of urate oxidase, a 135-kDa tetrameric assembly. Biochim Biophys Acta 1764:391–397. doi:10.1016/j.bbapap.2006.01.006

Cluzel P, Surette M, Leibler S (2000) An ultrasensitive bacterial motor revealed by monitoring signaling proteins in single cells. Science 287:1652–1655

Dzwolak W, Kato M, Taniguchi Y (2002) Fourier transform infrared spectroscopy in high-pressure studies on proteins. Biochim Biophys Acta 1595:131–144

Eloe EA, Lauro FM, Vogel RF, Bartlett DH (2008) The deep-sea bacterium *Photobacterium profundum* SS9 utilizes separate flagellar systems for swimming and swarming under high-pressure conditions. Appl Environ Microbiol 74:6298–6305. doi:10.1128/AEM.01316-08

Filabozzi A, Deriu A, Di Bari MT, Russo D, Croci S, Di Venere A (2010) Elastic incoherent neutron scattering as a probe of high pressure induced changes in protein flexibility. Biochim Biophys Acta 1804:63–67. doi:10.1016/j.bbapap.2009.08.025

Frey B, Hartmann M, Herrmann M, Meyer-Pittroff R, Sommer K, Bluemelhuber G (2006) Microscopy under pressure – an optical chamber system for fluorescence microscopic analysis of living cells under high hydrostatic pressure. Microsc Res Tech 69:65–72. doi:10.1002/jemt.20269

Fujisawa T, Inoue K, Oka T, Iwamoto H, Uruga T, Kumasaka T, Inoko Y, Yagi N, Yamamoto M, Ueki T (2000) Small-angle X-ray scattering station at the SPring-8 RIKEN beamline. J Appl Cryst 33:797–800. doi:10.1107/S002188980000131x

Horio T, Hotani H (1986) Visualization of the dynamic instability of individual microtubules by dark-field microscopy. Nature 321:605–607. doi:10.1038/321605a0

Hoshihara Y, Kimura Y, Matsumoto M, Nagasawa M, Terazima M (2008) An optical high-pressure cell for transient grating measurements of biological substance with a high reproducibility. Rev Sci Instrum 79:034101. doi:10.1063/1.2894331

Inoue S, Fuseler J, Salmon ED, Ellis GW (1975) Functional organization of mitotic microtubules. Physical chemistry of the *in vivo* equilibrium system. Biophys J 15:725–744. doi:10.1016/S0006-3495(75)85850-4

Inoue Y, Baker MA, Fukuoka H, Takahashi H, Berry RM, Ishijima A (2013) Temperature dependences of torque generation and membrane voltage in the bacterial flagellar motor. Biophys J 105:2801–2810. doi:10.1016/j.bpj.2013.09.061

Ishii A, Sato T, Wachi M, Nagai K, Kato C (2004a) Effects of high hydrostatic pressure on bacterial cytoskeleton FtsZ polymers *in vivo* and *in vitro*. Microbiology 150:1965–1972. doi:10.1099/mic.0.26962-0

Ishii Y, Nishiyama M, Yanagida T (2004b) Mechano-chemical coupling of molecular motors revealed by single molecule measurements. Curr Protein Pept Sci 5:81–87

Kamatari YO, Kitahara R, Yamada H, Yokoyama S, Akasaka K (2004) High-pressure NMR spectroscopy for characterizing folding intermediates and denatured states of proteins. Methods 34:133–143. doi:10.1016/j.ymeth.2004.03.010

Kawaguchi K, Ishiwata S (2000) Temperature dependence of force, velocity, and processivity of single kinesin molecules. Biochem Biophys Res Commun 272:895–899. doi:10.1006/bbrc.2000.2856

Kawarai T, Wachi M, Ogino H, Furukawa S, Suzuki K, Ogihara H, Yamasaki M (2004) SulA-independent filamentation of Escherichia coli during growth after release from high hydrostatic pressure treatment. Appl Microbiol Biotechnol 64:255–262. doi:10.1007/s00253-003-1465-6

Kimura T, Sakamoto K, Morishima I, Ishimori K (2006) Dehydration in the folding of reduced cytochrome c revealed by the electron-transfer-triggered folding under high pressure. J Am Chem Soc 128:670–671. doi:10.1021/ja055084w

Kitching JA (1957) Effects of high hydrostatic pressures on the activity of flagellates and ciliates. J Exp Biol 34:494–510

Kojima S, Blair DF (2004) The bacterial flagellar motor: structure and function of a complex molecular machine. Int Rev Cytol 233:93–134. doi:10.1016/S0074-7696(04)33003-2

Kojima H, Muto E, Higuchi H, Yanagida T (1997) Mechanics of single kinesin molecules measured by optical trapping nanometry. Biophys J 73:2012–2022. doi:10.1016/S0006-3495(97)78231-6

Kuwajima G (1988) Construction of a minimum-size functional flagellin of *Escherichia coli*. J Bacteriol 170:3305–3309

Macnab RM (1996) Flagella and Motility. In: Neidhardt FC (ed) *Escherichia coli* and *Salmonella*, 2nd edn. Amer Society for Microbiology, Washington, DC, pp 123–145

Manas P, Mackey BM (2004) Morphological and physiological changes induced by high hydrostatic pressure in exponential- and stationary-phase cells of *Escherichia coli*: relationship with cell death. Appl Environ Microbiol 70:1545–1554

Manson MD, Tedesco PM, Berg HC (1980) Energetics of flagellar rotation in bacteria. J Mol Biol 138:541–561

Meganathan R, Marquis RE (1973) Loss of bacterial motility under pressure. Nature 246:525–527

Minamino T, Imada K, Namba K (2008) Molecular motors of the bacterial flagella. Curr Opin Struct Biol 18:693–701. doi:10.1016/j.sbi.2008.09.006

Miyamoto Y, Muto E, Mashimo T, Iwane AH, Yoshiya I, Yanagida T (2000) Direct inhibition of microtubule-based kinesin motility by local anesthetics. Biophys J 78:940–949. doi:10.1016/S0006-3495(00)76651-3

Moussa M, Perrier-Cornet JM, Gervais P (2007) Damage in *Escherichia coli* cells treated with a combination of high hydrostatic pressure and subzero temperature. Appl Environ Microbiol 73:6508–6518. doi:10.1128/AEM.01212-07

Nagae T, Kawamura T, Chavas LMG, Niwa K, Hasegawa M, Kato C, Watanabe N (2012) High-pressure-induced water penetration into 3-isopropylmalate dehydrogenase. Acta Crystallogr D Biol Crystallogr 68:300–309. doi:10.1107/S0907444912001862

Namba K, Vonderviszt F (1997) Molecular architecture of bacterial flagellum. Q Rev Biophys 30:1–65

Nishiyama M (2013) High pressure sample container for optical microscopic observation. Japan Patent P5207300

Nishiyama M, Kojima S (2012) Bacterial motility measured by a miniature chamber for high-pressure microscopy. Int J Mol Sci 13:9225–9239. doi:10.3390/ijms13079225

Nishiyama M, Sowa Y (2012) Microscopic analysis of bacterial motility at high pressure. Biophys J 102:1872–1880. doi:10.1016/j.bpj.2012.03.033

Nishiyama M, Muto E, Inoue Y, Yanagida T, Higuchi H (2001) Substeps within the 8-nm step of the ATPase cycle of single kinesin molecules. Nat Cell Biol 3:425–428. doi:10.1038/35070116

Nishiyama M, Higuchi H, Yanagida T (2002) Chemomechanical coupling of the forward and backward steps of single kinesin molecules. Nat Cell Biol 4:790–797. doi:10.1038/ncb857

Nishiyama M, Kimura Y, Nishiyama Y, Terazima M (2009) Pressure-induced changes in the structure and function of the kinesin-microtubule complex. Biophys J 96:1142–1150. doi:10.1016/j.bpj.2008.10.023

Nishiyama M, Shimoda Y, Hasumi M, Kimura Y, Terazima M (2010) Microtubule depolymerization at high pressure. Ann N Y Acad Sci 1189:86–90. doi:10.1111/j.1749-6632.2009.05411.x

Nishiyama M, Sowa Y, Kimura Y, Homma M, Ishijima A, Terazima M (2013) High hydrostatic pressure induces counterclockwise to clockwise reversals of the *Escherichia coli* flagellar motor. J Bacteriol 195:1809–1814. doi:10.1128/JB.02139-12

Nogales E, Wang HW (2006) Structural intermediates in microtubule assembly and disassembly: how and why? Curr Opin Cell Biol 18:179–184. doi:10.1016/j.ceb.2006.02.009

Noji H, Yasuda R, Yoshida M, Kinosita K Jr (1997) Direct observation of the rotation of F_1-ATPase. Nature 386:299–302. doi:10.1038/386299a0

Okuno D, Nishiyama M, Noji H (2013) Single-molecule analysis of the rotation of F_1-ATPase under high hydrostatic pressure. Biophys J 105:1635–1642. doi:10.1016/j.bpj.2013.08.036

Otter T, Salmon ED (1985) Pressure-induced changes in Ca^{2+}-channel excitability in *Paramecium*. J Exp Biol 117:29–43

Porter SL, Wadhams GH, Armitage JP (2011) Signal processing in complex chemotaxis pathways. Nat Rev Microbiol 9:153–165. doi:10.1038/nrmicro2505

Ranatunga KW, Fortune NS, Geeves MA (1990) Hydrostatic compression in glycerinated rabbit muscle fibers. Biophys J 58:1401–1410. doi:10.1016/S0006-3495(90)82486-3

Rice S, Lin AW, Safer D, Hart CL, Naber N, Carragher BO, Cain SM, Pechatnikova E, Wilson-Kubalek EM, Whittaker M, Pate E, Cooke R, Taylor EW, Milligan RA, Vale RD (1999) A structural change in the kinesin motor protein that drives motility. Nature 402:778–784. doi:10.1038/45483

Ryu WS, Berry RM, Berg HC (2000) Torque-generating units of the flagellar motor of *Escherichia coli* have a high duty ratio. Nature 403:444–447. doi:10.1038/35000233

Salmon ED (1975a) Pressure-induced depolymerization of spindle microtubules. I. Changes in birefringence and spindle length. J Cell Biol 65:603–614

Salmon ED (1975b) Pressure-induced depolymerization of spindle microtubules. II. Thermodynamics of in vivo spindle assembly. J Cell Biol 66:114–127

Salmon ED (1975c) Pressure-induced depolymerization of brain microtubules in vitro. Science 189:884–886

Salmon ED, Ellis GW (1975) A new miniature hydrostatic pressure chamber for microscopy Strain-free optical glass windows facilitate phase-contrast and polarized-light microscopy of living cells. Optional fixture permits simultaneous control of pressure and temperature. J Cell Biol 65:587–602

Scharf BE, Fahrner KA, Turner L, Berg HC (1998) Control of direction of flagellar rotation in bacterial chemotaxis. Proc Natl Acad Sci U S A 95:201–206

Schnitzer MJ, Block SM (1997) Kinesin hydrolyses one ATP per 8-nm step. Nature 388:386–390. doi:10.1038/41111

Shioi JI, Matsuura S, Imae Y (1980) Quantitative measurements of proton motive force and motility in *Bacillus subtilis*. J Bacteriol 144:891–897

Silverman M, Simon M (1974) Flagellar rotation and the mechanism of bacterial motility. Nature 249:73–74

Sowa Y, Berry RM (2008) Bacterial flagellar motor. Q Rev Biophys 41:103–132. doi:10.1017/S0033583508004691

Stock D, Namba K, Lee LK (2012) Nanorotors and self-assembling macromolecular machines: the torque ring of the bacterial flagellar motor. Curr Opin Biotechnol 23:545–554. doi:10.1016/j.copbio.2012.01.008

Svoboda K, Schmidt CF, Schnapp BJ, Block SM (1993) Direct observation of kinesin stepping by optical trapping interferometry. Nature 365:721–727. doi:10.1038/365721a0

Taniguchi Y, Nishiyama M, Ishii Y, Yanagida T (2005) Entropy rectifies the Brownian steps of kinesin. Nat Chem Biol 1:342–347. doi:10.1038/nchembio741

Turner L, Caplan SR, Berg HC (1996) Temperature-induced switching of the bacterial flagellar motor. Biophys J 71:2227–2233. doi:10.1016/S0006-3495(96)79425-0

Usui K, Hiraki T, Kawamoto J, Kurihara T, Nogi Y, Kato C, Abe F (2012) Eicosapentaenoic acid plays a role in stabilizing dynamic membrane structure in the deep-sea piezophile *Shewanella violacea*: a study employing high-pressure time-resolved fluorescence anisotropy measurement. Biochim Biophys Acta 1818:574–583. doi:10.1016/j.bbamem.2011.10.010

Vass H, Black SL, Herzig EM, Ward FB, Clegg PS, Allen RJ (2010) A multipurpose modular system for high-resolution microscopy at high hydrostatic pressure. Rev Sci Instrum 81:053710. doi:10.1063/1.3427224

Vass H, Black SL, Flors C, Lloyd D, Ward FB, Allen RJ (2013) Single-molecule imaging at high hydrostatic pressure. Appl Phys Lett 102:154103. doi:10.1063/1.4802202

Walker RA, O'Brien ET, Pryer NK, Soboeiro MF, Voter WA, Erickson HP, Salmon ED (1988) Dynamic instability of individual microtubules analyzed by video light microscopy: rate constants and transition frequencies. J Cell Biol 107:1437–1448

Watanabe TM, Imada K, Yoshizawa K, Nishiyama M, Kato C, Abe F, Morikawa TJ, Kinoshita M, Fujita H, Yanagida T (2013) Glycine insertion makes yellow fluorescent protein sensitive to hydrostatic pressure. PLoS One 8:e73212. doi:10.1371/journal.pone.0073212

Welch TJ, Farewell A, Neidhardt FC, Bartlett DH (1993) Stress-response of *Escherichia-coli* to elevated hydrostatic-pressure. J Bacteriol 175:7170–7177

Winter R (2002) Synchrotron X-ray and neutron small-angle scattering of lyotropic lipid mesophases, model biomembranes and proteins in solution at high pressure. Biochim Biophys Acta 1595:160–184

Yasuda R, Noji H, Kinosita K Jr, Yoshida M (1998) F_1-ATPase is a highly efficient molecular motor that rotates with discrete 120 degree steps. Cell 93:1117–1124

Yasuda R, Noji H, Yoshida M, Kinosita K Jr, Itoh H (2001) Resolution of distinct rotational substeps by submillisecond kinetic analysis of F_1-ATPase. Nature 410:898–904. doi:10.1038/35073513

Yuan J, Berg HC (2010) Thermal and solvent-isotope effects on the flagellar rotary motor near zero load. Biophys J 98:2121–2126. doi:10.1016/j.bpj.2010.01.061

Zobell CE, Cobet AB (1964) Filament formation by *Escherichia coli* at increased hydrostatic pressures. J Bacteriol 87:710–719

Chapter 28
Ion Channels Activated by Mechanical Forces in Bacterial and Eukaryotic Cells

Masahiro Sokabe, Yasuyuki Sawada, and Takeshi Kobayashi

Abstract Since the first discovery of mechanosensitive ion channel (MSC) in non-sensory cells in 1984, a variety of MSCs has been identified both in prokaryotic and eukaryotic cells. One of the central issues concerning MSCs is to understand the molecular and biophysical mechanisms of how mechanical forces activate/open MSCs. It has been well established that prokaryotic (mostly bacterial) MSCs are activated exclusively by membrane tension. Thus the problem to be solved with prokaryotic MSCs is the mechanisms how the MSC proteins receive tensile forces from the lipid bilayer and utilize them for channel opening. On the other hand, the activation of many eukaryotic MSCs crucially depends on tension in the actin cytoskeleton. By using the actin cytoskeleton as a force sensing antenna, eukaryotic MSCs have obtained sophisticated functions such as remote force sensing and force-direction sensing, which bacterial MSCs do not have. Actin cytoskeletons also give eukaryotic MSCs an interesting and important function called "active touch sensing", by which cells can sense rigidity of their substrates. The contractile actin cytoskeleton stress fiber (SF) anchors its each end to a focal adhesion (FA) and pulls the substrate to generate substrate-rigidity-dependent stresses in the FA. It has been found that those stresses are sensed by some Ca^{2+}-permeable MSCs existing in the vicinity of FAs, thus the MSCs work as a substrate rigidity sensor that can transduce the rigidity into intracellular Ca^{2+} levels. This short review, roughly constituting of two parts, deals with molecular and biophysical mechanisms underlying the MSC activation process mostly based on our recent studies; (1) structure-function in bacterial MSCs activation at the atomic level, and (2) roles of actin cytoskeletons in the activation of eukaryotic MSCs.

Keywords Mechanosensitive channels • Membrane tension • Actin cytoskeleton • Stress fiber • Focal adhesion • Rigidity sensing

---

M. Sokabe (✉)
Mechanobiology Laboratory, Nagoya University Graduate School of Medicine, Nagoya, Japan

Mechanobiology Institute, National University of Singapore, Singapore, Singapore
e-mail: msokabe@med.nagoya-u.ac.jp

Y. Sawada • T. Kobayashi
Department of Physiology, Nagoya University Graduate School of Medicine, Nagoya, Japan

K. Akasaka, H. Matsuki (eds.), *High Pressure Bioscience*,
Subcellular Biochemistry 72, DOI 10.1007/978-94-017-9918-8_28

28.1 Introduction

Mechanosensing is an indispensable function to support the life of organisms throughout the phylogeny from bacteria to human. Not only specialized mechanoreceptors, which connect to the central nervous system, like inner ear hair cells, cutaneous tactile sensors and visceral baroreceptors but also every ordinary cell can sense mechanical stimuli and utilize the information to regulate their shape, motility, differentiation and cell cycle. The known major player in cell mechanosensing is the mechanosensitive ion channel (MSC) that can be activated primarily by increased membrane tension generated by externally applied mechanical forces to cells. Among MSCs, the bacterial MSCs MscL and MscS are the best studied ones owing to their resolved 3D protein structures of the closed state. They are activated simply by increased tension in the plasma membrane (lipid bilayer) and contribute to the cell volume regulation against hypotonic challenge, thus working as a safety bulb to protect cells from osmotic rupture.

On the other hand, eukaryotic MSCs, the majority of which are Ca^{2+} permeable, seem to be activated mainly by tension in the actin cytoskeleton stress fiber (SF) that connects to the focal adhesion (FA) adhering to the cell substrate. The supramolecular complex of SF/FA/MSC has an "active-touch sensing" capability, in which actively generated contractile forces in the SF pull the substrate via an FA and activate MSCs existing in the vicinity of the FA. As stresses in FAs generated by SF dragging of substrate would depend on rigidity of the substrate, the MSCs can transduce substrate rigidity into the amount of Ca^{2+} influx across the MSCs.

MSCs might have evolved from the simple bacterial ones, which directly respond to changes in the membrane tension, to the elaborated eukaryotic ones that can actively detect mechanical properties of their immediately surrounding environments. This short article summarizes our recent studies on the differential activation mechanisms between bacterial and eukaryotic MSCs.

28.2 Activation Mechanisms in Bacterial MSCs

Bacterial MscL and MscS are the only MSCs of which 3D protein structures in the closed state are resolved at the atomic level, and MscL has been better studied due to its simpler structure. As shown in Fig. 28.1, MscL is constituted of homopentamer of a hair-pin like structured subunit having two transmembrane α-helices called inner helix (TM1, white) and outer helix (TM2, pink), displaying fivefold radial symmetry around the central pore (right panel in Fig. 28.1a) (Steinbacher et al. 2007). Neighboring TM1s cross and interact each other near the cytoplasmic side through hydrophobic interactions, forming the most constricted pentagon-shaped hydrophobic part (called "gate") of the pore. The crossings stabilize the closed state of MscL. The issue is to understand the underlying molecular mechanism during the course of channel opening (gating) driven by tension increase in the membrane;

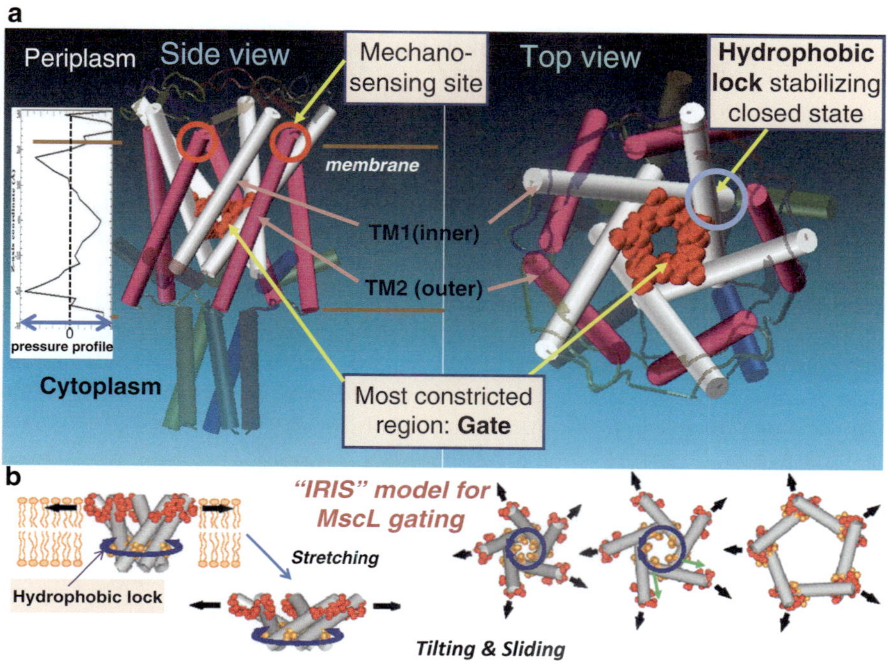

Fig. 28.1 3D structure and a gating model of *E. coli* MscL. (**a**) Schematized 3D structure of MscL based on a latest study (Steinbacher et al. 2007). Some hypothetical functional sites and transmembrane pressure profile (most *left*) are indicated. (**b**) IRIS model for MscL gating, where membrane stretch induces tilting of helices (*left*) and radially outward sliding of the crossings between inner TM1 helices (*right*) (Modified from (Sawada et al. 2012))

which amino acid in the channel protein senses tensile forces and how the sensed forces lead to the channel/gate opening of MscL.

We identified potential tension-sensing site in the MscL protein by using single channel current recordings in combination with site-directed mutagenesis. Amino acid (AA) residues facing the lipid bilayer were substituted one by one with a hydrophilic AA (asparagine, N) in order to weaken the hydrophobic interaction between the AAs and lipids to search which AA(s) is crucial in sensing tension in the membrane. The asparagine substitution of one of seven residues at the periplasmic end of transmembrane helices caused loss of MscL mechanosensitivity, whereas the substitution of the residues at the core of the bilayer did not affect the channel gating (Yoshimura et al. 2004; Yoshimura and Sokabe 2010). This result indicates that the hydrophobic interaction between lipids and one or some of the residues near the periplasmic surface may act as a tension sensor of MscL. This is consistent with the transmembrane tension (pressure) distribution with two peaks near the inner and outer membrane surfaces (most left graph in Fig. 28.1a).

Based on the above implication, we have proposed a model for MscL gating called IRIS model (Yoshimura et al. 2004). Upon membrane stretch, the AA residues near the outer surface of the membrane are dragged by lipids and TM1/TM2 helices are gradually tilted (left panel in Fig. 28.1b) accompanied by outwardly radial sliding of the crossings between TM1 helices, leading to gate expansion (right panel in Fig. 28.1b). This model has been supported by our recent single molecular imaging of closed and open structures of MscL by cryo-electron microscopy (Yoshimura et al. 2008).

To get insight into more detailed features of structural dynamics of MscL opening at the atomic detail, molecular dynamics (MD) simulations were performed (Sawada et al. 2012). We constructed a molecular model using the closed structure of MscL (Steinbacher et al. 2007) embedded in a lipid bilayer and immersed in water molecules. MD simulations were conducted under an increased membrane tension that was generated by reducing the lateral pressure only in the bilayer. Simulation time of all-atom MD of proteins is in general limited for a few tenths of nanoseconds (ns) to obtain reliable results, which is apparently much shorter than that (> a few hundred microseconds) required for actual MscL opening. Therefore, we focused on the two mechanisms critical for initiating MscL opening; (1) which residue(s) has the most potent and stable interaction with the surrounding lipids (identification of tension sensor) and (2) how the received force by the sensor induces expansion of the gate of MscL. Results principally succeeded in reproducing the initial process predicted by the "IRIS" model (Fig. 28.1b), that is, tilting of TM1/TM2 helices accompanied by sliding of the TM1 crossings towards channel opening.

To determine the major tension-sensing sites (tension sensor) in MscL, we estimated the interaction energy between individual AA residues from Gly76 to Ala89 in TM2 helix and surrounding lipids. Surprisingly, the interaction energy between Phe78 and lipids is conspicuously much lower than that of any other residues. Since the tension sensitivity at the level of individual AA residues depends on the mechanical strength of AA-lipids interaction, Phe78 was thought to act as the major tension sensor of MscL. Actually, TM2 was dragged by lipids specifically at Phe78 and tilted (Fig. 28.2a). Interestingly, aromatic ring of Phe78 was frequently facing the CH_2 residues of lipid acyl chains, implying that CH/Π interaction (Tsuzuki et al. 2000; Shanthi et al. 2010), as depicted in Fig. 28.2b, is the physicochemical mechanism for conspicuously strong interaction between Phe78 and lipids, by which Phe78 can act as the major tension sensor of MscL.

Next, in order to know the details how the gate is expanded, we sought AAs that stabilize the closed gate at the crossings between neighboring TM1 helices (see Fig. 28.1a), and found that the residues (Val16, Leu19 and Ala20) on a TM1 helix come into contact with Gly22 and Gly26 on the neighboring TM1 helix (left panel in Fig. 28.2c). Upon membrane stretching, Gly22 began to come into contact with its neighboring TM1 helix in $\sim$1 ns, fitting into a pocket formed by Val16, Leu19 and Ala20, then Gly26 instead of Gly22 began to come into contact with the pocket (middle to right panels in Fig. 28.2c). Time profile of interaction energies at those interacting AAs between neighboring TM1 helices revealed that an energy barrier exists during the 2 ns simulation. The potential energy difference between the two states separated by the barrier was calculated to be ca. 25 kcal mol^{-1} (42 k_BT), which value is comparable to the experimentally estimated free energy

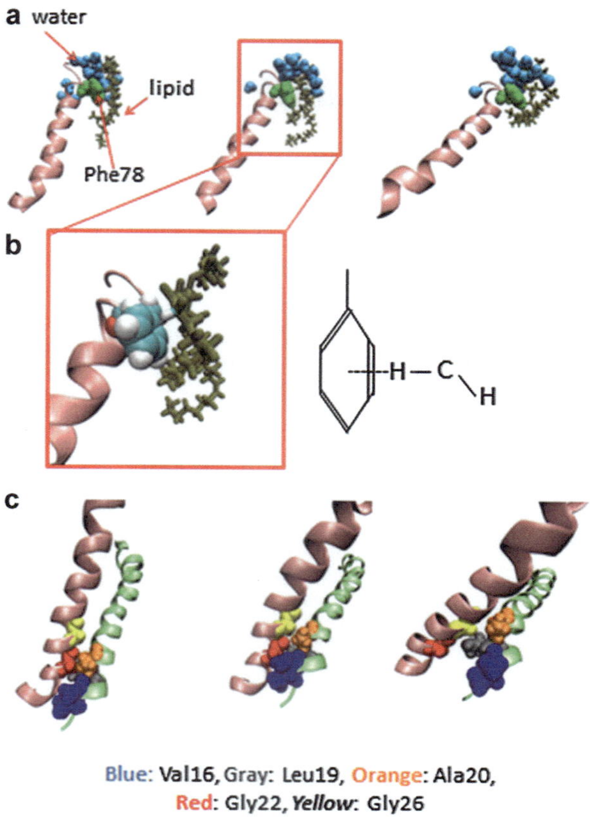

Fig. 28.2 Detailed structural changes at the putative tension sensor and gate at the atomic level, during the initial phase towards channel opening. (**a**) Snapshots of lipid dragging of Phe78 on a TM2 at 0, 2, 4 ns after membrane stretch, which causes TM2 (and TM1) tilting. (**b**) Interaction between aromatic ring of Phe78 and CH of a lipid acyl chain (CH/π interaction). (**c**) Snapshots of sliding between neighboring TM1s at 0, 3, 4 ns after stretch onset (Modified from (Sawada et al. 2012))

difference (~38 k_BT) between the closed and the first sub-conducting state of MscL (Sukharev et al. 1999). It is rather surprising that such a limited MD calculation gives a good number, the model might reflect an essence of initial process of MscL gating. Detailed calculations including other interactions that may contribute to the MscL gating are in progress.

28.3 Activation Mechanisms in Bacterial MscS

Another bacterial MS channel MscS has a smaller conductance (1 nS), a lower threshold for activation and a much more complicated structure in comparison with those of MscL. MscS is a homoheptamer of a subunit having three transmembrane α-helices called TM1, TM2, and TM3 (middle panel in Fig. 28.3a), displaying sevenfold radial symmetry around the central pore (right panel in Fig. 28.3a) with a large cage-shaped cytoplasmic domain having side windows (Bass et al. 2002) that are implicated to work as a molecular sieve against large sized molecules (left

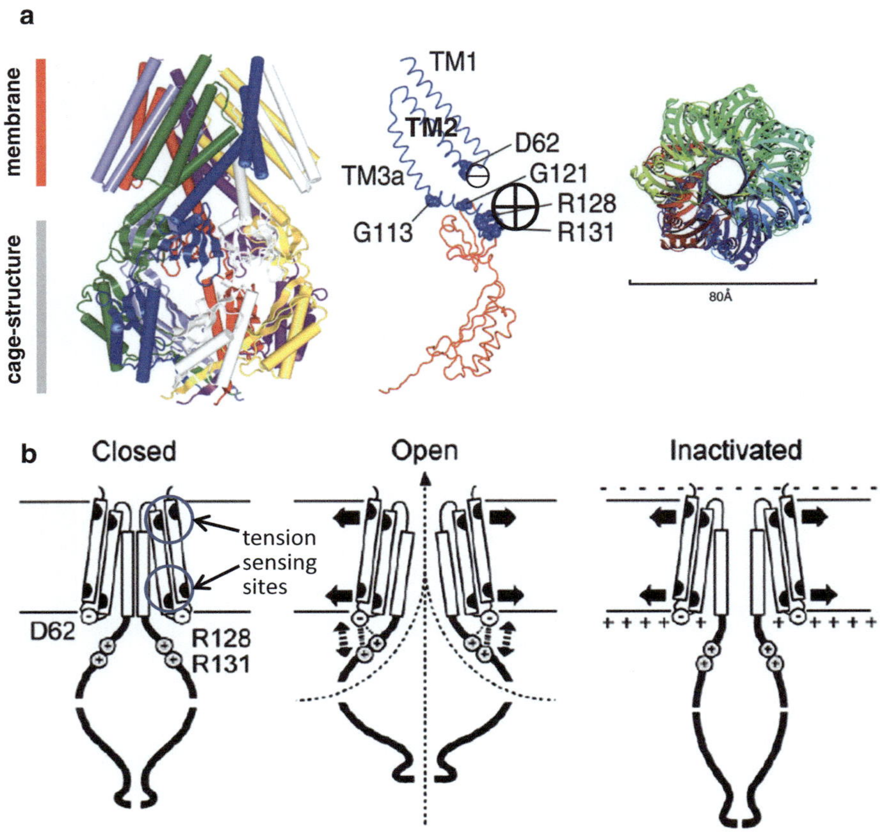

Fig. 28.3 3D structure of MscS and its gating model. (**a**) Side view of the whole (*left*) and subunit (*middle*) structures, and a top view of the whole structure, see text for details (Modified from (Bass et al. 2002)). (**b**) A model for the MscS activation (*opening*) and inactivation. Asp62 is located in the loop that connects TM1 and TM2. Arg128 and Arg131 are located in the upper surface of the cytoplasmic vestibule (*left panel*). When the channel opens, Asp62 forms a salt bridge with Arg128 and Arg131 (*middle panel*). When the cytoplasmic potential is positive, the electrostatic interaction is dissociated by the positive surface potential (*right panel*). The resulting shrinkage of the cytoplasmic vestibule brings about channel inactivation (Modified from (Nomura et al. 2008))

panel in Fig. 28.3a, b). TM3s line the pore and some AA residues of TM3 near the cytoplasmic surface of the membrane are postulated to form the main gate of MscS.

MscS as well as MscL can be activated exclusively by tension in the membrane. In contrast to the conformational changes during the MscL opening (tilting and sliding), MscS seems to open through a lateral expansion of TM helices along the membrane plain (Fig. 28.3b). How does membrane tension cause such a conformational change in MscS? As did in MscL, we tried toidentify the tension sensing site of MscS using the asparagine scanning method, and found that two separate tension sensing sites are located near both surfaces of the membrane (left panel in Fig. 28.3b) (Nomura et al. 2006), which conforms the lateral expansion model of TMs during MscS opening. This configuration of tension sensing sites may also partially account for the higher mechanosensitivity of MscS than MscL, since the number of mechanosensing sites of MscS can be estimated at 14 (2 sites/subunit × 7 subunits) a much larger number than that (five) of MscL. The sensed forces at the tension sensing sites may be transmitted through linkers between helices and/or lateral interactions among helices, eventually lead to a lateral expansion of TM3 helices. However, in the case of MscS opening, things are not so simple.

The large cytoplasmic domain has a crucial role in the MscL opening. As depicted in Fig. 28.3b, there is an electrostatic interaction between negatively charged Asp62 at the TM1-TM2 linker and positively charged Arg128/131 at the cytoplasmic domain. We found that destruction of this ionic interaction by site-directed mutagenesis caused loss of function (LOF) mutants (Nomura et al. 2008). Presumably, lateral expansion of TM1 and TM2 pulls up and distends the cytoplasmic domain (middle panel in Fig. 28.3b) (Machiyama et al. 2009), followed by stretching out of the linker directly connecting to the gate of TM1 to open the channel. Interestingly, under a large and/or long lasting hypotonic stress MscS transits to an inactivated state (right panel in Fig. 28.3b), where the channel never opens even with a very strong mechanical stimulus (membrane stretch). MscS can recover its closed state only when the membrane tension returns to the resting level. Inactivation of MscS can also be induced by membrane depolarization, which would occur when MscS opens for a long time due to a long exposure to hypotonic shock, because preferred anion permeability of MscS will cause anion out-flux from cells. This is a very smart way to prohibit excess loss of valuable cytoplasmic components. Another interesting potential function of the cytoplasmic domain is that it may work as a cytoplasmic osmometer. When MscS opens under hypotonic shock, cytoplasmic solutes and water will flow out through MscS, but large sized molecules that cannot permeate the side windows (14 Å in diameter) of the cytoplasmic domain will be condensed around the outside surface of the cage-domain, leading to an increased osmolarity there. This induces water outflow from the cage to the cytoplasm, leading to a shrinkage of the cage (right panel in Fig. 28.3b), which would uncouple the electrostatic interaction between TM3 and the cage to induce inactivation of MscS. This may be another protection mechanism to inhibit excess loss of cytoplasmic components. Based on the hypothetical model proposed here, inactivation as well as activation of MscS largely depends on the interaction of TM2 with TM3 and TM1. In other words, TM2 may work as a modulator to regulate the threshold for activation and inactivation of MscS.

While MscL has a large conductance (3 nS), a high activation threshold and a simple regulatory system, MscS has a smaller conductance (1 nS), a lower threshold, and a subtle regulatory system, by which they may play differential roles as an emergent device and a housekeeping one, respectively.

28.4 Activation Mechanisms in Eukaryotic MSCs

It has been proposed that eukaryotic MSCs are activated mainly by tension in the cytoskeleton based on pieces of indirect evidence (Sokabe et al. 1991; Sokabe and Sachs 1992). To test this hypothesis, we employed cultured human umbilical vein endothelial cells (HUVECs). HUVECs carry a 35 pS Ca^{2+} permeable MSC (Naruse and Sokabe 1993), which enabled us to monitor the channel activation through the measurement of intracellular Ca^{2+} increases as well as whole cell currents across the MSC. In addition, HUVECs have well developed F-actin based stress fibers (SFs) that anchor to FAs (Fig. 28.4 left). We hypothesized that SFs act as an intracellular force transmitter to activate the MSCs located in the vicinity of the integrins in the FA to which SFs are anchored (Fig. 28.4, right).

To verify this hypothesis we developed a sophisticated method by which we can apply localized mechanical stimuli onto FAs via SFs while monitoring Ca^{2+} influx in and near FAs by high speed near field microscopy (Hayakawa et al. 2008). Aseptically prepared HUVECs were plated on a fibronectin (FN) coated cover slip and a glass bead (5–10 μm) coated with FN was attached on the apical surface of a HUVEC. Focal proteins (such as integrin and vinculin) and F-actins were assembled in the membrane region attaching the bead to form FAs beneath the bead within 15 min. Each FA was connected via an SF to a preformed FA at the basal cell surface. A short pulse of stretch was applied through this structure to the basal FAs by displacing the bead with a piezo-driven glass pipette, while monitoring Ca^{2+} transient near the basal cell surface using the Ca^{2+} indicator fluo-3.

Immediately after a stimulation (0.1–2 μM displacement of the bead) an intracellular Ca^{2+} increase was observed around basal FAs, which was strongly

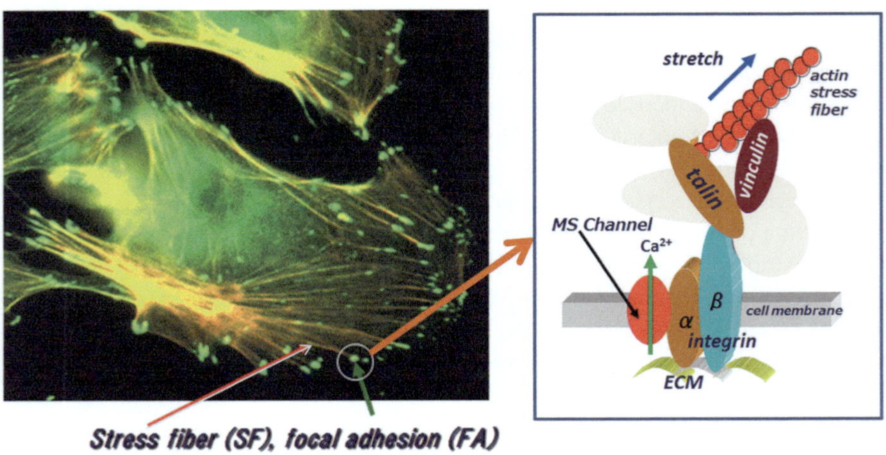

Fig. 28.4 Fluorescent image of HUVECs. Strand like SFs terminate at spot shaped FAs. *Right panel* shows hypothetical SF/FA/MSC complex, in which MSC is activated by force (tension) transmitted through SF and FA (integrin)

inhibited by external Ca^{2+} deprivation or by the potent MSC blocker gadolinium (Gd^{3+}), whereas thapsigalgin, which depletes intracellular Ca^{2+} stores, did not show any significant effect. As the F-actin depolymerizing drug cytochalasin D also inhibited the Ca^{2+} increase, it is highly likely that the stretch induced Ca^{2+} transient was mediated by the MSCs that are activated by the force transmitted along SFs.

To directly test the above idea, phalloidin coated fluorescent plastic beads, which selectively bind to actin filaments and SFs, were microinjected into the cell through a glass pipette for whole cell current recordings. A traction force was applied to a bead aggregate attached on an SF using the laser trapping technique while recording the resulting whole cell current and Ca^{2+} increase (Fig. 28.5a). An inward current and a Ca^{2+} increase were detected immediately after a force application to the bead aggregate, strongly supporting our hypothesis that tension in the SF activates MSCs. It was found that stretching a single SF could activate 20–40 MSCs and that force required to activate a single MSC was as small as 1 pN (Hayakawa et al. 2008).

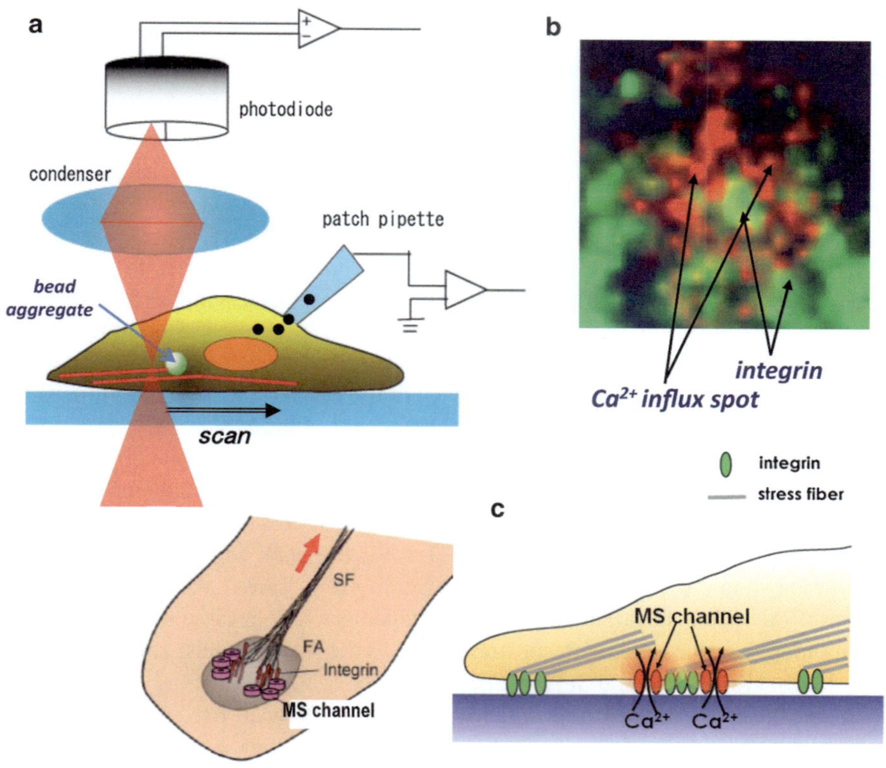

Fig. 28.5 Simultaneous measurements of whole cell currents and Ca^{2+} influx spots in response to an SF stretch. (**a**) Experimental setup. (**b**) A live fluorescent image of integrins and Ca^{2+} influx spots (Modified from (Hayakawa et al. 2008)). (**c**) Schematic diagram of hypothetical SF/FA/MSC complex and SF/integrin/MSC complex

We also could visualize, by ultra-fast near field microscopy, individual Ca^{2+}-influx spots that occurred near FAs within a few msec after a stretch. The Ca^{2+}-influx spots had a size as small as 300 nm and quantitative estimation suggested that the smallest spots reflect the images of the moment of Ca^{2+}-influx through single MSCs (Fig. 28.5b). Analysis of the spatial distribution of the Ca^{2+}-influx spots and integrins showed that MSCs are located as close as a few hundred nm apart from FAs (integrins). A hypothetical model for mechanosesning machinery composed of SF, FA and MSC is shown in Fig. 28.5c. The mechanism how the conveyed force via SF/integrins transmits (e.g., via membrane or submembraneous cytoskeletons) to MSCs awaits future studies. It should be noted that our results do not deny the role of membranes as a force transmitter but rather show up the actin stress fiber as an accessory system that gives new functions to MSCs, such as force concentration, remote force sensing, and force direction sensing. Unfortunately, as the molecular entity of the MSC in HUVECs has not been identified yet, we cannot discuss on the detailed supramolecular architecture of the SF/FA/MSC complex.

28.5 Active Mechanosensing: Rigidity Detection

Cellular motility, spreading, proliferation and differentiation are critically influenced by the substrate rigidity. To sense substrate rigidity, cells apply traction forces to cell–substrate adhesions via SFs and measure mechanical responses of the substrate. The supra-molecular complex of SF/FA/MSC shown in Fig. 28.5c may represent an ideal device to execute this task. In contrast to the Ca^{2+} transient induced by external forces, substrate rigidity sensing is induced by intrinsic mechanical stimuli, such as imposing traction forces generated in contractile SFs onto the ECM via SF/FA complex. MSCs located in or near FAs would convert the substrate-rigidity dependent stress generated in the FA into the level of cytoplasmic Ca^{2+} concentration ($[Ca^{2+}]_{cyt}$) via a local activation of the MSCs.

If this hypothesis stands and cells intermittently detect substrate rigidity, it is expected that cells will show spontaneous Ca^{2+} oscillations under no extrinsic mechanical stimulation. As the generated stress in FAs would larger in the cells on a rigid substrate than that on a soft substrate, the peak amplitude of the Ca^{2+} oscillation should depend on the substrate rigidity (Fig. 28.6a). Actually, HUVECs cultured on stiffer substrates showed spontaneous Ca^{2+} oscillations with larger amplitudes than those on soft substrates (top and bottom panels, respectively, in Fig. 28.6b) (Kobayashi and Sokabe 2010).

It has been implicated that rigidity sensing by cells is used for controlling cell migration. As can be seen in Fig. 28.6b (middle panel), the amplitudes of Ca^{2+} oscillations are much larger in the cells located at the boundary between soft and stiff substrates during durotaxis (mechano-taxis), strongly suggesting that spontaneous Ca^{2+} oscillations are used to search substrate with a stiffness proper for the cells during their migration. The rigidity sensing machinery SF/FA/MSC seems to be used more lively in the process of cell migration. Moving cells have

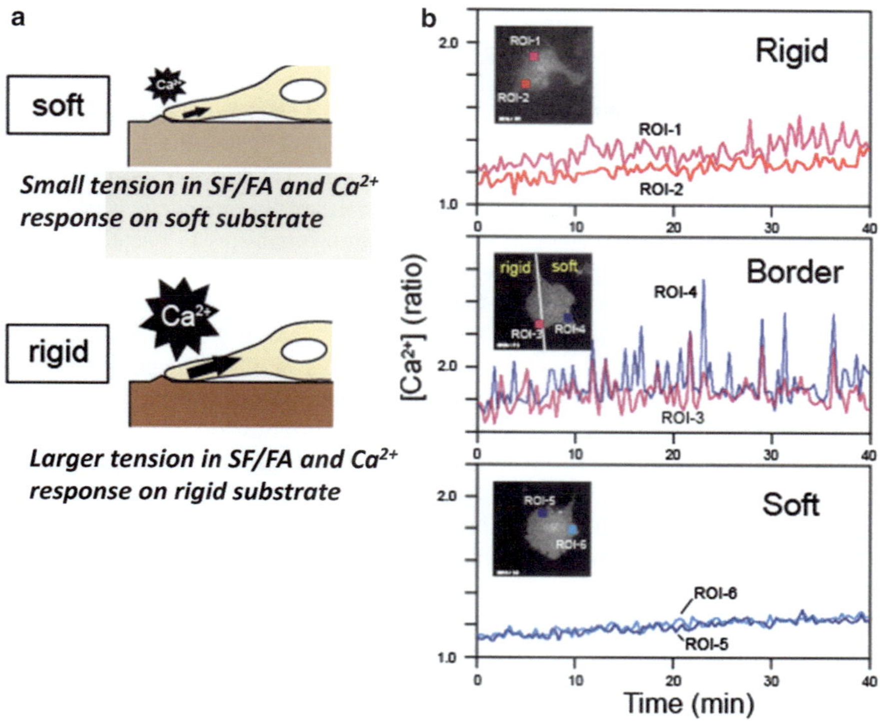

Fig. 28.6 Rigidity sensing by SF/FA/MSC complex. (**a**) Actively generated contractile forces in SF pull the substrate: tension in the SF/FA complex may depend on the substrate rigidity. (**b**) Spontaneous Ca^{2+} oscillations on a rigid (*upper*) and a soft (*lower*) substrate. Rigidity sensing is much more vigorous on the border between rigid and soft substrates during durotaxis (*middle*) (Modified from (Kobayashi and Sokabe 2010))

a rear-to-front calcium gradient with the lowest concentration at the front, and transient increases in the $[Ca^{2+}]_{cyt}$ have been thought to facilitate retraction at the rear by increasing SF contractility, or promoting FA disassembly (Ridley et al. 2003; Kiyoshima et al. 2011). Recently it has been reported that local increases in the $[Ca^{2+}]_{cyt}$ via TRP melastatin 7 (TRPM7) are required for cell migration or chemotaxis (Wei et al. 2009). As TRPM7 is a Ca^{2+}-permeable MSCs (Numata et al. 2007), it is critically important to examine the involvement of TRPM7 in sensing substrate rigidity by cells.

28.6 Summary and Perspectives

Bacterial MscL is activated exclusively by membrane tension and used to sense changes in the global tension in the cell membrane, followed by proper outflow of cytoplasmic components in order to adjust the intracellular osmolarity to the extracellular one. This function seems to be sufficient for MscL to work as an emergent safety valve protecting cells from lysis under a severe hypotonic downshock and the underlying activation mechanism in MscL is relatively simple. By contrast, MscS has a more sensitive and elaborated mechano-gating mechanism to deal with subtle environmental changes, working as a house-keeping device. Particularly, the inactivation mechanism of MscS, which is probably used to protect cells from losing excess amount of valuable cytoplasmic components, works in an ingenious manner. Further analyses of the gating mechanisms in bacterial MSCs would give the first complete showpiece, in terms of structure-function at the atomic level, among ion channels. Notwithstanding, MscS has not been inherited by eukaryotic cells; only limited species seem to inherit MscS homologues (Nakayama et al. 2007; Haswell et al. 2008). This may be because most eukaryotes have succeeded in developing various systems to protect the cells vulnerable to hypotonic challenge from direct exposure to such a dangerous stimulus.

Eukaryotes have evolved MSCs in several other directions. First, they utilize MSCs as a mechanosensor for specialized mechanoreceptors, such as inner ear hair cells, skin mechanoreceptors, visceral baroreceptors and muscle mechanoreceptors, to control their body reactions. Second, eukaryotic MSCs have been employed as a mechanosensor for nonsensory cells to detect mechanical forces generated by flows of body fluid, movements of organs and skeletal muscles, and gravity, making use of those mechanical cues for local (without nervous system) regulations of tissues and organs, including blood vessels, intestines, and bones. Third, eukyariotic MSCs are working as a mechanosensor to detect stresses created by intrinsic forces produced in cells, most of which are produced by polymelization-depolymerization dynamics of microtubules and microfilaments or contraction of actomyosin fibers like SFs. This function is critically important in the regulation of fundamental cell functions, such as growth, proliferation, differentiation and migration of cells. Finally, as a variation of the third type of cell mechanosensing, eukaryotic cells have evolved active touch sensing mechanisms, by which cells can detect mechanical properties (e.g., stiffness and viscosity) of their surrounding environments, including neighboring cells and substrates. This is realized by concerted operations among MSCs, contractile actin cytoskeletons and adhesive molecules. The active mechanosesning is used for cell migration towards the substrate with a proper rigidity, which may play crucial roles in embryonic development, wound healing, tissue regeneration and cancer invasion.

One of the characteristic features across eukaryotic MSCs is their association with a variety of accessory proteins such as cytoskeletons and adhesive molecules. This gives the MSCs, beyond the simple membrane tension sensing in bacterial MSCs, additional functions such as remote force sensing (via cytoskeletal linkage between membrane proteins and the nucleus), force direction sensing (e.g., tip-links between stereo-cilia in hair cells) and active mechanosensing as described above. One of the most important problems in the coming decade is to resolve the structure-function of supramolecular complex of MSC/accessory proteins, as well as molecular identification of eukaryotic MSCs. Moreover, identification and characterization of MSC-specific downstream signaling cascades remains challenging subjects to understand and to establish the biological and physiological roles of MSCs.

Acknowledgements This work was supported in part by Grant-in-Aid for Scientific Research from the Ministry of Education, Culture, Sports, Science, and Technology, Japan and from the Japan Society for the Promotion of Science (JSPS) (M.S., and T.K.), a grant ICORP/SORST Cell Mechanosensing project from Japan Science and Technology Agency (JST) (M.S.), and a grant from the Japan Space Forum (JSF) (M.S.).

References

Bass RB, Strop P, Barclay M, Rees DC (2002) Crystal structure of Escherichia coli MscS, a voltage-modulated and mechanosensitive channel. Science 298:1582–1587

Haswell ES, Peyronnet R, Barbier-Brygoo H, Meyerowitz EM, Frachisse JM (2008) Two MscS homologs provide mechanosensitive channel activities in the Arabidopsis root. Curr Biol 18:730–734

Hayakawa K, Tatsumi H, Sokabe M (2008) Actin stress fibers transmit and focus force to activate mechanosensitive channels. J Cell Sci 121:496–503

Kiyoshima D, Kawakami K, Hayakawa K, Tatsumi H, Sokabe M (2011) Force- and Ca^{2+}-dependent internalization of integrin in cultured endothelial cells. J Cell Sci 124:3859–3870

Kobayashi T, Sokabe M (2010) Sensing substrate rigidity by mechanosensitive ion channels with stress fibers and focal adhesions. Curr Opin Cell Biol 22:669–676

Machiyama H, Tatsumi H, Sokabe M (2009) Structural changes in the cytoplasmic domain of the mechanosensitive channel MscS during opening. Biophys J 97:1048–1057

Nakayama Y, Fujiu K, Sokabe M, Yoshimura K (2007) Mechanosensitive channel with a latch mechanism is present in the cytoplasm and chloroplast of Chlamydomonas. Proc Natl Acad Sci U S A 104:5883–5888

Naruse K, Sokabe M (1993) Involvement of stretch activated ion channels in Ca^{2+} mobilization to mechanical stretch in endothelial cells. Am J Physiol Cell Physiol 33:C1037–C1044

Nomura T, Sokabe M, Yoshimura K (2008) Interaction between the cytoplasmic and transmembrane domains of the mechanosensitive channel, MscS. Biophys J 94:1638–1645

Nomura T, Yoshimura K, Sokabe M (2006) Lipid-protein interaction of the MscS mechanosensitive channel examined by scanning mutagenesis. Biophys J 91:2874–2881

Numata T, Shimizu T, Okada Y (2007) TRPM7 is a stretch- and swelling-activated cation channel involved in volume regulation in human epithelial cells. Am J Physiol Cell Physiol 292:C460–C467

Ridley AJ, Schwartz AM, Burridge K, Firtel RA, Ginsberg MH, Borisy G, Parsons JT, Horwitz RA (2003) Cell migration: integrating signals from front to back. Science 302:1704–1709

Sawada Y, Murase M, Sokabe M (2012) The gating mechanism of the bacterial mechanosensitive channel MscL revealed by molecular dynamics simulations: from tension sensing to channel opening. Channels (Austin) 6:317–331

Shanthi V, Ramanathan K, Sethumadhavan R (2010) Exploring the role of C-H-....pi interactions on the structural stability of single chain "all-alpha" proteins. Appl Biochem Biotechnol 160:1473–1483

Sokabe M, Sachs F (1992) Towards molecular mechanism of the gating in stretch activated channel. In: Comparative aspects of mechanoreceptor systems. Springer, Berlin Heidelberg, pp 55–77

Sokabe M, Sachs F, Jing ZQ (1991) Quantitative video microscopy of patch clamped membranes stress, strain, capacitance, and stretch channel activation. Biophys J 59:722–728

Steinbacher S, Bass R, Strop P, Rees DC (2007) Structures of the prokaryotic mechanosensitive channels MscL and MscS. Curr Top Membr 58:1–24

Sukharev SI, Sigurdson WJ, Kung C, Sachs F (1999) Energetic and spatial parameters for gating of the bacterial large conductance mechanosensitive channel, MscL. J Gen Physiol 113:525–540
Tsuzuki S, Honda K, Uchimaru T, Mikami M, Tanabe K (2000) The magnitude of the CH/π interaction between benzene and some model hydrocarbons. J Am Chem Soc 122:3746–3753
Wei DC, Wang X, Chen M, Ouyang K, Song LS, Cheng H (2009) Calcium flickers steer cell migration. Nature 457:901–905
Yoshimura K, Nomura T, Sokabe M (2004) Loss-of-function mutations at the rim of the funnel of mechanosensitive channel MscL. Biophys J 86:2113–2120
Yoshimura K, Sokabe M (2010) Mechanosensitivity of ion channels based on protein–lipid interactions. J Roy Soc Interface 7:S307–S320
Yoshimura K, Usukura J, Sokabe M (2008) Gating-associated conformational changes in the mechanosensitive channel, MscL. Proc Natl Acad Sci U S A 105:4033–4038

Chapter 29
Gravitational Effects on Human Physiology

Yoriko Atomi

Abstract Physical working capacity decreases with age and also in microgravity. Regardless of age, increased physical activity can always improve the physical adaptability of the body, although the mechanisms of this adaptability are unknown. Physical exercise produces various mechanical stimuli in the body, and these stimuli may be essential for cell survival in organisms. The cytoskeleton plays an important role in maintaining cell shape and tension development, and in various molecular and/or cellular organelles involved in cellular trafficking. Both intra and extracellular stimuli send signals through the cytoskeleton to the nucleus and modulate gene expression via an intrinsic property, namely the "dynamic instability" of cytoskeletal proteins. αB-crystallin is an important chaperone for cytoskeletal proteins in muscle cells. Decreases in the levels of αB-crystallin are specifically associated with a marked decrease in muscle mass (atrophy) in a rat hindlimb suspension model that mimics muscle and bone atrophy that occurs in space and increases with passive stretch. Moreover, immunofluorescence data show complete co-localization of αB-crystallin and the tubulin/microtubule system in myoblast cells. This association was further confirmed in biochemical experiments carried out in vitro showing that αB-crystallin acts as a chaperone for heat-denatured tubulin and prevents microtubule disassembly induced by calcium. Physical activity induces the constitutive expression of αB-crystallin, which helps to maintain the homeostasis of cytoskeleton dynamics in response to gravitational forces. This relationship between chaperone expression levels and regulation of cytoskeletal dynamics observed in slow anti-gravitational muscles as well as in mammalian striated muscles, such as those in the heart, diaphragm and tongue, may have been especially essential for human evolution in particular. Elucidation of the intrinsic properties of the tubulin/microtubule and chaperone αB-crystallin protein complex systems is expected to provide valuable information for high-pressure bioscience and gravity health science.

Y. Atomi (✉)
204 Research Center for Science and Technology, Tokyo University of Agriculture and Technology, Koganei-shi, Tokyo 184-8588, Japan
e-mail: yatomi@cc.tuat.ac.jp

K. Akasaka, H. Matsuki (eds.), *High Pressure Bioscience*,
Subcellular Biochemistry 72, DOI 10.1007/978-94-017-9918-8_29

Keywords Cytoskeleton • Dynamic instability • Mechanical stress response • Molecular chaperone • Slow muscle

29.1 Overview

This chapter discusses the bases of physical adaptability as they relate to gravitational effects mediated through mechanical stresses on biological organisms that emerged on the earth, and how they relate to cellular systems as a unit of a life and the entire body as a multicellular organism. Cells can live autonomously by responding and adapting to their outer environment. Cellular responses to mechanical stresses are mainly mediated by the cytoskeleton, which governs cell survival systems through force development after connection to a fulcrum that consists of adhesion molecules secreted by the cell. The outer environment of some sixty trillion cells is the body itself. Therefore, physical activity or inactivity must play important roles to provide mechanical stimuli to the cells, especially cells found in force developing tissues that participate in activities/exercises, such as skeletal muscle, joints and bones. There are two types of muscle cells in the human body: fast- and slow-twitch cells (fibers). Compared with four-legged animals, the upright posture of humans promoted the development of slow twitch muscles cells. Anti-gravitational activity of the soleus muscle is important for maintaining posture with tonic contraction and is critical for overall health. The tubulin/microtubule system presents in three cytoskeletal systems and their molecular chaperone αB-crystallin is a key player in adaptation to physical activities and are also essential for good health.

29.2 Introduction to Physical Adaptability and Mechanical Stress and Gravity

Most biological organisms, including human beings, can adapt to various stresses from the environment. Recently, many studies have clarified the mechanisms at cellular and molecular levels through which this adaptability might be related to stress responses and stress tolerance. Such stresses may induce an "hormesis" effect, which often involves mild stress that is thought to be "beneficial stress", because of productive stress protein induction (for more in-depth discussion see Sect. 29.8). Mechanical stress associated with physical activities performed appropriately is usually beneficial for health because physical exercises affect various systems that have been acquired during evolution to ensure survival. Experimentally induced mental stress such as maternal separation in early life and/or restraint stress is negative stress (Kim et al. 2011; Niwa et al. 2013), which is thought to present a risk for the development of mental disorders later in life, particularly for individuals having specific genetic backgrounds (Niwa et al. 2013). However, these negative effects could be suppressed by an enriched environment (Niwa et al. 2013) or physical exercise (Kim et al. 2011). A recent research on exercising muscle, where mechanical contraction itself stimulates secretory mechanisms, has shown that skeletal muscle is in fact a secretory organ that can communicate with other organs such as the brain and adipose tissues so as to maintain overall health (Pedersen 2013).

Several groups of stress proteins are induced in response to stresses that allow cells to recover and withstand stress conditions. Many of these stress proteins function as chaperones, which assist the folding of newly synthesized and denatured proteins, the level of which are often increased by stresses. Since all organisms from prokaryotes to humans express conserved stress proteins, they are thus essential for the evolution of organisms. Although there are various stresses that induce stress protein synthesis, including heat, respiratory inhibitors and amino acid analogues, there are currently no studies showing the relationship between mechanical stress and stress protein induction/functions. For more than forty years, I have conducted studies relating to exercise physiology and have recently focused on the biological aspect of physical exercise. What happens in our exercising body (Fig. 29.1).

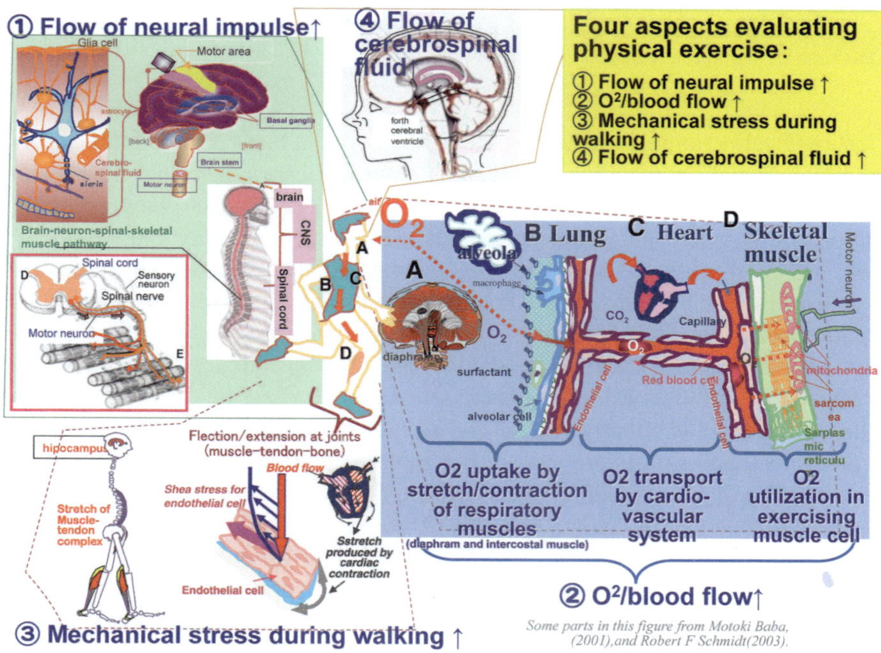

Fig. 29.1 What happens during exercise? The human body consists of some 60 trillion cells and the ECM molecules that those cells secrete. When we walk and run, coordinated skeletal muscle contractions develop joint movements and support our posture and body weight. Aerobic exercise promotes increased activity of four main aspects: (*1*) enhanced flow of neuronal impulses, (*2*) increased flow of oxygen to the bloodstream, (*3*) cellular activation induced by mechanical stimuli and (*4*) increased circulation of cerebrospinal fluid. All physical activities increase mechanical stresses on relevant cells in tissues and ECM

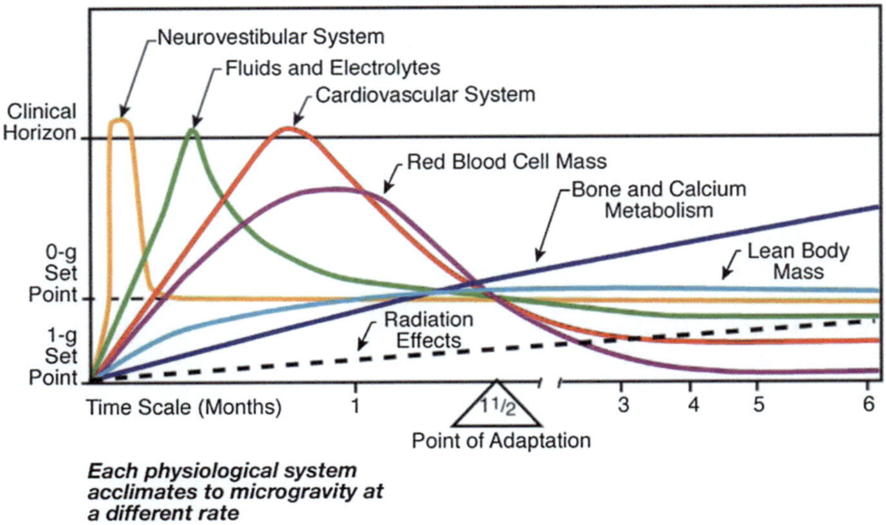

Fig. 29.2 Physiological responses of the human body in space. Each physiological system acclimates to microgravity at a different rate. (Nicogossian et al. 1989; American Society for Gravitational and Space Research 2013)

Biological organisms evolved on the earth. As such, many factors influence the evolution of biological organisms such as gravity, oxygen, light and magnetism. Gravity is the main factor that affects the morphogenesis of eukaryotes, especially multi-cellular organisms. Many studies concerning space biology and physiology have shown the importance of gravity in maintaining the functions of various organs in the human body. Physiological systems have been shown to acclimate to microgravity at different rates (Fig. 29.2) (Nicogossian et al. 1989). Changes in microgravity often result in the deterioration of physiological functions in the organism. Our studies using an experimental model of rat hindlimb suspension, which induces acute atrophy of bone and skeletal muscles of the hindlimb, demonstrated the suppressed expression of the molecular chaperone αB-crystallin and its substrate tubulin/microtubule (MT) system (one of three cytoskeletal filaments) under simulated microgravity conditions (Atomi et al. 1991a; Atomi et al. 1991b). This combination of a protein complex recognizing its own and other protein conformations in a living cell system seems to have co-evolved as a constitutive means to receive and respond to gravity. Interestingly, the MT system seems to have an intrinsic capacity to respond to gravity, as shown in in vitro studies by the Tabony group (Tabony and Job 1990; Tabony et al. 2002, 2007) and studies in plants by the Hoson group (Hoson et al. 2010).

All humans and animals inevitably age. With increasing age the number of cells in various tissues decreases, as do the functions of these tissues. Stress proteins that are induced by cells in response to various stresses are essential for the adaptation of biological systems. Continuous turnover of living materials occurring both inside and outside of cells is maintained through synthesis and degradation performed by cells that make up organs and tissues, and accordingly the biological organism can maintain a homeostasis of both cell number and mass. These physiological activities of multi-cellular organisms work under an integration of the brain-neural and endocrine systems as well as circulatory and respiratory and musculoskeletal systems. However, some cells that comprise connective tissues and joints exist in extracellular matrices (ECM), which consist of large protein and sugar molecules that are secreted by the cells themselves. Since some joints such as the knee have only neural innervation and do not have a direct blood supply, mechanical stress is essential for their survival. Mechanical stress is transduced via the ECM, and physical exercise increases these stresses to the benefit of many cells in the body.

Based on results of my studies concerning the mechanism of muscle atrophy, a key factor of mechanical stress is αB-crystallin and its cellular function. I will submit the hypothesis that the cytoskeleton (especially the tubulin/MT system) and the tension it generates are essential for living cells. An active lifestyle that includes physical activity may contribute to the maintenance of a basic environment of working functionality, especially for cells in tissues that are relevant to physical exercise.

29.2.1 *Decreased Working Capacity due to Inactivity and Aging as well as in Response to Microgravity but Increased Adaptability with Aerobic Muscle Activity: Hormesis Effect*

As the life expectancy continues to increase, so will the average age of the population, such that the age distribution will change from a pyramid shape to a bell-shaped or reverse triangle-shaped curve. Given this increase in the population of senior citizens, successful maintenance of the health and physical fitness of older people will be critical for Japan.

Biological changes with age, such as decreases in many physiological functions and capacity, appear to be a natural phenomenon. Development of various parameters during aging is maintained by the hormonal environment, such as human growth hormone, and growth factors secreted by somatic cells in the body. As age increases, the secretion of these growth-stimulating substances will decrease. As such, countermeasures to the consequences of this decreased functionality are needed.

Physical work capacity and physiological function deceases with inactivity as well as age. They also decrease in space. Although deterioration of various capabilities and function is likely inevitable, all decreases in ability could be induced by reduced activity. However, we also know that physical work capacity can increase with appropriate physical training, even in middle-aged and older people. Thirty years ago I found that aerobic work capacity (maximal oxygen uptake; VO_2 max)

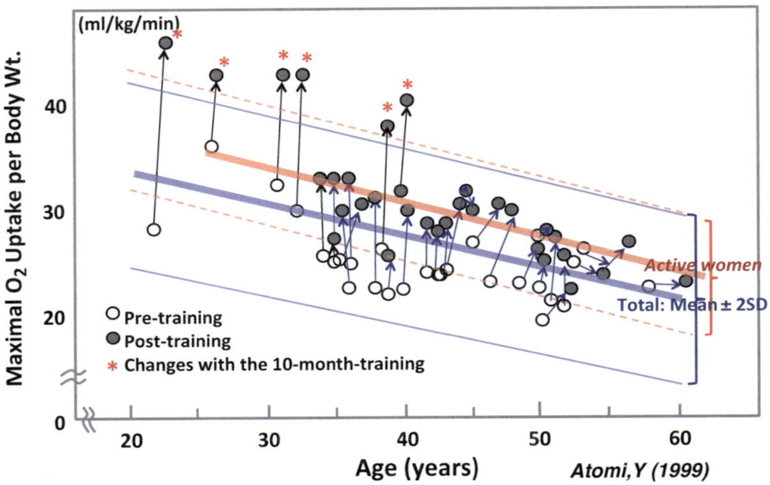

Fig. 29.3 The mean and range of improved maximal O_2 uptake with physical training in middle-aged and elderly women (Atomi and Miyashita 1974; Atomi 1999)

of Japanese women decreased with age, but after aerobic training the VO_2 max increased by 10–20 % above initial values (Atomi 1980) (Fig. 29.3). Although the working capacity gains made through training lessen with age, some effect could still be seen in older women. Tanaka's group reported that low intensity exercise for 1 h in older people maintains their working capacity and physiological function, but also variously affects blood pressure and immunological function (Tanaka et al. 1998). We believe that the effects of physical activity as well as balanced nutrition are important for health, although the precise reasons for these effects are still unclear. Furthermore, it is very difficult to differentiate whether functional decreases are due to age or inactivity.

29.3 Muscle Plasticity and Use/Disuse-Induced Muscle Volume Changes

Limb musculature in animals, especially slow muscles (slow-twitch fiber-dominant muscle) such as the soleus, which helps maintain posture and contracts against the earth's gravitational force, undergoes a significant reduction in mass during space flight or under unweighted conditions, such as those in the hindlimb suspension model. Removing passive tension on the hindlimb plantar flexor muscle through the hindlimb suspension method to induce a shortened state readily induces atrophy in the slow soleus but not in the fast muscles of the leg [see reviews in references (Musacchia et al. 1988; Thomason et al. 1989)]. In all experimentsin our studies,

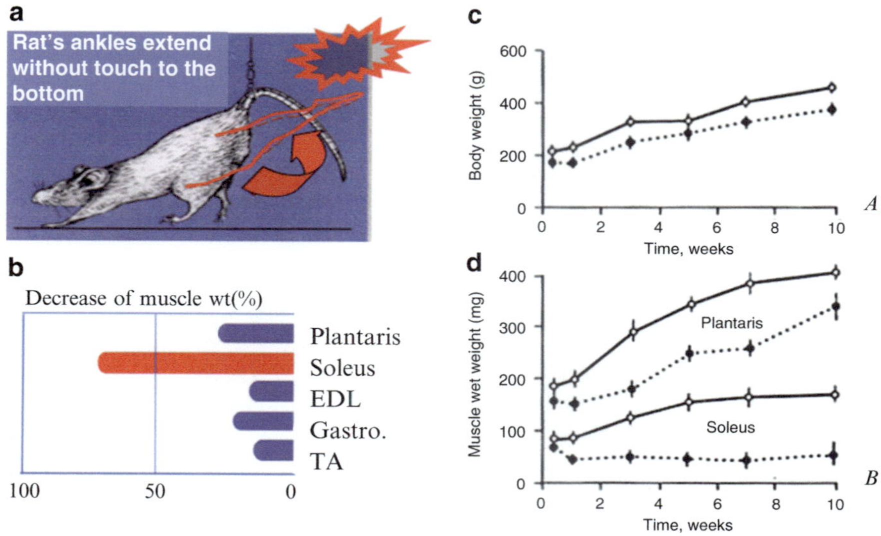

Fig. 29.4 Hindlimb suspension model (**a**), changes (Δ%) in soleus and four lower leg fast-twitch fiber-dominant muscles after one week of hindlimb suspension (HS) (**b**), changes in body weight, EDL: extensor digitorum longus, TA: tibialis anterior (**c**) and slow soleus and fast plantaris muscles (**d**) through a 10-week HS experiment. ○ Control, ● suspended

adult Wistar rats were used and muscle atrophy was induced by a standard hindlimb suspension method (Fig. 29.4) (Moray 1979), modified by Atomi et al. (1990, 1991a, b) that is known to produce rapid and reproducible levels of muscle atrophy (Musacchia et al. 1988; Thomason and Booth 1990). In our first hindlimb suspension experiment carried out over the course of 10 weeks, the mean weight of the slow soleus muscle declined rapidly, with a 40 % loss by the end of week one. By the third week, the muscle weight declined to 51 % of control values and remained at this reduced level during the subsequent 10 weeks of the suspension procedure. Meanwhile, the fast plantaris muscle of the atrophied hindlimb showed a 27 % decline during the first week, but soon recovered by 14 % before week three. The fast muscles (fast-twitch fiber-dominant muscle) then continued to grow throughout the 10-week experimental period so that by the end of the experiment they showed an average 50 % gain in weight. The total body weight of the experimental animals did not show radical differences when compared to control animals (Atomi 1992), but they did weigh roughly 10 % less than the controls after 10 weeks. According to Dart and Shultz (1989) this decrease in muscle mass seen with hindlimb suspension is associated with a reduction in the number and proliferation of satellite cells, and persisted for 30 days in young rats from 12 to 50 days of age. These observations show that weight-supporting activity is an essential factor in maintaining normal growth of the soleus muscle.

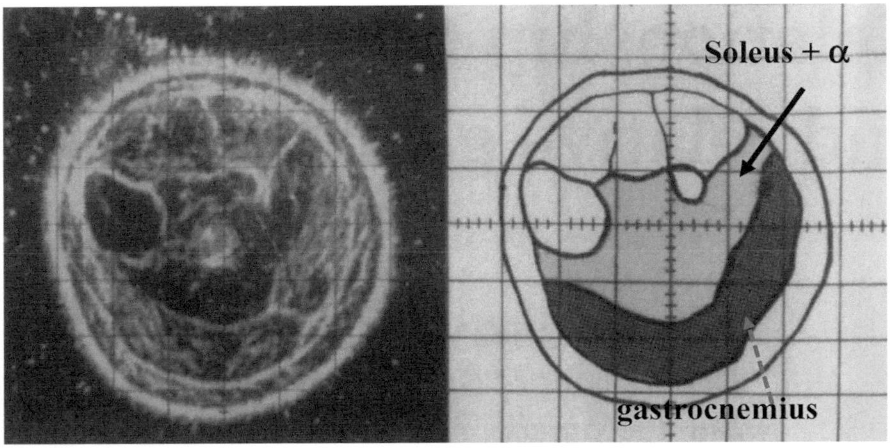

Fig. 29.5 Ultrasonic photograph (*left*) and traced figure (*right*) for leg muscles. *Shaded* area shows the total plantar flexor, including the gastrocnemius muscle (*darker color*) (Atomi et al. 1987)

29.3.1 Two Types of Muscle-Related Force, Energy, and Adaptability

Hindlimb plantar flexor muscles are composed of the gastrocnemius, soleus and some resting deep small muscles (Fig. 29.5). The gastrocnemius muscle is a "mixed" muscle, whereas the soleus muscle contains predominantly slow twitch fibers that have a high oxidative capacity. In humans the soleus is significantly developed and represents up to 40 % of the total plantar flexor and 56 % of the triceps surae muscle wet weight in cadavers (Endo 1937; Gollnick et al. 1974). In our data gathered from 18 cadavers, the soleus comprised 50–68 % of the triceps surae muscle (Satake et al. 1987). The relative development of these leg muscles, especially in children, is related to a higher lactate threshold during running (Atomi et al. 1987), whereas these muscles represent only 0.6 % of the total weight. The soleus muscle also plays an important role in the standing posture in humans. The limb musculature in animals, especially slow muscles like the soleus muscle that functions to maintain posture and counteract the gravitational force on earth, undergoes a significant reduction in mass during space flight or during unweighted conditions, such as hindlimb suspension. In humans, back muscles (erector supinae) are also slow muscles that are made up of about 70 % slow-twitch muscle fibers. There are about 400 muscles in the back that are involved in the support and regulation of erector bipedalism.

The soleus is a typical slow skeletal muscle that continuously maintains posture against gravity through neural and contractile activities. Compared with fast-twitch oxidative and fast-twitch oxidative glycolytic skeletal muscles (fast muscles), slow muscles are functionally characterized by slower contractile kinetics, larger oxidative metabolic capacity, and lower fatigability (Campbell et al. 2001). These

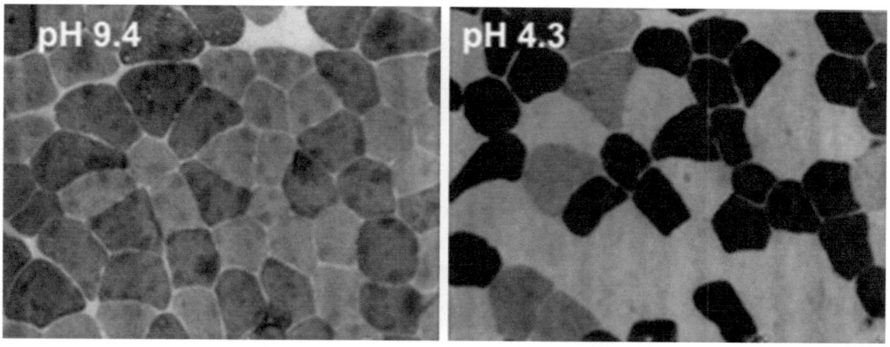

Fig. 29.6 Human skeletal muscles consist of cells with different fiber types and myosin ATPase activity, which is dependent on reaction buffer pH in the method used to differentiate fiber type (human vastus lateralis muscle; photographs kindly provided by Prof. Tomoo Ryushi). Fibers with *dark color* at pH 9.4 and *light color* at pH 4.3 correspond to fast-twitch fibers, and light at pH 9.4 and dark at pH 4.3 are slow-twitch cells

functional features make these muscles well-suited for their continuous activities that meet the mechanical demands arising from the environment.

Slow muscles are distinct from fast muscle both morphologically and biochemically. Morphologically, slow muscles are abundant in extracellular collagenous matrices (Zimmerman et al. 1993), capillary vessels, myoglobin, and mitochondria (Campbell et al. 2001). Their sarcomeres show wider Z-disks and diffuse M-lines on electron micrographs (Sjöström et al. 1982; Agarkova et al. 2004). Biochemically, slow muscle sarcomeres are composed of specific myoproteins. The myosin heavy chain (MHC) is predominantly MHC-I in slow muscles, while MHC-IIa, IId, or IIb are common in fast muscles (Eason et al. 2000; Campbell et al. 2001) (Fig. 29.6). M-lines are largely composed of embryonic heart (EH)-myomesin in slow muscles, and M-proteins in fast muscles (Sjöström et al. 1982; Schoenauer et al. 2008). The protein turnover rate is higher in slow muscles due to higher rates of protein synthesis (Atomi et al. 1991b) and higher chaperone contents have also been reported in slow muscle (Atomi et al. 2000). αB-crystallin is one of these chaperone proteins (Fujita et al. 2004; Ohto-Fujita et al. 2007) that is reported to be rich in the soleus muscle (Atomi et al. 1991b; Koh and Escobedo 2004; Sakurai et al. 2005).

29.3.2 Relationships Among MTs, αB-Crystallin, and Myosin Type in Mammalian Striated Muscles

Cellular cytoskeletons are generally considered to transduce mechanical environmental demands to drive morphological and metabolic adaptation of cells and organs (Lazarides 1981; Ingber 1997; Gundersen and Cook 1999). MTs form characteristically dynamic networks to transmit tensile forces, transport proteins and maintain mitochondria (Appaix et al. 2003), the Golgi apparatus (Ralston et al. 2001) and cell shape (Gundersen et al. 1989; Boudriau et al. 1993; Gundersen and Cook 1999; Sakurai et al. 2005). While many studies have followed "dynamic instability" during mitosis and in undifferentiated and proliferating cells, few have concerned differentiated cells or tissues. Stabilized MTs are thought to be enriched in detyrosinated tubulin after differentiation events such as myogenesis and neurogenesis as well as during directed cell motility (Bulinski and Gundersen 1991) due to the activity of a tubulin-specific carboxypeptidase that detyrosinates tubulin protomers within microtubules. On the other hand, the tubulin protomer itself has been reported to influence the voltage-dependent anion channel in the outer mitochondrial membrane (Sheldon et al. 2011; Tepp et al. 2014). The three cytoskeleton components, namely fibrous actin, MTs and intermediate filaments (IFs), are basically independently regulated, but they affect each other with different tensile and dynamic properties via interacting proteins. Co-localization of IF and MT networks has been reported in various cells (Gyoeva and Gelfand 1991). In fact, some specific proteins have been identified that connect actin and MTs (Leung et al. 1999; Karakesisoglou et al. 2000), or MTs and IFs (Liao and Gundersen 1998). MT disassembly has been suggested to be induced by actomyosin contraction of stress fibers in cultured cells (Waterman-Storer and Salmon 1997) in a manner that is similar to tonic contraction in heart and slow muscle.

In striated muscles MTs are essential for differentiating myotubes (Gundersen et al. 1989), although their significance in mature muscle remains unclear. Since passive and active mechanical stresses are well-documented to induce morphological as well as metabolic adaptations in striated muscles (Dulhunty and Franzini-Armstrong 1975; Atomi et al. 1991b; Sakurai et al. 2005), we can infer that MTs would play significant roles in such changes, especially in slow-twitch oxidative skeletal muscles (slow muscles) (Sakurai et al. 2005). To test our hypothesis over a wider range of muscle types in rats, we performed SDS-PAGE to analyze the correlation between α-tubulin and αB-crystallin in various skeletal and non-skeletal striated muscle tissues expressing different MHC isoforms (Fig. 29.7), and showed a good correlation between their protein levels ($r = 0.837$) (Fig. 29.8).

The tubulin and αB-crystallin contents (Fig. 29.8) are higher in MHC-I/α dominant muscles, intermediate in MHC-IId dominant muscles, and lower in MHC-IIb dominant muscles. As shown above, this strongly supports our hypothesis that dynamics between MTs and their constituent tubulin transduce mechanical stress to produce adaptive responses in mature muscles.

Persistently alternating passive lengthening and active shortening are the common features of MHC-I/α dominant muscles. Soleus muscles (MHC-I dominant) are stretched and shortened during walking and running, while heart muscles (MHC-α dominant) extend and contract during the pumping cycles of the heart. This alternating stress would elicit dynamic remodeling of MTs, which in turn may induce adaptive protein turnover in muscles as is seen in general cells.

MTs reportedly differentiate when cultured myotubes are stabilized by the replacement of tyrosine at the tip with glutamate (Gundersen et al. 1989).

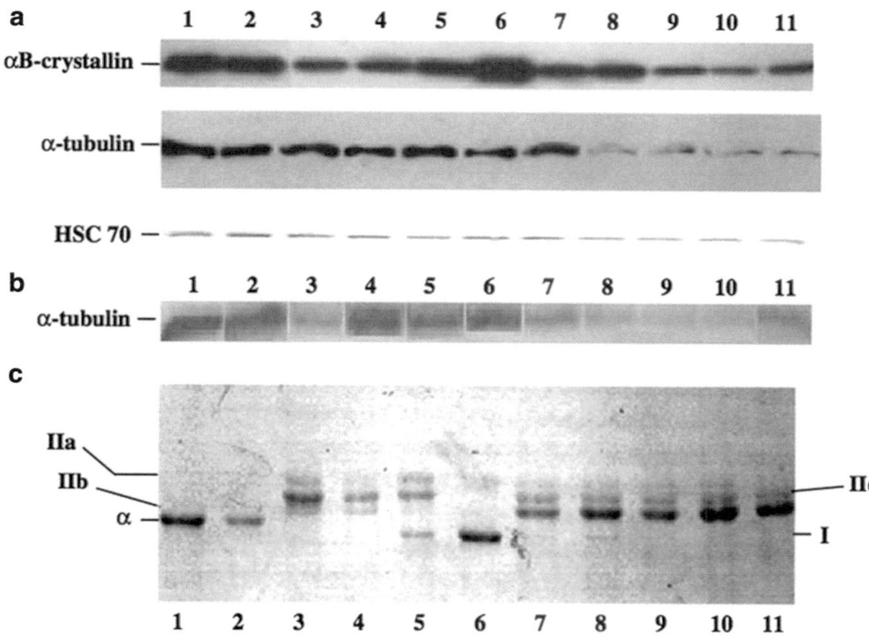

Fig. 29.7 Comparison of protein contents of MHC-isoforms, αB-crystallin and α-tubulin rat striated muscles. (**a**) Immunoblotting bands showing αB-crystallin, α-tubulin, and 70 kilodalton heatshock cognate protein (HSC70) (control) in various striated muscle tissues: atrium (*1*), ventricle (*2*), masseter (*3*), tongue (*4*), diaphragm (*5*), soleus (*6*), plantaris (*7*), gastrocnemius (*8*), extensor digitorum longus (*9*), tibialis anterior (*10*), and psoas (*11*) muscles. Protein quantities per lane were 1.6, 50, and 20 μg for αB-crystallin, α-tubulin, and HSC 70, respectively. (**b**) Immunoblotting bands showing α-tubulin corresponding to the same striated muscles as in (**a**). Protein quantities of total muscle extracts loaded per lane were lane (*1*) 8 μg, (*2*) 8 μg, (*3*) 30 μg, (*4*) 13 μg, (*5*) 15 μg, (*6*) 8 μg, (*7*) 28 μg, (*8*) 25 μg, (*9*) 25 μg, (*10*) 40 μg, (*11*) 50 μg. (**c**) Typical SDS-PAGE pattern showing MHC-isoform composition. IIa, IId, IIb, a, and I indicate MHC-IIa, MHC-IId, MHC-IIb, a-MHC, and MHC-I, respectively. The numbers correspond to those in (**a**). The amount of protein loaded per lane was 1.6 μg (Jee et al. 2009)

A considerable fraction of MTs in fast muscles may be stabilized with such a tip replacement. However, we infer that, at least in continuously stressed slow skeletal and heart muscles, a considerable fraction of MTs are dynamically remodeling.

A closer inspection of the correlation of protein levels suggests more details of the adaptive process in striated muscles (Fig. 29.8). The muscles with relatively higher tubulin/αB-crystallin ratios include the heart, tongue, and masseter muscles. These muscles work at higher alternating frequencies of passive lengthening and active shortening with relatively low loads. On the other hand, muscles having lower tubulin/αB-crystallin ratios include the soleus and diaphragm muscles, which work at lower alternating frequencies of passive stretching and active shortening with relatively high loads. Highly frequent switching of passive lengthening and

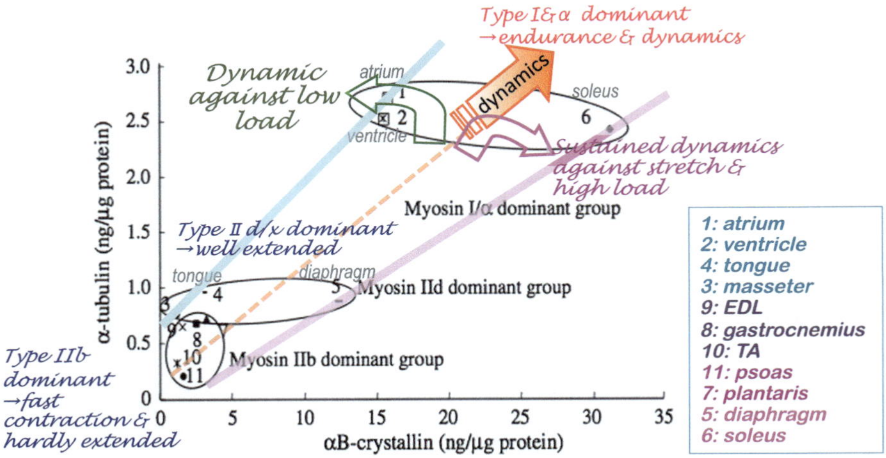

Fig. 29.8 Correlation between αB-crystallin and α-tubulin content in various muscle tissues. Muscles are grouped by dominant MHC-isoform. The numbers correspond to those in Fig. 29.7. There is a significant correlation ($P < 0.01$) between the quantity of αB-crystallin and α-tubulin ($n = 11$, Pearson correlation coefficient 0.837) (Jee et al. 2009, Modified by Atomi)

active shortening likely destabilizes microtubule networks, thus perturbing their rates of protein turnover. Therefore, there is likely an optimal switching frequency for the adaptive induction of αB-crystallin expression. The frequency of walking and breathing that governs the switching frequency of soleus muscle and diaphragm seems to be optimal for the adaptive function of MT networks.

Two other cytoskeletal protein systems are found in striated muscles. One is the IF system, including desmin, which maintains the integrity of the sarcomere structure and organizes cell shape (Agbulut et al. 1996). The mechanical character of desmin is similar to connecting, a gigantic protein filament that anchors thick filament lattices to Z-disks to transmit passive force in the sarcomere (Maruyama et al. 1976). The other system involves actin filaments. In cultured cells, β-actin filaments line the plasma membrane to form stress-fiber transmitting active forces as well as passive forces (Bray 2001). α-actin filaments constitute I-bands that transmit active forces of the thick filament lattice to Z-disks (Ingber 1997; Lele et al. 2006), which also include γ-actin (Nakata et al. 2001). The IFs for passive force balance in relaxed muscle, and actin filaments for active force transmission in contracting muscle are both specialized to sustain the stability of functioning muscle. However, our hypothesis suggested that MTs are specialized in transforming muscles in accordance with mechanical demands from the environment (Korfage and Van Eijden 1999). In cultured cells as well as in striated muscles, IFs and MTs frequently co-align (Gundersen et al. 1989; Boudriau et al. 1993), and αB-crystallin also works as a chaperone for the free forms of IFs (Nicholl and Quinlan 1994).

29.4 Response of Stress Protein αB-Crystallin in Muscles to Mechanical Stress and Stretch

The effects of muscle training on muscle volume and mass are readily apparent. Skeletal muscle has high plasticity such that training induces muscle hypertrophy/hyperplasia, while disuse induces muscle atrophy. In experiments using the rat hindlimb suspension model to mimic disuse atrophy experienced in space, the mass of soleus muscle exposed to anti-gravity conditions rapidly decreased to one half of initial values after one week. Shultz (Mozdziak et al. 1998) and Edgerton's groups (Allen et al. 1997) showed a decrease in muscle size and in the number of muscle cells, which likely reflects disuse-induced apoptosis of muscle cells.

Although muscle cells are known to undergo size changes, the mechanisms by which disuse induces decreases in muscle size and apoptosis are unclear. Therefore, I decided to elucidate a mechanism of skeletal muscle disuse atrophy using the rat hindlimb suspension model and a subtraction method to determine the protein compositions of suspended and control soleus muscles (Fig. 29.9). After one week

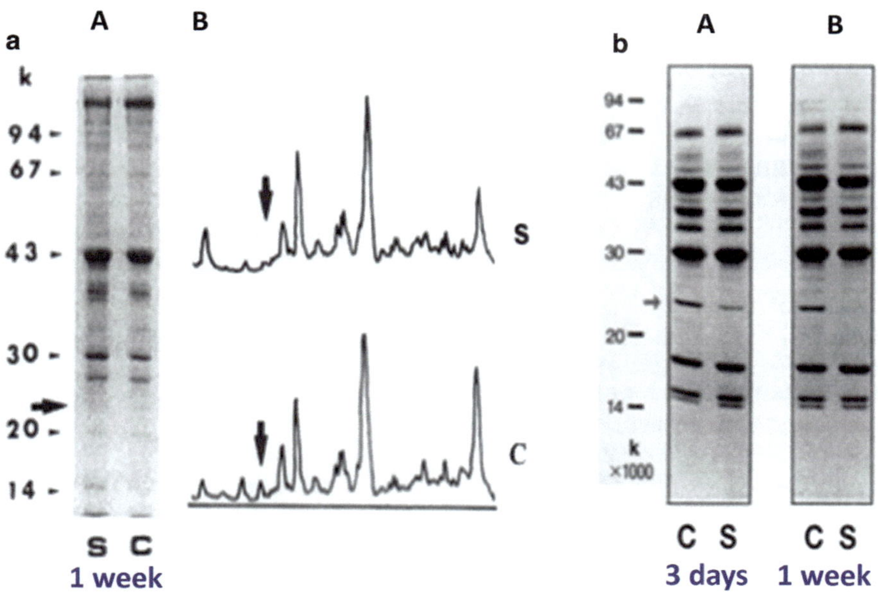

Fig. 29.9 Identification of a 22 kDa protein that is specifically decreased in atrophied soleus muscle after hindlimb suspension for 1 week. (**a**) Comparisons of total muscle homogenates from atrophied (S) and control (*C*) soleus muscle shown by SDS-PAGE (*A*) and densitometric analysis (*B*). *Arrows* indicate the 22 kDa protein. (**b**) In supernatants separated from the myofibrillar fraction following precipitation and centrifugation of total homogenates, the decrease in levels of the 22-kDa protein is greater for atrophied muscle after 3 days of suspension (*A*) and the protein is absent in atrophied muscles after 1 week of suspension (*B*). Molecular weight standards (Pharmacia) represent kilodaltons (Atomi 1992)

of suspension, we focused on a specific protein with levels that markedly decreased in both total homogenates (a) and soluble fractions (b) in atrophied soleus muscles in suspended rats. The decrease in a 22 kDa protein was previously detected in supernatants following the removal of the myofibrillar fraction.

When we move/exercise, relevant skeletal muscles contract and develop tension against load under regulation by innervation. Mechanical stresses such as stretch are induced by the weight of each step. Since the soleus muscle is attached to the calcaneus and the tibia, the degree of soleus stretching is determined only by the angle of the ankle joint, which differs from the other triceps surae muscles. Accordingly, an experiment to examine the stretch effect, particularly on the soleus muscle, can be easily done using the rat hindlimb suspension model. Continuous mechanical stretching or lengthening of muscle increases muscle mass, while shortening of muscle length results in augmentation of muscle atrophy (Goldspink 1980). On the other hand, denervation of skeletal muscle is known to aggravate muscle atrophy, and an experiment involving cross-innervation between slow and fast muscle reversed the physiological and biochemical characteristics depending on the fiber type (Close 1972). Therefore, we further examined the effects on muscle weight of continuous passive stretch and denervation of both slow and fast muscles under hindlimb suspension (Fig. 29.10). Innervated or denervated animals under hindlimb suspension were restrained in a fully dorsiflexed (stretched state) or plantarflexed (shortened state) position with tape (Atomi et al. 1990). Two weeks after initiating the experiment, passive stretch showed significant protective effects against muscle atrophy in both muscles and under both suspended and denervated/suspended conditions (Fig. 29.10). However, the effect of denervation is severe on the slow twitch soleus muscle compared with fast twitch extensor digitorum longus (EDL) muscle. These various effects may be due to differences in the function and design of muscles that receive and develop tension and from enduring mechanical stress derived from supporting the body's weight at each step during standing, walking and running.

αB-crystallin is highly expressed in soleus muscle as well as heart muscle, but to a lesser degree in fast muscle (Fig. 29.11). As described earlier (Fig. 29.2), αB-crystallin expression may be related to contractile character in muscles that do not tire easily. Slow twitch soleus muscles that are as large as those seen in humans are not found in other animal species, suggesting that bipedal endurance running may have contributed to the evolution of muscle structures in humans (Bramble and Lieberman 2004; Skoyles 2006). In mammals, sarcomeric myosin heavy chain (MHC) isoforms, which determine contraction properties such as slow- or fast-twitch, are encoded by two families of tandemly linked genes, where the process of isoform switching is under transcriptional control. In the human genome, six genes (chromosome 17p13) encode the skeletal MHC isoforms and two (chromosome 14q12) encode cardiac isoforms (Desjardins et al. 2002). Interestingly, since slow skeletal MHC is the same as cardiac β-magnesium / proton exchanger, this gene might have evolved differently than the fast MHC genes.

Slow muscle seems to be more sensitive to mechanical stress, because the Z-band that receives tension and transfers it to the ECM is wider and more well-developed in the MT network (Ralston et al. 2001). Such subtle differences in the structure

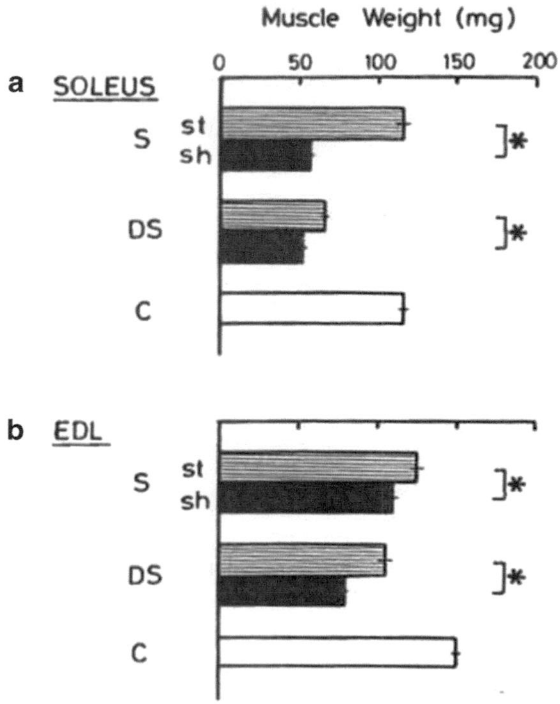

Fig. 29.10 Effect of passive stretching on weights of slow and fast rat muscle that are suspended or suspended/denervated. (**a**) Muscle weights of slow soleus and (**b**) fast EDL muscles in control (*C*), suspended (S) and denervated/suspended (DS) animals. *st* stretched muscle, *sh* shortened muscle. *$p < 0.05$ (Atomi 1992)

and function of muscles are relevant for physical activities and endurance exercise. Different activities of slow muscles are integrated by the brain-body balance axis, where they continue to drive circulation, and consume energy resulting in health benefits and enhanced muscle plasticity compared with physical activity involving fast muscles.

Although there are many studies that have analyzed the physiology and anatomy of muscle contraction, there are fewer studies on the cytoskeletal character and structures in skeletal muscle. Few papers have shown how αB-crystallin responds to mechanical stress derived from sarcomere damage occurring after eccentric contraction and movement of Z-bands (or precipitated fraction) to prevent rupture (Féasson et al. 2002; Paulsen et al. 2009; Frankenberg et al. 2014) that can be induced due to violent extension.

αB-crystallin levels did not decrease in the soleus muscle when its length was maintained with tape in a stretched position (Fig. 29.12). The muscle weight was also maintained after suspension with the soleus muscle stretched with tape. This protein was targeted and purified from rat slow muscle tissues and its amino acid sequence was measured and identified to be αB-crystallin (Atomi et al. 1991a). Both the protein levels and synthesis rates of αB-crystallin decreased after 24 h (Fig. 29.12a (B-SH) and b) despite the finding that levels of the soluble 30 kDa protein remained stable (Fig. 29.4b). Reversibly passive stretch increased the rate

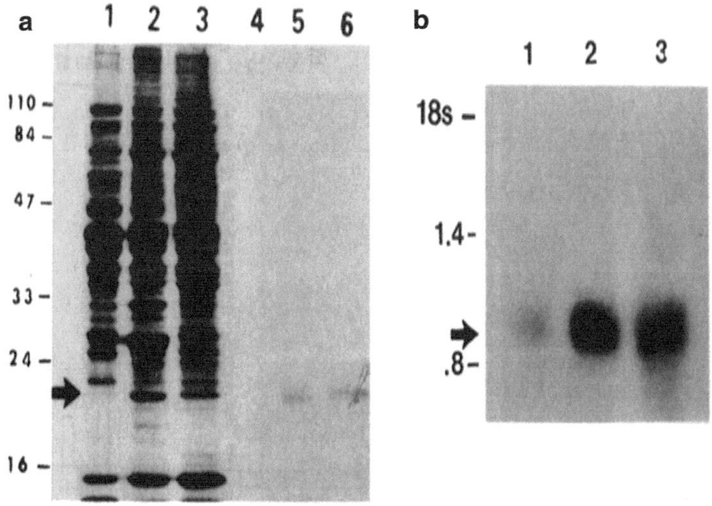

Fig. 29.11 Protein and mRNA expression levels of αB-crystallin in rat striated muscles, fast skeletal plantaris, slow skeletal soleus, and heart muscles. (**a**) Proteins analyzed by SDS-PAGE with Coomassie blue staining of 30 mg rat muscle soluble proteins (lane 1–3), immunoblotting (lane 4–6) of fast plantaris (lane I,4), slow soleus (lane 2,5) and heart (lane 3,6) muscles. The mRNA level of αB-crystallin is the same for striated muscles. (**b**) Northern blotting of 20 mg total RNA extracted from fast plantaris (lane 1), slow soleus (lane 2) and heart (lane 3) to αB-crystallin cDNA. Prestained protein molecular weight standard (Bio-Rad) in A are given in kDa. RNA molecular weight makers (BBL), size of α-actin 18 s rRNA in B is given in kb (Atomi et al. 1991a)

of protein synthesis. αB-crystallin mRNA also began to decrease after 36 h and to a significant degree after 3 days of suspension. After a 24 h recovery from suspension, αB-crystallin mRNA levels recovered (Atomi et al. 1991b). These experiments suggest that the constitutive expression of αB-crystallin in slow soleus muscle is induced by muscle contraction and mechanical stretching on the ground, but the expression decreases with disuse.

29.5 Adaptive Responses to Mechanical Stress: Stress Protein αB-Crystallin Localizes along the Cytoskeleton

While the bulk of αB-crystallin in muscle can be extracted in the soluble fraction, its exact localization is difficult to determine and whether it truly resides in the soluble pool or is present in a loosely bound state is unclear. Furthermore, small heat shock proteins (sHSPs), including αB-crystallin, may recognize very early conformational changes of substrate proteins under stress. Figure 29.13 shows the ordered striation of fluorescent bands reacting to affinity-purified anti-αB-crystallin as observed by immunofluorescence microscopy. This striation corresponds to the

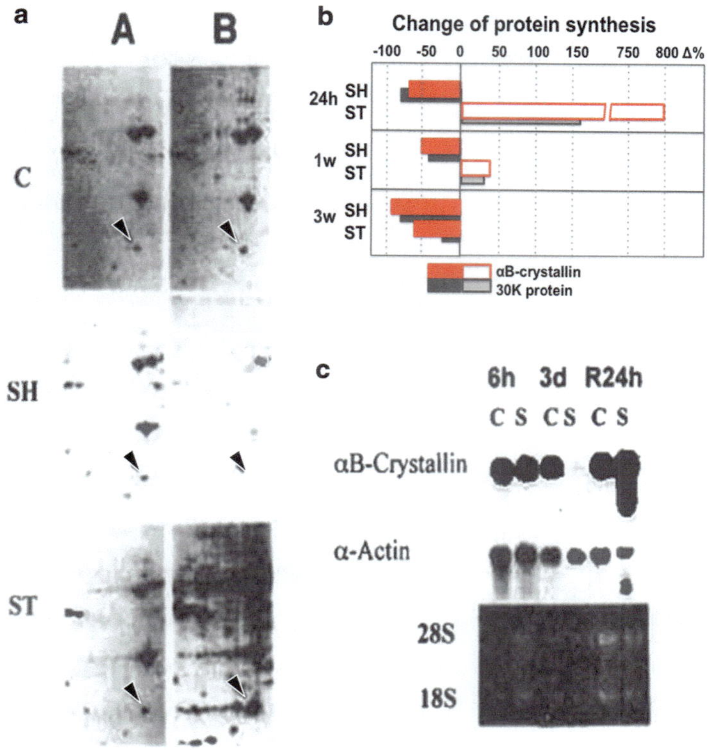

Fig. 29.12 (**a**) Effects of soleus muscle shortening (SH) and lengthening (passive stretch: ST) on muscle protein synthesis (*A*: A-coomassie brilliant blue (CBB) staining, B-fluorography). *Arrowhead* shows αB- crystalline. (**b**) Percent change of protein synthesis after suspension experiment with shortening and lengthening. (**c**) The change for mRNA (*S* suspension, *C* control) (Atomi et al. 1991b; Atomi 1992)

Z-band in myofibrils prepared from glycerinated fibers as determined by both light and electron microscopy (Atomi et al. 1991a). Accordingly, αB-crystallin localizes around the Z-bands of the sarcomere structure. Z-bands in muscle correspond to structures that are similar to focal adhesions and/or the cytoskeleton in cultured cells. This image of the colocalization of αB-crystallin was reproduced in vivo time-lapse images of beating hearts using geen fluorescent protein (GFP)-αB-crystallin (unpublished data), thus it could be said that αB-crystallin may play a dynamic role as a chaperone of unstable proteins in muscle cells.

Examination of αB-crystallin is simpler in undifferentiating cultured cells, which express three cytoskeletal proteins of actin, IFs and tubulin/MTs that are each differently regulated, and change the cell shape depending on the cell cycle. Living cells have "dynamic instability" that occurs due to the assembly/disassembly of cytoskeleton protein molecules, which is an intrinsic property of the tubulin/MT system (Mitchison and Kirschner 1984). According to comparisons of colocalizations

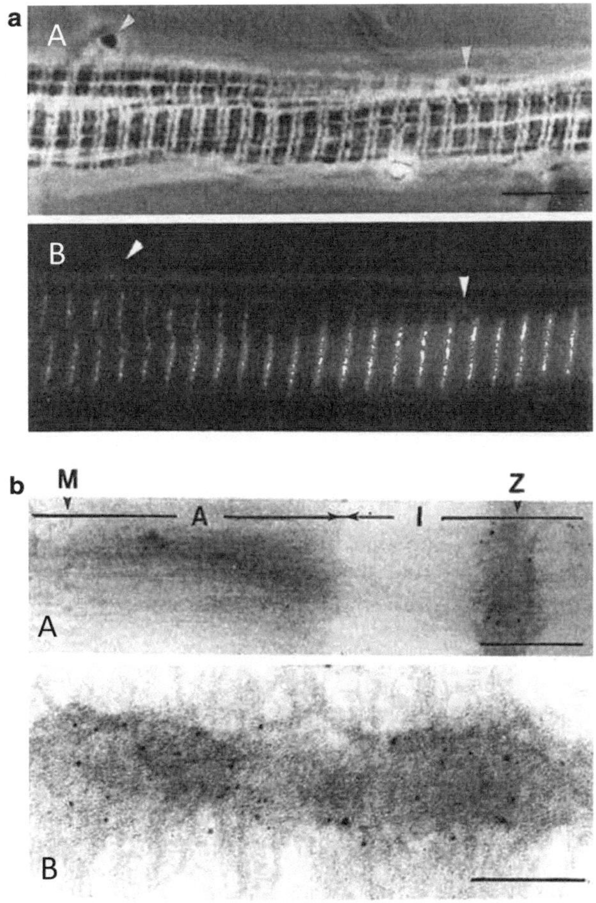

Fig. 29.13 Immunohisto chemical and electron microscopic localization of αB-crystallin myofibrils in soleus muscle (Atomi et al. 1991a). (**a**) Indirect immunofluorescence localization of αB-crystallin in myofibrils incubated with affinity-purified anti-αB-crystallin antibody visualized by phase contrast (*A*) and fluorescence (*B*) microscopy. Bar = 10 μm. (**b**) Electron micrographs of negatively stained myofibrils decorated with affinity-purified anti-αB-crystallin antibody, followed by gold-labeled protein A: Bar = 500 nm, B: Bar = 200 nm. Both immunofluorescence and electron micrographs show the localization of αB-crystallin at the Z bands of myofibrils (Atomi et al. 1991a)

among αB-crystallin and these three types of cytoskeletal proteins, αB-crystallin unexpectedly co-localized with MTs (Fig. 29.14) and with actin filaments at focal adhesions. Indeed, previous papers showed that αB-crystallin works as a molecular chaperone for actin (Singh et al. 2007), and IFs (Nicholl and Quinlan 1994; Houck et al. 2011). We showed that αB-crystallin prevents tubulin aggregation in vitro (Arai and Atomi 1997; Ohto-Fujita et al. 2007) and contributes to MT stability in αB-crystallin overexpressing cultured cells (unpublished data). αB-crystallin can function as a chaperone for the maintenance of cytoskeleton molecules. αB-crystallin-overexpressing cells are resistant to MT disassembly-inducing reagents as well as high calcium ion concentrations that promote MT disassembly in vitro and in cultured cells (myoblast and glia cells). sHSPs, including αB-crystallin, are believed to recognize early conformational changes that occur upon protein denaturation. Tubulin molecules assemble above 20 °C into MTs and their turnover is rapid in cells. αB-crystallin appears to recognize such tubulin/MT conformational changes and maintains that dynamic stability in a cell.

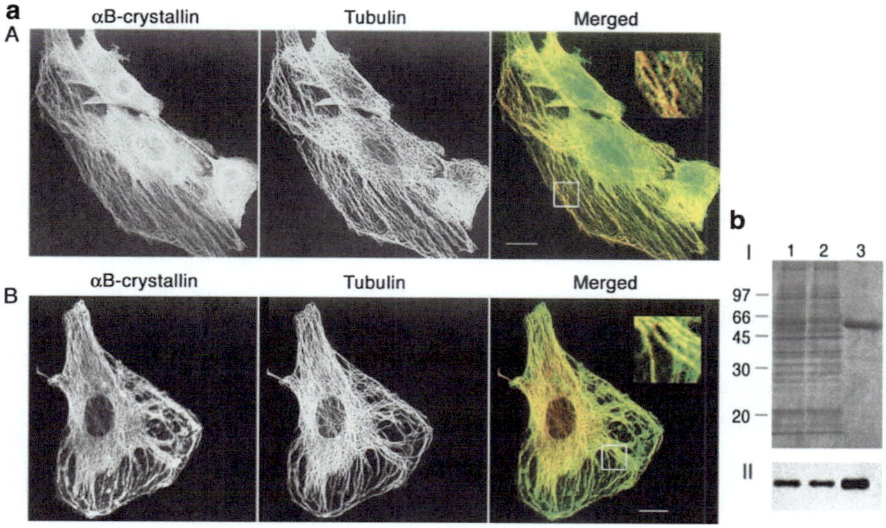

Fig. 29.14 Association of αB-crystallin with MTs. (**a**) Co-localization of αB-crystallin and MTs in formaldehyde fixed L6E9 cells (*A*) Co-localization of αB-crystallin and MT in methanol-fixed L6E9 cells (*B*). Merged immunofluorescence micrographs (*right*) of the same cells stained for αB-crystallin (*left*) and tubulin (*middle*). Insets show an enlargement (threefold) of the boxed area in the right-hand panels. (**b**) Co-precipitation of αB-crystallin and reconstituted MTs from L6E9 cell lysates. SDS-PAGE of total proteins (lane 1), the supernatant (lane 2), the precipitate (lane 3) in I and their immunoblots with the C1 antibody in II. Scale bars: 10 mm (Fujita et al. 2004)

29.6 Chaperone for Tubulin/Microtubules Responding to Mechanical Stress

"Dynamic instability" (Mitchison and Kirschner 1984) is a necessary feature of MTs to realize their adaptive transforming function. The heterodimers of α- and β-tubulin self-assemble from rigid MTs at the expense of free energy derived from GTP, whereupon they are capable of transmitting passive tensile forces, and also producing active pushing force to create cell polarity (Li and Gundersen 2008). In proliferating cells, assembled MTs, together with motor proteins, can function as a railroad for the shuttling of newly synthesized and denatured proteins, as well as for transmitting signals to regulate metabolism (Gundersen and Cook 1999). We infer that MTs in mature striated muscle cells have functions that are similar to those in proliferating cells. In our hypothesis, αB-crystallin expression would be induced by the MT network under such optimal dynamic instability due to the switching of passive and active tensile stresses. As a general rule in biochemical regulation, equilibrium between MTs and tubulin dimers may receive feedback from αB-crystallin, which is supported by several recent reports. For instance, αB-crystallin in striated muscles is reported to interact with both tubulin dimers (Arai and Atomi 1997; Ohto-Fujita et al. 2007) and MAPs (microtubule-associated proteins) of MTs

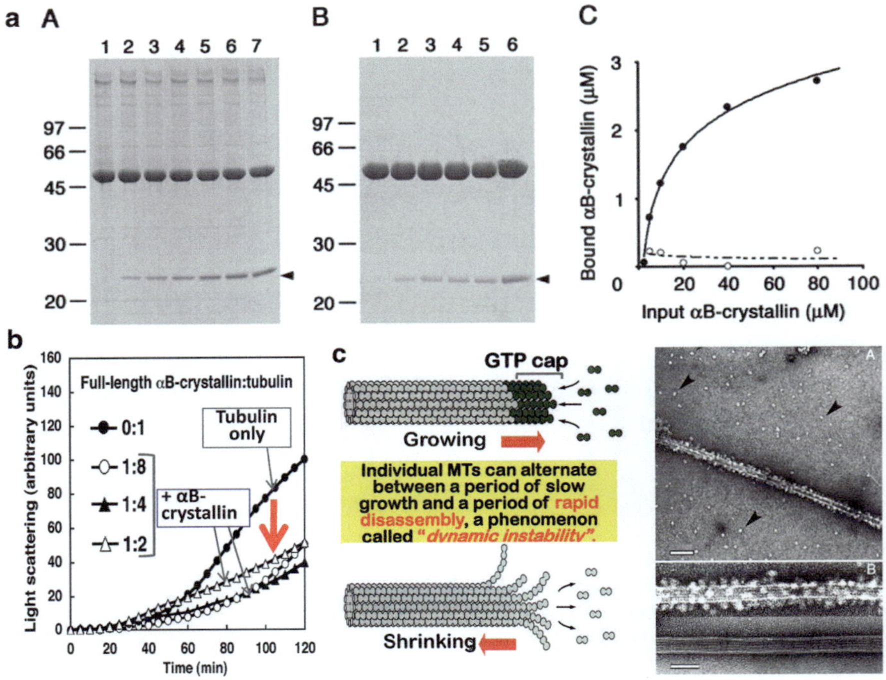

Fig. 29.15 Binding or association of αB-crystallin to reconstituted MTs (**a**) and tubulin dimers under stress conditions (42 °C) (**b**) αB-crystallin-coated MT observed by electron microscopy (**c**) schematic representation of "MT dynamic instability". (**a**) CBB-stained gel of co-precipitated αB-crystallin and MAP-MTs (*A*). All lanes contain 1 mg ml^{-1} MAP-MTs. Lanes 1–7 contain 0, 2.5, 5, 10, 20, 40 and 80 μM αB-crystallin, respectively. *Arrowhead* indicates αB-crystallin band. CBB-stained gel of co-precipitated αB-crystallin and proximal convoluted tubule (PCT)-MT (*B*). All lanes contain 10 μM PCT-MT. Lanes 1–7 contain 0, 5, 10, 20, 40 and 80 μM 5 μM αB-crystallin, respectively. The amount of co-precipitated αB-crystallin was plotted for input αB-crystallin (*C*). Closed circles represent microsomal triglyceride transfer protein (MTP)-MTs and open circles indicate PCT-MTs. (**b**) αB-crystallin protects tubulin from aggregation following 42 °C heat stress in vitro as observed by changes in turbidity (Ohto-Fujita et al. 2007). (**c**) αB-crystallin oligomer binds to MAP-microtubules in vitro but does not bind to PCT-tubulin (Fujita et al. 2004)

(Fujita et al. 2004) (Fig. 29.15). αB-crystallin also contributes MT resistance to nocodazole and calcium ions in vitro (Fig. 29.16) (Fujita et al. 2004).

In oligodendroglia, Bauer and Richter-Landsberg (2006) suggested that the stress-induced binding of αB-crystallin prevents MTs from aggregating. Our hypothesis as supported by the present correlation analysis thus further suggests that MTs and αB-crystallin constitute a conducting system that links mechanical environmental demands to morphological and biochemical processes, which eventually evolve into adaptive transformation in the function and metabolism of mature striated muscle cells. Some studies to test the hypothesis showed a close relationship between αB-crystallin and MTs, both in cells and in vitro,

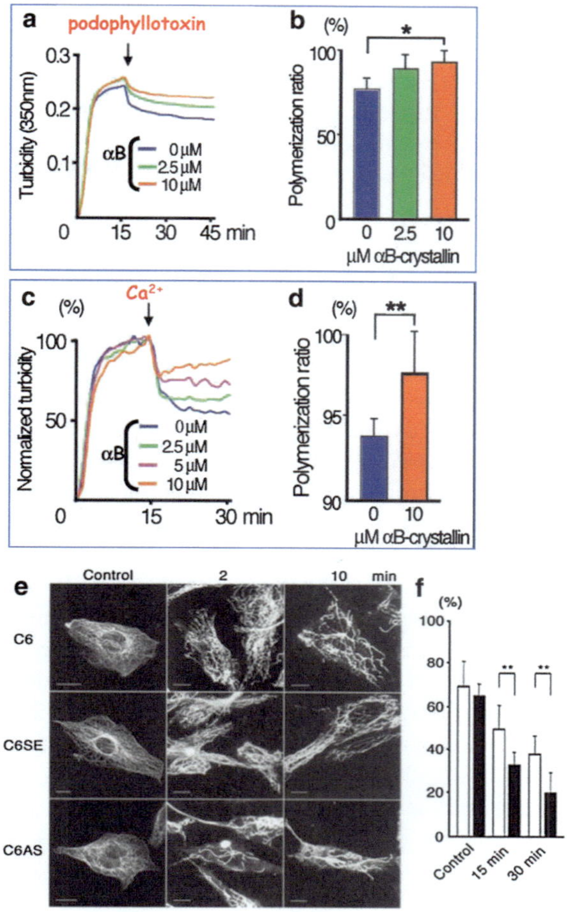

Fig. 29.16 αB-crystallin increases MT resistance to disassembly in vitro. (*A*) Changes in turbidity following MT assembly over 15 min, and disassembly after the addition of 10 μM podophyllotoxin (indicated by *arrow*) with (2.5 μM, 10 μM) or without αB-crystallin. (*B*) Mean levels of remaining MTs and standard deviations for each of three trials calculated from the ratio of the absorbance at 44–45 min to the level at 14–15 min, (at polymerized peak) with different amounts of αB-crystallin. αB-crystallin (10 μM) significantly increased the amount of remaining MT compared with that in the absence of αB-crystallin. Values are means ± s.d. (n = 3, *significantly different at $P < 0.05$). (*C*) Changes in turbidity upon MT protein assembly over 15 min and disassembly after the addition of 1 mM calcium (indicated by *arrow*) with (2.5 μM, 5.0 μM, 10.0 μM) or without αB-crystallin. (*D*) The ratio of the attained polymerization as a mean value at 14–15 min after 113 μM calcium addition with αB-crystallin (10 μM) was significantly higher than that without αB-crystallin. Values are means ± s.d. (n = 6, **significantly different at $P < 0.01$). (*E*) Remaining MT after 0, 2, and 10 min nocodazole treatment in αB-crystallin-expressing C6 glioma cells (*C*), overexpressing C6 cells (C6SE), lower expressing cells (C6AS). (*F*) Comparison of the polymer ratio to the amount of total tubulin in C6 and C6AS cells, as quantified by immunoblotting of the isolated dimers and polymers with (for 15 min and 30 min, respectively) and without (control) nocodazole treatment. The mean polymer ratio and the standard deviations of six samples are shown for C6 cells (*white* columns) and αB-crystallin- antisense-expressing C6AS cells (*black* columns). **Significantly different at $P < 0.01$. C6 cells = rat glioma cells, C6 SE = C6 cells to overexpress αB-crystallin protein (the cell that artificially (genetically) overexpresses sense strand of αB-crystallin gene), C6 AS = C6 cells to inhibit αB-crystallin protein (the cell that artificially (genetically) overexpress anti-sense strand of αB-crystallin gene)

and a resistant effect of αB-crystallin towards nocodazole- and calcium-induced disassembly of MT. This suggests that the role of αB-crystallin in MT function might be that of a possible modulator for MT disassembly as mediated by MAPs.

These results remind us of a hypothesis by Carver et al. (2002) to explain the ability of sHSPs to protect other proteins from denaturation in that the oligomer form of sHSP can supply interactive sites for the substrates on the surface of the sHSP spheres. As such, the surface of αB-crystallin oligomers, when attached to MTs by MAPs, may thus be accessible as a chaperone bed for denatured tubulin dimers and/or other substrates. Taken together, there is the possibility that one important role of αB-crystallin is to protect cytoskeletal proteins from denaturation while keeping monomers close to polymeric cytoskeletons, with the transformation between the two forms being dynamically regulated (Kirschner 1980; Mitchison and Kirschner 1984; Goldman et al. 1999) under both normal and stressed conditions. Together MTs are essential for muscle differentiation (Gundersen et al. 1989) while αB-crystallin is also necessary for myotube differentiation (Fig. 29.17) (Sakurai et al. 2005).

Otherwise, the possible additive functions of MTs and αB-crystallin might be closely related to the maintenance of cytoskeletal integrity (Ingber 1993) in the face of long-lasting tonic contraction. Because all sHSPs are highly expressed in muscle cells, especially in heart (Dubin et al. 1989) and slow skeletal muscle (fiber-type-dependent expression) (Atomi et al. 2000), the function of these tissues might be related to tonic contraction, and are maintained in a continuous energy-consuming fashion. In addition, a close correlation has been reported between oxidative metabolism and the expression level of αB-crystallin in various cells in

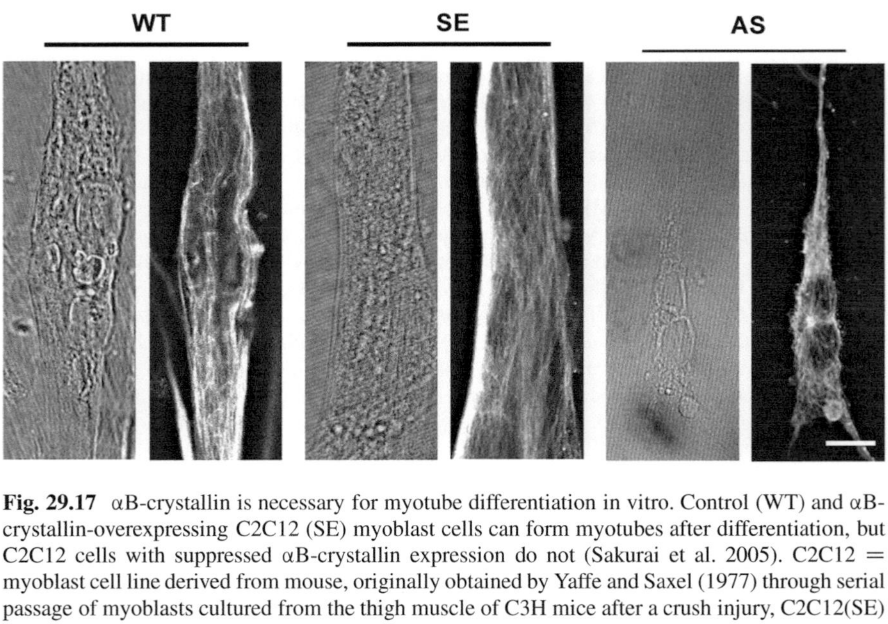

Fig. 29.17 αB-crystallin is necessary for myotube differentiation in vitro. Control (WT) and αB-crystallin-overexpressing C2C12 (SE) myoblast cells can form myotubes after differentiation, but C2C12 cells with suppressed αB-crystallin expression do not (Sakurai et al. 2005). C2C12 = myoblast cell line derived from mouse, originally obtained by Yaffe and Saxel (1977) through serial passage of myoblasts cultured from the thigh muscle of C3H mice after a crush injury, C2C12(SE) = C2C12 mouse myoblast cell line that overexpress sense strand of αB-crystallin gene

rat tissues (Iwaki et al. 1990). Furthermore, slow muscle with higher oxidative metabolism in mitochondria is regulated by the coactivator peroxisome proliferator-activated receptor-gamma coactivator 1α (Lin et al. 2002), which is induced by glucocorticoid and MT-related drugs such as colchicine and taxol that influence dynamic instability (Arany et al. 2008).

To determine the functional contribution of the interaction of αB-crystallin and tubulin/MT in skeletal muscle in greater detail, the dynamic phase of myotube formation was compared in the presence and the absence of αB-crystallin using cultured myoblast cells (Fig. 29.17). An immunostaining study using cultured myoblast C2C12SE and C2C12AS cells, which have up- and downregulated αB-crystallin expression, respectively, showed that C2C12SE cells had dense MT networks while C2C12AS cells had thin MT networks. Although αB-crystallin may have other roles given the reduced capacity of C2C12AS cells to fuse into myotubes, these observations imply that αB-crystallin is essential for the formation of dense MT networks and may maintain dynamic MT reorganization during differentiation. Together these results are in keeping with the finding that MT stabilization is essential for the fusion of myoblast cells into myotubes and in turn preserving the orientation of the myotube (Gundersen et al. 1989; Chang et al. 2002). Recently the essential role of MT stabilization for myotube formation was found to involve the ubiquitin ligase enzyme 3 Muscle Ring Finger (MURF) (Spencer et al. 2000; McElhinny et al. 2004). This possible new role for αB-crystallin raised another question that was addressed in our previous paper showing that a decrease in αB-crystallin expression increased the rate of nocodazole-induced MT disassembly (Fujita et al. 2004). This study also examined whether or not free tubulin molecules produced after nocodazole-induced disassembly could reassemble into MTs. A decreased capacity of MT reassembly could be due to tubulin denaturation that in turn results in decreased levels of MTs. On the other hand, increases in αB-crystallin expression may decrease the rate of nocodazole-induced MT disassembly and promote tubulin refolding. Reassembled or disassembled tubulin may be degraded with the assistance of αB-crystallin acting as a ligase enzyme 3 in the ubiquitin proteasome system (Lin et al. 2006). This degradation may be related to phosphorylation of αB-crystallin, particularly at Ser59, which was shown to inhibit apoptosis (Webster 2003; Hishiya and Takayama 2008; Steinman 2009). Three serines found in the N-terminal one-third of αB-crystallin were shown to be phosphorylated by various kinases (Fujita et al. 2004) that may contribute to MT dynamic stability via MAPs while also preserving dynamic instability through a continuous supply of free tubulins mediated by the chaperone activity of the C-terminal "α-crystallin domain" of αB-crystallin. In a chaperone experiment using αB-crystallin deletion mutants (Ohto-Fujita et al. 2007), tubulin aggregation that followed heat-denaturation was almost completely suppressed by the C-terminal α-crystallin domain, where phosphorylation is not a factor. Many new data showing good effects of αB-crystallin in αB-crystallin knockout mice that have reduced immunological capacity are also available (Ousman et al. 2007; Steinman 2009; Shao et al. 2013). As such, additional molecular mechanisms of αB-crystallin should emerge in the near future.

29.7 Applications

Recent studies showed that some peptides from αB-crystallin as well as αA-crystallin could protect against the aggravation of inflammation (Hochberg et al. 2014; Wang et al. 2014; Healy et al. 2014). sHSPs such as αB-crystallin are thought to recognize conformation changes at the very early phases of protein denaturation. Therefore these hormesis effects, in which stress proteins are induced and proteostasis is maintained (Labbadia and Morimoto 2014), are expected to be promoted by sHSP activity. Only mild stress is effective for this induction that results in beneficial effects.

29.8 Stretching is Good for Both Cells and Humans

In 1997, the paper "Stretching is good for a cell" with the subtitle "the shape is the thing" was published in *Science* (Ruoslahti 1997). Humans are able to stand due to the development of large soleus muscles that allow an erect posture while falling is prevented by the stretch-contraction cycle of the soleus muscle (Fig. 29.18, left). As with humans when they stand, cells need substrates for attachment. Most cells can form a fulcrum for adhesion, whereupon the cell can develop tension

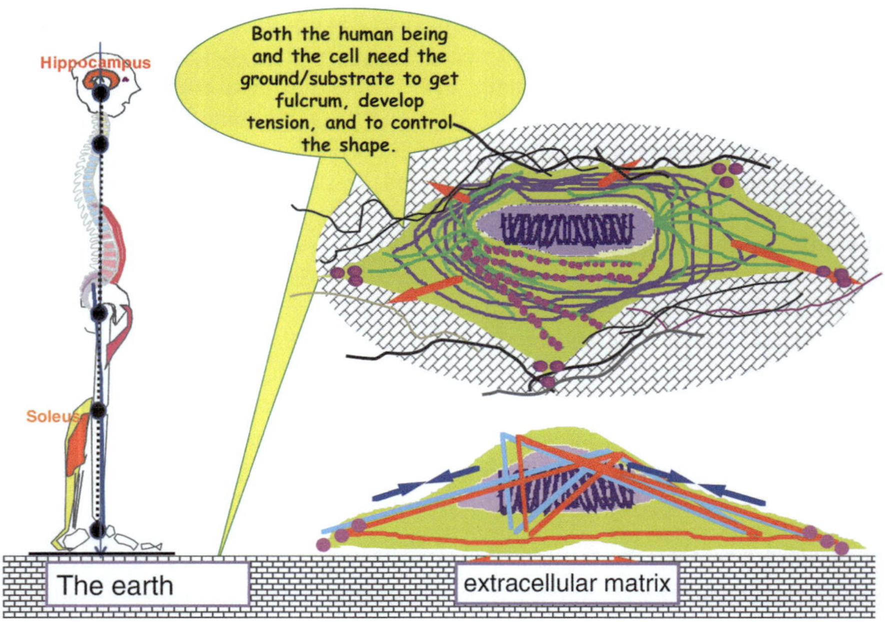

Fig. 29.18 Stretch is good for humans and cells, which can be maintained by the cytoskeleton with tensile and dynamic properties supported by αB-crystallin

and stretch (Fig. 29.18, right). Ruoslahti likely imagined this idea for endothelial cells in blood vessels. As myoblast cells differentiate into myotubes that make up skeletal muscle, fibroblast cells found in joints and their surrounding tendons receive tension signals during walking and running (Fig. 29.1). Most cells of the body function independently via an adhering substrate, while secreting ECM and growth factors to maintain and form shape-developing tension through dynamically regulated processes.

Usually we give little thought to standing, although humans can stretch against the ground due to gravity, which is similar to enhanced cellular function following connection to a substrate. If the cell cannot attach to a substrate, it maintains a rounded shape and eventually undergoes apoptosis due to the loss of cell adhesion. Stretch induces the gene expression of the stress protein αB-crystallin, which is reported to move stretched Z-bands or myofibril fractions by eccentric contraction (Féasson et al. 2002; Paulsen et al. 2009; Frankenberg et al. 2014). Skeletal muscles connect to the skeleton (bone), which provides a fulcrum through which tension can be developed. When stretching before or after exercise, a similar structure is activated in the same manner. As I discussed at the beginning of this chapter, the cells of our body are structured so that they can pass signals to genes through mechanical stretch.

Stretching is especially essential for maintaining muscle mass. Cell adhesion and cell stretching organize cytoskeletal networks dynamically in the exercising body and result in production of tension in cells and skeletal muscle. Muscle shortening results in decreased tension and stops the development of tension completely if the body remains relaxed for a sufficiently long period of time, as would be experienced by bed-redden patients. Tension developed with actin filaments and actin dynamics is related by c-fos-dependent gene expression followed by translocation of the serum response factor (SRF) transcription factor and related proteins from the cell surface to the nucleus (Treisman 2013; Esnault et al. 2014). There is a stretch-response element upstream of several genes that are relevant to muscle function, including αB-crystallin, c-fos, vinculin, and actin, which are immediate early genes (Treisman 1995). αB-crystallin colocalizes to the cytoskeleton and functions as a chaperone for actin, MTs and IFs. Most cells in our body live independently via adhesion to a substrate, secreting ECM and growth factors, while maintaining their shape by developing tension. This process is dynamically regulated.

29.9 Stress Proteins and Molecular Chaperones in Cells and in Life

This paper describes important aspects of the sHSP molecular chaperone αB-crystallin, which is constitutively expressed not only in striated muscles but also in brain cells. sHSPs are perhaps under-appreciated in the chaperone research field, although the review by Quinlan and Ellis (2013) stated: "we return to first principles, and shine the spotlight on the role of chaperones in biomolecular assemblies, as first defined for nucleosomes by Laskey et al. (1978)". This initial concept is now widened to include the modulation of many assembled polymers during normal cell growth and development. In addition, assembly chaperones have a pivotal role in all adaptive responses to environmental and ecological stresses that can shape species evolution. For example, the potential impact of the cellular chaperone complement on the preservation and evolution of the species by the moderation of genetic mutation (Mendez et al. 2010; Bandyopadhyay et al. 2012; Bogumil and Dagan 2012; Gros and Tenaillon 2009) is now emerging. Consequently, chaperones are at the forefront of adaptive responses of microbial (Tsai et al. 2010; Ladner et al. 2012; Li et al. 2012), plant (Iki et al. 2010; Ladner et al. 2012; Brígido and Oliveira 2012; Goyal et al. 2012) and animal (Stecyk et al. 2012) life to different environments. Quinlan focuses on αB-crystallin and intermediate filaments, and most of his studies pertain to lens-related cells, which express abundant intermediate filaments (Song et al. 2009). Molecular chaperones thus are essential guardians for several cell systems over species. sHSPs can form high molecular weight oligomers to various extents, which can affect protein processing and has rather broad target specificities. Kenyon, a leading scientist of longevity, used *Caenorhabditis elegans* mutants as an experimental model to determine that the transcriptional heat shock factor (HSF), which controls HSP family gene expression, has a strong role in longevity (Garigan et al. 2002). sHSPs also work well to suppress polyglutamine-expansion protein aggregation in a *Caenorhabditis elegans* model (Hsu et al. 2003). To avoid conformational disorders such as Alzheimer's disease, which produces fiber protein deposits in the brain of aged people, HSP induction and maintenance of proteostasis are of significant importance (Labbadia and Morimoto 2014).

Anfinsen's dogma, also known as the thermodynamic hypothesis, is well-known to protein scientists. Anfinsen purported that all the necessary information for a protein to adopt its final conformation is encoded in the amino acid sequence. However, we now know that molecular chaperones are essential for protein folding and refolding in cells. These chaperones maintain protein conformation over a small range of homeostasis for cell survival. Slow muscle tissues constitutively express abundant stress proteins for proteostasis because muscle cells constantly produce abundant amounts of superoxides in the mitochondria.

Deterioration of the body with aging could be decreased by induced stress proteins. This effect is referred to as "hormesis", which is the response of organisms to mild stressors that result in improved health and longevity. Mild heat shock has been thought to induce hormetic responses because it promotes the activation of HSPs, which may extend lifespan (Lagisz et al. 2013). Calabrese et al. (2011) termed genes for stress proteins as vitagenes. Through an active lifestyle, living cells in our body are constantly stimulated and thus they constitutively express stress proteins and various other housekeeping proteins. Mechanical stress is essential for the maintenance of other tissues that participate in exercise, such as the heart, the endothelial cells of blood vessels, bone cells and fat cells, as well muscle tissues (Fig. 29.2). Our body has a system to express stress proteins not only inductively but also constitutively. If we consider an active lifestyle at the molecular level, muscle contraction produces various mechanical stimuli in the body. These appropriate stresses may be essential for cellular maintenance, because stretching is also good for a cell. Appropriate stress is produced in standing and moving the body against the earth's gravity, which induces cellular expression of appropriate stress proteins. Both life units - the cell at a micro level, and the body as a whole at the macro level - have similar relationships to their environment. As such, exercise is essential for both cells and particularly for humans to have a long life span. An active lifestyle is thus important to maintain optimal health.

29.10 Summary

An active lifestyle could increase not only a person's working capacity but also his/her overall health, regardless of age. Although determining whether an active lifestyle is similarly beneficial for all cells in the body is difficult, each cell does have common basic systems even if they have already differentiated and are functioning as a component of specific tissues. Functions of the chaperone system to induce dynamic stability and maintain critical systems should be essential for all cells in the body, because of the principle of "activity dependence" wherein survival mechanisms that might have been put into place during life's early origins are important for continued survival and evolution on earth. As such, the 'adaptability' of living organisms is an important consideration. Here I put forward that gravity health science will be an important line of research to improve the overall health of aging societies (Fig. 29.19). To elucidate the mechanism and support our hypothesis

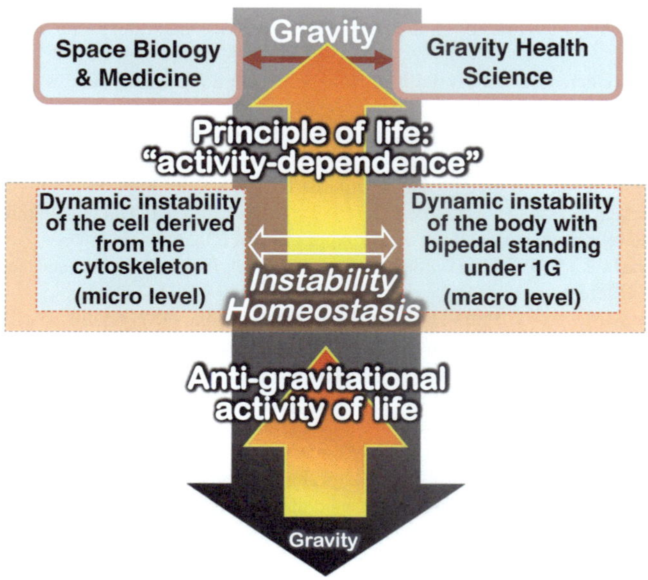

Fig. 29.19 Hypothesis of gravity health science as a countermeasure to maintain dynamic instability in organisms to prevent deterioration in space and during aging. Principle of "activity dependence" working at various levels from micro to macro units that must be acquired during the emergence of life under gravity

using high pressure bioscience is thus important for proceeding in the frontier of life sciences. Only from basic science can normal applications be developed.

References

Agarkova I, Schoenauer R, Ehler E, Carlsson L, Carlsson E, Thornell LE, Perriard JC (2004) The molecular composition of the sarcomeric M-band correlates with muscle fiber type. Eur J Cell Biol 83:193–204

Agbulut O, Li Z, Mouly V, Butler-Browne GS (1996) Analysis of skeletal and cardiac muscle from desmin knock-out and normal mice by high resolution separation of myosin heavy-chain isoforms. Biol Cell 88:131–135

Allen DL, Linderman JK, Roy RR, Bigbee AJ, Grindeland RE, Mukku V, Edgerton VR (1997) Apoptosis: a mechanism contributing to remodeling of skeletal muscle in response to hindlimb unweighting. Am J Physiol 273(2 Pt 1):C579–C587

American Society for Gravitational and Space Research (2013) Physiological slides. American Society for Gravitational and Space Research Website. https://www.asgsr.org/index.php/education/slide-sets

Appaix F, Kuznetsov AV, Usson Y, Kay L, Andrienko T, Olivares J, Kaambre T, Sikk P, Margreiter R, Saks V (2003) Possible role of cytoskeleton in intracellular arrangement and regulation of mitochondria. Exp Physiol 88:175–190

Arai H, Atomi Y (1997) Chaperone activity of αB-crystallin suppresses tubulin aggregation through complex formation. Cell Struct Funct 22:539–544

Arany Z, Wagner BK, Ma Y, Chinsomboon J, Laznik D, Spiegelman BM (2008) Gene expression-based screening identifies microtubule inhibitors as inducers of PGC-1α and oxidative phosphorylation. Proc Natl Acad Sci U S A 105:4721–4726

Atomi Y (1980) Aerobic capacity of Japanese adult women. Doctoral Thesis, The University of Tokyo

Atomi Y (1992) Decreased αB-crystallin in soleus muscle atrophy and role of αB-Crystallin in muscle. Med Sport Sci 37:171–192

Atomi Y (1999) Why active life is important for health and longevity? From the studies of stress protein and mechanical stress. In: Current reviews of medical science by women, International Conference of Women Engineers and Scientists, Tokyo, 150–153

Atomi Y, Fukunaga T, Hana H, Yamamoto Y (1987) Relationship between lactate threshold during running and relative gastrocnemius area. J Appl Physiol 63:2343–2347

Atomi Y, Miyashita M (1974) Maximal aerobic power of Japanese active and sedentary adult females of different ages (20 to 62 years). Med Sci Sports 6:223–225

Atomi Y, Toro K, Masuda T, Hatta H (2000) Fiber-type-specific αB-crystallin distribution and its shifts with T(3) and PTU treatments in rat hindlimb muscles. J Appl Physiol 88:1355–1364

Atomi Y, Yamada S (1986) Mechanism of muscle atrophy. III. A model of muscle atrophy with the tail suspension of rat. Dept Sports Sci Univ Tokyo 20:8–12

Atomi Y, Yamada S, Hong Y-M (1990) Dynamic expression of αB-crystallin in skeletal muscle -effects of unweighting, passive stretch and denervation. Proc Japan Acad 66B:203–208

Atomi Y, Yamada S, Nishida T (1991a) Early changes of αB-crystallin mRNA in rat skeletal muscle to mechanical tension and denervation. Biochem Biophys Res Commun 181: 1323–1330

Atomi Y, Yamada S, Strohman R, Nonomura Y (1991b) αB-crystallin in skeletal muscle: purification and localization. J Biochem 110:812–822

Baba M (2001) Brain and nerves in pictures: their system and mechanism of the disorder, 2nd Japanese edn. Igaku Shoin, Tokyo, 1–233

Bandyopadhyay A, Saxena K, Kasturia N, Dalal V, Bhatt N, Rajkumar A, Maity S, Sengupta S, Chakraborty K (2012) Chemical chaperones assist intracellular folding to buffer mutational variations. Nat Chem Biol 8:238–245

Bauer NG, Richter-Landsberg C (2006) The dynamic instability of microtubules is required for aggresome formation in oligodendroglial cells after proteolytic stress. J Mol Neurosci 29: 153–168

Bogumil D, Dagan T (2012) Cumulative impact of chaperone-mediated folding on genome evolution. Biochemistry 51:9941–9953

Boudriau S, Vincent M, Côté CH, Rogers PA (1993) Cytoskeletal structure of skeletal muscle: identification of an intricate exosarcomeric microtubule lattice in slow- and fast-twitch muscle fibers. J Histochem Cytochem 41:1013–1021

Bramble DM, Lieberman DF (2004) Endurance running and the evolution of Homo. Nature 432:345–352

Bray D (2001) Chapter 5: Actin filaments. In: Bray D (ed) Cell movements: from molecules to motility, 2nd edn. Garland, New York, 65–80

Brígido C, Oliveira S (2012) Most acid-tolerant chickpea mesorhizobia show induction of major chaperone genes upon acid shock. Microb Ecol 65:145–153

Bulinski JC, Gundersen GG (1991) Stabilization of post-translational modification of microtubules during cellular morphogenesis. Bioessays 13:285–293

Calabrese V, Cornelius C, Cuzzocrea S, Iavicoli I, Rizzarelli E, Calabrese EJ (2011) Hormesis, cellular stress response and vitagenes as critical determinants in aging and longevity. Mol Aspects Med 32:279–304

Campbell WG, Gordon SE, Carlson CJ, Pattison JS, Hamilton MT, Booth FW (2001) Differential global gene expression in red and white skeletal muscle. Am J Physiol Cell Physiol 280: C763–C768

Carver JA, Lindner RA, Lyon C, Canet D, Hernandez H, Dobson CM, Redfield C (2002) The interaction of the molecular chaperone α-crystallin with unfolding α-lactalbumin: a structural and kinetic spectroscopic study. J Mol Biol 318:815–827

Chang W, Webster DR, Salam AA, Gruber D, Prasad A, Eiserich JP, Bulinski JC (2002) Alteration of the C-terminal amino acid of tubulin specifically inhibits myogenic differentiation. J Biol Chem 277:30690–30698

Close RI (1972) Dynamic properties of mammalian skeletal muscles. Physiol Rev 52:129–197

Dart KC, Schultz E (1989) Hindlimb suspension suppresses muscle growth and satellite cell proliferation. J Appl Physiol 67:1827–1834

Desjardins PR, Burkman JM, Shrager JB, Allmond LA, Stedman HH (2002) Evolutionary implications of three novel members of the human sarcomeric myosin heavy chain gene family. Mol Biol Evol 19:375–393

Dubin RA, Wawrousek EF, Piatigorsky J (1989) Expression of the murine αB-crystallin gene is not restricted to the lens. Mol Cell Biol 9:1083–1091

Dulhunty AF, Franzini-Armstrong C (1975) The relative contributions of the folds and caveolae to the surface membrane of frog skeletal muscle fibres at different sarcomere lengths. J Physiol 250:513–539

Eason JM, Schwartz GA, Pavlath GK, English AW (2000) Sexually dimorphic expression of myosin heavy chains in the adult mouse masseter. J Appl Physiol 89:251–258

Endo M (1937) Muscle weights of human lower legs related to the development. J Tokyo Med Sci 51:1177–1185

Esnault C, Stewart A, Gualdrini F, East P, Horswell S, Matthews N, Treisman R (2014) Rho-actin signaling to the MRTF coactivators dominates the immediate transcriptional response to serum in fibroblasts. Genes Dev 28:943–958

Féasson L, Stockholm D, Freyssenet D, Richard I, Duguez S, Beckmann JS, Denis C (2002) Molecular adaptations of neuromuscular disease-associated proteins in response to eccentric exercise in human skeletal muscle. J Physiol 543(Pt 1):297–306

Frankenberg NT, Lamb GD, Overgaard K, Murphy RM, Vissing K (2014) Small heat shock proteins translocate to the cytoskeleton in human skeletal muscle following eccentric exercise independently of phosphorylation. J Appl Physiol 116:1463–1472

Fujita Y, Ohto E, Katayama E, Atomi Y (2004) αB-Crystallin-coated MAP microtubule resists nocodazole and calcium-induced disassembly. J Cell Sci 117:1719–1726

Garigan D, Hsu AL, Fraser AG, Kamath RS, Ahringer J, Kenyon C (2002) Genetic analysis of tissue aging in *Caenorhabditis elegans*: a role for heat-shock factor and bacterial proliferation. Genetics 161:1101–1112

Goldman RD, Chou YH, Prahlad V, Yoon M (1999) Intermediate filaments: dynamic processes regulating their assembly, motility, and interactions with other cytoskeletal systems. FASEB J 13(Suppl 2):S261–S265

Goldspink DF (1980) Physiological factors influencing protein turnover and muscle growth in mammals. In: Goldspink DF (ed) Development and specialization of skeletal muscle. Cambridge University Press, London, pp 65–89

Gollnick PD, Sjödin B, Karlsson J, Jansson E, Saltin B (1974) Human soleus muscle: a comparison of fiber composition and enzyme activities with other leg muscles. Pflugers Arch 348:247–255

Goyal RK, Kumar V, Shukla V, Mattoo R, Liu Y, Chung SH, Giovannoni JJ, Mattoo AK (2012) Features of a unique intronless cluster of class I small heat shock protein genes in tandem with box C/D snoRNA genes on chromosome 6 in tomato (*Solanum lycopersicum*). Planta 235: 453–471

Gros PA, Tenaillon O (2009) Selection for chaperone-like mediated genetic robustness at low mutation rate: impact of drift, epistasis and complexity. Genetics 182(2):555–564

Gundersen GG, Cook TA (1999) Microtubules and signal transduction. Curr Opin Cell Biol 11: 81–94

Gundersen GG, Khawaja S, Bulinski JC (1989) Generation of a stable, posttranslationally modified microtubule array is an early event in myogenic differentiation. J Cell Biol 109:2275–2288

Gyoeva FK, Gelfand VI (1991) Coalignment of vimentin intermediate filaments with microtubules depends on kinesin. Nature 353:445–448

Healy EF, Little C, King PJ (2014) A model for small heat shock protein inhibition of polyglutamine aggregation. Cell Biochem Biophys 69:275–281

Hishiya A, Takayama S (2008) Molecular chaperones as regulators of cell death. Oncogene 27:6489–6506

Hochberg GK, Ecroyd H, Liu C, Cox D, Cascio D, Sawaya MR, Collier MP, Stroud J, Carver JA, Baldwin AJ, Robinson CV, Eisenberg DS, Benesch JL, Laganowsky A (2014) The structured core domain of αB-crystallin can prevent amyloid fibrillation and associated toxicity. Proc Natl Acad Sci U S A 111:E1562–E1570

Hoson T, Matsumoto S, Soga K, Wakabayashi K (2010) Cortical microtubules are responsible for gravity resistance in plants. Plant Signal Behav 5:752–754

Houck SA, Landsbury A, Clark JI, Quinlan RA (2011) Multiple sites in αB-crystallin modulate its interactions with desmin filaments assembled in vitro. PLoS One 6:e25859

Hsu AL, Murphy CT, Kenyon C (2003) Regulation of aging and age-related disease by DAF-16 and heat-shock factor. Science 300:1142–1145

Iki T, Yoshikawa M, Nishikiori M, Jaudal MC, Matsumoto-Yokoyama E, Mitsuhara I, Meshi T, Ishikawa M (2010) In vitro assembly of plant RNA-induced silencing complexes facilitated by molecular chaperone HSP90. Mol Cell 39:282–291

Ingber DE (1993) Cellular tensegrity: defining new rules of biological design that govern the cytoskeleton. J Cell Sci 104:613–627

Ingber DE (1997) Tensegrity: the architectural basis of cellular mechanotransduction. Annu Rev Physiol 59:575–599

Iwaki T, Kume-Iwaki A, Goldman JE (1990) Cellular distribution of αB-crystallin in non-lenticular tissues. J Histochem Cytochem 38:31–39

Jee H, Sakurai T, Kawada S, Ishii N, Atomi Y (2009) Significant roles of microtubules in mature striated muscle deduced from the correlation between tubulin and its molecular chaperone αB-crystallin in rat muscles. J Physiol Sci 59:149–155

Karakesisoglou I, Yang Y, Fuchs E (2000) An epidermal plakin that integrates actin and microtubule networks at cellular junctions. J Cell Biol 149:195–208
Kim BS, Kim MY, Leem YH (2011) Hippocampal neuronal death induced by kainic acid and restraint stress is suppressed by exercise. Neuroscience 194:291–301
Kirschner MW (1980) Implications of treadmilling for the stability and polarity of actin and tubulin polymers *in vivo*. J Cell Biol 86:330–334
Koh TJ, Escobedo J (2004) Cytoskeletal disruption and small heat shock protein translocation immediately after lengthening contractions. Am J Physiol Cell Physiol 286:C713–C722
Korfage JA, Van Eijden TM (1999) Regional differences in fibre type composition in the human temporalis muscle. J Anat 194:355–362
Labbadia J, Morimoto RI (2014) Proteostasis and longevity: when does aging really begin? F1000Prime Rep 6:7
Ladner JT, Barshis DJ, Palumbi SR (2012) Protein evolution in two co-occurring types of Symbiodinium: an exploration into the genetic basis of thermal tolerance in Symbiodinium clade D. BMC Evol Biol 12:217
Lagisz M, Hector KL, Nakagawa S (2013) Life extension after heat shock exposure: assessing meta-analytic evidence for hormesis. Ageing Res Rev 12:653–660
Laskey RA, Honda BM, Mills AD, Finch JT (1978) Nucleosomes are assembled by an acidic protein which binds histones and transfers them to DNA. Nature 275:416–420
Lazarides E (1981) Intermediate filaments – chemical heterogeneity in differentiation. Cell 23:649–650
Lele TP, Thodeti CK, Ingber DE (2006) Forced meets chemistry: analysis of mechanochemical conversion in focal adhesions using fluorescence recovery after photobleaching. J Cell Biochem 97:1175–1183
Leung CL, Sun D, Zheng M, Knowles DR, Liem RK (1999) Microtubule actin cross-linking factor (MACF): a hybrid of dystonin and dystrophin that can interact with the actin and microtubule cytoskeletons. J Cell Biol 147:1275–1286
Li K, Jiang T, Yu B, Wang L, Gao C, Ma C, Xu P, Ma Y (2012) Transcription elongation factor GreA has functional chaperone activity. PLoS One 7:e47521
Li R, Gundersen GG (2008) Beyond polymer polarity: how the cytoskeleton builds a polarized cell. Nat Rev Mol Cell Biol 9:860–873
Liao G, Gundersen GG (1998) Kinesin is a candidate for cross- bridging microtubules and intermediate filaments. Selective binding of kinesin to detyrosinated tubulin and vimentin. J Biol Chem 273:9797–9803
Lin J, Wu H, Tarr PT, Zhang CY, Wu Z, Boss O, Michael LF, Puigserver P, Isotani E, Olson EN, Lowell BB, Bassel-Duby R, Spiegelman BM (2002) Transcriptional co-activator PGC-1α drives the formation of slow-twitch muscle fibres. Nature 418:797–801
Lin DI, Barbash O, Kumar KG, Weber JD, Harper JW, Klein-Szanto AJ, Rustgi A, Fuchs SY, Diehl JA (2006) Phosphorylation-dependent ubiquitination of cyclin D1 by the SCF (FBX4-αB crystallin) complex. Mol Cell 24:355–366
McElhinny AS, Perry CN, Witt CC, Labeit S, Gregorio CC (2004) Muscle-specific RING finger-2 (MURF-2) is important for microtubule, intermediate filament and sarcomeric M-line maintenance in striated muscle development. J Cell Sci 117:3175–3188
Maruyama K, Natori R, Nonomura Y (1976) New elastic protein from muscle. Nature 262:58–60
Mendez R, Fritsche M, Porto M, Bastolla U (2010) Mutation bias favors protein folding stability in the evolution of small populations. PLoS Comput Biol 6:e1000767
Mitchison T, Kirschner M (1984) Dynamic instability of micro-tubule growth. Nature 312: 237–242
Morey ER (1979) Spaceflight and bone turnover: correlation with a new rat model of weightlessness. BioSci 29:168–172
Mozdziak PE, Troung Q, Macius A, Schultz E (1998) Hindlimb suspension reduces muscle regeneration. Eur J Appl Physiol Occup Physiol 78:136–140
Musacchia XJ, Steffen JM, Fell RD (1988) Disuse atrophy of skeletal muscle: animal models. Exerc Sport Sci Rev 16:61–87

Nakata T, Nishina Y, Yorifuji H (2001) Cytoplasmic gamma actin as a Z-disc protein. Biochem Biophys Res Commun 286:156–163
Nicholl RA, Quinlan RA (1994) Chaperone activity of α-crystallins modulates intermediate filament assembly. EMBO J 13:945–953
Nicogossian AE, Huntoon CL, Pool SL (1989) Space physiology and medicine, 2nd edn. Lea and Febiger, Philadelphia
Niwa M, Jaaro-Peled H, Tankou S, Seshadri S, Hikida T, Matsumoto Y, Cascella NG, Kano S, Ozaki N, Nabeshima T, Sawa A (2013) Adolescent stress-induced epigenetic control of dopaminergic neurons via glucocorticoids. Science 339:335–339
Ohto-Fujita E, Fujita Y, Atomi Y (2007) Analysis of the αB-crystallin domain responsible for inhibiting tubulin aggregation. Cell Stress Chaperones 12:163–171
Ousman SS, Tomooka BH, van Noort JM, Wawrousek EF, O'Connor KC, Hafler DA, Sobel RA, Robinson WH, Steinman L (2007) Protective and therapeutic role for αB-crystallin in autoimmune demyelination. Nature 448:474–479
Papaseit C, Pochon N, Tabony J (2000) Microtubule self-organization is gravity-dependent. Proc Natl Acad Sci U S A 97:8364–8368
Paulsen G, Lauritzen F, Bayer ML, Kalhovde JM, Ugelstad I, Owe SG, Hallén J, Bergersen LH, Raastad T (2009) Subcellular movement and expression of HSP27, αB-crystallin, and HSP70 after two bouts of eccentric exercise in humans. J Appl Physiol 107:570–582
Pedersen BK (2013) Muscle as a secretory organ. Compr Physiol J 3:1337–1362
Quinlan JA, Ellis RJ (2013) Chaperones: needed for both the good times and the bad times. Philos Trans R Soc Lond B Biol Sci 368:e0091
Ralston E, Lu Z, Ploug T (1999) The organization of the Golgi complex and microtubules in skeletal muscle is fiber type-dependent. J Neurosci 19:10694–10705
Ralston E, Ploug T, Kalhovde J, Lomo T (2001) Golgi complex, endoplasmic reticulum exit sites, and microtubules in skeletal muscle fibers are organized by patterned activity. J Neurosci 21:875–883
Ruoslahti E (1997) Stretching is good for a cell. Science 276:1345–1346
Sakurai T, Fujita Y, Ohto E, Oguro A, Atomi Y (2005) The decrease of the cytoskeleton tubulin follows the decrease of the associating molecular chaperone αB-crystallin in unloaded soleus muscle atrophy without stretch. FASEB J 19:1199–1201
Satake T, Atomi Y, Okajima Y (1987) Effect of physical activity in daily life on muscle weight and relative weight of the M. triceps surae. J Phys Fitness Jpn 36:25–30
Schmidt RF ed., Uchizono K et al. transl. (1988) Fundamentals of neurophysiology, 2nd Japanese edn. Kinpodo, Kyoto, pp 1–325
Schoenauer R, Lange S, Hirschy A, Ehler E, Perriard JC, Agarkova I (2008) Myomesin 3, a novel structural component of the M-band in striated muscle. J Mol Biol 376:338–351
Shao W, Zhang SZ, Tang M, Zhang XH, Zhou Z, Yin YQ, Zhou QB, Huang YY, Liu YJ, Wawrousek E, Chen T, Li SB, Xu M, Zhou JN, Hu G, Zhou JW (2013) Suppression of neuroinflammation by astrocytic dopamine D2 receptors via αB-crystallin. Nature 494:90–94
Sheldon KL, Maldonado EN, Lemasters JJ, Rostovtseva TK, Bezrukov SM (2011) Phosphorylation of voltage-dependent anion channel by serine/threonine kinases governs its interaction with tubulin. PLoS One 6:e25539
Singh BN, Rao KS, Ramakrishna T, Rangaraj N, Rao CM (2007) Association of αB-crystallin, a small heat shock protein, with actin: role in modulating actin filament dynamics *in vivo*. J Mol Biol 366:756–767
Sjöström M, Kidman S, Larsén KH, Angquist KA (1982) Z- and M-band appearance in different histochemically defined types of human skeletal muscle fibers. J Histochem Cytochem 30:1–11
Skoyles JR (2006) Human balance, the evolution of bipedalism and dysequilibrium syndrome. Med Hypotheses 66:1060–1068
Song S, Landsbury A, Dahm R, Liu Y, Zhang Q, Quinlan RA (2009) Functions of the intermediate filament cytoskeleton in the eye lens. J Clin Invest 119:1837–1848

Spencer JA, Eliazer S, Ilaria RL, Richardson JA, Olson EN (2000) Regulation of microtubule dynamics and myogenic differentiation by MURF, a striated muscle RING-finger protein. J Cell Biol 150:771–784

Stecyk JA, Couturier CS, Fagernes CE, Ellefsen S, Nilsson GE (2012) Quantification of heat shock protein mRNA expression in warm and cold anoxic turtles (Trachemys scripta) using an external RNA control for normalization. Genom Proteom 7:59–72

Steinman L (2009) A molecular trio in relapse and remission in multiple sclerosis. Nat Rev Immunol 9:440–447

Tabony J, Job D (1990) Spatial structures in microtubular solutions requiring a sustained energy source. Nature 346:448–451

Tabony J, Glade N, Papaseit C, Demongeot J (2002) Microtubule self-organisation and its gravity dependence. Adv Space Biol Med 8:19–58

Tabony J, Rigotti N, Glade N, Cortès S (2007) Effect of weightlessness on colloidal particle transport and segregation in self-organising microtubule preparations. Biophys Chem 127:172–180

Tanaka H, DeSouza CA, Seals DR (1998) Arterial stiffness and hormone replacement use in healthy postmenopausal women. J Gerontol 53:M344–M346

Tepp K, Mado K, Varikmaa M, Klepinin A, Timohhina N, Shevchuk I, Chekulayev V, Kuznetsov AV, Guzun R, Kaambre T (2014) The role of tubulin in the mitochondrial metabolism and arrangement in muscle cells. J Bioenerg Biomembr 46:421–434

Thomason DB, Booth FW (1990) Atrophy of the soleus muscle by hindlimb unweighting. J Appl Physiol 68:1–12

Thomason DB, Riggs RB, Booth FW (1989) Protein metabolism and β-myosin heavy-chain mRNA in unweighted soleus muscle. Am J Physiol 257:R300–R305

Treisman R (1995) Journey to the surface of the cell: Fos regulation and the SRE. EMBO J 14:4905–4913

Treisman R (2013) Shedding light on nuclear actin dynamics and function. Trends Biochem Sci 38:376–377

Tsai YL, Chiang YR, Narberhaus F, Baron C, Lai EM (2010) The small heat-shock protein HspL is a VirB8 chaperone promoting type IV secretion-mediated DNA transfer. J Biol Chem 285:19757–19766

Wang W, Sreekumar PG, Valluripalli V, Shi P, Wang J, Lin YA, Cui H, Kannan R, Hinton DR, MacKay JA (2014) Protein polymer nanoparticles engineered as chaperones protect against apoptosis in human retinal pigment epithelial cells. J Control Release 191:4–14

Waterman-Storer CM, Salmon ED (1997) Actomyosin-based retrograde flow of microtubules in the lamella of migrating epithelial cells influences microtubule dynamic instability and turnover and is associated with microtubule breakage and treadmilling. J Cell Biol 139:417–434

Webster KA (2003) Serine phosphorylation and suppression of apoptosis by the small heat shock protein αB-crystallin. Circ Res 92:130–132

Yaffe D, Saxel O (1977) Serial passaging and differentiation of myogenic cells isolated from dystrophic mouse muscle. Nature 270:725–727

Zimmerman SD, Mccormick RJ, Vadlamudi RK, Thomas DP (1993) Age and training alter collagen characteristics in fast- and slow-twitch rat limb muscle. J Appl Physiol 75:1670–1674

Part IX
Methodology

Editors' Foreword of Part IX

Technology is by far the importance in performing any measurements under high pressure. In the high-pressure bio-science field, several new high pressure techniques have come into play in recent years. Among these, here we choose the following four techniques, as fundamental ones to reveal structural changes of bio-macromolecules under pressure. For High Pressure Microscopy, see Chap. 27 by Nishiyama. For High Pressure Macromolecular Crystallography, see also Chap. 11 by Dhaussy and Girard. For other well-established techniques, please consult standard text books for high-pressure experiments.

Chapter 30	High Pressure Small-Angle X-Ray Scattering	Tetsuro Fujisawa
Chapter 31	High Pressure Macromolecular Crystallography	Nobuhisa Watanabe
Chapter 32	High-Pressure Fluorescence Spectroscopy	Akihiro Maeno and Kazuyuki Akasaka
Chapter 33	High Pressure NMR Spectroscopy	Kazuyuki Akasaka

Chapter 30
High Pressure Small-Angle X-Ray Scattering

Tetsuro Fujisawa

Abstract Small-angle scattering, solution scattering from proteins in solution, reflects the shape of the scatter as a spread of electron density, which is common to protein crystallography. Although the obtained resolution of small-angle scattering is inferior to that of crystallography, it shows the global image of protein structure *in solution* without constraints of neighboring molecules in crystal lattice. At ambient pressure, data collection technology and analyses of small-angle scattering method developed so greatly in recent 10 years that it is recognized as one of the powerful method of structural biology. In parallel, many efforts have been made to apply this technique under high pressure. The instrumentation and interpretation of small-angle scattering under pressure, however, requires special considerations. The present chapter reviews the technological aspect of scattering from protein solution especially optimized for synchrotron X-ray sources.

Keywords Contrast effect • Cell design • Data calibration • Synchrotron X-ray

30.1 Introduction

Pressure effectso n biological system have been studied extensively up to several giga Pascal. Among them, the interaction between subunits is known to be perturbed by high pressure which occurs around a few hundred mega Pascal (Silva and Weber 1988; Gross and Jaenicke 1994; Heremans and Smeller 1998; Silva and Foguel 2009). It is the pressure range that HP-SAXS contributed to the probing of protein under pressures. In the last decade main breakthroughs on protein research under high pressure were achieved by protein crystallography or high-resolution NMR (Fourme and Akasaka 2012). Compared to those techniques, small-angle X-ray scattering (SAXS) has a limited resolution, in which it observes not fine structural change such as protein cavity but global structural change such as reorientation of domains and subunits or reorganization of fluctuating portion governed by weak

T. Fujisawa (✉)
Department of Chemistry and Biomolecular Science, Gifu University, 1-1 Yanagido, Gifu 501-1193, Japan
e-mail: fujisawa@gifu-u.ac.jp

K. Akasaka, H. Matsuki (eds.), *High Pressure Bioscience*,
Subcellular Biochemistry 72, DOI 10.1007/978-94-017-9918-8_30

interaction forces (Koch et al 2003; Svergun 2010). SAXS under high pressure (HP-SAXS), therefore has provided unique structural information: Protein folding of small globular proteins (Panick et al. 1998; Spinozzi et al. 2007), subunit interaction of oligomeric proteins (Fujsiawa et al. 1999; Skouri-Panet et al. 2006), stability of filamentous protein (Popp et al. 2008) and more recently protein-protein interactions (Möller et al. 2012; Ortore et al. 2009; Schroer et al. 2012). However, the number of applications seems still limited, compared to those at ambient pressures. This mainly comes from the difficulties of HP-SAXS instrumentation.

HP-SAXS on diffracting system such as lipid membrane and polymers started rather earlier and mostly used neutron sources and anvil cell (Winter 2002). Initial data collection of HP-SAXS from protein solution started with anvil cell (Czeslik et al 1996), and later shifted to high pressure cell (HP-cell) similar to optical cell (Kato and Fujisawa 1998; Pressl et al. 1997). Since 1990s several designs for HP-SAXS including our group have been reported (Brooks et al. 2010; Möller et al. 2012; Nishikawa et al. 2001; Skouri-Panet et al. 2006) and most of all, Ando et al. well documented the design of HP-cell (Ando et al. 2008). The present chapter focuses on the key consideration for HP-cell design and basic manipulation of HP-SAXS data that are rarely documented.

30.2 Instrumentation for HP-SAXS

30.2.1 Overview of HP-SAXS Experiment and Experimental Setup

HP-SAXS measurement usually uses synchrotron facility with 8–14 keV X-rays because of the beam size and brilliance. Lab X-ray sources are scarcely used. In practice, HP-SAXS experiments were undertaken by replacing standard sample cell holder by HP-cell at beamline station. In any facilities, the workflow of HP-SAXS experiment is similar to that of standard SAXS at ambient pressure. Sample and buffer data collection is made by exchanging the solutions of the cell. Sample change is mostly done inside the radiation shielding hutch, dismounting a high pressure cell from the stage, disassembling the parts, washing and inserting sample solutions, reassembling the cell, then setting back to the position exactly the same as before (Fujisawa et al. 2000). In HP-SAXS, there are two ways of sample exchange procedures depending on cell design. One is direct injection of sample solution into the cell (Nishikawa et al. 2001; Pressl et al. 1997) and the other is replacing inner sample cell inside cell chamber (Ando et al. 2008; Brooks et al. 2010; Skouri-Panet et al. 2006). In either case, size and weight of the HP-cell should not far exceed those of standard cell and the workflow described above should apply. The high-pressure network, therefore, results in a system consisting of high-pressure control part and sample observing part. Figure 30.1 shows the experimental setup of BL45XU-SAXS at SPring-8 (Fujisawa et al. 2000). The HP-cell is connected by high-pressure tube, remote controlled valve, and pressure generating system

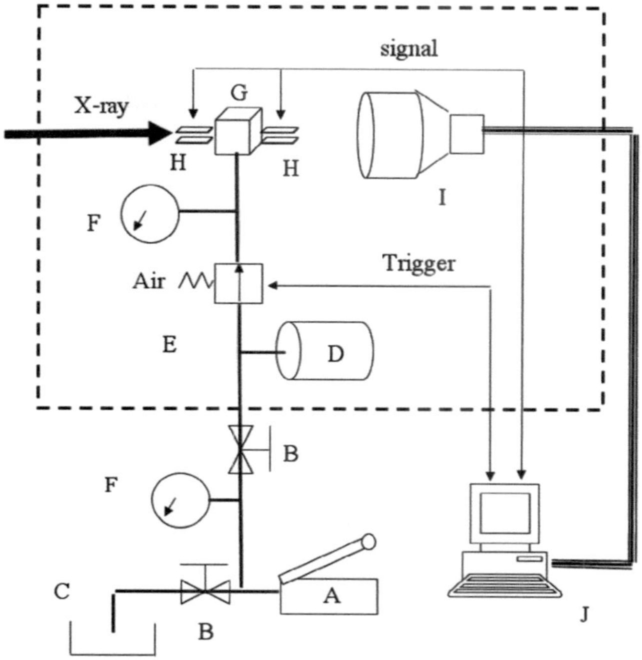

Fig. 30.1 Pressure network for HP-SAXS experiments at BL45-SAXS, SPring-8

outside radiation shielding hutch. Initially the author used HP-cell with pressure intensifier (×20) (Nishikawa et al. 2001), which reduces the pressure supply less than 25 MPa. This is economical, and HP tubes are more flexible, reducing troubles related to tube connections. At present we use more conservative setup without pressure intensifier (×1) in order to satisfy the precision of pressure setting (*ca.* 2 MPa). As for pressurizing fluid, we use water. Considering the corrosives of pump and other high-pressure parts, fluorinert or kerosene may be preferred, but it causes troubles when leakage occurs inside the hutch. Monitoring transmission with ion chambers before and after sample cell is very effective to detect problems *in situ* without processing data. The problems include leakage from window or mismatch of data collection pressure between sample and buffer.

Schematic of HP-SAXS experiment at BL45XU, SPring-8 (Fujisawa et al. 2000) is shown. Each symbols represent as follows, A: pump; B: valve; C: drain; D: HP-reservoir; E: remote valve; F: pressure gauge; G: X-ray HP chamber; H: ion chamber; I: x-ray detector; J: controller.

30.2.2 Window of HP-SAXS Cell

Sample chamber is the most difficult part of HP-SAXS instrumentation. Figure 30.2 shows our HP-SAXS cell (Minatogawa Steel Works, Kobe, Japan). The main design

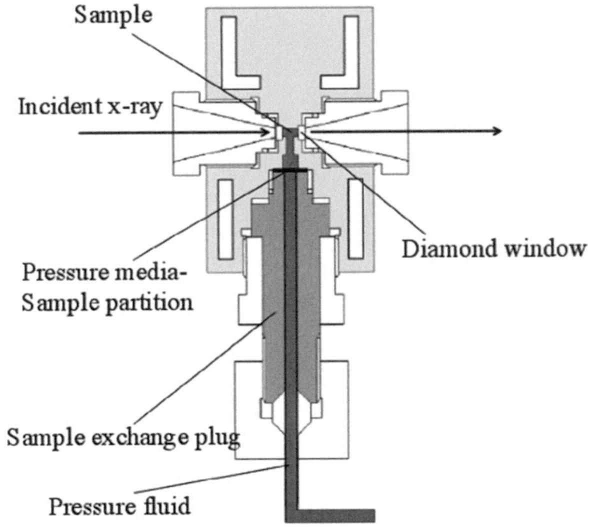

Fig. 30.2 Structure of HP-cell

issue is the cell window. The materials used for optical or neutron HP-cell, berrium and sapphire have high parasitic scattering or absorption. For SAXS from protein solutions, it is essential to reduce parasitic scattering and guarantee the identical condition in between sample and buffer data collection. At present almost all HP-cells use diamond. To obtain good size and quality of diamond is difficult, which also limits the cell design. Our cell has two synthetic diamond windows of 1 mm in thickness and 4.5 mm in diameter (synthetic diamond Type Ib, 100 plane).

30.2.3 Typical HP-Cell for HP-SAXS

The parasitic scattering of diamond window depends on two factors. The first is the types of diamond (Ando et al. 2008; Wang et al. 2012) and the second is how beam traverses the window. As for the latter, hyperbolic streaks composed of light and dark lines known as Kossel line can be seen. Figure 30.3 shows that Kossel line changed the position as stress destroyed the perfection of crystals in our case (Nishikawa et al. 2001). Adjustment of the tilt angle of the cell (up to 10 mrad) was mandatory to change the position of Kossel lines from the beam stop. By using two tapered pins on the cell stage realized the positional reproducibility within 10 μm and the tilt angle within 0.3 mrad. The introduction of adjustment design into HP-cell may be skipped when using more good diamond type such as CVD diamond (Wang et al. 2012).

The other overlooked agenda concerning cell window is the sealing. The diamond window is supported by a Poulter type (Poulter 1932). The seal between plug and widows is realized by mirror polished head of the window plug. Once the good contact between the diamond and window plug was established, the

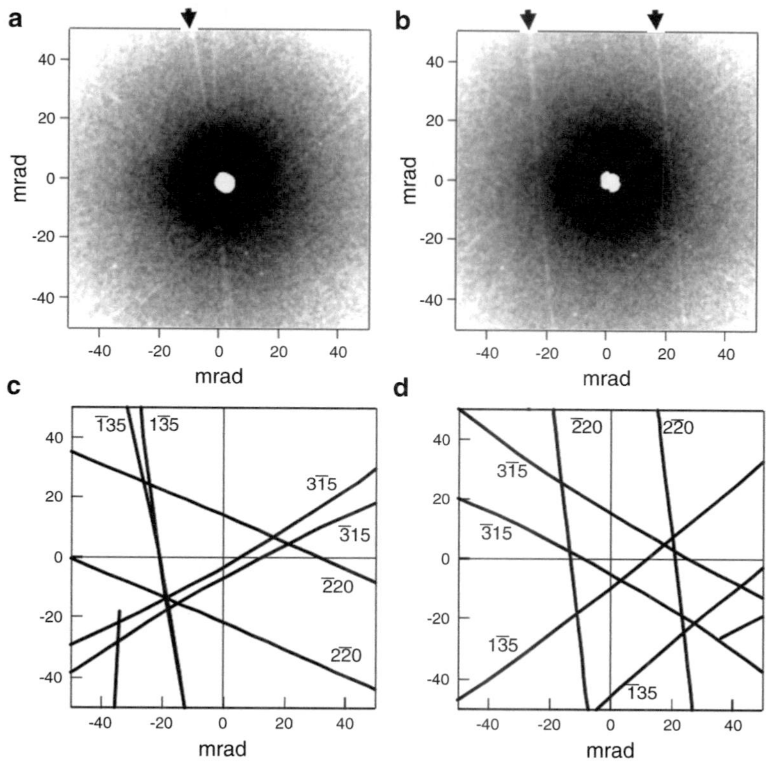

Fig. 30.3 Kossel line from diamond window material

sealing is complete. Although this is a standard technique for optical cell, the window plug is much smaller than optical cell so that the fabrication of contact face seems sometimes insufficient. Besides, HP-SAXS sample exchange duty cycle accompanies frequent compression and decompression cycle which is achieved by small pressure steps (*ca.* 2 MPa). This experimental procedure hinders the use of a seal bracket like anvil cell and sometimes resulted in pressure leaks from the window seal. The author therefore designed the cell not to touch window part when loading samples.

Stress on the window perturbs the perfection of diamond crystals, which resulted in Kossel lines. When the line cross the beam stop, parasitic scattering greatly increases. The figure was modified by Nishikawa et al. (Nishikawa et al. 2001).

30.2.4 Isolation of HP-SAXS Cell

The final agenda on the cell design is the method of sample isolation. It is always necessary to separate sample solution and pressurizing fluid. Our former high-pressure cell has a pressure intensifier (×20) inside the cell, which was realized by a free piston, whose inner and outer area are asymmetric. That piston also serves as a separator for sample and pressurizing fluid (Nishikawa et al. 2001). The design was effective, when the measurement is done at single high pressure (over 100 MPa) for long duration, but not suitable for setpwise increase of pressure with small pressure increment. The control of pressure in low-pressure range was deteriorated by the time since packing friction of the free piston changed. The author therefore currently uses a membrane partition (parafilm) between the sample and pressure medium placed on the top of the sample exchange plug pushed by O-ring (Skouri-Panet et al. 2006). When glue is used, its selection should be cared because most glues are easy to come unglued under high pressure condition.

30.2.5 Specification of Our HP-SAXS Cell

A few additional remarks on cell design are described below. We set our temperature range of the cell to be biologically relevant, from −20 °C to 60 °C. This widens the selection of seal materials. Maximum pressure is 400 MPa. In order to satisfy the identical condition, at least 2 MPa of pressure precision is required so that plasticity of the seal becomes prerequisite. The outer dimension of the HP-cell is roughly 7 cm × 4 cm × 4 cm and its weight is about 1 kg. The path length was 4 mm. The sample cell volume is 50 μL, which is used with a spacer. The cell is made of stainless, SUS630 alloy, which has a high yield strength and highly machinable feature. Inconel 725 alloy was also used by other group (Ando et al. 2008). The structure of our cell is so simple that several minutes are sufficient to exchange sample by experienced hands. But there are some drawbacks. First, to make a compact and one body HP-cell requires a well-skilled craftsperson. Second, in our design metal seals such as anti-extrusion metal rings are prohibited because protein buffers are often corrosive. Third, HP-SAXS experiments sometimes end up with irreversible aggregative deposition, which follows troubles of cleaning. Alternatively, the use of inner cell becomes more popular as described above (Ando et al. 2008; Brooks et al. 2010; Skouri-Panet et al. 2006). The inner cell certainly eliminates sample volume (less than 20 μL) and realizes faster sample change. It also protects corrosion from biological buffers. Its drawbacks may be more window materials to be traversed by beam. Careful inspection of the identity of parasitic scattering from window materials among inner cells should be undertaken (Wang et al. 2012).

30.3 Preliminary Data Manipulation of HP-SAXS Data

30.3.1 Change of Intensity Inherit to Compression

HP-SAXS intensities depend on pressure because of the contrast change (Ando et al. 2008; Kato and Fujisawa 1998). The origin of SAXS arises from the density

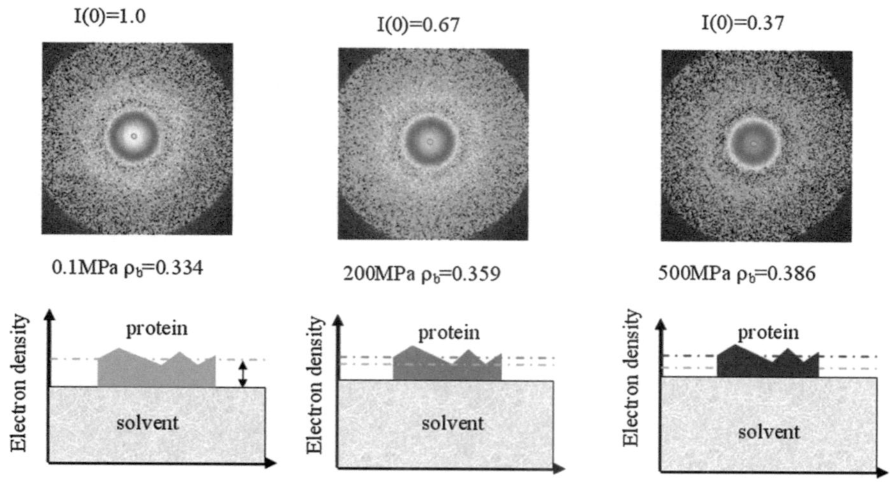

Fig. 30.4 Change of SAXS intensities upon compression

difference between protein and solvent, *i.e.*, contrast. When the protein solution is compressed, the compressibility of water (*ca.* 4.5×10^{-4} MPa^{-1}) is much larger than protein (*ca.* 10^{-4} MPa^{-1}) so that the contrast decreases as pressure increases (Fig. 30.4) (Fujisawa et al. 2002; Gekko and Noguchi 1979; Spinozzi et al. 2007). It should be remarked that the contrast decrease should also occur in protein crystals. Upon compression the intensities of SAXS decreases even if the shape of scattering object is unchanged. In order to study the change of protein shape by high pressure, it is necessary to calibrate the contrast effect. Spinozzi et al. used global fitting technique, in which the compressibilities of protein and water were determined for a series of HP-SAXS data under deliberate assumptions (Spinozzi et al. 2007). Some HP-SAXS data were simply calibrated by matching the high q region of HP-SAXS (Popp et al. 2008), but mostly not clearly documented. We designed a conventional calibration procedure for the contrast change as follows.

Upper panels show the real buffer subtracted images of HP-SAXS data from cytochrome-c (Fujisawa et al. 2002). The electron densities of water at corresponding pressures are listed. Lower panels depict how contrast change occurs.

30.3.2 Contrast Change by High Pressure

In order to characterize the contrast problem quantitatively, we first exemplify HP-SAXS of BSA where the structural change of BSA by high pressure is so small that the observed decrease of intensity can be solely ascribe to contrast change. In this case, the shape of SAXS profile is conserved, while the total intensity decreases. Figure 30.5 shows how HP-SAXS from BSA solution changes with pressures. For clarity, Kratky plot (q vs. $q^2I(q)$ plot) was employed (q is the scattering vector

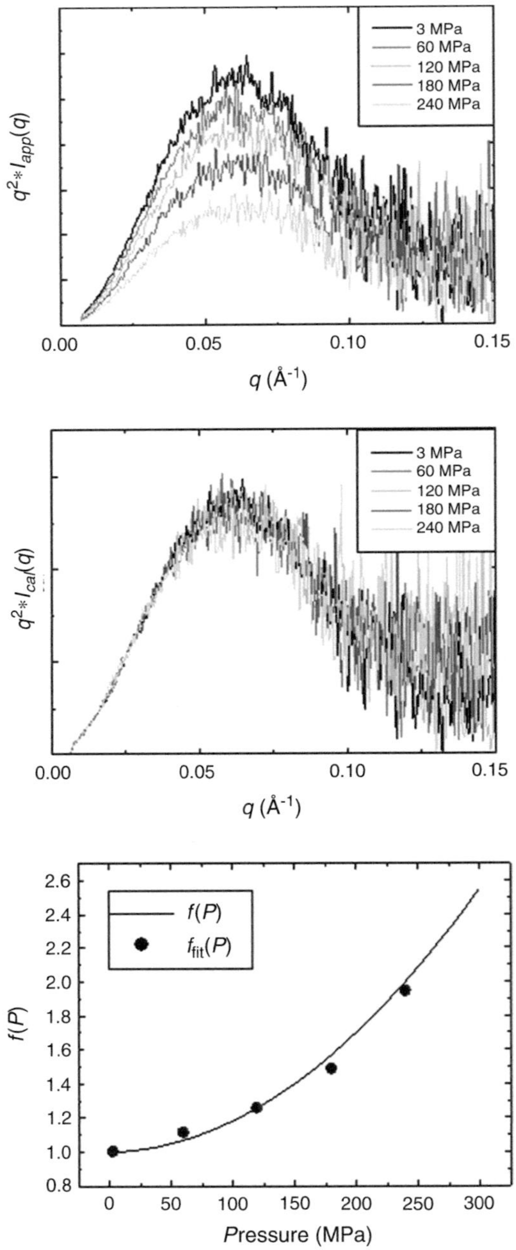

Fig. 30.5 Example of contrast change of HP-SAXS data

defined by $q = 4\pi \sin\theta/\lambda$, 2θ; scattering angle, λ; wave length). The height of the peak of Kratky plot decreases upon compression, while they coincides with each other by multiplying coefficient $f_{\mathrm{fit}}(P)$ at pressure P, as defined below.

$$f_{\mathrm{fit}}(P) = \frac{\sum_{k-1}^{N} I\left(q_k, P\right) I\left(q_k, 1\right)}{\sum_{k-1}^{N} I^2\left(q_k, P\right)}, \quad (30.1)$$

where $I(q,P)$ and $I(q,1)$ are $I(q)$ at P and 0.1 MPa, respectively. Figure 30.5 clearly points that the shape of $I(q)$ is conserved while its intensity decreased and the $f_{\mathrm{fit}}(P)$ can be a measure for calibration in this case. We set the correction coefficient $f(P)$ to accommodate with $f_{\mathrm{fit}}(P)$ as follows,

$$\begin{aligned} f(P) &= \frac{V_{\mathrm{Porod},P} / V_{\mathrm{Porod,1bar}}}{I\left(0, P\right) / I\left(0, 1\right)} \\ V_{\mathrm{Porod}} &= \frac{2\pi^2 I(0)}{\int_0^{\pi} q^2 I(q) dq} \end{aligned} \quad (30.2)$$

where $I(0,P)$ is the scattering intensity of zero angles at P pressure and V_{porod} is the scattering volume, so called, Porod volume (Guinier and Fournet 1955; Koch et al. 2003; Svergun 2010). $I(0,P)$ is proportional to square of excess scattering density, *i.e.*, contrast and easily calculated by above equation. Since protein volume is directly correlated with molecular weight, the numerator of Eq (30.2) is the molecular weight correction term which compensates the change of $I(0)$ arisen from the molecular weight. This contrast effect of high pressure depends on the solvent composition (data not shown) so that the correction should be made for every buffer conditions.

Typical example of HP-SAXS data from a single peptide globular protein, which is supposed to be unchanged by this pressure range. From top to the bottom: raw HP-SAXS data from BSA solution (2 mg ml^{-1}, Exposure time is 1 s); Inspection on the shape change of HP-SAXS data by multiplying $f_{\mathrm{fit}}(P)$; correlation between $f(P)$ and $f_{\mathrm{fit}}(P)$.

30.3.3 Performance of Calibration on Contrast Change by High Pressure

The demonstration of calibration based on Eq. (30.2) is presented for association/dissociation system of protein complex. Microbial nitrilase is known to homo-oligomerize upon heating (Nagasawa et al. 2000). Similarly, pressure perturbed association state, *i.e.*, molecular weight of nitrilase oligomers. Figure 30.6 shows both raw and calibrated HP-SAXS data from nitrilase. In raw HP-SAXS data (Fig. 30.6, left), both shape and intensity of $I(q)$ changed with pressure. It should be noted that $I(q)$ in $q > 0.125$ Å^{-1} differs with pressures. The high q region of SAXS

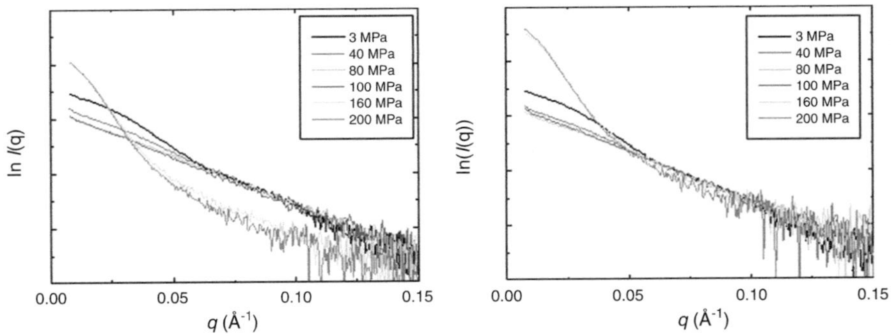

Fig. 30.6 Performance of calibration on contrast change by high pressure

profiles reflects the density fluctuation inside the molecule (Guinier and Fournet 1955). The level of $I(q)$ in that q region was constant with temperature induced association of nitrilase (data not shown), because the protein fold generally does not change drastically in most association/dissociation system. The matching of $I(q)$ at high q region, therefore, could be a measure for correct calibration. Figure 30.6 demonstrates the calibrated curves meet this measure. Ando et al. collected wide q range and calibrated curves by direct matching of high q region, rather than using $I(0)$ or V_{Porod} as described in Eq (30.2). The main advantages of the present method are as follows. First, the calibration can be made even for limited q range, which easily accommodates the beamline setup. Second, using integral value such as V_{Porod} rather than matching noisy part of high q data, seems statistically favorable. The present calibration procedure opens the limit of scattering range and enables a conventional data check during data collection.

The performance of calibration was checked by an association/dissociation system of microbial nitrilase oligomers (Nagasawa et al. 2000). Left; raw $I(q)$s have a mismatch at $0.125 < q < 0.15$ Å^{-1}, Right; calibrated $I(q)$s have the same intensity level at $0.125 < q < 0.15$ Å^{-1}.

30.4 Summary

HP-SAXS appears very attractive technique, since this is the sole methodology to obtain quaternary structure information of a large size protein complex in solution under pressure. In these decades, many designs have been developed in order to satisfy the HP-SAXS data collection. Among all, a cell designed by Ando et al. (2008) is noteworthy. Their cell consists of several assemblies of different parts, which looks easy to fabricate compared to one cell body like us. Emerging new type of diamond would improve this type of cell.

When the instrumentation becomes matured, it is more and more important to decide what sample to be measured with HP-SAXS. Obtaining beamtime at synchrotron facilities is very competitive so that lab-level experiment in advance is indispensable. Size information obtainable at ambient pressure such as size exclusion chromatography, dynamic light scattering and analytical ultracentrifugation are not easily accessible under high pressure environment. The author proposed the use of native poly acrylamide gel electrophoresis under high pressure for that purpose (Fujisawa et al. 2013) which reflect quantitative behavior of oligomeric proteins under pressure (Paladini et al. 1994).

The development of HP-SAXS technique is not only for protein research under extreme conditions but also helpful for standard measurement. For example, NMR researchers recently used high pressure as a practical tool for refolding and solubilizing sample proteins (Ogura et al. 2013). Similarly, high pressure should work improving the quality of SAXS data by which nonspecific binding of proteins is hindered. It should be noted that contrast effect by pressure affects radius of gyration only by ca. 2 % at 500 MPa (Kato and Fujisawa 1998). HP-SAXS appears to have a potential to extend the possibilities of SAXS technique and may be integrated into standard setup at beamline in future.

Acknowledgements The author thanks to Drs. Masato Moritoki and Hiroshi Matsuo for their help in designing HP-cells. Drs. Ryo Ishiguro and Takaaki Hikima and Mr. Ryosuke Noda contributed to the data collection of nitrilase. All HP-SAXS experiments were done at BL45XU-SAXS, SPring-8, Harima with the approval of RIKEN (Proposal Nos. 20080035, 20100104, 20110050, 20120089, 20130092).

References

Ando N, Chenevier P, Novak M et al (2008) High hydrostatic pressure small-angle X-ray scattering cell for protein solution studies featuring diamond windows and disposable sample cells. J Appl Crystallogr 41:167–175. doi:10.1107/S0021889807056944

Brooks NJ, Gauthe BLLE, Terrill NJ et al (2010) Automated high pressure cell for pressure jump x-ray diffraction. Rev Sci Instrum 81:064103. doi:10.1063/1.3449332

Czeslik C, Malessa R, Winter R, Rapph G (1996) High pressure synchrotron X-ray diffraction studies of biological molecules using the diamond anvil technique. Nucl Instrum Methods Phys Res Sect A 368:847–851

Fourme R, Girard E, Akasaka K (2012) High-pressure macromolecular crystallography and NMR: status, achievements and prospects. Curr Opin Struct Biol 22:636–642. doi:10.1016/j.sbi.2012.07.007

Fujisawa T, Inoue K, Oka T et al (2000) Small-angle X-ray scattering station at the SPring-8 RIKEN beamline. J Appl Crystallogr 33:797–800

Fujisawa T, Nishikawa Y, Inoko Y (2002) The use of a small-angle X-ray scattering technique with a third-generation synchrotron X-ray source in high-pressure biochemistry. In: Hayashi R (ed) Trends in high pressure bioscience and biotechnology: Prog Biotechnol, vol 19. Elsevier, Amsterdam, pp 615–620, Section: High pressure methodology

Fujisawa T, Ueda T, Kameyama K et al (2013) Analysis of oligomeric transition of silkworm small heat shock protein sHSP20.8 using high hydrostatic pressure native PAGE. High Pressure Res 33:258–264

Fujsiawa T, Kato M, Inoko Y (1999) Structural characterization of lactate dehydrogenase dissociation under high pressure studied by synchrotron high-pressure small-angle X-ray scattering. Biochemistry 38:6411–6418

Gekko K, Noguchi H (1979) Compressibility of globular proteins in water at 25 °C. J Phys Chem 83:2706–2714. doi:10.1021/j100484a006

Gross M, Jaenicke R (1994) Proteins under pressure. The influence of high hydrostatic pressure on structure, function and assembly of proteins and protein complexes. Eur J Biochem 221:617–630

Guinier A, Fournet G (1955) Small-angle scattering of X-rays. Wiley, New York

Heremans K, Smeller L (1998) Protein structure and dynamics at high pressure. Biochim Biophys Acta 1386:353–370

Kato M, Fujisawa T (1998) High-pressure solution x-ray scattering of protein using a hydrostatic cell with diamond windows. J Synchrotron Radiat 5:1282–1286

Koch MH, Vachette P, Svergun DI (2003) Small-angle scattering: a view on the properties, structures and structural changes of biological macromolecules in solution. Q Rev Biophys 36:147–227

Möller J, Schroer MA, Erlkamp M et al (2012) The effect of ionic strength, temperature, and pressure on the interaction potential of dense protein solutions: from nonlinear pressure response to protein crystallization. Biophys J 102:2641–2648. doi:10.1016/j.bpj.2012.04.043

Nagasawa T, Wieser M, Nakamura T et al (2000) Nitrilase of Rhodococcus rhodochrous J1. Conversion into the active form by subunit association. Eur J Biochem 267:138–144

Nishikawa Y, Fujisawa T, Inoko Y, Moritoki M (2001) Improvement of a high pressure cell with diamond windows for solution X-ray scattering of proteins. Nucl Instrum Methods Phys Res Sect A 467–468:1384–1387

Ogura K, Kobashigawa Y, Saio T et al (2013) Practical applications of hydrostatic pressure to refold proteins from inclusion bodies for NMR structural studies. Protein Eng Des Sel 26:409–416. doi:10.1093/protein/gzt012

Ortore MG, Spinozzi F, Mariani P et al (2009) Combining structure and dynamics: non-denaturing high-pressure effect on lysozyme in solution. J R Soc Interface 6(Suppl 5):S619–S634. doi:10.1098/rsif.2009.0163.focus

Paladini AA, Weber G, Erijman L (1994) Analysis of dissociation and unfolding of oligomeric proteins using a flat bed gel electrophoresis at high pressure. Anal Biochem 218:364–369. doi:10.1006/abio.1994.1193

Panick G, Malessa R, Winter R et al (1998) Structural characterization of the pressure-denatured state and unfolding/refolding kinetics of staphylococcal nuclease by synchrotron small-angle X-ray scattering and Fourier-transform infrared spectroscopy. J Mol Biol 275:389–402. doi:10.1006/jmbi.1997.1454

Popp D, Narita A, Oda T et al (2008) Molecular structure of the ParM polymer and the mechanism leading to its nucleotide-driven dynamic instability. EMBO J 27:570–579

Poulter TC (1932) Apparatus for optical studies at high pressure. Phys Rev 860–871

Pressl K, Kriechbaum M, Steinhart M, Laggner P (1997) High pressure cell for small- and wide-angle x-ray scattering. Rev Sci Instrum 68:4588. doi:10.1063/1.1148436

Schroer M, Tolan M, Winter R (2012) Exploring the thermodynamic derivatives of the structure factor of dense protein solutions. Phys Chem Chem Phys 14:9486–9491. doi:10.1039/c2cp41041a

Silva JL, Foguel D (2009) Hydration, cavities and volume in protein folding, aggregation and amyloid assembly. Phys Biol 6:015002. doi:10.1088/1478-3975/6/1/015002

Silva JL, Weber G (1988) Pressure-induced dissociation of brome mosaic virus. J Mol Biol 199:149–159

Skouri-Panet F, Quevillon-Cheruel S, Michiel M et al (2006) sHSPs under temperature and pressure: the opposite behaviour of lens alpha-crystallins and yeast HSP26. Biochim Biophys Acta 1764:372–383. doi:10.1016/j.bbapap.2005.12.011

Spinozzi F, Mariani P, Saturni L et al (2007) Met-myoglobin association in dilute solution during pressure-induced denaturation: an analysis at pH 4.5 by high-pressure small-angle X-ray scattering. J Phys Chem B 111:3822–3830. doi:10.1021/jp063427m

Svergun DI (2010) Small-angle X-ray and neutron scattering as a tool for structural systems biology. Biol Chem 391:737–743. doi:10.1515/BC.2010.093

Wang S, Meng Y-F, Ando N et al (2012) Single-crystal CVD diamonds as small-angle X-ray scattering windows for high-pressure research. J Appl Crystallogr 45:453–457. doi:10.1107/S0021889812010722

Winter R (2002) Synchrotron X-ray and neutron small-angle scattering of lyotropic lipid mesophases, model biomembranes and proteins in solution at high pressure. Biochim Biophys Acta 1595:160–184

Chapter 31
High Pressure Macromolecular Crystallography

Nobuhisa Watanabe

Abstract X-ray crystallography is a powerful tool for the high resolution structural study of biomacromolecules. One of the most important beneficial feature of the method is the possibility of the direct observation of hydration water structure. In recent years, significant development in high-pressure macromolecular crystallography (HPMX) using a diamond anvil cell (DAC) has been performed in combination with shorter wavelength X-ray of synchrotron radiation. The number of protein structures determined by HPMX at pressure ranging from several hundred MPa to 1 GPa is gradually growing. In this chapter, we describe DAC with its usage, and then an example of HPMX study on hydration structure of 3-isopropylmalate dehydrogenase (IPMDH).

Keywords Diamond anvil cell • DAC • 3-Isopropylmalate dehydrogenase • IPMDH

31.1 Historical Background

High-pressure study uses various experimental techniques including spectroscopies, NMR, a small angle scattering, etc. Among them X-ray crystallography is a unique method which provides high resolution structural informations including structure of hydration waters. Therefore, it is challenging to develop high-pressure crystallography, HPMX, and use it to study protein structures with pressure perturbation (Fourme et al., 2001; Kurpiewska and Lewinski, 2010). However, there are some specific problems which should be considered using HPMX. For example, protein crystals are fragile and susceptive to dehydration and should be kept in the buffer where the crystal was grown with precipitant. The number of high-pressure structures deposited in the Protein Data Bank is still small, but

N. Watanabe (✉)
Synchrotron Radiation Research Center, Nagoya University, Furo-cho Chikusa-ku, Nagoya 4648603, Japan
e-mail: nobuhisa@nagoya-u.jp

K. Akasaka, H. Matsuki (eds.), *High Pressure Bioscience*, Subcellular Biochemistry 72, DOI 10.1007/978-94-017-9918-8_31

several macromolecular structures, from short DNA (Girard et al., 2007b) to large biological molecular assemblies such as viruses capsid (Girard et al., 2005), have been successfully determined at high pressure.

Historically the pressure cells for X-ray diffraction studies were made of beryllium, which allows x-rays to penetrate. Pioneering study using the beryllium cell was done by Kundrot and Richards in 1980, and the crystal structure of hen egg-white lysozyme was observed at up to 100 MPa (Kundrot and Richards, 1987). There is a limit of the upper pressure at about 200 MPa and the sample crystal in the cell is not visible, nonetheless, variants of the beryllium cells are used by several groups (Collins et al., 2005; Suzuki et al., 2010; Urayama et al., 2002).

31.2 Diamond-Anvil Cell (DAC)

Another high-pressure device, mostly used in geological research nowadays, is a diamond anvil cell (DAC). Various types of DAC's are optimized for the respective measurement techniques. In 1996, Katrusiak and Dauter used DAC for HPMX for the first time (Katrusiak and Dauter, 1996). A large variety of them for X-ray studies are commercially available, each cell being specifically designed for particular applications.

31.2.1 DAC for HPMX

Many improvements and optimizations of the DAC for protein crystallography have been performed by Fourme and co-workers (Girard et al., 2007a). The DAC's used in HPMX consist of hard metal body, two opposed diamonds with support plates, and the metal gasket, see Fig. 31.1.

Important development of the DAC is a gasket method to hold liquid materials in a sample chamber. The gasket is a thin metal sheet with a small hole. Two opposed diamonds and the gasket form the sample chamber in which the protein crystal is kept in the buffer. The solvent content of protein crystals is a ranges between 20 % and 80 % and susceptible to dehydration. By placing a buffer with the crystal in the sample chamber, dehydration will be avoided and the buffer also acts as a pressure medium that transmits the compression.

The protein crystal is compressed with increasing pressure generated by the squeezing of the chamber volume between the two diamond's culets. The principles of the DAC's for MPMX can be found in preceding reviews (Fourme et al., 2009; Kurpiewska and Lewinski, 2010). Diamond anvils are polished single crystal diamonds. They are usually glued to the metal or hard ceramic support discs. Typical diamonds for HPMX are about 3 mm height with a large culet diameter. A conical mount shape of the diamond (Boehler-Almax design anvil) (Boehler, 2006), which provide a useful aperture larger than the conventional mount shape, is used for HPMX (Girard et al., 2007a). The design allows for collecting data over a large rotation angle up to approximately 70°.

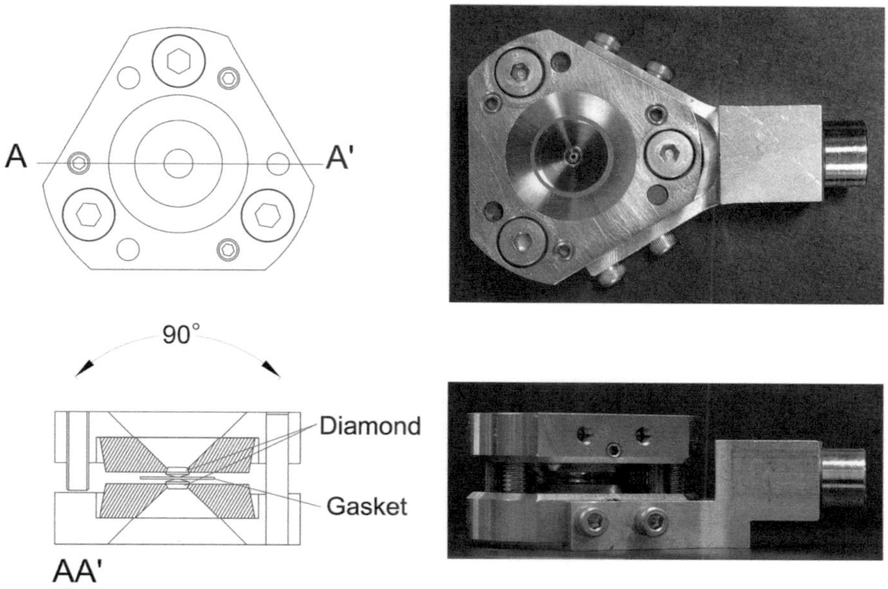

Fig. 31.1 Diagrams and photographs of the Merrill-Bassett type DAC. With the Boehler-Almax anvil, typical optical aperture angles of the DAC are 90°, and effective rotation angle for the data collection is up to approximately 70°

The gasket is a metallic foil placed between culets of two diamonds. The sample chamber is made by drilling a hole at the center of the gasket, and the volume must be large enough to accommodate protein crystal. Since the diameter of the gasket hole decreases slightly with compressing, the thickness of the gasket and the size of the hole confine the dimensions of the studied protein crystals. We usually use the Merrill-Bassett type DAC (Merrill and Bassett, 1974), and the anvil has a large culet diameter of 1 mm. Two gasket types, 0.3 mm thick stainless steal with 0.7 mm hole or 0.5 mm thick copper with 0.6 mm hole, are selected depending to sample crystal sizes.

Pressure inside the sample chamber at data collection must be monitored. The standard method to monitor pressure is a wavelength shift of fluorescence of ruby R_1-line (694.25 nm at atmospheric pressure). The shift of R_1-line is linearly proportional to the pressure as 2.74 GPa nm^{-1} (Piermarini et al., 1975). Unfortunately, however, it remains a challenge to determine low pressures in the 0.1 MPa–1 GPa pressure domain for lack of appropriate calibrants.

31.2.2 Crystal Mounting

To perform the HPMX experiment, the protein crystal is transferred from the crystallization drop, where the crystal has grown, to the DAC sample chamber.

The crystal must be transported and encapsulated in the chamber as quickly as possible in order to prevent crystal dehydration. In addition, protein crystals are only stable at specific crystallization conditions (buffer pH, precipitant contents and concentrations, etc.). The buffer solution used for pressure medium should be better suited for the crystal. A potential solution is using the crystallization mother liquor as-is. However, the solubility of protein crystals depends on pressure, and decrease or increase in solubility with pressure is unpredictable (Sazaki et al., 1999; Waghmare et al., 2000). Usually, a buffer with higher precipitant concentrations will be used, but the conditions should be tested and confirmed before HPMX experiment.

Some salts generally used as precipitant, for example ammonium sulfate, will crystallize under high pressure and causes trouble in the sample chamber (Kurpiewska and Lewinski, 2010). Moreover, the pH of buffers will be shifted with pressure (Kitamura and Itho, 1987). Searching crystallization conditions appropriate to the high-pressure experiment is necessary.

Since the protein crystal in the sample chamber is floating in the buffer solution which acts as a pressure medium, some sort of means for immobilization of the crystal is necessary. Katrusiak and Dauter had used cotton wool fiber (Katrusiak and Dauter, 1996). As shown in Fig. 31.2, we use fibers in a cigarette filter for this purpose (Nagae et al., 2012).

One of the unfavorable aspect of high pressure data collection using the DAC is the limited aperture. It is not possible to fully rotate the crystal to get diffraction images. Only a crystal with higher symmetry space group can be measured for a high completeness data set with one crystal with less than 90° rotation. For crystals of lower symmetry, it is necessary to use several differently oriented crystals and merge the data to make them a complete data set. In such cases, the use of splinters made by boron nitride or diamond, which help to change the sample crystal orientation, is proposed (Girard et al., 2007a). The pressure setting of the DAC is poorly-reproducible, and it is not possible to collect diffraction data again at the same pressure within a few MPa difference. However, the large sample chamber

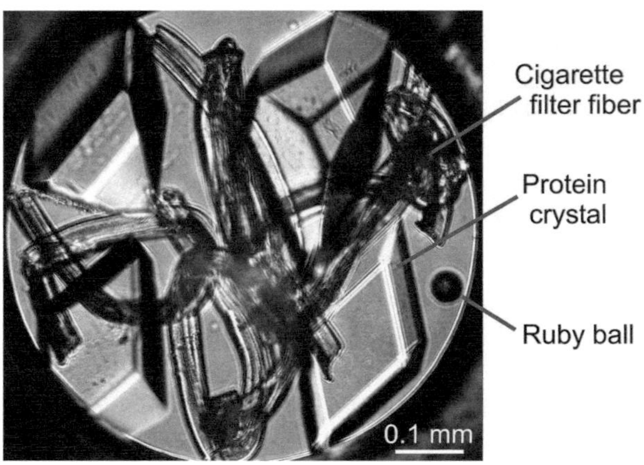

Fig. 31.2 Photograph of the crystals in the sample chamber of DAC. Four protein crystals are tiled with different orientations from each other, and fixed in position using the knotted cigarette-filter fibers. A small ruby ball for the pressure standard is seen at the right side of the chamber. In this case, the diameter of the sample chamber is 0.7 mm

of the DAC allows accommodating several crystals together. As shown in Fig. 31.2, our cigarette-filter fiber technique is also effective for keeping those crystals with different orientations.

31.2.3 Data Collection

In the HPMX experiment, a shorter wavelength X-ray is used for reducing absorption by the diamonds. Using shorter wavelength is also beneficial for high resolution data collection with the limited aperture angle of the DAC, because the diffraction angle depends on resolution after the Bragg's Law. Fourme and co-workers developed instrumentations for HPMX using synchrotron radiation and reported in several papers (Fourme et al., 2001, 2009, 2011). They use ultra-short-wavelength X-rays of 0.3–0.4 Å range at the ESRF (European Synchrotron Radiation Facility). However, the diffraction intensity of crystals decline proportional to the cube of wavelength (λ^3). If we use shorter wavelength, the reduced scattering power should be compensated by using a highly collimated intense beam. Since the HPMX experiment uses a hydrostatic compression, the data collection will be performed at room temperature. In such condition, rapid crystal degradation is unavoidable. The damage at the irradiated regions of the crystal are even visually recognized (Fourme et al., 2001, 2009). For such occasions, they expose different part of the sample crystal successively during data collection to alleviate the degradation of the crystal. Using more than one crystal for collecting the complete data set is also suitable to alleviate the degradation of crystals induced by X-ray irradiation at room temperature.

While on the other hand, our experimental setup at Beamline NW12A at the Photon Factory (PF) is unique (Chavas et al., 2013). At NW12A, the shortest wavelength available is 0.7 Å, limited by the cut-off of the focusing mirror, rather than by the X-ray undulator source. We usually use 0.75 Å with $100 \times 100\,\mu$m to $200 \times 200\,\mu$m beam sizes. Using a larger crystal becomes essential, but data collection with a relatively larger size lower density beam should quite effective to increase the crystal lifetime at room temperature. Advancing of the radiation damage of the crystals will be checked easily as an increasing mosaicity at the data processing step.

31.3 Applications: Hydration Structure of Proteins

Applications of HPMX to investigate functional sub-states of biological macromolecules are set out in Chap. 12 of this monograph. As described above, HPMX is a unique method which provides high resolution structural informations including hydration waters at a molecular surface and an internal cavity (Collins et al., 2005; Nagae et al., 2012). These hydration structure affects activity of the enzymatic functions, and study on the pressure-induced hydration may help to explain the pressure dependence of the activity of proteins (Mozhaev et al., 1996). Here we briefly describe our results on 3-isopropylmalate dehydrogenase (IPMDH) as the ones of the examples (Nagae et al., 2012).

The organisms living in the deep sea such as the Mariana Trench should be adapted to extremely high-pressure environment. Protein, for example, IPMDH from deep-sea bacteria *Shewanella benthica* DB21MT2 (SbIPMDH) remains active under such extreme conditions, while that from land bacteria *S. oneidensis* MR-1 (SoIPMDH) is inactivated, even though these two enzymes share 85 % amino-acid identity (Kasahara et al., 2009). Crystal structures of SoIPMDH have been determined at about 2 Å resolution under pressures ranging from 0.1 to 650 MPa using the HPMX technique. (PDB ID's are 3vkz, 0.1 MPa; 3vl2, 160 MPa; 3vl3, 340 MPa; 3vl4, 410 MPa; 3vl6, 580 MPa; 3vl7, 650 MPa.)

31.3.1 Water Penetration into the Internal Cavity

As shown in Fig. 31.3a, b, most of internal cavities of SoIPMDH are compressed and decrease their volumes as pressure increases. The volume of one relatively large cavity at the dimer interface also shows an initial decrease with pressure from 0.1 and 160 MPa, however, followed by a constant volume increase with further pressurization. The cavity is surrounded by hydrophobic residues belonging to both subunits of the SoIPMDH dimer (Pro120, Leu121, Ile125, Leu232, and Leu258). In parallel with the volume increase, penetration of water molecules into the cavity is observed. The difference electron density maps around the cavities at 0.1 and 580 MPa are shown in Fig. 31.3c, d. No electron density in the cavity could be observed at 0.1 MPa, while at 580 MPa two peaks of water molecules are clearly visible. The water molecules are at the hydrogen bonding distances from main-chain carbonyl oxygen atoms of Pro120 and Leu232, and the local structural changes around the cavity are induced by them. A quick estimation of the occupancy of the water molecules was performed and the estimated occupancy values of the penetrated water are 0.42 at 410 MPa, and increased to 0.51 and 0.76 at 580 and 650 MPa, respectively.

In the previously published work on L99A mutated T4 lysozyme, water penetration into the large hydrophobic cavity created by the mutation was observed at high pressure (Collins et al., 2005). In the case of T4 lysozyme, there is no need to increase the volume of the cavity in order to accommodate water molecules, because the cavity is large enough as 160 $Å^3$. While on the other hand, the volume of the cavity at the dimer interface of SoIPMDH is only 56 $Å^3$ at 0.1 MPa, and slightly reduces to 50 $Å^3$ at 160 MPa. As a consequence, there is a need to increase the volume of the cavity at higher pressures (over 340 MPa) in order to accommodate additional water molecules. At 650 MPa, the destabilization of the dimer itself might be initiated because of the penetrated water molecules in the cavity.

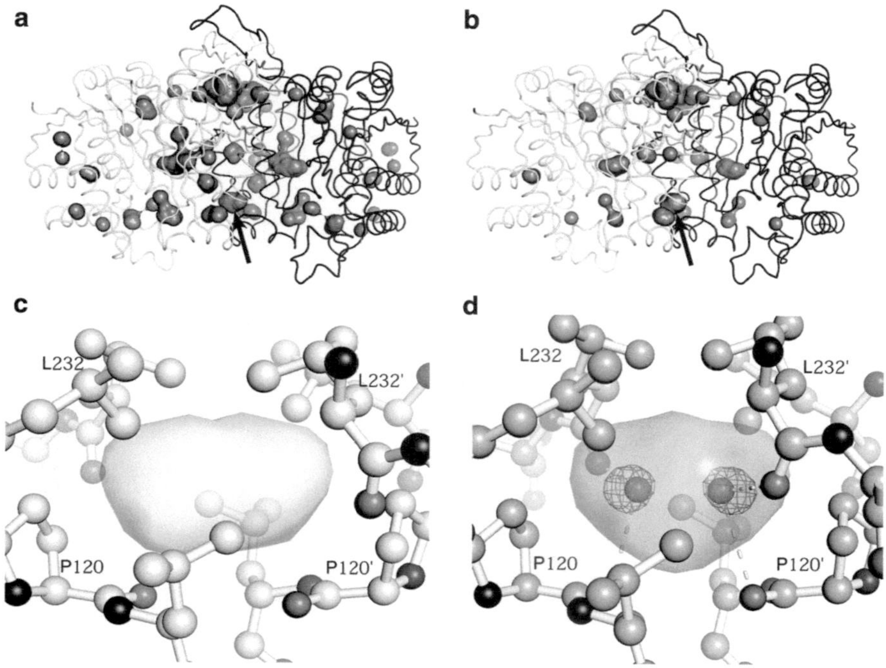

Fig. 31.3 Internal cavities of the SoIPMDH dimer and observed water penetration. (**a**) and (**b**) show internal cavities of the SoIPMDH dimer as surface representations at 0.1 and 580 MPa, respectively. The cavity at the dimer interface is indicated by an *arrow*. (**c**) and (**d**) are magnified views around the cavity at 0.1 and 580 MPa, respectively. No positive peaks are observed in (**c**) at 0.1 MPa. The transparent surfaces in (**c**) and (**d**) represent the cavities; hydrogen bonds are also shown in (**d**) (Figures are modified from Nagae et al. 2012)

31.3.2 Waters Squeezing into a Molecular Surface Cleft

The numbers of the hydration waters are increased with pressure, and the characteristic structural change is the appearance of three waters at a cleft of SoIPMDH surface opposite the active site (Fig. 31.4). At 580 MPa, the width of the cleft is expanded by the penetration of water molecules. Two water molecules O697 and O698 appeared inside the cleft, together with another water molecule O699 near there. These three waters were not observed in the structures at lower pressures ranging from 0.1 to 410 MPa. The widening of the cleft together with the water penetration can be considered as an initial step in the pressure denaturation process of SoIPMDH.

In detail, Ser266 forms a hydrogen bond with O697, allowing anchoring of the water into the cleft. Additionally, the three new water molecules form a hydrogen bond network among each other as well as with atoms of the protein. In the structures at pressures ranging from 0.1 to 410 MPa, the surface at the bottom of the

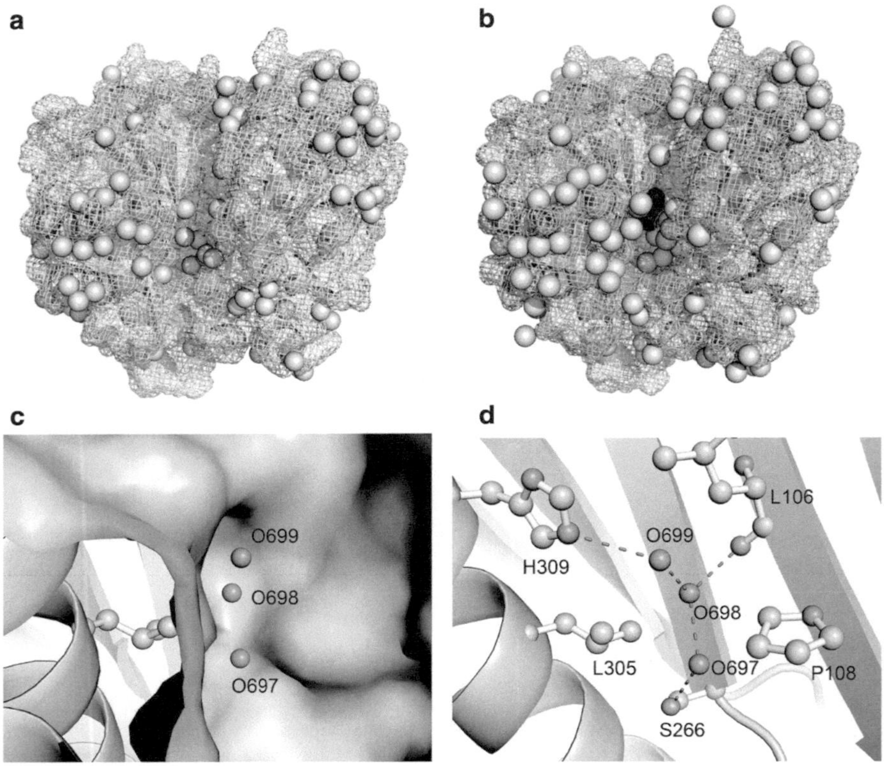

Fig. 31.4 Waters squeezing into a molecular surface cleft of SoIPMDH. (**a**) Surface waters observed at 0.1 MPa. (**b**) That of 580 MPa. The penetrating waters are drawn in *black*. (**c**) Surface representation around Pro108 and Leu305 at 580 MPa. (**d**) Hydrogen-bond network of the penetrating water molecules O697, O698 and O699 (Figures (**c**) and (**d**) are modified from Nagae et al. 2012)

cleft is non-polar since it is mostly composed of Pro108 and Leu305 side-chains. As a consequence, the penetrating water molecules, especially O697 and O698, force this non-polar surface to open. Generation of the cleft accompanied with water penetration observed in this study can be a representation of the conformational change driven by the partial molar volume reduction induced by the pressure increase. The hydrogen bond between O697 and Ser266 seems to play an important role in this structural change. In land bacterium's SoIPMDH, the bottom of the cleft is characterized by the presence of Ser266, while the deep-sea bacterium's SbIPMDH favors an alanine at the same position. A possible explanation of the pressure resistance of SbIPMDH could be that a water molecule cannot be stabilized at the same position of O697 because of the lack of a side chain in Ala266.

Acknowledgements This work was partially supported by a KAKENHI Grant-in-Aid for Challenging Exploratory Research (21657027), and has been performed with Dr. Takayuki Nagae, Nagoya University.

References

Boehler R (2006) New diamond cell for single-crystal x-ray diffraction. Rev Sci Instrum 77:115103

Chavas LMG, Nagae T, Yamada H, Watanabe N, Yamada Y, Hiraki M, Matsugaki N (2013) New methodologies at PF AR-NW12A: implementation of high-pressure macromolecular crystallography. J Synchrotron Rad 20:838–842

Collins MD, Hummer G, Quillin ML, Matthews BW, Gruner SM (2005) Cooperative water filling of a nonpolar protein cavity observed by high-pressure crystallography and simulation. Proc Natl Acad Sci USA 102:16668–16671

Fourme R, Ascone I, Kahn R, Girard E, Hoerentrup C et al (2001) High pressure protein crystallography: instrumentation, methodology and results of data collection on lysozyme crystals. J Synchrotron Radiat 8:1149–1156

Fourme R, Girard E, Dhaussy A-C, Medjoubi K, Prangé T, Ascone I, Mézouar M, Kahn R (2011) A new paradigm for macromolecular crystallography beamlines derived from high-pressure methodology and results. J Synchrotron Rad 18:31–36

Fourme R, Girard E, Kahn R, Dhaussy A-C, Ascone I (2009) Advances in high-pressure biophysics: status and prospects of macromolecular crystallography. Annu Rev Biophys 38:153–171

Girard E, Dhaussy A-C, Couzinet B, Chervin J-C, Mezouar M, Kahn R, Ascone I, Fourme R (2007a) Toward fully fledged high-pressure macromolecular crystallography. J Appl Crystallogr 40:912–918

Girard E, Kahn R, Mezouar M, Dhaussy A-C, Lin T, Johnson JE, Fourme R (2005) The first crystalstructure of a macromolecular assembly under 100 high pressure: CpMV at 330 MPa. Biophys J 88:3562–3571

Girard E, Prange T, Dhaussy A-C, Migianu-Griffoni E, Lecouvey M, Chervin J-C, Mezouar M, Kahn R, Fourme R (2007b) Adaptation of the base-paired double-helix molecular architecture to extreme pressure. Nucleic Acids Res 35:4800–4808

Kasahara R, Sato T, Tamegai H, Kato C (2009) Piezo-adapted 3-isopropylmalate dehydrogenase of the obligate piezophile *Shewanella benthica* DB21MT-2 isolated from the 11,000-m depth of the Mariana Trench. Biosci Biotech Biochem 3:2541–2543

Katrusiak A, Dauter Z (1996) Compressibility of lysozyme protein crystals by X-ray diffraction. Acta Crystallogr D52:607–608

Kitamura Y, Itho T (1987) Reaction volume of protonic ionization for buffering agents. Prediction of pressure dependence of pH and pOH. J Solut Chem 16:715–725

Kundrot CE, Richards FM (1987) Crystal structure of hen egg-white lysozyme at a hydrostatic pressure of 1000 atmospheres. J Mol Biol 193:157–170

Kurpiewska K, Lewinski K (2010) High pressure macromolecular crystallography for structural biology: a review. Cent Eur J Biol 5:531–542

Merrill L, Bassett WA (1974) Miniature diamond anvil pressure cell for single crystal x-ray diffraction studies. Rev Sci Instrum 45:290–294

Mozhaev VV, Heremans K, Frank J, Masson P, Balny C (1996) High pressure effects on protein structure and function. Proteins Struct Funct Genet 24:81–91

Nagae T, Kawamura T, Chavas LMG, Niwa K, Hasegawa M, Kato C, Watanabe N (2012) High pressure induced water penetration into 3-isopropylmalate dehydrogenase. Acta Crystallogr D68:300–309

Piermarini GJ, Block S, Barnett JD, Forman RA (1975) Calibration of the pressure dependence of the R1 Ruby fluorescence line to 195 kbar. J Appl Phys 46:2774–2780

Sazaki G, Nagatoshi Y, Suzuki Y, Durbin SD, Miyashita S, Nakada T et al (1999) Solubility of tetragonal and orthorhombic lysozyme crystals under high pressure. J Crystallogr Growth 196:204–209

Suzuki Y, Tsukamoto M, Sakuraba H, Matsumoto M, Nagasawa M, Tamura K (2010) Design of a standalone-type beryllium vessel for high-pressure protein crystallography. Rev Sci Instrum 81:084302

Urayama P, Phillips GN, Gruner SM (2002) Probing substates in sperm whale myoglobin using high-pressure crystallography. Structure 10:51–60

Waghmare RY, Pan XJ, Glatz CE (2000) Pressure and concentration dependence of nucleation kinetics for crystallization of subtilisin. J Crystallogr Growth 210:746–757

Chapter 32
High-Pressure Fluorescence Spectroscopy

Akihiro Maeno and Kazuyuki Akasaka

Abstract The combination of fluorescence and pressure perturbation is a widely used technique to study the effect of pressure on a protein system to obtain thermodynamic, structural and kinetic information on proteins. However, we often encounter the situation where the available pressure range up to 400 MPa of most commercial high-pressure fluorescence spectrometers is insufficient for studying highly pressure-stable proteins like inhibitors and allergenic proteins. To overcome the difficulty, we have recently developed a new high-pressure fluorescence system that allows fluorescence measurements up to 700 MPa. Here we describe the basic design of the apparatus and its application to study structural and thermodynamic properties of a couple of highly stable allergenic proteins, hen lysozyme and ovomucoid, using Tryptophan and Tyrosine/Tyrosinate fluorescence, respectively. Finally, we discuss the utility and the limitation of Trp and Tyr fluorescence. We discuss pitfalls of fluorescence technique and importance of simultaneous use of other high-pressure spectroscopy, particularly high-pressure NMR spectroscopy.

Keywords Fluorescence spectroscopy • Highly stable protein • Intermediate conformer • Pressure • *p-T* phase diagram • Tryptophan • Tyrosine/Tyrosinate

32.1 Introduction

Most proteins have several fluorescent amino acid residues that are tryptophan, tyrosine and phenylalanine. These fluorophores provide the intrinsic emission

A. Maeno (✉)
High Pressure Protein Research Center, Institute of Advanced Technology, Kinki University, 930 Nishimitani, Kinokawa 649-6493, Japan
e-mail: maeno@waka.kindai.ac.jp

K. Akasaka
High Pressure Protein Research Center, Institute of Advanced Technology, Kinki University, 930 Nishimitani, Kinokawa 649-6493, Japan

Present Affiliation: Graduate School of Life and Environmental Sciences, Kyoto Prefectural University, 1-5 Hangi-cho, Shimogamo, Sakyo-ku, Kyoto 606-8522, Japan
e-mail: akasaka@kpu.ac.jp

K. Akasaka, H. Matsuki (eds.), *High Pressure Bioscience*,
Subcellular Biochemistry 72, DOI 10.1007/978-94-017-9918-8_32

spectrum of a protein by excitation at different absorption wavelengths. As an advantage, intrinsic protein fluorescence is quite sensitive to chemical environment of fluorophores and can be acquired very quickly in nanosecond time range, though it only gives limited information on protein conformation based on the change in Trp environment. Therefore fluorescence spectroscopy has been widely used as a convenient approach to study conformational transition, folding, self-association and dissociation mechanism of proteins, but normally limited up to 400 MPa (Royer 2002, 2006; Munishkina and Fink 2007; Shah et al. 2012). Pressure shifts equilibria within various conformational states of a protein in solution through volume differences (ΔV) in contrast to temperature perturbing conformational equilibria through heat capacity differences (ΔC_p), which cause both enthalpy and entropy changes. Therefore we can utilize pressure perturbation to investigate the intrinsic conformational fluctuation of a protein by increasing the population of lower volume conformers or sub-states by increasing pressure among the multiple conformational sub-states with different partial molar volumes, provided that the multiple sub-states exist in marginal balance of stability and that the volume difference (ΔV) among them are sufficiently large. For some proteins in nature, these conditions are rather hard to realize, in which case pressure denaturation cannot be realized in the pressure range of 0.1–400 MPa, which is the range of pressure for most commercial high-pressure fluorescence spectrometers. We, therefore, constructed a high-pressure fluorescence spectrometer system that spans a pressure range of 0.1–700 MPa.

In this review, we illustrate the use of high-pressure fluorescence spectrometer for analyzing a thermodynamic stability of a protein (a phase diagram) on pressure (p)-temperature (T) axes and for studying a highly stable conformational intermediate. Here, we choose two proteins, hen egg white lysozyme and ovomucoid, both being well-known as strong allergenic proteins in hen egg. As probes, we adopt the maximum wavelength of fluorescence emission from naturally occurring Trp residues for hen lysozyme, but the maximum wavelength of fluorescence emission from tyrosine and tyrosinate for ovomucoid, which lacks Trp residue in its amino acid sequence.

32.2 Intrinsic Fluorophores in a Protein

32.2.1 Tryptophan Fluorescence

Aromatic amino acid residues that are contained in most proteins work as inherent fluorophores: tryptophan (Trp), tyrosine (Tyr) and phenylalanine (Phe). Among these, Trp becomes dominant probe in a protein because its extinction coefficient is by far larger than those of Tyr and Phe, though the quantum yields of Trp and Tyr are approximately the same ($\Phi = \sim 0.14$) in the neutral solution at room temperature (Chen 1967). In the case of Phe, since both the absorption coefficient

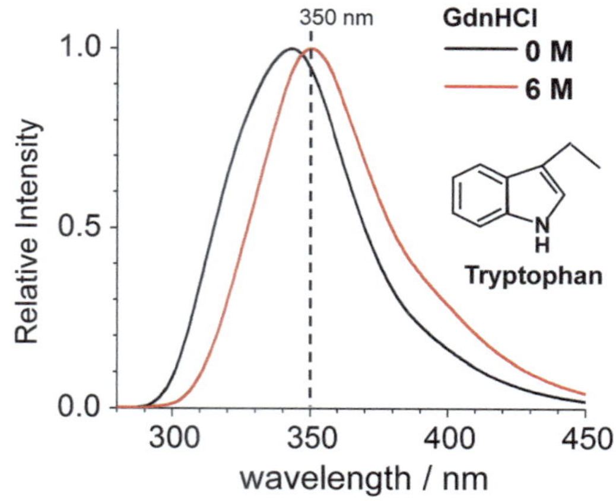

Fig. 32.1 The fluorescence spectra of tryptophan residues in the protein (hen lysozyme) at pH 7.0 at 25 °C. Tryptophan fluorescence sensitively reflects the polarity around the ring. The maximum wavelength of tryptophan emission (λ_{max}) shows *red*-shift, depending upon the exposure to the solvent. The degree of protein-unfolding induced by chemical denaturant such as guanidine hydrochloride (Gdn HCl)

and the quantum yield are both quite small, the emission from this residue is rarely observed in a protein sample. Another advantage of using Trp fluorescence is that most globular proteins have a relatively restricted number (a few to several) of Trp residues, and therefore, site-specific conformational information on the local environment may be obtained from a selective excitation of Trp residues at approximately 295 nm. Because, in the excited state, the dipole moment of the Trp ring is quite large (Pierce and Boxer 1995), the emission energy is highly sensitive to the change in polarity. Therefore, the shift in the maximum wavelength (λ_{max}) of Trp emission is considered to represent sensitively the change in the microenvironment of the Trp ring (Royer 2006). The blue-shifted wavelength (λ_{max} ~330 nm) shows that the Trp ring is located in the non-polar environment or buried in the hydrophobic core of the protein, while the red-shifted wavelength (λ_{max} ~350–355 nm) indicates that the Trp ring is located in the polar environment or exposed to the aqueous environment (Fig. 32.1) (Lakowicz 2006). Based on the variation of the emission wavelength induced by a perturbation, we can utilize Trp fluorescence as a probe of conformational transitions from the folded state to the intermediate or to the globally unfolded state in many globular proteins.

32.2.2 Tyrosine Fluorescence

We occasionally encounter proteins containing no tryptophan residue in their primary sequence. In this case, tyrosine (Tyr) is available as the dominant fluorescence probe in proteins. Figure 32.2a is the schematic view of the electronic states of a molecule and the transitions between them, called *Jablonski diagram*, drawn for the absorption and the emission of tyrosine. On the ground state S_0, under acidic or neutral pH (pH < 10.4 : pK_a), Tyr is dominantly excited at 280 nm and gives intrinsic emission at ~304 nm (Fig. 32.2a, transition *a*). For the pH of the solution

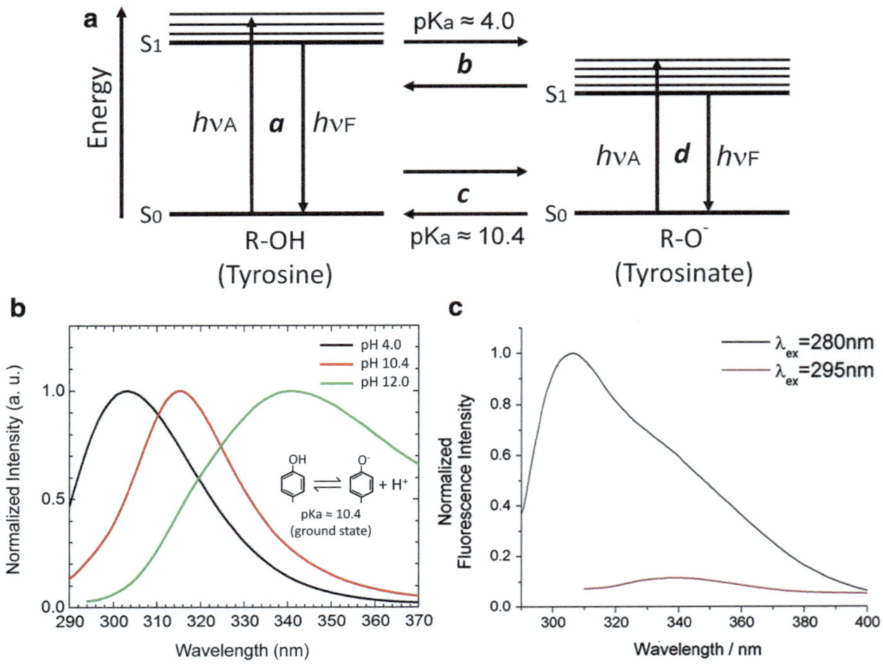

Fig. 32.2 The feature of tyrosine (Tyr) fluorescence. (**a**) Schematic view of energy-level diagram for tyrosine and tyrosinate fluorescence. (**b**) The fluorescence spectra of amino acid tyrosine at three pH values, at 0.1 MPa at 25 °C. All the excitations were made at 280 nm and the emission intensity is normalized to unity at its maximum. The emission with λ_{max} at ~304 nm is assignable to tyrosine, the emission with λ_{max} at ~350 nm to tyrosinate, and the emission with λ_{max} at ~315 nm to a tyrosine/tyrosinate mixture. (**c**) The emission spectra of tyrosine (*black line*) and tyrosinate (*red line*) in the presence of proton acceptor. Tyrosine fluorescence excited at 280 nm shows the emission with a maximum at ~304 nm and a shoulder at approximately 340 nm, while that of tyrosinate excited at 295 nm gives lower emission with maximum wavelength (λ_{max}) at 340 nm

is larger than the pK_a(10.4), the side-chain turns from tyrosine to tyrosinate due to deprotonation in the ground state (Fig. 32.2b, *inset*), and the ionized tyrosinate gives emission at ~340 nm by excitation at 280 nm (Fig. 32.2b, *green*) (O'neil and Hofmann 1987; Ross et al. 1992) (Fig. 32.2a, reaction *c* and *d*).

Interestingly, the pK_a value of the Tyr residue decreases from 10.4 to approximately 4.0 in the excited state (Fig. 32.2a, reaction *b*), provided that its life time is long. However, since the transition (emission) a is fast, only a fraction of the excited state of tyrosine goes to the excited state of tyrosinate (Fig. 32.2a, reaction *b*) (O'neil and Hofmann 1987). This means that intrinsic tyrosinate emission at 340 nm is difficult to observe under neutral pH conditions. Therefore, when the pH of the solution is below its pK_a(10.4), the fluorescence emission of Tyr occurs at λ_{max} ~304 nm by excitation at 280 nm (Fig. 32.2b, *black*) (Martin et al. 1958).

As a result, the intrinsic emission of tyrosine is insensitive to the polarity around the Tyr ring under pH conditions below its pK_a(10.4), namely λ_{max} does not vary irrespective of whether Tyr residue is buried in the interior of the protein or exposed to the solvent. On the other hand, the presence of proton-acceptor (e.g. acetic acid, phosphate, aspartic acid and glutamic acid) accelerates the deprotonation of phenolic hydroxyl group of tyrosine in the excited state (Rayner et al. 1978; Szabo et al. 1978). When the Trp-lacking protein coexists with the sufficient amount of proton acceptors, tyrosinate emission excited at 280 nm is often observed as a shoulder at near 340 nm even under neutral pH (Fig. 32.2c, *black line*) (Rayner et al. 1978). Alternatively, we can obtain tyrosinate emission independently by excitation at 295 nm (Fig. 32.2c, *red line*). The combination of tyrosine and tyrosinate fluorescence will be useful as probes for studying thermodynamics, conformational transition and kinetic reaction in Trp-lacking proteins.

32.3 High Pressure Fluorescence Apparatus Operating at 3–700 MPa: Innovated Design of Pressure-Resistive Optical Vessel and Inner Cell

The combination of fluorescence spectroscopy and pressure perturbation has been utilized as a convenient means to trap and characterize structures and thermodynamic properties of folding intermediates (Royer 2006; Mitra et al. 2007), and to discuss the contributions of volume changes of conformational transition among protein structures (Maeno et al. 2009), as well as to study pressure-induced dissociation of amyloid fibrils (Abdul Latif et al. 2007; Shah et al. 2012). While many proteins respond sensitively to pressure and often unfold below 400 MPa, some proteins are exceptionally stable and need higher pressures to denature. A class of such proteins is commonly found in proteinase inhibitors (Hiromi et al. 1985) and allergenic proteins (Astwood et al. 1996). In order to resolve such experimental difficulty, we have recently developed a high-pressure fluorescence spectrometer, for which the pressure limit is extended from 400 to 700 MPa. This methodological advance allows pressure-resistive protein to become a target of investigation. Here, we show some points of improvements in our high-pressure fluorescence system, the detail of which will be published elsewhere.

We have developed a high-pressure fluorescence spectrometer system in collaboration with Syn-Corporation. Co. Ltd. (Kyoto, Japan) (Fig. 32.3). The system consists of a pressure-resistive optical vessel (Fig. 32.3b-2 and 32.3c) and the inner cell parts (Fig. 32.3d), and was added to a conventional high-pressure fluorescence system (FP-6500, JASCO Inc., Japan) (Fig. 32.3a). Though the detailed design of the body of pressure-resistive optical vessel is not shown here, the material of the body part is stainless steel (JIS SUS630). This vessel has three sapphire windows for excitation light or emission light to penetrate, each attached to an optical plug.

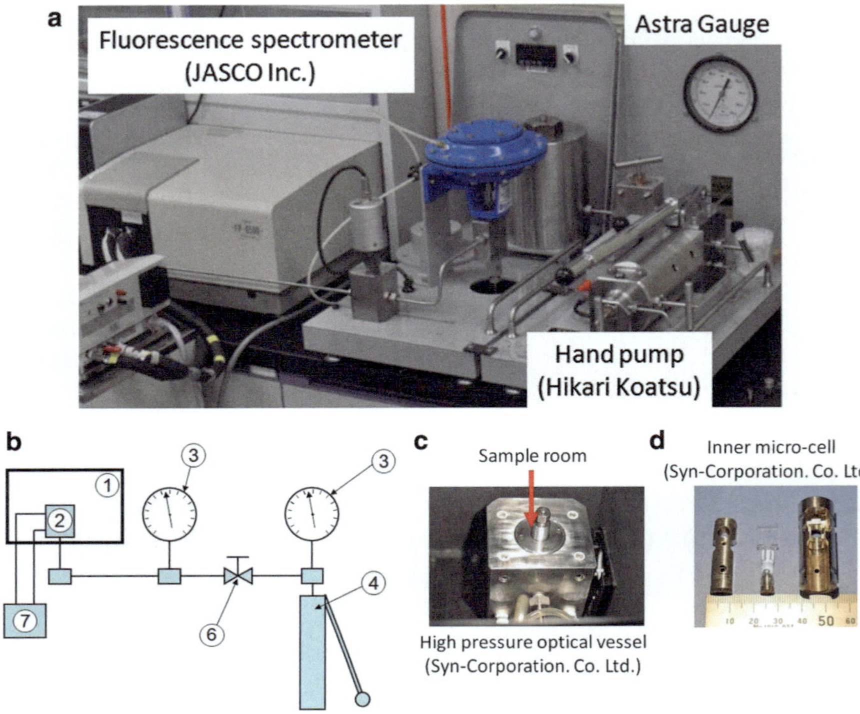

Fig. 32.3 (**a**) and (**b**) Overview of our improved high-pressure fluorescence system available between 3 and 700 MPa. (**c**) High pressure optical vessel made by SUS 630 and (**d**) the optical inner micro-cell made by quartz (*center*) and its holder parts (*right* and *left*)

When the internal pressure of vessel reaches 700 MPa, these optical plugs are sustained by using three grand nuts since considerable force is applied to each plug. Upper portion plug is the lid of the high-pressure optical vessel, and the holder to fix an inner optical cell is attached to this plug in the vessel (arrow in Fig. 32.3c). The optical inner micro-cell corresponding to sample chamber consists of two parts. One is an optical cell made of doped fused silica glass and another is a diaphragm which was made of polytetrafluoroethylene (PTFE) (Fig. 32.3d, center part). The total volume of sample chamber is about 0.2 cm^3. When the inner optical cell is compressed by hydrostatic pressure, the cylinder of a diaphragm of inner cell will make a dent. Therefore, the wall of this cylinder is made very thin, the thickness of which is about 0.25 mm. Importantly, in order to maintain the reproducibility of the cell attachment position, this inner optical cell is tightly secured in an inner cell holder (Syn-Corporation. Co. Ltd., Japan) (Fig. 32.3d, left and right metal parts).

32.4 Application to Highly Stable Proteins

Knowledge on thermodynamic stability and conformational state of a globular protein forms a basis for understanding its function, folding as well as misfolding into amyloid fibrils (Moriarty and Colon 2005). In addition to the characterization of them on the temperature axis (Privalov 1989, 1990), the characterization of the same on the pressure axis is increasingly important in basic protein science (Zipp and Kauzmann 1973; Nash and Jonas 1997; Akasaka 2006) as well as in the applied protein science (Hayashi 2002). Here, we introduce the methodology to characterize thermodynamic property and/or high-energy intermediary state of highly stable allergenic proteins in hen egg white, using tryptophan and tyrosine/tyrosinate fluorescence spectroscopy in the pressure range from 3 to 700 MPa. Fluorescence spectroscopy up to 700 MPa was pioneered by Hoa et al. (1982, Hoa and Marden 1982) to study heme proteins. Hen egg is well known as one of food allergen, and its allergenicity usually arises because of the excessive immune response to "allergenic protein" as antigen. As common feature, they have quite stable folded conformers so that its allergenic potency remains even after perturbations of heat, pressure and chemical denaturants. Therefore, so far, basic information such as thermodynamic and conformational stability in these proteins has rarely been reported.

32.4.1 p-T Phase Diagram of Hen Lysozyme at pH 2 Using Trp Fluorescence

Hen egg white lysozyme (HEWL), consisting of 129 amino acid residues (approximately 14.4 kDa), is one of the most well characterized globular proteins as a model system both for structure and function, and is also known as a source of allergenicity in egg white. There are six tryptophan residues in the molecule, two of which (Trp62 and Trp63) are found in the β domain and the rest are found in the α domain, of which Trp28, Trp108 and Trp111 are close to the large water-containing cavity (navy spheres in Fig. 32.4c, left). Therefore, we can utilize tryptophan fluorescence averaged from them as a probe of thermodynamic analysis for global unfolding. However, in the previous study, the global unfolding of HEWL has been hardly attainable even at acidic pH and at 400 MPa which is the highest pressure available in a traditional high pressure fluorescence system, because of the high pressure-resistance of this protein. With the improved fluorescence system, the study of the pressure-induced unfolding equilibrium of HEWL was carried out at pH 2 in the pressure range of 3–700 MPa. To depict a rather complete free energy landscape of HEWL on the pressure-temperature plane, temperature was widely varied between −10 °C and 50 °C.

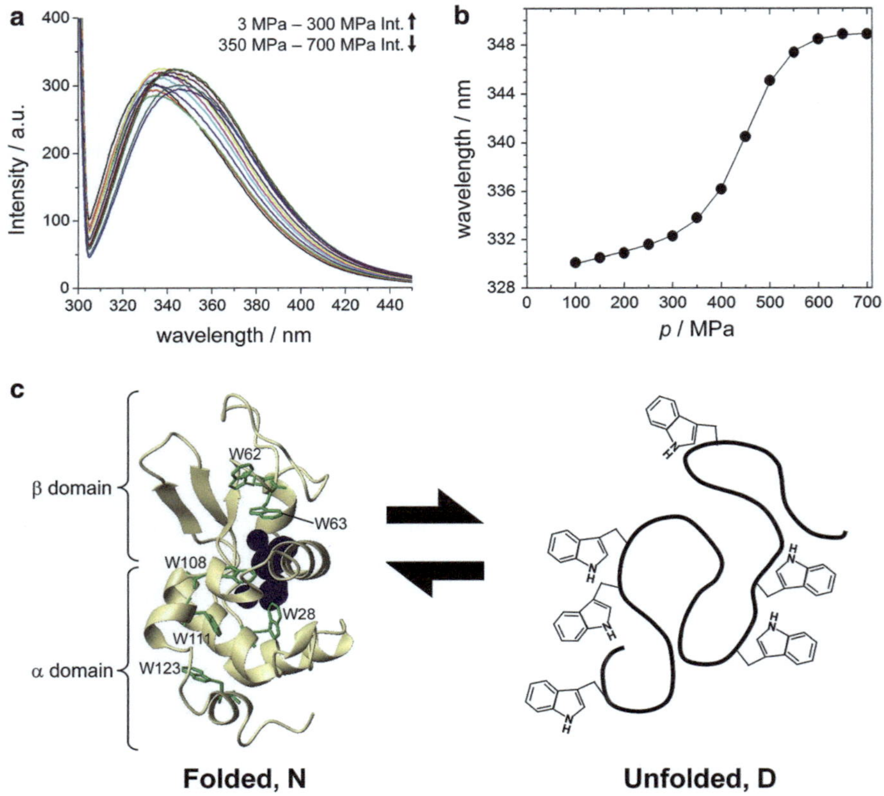

Fig. 32.4 Tryptophan fluorescence on hen egg white lysozyme at pH 2. (**a**) Trp fluorescence changes as a function of pressure at 15 °C. The upward and downward *arrows* indicate whether the fluorescence intensity is increased or decreased with increasing pressure in the pressure range indicated. (**b**) Plots of the wavelength of maximum fluorescence intensity (λ_{max}) as a function of pressure at 15 °C. The *solid curves* are the best-fit of eq. 32.2 to λ_{max}, giving ΔG_0 and ΔV (Eq. 32.3) values. Obtained thermodynamic parameters at various temperatures (−10 °C to 50 °C) are listed in Table 32.1. (**c**) (*Left*) The folded structure N of hen lysozyme in crystal (PDB ID; 135 L), which consists of two domains, α and β. The six tryptophan residues are colored *green* and the water-accessible internal cavity is shown by navy-spheres. (*Right*) The expected fully unfolded structure D of hen lysozyme including 6 Trp residues which exposed to solvent

32.4.1.1 Pressure-Induced Changes in Trp Fluorescence Spectra of HEWL at 0.1–700 MPa

Figure 32.4a gives the representative averaged fluorescence spectra obtained from six Trp residues contained in HEWL as a function of pressure from 3 to 700 MPa at 15 °C, and the rest were measured at – 5 °C, −10 °C, 5 °C, 25 °C, 40 °C and 50 °C, respectively. For details of pressure-induced spectral changes at respective temperatures, see a reference of Maeno et al. (2009). At all measured-temperatures, the fluorescence spectral changes in λ_{max} are fully reversible with pressure, but the intensity of emission shows less reversibility (~80 %). The lack of full reversibility is often encountered in high-pressure fluorescence experiments owing to some technical reasons. The shift in λ_{max} of Trp fluorescence is considered to represent correctly the change in the microenvironment of the tryptophan ring, which has already described in Sect. 32.2.1. From the plot of λ_{max} against pressure (Fig. 32.4b), we note that λ_{max} stayed initially within 330–335 nm at all temperatures (−10 °C to 50 °C), showing that Trp residues are almost fully buried in the folded conformer of the protein. Finally at 700 MPa, λ_{max} shifted to 349–350 nm, indicating that all the six Trp residues become exposed to the solvent. Therefore, we conclude that this protein is totally unfolded at 700 MPa at all measured temperature points. These results presumably represent pressure-induced two-state equilibrium transitions of HEWL from the folded conformer (N; "native" conformer) to the unfolded conformer (D; denatured conformer) (Fig. 32.4c).

32.4.1.2 Thermodynamic Analysis: Volume, Expansivity, Stability and Heat Capacity Changes

We assume the two-state transition between N and D, N ↔ D, and best-fit the observed changes in the wavelength of maximum fluorescence intensity (λ_{max}) in Fig. 32.4b with Eq. 32.2, which gives ΔV and ΔG^0 values listed in Table 32.1, together with pressures at the midpoint of transition p_m (Eq. 32.4). (For the detail of thermodynamic analysis, see a reference of Maeno et al. (2009)).

$$\Delta G = G_D - G_N = -RT \ln K \tag{32.1}$$

$$\lambda_{max} = \frac{\lambda_{max} + \lambda_{max} \exp\left[-\left\{\Delta G^0 + \Delta V^0 \left(p - p^0\right) + \Delta\alpha \left(p - p^0\right)\left(T - T^0\right)\right\}/RT\right]}{1 + \exp\left[-\left\{\Delta G^0 + \Delta V^0 \left(p - p^0\right) + \Delta\alpha \left(p - p^0\right)\left(T - T^0\right)\right\}/RT\right]} \tag{32.2}$$

Table 32.1 Thermodynamic parameters for unfolding of hen lysozyme determined from Trp fluorescence experiments (Maeno et al. 2009)

T (°C)	ΔG^0 (kJ mol^{-1})[a]	ΔV (ml mol^{-1})[b]	p_m (MPa)[c]
−10	31.5 ± 3.8	−85.9 ± 3.8	366.7
−5	22.1 ± 2.6	−58.2 ± 6.7	379.7
5	23.1 ± 1.5	−51.0 ± 3.5	452.9
15	25.7 ± 1.1	−56.2 ± 2.4	357.3
25	9.5 ± 3.4	−22.4 ± 9.3	424.1
40	7.6 ± 2.8	−23.4 ± 8.2	324.8
50	4.2 ± 3.1	−16.8 ± 11.8	250.0

[a]Gibbs free energy change at 0.1 MPa calculated with Eq. 32.2
[b]Partial molar volume change ($\Delta V = \Delta V^0 + \Delta\alpha\ (T\text{-}T^0)$) calculated with Eq. 32.2
[c]The denatured pressure

$$\Delta V = V_{\rm D} - V_{\rm N} = \Delta V^0 + \Delta\alpha\left(T - T^0\right) \tag{32.3}$$

$$p_{\rm m} = \Delta G^0/\Delta V \tag{32.4}$$

$$\Delta G = \Delta H_{\rm m}\left(1 - \frac{T}{T_{\rm m}}\right) + \Delta C_{\rm p}\left\{T - T_{\rm m} - T\ln\left(\frac{T}{T_{\rm m}}\right)\right\} \tag{32.5}$$

$$\Delta S_{\rm m} = \frac{\Delta H_{\rm m}}{T_{\rm m}} \tag{32.6}$$

where ΔG is the Gibbs free energy difference between N and D conformer, K is the equilibrium constant, ΔV is the change in partial molar volume of conformer D relative to N and $\Delta\alpha$ is the change in expansivity, respectively. In addition, $T_{\rm m}$, $p_{\rm m}$, $\Delta H_{\rm m}$, $\Delta S_{\rm m}$ and $\Delta C_{\rm p}$ represent the temperature and pressure at the midpoint of transition between N and D state, the unfolding enthalpy change at $T_{\rm m}$, the unfolding entropy change at $T_{\rm m}$ and the unfolding heat capacity change, respectively.

In Table 32.1, we note that $p_{\rm m}$ ranges from 250 (50 °C) to 453 MPa (5 °C), revealing a relatively high stability of HEWL against pressure unfolding even at low pH. The stability extrapolated to 0.1 MPa (ΔG^0) shows an increasing trend by lowering temperature, but the values may not be as reliable as ΔV in the present case, because 0.1 MPa is far from $p_{\rm m}$ and no reliable data points are available to 0.1 MPa. Figure 32.5a gives plots of ΔV against temperature (Eq. 32.3), which are all negative within the temperature range from −10 °C to 50 °C, but with a significant temperature dependence. From the slope of the dashed line in Fig. 32.5a, we estimate the expansivity change upon unfolding $\Delta\alpha$ (=1.07 ml mol^{-1} deg^{-1}) in Eq. 32.2. This value is comparable to those in staphylococcal nuclease (1.33 ml mol^{-1} deg^{-1}) (Lassalle et al. 2000) as well as in metmyoglobin (1.8 ml mol^{-1} deg^{-1}) (Zipp and Kauzmann 1973) and in ribonuclease A (1.32 ml mol^{-1} deg^{-1}) (Brandts et al. 1970). The positive value of $\Delta\alpha$ is taken to indicate the increased thermal volume due to the increased exposure of the polypeptide chain upon unfolding (Royer 2006). Figure 32.5b shows the plot of the experimentally determined values of ΔG (Eq. 32.1) against temperature at constant pressures, which depict concave features at all pressures studied. Data are limited to above 200 MPa, as no significant fraction unfolds in the lower pressure range to give sufficiently reliable ΔG values below 200 MPa. The plots were best fitted to Eqs. 32.5 and 32.6, giving parameters of $\Delta C_{\rm p}$, $T_{\rm s}$ (the temperature of maximum stability: the temperature for $\Delta S = 0$), $T_{\rm m}$, $\Delta H_{\rm m}$ and $\Delta S_{\rm m}$ (for both heat and cold denaturations) as summarized in Table 32.2. The ΔG values are fitted reasonably well with a single positive $\Delta C_{\rm p}$ value at each pressure, covering both the cold denaturation and heat denaturation ranges. In general, a positive $\Delta C_{\rm p}$ upon unfolding is accepted as due to the exposure of nonpolar amino acid groups to the solvent water (Royer 2006). In accordance with this, $\Delta H_{\rm m}$ for heat denaturation increases with increasing $T_{\rm m}$. Interestingly, while the stability is found to decrease with pressure, $T_{\rm s}$ increases with increasing pressure.

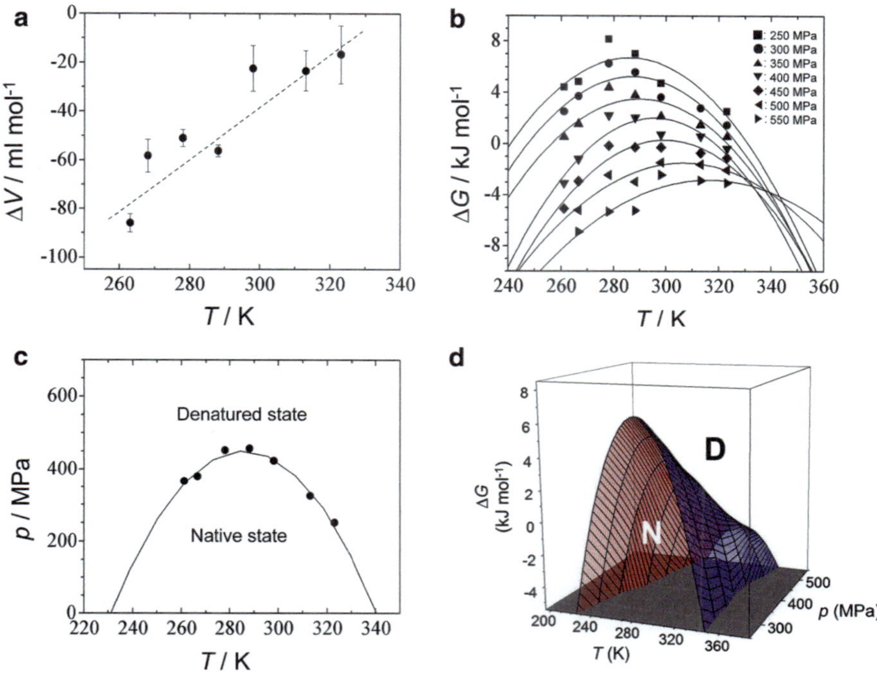

Fig. 32.5 Thermodynamic analysis of HEWL at pH 2. (**a**) Plot of the change in partial molar volume ΔV on unfolding against temperature. Best-fit to Eq. 32.3 gives a change in expansivity $\Delta\alpha$ on unfolding (260–320 K, pH 2) to be 1.07 ml mol^{-1} deg^{-1}. (**b**) Plots of the change in free energy ΔG on unfolding at various pressures. The *solid lines* show best-fit of eq. 32.5 to the experimental points, with melting temperature (T_m), enthalpy change at T_m (ΔH_m) and heat capacity change on unfolding (ΔC_p) as fitting parameters listed in Table 32.2. (**c**) The phase diagram of hen lysozyme at pH 2 as obtained by fluorescence measurements. Plot of p_m (pressure at the midpoint of transition between N and U or half-denaturation) obtained from Eq. 32.4 against T_m. The *solid curve* represents the best-fit of Eq. 32.5 for $\Delta G = 0$ to the experimental points. (**d**) Three-dimensional free-energy landscape of hen lysozyme at pH 2 in the range of pressure from 250 to 550 MPa and temperature from 200 to 380 K, drawn as overlay at various pressures

32.4.1.3 Pressure-Temperature Phase Diagram and Free Energy Landscape

Based on the experimentally obtained thermodynamic parameters, we depicted the phase diagram (for $\Delta G = 0$) of HEWL for the first time in aqueous environment on the pressure-temperature plane (Fig. 32.5c). So far, the thermodynamic stability on the temperature-pressure plane in aqueous environment has been reported for a limited number of proteins, including metmyoglobin (Zipp and Kauzmann 1973), ribonuclease A (Brandts et al. 1970), chymotrypsinogen (Hawley 1971) and staphylococcal nuclease (Lassalle et al. 2000; Panick et al. 1998; Panick et al. 1999). In these proteins, except for chymotrypsino-

Table 32.2 Thermodynamic parameters for heat and cold denaturation of hen lysozyme (Maeno et al. 2009)

P (MPa)		ΔH_m (kJ mol^{-1})[a]	ΔS_m (J mol^{-1} K^{-1})[b]	ΔC_p (kJ mol^{-1} K^{-1})[c]	T_m (K)[d]	T_s (K)[e]
250	Heat	95.1 ± 14.2	287.0 ± 14.5	1.9 ± 0.8	331.4 ± 3.0	286
	Cold	-77.4 ± 10.6	-319.4 ± 14.3		242.3 ± 9.6	
300	Heat	83.9 ± 7.5	256.3 ± 8.6	1.9 ± 0.6	327.4 ± 4.2	287
	Cold	-70.2 ± 8.3	-282.8 ± 8.5		248.2 ± 2.0	
350	Heat	66.3 ± 5.6	204.7 ± 6.0	1.8 ± 0.5	323.9 ± 2.2	290
	Cold	-56.4 ± 13.0	-219.6 ± 13.8		256.8 ± 4.5	
400	Heat	54.2 ± 14.0	170.0 ± 14.6	1.8 ± 0.6	318.8 ± 4.1	296
	Cold	-48.5 ± 6.0	-178.0 ± 6.3		272.5 ± 1.8	

[a]Unfolding enthalpy at T_m
[b]Unfolding entropy at T_m
[c]Unfolding heat capacity at T_m
[d]Denaturation temperature
[e]Temperature for maximum stability ($(\partial \Delta G/\partial T) = \Delta S = 0$)

gen which clearly gives a region of pressure-induced folding at elevated temperature (Hawley 1971), an ellipsoid type pattern like that in Fig. 32.5c with no regions of pressure-induced folding has been commonly observed, although the individual pattern is characteristic of each protein (Ravindra and Winter 2003). Finally, we have been successful in drawing the energy landscape of hen lysozyme at low pH on pressure and temperature axes on the approximation of null isothermal compressibility change ($\Delta\kappa = 0$) within the pressure range studied. The knowledge and the method presented here will serve as a basis for studying dynamics and folding of HEWL as a whole, but will also provide useful information as to the condition for the formation of amyloid fibrils or insoluble aggregates in a wider perspective (Abdul Latif et al. 2007; Shah et al. 2012).

32.4.2 Characterization of an Intermediate Conformer of Chicken Ovomucoid at pH 12 Using Tyrosine/Tyrosinate Fluorescence

Ovomucoid (OVM) from chicken egg white, consisting of 186 amino acid residues (approximately 28 kDa), is well known as an allergenic glycoprotein (Kato et al. 1976) as well as a trypsin inhibitor (Feeney et al. 1963; Hiromi et al. 1985). This protein consists of three sub-domains (domain 1 – domain 3), each having three intra-domain disulfide bonds and two Tyr residues, but no Trp residue is included in the primary sequence (Beeley 1976; Kato et al. 1987). Investigation of its thermodynamic stability is important from the practical viewpoint, as it consists of

11 % of the total protein content and is the most potent allergen in egg white. Though the unfolding of OVM that proceeds in two distinctive steps with a intermediary formation in between them has been shown using urea and guanidine hydrochloride (Waheed et al. 1977; Mustafa et al. 2011), studies on pressure axis, which can give the structural as well as thermodynamic information on semi-stable high energy conformers of this protein located higher on the folding funnel (Akasaka 2006), have not been carried out. Such information would be crucial for understanding the inhibitory function, enzymatic degradability, allergenicity and gel formation (Akasaka 2010). We show the results of high-pressure fluorescence measurements on OVM using improved high-pressure fluorescence spectrometer that can operate in the pressure range up to 700 MPa. To detect unfolding transitions of OVM, we utilize the change in fluorescence emission from six Tyr residues containing two each per domain. A highly alkaline solution condition, pH 12.0, was chosen to ensure that a Tyr residue of OVM will automatically give distinct tyrosinate fluorescence due to the side-chain ionization of tyrosine upon exposure to the alkaline environment.

32.4.2.1 Fluorescence of OVM by Excitation of Tyrosines

Figure 32.6a shows an overlay of the fluorescence spectra from six Tyr residues of OVM in the solution condition of pH 12.0 measured at pressures from 5 to 700 MPa at 25 °C by excitation of tyrosine absorption at 280 nm. At 5 MPa, the fluorescence emission is dominated by the emission with λ_{max} at 306 nm (Fig. 32.6a) assignable to tyrosine (R-OH) (cf. Fig. 32.2b), showing that even at 25 °C in the solution of pH 12.0 the effective pK_a values of all six Tyr of OVM are larger than 12. This finding indicates, in turn, that all the three domains of OVM are surely folded and that the phenolic OH groups of the Tyr residues are largely protected from the solvent, even at pH 12.0 at 25 °C. With increasing pressure, the decrease of the emission at 306 nm and a concomitant increase in the emission around 350 nm (assignable to tyrosinate (R-O$^-$), see inset) were observed, with a clear isoemissive point at 335 nm (*red circle* in Fig. 32.6a). This result strongly suggests that a two-state transition takes place between the tyrosine (R-OH) and tyrosinate (R-O$^-$) forms of Tyr residues of OVM in the wide pressure range from 5 to 700 MPa (Fig. 32.6b). Quantitatively, the fraction (47/75 $\approx$ 2/3) of the original emission at 306 nm at 5 MPa is decreased at 700 MPa, while a significant fraction (28/75 $\approx$ 1/3) of the original emission still remains even after the transition is nearly complete at 700 MPa (Fig. 32.6a). These results indicate that 4 Tyr residues out of six corresponding to two domains have transformed into tyrosinate (R-O$^-$), while two Tyr residues corresponding to one domain remain as tyrosine (R-OH) even at 700 MPa, strongly suggesting that the folding state of one domain of OVM is strongly resistive against denaturation, even to the extreme condition of pH 12.0 at 700 MPa.

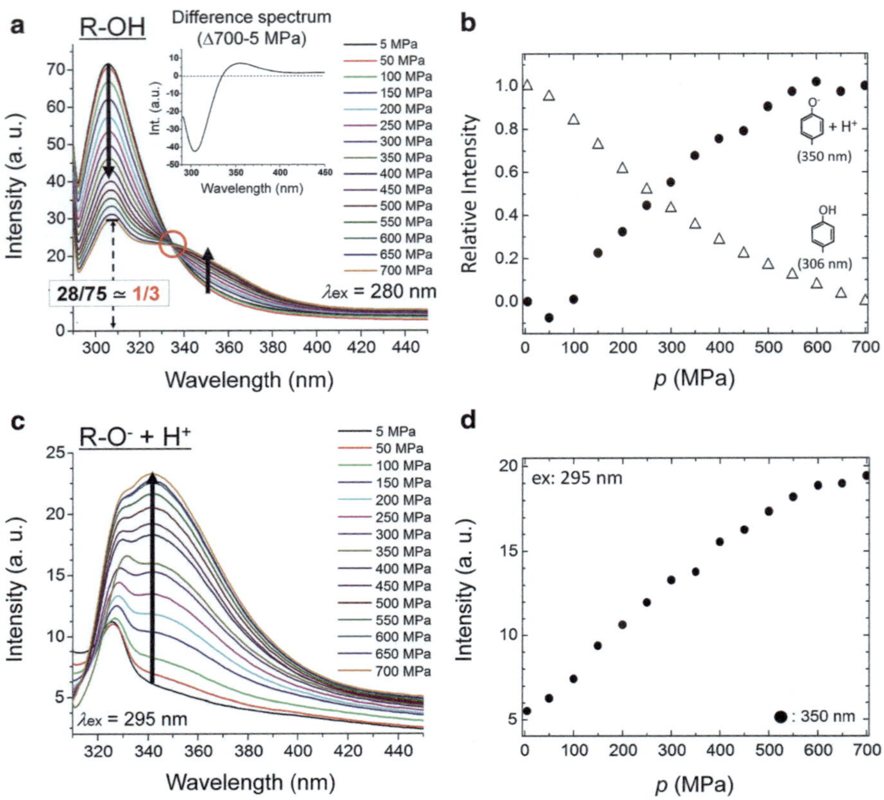

Fig. 32.6 Pressure-dependent changes of the fluorescence emission spectra of OVM at pH 12.0 at 25 °C between 5 and 700 MPa. (**a**) Spectral changes by excitation at 280 nm (tyrosine absorption). An isoemissive point is observed at 335 nm. The *inset* shows the difference spectrum obtained by subtracting the spectrum at 700 MPa from that at 5 MPa. (**b**) Plot of emission intensities at 306 nm (tyrosine) and at 350 nm (tyrosinate) (from panel A), after correcting the measured intensity by taking the solvent contraction into account (for example, ∼9 % by 300 MPa at 25 °C) (Bridgman 1912). *Open triangle*: 306 nm; *filled circle*: 350 nm. (**c**) Spectral changes by excitation at 295 nm (tyrosinate absorption). (**d**) Plot of emission intensity at 350 nm (tyrosinate) (from panel C) (Adapted from Fig. 2 in Maeno et al. (2013) by permission of Elsevier Inc.)

32.4.2.2 Fluorescence of OVM by Excitation of Tyrosinate

To confirm the production of tyrosinates (R-O$^-$) in ground state of OVM at high pressure, next we excite directly the tyrosinate absorption at 295 nm to observe its fluorescence at pH 12.0 both at 25 °C (Fig. 32.6c). Initially, we observe emission with λ_{max} around 325 nm at 5 MPa (*black line* in the spectrum). By increasing pressure, the fluorescence emission at λ_{max} ∼350 nm, assignable to tyrosinates (R-O$^-$) (cf. Fig. 32.2b), increased dramatically to the final value at 700 MPa (Fig. 32.6c). Thus, we have at least two fluorescent species, one giving the emission with λ_{max} around 325 nm existent already at 5 MPa and another giving a broad emission with λ_{max} at ~350 nm that increases with pressure. First, the species giving the emission with λ_{max} at around 350 nm can be identified with fully exposed tyrosinates (R-O^-), and the result of Fig. 32.6d indicates that the fraction of tyrosinates (R-O^-) increases dramatically with the increase in pressure. These results confirm those from Fig. 32.6a that a majority of tyrosine (R-OH) residues turn into tyrosinates (R-O^-) with increasing pressure. Figure 32.6d shows plots of the emission intensities at 350 nm at various pressures at 25 °C, obtained from Fig. 32.6c. Second, the relatively weak emission with λ_{max} around 325 nm needs attention (Fig. 32.6c). This emission appears already at 5 MPa when OVM is in the basic folded state N and stays as shoulder even at 700 MPa when OVM is in the intermediate state I in which only the most stable domain is folded. Therefore, the most reasonable interpretation seems that this emission belongs to a Tyr residue in the most stable domain.

32.4.2.3 Characterization of Intermediate Conformer of OVM

With chicken ovomucoid at pH 12.0 as target, we found a distinct intermediate state I at 700 MPa, in which two domains are unfolded, leaving one domain intact. So far, we have not specifically identified the intermediate conformer, under our specific experimental conditions, as to which one of the three domains of chicken OVM is the domain that remains folded even at 700 MPa at 25 °C. However, previous urea titration experiments have been carried out on isolated domains or domain combinations of chicken OVM at pH 7.0, and showed that domain 3 has by far the highest stability ([urea] $1/2 \approx 6.4$ M) compared to two other domains ([urea] $1/2 \approx 2.7$ M) (Das et al. 1991). Based on this result, it appears rather certain that the single domain remaining intact at an extreme condition of pH 12.0 and 700 MPa be assigned to domain 3, giving the conformational equilibrium of OVM as depicted in Fig. 32.7. In this regard, we plan to carry on an independent assignment of domain 3 using high-pressure NMR in later studies. The outstanding stability of domain 3 is considered responsible for the strong allergenicity of this protein. In the present work, we have restricted our solution condition to a highly alkaline pH, pH 12.0, to study conformational fluctuations of a Trp-lacking protein by utilizing the strong tyrosinate fluorescence at high pH. The experiment has been successful in monitoring differential stabilities of domains and to find a locally unfolded intermediate state of OVM in a clear-cut manner (Fig. 32.7). This work can be extended to the study of conformational fluctuations in Trp-lacking proteins in the neutral pH region by using tyrosine/tyrosinate fluorescence in combination with various proton acceptors to facilitate tyrosinate production (O'neil and Hofmann 1987). We have already carried out same experiments in OVM at neutral pH, and the thermodynamic feature of this allergenic protein will be reported elsewhere. Thermodynamic properties of the intermediate state disclosed here are expected to be a key to establish a new strategy to eliminate an allergenic potency from allergens.

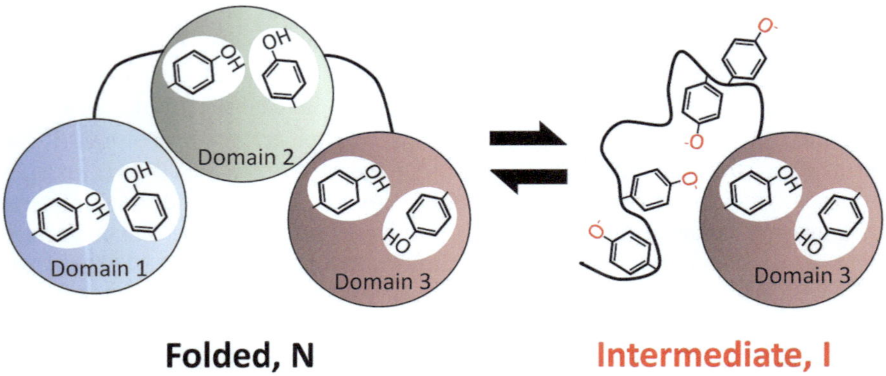

Fig. 32.7 Schematic view of conformational fluctuation of OVM between the folded conformer N and the intermediate conformer I deduced from the experiments at pH 12.0 at 25 °C. All the tyrosine side-chains of OVM are buried at 0.1 MPa at 25 °C at pH 12.0, but those in domains 1 and 2 become exposed to the solvent in the form of tyrosinate at 700 MPa, while those in domain 3 remain buried (Adapted from Fig. 6 in Maeno et al. (2013) by permission of Elsevier Inc.)

32.5 Caution in the Use of High-Pressure Fluorescence Spectroscopy for Studying Proteins

As we have shown here, fluorescence is very sensitive but provides only the limited information around the ring of inherent fluorophores (tryptophan and tyrosine) in a protein. Therefore, we recommend the combination of high-pressure fluorescence spectroscopy and other high-pressure methodologies that can compensate mutually, particularly high-pressure NMR spectroscopy (Maeno et al. 2009; Kitahara et al. 2012; Sunilkumar et al. 2012), when we study pressure-induced local and/or global unfolding of a protein. Otherwise we may erroneously take the site-specific internal hydration around the intrinsic fluorophores with the global unfolding of the protein. For instance, there is a case such that, while the λ_{max} value of Trp fluorescence tells that the protein is extensively hydrated, the high-pressure NMR spectrum shows that the protein is still folded (Sunilkumar et al. 2012). This could arise because the Trp fluorescence reports essentially the change in the local environment of the Trp residue, which does not necessarily coincide with the change in the local environment of other parts of the protein structure. This problem can be avoided if one monitors pressure-induced conformational changes also by NMR, preferably at least at some points of measurement. When the extra-high pressure experiment (e.g. up to 700 MPa) is to be carried out below the freezing temperature of the buffer, we should pay attention to the phase diagram of water. Otherwise, during the experiment at high pressure and low temperature, the buffer might be iced, and then the inner-cell might get a crack due to the volume expansion of the iced water. For instance, the pressure limit under freezing temperature is 200 MPa at −20 °C and 630 MPa at 0 °C.

32.6 Conclusion

The fluorescence spectroscopy combined with pressure perturbation is a well-established technique, for studying conformational transition, folding, dynamics of proteins and amyloid-forming systems. Use of the natural fluorescent probe Trp is particularly convenient, but caution is also necessary to differentiate the local structural change from the overall structural change. The improved high-pressure fluorescence system (0.1–700 MPa) introduced here is particularly; useful to extend the technique to cover a wide range of pressure-resistant proteins such as highly stable allergenic proteins for characterizing their thermodynamic stability and high-energy intermediates. The advancement would provide critical information and strategy for decreasing allergenicity, finding efficient pasteurization and improving food quality and other utility of pressure.

References

Abdul Latif AR, Kono R, Tachibana H, Akasaka K (2007) Kinetic analysis of amyloid protofibril dissociation and volumetric properties of the transition state. Biophys J 92:323–329

Akasaka K (2006) Probing conformational fluctuation of proteins by pressure perturbation. Chem Rev 106:1814–1835

Akasaka K (2010) High pressure NMR study of proteins seeking roots for function, evolution, disease and food applications. High Pressure Res 30:453–457

Astwood JD, Leach JN, Fuchs RL (1996) Stability of food allergens to digestion in vitro. Nat Biotechnol 14:1269–1273

Beeley JG (1976) Location of the carbohydrate groups of ovomucoid. Biochem J 159:335–345

Brandts JF, Olivera RJ, Westort C (1970) Thermodynamics of protein denaturation. Effect of pressure on the denaturation of ribonuclease A. Biochemistry 9:1038–1047

Bridgman PW (1912) Water, in the liquid and five solid forms, under pressure. Proc Am Acad Arts Sci 47:441–558

Chen RF (1967) Fluorescence quantum yields of tryptophan and tyrosine. Anal Lett 1:35–42

Das BK, Agarwal SK, Khan MY (1991) Unfolding–refolding behavior of chicken egg white ovomucoid and its correlation with the three domain structure of the protein. Biochim Biophys Acta 1076:343–350

Feeney RE, Stevefs FC, Osuga DT (1963) The specificities of chicken ovomucoid and ovoinhibitor. J Biol Chem 238:1415–1418

Hawley SA (1971) Reversible pressure-temperature denaturation of chymotrypsinogen. Biochemistry 10:2436–2442

Hayashi R (2002) Trends in high pressure bioscience and biotechnology, vol 19, Progress in biotechnology. Elsevier, Amsterdam

Hiromi K, Akasaka K, Mitsui Y, Tonomura B, Murao S (eds) (1985) Protein protease inhibitor – the case of Streptomyces subtilisin inhibitor (SSI). Elsevier, Amsterdam, pp 1–14, Chapter 1

Hoa HB, Marden MC (1982) The pressure dependence of the spin equilibrium in camphor-bound ferric cytochrome P450. Eur J Biochem 124:311–315

Hoa HB, Douzou GP, Dahan N, Balny C (1982) High pressure spectrometry at subzero temperatures. Anal Biochem 120:125–135

Kato I, Schrode J, Wilson KA, Laskowski M (1976) Evolution of proteinase inhibitors. Protect Biol Fluids 23:235–243

Kato I, Schrode J, Kohr WJ, Laskowski M (1987) Chicken ovomucoid: determination of its amino acid sequence, determination of the trypsin reactive site, and preparation of all three of its domains. Biochemistry 26:193–201

Kitahara R, Simorellis AK, Hata K, Maeno A, Yokoyama S, Koide S, Akasaka K (2012) A delicate interplay of structure, dynamics and thermodynamics for function: a high pressure NMR study of outer surface protein A. Biophys J 126:1225–1236

Lakowicz JR (2006) Principles of fluorescence spectroscopy, 3rd edn. Springer, New York

Lassalle MW, Yamada H, Akasaka K (2000) The pressure-temperature free energy-landscape of staphylococcal nuclease monitored by ^{1}H NMR. J Mol Biol 298:293–302

Maeno A, Matsuo H, Akasaka K (2009) The pressure-temperature phase diagram of hen lysozyme at low pH. Biophysics 5:1–9

Maeno A, Matsuo H, Akasaka K (2013) Tyrosine/tyrosinate fluorescence at 700 MPa: a pressure unfolding study of chicken ovomucoid at pH 12. Biophys Chem 183:57–63

Martin RB, Edsall JT, Wetlaufer DB, Hollingworth BR (1958) A complete ionization scheme for tyrosine and the ionization constants of some tyrosine derivatives. J Biol Chem 233:1429–1435

Mitra L, Hata K, Kono R, Maeno A, Isom D, Rouget JB, Winter R, Akasaka K, García-Moreno B, Royer CA (2007) *V*(i)-value analysis: a pressure-based method for mapping the folding transition state ensemble of proteins. J Am Chem Soc 129:14108–14109

Moriarty DF, Colon W (2005) Structural intermediates of globular proteins as precursors to amyloid formation. In: Sipe JD (ed) Amyloid proteins. Wiley-VCH Verlag GmbH & Co. KGaA, Weinheim, pp 275–300

Munishkina LA, Fink AL (2007) Fluorescence as a method to reveal structures and membrane-interactions. Biochim Biophys Acta 1768:1862–1885

Mustafa MG, Dar TA, Ali S (2011) Polyols stabilize the denatured states of multidomain protein ovomucoid. Int J Biol Chem 5:327–341

Nash DP, Jonas J (1997) Structure of pressure-assisted cold denatured lysozyme and comparison with lysozyme folding intermediates. Biochemistry 36:14375–14383

O'neil JDJ, Hofmann T (1987) Tyrosine and tyrosinate fluorescence of pig intestinal Ca^{2+}-binding protein. Biochem J 243:611–615

Panick G, Malessa R, Winter R, Rapp G, Frye K, Royer CA (1998) Structural characterization of the pressure-denatured state and unfolding/refolding kinetics of staphylococcal nuclease by synchrotron small-angle X-ray scattering and Fourier-transform infrared spectroscopy. J Mol Biol 16:389–402

Panick G, Vidugiris GJA, Malessa R, Rapp G, Winter R, Royer CA (1999) Exploring the temperature-pressure phase diagram of staphylococcal nuclease. Biochemistry 38:4157–4164

Pierce DW, Boxer SG (1995) Stark effect spectroscopy of tryptophan. Biophys J 68:1583–1591

Privalov PL (1989) Thermodynamic problems of protein structure. Annu Rev Biophys Chem 18:47–69

Privalov PL (1990) Cold denaturation of proteins. Crit Rev Biochem Mol Biol 25:281–305

Ravindra R, Winter R (2003) On the temperature-pressure free-energy landscape of protein. Chem Phys Chem 4:359–365

Rayner DM, Krajcarski DT, Szabo AG (1978) Excited state acid-base equilibrium of tyrosine. Can J Chem 56:1238–1245

Ross JBA, Laws WR, Rousslang KW, Wyssbrod HR (1992) Tyrosine fluorescence and phosphorescence from proteins and polypeptides. In: Lacowicz JR (ed) Topics in fluorescence spectroscopy, vol 3. Plenum Press, New York, pp 1–63

Royer CA (2002) Revisiting volume changes in pressure-induced protein unfolding. Biochim Biophys Acta 1595:201–209

Royer CA (2006) Probing protein folding and conformational transitions with fluorescence. Chem Rev 106:1769–1784

Shah BR, Maeno A, Matsuo H, Tachibana H, Akasaka K (2012) Pressure-accelerated dissociation of amyloid fibrils in wild-type hen lysozyme. Biophys J 102:121–126

Sunilkumar PN, Maeno A, Matsuo H, Oda M, Morii H, Akasaka K (2012) Extensively hydrated but folded: a novel state of globular proteins stabilized at high pressure and low temperature. Biophys J 102:L8–L10

Szabo AG, Lynn KR, Krajcarski DT, Rayner DM (1978) Tyrosinate fluorescence maxima at 345 nm in proteins lacking tryptophan at pH 7. FEBS Lett 94:249–252

Waheed A, Qasim MA, Salahuddin A (1977) Characterization of stable conformational states in urea-induced transition in ovomucoid. Eur J Biochem 76:383–390

Zipp A, Kauzmann W (1973) Pressure denaturation of metmyoglobin. Biochemistry 12:4217–4228

Chapter 33
High Pressure NMR Spectroscopy

Kazuyuki Akasaka

Abstract The combination of high-resolution NMR spectroscopy with pressure perturbation, known as variable-pressure NMR spectroscopy or simply high pressure NMR spectroscopy, is a relatively recent accomplishment, but is a technique expanding rapidly with high promise in future. The importance of the method is that it allows, for the first time in history, a systematic means of detecting and analyzing the structures and thermodynamic stability of high-energy sub-states in proteins. High-energy sub-states have been only vaguely known so far, as normally their populations are too low to be detected by conventional spectroscopic techniques including NMR spectroscopy. By now, however, high pressure NMR spectroscopy has established unequivocally that high-energy conformers are universally present in proteins in equilibrium with their stable folded counterparts. This chapter describes briefly the techniques of high pressure NMR spectroscopy and its unique and novel aspects as a method to explore protein structure in its high-energy paradigm with illustrative examples. It is now well established that high pressure NMR spectroscopy is a method to study *intrinsic* fluctuations of proteins, rather than those forced by pressure, by detecting structural changes amplified by pressure. Extension of the method to other bio-macromolecular systems is considered fairly straightforward.

Keywords High-energy sub-states • Pressure-resisting cells • Pressure-dependent structural changes • Structural fluctuations • Thermodynamic analysis

33.1 Introduction

The NMR spectroscopy is the only spectroscopy that can give structural information in all atom basis on a protein in solution. However, the target of the method so far has

K. Akasaka (✉)
High Pressure Protein Research Center, Institute of Advanced Technology, Kinki University, 930 Nishimitani, Kinokawa 649-6493, Japan

Present Affiliation: Graduate School of Life and Environmental Sciences, Kyoto Prefectural University, 1-5 Hangi-cho, Shimogamo, Sakyo-ku, Kyoto 606-8522, Japan
e-mail: akasaka@kpu.ac.jp

K. Akasaka, H. Matsuki (eds.), *High Pressure Bioscience*,
Subcellular Biochemistry 72, DOI 10.1007/978-94-017-9918-8_33

been to elucidate a "single" structure that exists stably in solution, which is a major conformer conventionally called the "native" conformer N. The NMR spectrum is insensitive to conformational fluctuations occurring in the time range << ms, nor can it detect conformers that may exist in rare population (<<10 %), while such rare conformers may be crucial for the reaction or the function of the protein. Information about these rare conformers and associated fluctuations are not fully recovered by the use of various spin relaxation measurements in the laboratory frame (T_1, T_2, NOE) or in the rotating frame (μs-ms).

The major role of the high pressure NMR spectroscopy is to make such rare conformers and associated conformational fluctuations detectable by NMR spectroscopy. For this purpose, we need, as much as possible, a non-invasive perturbation of the system in intrinsic fluctuations. We found that this is the pressure perturbation in the range of a few to several hundred MPa or up to several kbar (1 bar $= 10^5$ Pa $= 0.9869$ atm). Such a pressure causes a shift of equilibrium population among ensembles of conformers because their partial molar volumes are intimately coupled with their conformations. The resultant spectral difference between 0.1 MPa and pressure p represents the fluctuating part of the macromolecular structure.

If the pressure is varied sufficiently (up to several kbar), a considerable range of the conformational space of a protein may be explored from within the basic folded ground state, to a larger part of the biologically relevant conformational space even close to that of full unfolding. Here, all NMR parameters, particularly chemical shifts and signal intensities along with J-coupling, linewidth/line shape, spin relaxation and NOE are used in one-dimensional and multi-dimensional spectroscopy. Preferably (and it is often essential), all or the relevant signals in question must have been assigned to specific residues or atoms at ambient pressure, usually using $^{15}N/^{13}C$ double-labeled protein samples. Independent assignments at elevated pressures may not be required if the spectral changes with pressure are continuous, but may be required in special cases. Once the assignments are complete, regions of conformational change, high mobility or local disorder/unfolding can be directly identifiable in the spectrum.

33.2 High Pressure NMR Techniques

There are basically two designs for high-pressure NMR apparatus on solution samples; the so-called *high-pressure probe method* and the so-called *pressure-resisting cell method.* In the former method, which dates back to as early as 1954 (Benedek and Purcell 1954), and more recently developed further by Jonas for bio-macromolecular studies (Jonas 2002), the radiofrequency transmitter/detection coils are placed within a high-pressure vessel, the entirety of which is placed in the static magnetic field. In the latter method, only the sample solution is kept under pressure in a pressure-resisting cell, while the transmitter/detection coil itself is kept at ambient pressure (Akasaka and Yamada 2001). The former has a merit of covering a relatively high pressure range up to ∼GPa, while its utility to bio-macromolecular studies is limited because of the lack of the sufficient homogeneity in static magnetic field and the capability in various pulse sequences for one-dimensional and two-dimensional measurements. Thus the pressure-resisting cell method, in combination with a commercially available spectrometer with versatility of pulse sequence, is currently the method of choice for most purposes of bio-macromolecular studies (Akasaka and Yamada 2001).

33.2.1 The Pressure-Resisting Cell Method

The pressure-resisting cell method dates back to 1974 when Yamada introduced a hand-made glass capillary cell for studying pressure-dependent structural change in small organic molecules in solution (Yamada 1974). Since then, a variety of non-magnetic cell materials with high tensile strength has been introduced, including synthetic quartz (Yamada et al. 2001), sapphire (Kremer et al. 2004), and Zirconia (Peterson et al. 2005). The advantage common to all the pressure-resisting cell methods is that no modification of NMR hardware including the probe is required, so that any commercially available NMR spectrometer can be employed for high-pressure experiments with all their advanced capabilities with multiple pulse sequences. (See Fig. 33.1 for the conceptual view of the pressure-resisting cell method). Usually, the pressure-resisting cells are used repeatedly with sample changes.

On the other hand, the high-pressure probe method retains the advantage of working in a higher pressure range (up to ∼GPa) than the pressure-resisting cell method (currently up to 3∼4 kbar), and will continue to be useful for specific needs.

33.2.2 The Advantage and Disadvantage of the Synthetic Quartz Cell

Extensive works on proteins using one-dimensional and two-dimensional NMR with variable pressure have been carried out using the synthetic quartz cell in the pressure range up to 2–4 kbar (Akasaka and Yamada 2001; Akasaka 2003, 2006; Kamatari et al. 2004; Kalbitzer et al. 2009; Kitahara et al. 2013), which have contributed to the opening of the new field of protein structure study. Its excellent resolution as well as its relative high-resistance to pressure (up to 4 kbar (for example, see Maeno et al. 2009)), stands out among the cell materials used so far. It consists of a long capillary part and the body part at the end, which goes into the detection coil (Fig. 33.1, left) (Yamada et al. 2001; Akasaka and Yamada 2001). The cell endures repeated use and its accidental break during measurement causes no damage to the probe, partly because the cell is protected by a Teflon horse, which shrinks automatically upon pressure-release and a Teflon protection tube surrounding the entire quartz cell, which prohibits the leakage of the sample solution into the probe.

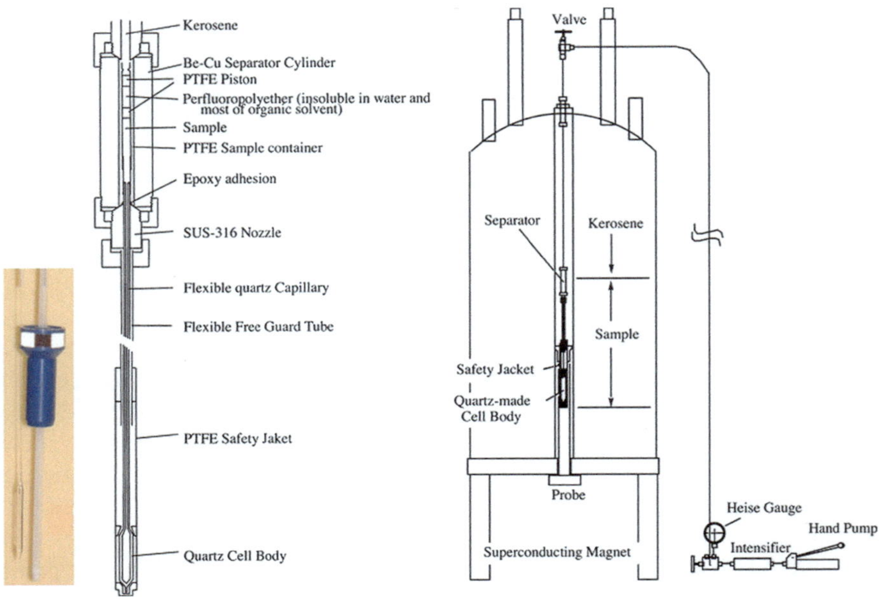

Fig. 33.1 Schematic illustration of the high-pressure NMR apparatus using a pressure-resisting cell. On the *left*, a photo is shown of a quartz cell with a sample-solution-containing body part (O.D/I.D = 3.5 mm/1 mm) housed in a Teflon protecting tube and a long capillary stem for mediating hydrostatic pressure from an external pump (See Yamada et al. (2001) for details of the cell design). The maximum operating pressure can go up to 4 kbar or higher (Akasaka and Yamada 2001)

However, despite the great advantage of this cell technique and its success, there have been several drawbacks and limitations as well; these include the requirement of special techniques for its manufacture (Yamada et al. 2001), the requirement of special equipment for handling sample solution into and out of the cell, and the small sample volumes in the body part (with inner diameter ~1 mm/outer diameter ~3.5 mm for the 5 mm probe and with inner diameter ~1.5 mm/outer diameter ~6.0 mm for the 8 mm probe), which prohibits its use for proteins with low solubility. For these reasons, the quartz cell technique has gained limited popularity, so far, in bio-macromolecular application. However, these drawbacks may be overcome in future by improved cell design and the use of a higher sensitivity spectrometer with a cryo-probe setup.

33.2.3 The Advantage and Disadvantage of the Ceramic Cell

Recently, a cylindrically shaped cell made out of the aluminum-toughened zirconia has been introduced commercially (Daedalus Innovations, See Fig. 33.4 of Chapter

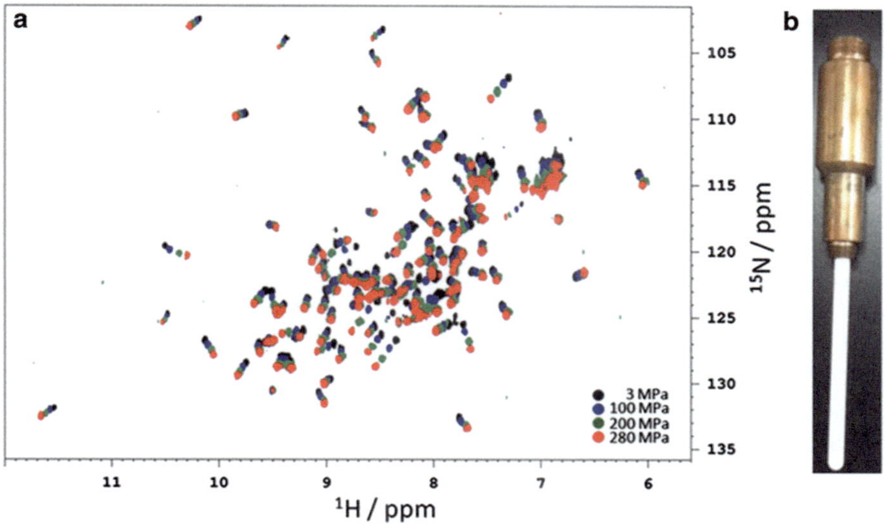

Fig. 33.2 ^{15}N/^{1}H HSQC spectra of ^{15}N-labeled *Staphylococcal* nuclease (1.3 mM, pH 7.5) measured with a zirconia cell at pressures from 3 to 280 MPa at 25 °C. The spectral changes are fully reversible with pressure. A home-made zirconia cell of O.D/I.D = 6.0 mm/2.0 mm (a photo on the *right*) is used with a protective measure for a cell burst (not shown) on a 8 mm probe of a Bruker AVANCE-600 spectrometer

by Roche et al.) and independently by us, which can endure pressure up to 3 kbar (See Fig. 33.2). The great advantages of these cells are their toughness against mechanical shock, the cylindrical shape easy for sample filling and its relatively large inner diameter assuring a high signal-to-noise ratio (currently up to 3 mm is realized for the outer diameter of 5 mm).

The cell can be used repeatedly with sample changes. However, while, in the quartz cell method, the plastic hose and the safety jacket prevents damage to the probe in time of a cell burst (Yamada et al. 2001), commercially available ceramic cells do not have a protection against an accidental cell burst. Thus one must make an extreme care that an accidental cell burst would not cause a serious damage to the probe. At this stage, one is required to make a careful trade between the handiness of the experiment and the danger of damaging the probe. Nonetheless, because of the handiness and relatively high sensitivity, the development of zirconia cells is rapidly widening the utility of high pressure NMR spectroscopy to a larger community in protein science. Further development of pressure-resisting cell techniques to a higher pressure range, preferably with a protective measure for a cell burst, is anticipated for extending its utility in the exploration of the higher-energy sub-states of proteins.

33.2.4 Experimental Procedures

Choice of Buffers Buffers to be used for pressure studies including high pressure NMR are preferred to be those having little variation of pK_a with pressure. There is no perfect buffer in this sense, but one should choose those buffers with ionization volume as small as possible. We find that the following buffers are relatively inert to pressure variation: maleic acid ($\Delta V° = -5.1$ cm^3 mol^{-1}), MES (2-(N-morpholino)ethanesulfonic acid, −3.9 cm^3 mol^{-1}), Tris (2-Amino-2-hydroxymethyl-propane-1,3-diol, −4.3 cm^3 mol^{-1}), or imidazole buffers (−1.8 cm^3 mol^{-1}) (Kitamura and Itoh 1987). One must be aware, however, that those buffers inert to pressure variation tend to be more susceptible to temperature variation and are not suited for temperature variation studies. For ^{1}H NMR measurements, appropriately deuterium-substituted version of the buffering agents (commercially obtainable) are to be used. The final buffer solution must contain 5–10 % of ^2H_2O to maintain the field-frequency lock during the measurement.

Chemical Shift Reference As a chemical shift standard the methyl signal of DSS (4,4-dimethyl-4-silapentane-1-sulfonic acid) is to be set at origin ($\delta = 0$) as usual. However, when this overlaps with some methyl proton signals in the ^{1}H NMR spectrum, one may use an aliquot of dioxane, which gives a sharp singlet at $\delta = 3.75$ ppm downfield from DSS nearly independently of pressure, as alternate internal reference of chemical shift.

Pulse Sequences Because the pressure-resisting cell method can be combined with any commercial spectrometer and probe, practically any pulse sequence that the hardware of the spectrometer permits can be used for high-pressure NMR experiments, limited only by the signal-to-noise ratio due to the limited sample volume of the pressure-resisting cell.

33.3 Pressure-Dependent Structural Changes in Bio-macromolecules

All the NMR parameters can be used to detect structural changes of proteins and other bio-macromolecules with NMR spectroscopy under pressure. Here we focus on the characteristic features of structural change in proteins, using mainly chemical shifts and signal intensities.

33.3.1 Pressure-Dependent Changes Within the Folded Protein Structure

Our first paper came in 1997 using home-developed high-pressure NMR facility, when we started ^{1}H NMR measurements under variable pressure on a solution of

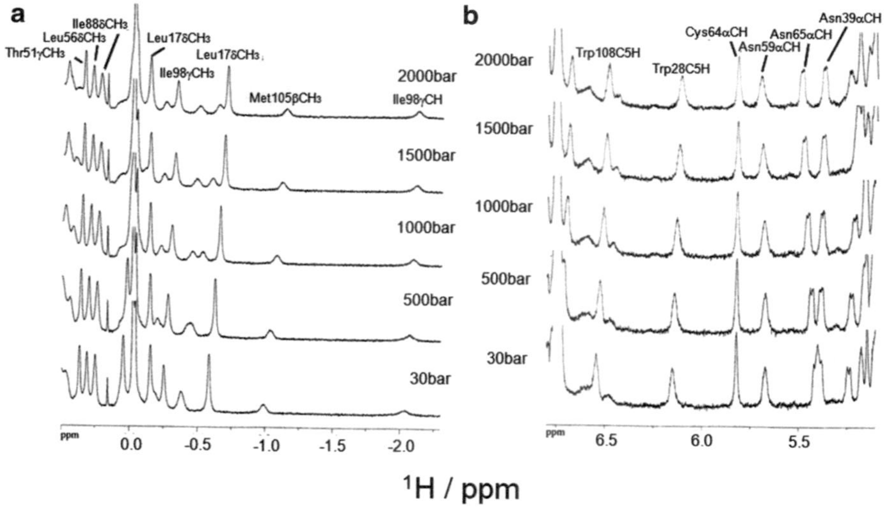

Fig. 33.3 Selected regions of 750 MHz ^{1}H NMR spectra of hen lysozyme in the folded state (3.5 mM in 100 mM glycine buffer, 99.96 % ^2H_2O, pH 2.0), recorded with a quartz cell at 45 °C at 500 bar intervals up to 2,000 bar. (**a**) The region for high-field shifted side-chain protons. (**b**) The region for low-field shifted Cα protons and high-field shifted Trp C_5 protons. The spectral changes are fully reversible with pressure (Akasaka et al. 1997)

hen lysozyme. The abstract of this paper in the following describes the surprise of observing an overall conformational change of the protein hen lysozyme in its thermodynamically stable folded state in solution (Akasaka et al. 1997) (Fig. 33.3): *We demonstrate, for the first time in solution, that pressure induces changes in the overall folded structure of a protein (hen lysozyme). This was made possible by using a home-developed, on-line continuously variable pressure cell on a high resolution NMR spectrometer operating at 750 MHz. We could follow pressure-induced diamagnetic chemical shifts of more than 26 protons of hen lysozyme at variable pressure in the range of 1 to 2000 bar. The results indicate that the main effect of the pressure is a compaction of the hydrophobic core part of the protein consisting of bulky side-chains. The technique introduced here provides a general method with which one can probe microscopic internal flexibility of a protein in solution.*

Since the spectrum remains that of folded throughout the pressure range studied, the observed spectral changes represent conformational changes of hen lysozyme within the folded ensemble, site-specific details of which are disclosed later by high pressure ^{15}N NMR spectroscopy (Kamatari et al. 2001) and further in coordinate changes such that both compression and expansion are taking place preferentially around the water-containing cavities besides other regions of the protein (Refaee et al. 2003) see Fig. 6.2 in Chap. 6 (Williamson). In a still later work, the dynamic hydration of the pre-existing cavities is suggested to be the initial step for unfolding (Kamatari et al. 2011) (cf. Chap. 12 by Li and Kamatari).

We also found that pressure induces average changes in hydrogen bonds (Li et al. 1998) as well as in ϕ, ψ torsion angles in the polypeptide chain within the folded manifold (Akasaka et al. 1999). In BPTI, the average shrinking of hydrogen bonds at 200 MPa is estimated to be 0.020 Å for the NH—O = C (Li et al. 1998) and 0.029 ± 0.117 Å for those including hydrogen bonds to internal buried water molecules (Williamson et al. 2003), and similar values are obtained in other proteins (Kitahara et al. 2013).

33.3.2 Pressure-Induced Transition into Higher-Energy Sub-states

Another major surprise came in the study of the Ras-binding domain of Ral guanine dissociation stimulator (RalGDS), when some cross peaks disappear and new cross peaks appear in the central region of the spectrum above 1,000 bar, showing that a local unfolding event takes place in this protein (Fig. 33.4). The result clearly indicates the presence of an intermediate (I) undergoing slow exchange (>> ms) with the basic folded conformer N, as two separate signals are found for the same

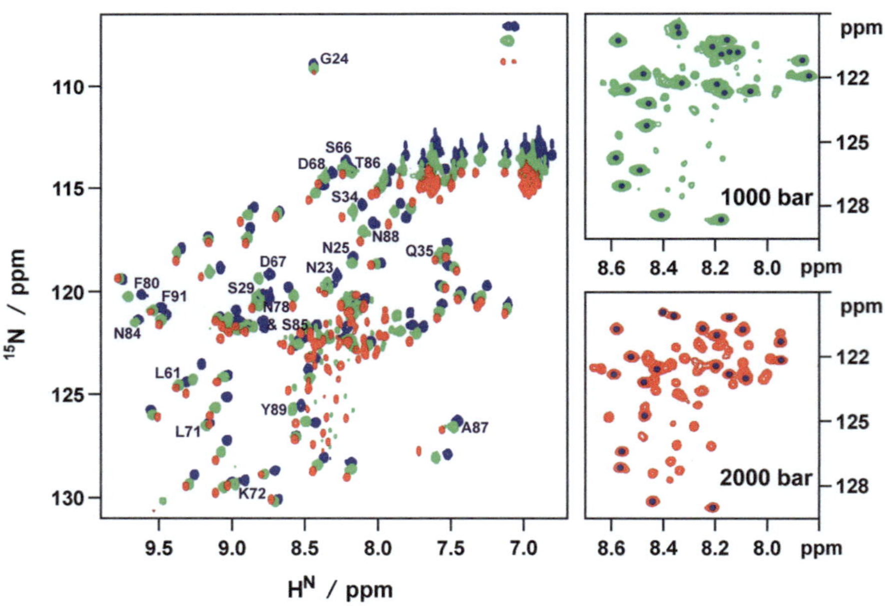

Fig. 33.4 Superposition of ^{15}N/^{1}H HSQC spectra of the Ras-binding domain of RalGEF (11–97) measured at 297 K at 30 bar (*blue*), 1,000 bar (*green*) and 2,000 bar (*red*). Data were recorded at a proton frequency of 750.13 MHz and a nitrogen frequency of 76.01 MHz for 1.2 mM RalGEF-RBD in 15 mM Tris-HCl, 150 mM NaCl, 10 mM DTE, pH 7.3 in 95 % $^{1}H_2O$ and 5 % $^{2}H_2O$. The spectral changes are fully reversible with pressure (Inoue et al. 2000)

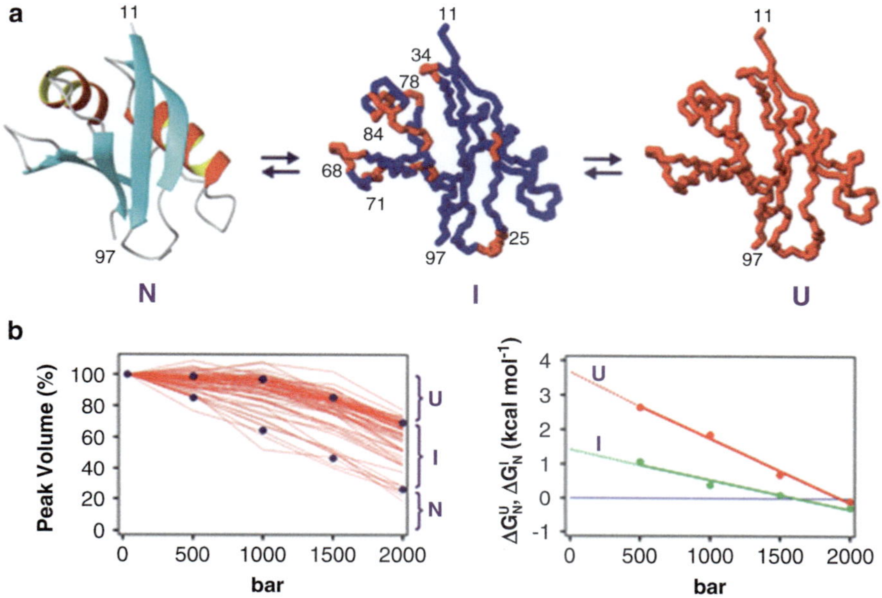

Fig. 33.5 Conformational equilibrium in the Ras-binding domain of RalGEF. (**a**) N, I and U are found to exist in equilibrium at all pressures. Only the native structure (N) is shown by a ribbon model from the NMR structure at 1 bar. The intermediate structure (I) is shown with locally disordered regions depicted in *red*. The unfolded state (U) is depicted as completely disordered. (**b**) Population (*left*) and stability (*right*) of *N*, *I* and *U* plotted against pressure. (*Left*) Plot of cross peak intensities (or peak volumes) of all separately observable individual "native" signals in the $^{15}N/^{1}H$ HSQC spectra against pressure. The *dots* show the average values of the curves from which relative populations of N, I and U are estimated as shown. (*Right*) Plot of Gibbs energy differences between I and N and between U and N (calculated from Eq. 33.1) against pressure. The extrapolation to 1 bar following Eq. 33.3 yields standard Gibbs energy differences (at 1 bar, 297 K) $\Delta G_0 = 1.4 \pm 0.16$ kcal mol^{-1} between I and N and 3.6 ± 0.14 kcal mol^{-1} between U and N, The lopes give $\Delta V_{IN} = -36$ ml mol^{-1} and $\Delta V_{UN} = -78$ ml mol^{-1} (Inoue et al. 2000)

residues in Fig. 33.4 (Inoue et al. 2000). In conformer I, part of the protein structure is unfolded, while the rest remains folded. With further increase in pressure, all the cross peak intensities begin decreasing, indicating the onset of unfolding of the entire protein molecule into U. Namely, there are at least three conformational species N, I and U are involved (N $\leftrightarrows$ I $\leftrightarrows$ U) within the pressure range studied (Fig. 33.5a).

33.3.3 Thermodynamic Analysis of High-Energy Sub-states

Equilibrium populations of N, I and U can be determined experimentally as a function of pressure with cross peak intensities as representing respective populations of N, I and U (Fig. 33.5b, left), thereby the equilibrium constant K, for example, between N and I, by Eq. 33.1 and the Gibbs energy difference ΔG by Eq. 33.2 or more simply by Eq. 33.3 as a function of pressure.

$$K = [I]/[N] = \exp(-\Delta G/RT) \tag{33.1}$$

$$\Delta G = G_{\mathrm{I}} - G_{\mathrm{N}} = \Delta G\,(p_0) + \Delta V^\circ\,(p - p_0) - (1/2)\,\Delta\beta V^\circ(p - p_0)^2 \tag{33.2}$$

An approximation by neglecting the second order term gives

$$\Delta G = G_{\mathrm{I}} - G_{\mathrm{N}} = \Delta G\,(p_0) + \Delta V^\circ\,(p - p_0) \tag{33.3}$$

Here, ΔG and $\Delta G(p_0)$ are the Gibbs energy changes from N to I at pressure p and p_0 (1 bar or other low pressures chosen in the experiment), respectively; ΔV° is the partial molar volume change at p_0 (1 bar); $\Delta\beta$ is the change in compressibility coefficient; R is the gas constant, and T is the absolute temperature (Akasaka 2006).

Plotting ΔG as a function of pressure according to Eq. 33.3 gives Fig. 33.5b, right. Here, the extrapolation of ΔG to 1 bar gives $\Delta G(p_0)$, the Gibbs energy difference at 1 bar, showing that I and U exist by 1.4 and 3.6 kcal mol^{-1} above N in Gibbs energy, respectively, at 1 bar with ΔV_{IN} of −36 ml mol^{-1} and ΔV_{UN} of −78 ml mol^{-1} (Inoue et al. 2000). Conformers I and U may therefore be called high-energy conformers of RalGDS-RBD.

This example shows a typical event occurring for a protein in solution under pressure. The Gibbs energy levels of I and U relative to that of N ($\Delta G(p_0)$) are sufficiently high so that I and U are spectroscopically invisible at 1 bar, but are sufficiently low so that their $\Delta G(p)$ may become negative at 1–2 kbar, making them clearly visible in the NMR spectra (Fig. 33.6, right), thanks to their sufficiently

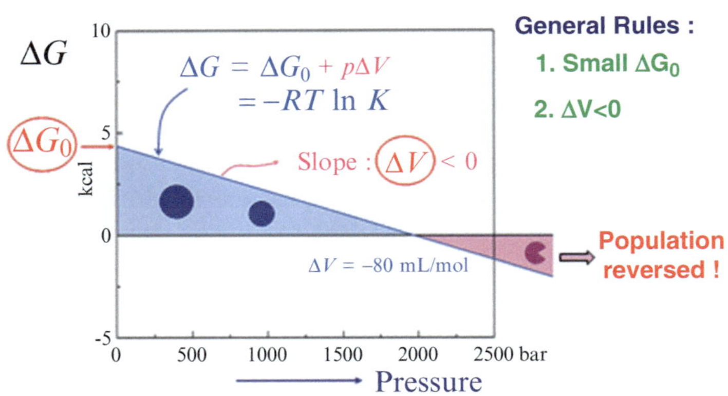

Fig. 33.6 Schematic expression of Eq. 33.3: $\Delta G = G_{\mathrm{I}} - G_{\mathrm{N}} = \Delta G(p_0) + \Delta V^\circ(p - p_0)$, with N representing the folded ground state and I representing any high-energy state. A specific case with $\Delta V = -80$ mL mol^{-1} and $\Delta G^\circ = 4$ kcal mol^{-1} is drawn

large negative ΔV values. The finding of a locally unfolded intermediate I in the Ras-binding domain of RalGDS is just an example, which is followed by similar local unfolded conformers or other high-energy conformers in many other globular proteins (Akasaka 2003, 2006; Kamatari et al. 2004; Fourme et al. 2012), showing that semi-stable intermediates occur rather generally in globular proteins. The partial molar volume of a protein becomes less and less, as the protein looses its conformational order (or folding contacts) toward unfolding ("Volume rule"). The volume rule is considered to arise as the space occupied by "voids" within a folded protein becomes occupied by bulk water molecules as unfolding proceeds (cf. Fourme et al. 2012).

33.4 Exploration of High-Energy Paradigm of Protein Structure with High Pressure NMR

Besides the locally unfolded intermediates shown above, there are many forms of high-energy sub-states between the basic folded conformer and the fully unfolded conformer, which constitute the high-energy paradigm of proteins (Fig. 33.7) (Akasaka 2010). These include low-lying excited states detectable typically by non-linear pressure shifts (Akasaka and Li 2001), alternately folded conformers (Kitahara et al. 2005), locally or partially unfolded conformers (Lassalle et al. 2003; Kamatari et al. 2004; Akasaka 2006), molten globule-like conformers (Kitahara et al. 2002), fully hydrated unfolded conformers and reversibly aggregated species (Niraula et al. 2004; Kamatari et al. 2005; Abdul Latif et al. 2007; Sasaki et al. 2008; Akasaka et al. 2014).

Detection and characterization of high-energy sub-states including alternately fold, locally unfolded intermediates or low-lying excited states, represented by Fig. 33.5b, are among the highlights of high pressure NMR spectroscopy of proteins. When the exchange between a high-energy sub-state and the ground state N is rapid with respect to the NMR time scale (<< ms), the signals of the two conformers would merge into one, in which case the existence of and the transition into a higher-

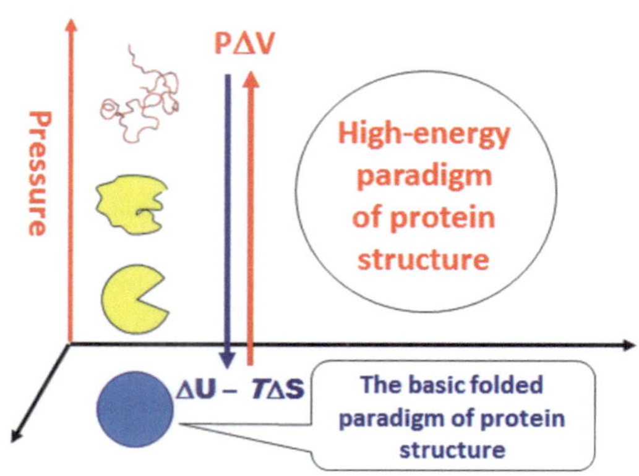

Fig. 33.7 The high-energy paradigm of protein structure to be explored with high pressure NMR spectroscopy. This may be performed by taking advantage of the "Volume Rule" of proteins such that the partial molar volume of a protein decreases in parallel with the loss of conformational order (cf. Fourme et al. 2012)

energy state is detected as non-linearity in pressure-induced chemical shift. Such phenomena are considered to occur when the excited states are only slightly higher in Gibbs energy than the state N, namely for low-lying excited states. Non-linear pressure shift is quite a sensitive measure of the presence of low-lying excited states in proteins (Akasaka and Li 2001). This topic is discussed separately by Li and Kamatari (Chap. 12) in detail in their chapter.

Actual examples of high-energy sub-states in proteins disclosed by high pressure NMR studies are found in Chaps. 6 (Williamson), 9 (Kalbitzer), 10 (Kitahara), 11 (Dhaussy and Girard) and 12 (Li and Kamatari). There we will see that high-energy sub-states are often correlated with naturally occurring cavities and possibly with their hydration. In Chaps. 6 and 10, structures of high-energy conformers within the folded manifold or outside the folded manifold are expressed in average atomic coordinates, the former using pressure-induced chemical shifts as constraints and the latter using NOE-based distances as constraints.

Structures of high-energy sub-states for larger protein molecules can be studied with the recently developed alternative technique, high-pressure macromolecular crystallography (See Chap. 11 (Dhaussy and Girard) and Chap. 31 (Watanabe)). Here, the concept of conformational equilibrium in Fig. 33.6 is considered to be valid (Fourme et al. 2012) within a limited range of pressure (See Chap. 11 (Dhaussy and Girard)).

33.5 Relationship Between the Pressure-Induced Structural Change and the Conformational Fluctuation

In Sect. 33.4, we have observed that various structural changes take place in a protein with increasing pressure, which are all reversible with pressure. This would mean that all conformers detected exist in dynamic equilibrium in the entire pressure range studied. Furthermore, the structural changes by pressure (in the pressure range so far employed), as depicted in Fig. 33.7a are actually quite small and are considered to lie within the range or the amplitude of fluctuation occurring intrinsically at 1 bar; for example, the average shrinking of a NH—O = C hydrogen bond is ~0.020 Å at 2 kbar (Li et al. 1998) and the estimated average changes in ϕ, ψ torsion angles are estimated to be 1–2° at 2 kbar for the β-sheet part and slightly larger for the α-helix and loop, which are considered well within the ranges of fluctuation at 1 bar (Akasaka et al. 1999). These findings have led us to the concept that what we observe as structural changes induced by pressure is an amplification of the conformational fluctuation intrinsically present at ambient pressure (Akasaka 2003, 2006, 2010, 2014).

General procedures are available from statistical thermodynamics to estimate the magnitude of these fluctuations in any given system (Cooper 1976). Namely, the isothermal compression of the volume $(\beta_T V)$ of a protein under pressure is correlated with the mean-square fluctuation of its volume $\langle(\delta V)^2\rangle$ through the relation

$$\langle(\delta V)^2\rangle = \beta_T VkT \tag{33.4}$$

$$\beta_T = -(1/V)\ (\partial V/\partial p)_T \tag{33.5}$$

where V is the volume of a macromolecular system, β_T is the isothermal compressibility coefficient, k is Boltzmann's constant and T is the absolute temperature. Equation 33.4 states that we acquire information about the fluctuation of a partial molar volume of a protein by measuring its macroscopic compressibility, i.e., compression per unit pressure (See Chap. 5 (Gekko)). This is basically a macroscopic equation relating the isothermal compressibility with the fluctuation of the partial volume of a macromolecule (e.g., protein) in solution.

A fundamental question here is if this concept is assured also for the microscopic structural changes and the conformational fluctuations in the atomic level. Pressure-dependent structural changes are observed by high pressure NMR or high-pressure macromolecular crystallography, in principle as changes in coordinates of individual atoms, rather than changes in macroscopic quantities such as volume and compressibility. How the changes in coordinates of individual atoms are related to their positional fluctuations are the questions to be answered for the high-pressure experiment beyond the macroscopic relationship above.

A general answer to this question is given in Chap. 7 (Hirata), based on the generalized Langevin equation and the three-dimensional interaction site model (3D-RISM) theory under the regime of the linear response theory. As the consequence, we obtain the following relationship (Eq. 33.6), by the combination of Eqs. 7.26 and 7.27. Equation 33.6 gives the *microscopic* relation, in place of the macroscopic relation Eq. 33.4, correlating the pressure-induced coordinate changes with the intrinsic fluctuations of the coordinates of the protein-water system under ambient conditions.

$$\Delta \mathbf{R}_\alpha = (k_B T)^{-1} \Sigma_\beta < \Delta \mathbf{R}_\alpha \Delta \mathbf{R}_\beta > 0 \cdot \left(\partial \Delta V/\partial \mathbf{R}_\beta\right)_{T,P} \cdot P \tag{33.6}$$

Here the left hand side gives the actual coordinate changes observed by high pressure NMR, while in the right hand side the intrinsic fluctuations of coordinates under ambient conditions represented by $\Sigma_\beta < \Delta\mathbf{R}_\alpha \Delta\mathbf{R}_\beta >_0$ is amplified by the partial molar volume change caused by coordinate changes $(\partial\Delta V/\partial\mathbf{R}_\beta)_{T,P}$ induced by pressure P.

Thus the whole relation represented by Eq. 33.6 assures what we observe experimentally by coordinate changes in high pressure NMR represents the intrinsic fluctuations in coordinates that are amplified by pressure via the coordinate-dependent partial molar volume change. They identified those different conformations under pressure as fluctuating states of protein when the solution is in the native condition, or ambient pressure. The expression is to put the concept by Akasaka that the structural changes caused by pressure are the manifestation of intrinsic conformational fluctuations of a protein (Akasaka 2003, 2006, 2010, 2014) onto a firm theoretical basis (Hirata and Akasaka 2015).

33.6 Concluding Remark

High pressure NMR spectroscopy provides a general and powerful means to explore the high-energy paradigm of protein structures, which so far has been left largely untouched. It is a method to detect intrinsic fluctuations of proteins site-specifically by amplifying the fluctuations and detecting them sensitively and clearly as structural changes under pressure. The general significance of the high-energy sub-states in biology is inferred by noting that they are generally more reactive than their ground-state counterparts and are more specific in structure to individual proteins so that they are likely to be the crucial states of proteins for function (Akasaka 2014). The general significance of the high-energy sub-states in bio-technological application is that we gain the possibility of choosing different functional sub-states of a protein according to our specific needs by finely tuning pressure (Akasaka 2010). For these reasons and with further development of the technique, high pressure NMR will serve as an indispensable tool in biological science and bio-technology in years to come.

References

Abdul Latif AR, Kono R, Tachibana H, Akasaka K (2007) Kinetic analysis of amyloid protofibril dissociation and volumetric properties of the transition state. Biophys J 92:323–329

Akasaka K (2003) Highly fluctuating protein structures revealed by variable-pressure nuclear magnetic resonance. Biochemistry 42:10875–10885

Akasaka K (2006) Probing conformational fluctuation of proteins by pressure perturbation. Chem Rev 106:1814–1835

Akasaka K (2010) High pressure NMR study of proteins – seeking roots for function, evolution, disease and food applications. High Pressure Res 30:453–457

Akasaka K (2014) Pressure and protein dynamism. High Pressure Res 34:222–235

Akasaka K, Li H (2001) Low-lying excited states of proteins revealed from nonlinear pressure shifts in ^{1}H and ^{15}N NMR. Biochemistry 40:8665–8671

Akasaka K, Yamada H (2001) On-line cell high-pressure nuclear magnetic resonance technique: application to protein studies. Methods Enzymol 338:134–158

Akasaka K, Tezuka T, Yamada H (1997) Pressure-induced changes in the folded structure of lysozyme. J Mol Biol 271:671–678

Akasaka K, Li H, Yamada H, Li R, Thoresen T, Woodward CK (1999) Pressure response of protein backbone structure. Pressure-induced amide ^{15}N chemical shifts in BPTI. Protein Sci 8:1946–1953

Akasaka K, Maeno A, Murayama T, Tachibana H, Fujita Y, Yamanaka H, Nishida N, Atarashi R (2014) Pressure-assisted dissociation and degradation of "proteinase K-resistant" fibrils prepared by seeding with scrapie-infected hamster prion protein. Prion 8:314–318

Benedek GB, Purcell EM (1954) Nuclear magnetic resonance in liquids under high pressure. J Chem Phys 22:2003–2012

Cooper A (1976) Thermodynamic fluctuations in protein molecules. Proc Natl Acad Sci U S A 73:2740–2741

Fourme R, Girard E, Akasaka K (2012) High-pressure macromolecular crystallography and NMR: status, achievements and prospects. Curr Opin Struct Biol 22:636–642

Hirata F, Akasaka K (2015) Structural fluctuation of proteins induced by thermodynamic perturbation. J Chem Phys 142, 044110.

Inoue K, Yamada H, Akasaka K, Herrmann C, Kremer W, Maurer T, Döker R, Kalbitzer HR (2000) Pressure-induced local unfolding of the Ras binding domain of RalGDS. Nat Struct Biol 7:547–550

Jonas J (2002) High-resolution nuclear magnetic resonance studies of proteins. Biochim Biophys Acta 1595:145–159

Kalbitzer HR, Spoerner M, Ganser P, Hosza C, Kremer W (2009) Fundamental link between folding states and functional states of proteins. J Am Chem Soc 131:16714–16719

Kamatari YO, Yamada H, Akasaka K, Jones JA, Dobson CM, Smith LJ (2001) Response of native and denatured hen lysozyme to high pressure studied by $^{15}N/^{1}H$ NMR spectroscopy. Eur J Biochem 268:1782–1793

Kamatari YO, Kitahara R, Yamada H, Yokoyama S, Akasaka K (2004) High-pressure NMR spectroscopy for characterizing folding intermediates and denatured states of proteins. Methods 34:133–143

Kamatari YO, Yokoyama S, Tachibana H, Akasaka K (2005) Pressure-jump NMR study of dissociation and association of amyloid protofibrils. J Mol Biol 349:916–921

Kamatari YO, Smith LJ, Dobson CM, Akasaka K (2011) Cavity hydration as a gateway to unfolding: an NMR study of hen lysozyme at high pressure and low temperature. Biophys Chem 156:24–30

Kitahara R, Yamada H, Akasaka K, Wright PE (2002) High pressure NMR reveals that apomyoglobin is an equilibrium mixture from the native to the unfolded. J Mol Biol 320:311–319

Kitahara R, Yokoyama S, Akasaka K (2005) NMR snapshots of a fluctuating protein structure: ubiquitin at 30 bar–3 kbar. J Mol Biol 347:277–285

Kitahara R, Hata K, Li H, Williamson M, Akasaka K (2013) Pressure-induced chemical shifts as probes for conformational fluctuations in proteins. Prog Nucl Mag Res Sp 71:35–58

Kitamura Y, Itoh T (1987) Reaction volume of protonic ionization for buffering agents. Prediction of pressure dependence of pH and pOH. J Solution Chem 16:715–725

Kremer W, Arnold MR, Kachel N, Kalbitzer HR (2004) The use of high-sensitivity sapphire cells in high pressure NMR spectroscopy and its application to proteins. Spectroscopy 18:271–278

Lassalle MW, Li H, Yamada H, Akasaka K, Redfield C (2003) Pressure-induced unfolding of the molten globule of all-Ala alpha-lactalbumin. Protein Sci 12:66–72

Li H, Yamada H, Akasaka K (1998) Effect of pressure on individual hydrogen bonds in proteins. Basic pancreatic trypsin inhibitor. Biochemistry 37:1167–1173

Maeno A, Matsuo H, Akasaka K (2009) The pressure-temperature phase diagram of hen lysozyme at low pH. Biophysics 5:1–9

Niraula TN, Konno T, Li H, Yamada H, Akasaka K, Tachibana H (2004) Pressure-dissociable reversible assembly of intrinsically denatured lysozyme is a precursor for amyloid fibrils. Proc Natl Acad Sci U S A 101:4089–4093

Peterson RW, Lefebvre BG, Wand J (2005) High resolution NMR studies of encapsulated proteins in liquid ethane. J Am Chem Soc 127:10176–10177

Refaee M, Tezuka T, Akasaka K, Williamson MP (2003) Pressure-dependent changes in the solution structure of hen egg-white lysozyme. J Mol Biol 327:857–865

Sasaki K, Gaikwad J, Hashiguchi S, Kubota T, Sugimura K, Kremer W, Kalbitzer HR, Akasaka K (2008) Reversible monomer-oligomer transition in human prion protein. Prion 2:118–122

Williamson MP, Akasaka K, Refaee M (2003) The solution structure of bovine pancreatic trypsin inhibitor at high pressure. Protein Sci 12:1971–1979

Yamada H (1974) Pressure-resisting glass cell for high pressure, high resolution NMR measurement. Rev Sci Instrum 45:640–642

Yamada H, Nishikawa K, Honda N, Shimura T, Aksaka K, Tabayashi K (2001) Pressure-resisting cell for high-pressure, high-resolution nuclear magnetic resonance measurement at very high magnetic fields. Rev Sci Instrum 72:1463–1471

ERRATUM

Chapter 5
Volume and Compressibility of Proteins

Kunihiko Gekko

K. Akasaka, H. Matsuki (eds.), *High Pressure Bioscience*,
Subcellular Biochemistry 72, DOI 10.1007/978-94-017-9918-8

DOI 10.1007/978-94-017-9918-8_34

In Table 5.1 (page 81), the solvent name is mentioned incorrectly as *Chymotrypsinogen A*. The correct solvent name is *α-Chymotrypsinogen A*.

In Tables 5.2 (page 86) and 5.3 (page 97), the protein name is mentioned incorrectly as *Chymotrypsinogen A*. The correct protein name is *α-Chymotrypsinogen A*.

The online version of the original chapter can be found at
http://dx.doi.org/10.1007/978-94-017-9918-8_5

Appendix

1. Units of Pressure

Pa	bar	kgf cm^{-2}	atm	Torr	psi
1	1×10^{-5}	1.020×10^{-5}	9.869×10^{-6}	7.501×10^{-3}	1.450×10^{-4}
1×10^5	1	1.020	9.869×10^{-1}	7.501×10^2	1.450×10
9.807×10^4	9.807×10^{-1}	1	9.678×10^{-1}	7.356×10^2	1.422×10
1.013×10^5	1.013	1.033	1	7.600×10^2	1.470×10
1.333×10^2	1.333×10^{-3}	1.360×10^{-3}	1.316×10^{-3}	1	1.934×10^{-2}
6.895×10^3	6.895×10^{-2}	7.301×10^{-2}	6.805×10^{-2}	5.172×10	1

K. Akasaka, H. Matsuki (eds.), *High Pressure Bioscience*,
Subcellular Biochemistry 72, DOI 10.1007/978-94-017-9918-8

2. Typical Values of Pressure (Approximate Values)

Lowest pressure obtained in laboratory	1 p (10^{-12}) Pa
Atmospheric pressure of the Earth	
100 km altitude	0.1 Pa
10 km altitude	2.6 kPa
Blood pressure	1–15 kPa
Water pressure of a garden hose	300 kPa
Normal atmospheric air pressure	1 atm = 0.1013 MPa = 760 Torr = 14.7 psi
Under the foot of an elephant	1 M (10^6) Pa
Ballet dancer on one foot	2 MPa
Atmosphere of Venus	9.2 MPa
Deep trenches in pacific ocean	100 MPa
Center of the Earth	360 G (10^9) Pa
Center of the Sun	25 P (10^{15}) Pa

3. Historical Manuscript of PW Bridgman (J Biol Chem 19: 511–512 (1914))

THE COAGULATION OF ALBUMEN BY PRESSURE.

By P. W. BRIDGMAN.

(From the Jefferson Physical Laboratory of Harvard University, Cambridge.)

(Received for publication, October 20, 1914.)

The purpose of this note is to state a fact of possible biological interest which I have discovered incidentally in the course of other work. If the white of an egg is subjected to hydrostatic pressure at room temperature, it becomes coagulated, presenting an appearance much like that of a hard boiled egg.

The albumen was enclosed in a nickel-steel case, and pressure transmitted to it by mercury. The high pressure apparatus was that which I have already described.[1] Pressure may be applied so slowly that the rise of temperature due to the compression is inappreciable. At room temperature (20°) the limits of pressure necessary to produce the coagulation were fairly well marked. A pressure of 5000 atmospheres (75,000 pounds per square inch) applied for thirty minutes produced a perceptible stiffening of the white, but little more; 6000 atmospheres for thirty minutes produced a coagulation in appearance like curdled milk; while 7000 for thirty minutes resulted in apparently complete coagulation, the white being capable of standing under its own weight. If the duration of the pressure of 5000 was increased to one hour, the coagulation was only slightly increased in amount. 3000 atmospheres applied for sixteen hours produced a barely perceptible thickening of the white. The effect of temperature, which is not large, seems to be such that the ease of coagulation increases at low temperatures, contrary to what one might expect. 6000 atmospheres applied at 0° for one hour produced a somewhat greater stiffening than would have been produced at 20°. Pressures considerably higher than 7000 did not alter the effect. The first time I tried the experiment was with a pressure of 12,000

[1] *Proc. Amer. Acad. of Arts and Sciences*, xlix, p. 627, 1914.

atmospheres for twenty minutes; the resulting product was indistinguishable in appearance from that produced by 7000. A pressure of 12,000 at 20° was high enough to compel the water to freeze to a modification of ice, ice VI, denser than water. It is interesting that the coagulated white had not apparently been affected by this freezing.

I have made no attempt to determine whether the nature of the coagulation produced by pressure is the same as that produced by heat. If one can judge by appearances, the two may be different. In the course of twenty-four hours there separates from the pressure-coagulated white a small quantity of some watery fluid, in which the coagulated part remains insoluble.

Index

K. Akasaka, H. Matsuki (eds.), *High Pressure Bioscience*,
Subcellular Biochemistry 72, DOI 10.1007/978-94-017-9918-8